Texts and Monographs in Physics

Texts and Monographs in Physics

S. Albeverio, F. Gesztesy, R. Høegh-Krohn, and H. Holden: **Solvable Models in Quantum Mechanics** (1988).

R. Bass: **Nuclear Reactions with Heavy Ions** (1980).

A Bohm: **Quantum Mechanics: Foundations and Applications,** Second Edition (1986).

O. Bratelli and D.W. Robinson: **Operator Algebras and Quantum Statistical Mechanics.** Volume I: C*- and W*-Algebras. Symmetry Groups. Decomposition of States (1979). Volume II: Equilibrium States. Models in Quantum Statistical Mechanics (1981).

K. Chadan and P.C. Sabatier: **Inverse Problems in Quantum Scattering Theory** (1977).

M. Chaichian and N.F. Nelipa: **Introduction to Gauge Field Theories** (1984).

G. Gallavotti: **The Elements of Mechanics** (1983).

W. Glöckle: **The Quantum Mechanical Few-Body Problem** (1983).

W. Greiner, B Müller, and J. Rafelski: **Quantum Electrodynamics of Strong Fields** (1985).

J.M. Jauch and F. Rohrlich: **The Theory of Photons and Electrons: The Relativistic Quantum Field Theory of Charged Particles with Spin One-half,** Second Expanded Edition (1980).

J. Kessler: **Polarized Electrons (1976).** Out of print. (Second Edition available as Springer Series in Atoms and Plasmas, Vol. 1.)

G. Ludwig: **Foundations of Quantum Mechanics I** (1983).

G. Ludwig: **Foundations of Quantum Mechanics II** (1985).

O. Nachtmann: **Elementary Particle Physics: Concepts and Phenomena** (1990).

R.G. Newton: **Scattering Theory of Waves and Particles,** Second Edition (1982).

A. Perelomov: **Generalized Coherent States and Their Applications** (1986).

H. Pilkuhn: **Relativistic Particle Physics** (1979).

R.D. Richtmyer: **Principles of Advanced Mathematical Physics.** Volume I (1978). Volume II (1981).

W. Rindler: **Essential Relativity: Special, General, and Cosmological,** Revised Second Edition (1980).

P. Ring and P. Schuck: **The Nuclear Many-Body Problem** (1980).

R.M. Santilli: **Foundations of Theoretical Mechanics.** Volume I: The Inverse Problem in Newtonian Mechanics (1978). Volume II: Birkhoffian Generalization of Hamiltonian Mechanics (1983).

M.D. Scadron: **Advanced Quantum Theory and Its Applications Through Feynman Diagrams** (1979).

N. Straumann: **General Relativity and Relativistic Astrophysics** (1984).

C. Truesdell and S. Bharatha: **The Concepts and Logic of Classical Thermodynamics as a Theory of Heat Engines: Rigourously Constructed upon the Foundation Laid by S. Carnot and F. Reech** (1977).

F.J. Yndurâin: **Quantum Chromodynamics: An Introduction to the Theory of Quarks and Gluons** (1983).

Otto Nachtmann

Elementary Particle Physics
Concepts and Phenomena

Translated by A. Lahee and W. Wetzel

With 171 Illustrations

Springer-Verlag
Berlin Heidelberg New York
London Paris Tokyo

Otto Nachtmann
Institut für Theoretische Physik
Universität Heidelberg
6900 Heidelberg
Federal Republic of Germany

Library of Congress Cataloging in Publication Data
Nachtmann, Otto.
[Phänomena und Konzepte der Elementareilchenphysik. English]
Elementary particle physics : concepts and phenomena / Otto Nachtmann.
p. cm. — (Texts and monographs in physics)
Translation of: Phänomena und Konzepte der Elementareilchenphysik.
Includes index.
ISBN-13:978-3-540-51647-7 e-ISBN-13:978-3-642-61281-7
DOI: 10.1007/978-3-642-61281-7

1. Particles (nuclear physics) I. Title. II. Series.
QC793.2.N3313 1989
539.7'21—dc19

88-27328
CIP

Reprint of the original edition 1990

Typesetting: Asco Trade Typesetting Ltd., Hong Kong; printing and binding: R.R. Donnelley & Sons, Harrisonburg, Virginia, U.S.A.
2156/NY30-543210–Printed on acid-free paper

Preface

This book grew—how could it be otherwise?—out of a series of lectures which the author held at the University of Heidelberg. The purpose of these lectures was to give an introduction to the phenomenology of elementary particles for students both of theoretical and experimental orientation. With the present book the author has set himself the same aim. The reader is assumed to be familiar with ordinary nonrelativistic quantum mechanics as presented, e.g., in the following books: *Quantum Mechanics*, by L.I. Schiff (McGraw-Hill, New York, 1955); *Quantum Mechanics*, Vol. I, by K. Gottfried (W.A. Benjamin, Reading, Ma., 1966).

The setup of the present book is as follows. In the first part we present some basic general principles and concepts which are used in elementary particle physics. The reader is supposed to learn here the "language" of particle physics. An introductory chapter deals with special relativity, of such fundamental importance for particle physics, which most of the time is high energy, i.e., highly relativistic physics. Further chapters of this first part deal with the Dirac equation, with the theory of quantized fields, and with the general definitions of the scattering and transition matrices and the cross-sections.

The phenomenology of elementary particles is today dominated by gauge theories. The central issue of the present book is to make the reader familiar with the corresponding physical principles. We have tried to put a certain order into the presentation of these concepts, going from the simple to the more complicated cases. Thus, in the second part of the book we deal with quantum electrodynamics (QED), the most simple gauge theory, where the gauge group is Abelian and unbroken. The third part of the book is dedicated to the phenomena of strong interactions, to the world of quarks and gluons. There we encounter for the first time a non-Abelian gauge theory, quantum chromodynamics (QCD), where the gauge group is more complicated than in the QED case, but still unbroken. In the fourth part we deal with the theory of electroweak interactions, where we also have a non-Abelian gauge group but as a new phenomenon the spontaneous breaking of the gauge symmetry. In the final chapter we make some cursory remarks concerning developments in the theory of elementary particle physics going beyond the so-called standard model.

Everywhere in this book we have tried to emphasize physical points of view and considerations. For these reasons we have chosen the interaction picture to describe scattering processes. This is legitimate also in quantum field theory if all intermediate calculations are thought to be done in the regularized theory. Our special aim was to provide the reader with all necessary knowledge enabling him to do actual calculations related to problems in elementary particle phenomenology. On the other hand, students not specializing in elementary particle physics should be able to get an overview of the field by reading a few chapters easily picked out by their titles. It was not our aim to give a detailed presentation of quantum field theory. For this we can refer the reader, e.g., to the following text books: *Relativistic Quantum Fields*, by J.D. Bjorken and S.D. Drell (McGraw-Hill, New York, 1965); *Quantum Field Theory*, by C. Itzykson and J.-B. Zuber (McGraw-Hilll, New York, 1980). Readers wanting to broaden their knowledge on the experimental aspects of high energy physics are referred, e.g., to the book: *Introduction to High Energy Physics* by D.H. Perkins (Addison-Wesley, Reading, Ma., 1982). Finally, we would like to mention a few books where the reader can find presentations of topics in particle phenomenology which were of central interest some years ago and which could not be dealt with in too much detail in the present book: *High Energy Hadron Physics* by M.L. Perl (Wiley, New York, 1974); *An Introduction to Regge Theory and High Energy Physics* by P.D.B. Collins, (Cambridge University Press, Cambridge, 1977); *Elementary Particle Physics* by G. Källen (Addison-Wesley, Reading, Ma., 1964).

References to original literature are indicated in the text by the name of the first author and the year of publication. The corresponding complete references are collected at the end of the book. References to experimental results are always understood as giving examples illustrating theoretical developments, not in the sense that the author wants to ascribe any priorities. Numbers for physical quantities are, if not otherwise indicated, taken from "Review of Particle Properties" (Particle Data Group, *Phys. Lett.* **170B**, 1, 1986).

This book was originally published in German. The author would like to thank first of all the colleagues who helped to put together the original version, in particular, B. Stech, D. Gromes, I. Bender, W. Wetzel, M. Wirbel, and A. Reiter. Such a book would have been impossible without the kind help of many colleagues from experimental physics. The author expresses his gratitude to B. Naroska, J. Drees, C. Geweniger, J. Heintze, J. von Krogh,

A. Putzer, K. Schubert, and A. Wagner, as well as their collaborators. Special thanks are due to J. Heintze for allowing the use of one of his original drawings. The author would also like to thank the students attending his lecture courses for posing many questions which helped to clarify the ideas and the presentation. The English translation would have been impossible without the untiring efforts of A. Lahee and W. Wetzel, and the author would like to express his deep gratitude to them. Thanks are also due to all collaborators involved in producing this book, first at Vieweg-Verlag for the German version and then at Springer-Verlag for the English version. Finally, the author would like to thank his wife for her understanding and encouragement during this whole enterprise.

Abbreviations

CERN	Centre Europeén de la Recherche Nucléaire; European Nuclear Research Center, Geneva, Switzerland
CESR	Cornell Electron–Positron Storage Ring, Ithaca, NY, U.S.A. (Center of mass energy ca. 10 GeV)
DESY	Deutsches Elektronen–Synchrotron; German Electron Synchrotron, Hamburg, Federal Republic of Germany
DORIS	Electron–Positron Storage Ring at DESY (Center of mass energies initially 2–7 GeV, now ca. 10 GeV)
FNAL	Fermi National Accelerator Laboratory, Batavia, Illinois, U.S.A.
HERA	Electron–Proton Storage Ring (Currently under construction at DESY. Center of mass energy ca. 300 GeV)
ISR	Proton–Proton Storage Ring at CERN (Closed 1984. Center of mass energy ca. 20–60 GeV)
KEK	National Laboratory for High Energy Physics, Tsukuba, Japan
LEP	Electron–Positron Storage Ring (Currently under construction at CERN. Center of mass energy ca. 100 GeV)

PEP	Electron–Positron Storage Ring at SLAC (Center of mass energy ca. 30 GeV)
PETRA	Electron–Positron Storage Ring at DESY (Closed 1986. Center of mass energies ca. 14–45 GeV)
SLAC	Stanford Linear Accelerator Center, Stanford, Ca. U.S.A.
SLC	Electron–Positron Collider (Near completion at SLAC. Center of mass energies ca. 100 GeV)
SPEAR	Electron–Positron Storage Ring at SLAC (Center of mass energies ca. 2–7 GeV)
$Sp\bar{p}S$	Proton–Antiproton Storage Ring at CERN (Center of mass energy ca. 600 GeV)
TEVATRON	Proton–Antiproton Storage Ring at FNAL (Center of mass energy ca. 1800 GeV)
TRISTAN	Electron–Positron Storage Ring at KEK (Center of mass energy ca. 50 GeV)
UNK	Proton Accelerator under construction at Serpukhov, USSR (Proton energies up to 3000 GeV)

Notation and Symbols

$=$	equals
$\equiv$	identically equals
$\cong$	approximately equals
$\sim$	essentially equal to or equivalent
$\propto$	proportional
$\in$	is contained in
T	transpose
$\dagger$	Hermitian conjugate
sr	steradian
eV	electron volt; 1 keV = 10^3 eV, 1 MeV = 10^6 eV, 1 GeV = 10^9 eV, 1 TeV = 10^{12} eV
f	Fermi; 1 f = 10^{-13} cm
b	Barn; 1 b = 10^{-24} cm^2
s	second
Hz	Hertz; 1 Hz = 1 s^{-1}, 1 MHz = 10^6 s^{-1}
$\sum$	summation
$\sum'$	summation over final states and average over initial states
c number	complex number
c.c.	complex conjugate
h.c.	Hermitian conjugate
$\boldsymbol{x} = (x^j)$	three-vector. The components are labeled with Latin indices ($j = 1, 2, 3$)

$x = (x^\mu)$	four-vector. The components are labeled with Greek indices ($\mu = 0, 1, 2, 3$)
$d^3x = dx^1\, dx^2\, dx^3$	volume element in ordinary three-dimensional space
$dx = dx^0\, dx^1\, dx^2\, dx^3$	volume element in Minkowski space
$\mathbf{A} = (A_{ij})$	matrix or operator with matrix elements A_{ij}
$[\mathbf{A}, \mathbf{B}] = \mathbf{AB} - \mathbf{BA}$	commutator
$\{\mathbf{A}, \mathbf{B}\} = \mathbf{AB} + \mathbf{BA}$	anticommutator
$\det \mathbf{A}$	determinant of matrix $\mathbf{A}$
$\mathrm{Tr}\{\mathbf{A}\}$	trace of matrix $\mathbf{A}$
ε^{ijk}	totally antisymmetric symbol in three dimensions ($\varepsilon^{123} = 1$)
$\varepsilon_{\mu\nu\rho\sigma}$	totally antisymmetric symbol in four dimensions ($\varepsilon_{0123} = 1$)
δ_{ij}	Kronecker delta; $\delta_{ij} = 1$ for $i = j$; $\delta_{ij} = 0$ for $i \neq j$
$\delta(x)$	Dirac delta function ($-\infty < x < \infty$); $\delta(x) = 0$ for $x \neq 0$, $\int_{-\infty}^{\infty} dx\, \delta(x) = 1$
$\theta(x)$	theta function ($-\infty < x < \infty$); $\theta(x) = 0$ for $x < 0$, $\theta(x) = 1$ for $x > 0$
$O(E)$	terms of order E

Contents

PART I

RELATIVISTIC KINEMATICS AND QUANTUM FIELDS

CHAPTER 1

Introduction

Our intention in this book is to work through the theoretical principles of elementary particle physics. Today we know of the existence of a whole series of elementary particles. Examples are the electron (e), the photon (γ), the proton (p), the neutron (n), the pi-mesons or pions (π), the K-mesons (K), and the hyperons (Λ, Σ, etc.). Of these, the electron and the proton have been known for the longest time. At the time when the electron was discovered (Thomson 1897), there was still a general belief in the validity of classical physics in which particles and waves were totally separate concepts. It is really no more than a historical accident that electrons were first observed in an experiment in which they behaved like classical particles, whilst experiments in which electron beams show a wave-like character were not carried out until much later (Davisson 1927; Thomson 1927). In the case of light quanta, the situation was exactly the reverse. The wave nature of light was a dogma of the nineteenth century. It was only with the discovery of the photoelectric effect that physicists were forced to think again. The revolutionary idea of Einstein—the hypothesis of light quanta, in which light is also assumed to possess a particle nature (Einstein 1905)—at first met with great scepticism. It took approximately 20 years before the γ-quantum was accepted as a "particle" (see the historical survey of Pais (1979)). In the meantime, with the rise of quantum mechanics, it had been realized that particles and waves cannot be regarded as strictly separate phenomena, but must instead be considered as two aspects of the same phenomenon.

In historical sequence, the next elementary particles to be discovered were the following: The proton (Rutherford 1911, 1919), the neutron (Chadwick

1932), the positron, the positive partner of the electron (Anderson 1932, 1933), the muon (Anderson 1937, 1938; Street 1937), the pi-meson (or pion) (Lattes 1947), and the K-meson (Rochester 1947). Since the beginning of the 1950s large accelerators have become available, built for the purpose of studying reactions between elementary particles and for discovering new particles. Since then, experimental physicists have been extremely successful in the search for new particles, and today we know of several hundred "elementary" particles (Particle Data Group 1986). It is the aim of the theory of elementary particles to explain all phenomena concerning these particles in terms of one physical principle which should be as simple and as general as possible. The ideal example of a successful theory may be found in Maxwell's electrodynamics. By means of the Maxwell equations we are able to describe all magnetic and electrical phenomena! In elementary particle physics we have not yet reached the stage where we can write down a set of basic equations and derive from these all the observed phenomena. In the last 15 years, however, major progress has been made in this direction.

For a long time now it has been known that there are various forces which can act between elementary particles. The gravitational and electromagnetic forces were already known from classical physics. New additions to these were the weak interaction, responsible, for example, for radioactive β-decay, and the strong interaction responsible for α-decay. Thus we have four forces:

(i) Gravitation.
(ii) Weak interaction (e.g., β-decay, μ-decay).
(iii) Electromagnetic interaction (e.g., Coulomb scattering).
(iv) Strong interaction (e.g., α-decay, nuclear forces, pion–nucleon scattering).

Gravitation is currently believed to play a negligible role in elementary particle physics on the phenomenological level, since the gravitational forces between individual particles are extremely weak. To illustrate this consider the Bohr radius for a system of two neutrons subjected (hypothetically) only to gravitational attraction

$$r_B(\text{grav.}) = \frac{2\hbar^2}{G_N m_n^3} \cong 7 \cdot 10^{24}\,\text{cm}. \tag{1.1}$$

Here G_N is Newton's gravitational constant, m_n is the neutron mass, and $\hbar$ is Planck's constant divided by 2π.

The electromagnetic force is considerably stronger. The Bohr radius of the hydrogen atom, which is held together by the Coulomb force between the electron and the proton, has, as is well known, a value of about 10^{-8} cm:

$$r_B(\text{em}) = \frac{1}{\alpha m_e}\frac{\hbar}{c} \cong 0.5 \cdot 10^{-8}\,\text{cm}, \tag{1.2}$$

where c is the velocity of light, m_e is the electron mass, and α the fine-structure constant

$$\alpha \cong \frac{1}{137}. \tag{1.3}$$

Whereas the gravitational and electromagnetic forces have an infinite range, i.e., they produce potentials which fall off as $1/r$ with distance r, the weak and strong forces are very short range. The strong force has a range of about 10^{-13} cm, and the weak force a range of about 10^{-16} cm. With the presently available experimental methods it has been possible, in recent years, to reach a spatial resolution of this order of magnitude.

Very roughly, we can classify the elementary particles according to the type of interaction in which they take part. All of the known elementary particles are subject to gravitation and to the weak interaction. The strong interaction, on the other hand, is only effective between particles belonging to the class known as hadrons. Examples of hadrons are protons, neutrons, and pi-mesons (pions). Particles that are not influenced by the strong force are called leptons. Members of this group include the electron, the muon, and the neutrino (ν). The latter is produced, for example, in the β-decay of a neutron,

$$\mathrm{n} \to \mathrm{p} + \mathrm{e}^- + \bar{\nu}_\mathrm{e}. \tag{1.4}$$

All hadrons and leptons, apart from the neutrinos, take part in the electromagnetic interaction. Note that even hadrons with zero total charge, such as the neutron, feel the electromagnetic interaction, for example, through their charge distribution, their magnetic moment, or in inelastic reactions.

It is the hadrons in particular that have contributed to the population explosion among the elementary particles. In Tables 1.1 and 1.2 we list the presently known hadrons, and have further subdivided them into mesons and baryons. Mesons are particles with integral spin whereas baryons possess half-integral spin.

The mere number of hadrons suggests that not all of them, if any, can be really elementary. It is currently believed that, in reality, all hadrons are composed of further constituents—the so-called *quarks*. We now know of the existence of a whole series of quarks:

u (up)
d (down)
s (strange)
c (charm)
b (bottom)
t (top)?

The t-quark is indicated with a question mark because up to now only indirect experimental evidence for its existence was obtained. A proton is presently believed to be a bound state of two u-quarks and a d-quark

$$\mathrm{p} \sim \mathrm{uud}. \tag{1.5}$$

These quarks, however, have rather peculiar properties, such as fractional

Table 1.1 List of all known meson isospin multiplets, arranged according to their quantum numbers strangeness, S, charm, C, and beauty, B. Also given are the masses (rounded to the nearest MeV), the isospin and G-parity, I^G, and the spin, parity, and charge-conjugation quantum number, J^{PC}. For states with no dot, only weak experimental evidence exists (after Particle Data Group 1986).

Nonstrange ($S = 0$; $C, B = 0$)						Strange ($\vert S\vert = 1$; $C, B = 0$)	
Entry	$I^G(J^{PC})$	Entry	$I^G(J^{PC})$	Entry	$I^G(J^{PC})$	Entry	$I(J^P)$
● π	$1^-(0^{-+})$	● ρ (1600)	$1^+(1^{--})$	e^+e^- (1100–2200)	(1^{--})	● K	$1/2(0^-)$
● η	$0^+(0^{-+})$	● ω_3 (1670)	$0^-(3^{--})$	$\bar{N}N$ (1200–3600)		● K^* (892)	$1/2(1^-)$
● ρ (770)	$1^+(1^{--})$	● π_2 (1680)	$1^-(2^{-+})$	X (1900–3600)		● K_1 (1280)	$1/2(1^+)$
● ω (783)	$0^-(1^{--})$	● ϕ (1680)	$0^-(1^{--})$	● η_c (2980)	$0^+(0^{-+})$	● K_0^* (1350)	$1/2(0^+)$
● η' (958)	$0^+(0^{-+})$	● ρ_3 (1690)	$1^+(3^{--})$	● J/ψ (3097)	$0^-(1^{--})$	● K_1 (1400)	$1/2(1^+)$
● f_0 (975)	$0^+(0^{++})$	X (1700)	$^+$	● χ_0 (3415)	$0^+(0^{++})$	K^* (1410)	$1/2(1^-)$
● a_0 (980)	$1^-(0^{++})$	● f_2 (1720)	$0^+(2^{++})$	● χ_1 (3510)	$0^+(1^{++})$	● K_2^* (1430)	$1/2(2^+)$
● ϕ (1020)	$0^-(1^{--})$	f_0 (1730)	$0^+(0^{++})$	● χ_2 (3555)	$0^+(2^{++})$	K (1460)	$1/2(0^-)$
● h_1 (1190)	$0^-(1^{+-})$	π (1770)	$1^-(0^{-+})$	η_c (3590)	$(\ ^+)$	K_2 (1580)	$1/2(2^-)$
● b_1 (1235)	$1^+(1^{+-})$	f_2 (1810)	$0^+(2^{++})$	● ψ (3685)	$0^-(1^{--})$	● K_2 (1770)	$1/2(2^-)$
f_0 (1240)	$0^+(0^{++})$	● ϕ_J (1850)	0	● ψ (3770)	(1^{--})	● K_3^* (1780)	$1/2(3^-)$
ρ (1250)	$1^+(1^{--})$	X (1935)		● ψ (4030)	(1^{--})	K^* (1790)	$1/2(1^-)$
● f_2 (1270)	$0^+(2^{++})$	● f_4 (2030)	$0^+(4^{++})$	● ψ (4160)	(1^{--})	K (1830)	$1/2(0^-)$
● a_1 (1270)	$1^-(1^{++})$	a_4 (2040)	$1^-(4^{++})$	● ψ (4415)	(1^{--})	● K_4^* (2060)	$1/2(4^+)$
η (1275)	$0^+(0^{-+})$	a_3 (2050)	$1^-(3^{++})$	● Υ (9460)	(1^{--})	K_2 (2250)	$1/2(2^-)$
● f_1 (1285)	$0^+(1^{++})$	π_2 (2100)	$1^-(2^{-+})$	● χ_{b0} (9860)	$(\ ^{++})$	K_3 (2320)	$1/2(3^+)$
● f_0 (1300)	$0^+(0^{++})$	ρ (2150)	$1^+(1^{--})$	● χ_{b1} (9895)	$(\ ^{++})$	K_4 (2500)	$1/2(4^-)$
● π (1300)	$1^-(0^{-+})$	f_2 (2150)	$0^+(2^{++})$	● χ_{b2} (9915)	$(\ ^{++})$	Charmed ($\vert C\vert = 1$)	
● a_2 (1320)	$1^-(2^{++})$	X (2220)	$0\ (\ ^{++})$	● Υ (10023)	(1^{--})	● D	$1/2(0^-)$
f_2 (1410)	$0^+(2^{++})$	f_2 (2240)	$0^+(2^{++})$	χ_{b0} (10235)	$(\ ^{++})$	● D^* (2010)	$1/2(1^-)$
● f_1 (1420)	$0^+(1^{++})$	ρ_3 (2250)	$1^+(3^{--})$	● χ_{b1} (10255)	$(\ ^{++})$	D^* (2420)	$1/2(\ \ )$
● η (1440)	$0^+(0^{-+})$	f_4 (2300)	$0^+(4^{++})$	● χ_{b2} (10270)	$(\ ^{++})$	● D_s	$0\ (0^-)$
● f_2' (1525)	$0^+(2^{++})$	ρ_5 (2350)	$1^+(5^{--})$	● Υ (10355)	(1^{--})	D_s^* (2110)	
f_1 (1530)	$0^+(1^{++})$	a_6 (2450)	$1^-(6^{++})$	● Υ (10575)	(1^{--})	Bottom ($\vert B\vert = 1$)	
● f_0 (1590)	$0^+(0^{++})$	f_6 (2510)	$0^+(6^{++})$	● Υ (10860)	(1^{--})	● B	
				● Υ (11020)	(1^{--})	B^* (5325)	
						Exotics	

charges for example. Furthermore, they have never to date been observed for certain as isolated particles. This fact has led the theoreticians to propose that quarks can only exist in bound states (the *confinement* hypothesis).

According to this picture of the structure of matter, the quarks and the spin-1/2 leptons are the fundamental fermions. We group them into three "families" (Table 1.3). The four forces mentioned previously are believed to be exerted by means of the exchange of bosons. These processes may be represented pictorially in terms of Feynman diagrams (Fig. 1.1). The quanta corresponding to the electromagnetic field, the photons (γ), are well known. The

Table 1.2 List of all known baryons with their masses (rounded to the nearest MeV). The number of stars represents the quality of the experimental evidence for the particles with one star being very weak, and four stars being completely certain (after Particle Data Group 1986).

N(939) *P*11	****	Δ(1232) *P*33	****	Z_0(1780) *P*01	*	Σ(1193) *P*11	****	Ξ(1318) *P*11	****
N(1440) *P*11	****	Δ(1550) *P*31	*	Z_0(1865) *D*03	*	Σ(1385) *P*13	****	Ξ(1530) *P*13	****
N(1520) *D*13	****	Δ(1600) *P*33	**	Z_1(1725) *P*11	*	Σ(1480)	*	Ξ(1630)	*
N(1535) *S*11	****	Δ(1620) *S*31	****	Z_1(1900) *P*13	*	Σ(1560)	**	Ξ(1680)	**
N(1540) *P*13	*	Δ(1700) *D*33	****	Z_1(2150)	*	Σ(1580) *D*13	**	Ξ(1820) 13	***
N(1650) *S*11	****	Δ(1900) *S*31	***	Z_1(2500)	*	Σ(1620) *S*11	**	Ξ(1940)	**
N(1675) *D*15	****	Δ(1905) *F*35	****			Σ(1660) *P*11	***	Ξ(2030) 1	***
N(1680) *F*15	****	Δ(1910) *P*31	****	Λ(1116) *P*01	****	Σ(1670) *D*13	****	Ξ(2120)	*
N(1700) *D*13	***	Δ(1920) *P*33	***	Λ(1405) *S*01	****	Σ(1690)	**	Ξ(2250)	**
N(1710) *P*11	***	Δ(1930) *D*35	***	Λ(1520) *D*03	****	Σ(1750) *S*11	***	Ξ(2370) 1	**
N(1720) *P*13	****	Δ(1940) *D*33	*	Λ(1600) *P*01	***	Σ(1770) *P*11	*	Ξ(2500)	*
N(1960) ?	*	Δ(1950) *F*37	****	Λ(1670) *S*01	****	Σ(1775) *D*15	****		
N(1990) *F*17	**	Δ(2000) *F*35	**	Λ(1690) *D*03	****	Σ(1840) *P*13	*	Ω(1672) *P*03	****
N(2000) *F*15	**	Δ(2150) *S*31	*	Λ(1800) *S*01	***	Σ(1880) *P*11	**		
N(2080) *D*13	**	Δ(2200) *G*37	*	Λ(1800) *P*01	***	Σ(1915) *F*15	****	Λ_c(2281)	****
N(2090) *S*11	*	Δ(2300) *H*39	**	Λ(1820) *F*05	****	Σ(1940) *D*13	***	Σ_c(2450)	**
N(2100) *P*11	*	Δ(2350) *D*35	*	Λ(1830) *D*05	****	Σ(2000) *S*11	*	Ξ_c(2460)	*
N(2190) *G*17	****	Δ(2390) *F*37	*	Λ(1890) *P*03	****	Σ(2030) *F*17	****	Ω_c(2740)	*
N(2200) *D*15	**	Δ(2400) *G*39	**	Λ(2000)	*	Σ(2070) *F*15	*	Λ_b(5500)	*
N(2220) *H*19	****	Δ(2420) *H*311	****	Λ(2020) *F*07	*	Σ(2080) *P*13	**		
N(2250) *G*19	****	Δ(2750) *I*313	**	Λ(2100) *G*07	****	Σ(2100) *G*17	*	*Dibaryons*	
N(2600) *I*111	***	Δ(2950) *K*315	**	Λ(2110) *F*05	***	Σ(2250)	***	*NN*(2170) 1*D*2	**
N(2700) *K*113	**	Δ(~3000)		Λ(2325) *D*03	*	Σ(2455)	**	*NN*(2250) 3*F*3	**
N(~3000)				Λ(2350)	***	Σ(2620)	**	*NN*(?)	*
				Λ(2585)	**	Σ(3000)	*	*NN*(2130) 3*S*1	**
						Σ(3170)	*	Ξ*N*(?)	*

Table 1.3 The three known families of fundamental fermions and their electrical charges. The question-mark by the t-quark is to indicate that its existence is not yet experimentally proven.

Families			Charge
ν_e	ν_μ	ν_τ	0
e	μ	τ	-1
u	c	t (?)	$\frac{2}{3}$
d	s	b	$-\frac{1}{3}$

quanta of the strong field are today regarded as being the so-called *gluons* (G). For the existence of these particles we have indirect, but nonetheless, convincing evidence. It is also believed that the gluons, in the same way as the quarks, cannot be observed as free particles. The quanta of the weak field, the W- and the Z-boson, are 80–90 times heavier than a proton. They were discovered in 1983 at the proton–antiproton storage ring (Sp$\bar{\text{p}}$S) at CERN.

One of the objectives of this book is to acquaint the reader with the modern ideas about the structure of matter and the fundamental forces of nature. The appropriate framework for the theoretical description of elementary-particle phenomena is the quantum field theory whose basic principles we must thus address. Before doing this, we will briefly describe the units which are generally used in elementary particle physics.

As the unit of velocity we take the speed of light c, and all quantities which have the dimension of action are measured in units of $\hbar$ where $\hbar = h/2\pi$ and h is Planck's constant. Thus in our system of units we write

$$c = \hbar = 1. \tag{1.6}$$

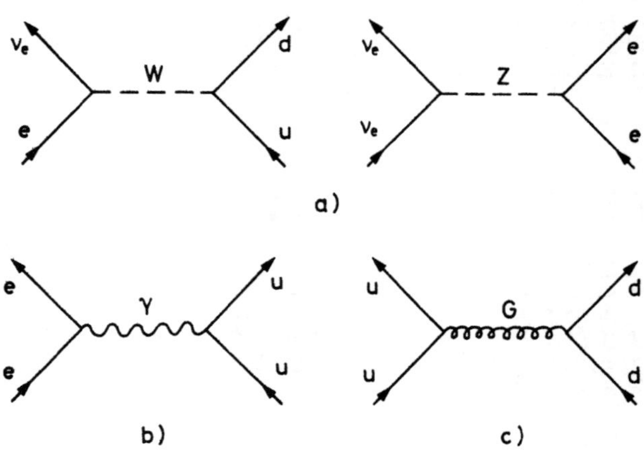

Figure 1.1 Examples of Feynman diagrams for the weak (a), the electromagnetic (b), and the strong interaction (c).

As a result of setting $c = 1$, lengths and times now have the same dimensions and, due to the Einstein relation

$$E = mc^2, \tag{1.7}$$

the same is true for energy and mass. The relationship, also discovered by Einstein, between frequency and energy

$$E = \hbar\nu = \hbar\omega \tag{1.8}$$

where $\omega = 2\pi\nu$, then endows energy and reciprocal time with the same dimensions since $\hbar = 1$. A single arbitrary unit remains, and this may be chosen to be length, time, or energy. A frequently used unit of energy is the MeV = 10^6 eV or alternatively the GeV = 10^3 MeV. The experimental values of c and $\hbar$ (Particle Data Group 1986), together with a few useful conversion formulas, are given in (1.9)–(1.11) below.

$$\begin{aligned} c &= 2.997\,924\,58\,(1, 2)\cdot 10^{10}\,\mathrm{cm\,s^{-1}}, \\ \hbar &= 1.054\,588\,7\,(57)\cdot 10^{-27}\,\mathrm{erg\,s}, \end{aligned} \tag{1.9}$$

$$1\,\mathrm{MeV} = 1.602\,189\,2\,(46)\cdot 10^{-6}\,\mathrm{erg}, \tag{1.10}$$

$$\begin{aligned} 1\,\mathrm{MeV}^{-1} &\mathrel{\hat{=}} 197.328\,58\,(51)\cdot 10^{-13}\,\mathrm{cm} \\ &\mathrel{\hat{=}} 6.582\,173\,(17)\cdot 10^{-22}\,\mathrm{s}. \end{aligned} \tag{1.11}$$

The numbers in brackets represent the uncertainty corresponding to one standard deviation in the last figures of the main number.

CHAPTER 2

The Theory of Special Relativity and Relativistic Kinematics

A large part of our knowledge about elementary particles stems from scattering experiments. As an example we shall consider proton–nucleon scattering as it occurs when a beam of protons from an accelerator impinges on a target (Fig. 2.1.). If the energy of the incident protons is large enough, then as a result of Einstein's equivalence relationship between mass and energy (Einstein 1905b), it is possible for the kinetic energy of the original protons to be converted into rest mass of new particles. A typical reaction might be

$$p + p \rightarrow p + p + \pi^+ + \pi^- . \tag{2.1}$$

The greater the energy of the protons, the larger is the number and the mass of the particles that can be produced. Thus it is clear why particle physicists are mainly interested in collisions between fast, relativistic particles. To begin with we will therefore discuss the basic principles of special relativity, and the relativistic kinematics of a collision between two particles.

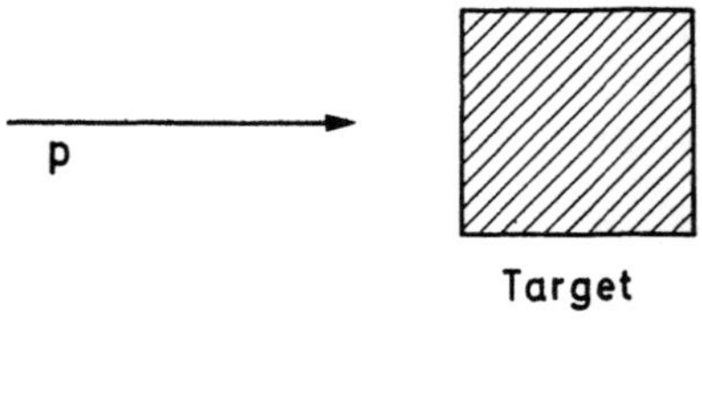

Figure 2.1 Scattering of a proton on a target.

2.1 The Basic Principles of Special Relativity

Let us take as our frame of reference an inertial system. For our purposes we may assume that the system of fixed stars provides (with sufficient accuracy) such an inertial frame of reference. In this system, time will be measured with ordinary clocks and the position of an event will be described in terms of Cartesian coordinates. We impose the further condition that the clocks are at rest relative to the reference system. Clocks at different points in space will be synchronized by means of light signals, whereat we set the speed of light $c = 1$. To any event, i.e., something taking place at a particular time and spatial point, we attribute a four-vector x where the uppermost component is the time of the event $x^0 = t$, and the remaining three components are the spatial Cartesian coordinates x^1, x^2, x^3:

$$x = (x^\mu) = \begin{pmatrix} x^0 \\ x^1 \\ x^2 \\ x^3 \end{pmatrix} = \begin{pmatrix} t \\ \boldsymbol{x} \end{pmatrix}, \qquad \mu = 0, 1, 2, 3. \tag{2.2}$$

Here, and in the following, we use the Greek letters (μ, ν, ...) as indices for the components of four-vectors and four-tensors, examples of which we shall be encountering shortly.

The set of four-dimensional vectors of (2.2) forms the Minkowski space[1] (Minkowski 1909). This can be visualized by taking a two-dimensional section as shown in Fig. 2.2.

We will now assume that an observer at point $\boldsymbol{y}$ and at time y^0 sends out a light signal in all directions. The spatial coordinates $\boldsymbol{x}$ of the wavefront of the light at a time $x^0 > y^0$ then satisfy (with $c = 1$) the equation

$$|\boldsymbol{x} - \boldsymbol{y}| = (x^0 - y^0),$$

or alternatively

$$(x^0 - y^0)^2 - (\boldsymbol{x} - \boldsymbol{y})^2 = 0, \qquad x^0 > y^0. \tag{2.3}$$

The set of points x in Minkowski space which satisfy this condition is termed the *forward light cone* of the point y. Analogously, the backward light cone is formed by a contracting spherical light wave which at time y^0 reaches the spatial point $\boldsymbol{y}$. Thus we define the light cone of the point y as the set of points x satisfying

$$(x^0 - y^0)^2 - (\boldsymbol{x} - \boldsymbol{y})^2 = 0 \tag{2.4}$$

[1] Strictly speaking, we should distinguish between Minkowski point space formed by the set of all events, and Minkowski vector space formed by all four-vectors x assigned to the events relative to a given space–time coordinate frame. We will, however, in the following always speak of Minkowski space and the context should make it clear whether the point or vector space is meant. In the same spirit we will often denote events, i.e., points in Minkowski point space, by their corresponding four-vectors.

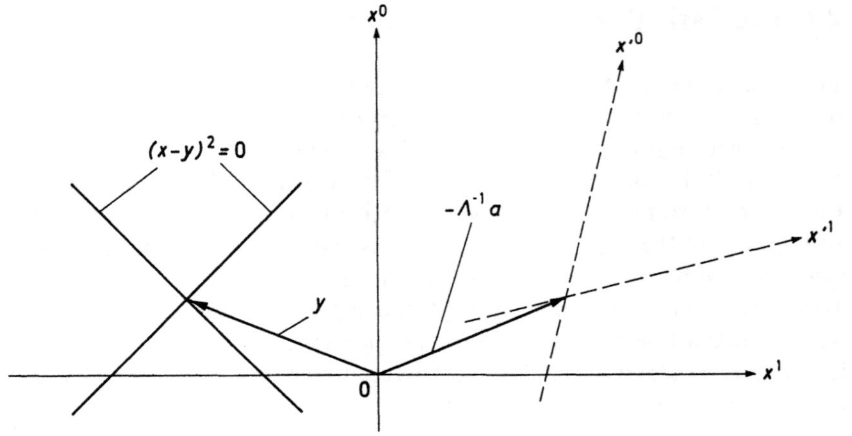

Figure 2.2 A two-dimensional section through Minkowski space displaying the light cone of the point y and a transition to new coordinates.

with its two parts additionally satisfying:

Forward light cone	$x^0 > y^0$.
Backward light cone	$x^0 < y^0$.

In (2.4) we encounter a frequently appearing quadratic form which is used to introduce *a pseudometric* in Minkowski space. Let x and y be two arbitrary four-vectors. We define their scalar product (in the Minkowski sense) as

$$(x, y) = x^0 y^0 - \boldsymbol{x} \cdot \boldsymbol{y}. \tag{2.5}$$

For $x = y$ we obtain from this the Minkowski quadratic form, which however, as can easily be seen, is not positive definite:

$$(x, x) \equiv x^2 = (x^0)^2 - (\boldsymbol{x})^2 \begin{cases} > 0, \\ = 0, \\ < 0. \end{cases} \tag{2.6}$$

It is also convenient to introduce a matrix and tensor notation. For this we define the *metric tensor* $g_{\mu\nu}$ in Minkowski space and its corresponding matrix $\mathbf{g}$ via

$$\mathbf{g} = (g_{\mu\nu}) = \begin{pmatrix} 1 & 0 & 0 & 0 \\ 0 & -1 & 0 & 0 \\ 0 & 0 & -1 & 0 \\ 0 & 0 & 0 & -1 \end{pmatrix}. \tag{2.7}$$

The components of the inverse matrix $\mathbf{g}^{-1}$ will be denoted by $g^{\mu\nu}$. The following relation then holds:

$$g^{\mu\nu} g_{\nu\rho} = \delta^{\mu}_{\rho}, \tag{2.8}$$

where δ^μ_ρ is the Kronecker delta. Here the Einstein summation convention over repeated indices is used. From (2.7) it follows, of course that $\mathbf{g}$ and $\mathbf{g}^{-1}$ are given by the same matrix

$$\mathbf{g}^{-1} = (g^{\mu\nu}) = \begin{pmatrix} 1 & 0 & 0 & 0 \\ 0 & -1 & 0 & 0 \\ 0 & 0 & -1 & 0 \\ 0 & 0 & 0 & -1 \end{pmatrix}. \tag{2.9}$$

This is valid, however, only for our choice of a Cartesian basis and not for the general case. We will use the metric tensor for the purpose of changing superscripts to subscripts and vice versa. For example, for a four-vector x, we define the components with subscripts as follows:

$$x_\mu = g_{\mu\nu} x^\nu = \begin{cases} x^0 & \text{for} \quad \mu = 0, \\ -x^\mu & \text{for} \quad \mu = 1, 2, 3. \end{cases} \tag{2.10}$$

Then for the scalar product of (2.5) we can write

$$(x, y) = g_{\mu\nu} x^\mu y^\nu = x^\mu y_\mu = x_\mu y^\mu. \tag{2.11}$$

In choosing a metric tensor in Minkowski space one has a certain amount of freedom. We could, for example, have taken $-g_{\mu\nu}$ instead of $g_{\mu\nu}$ or alternatively the Pauli metric, which makes use of an imaginary time unit. The convention used here follows that of Bjorken (1965).

Let us now consider a second inertial system, one which is in uniform motion with respect to our original system. The basic postulate of the theory of special relativity is that the laws of nature must have the same form in every inertial system, a postulate which was confirmed by the Michelson–Morley experiment (Michelson 1881, 1887; see also the discussion by von Laue 1952).

An event with the four-vector x in the first reference system corresponds in the second system to a four-vector x':

$$x' = \begin{pmatrix} x'^0 \\ x'^1 \\ x'^2 \\ x'^3 \end{pmatrix}. \tag{2.12}$$

Without loss of generality, we can write the relation between the old and new coordinates in the following form:

$$\begin{aligned} x'^\mu &= a^\mu + \mathrm{f}^\mu(x), \\ \mathrm{f}^\mu(x = 0) &= 0. \end{aligned} \tag{2.13}$$

where the $\mathrm{f}^\mu(x)$ are as yet undetermined functions of x and $a = (a^\mu)$ is a constant four-vector. In special relativity one assumes homogeneity of time and a Euclidian structure of space. Thus we may write

$$x''^\mu = x'^\mu - a^\mu \tag{2.14}$$

as the coordinates of the event in a further inertial system, which is related to the primed system by a simple translation of the coordinate origin. The double-primed and unprimed (original) systems then have the same space–time coordinate origin.

$$x''^{\mu} = f^{\mu}(x),$$
$$f^{\mu}(x = 0) = 0. \tag{2.15}$$

It is now easy to see that the $f^{\mu}(x)$ must be homogeneous, linear functions. Indeed, two events with coordinates x^{μ} and $2x^{\mu}$ in the unprimed system must have coordinates x''^{μ} and $2x''^{\mu}$ in the double-primed system. If this were not the case, then the transformation formulas from the unprimed to the double-primed system would differ according to whether we chose the meter or the centimeter as our unit of length. However, this state of affairs is impossible so long as no one length scale is assigned particular physical significance over the others in empty space–time.

Thus we obtain

$$x''^{\mu} = \Lambda^{\mu}{}_{\nu}x^{\nu},$$
$$x'^{\mu} = \Lambda^{\mu}{}_{\nu}x^{\nu} + a^{\mu}, \tag{2.16}$$

where the $\Lambda^{\mu}{}_{\nu}$ are the constants which can be conveniently written in the form of a 4×4 matrix $\Lambda = (\Lambda^{\mu}{}_{\nu})$. Note that in the framework of the general theory of relativity, the above considerations are no longer valid since a nonzero curvature of the space–time continuum determines a natural length scale at each point.

The postulate of the constancy of the speed of light in all inertial frames now leads us to a condition for Λ. Let us again consider the light signal which, in the unprimed system, is emitted from the spatial position $\boldsymbol{y}$ at time y^0. For the points x on the wavefront we have, according to (2.3)

$$(x - y)^2 = 0. \tag{2.17}$$

In the primed system we have for the place and time of emission and for the points on the wavefront

$$y'^{\mu} = \Lambda^{\mu}{}_{\nu}y^{\nu} + a^{\mu},$$
$$x'^{\mu} = \Lambda^{\mu}{}_{\nu}x^{\nu} + a^{\mu}, \tag{2.18}$$
$$x'^{\mu} - y'^{\mu} = \Lambda^{\mu}{}_{\nu}(x^{\nu} - y^{\nu}),$$

$$(x' - y')^2 = g_{\mu\nu}(x'^{\mu} - y'^{\mu})(x'^{\nu} - y'^{\nu})$$
$$= g_{\mu\nu}\Lambda^{\mu}{}_{\rho}\Lambda^{\nu}{}_{\sigma}(x^{\rho} - y^{\rho})(x^{\sigma} - y^{\sigma}). \tag{2.19}$$

In the primed system the speed of light must also be equal to 1 and thus, if (2.17) is valid, we also should have the relation

$$(x' - y')^2 = 0. \tag{2.20}$$

A sufficient and, as can easily be seen, also a necessary condition for this to be true is that the matrix Λ satisfies the condition

$$\Lambda^{\mu}{}_{\rho}\Lambda^{\nu}{}_{\sigma}g_{\mu\nu} = \kappa(\Lambda)g_{\rho\sigma} \tag{2.21}$$

with some factor $\kappa(\Lambda)$ whose significance remains to be discussed. We then have

$$(x' - y')^2 = \kappa(\Lambda)(x - y)^2 = 0. \tag{2.22}$$

Sylvester's "inertia of signature" theorem for quadratic forms (see Hohn 1958) demands

$$\kappa(\Lambda) > 0. \tag{2.23}$$

If $\kappa \neq 1$, then by applying a dilatation, i.e., a change of scale such that

$$x'^\mu \to \sqrt{\kappa} x'^\mu, \tag{2.24}$$

$$\Lambda^\mu{}_\nu \to \sqrt{\kappa} \Lambda^\mu{}_\nu, \tag{2.25}$$

it is always possible to arrive at $\kappa = 1$. The dilatations (2.24) are physical invariance transformations provided we confine ourselves to pure electromagnetic-wave phenomena. The dilatation invariance is broken, however, as soon as we allow for the atomic structure of matter. In the following we shall consider this broken dilatation invariance no further, although it should be mentioned that it plays an important role in more theoretical aspects of elementary particle physics. For our purposes we will thus be considering only matrices Λ which satisfy the condition

$$\Lambda^\mu{}_\rho \Lambda^\nu{}_\sigma g_{\mu\nu} = g_{\rho\sigma}. \tag{2.26}$$

We have already given a schematic illustration of the transition to new coordinates in Fig. 2.2. For such transitions the relativistic squared interval between any two points x and y in Minkowski space is an invariant quantity:

$$\begin{aligned}(x' - y')^2 &= (x'^\mu - y'^\mu)(x'_\mu - y'_\mu) \\ &= (x^\mu - y^\mu)(x_\mu - y_\mu) = (x - y)^2. \end{aligned} \tag{2.27}$$

The usefulness of the tensor notation is demonstrated here. It can easily be seen that the contraction of a four-vector superscript with a four-vector subscript leads quite generally to a result which is independent of the frame of reference.

The transformations that correspond to a transition from one inertial frame to another form a group known as the *inhomogeneous Lorentz group* or *Poincaré group*. An element of this group will be denoted by (Λ, a):

$$(\Lambda, a)\colon x^\mu \to x'^\mu = \Lambda^\mu{}_\nu x^\nu + a^\mu, \tag{2.28}$$

where Λ must satifsy (2.26) and $a = (a^\mu)$ is an arbitrary constant four-vector. As can readily be shown, the product of two transformations satisfies

$$(\Lambda, a)(\Lambda', a') = (\Lambda\Lambda', \Lambda a' + a). \tag{2.29}$$

An important subgroup of the Poincaré group is the group of translations in space and time that consists of all transformations of the form

$$(1, a)\colon x^\mu \to x'^\mu = x^\mu + a^\mu. \tag{2.30}$$

As we shall see later, this group is related to the conservation of energy and momentum.

The transformations which leave the coordinate origin unaltered likewise form a subgroup of the Poincaré group, which is called the (homogeneous) *Lorentz group*. A general element of this group has the form $(\Lambda, 0)$

$$(\Lambda, 0)\colon x^\mu \to x'^\mu = \Lambda^\mu{}_\nu x^\nu . \tag{2.31}$$

Examples of Lorentz transformations are the usual rotations in space. The corresponding matrices $\Lambda = (\Lambda^\mu{}_\nu)$ in (2.31) have the form

$$\Lambda = \left(\begin{array}{c|c} 1 & 0 \\ \hline 0 & \mathbf{R} \end{array}\right), \tag{2.32}$$

where $\mathbf{R}$ may be any orthogonal 3×3 matrix ($\mathbf{R}^\mathrm{T}\mathbf{R} = 1$). Further examples are the special Lorentz transformations (boosts) which, for the transition to a system moving in the x^1 direction (Fig. 2.3), correspond to a Λ matrix in (2.31) of the form

$$\Lambda(v) = \begin{pmatrix} \dfrac{1}{\sqrt{1-v^2}} & \dfrac{v}{\sqrt{1-v^2}} & 0 & 0 \\ \dfrac{v}{\sqrt{1-v^2}} & \dfrac{1}{\sqrt{1-v^2}} & 0 & 0 \\ 0 & 0 & 1 & 0 \\ 0 & 0 & 0 & 1 \end{pmatrix}. \tag{2.33}$$

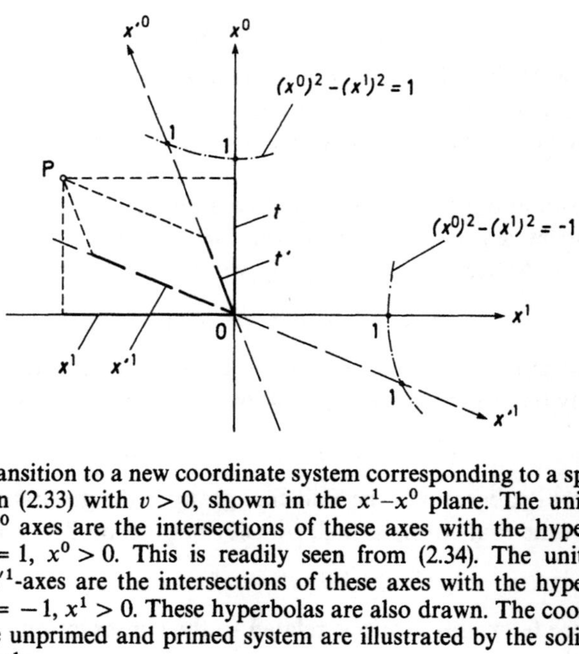

Figure 2.3 Transition to a new coordinate system corresponding to a special Lorentz transformation (2.33) with $v > 0$, shown in the x^1–x^0 plane. The unit markers on the x^0 and x'^0 axes are the intersections of these axes with the hyperbola branch $(x^0)^2 - (x^1)^2 = 1$, $x^0 > 0$. This is readily seen from (2.34). The unit markers on the x^1- and x'^1-axes are the intersections of these axes with the hyperbola branch $(x^0)^2 - (x^1)^2 = -1$, $x^1 > 0$. These hyperbolas are also drawn. The coordinates of an event P in the unprimed and primed system are illustrated by the solid and dashed lines, respectively.

Here v is the relative velocity between the two inertial frames; its modulus must of course be less than the speed of light ($|v| < 1$). The relationship between unprimed and primed coordinates in this example is given by

$$x'^{\mu} = \Lambda^{\mu}{}_{\nu}(v)x^{\nu},$$

$$\begin{pmatrix} t' \\ x'^1 \\ x'^2 \\ x'^3 \end{pmatrix} = \begin{pmatrix} \dfrac{t + vx^1}{\sqrt{1 - v^2}} \\ \dfrac{vt + x^1}{\sqrt{1 - v^2}} \\ x^2 \\ x^3 \end{pmatrix}. \tag{2.34}$$

If we now allow v in (2.33) to vary continuously, we obtain a set of Lorentz transformations which also includes the identity transformation ($v = 0$). All transformations in the Lorentz group which can be reached via a continuous variation of the matrices Λ from the identity transformation, form a subgroup of the Lorentz group: the proper orthochronous Lorentz group or identity component.

We would now like to characterize the Lorentz transformations of the identity component more explicitly. For this purpose we write (2.26) in matrix form

$$\Lambda^{\mathrm{T}}\mathbf{g}\Lambda = \mathbf{g}$$

and form the determinants. It follows that

$$\begin{aligned} (\det \Lambda)^2 &= 1, \\ \det \Lambda &= \pm 1. \end{aligned} \tag{2.35}$$

Clearly all transformations which can be reached continuously from the identity transformation, i.e., the unit matrix, must have $\det \Lambda = +1$. However, this condition is not enough. To see this we consider the four-vector

$$x = \begin{pmatrix} 1 \\ 0 \\ 0 \\ 0 \end{pmatrix}, \tag{2.36}$$

$$(x, x) = 1.$$

By means of a Lorentz transformation, x is transformed into the vector x' which in the Minkowski sense has the same length:

$$x' = \Lambda x = \begin{pmatrix} \Lambda^0{}_0 \\ \Lambda^1{}_0 \\ \Lambda^2{}_0 \\ \Lambda^3{}_0 \end{pmatrix}, \tag{2.37}$$

$$(x', x') = (x'^0)^2 - (x'^1)^2 - (x'^2)^2 - (x'^3)^2 = 1.$$

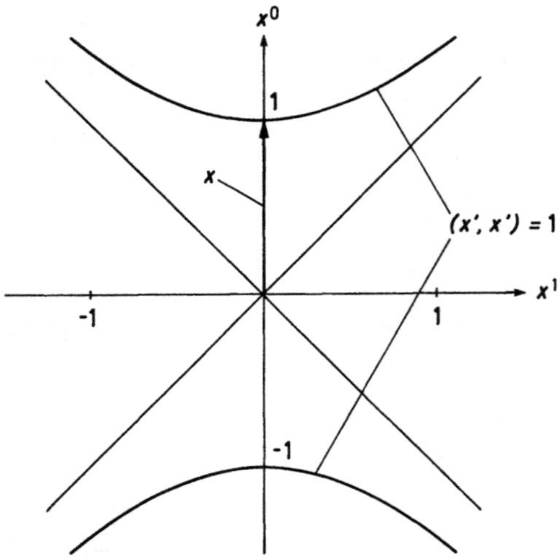

Figure 2.4 Intersection of the hyperboloid $(x', x') = 1$ (2.37) with the x^1–x^0 plane in Minkowski space.

All of the vectors x' lie on a hyperboloid of two sheets in Minkowski space (Fig. 2.4).

One sheet of the hyperboloid lies completely within the forward light cone and the other within the backward light cone with respect to the origin. It is clear that through transformations obtained by çontinuous transitions from the unit transformation one can only reach the upper sheet of the hyerboloid. For the transformations belonging to the identity component it must thus hold that

$$x'^0 = \Lambda^0{}_0 > 0. \tag{2.38}$$

The conditions

$$\begin{aligned} \det \Lambda &= +1, \\ \Lambda^0{}_0 &> 0, \end{aligned} \tag{2.39}$$

are thus necessary in order for a transformation Λ to belong to the proper orthochronous Lorentz group. As can easily be seen, these are also sufficient conditions.

We mention now two Lorentz transformations which do not belong to the identity component. First we consider the parity operation, i.e., the transformation from a spatial Cartesian right-handed system to a left-handed system (Fig. 2.5). The 4×4 matrix Λ_P which describes the parity operation is given by

$$\Lambda_P = \begin{pmatrix} 1 & 0 & 0 & 0 \\ 0 & -1 & 0 & 0 \\ 0 & 0 & -1 & 0 \\ 0 & 0 & 0 & -1 \end{pmatrix}. \tag{2.40}$$

$$\begin{pmatrix} t' \\ \boldsymbol{x}' \end{pmatrix} = \Lambda_P \begin{pmatrix} t \\ \boldsymbol{x} \end{pmatrix} = \begin{pmatrix} t \\ -\boldsymbol{x} \end{pmatrix}.$$

Another interesting transformation that does not belong to the unit com-

Figure 2.5 Transition from a spatially right-handed system to a left-handed system.

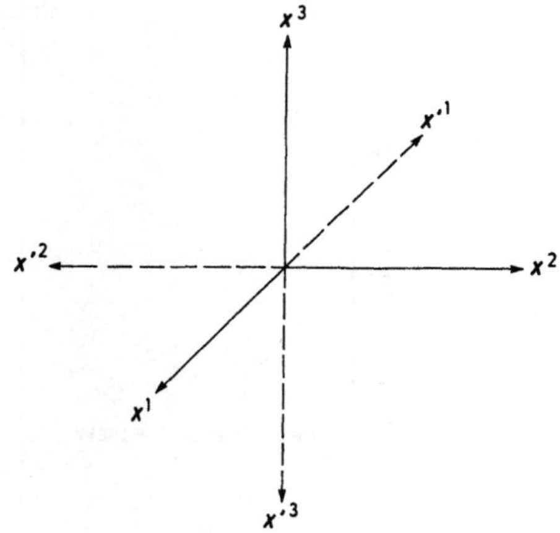

ponent is the reflection of the time axis, i.e., time reversal (Fig. 2.6)

$$\Lambda_T = \begin{pmatrix} -1 & 0 & 0 & 0 \\ 0 & 1 & 0 & 0 \\ 0 & 0 & 1 & 0 \\ 0 & 0 & 0 & 1 \end{pmatrix}. \tag{2.41}$$

$$\begin{pmatrix} t' \\ \boldsymbol{x}' \end{pmatrix} = \Lambda_T \begin{pmatrix} t \\ \boldsymbol{x} \end{pmatrix} = \begin{pmatrix} -t \\ \boldsymbol{x} \end{pmatrix}.$$

The transformations Λ_P and Λ_T both play an important role in particle physics.

At this point we will close our discussion of special relativity. For further information the reader is referred to the textbooks on the subject, e.g., Landau (1975), von Laue (1952), Weinberg (1972b).

According to our present knowledge, special relativistic invariance is valid in particle physics provided we neglect gravitation. The equations of motion of the particles appear the same in all systems that are related to one another by Poincaré transformations as given in (2.28), with the restriction that Λ

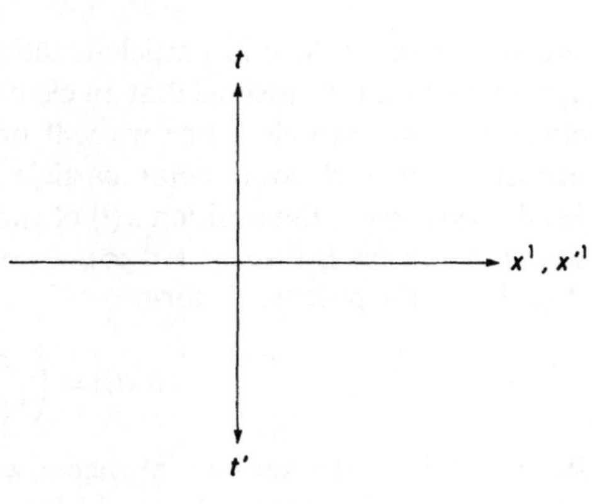

Figure 2.6 The time-reversal transformation of the coordinates.

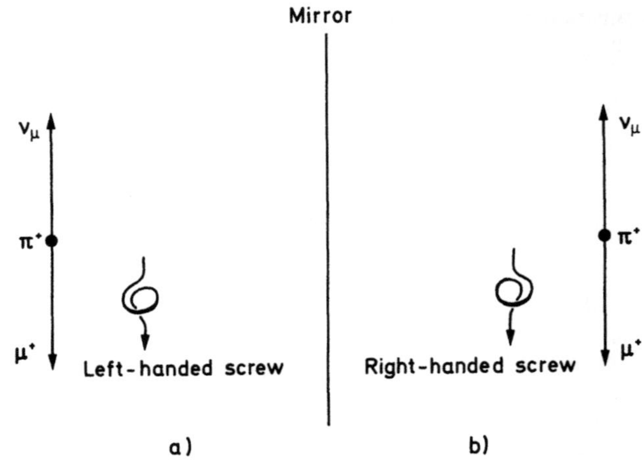

Figure 2.7 The decay $\pi^+ \to \mu^+ + \nu_\mu$ of a pion at rest for which the muon helicity always corresponds to a left-handed screw (a). The mirror image of this reaction (b) is not observed in nature.

belongs to the identity component. For a long time it was believed that the parity and time reversal transformations (2.40), (2.41) also leave the equations of motion unaltered. In fact experiments in particle physics demonstrated that neither parity invariance nor time reversal invariance are valid in nature. For example, it was found that the polarization the positive muon produced in the decay of a π^+-meson at rest

$$\pi^+ \to \mu^+ + \nu_\mu \tag{2.42}$$

always has the sense of a left-handed screw and never that of a right-handed one. If we allow this decay process to take place in front of a mirror, then the process in the mirror is one that never occurs in nature (Fig. 2.7).

2.2 Energy and Momentum of Relativistic Particles

We are interested here in particle reactions. A rough, but very often useful, approximation is to assume that an elementary particle can be regarded as a classical point particle. Thus we will briefly consider the most important properties of a classical point particle with mass $m > 0$. If we mark in Minkowski space, the position $\boldsymbol{x}(t)$ of such a particle and the corresponding time t for all times $(-\infty < t < \infty)$, we obtain the *world line* of the particle (Fig. 2.8) in the parametric form

$$x(t) = \begin{pmatrix} t \\ \boldsymbol{x}(t) \end{pmatrix}. \tag{2.43}$$

Instead of the time in the inertial system, we will introduce the particle's proper time τ as a parameter on the world line; this time is the one indicated by a

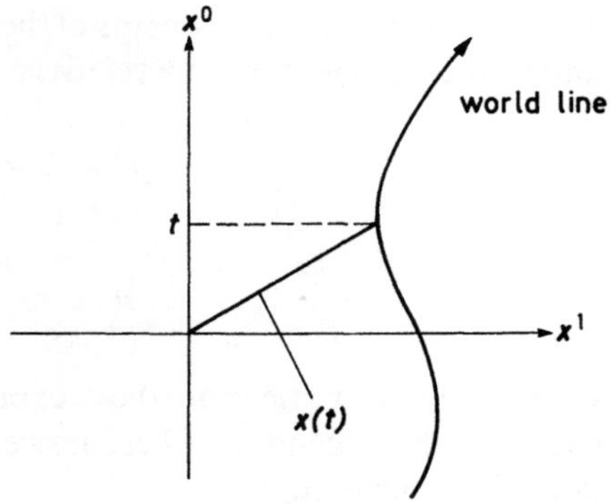

Figure 2.8 The world line (2.43) of a classical point particle in Minkowski space.

clock which moves with the particle. In order to relate the proper time to the normal time t, we consider an inertial frame in which the particle is momentarily at rest, i.e., the instantaneous rest frame of the particle. In this system the differential element of the world line of the particle is given, according to the definition of proper time by

$$\mathrm{d}x' = \begin{pmatrix} \mathrm{d}\tau \\ 0 \\ 0 \\ 0 \end{pmatrix}. \tag{2.44}$$

As a result of the invariance of the relativistic squared interval it then holds that

$$\begin{aligned}(\mathrm{d}x')^2 = \mathrm{d}\tau^2 = (\mathrm{d}x)^2 &= \mathrm{d}t^2 - (\mathrm{d}\boldsymbol{x})^2 \\ &= \mathrm{d}t^2\,(1 - \mathbf{v}^2(t)),\end{aligned} \tag{2.45}$$

where

$$\mathbf{v}(t) = \frac{\mathrm{d}\boldsymbol{x}}{\mathrm{d}t}(t).$$

This enables us to arrive at the desired relationship between τ and t:

$$\tau = \int_{t_0}^{t} \mathrm{d}t'\,\sqrt{1 - \mathbf{v}^2(t')}. \tag{2.46}$$

The time t_0 may be chosen at will since it simply determines the starting point ($\tau = 0$) of the proper time.

The four-velocity u and four-momentum p are defined by

$$u = \frac{\mathrm{d}x}{\mathrm{d}\tau}(\tau) = \begin{pmatrix} \dfrac{\mathrm{d}t}{\mathrm{d}\tau} \\[2mm] \dfrac{\mathrm{d}\boldsymbol{x}}{\mathrm{d}\tau} \end{pmatrix} = \begin{pmatrix} \dfrac{1}{\sqrt{1-\mathbf{v}^2}} \\[2mm] \dfrac{\mathbf{v}}{\sqrt{1-\mathbf{v}^2}} \end{pmatrix},$$

$$p = m\frac{\mathrm{d}x}{\mathrm{d}\tau}. \tag{2.47}$$

The time and space components of the four-momentum p are called the energy and (three-)momentum of a relativistic particle

$$p^0 = E = \frac{m}{\sqrt{1-\mathbf{v}^2}},\qquad \mathbf{p} = \frac{m\mathbf{v}}{\sqrt{1-\mathbf{v}^2}}. \tag{2.48}$$

In the nonrelativistic limit these expressions become the usual formulas for energy and momentum, where however, the rest mass energy must also be taken into account:

$$E \cong m + \tfrac{1}{2}m\mathbf{v}^2,\qquad \mathbf{p} \cong m\mathbf{v} \quad \text{for} \quad |\mathbf{v}| \ll 1. \tag{2.49}$$

Empirically, we find that the conservation of four-momentum is valid in all reactions. In other words, the sum of the four-momenta of all particles that take part in the reaction is the same after the reaction as before:

$$\mathrm{a}_1 + \mathrm{a}_2 + \cdots + \mathrm{a}_n \to \mathrm{b}_1 + \mathrm{b}_2 + \cdots + \mathrm{b}_n,\qquad p^\mu_{\mathrm{a}_1} + p^\mu_{\mathrm{a}_2} + \cdots + p^\mu_{\mathrm{a}_n} = p^\mu_{\mathrm{b}_1} + p^\mu_{\mathrm{b}_2} + \cdots + p^\mu_{\mathrm{b}_n}. \tag{2.50}$$

This also includes, of course, the case where some particles are nonrelativistic and others highly relativistic. The validity of the conservation law for energy and momentum can thus only be maintained if we define precisely the expressions in (2.48) as the energy and momentum of a relativistic particle. Out of historical interest it should be mentioned that equation (2.48) is, in essence, due to Einstein (Einstein 1905b). He considered the energy of a system of light waves interacting with matter within the framework of Maxwellian electrodynamics, and applied to this the basic postulate of special relativity which we discussed in Sect. 2.1.

We will now study the behavior of the four-momentum under the action of Poincaré transformations (2.28). Let us go over to a primed reference frame

$$x'^\mu = \Lambda^\mu{}_\nu x^\nu + a^\mu. \tag{2.51}$$

Since $\mathrm{d}\tau$, according to (2.45), is an invariant quantity, it follows that the four-momentum transforms like a four-vector:

$$p'^\mu = m\frac{\mathrm{d}x'^\mu}{\mathrm{d}\tau} = m\Lambda^\mu{}_\nu\frac{\mathrm{d}x^\nu}{\mathrm{d}\tau} = \Lambda^\mu{}_\nu p^\nu. \tag{2.52}$$

Furthermore, we find that the relativistic square of the four-momentum is equal to the square of the mass

$$p'^\mu p'_\mu = p^\mu p_\mu = m^2\left(\frac{\mathrm{d}x}{\mathrm{d}\tau}\right)^2 = m^2. \tag{2.53}$$

Thus when we take into account the fact that the energy of a particle is always

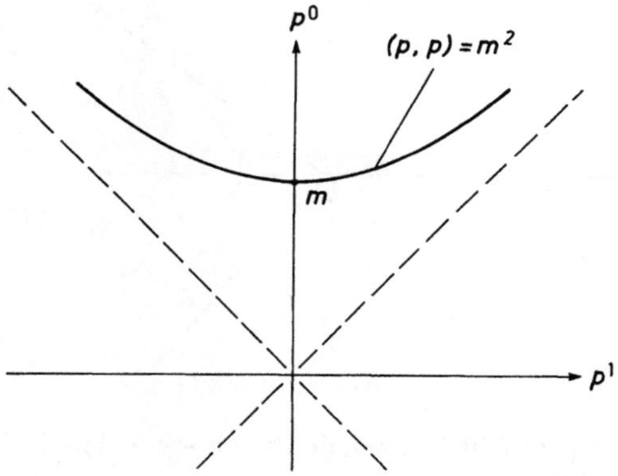

Figure 2.9 Two-dimensional section of momentum space with the mass shell of a particle of mass m (2.54).

greater than zero we find

$$\begin{aligned} E^2 - \boldsymbol{p}^2 &= m^2, \\ E &> 0. \end{aligned} \tag{2.54}$$

We proceed now to represent energy and momentum in a four-dimensional space, the so-called momentum space. The possible values for the four-momentum of a particle with mass m then lie, according to (2.54), on one sheet of a hyperboloid known as the *mass shell* (Fig. 2.9).

For massless particles such as the photon γ, the model of classical point particles clearly fails. In this case Einstein's hypothesis of light quanta (Einstein 1905a) helps us to make further progress. According to this hypothesis a plane wave

$$e^{i(\boldsymbol{kx} - \omega t)} \tag{2.55}$$

with wave vector $\boldsymbol{k}$ and angular frequency ω describes a photon with momentum $\hbar\boldsymbol{k}$ and energy $\hbar\omega$. In our system of unit ($\hbar = 1$), the four-momentum of a photon is thus equal to the four-wave vector

$$k = (k^\mu) = \begin{pmatrix} \omega \\ \boldsymbol{k} \end{pmatrix}. \tag{2.56}$$

An analagous relationship holds for the massless neutrinos. In momentum space the four-momenta of massless particles form the forward light cone of the origin:

$$\begin{aligned} k^2 = \omega^2 - \boldsymbol{k}^2 &= 0, \\ \omega &> 0. \end{aligned} \tag{2.57}$$

2.3 The Relativistic Kinematics of a Collision Between Two Particles

We now wish to discuss the kinematics of a two-body reaction

$$\mathrm{a} + \mathrm{b} \rightarrow \mathrm{c} + \mathrm{d}. \tag{2.58}$$

Consider for instance the case in which particle b is initially at rest in the

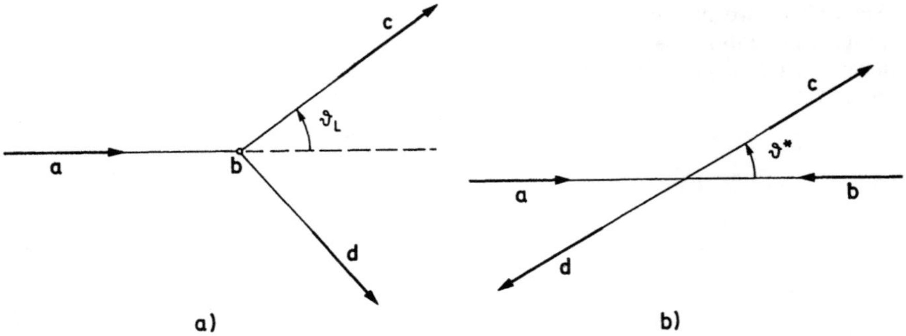

Figure 2.10 The reaction a + b → c + d in the laboratory system (a) and in the center of mass system (b).

laboratory frame (Fig. 2.10a)). The four-momenta of the particles in the laboratory system will be denoted by p_a, p_b, p_c, and p_d:

$$p_a = \begin{pmatrix} E_L \\ \boldsymbol{p}_L \end{pmatrix}, \quad p_b = \begin{pmatrix} m_b \\ \mathbf{0} \end{pmatrix}, \quad p_c = \begin{pmatrix} E_c \\ \boldsymbol{p}_c \end{pmatrix}, \quad p_d = \begin{pmatrix} E_d \\ \boldsymbol{p}_d \end{pmatrix}. \tag{2.59}$$

Here E_L and $\boldsymbol{p}_L$ are the energy and momentum of the projectile, a, in the laboratory frame. Let the scattering angle in the laboratory frame be the angle between the momenta of particles a and c:

$$\cos \vartheta_L = \frac{\boldsymbol{p}_a \cdot \boldsymbol{p}_c}{|\boldsymbol{p}_a||\boldsymbol{p}_c|}. \tag{2.60}$$

Then by the conservation of four-momentum we have

$$p_a + p_b = p_c + p_d. \tag{2.61}$$

It is often useful to consider the same reaction in different reference frames. Today this can also be realized experimentally. One can, for example, perform an antiproton–proton scattering experiment with a fixed target machine, or in a storage ring such as Sp$\bar{\text{p}}$S at CERN, where two particle beams are directed towards each other.

We now consider the reaction (2.58) in the center of mass frame (Fig. 2.10b) where the four-momenta are denoted by p_a^*, p_b^*, p_c^* and p_d^*. As a consequence of momentum conservation and according to the definition of the center of mass frame we have

$$\boldsymbol{p}_a^* + \boldsymbol{p}_b^* = \boldsymbol{p}_c^* + \boldsymbol{p}_d^* = 0. \tag{2.62}$$

The center of mass energy is denoted by $\sqrt{s}$ and by conservation of energy we have

$$E_a^* + E_b^* = E_c^* + E_d^* = \sqrt{s}. \tag{2.63}$$

From (2.62) we see that in the center of mass frame the two particles emerge from the scattering in opposite directions with momenta of equal magnitude

$$|\boldsymbol{p}_c^*| = |\boldsymbol{p}_d^*|. \tag{2.64}$$

The scattering angle in the center of mass system is

$$\cos \vartheta^* = \frac{\boldsymbol{p}_a^* \cdot \boldsymbol{p}_c^*}{|\boldsymbol{p}_a^*||\boldsymbol{p}_c^*|}. \tag{2.65}$$

We now wish to find the relationship between the variables that describe the scattering in the center of mass frame and those in the laboratory frame. The simplest way to do this is to use relativistic invariants, as was first done by Møller (1932). The notation commonly used today is that of Mandelstam (1958).

As one of these invariant variables we choose s, the square of the center of mass energy, which according to (2.62), (2.63) can also be written as the square of the four-vector $p_a^* + p_b^*$. Since this quadratic form is relativistically invariant the same result is obtained in the laboratory frame

$$s = (p_a^* + p_b^*)^2 = (p_a + p_b)^2. \tag{2.66}$$

The conservation of four-momentum further implies

$$s = (p_c^* + p_d^*)^2 = (p_c + p_d)^2. \tag{2.67}$$

Equation (2.66) then gives us the relationship between the center of mass energy and the energy E_L of the projectile in the laboratory frame (2.59):

$$\begin{aligned} s &= (p_a + p_b)^2 = p_a^2 + 2(p_a, p_b) + p_b^2, \\ s &= m_a^2 + 2E_L m_b + m_b^2, \\ E_L &= \frac{s - m_a^2 - m_b^2}{2m_b}. \end{aligned} \tag{2.68}$$

We now consider the example of protons and antiprotons in a storage ring. The particles are assumed to fly collinearly towards one another, each having an energy of 270 GeV (particles a and b in Fig. 2.10(b)). The center of mass energy is then

$$\sqrt{s} = 540 \text{ GeV}. \tag{2.69}$$

What energy would an antiproton need when scattering from a proton at rest in order to produce the same center of mass energy in the collision? According to (2.68), and using the fact that proton and antiproton both have the same mass of about 1 GeV, the energy required is given by

$$E_L = \frac{(540)^2 - 2}{2} \text{GeV} \cong 1.5 \cdot 10^5 \text{ GeV}. \tag{2.70}$$

Even in the most modern accelerators it is quite impossible to produce antiprotons of this energy. However, using the proton antiproton storage ring Sp$\bar{\text{p}}$S at CERN in Geneva the equivalent center of mass energy can be reached.

It has thereby been possible to produce the heavy W- and Z-bosons whose rest masses are approximately 80 and 90 GeV. The demand for high center of mass energies was, and remains, one of the main incentives for the construction of such storage rings.

We next calculate the laboratory momentum of particle a. Using (2.68) we find

$$
\begin{aligned}
|\boldsymbol{p}_\mathrm{L}| &= \sqrt{E_\mathrm{L}^2 - m_\mathrm{a}^2} \\
&= \frac{1}{2m_\mathrm{b}}\{(s - m_\mathrm{a}^2 - m_\mathrm{b}^2)^2 - 4m_\mathrm{a}^2 m_\mathrm{b}^2\}^{1/2} \\
&= \frac{1}{2m_\mathrm{b}}\{s^2 + m_\mathrm{a}^4 + m_\mathrm{b}^4 - 2sm_\mathrm{a}^2 - 2sm_\mathrm{b}^2 - 2m_\mathrm{a}^2 m_\mathrm{b}^2\}^{1/2}.
\end{aligned} \tag{2.71}
$$

The particular combination of variables encountered here frequently shows up in kinematic calculations. It is thus common to introduce a special notation for this, and define the function

$$
\begin{aligned}
w(x, y, z) &= (x^2 + y^2 + z^2 - 2xy - 2xz - 2yz)^{1/2} \\
&= [x - (\sqrt{y} + \sqrt{z})^2]^{1/2}[x - (\sqrt{y} - \sqrt{z})^2]^{1/2}.
\end{aligned} \tag{2.72}
$$

The alternative representation of w given here is also very useful. It is obvious from the definition that w is totally symmetric in its three variables.

The energy and momenta of the particles in the center of mass frame can also be expressed in terms of the variable s. Since the four-vector $p_\mathrm{a}^* + p_\mathrm{b}^*$ has only an energy component, it follows from (2.63) that

$$
\begin{aligned}
E_\mathrm{a}^* &= \frac{(p_\mathrm{a}^*, p_\mathrm{a}^* + p_\mathrm{b}^*)}{\sqrt{s}} = \frac{(p_\mathrm{a}, p_\mathrm{a} + p_\mathrm{b})}{\sqrt{s}} \\
&= \frac{1}{\sqrt{s}}(p_\mathrm{a}^2 + (p_\mathrm{a}, p_\mathrm{b})) \\
&= \frac{1}{\sqrt{s}}(p_\mathrm{a}^2 + \tfrac{1}{2}(p_\mathrm{a} + p_\mathrm{b})^2 - \tfrac{1}{2}p_\mathrm{a}^2 - \tfrac{1}{2}p_\mathrm{b}^2), \\
E_\mathrm{a}^* &= \frac{s + m_\mathrm{a}^2 - m_\mathrm{b}^2}{2\sqrt{s}}.
\end{aligned} \tag{2.73}
$$

Here we have used the invariance of the relativistic scalar product. Analogously we find:

$$
\begin{aligned}
E_\mathrm{b}^* &= \frac{s + m_\mathrm{b}^2 - m_\mathrm{a}^2}{2\sqrt{s}}, \\
E_\mathrm{c}^* &= \frac{s + m_\mathrm{c}^2 - m_\mathrm{d}^2}{2\sqrt{s}}, \\
E_\mathrm{d}^* &= \frac{s + m_\mathrm{d}^2 - m_\mathrm{c}^2}{2\sqrt{s}}.
\end{aligned} \tag{2.74}
$$

For the magnitude of the particle momenta in the center of mass frame we then obtain

$$|\boldsymbol{p}_{\mathrm{b}}^*| = |\boldsymbol{p}_{\mathrm{a}}^*| = \sqrt{(E_{\mathrm{a}}^*)^2 - m_{\mathrm{a}}^2}$$

$$= \left(\frac{(s + m_{\mathrm{a}}^2 - m_{\mathrm{b}}^2)^2}{4s} - m_{\mathrm{a}}^2\right)^{1/2}$$

$$= \frac{1}{2\sqrt{s}} w(s, m_{\mathrm{a}}^2, m_{\mathrm{b}}^2), \tag{2.75}$$

$$|\boldsymbol{p}_{\mathrm{c}}^*| = |\boldsymbol{p}_{\mathrm{d}}^*| = \frac{1}{2\sqrt{s}} w(s, m_{\mathrm{c}}^2, m_{\mathrm{d}}^2). \tag{2.76}$$

We now introduce two further invariant variables

$$\begin{aligned} t &= (p_{\mathrm{a}} - p_{\mathrm{c}})^2 = (p_{\mathrm{b}} - p_{\mathrm{d}})^2, \\ u &= (p_{\mathrm{a}} - p_{\mathrm{d}})^2 = (p_{\mathrm{b}} - p_{\mathrm{c}})^2. \end{aligned} \tag{2.77}$$

The three invariants s, t, and u are however not independent of one another. In a two-particle reaction of the type (2.58) there are (for particles of zero spin at least) only two independent kinematic variables. To show this, let us consider our reaction in the center of mass frame. In this system all energies and the magnitudes of all momenta are determined by the center of mass energy $\sqrt{s}$ (2.73)–(2.76). The only remaining independent variable is the scattering angle in the center of mass system ϑ^* (2.65). The invariant variable t is proportional to $\cos \vartheta^*$. More precisely, we have

$$\begin{aligned} t &= (p_{\mathrm{a}}^* - p_{\mathrm{c}}^*)^2 \\ &= m_{\mathrm{a}}^2 + m_{\mathrm{c}}^2 - 2(p_{\mathrm{a}}^*, p_{\mathrm{c}}^*) \\ &= m_{\mathrm{a}}^2 + m_{\mathrm{c}}^2 - 2(E_{\mathrm{a}}^* E_{\mathrm{c}}^* - |\boldsymbol{p}_{\mathrm{a}}^*||\boldsymbol{p}_{\mathrm{c}}^*| \cos \vartheta^*) \\ &= m_{\mathrm{a}}^2 + m_{\mathrm{c}}^2 \\ &\quad - \frac{1}{2s}(s + m_{\mathrm{a}}^2 - m_{\mathrm{b}}^2)(s + m_{\mathrm{c}}^2 - m_{\mathrm{d}}^2) \\ &\quad + \frac{1}{2s} w(s, m_{\mathrm{a}}^2, m_{\mathrm{b}}^2) w(s, m_{\mathrm{c}}^2, m_{\mathrm{d}}^2) \cos \vartheta^*. \end{aligned} \tag{2.78}$$

The variable u differs from t only in that the roles of the particles c and d are exchanged. Since in the center of mass system these particles fly apart in opposite directions, the corresponding scattering angle between particles a and d is equal to $\pi - \vartheta^*$. Therefore we have

$$\begin{aligned} u &= (p_{\mathrm{a}}^* - p_{\mathrm{d}}^*)^2 \\ &= m_{\mathrm{a}}^2 + m_{\mathrm{d}}^2 \\ &\quad - \frac{1}{2s}(s + m_{\mathrm{a}}^2 - m_{\mathrm{b}}^2)(s + m_{\mathrm{d}}^2 - m_{\mathrm{c}}^2) \\ &\quad - \frac{1}{2s} w(s, m_{\mathrm{a}}^2, m_{\mathrm{b}}^2) w(s, m_{\mathrm{c}}^2, m_{\mathrm{d}}^2) \cos \vartheta^*. \end{aligned} \tag{2.79}$$

Addition of equations (2.78) and (2.79) yields the important relationship

$$s + t + u = m_a^2 + m_b^2 + m_c^2 + m_d^2. \tag{2.80}$$

We now investigate more closely the physically allowed range of these kinematic variables. It is clear that the reaction (2.58) can only take place if the center of mass energy is larger than the sum of the masses of the particles in the initial (respectively the final) state:

$$\begin{aligned} \sqrt{s} &\geq m_a + m_b, \\ \sqrt{s} &\geq m_c + m_d. \end{aligned} \tag{2.81}$$

On the other hand, the scattering angle in the center of mass frame can take any value between 0 and π

$$0 \leq \vartheta^* \leq \pi, \qquad -1 \leq \cos\vartheta^* \leq 1. \tag{2.82}$$

From these conditions we can derive the kinematic limits for the variables t and u. Since the resulting formulas are rather lengthy, we shall discuss only the high energy limit

$$s \gg m_a^2, m_b^2, m_c^2, m_d^2. \tag{2.83}$$

For a given value of s we then obtain from (2.78), (2.79) the following allowed ranges for t and u:

$$t \cong -\frac{s}{2}(1 - \cos\vartheta^*), \qquad 0 \gtrsim t \gtrsim -s, \tag{2.84}$$

$$u \cong -\frac{s}{2}(1 + \cos\vartheta^*), \qquad 0 \gtrsim u \gtrsim -s. \tag{2.85}$$

A brief example will help to familiarize us with the variables. We consider elastic proton–proton scattering

$$\mathrm{p} + \mathrm{p} \to \mathrm{p} + \mathrm{p}, \tag{2.86}$$

which has been studied in detail at the ISR storage rings at a center of mass energy of $\sqrt{s} \cong 53$ GeV. As can be seen in Fig. 2.11, the differential cross-section has a pronounced minimum at

$$t \simeq 1.34\ \mathrm{GeV}^2. \tag{2.87}$$

The questions is: To what scattering angle in the center of mass system does this value of t correspond? From (2.84) we obtain

$$1 - \cos\vartheta^* \cong \frac{\vartheta^{*2}}{2} \cong -\frac{2t}{s},$$

and thus

$$\vartheta^* \cong 2\sqrt{\frac{-t}{s}} \cong 44\ \mathrm{mrad} \mathrel{\hat=} 2.5^\circ. \tag{2.88}$$

We move on now to a discussion of reactions related to (2.58) by crossing. By this we mean processes in which one or more particles are replaced by their corresponding antiparticles, which then appear on the other side of the equation describing the reaction. By crossing particles over from one side to the

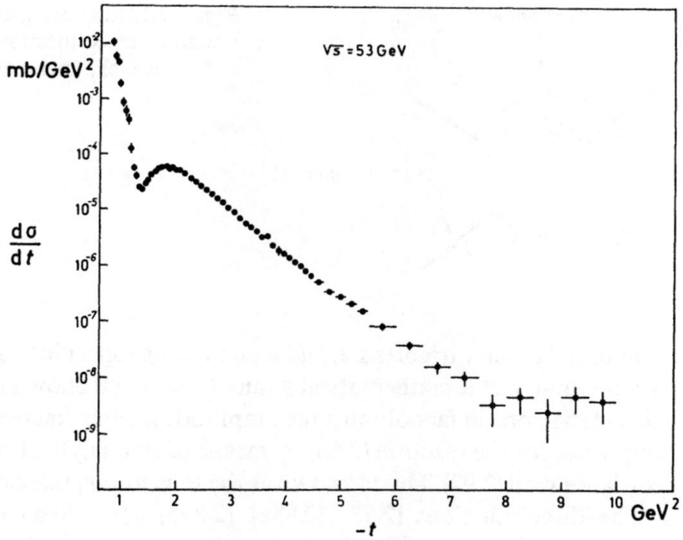

Figure 2.11 The differential cross-section for proton–proton elastic scattering at a center of mass energy of 53 GeV (after de Kerret 1977).

other of the reaction (2.58) we may, for example, obtain the following reactions:

$$a + \bar{c} \rightarrow \bar{b} + d, \tag{2.89a}$$

$$a + \bar{d} \rightarrow \bar{b} + c, \tag{2.89b}$$

$$a \rightarrow \bar{b} + c + d. \tag{2.89c}$$

The reaction (2.89c) represents the decay of particle a into three particles. In order that this reaction can take place, the masses must of course satisfy the condition

$$m_a \geq m_b + m_c + m_d. \tag{2.90}$$

Let us consider the reaction (2.89a) in more detail. We denote the invariant variables for this case by s', t', and u':

$$s' = (p_a + p_{\bar{c}})^2,$$
$$t' = (p_a - p_{\bar{b}})^2, \tag{2.91}$$
$$u' = (p_a - p_d)^2.$$

We can obtain these variables as a permutation of the variables s, t, and u (2.66), (2.77) if we make the following substitutions

$$p_c \rightarrow -p_{\bar{c}},$$
$$p_b \rightarrow -p_{\bar{b}},$$
$$t \rightarrow s', \tag{2.92}$$
$$s \rightarrow t',$$
$$u \rightarrow u'.$$

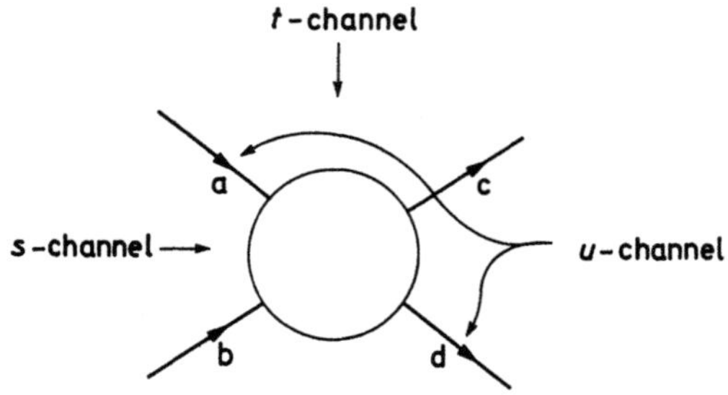

Figure 2.12 Scattering diagram to visualize the connection between s-, t-, and u-channel reactions.

The "continuation" of the variables s, t, and u into a range specific to a crossed reaction is more than just a mathematical game. As was first shown by Jauch and Rohrlich (1955), one in fact obtains the amplitude for the reaction (2.89a) from the amplitude for the reaction (2.58) by means of an analytical continuation in accordance with (2.92). This is known as the *substitution rule* or *crossing symmetry*. The three reactions (2.58), (2.89a), (2.89b) are called s-, t-, and u-channel reactions, due to the fact that s, t, and u, respectively, take on the role of the square of the center of mass energy. It is useful to visualize the connection between crossed reactions by means of a scattering diagram (Fig. 2.12) in which lines for outgoing particles can also be interpreted as lines for incoming antiparticles.

To end this chapter, we introduce the reader to a commonly used graphical representation of the variables s, t, and u. This is useful, for example, when one wishes to study the distribution of events in a two-dimensional scatter plot. For this purpose we could simply choose s and t as orthogonal coordinates. It is more useful, however, to take the coordinate axes $s = 0$ and $t = 0$ at an angle of 60° to one another (Fig. 2.13). The values of s and t are given by the perpendicular distances to the respective coordinate axes. Positive values appear on the side indicated by arrows. The three straight lines

$$
\begin{aligned}
s &= 0,\\
t &= 0,\\
u &= m_a^2 + m_b^2 + m_c^2 + m_d^2 - s - t = 0,
\end{aligned}
\tag{2.93}
$$

can then easily be shown to enclose an equilateral triangle of height

$$h = m_a^2 + m_b^2 + m_c^2 + m_d^2. \tag{2.94}$$

Every point P in the plane corresponds to a possible set of values of s, t, and u which, together with their signs, are given by the distances and directions to the coordinate axes $s = 0$, $t = 0$, and $u = 0$. This can most easily be seen by comparing the areas A of the triangles with vertices Q_1, Q_2, Q_3; Q_1, Q_3, P, etc., as shown in Fig. 2.13:

$$A(Q_1Q_3P) - A(Q_1Q_2P) - A(Q_2Q_3P) = A(Q_1Q_2Q_3). \tag{2.95}$$

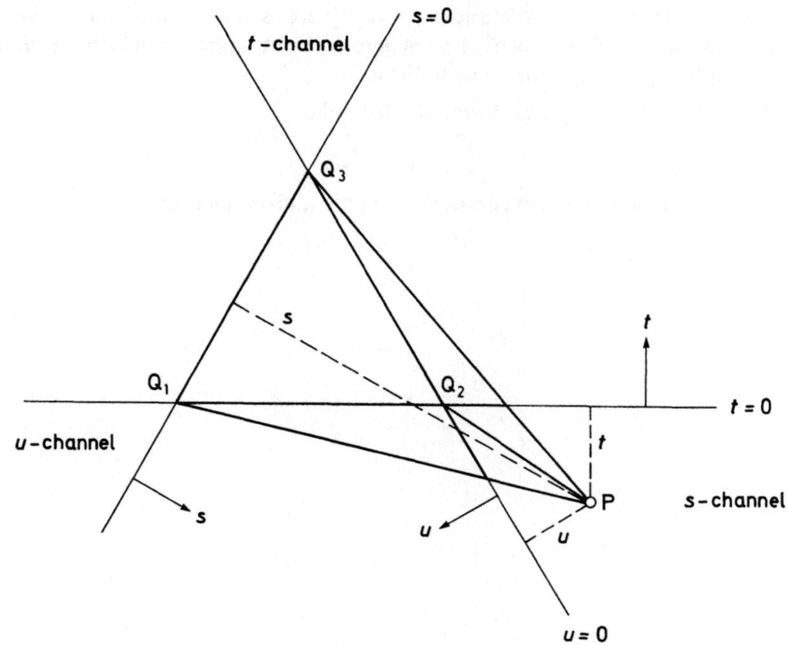

Figure 2.13 The s-, t-, u-plane. Positive values of s, t, and u are to be drawn in the direction of the corresponding arrows. The point marked P represents an s-channel reaction ($s > 0, t < 0, u < 0$).

The heights of the triangles Q_1Q_3P, Q_1Q_2P, and $Q_1Q_2Q_3$ are, by construction, s, $-t$, and h, respectively. From elementary rules of geometry it now follows from (2.95) that the height $-u$ of the triangle Q_2Q_3P fulfils the desired condition

$$s + t + u = h = m_a^2 + m_b^2 + m_c^2 + m_d^2.$$

With this we close our discussion of kinematic variables.

Exercises

2.1 Show that every homogeneous Lorentz transformation Λ that satisfies (2.39) can be reached continuously from the identity transformation.

2.2 Consider proton–proton scattering with a center of mass energy of $\sqrt{s} = 53$ GeV (2.86). Calculate the equivalent energy in the laboratory system, E_L, and also the values of u and ϑ_L corresponding to the minimum in the elastic cross-section at $t = -1.34\ \text{GeV}^2$ (Fig. 2.11).

2.3 What is the allowed phase space region in the s-, t-, u-plane for the reaction

$$e^-(p_1) + e^+(p_2) \to Z(p_3) + \gamma(p_4)?$$

Use $m_z \cong 90$ GeV, $m_e \cong 0$, and $m_\gamma = 0$. Where is the physical region for the crossed reaction $Z \to e^- + e^+ + \gamma$? Where do events appear in which the photon γ is emitted parallel to the e^- or to the e^+?

2.4 Derive from (2.52) the transformation formula

$$p_\nu = \Lambda^\mu{}_\nu p'_\mu$$

for the covariant or lower components of the four-momentum.

CHAPTER 3

Particles and Fields

In the previous chapter we discussed classical relativistic point particles. Physical particles must of course obey the rules of quantum mechanics. The appropriate formalism for describing the quantum behavior of elementary particles is the *quantum field theory* founded by Dirac, Jordan, Pauli, and Heisenberg (Dirac 1927; Jordan 1928; Heisenberg 1929, 1930). An important feature of this theory is that it unites wave- and particle-like aspects. For example, quantum field theory can describe both the propagation of a light wave and the photoelectric effect.

3.1 The Schrödinger, Dirac, and Heisenberg Pictures of Quantum Mechanics

To begin with, we shall recap on a few of the important concepts of ordinary quantum mechanics. To the states of a system we assign vectors in Hilbert space. To observable physical quantities such as the position and momentum of a particle, there correspond operators in this space. We shall make use of the Dirac notation (see Dirac 1958) in which state vectors are denoted by the symbol $|\ \rangle$. Within these brackets we can write the symbols characteristic of the state to be described. The scalar product of two state vectors $|a\rangle$ and $|b\rangle$ is denoted by $\langle a|b\rangle$. In ordinary quantum mechanics it is usual to adopt the so-called Schrödinger picture in which operators are time independent. Here, the state vector is time dependent and it satisfies the

Schrödinger equation:

$$\mathrm{i}\frac{\partial}{\partial t}|t\rangle = \mathbf{H}|t\rangle, \tag{3.1}$$

where $\mathbf{H}$ is the Hamilton operator and $|t\rangle$ is the state vector at time t.

As an example, let us consider an ordinary Schrödinger particle whose state vector $|t\rangle$ is given by the usual wave function $\psi(\boldsymbol{x}, t)$. The scalar product of this state vector with itself is the normalization integral

$$\langle t|t\rangle = \int \mathrm{d}^3x\, \psi^*(\boldsymbol{x}, t)\psi(\boldsymbol{x}, t) = 1.$$

The expectation value of an operator $\mathbf{A}$, for example, the position operator $\vec{\mathbf{X}}$, in the state $|t\rangle$, is

$$\begin{aligned} a(t) &= \langle t|\mathbf{A}|t\rangle, \\ \boldsymbol{x}(t) &= \langle t|\vec{\mathbf{X}}|t\rangle = \int \mathrm{d}^3y\, \psi^*(\boldsymbol{y}, t)\boldsymbol{y}\psi(\boldsymbol{y}, t). \end{aligned} \tag{3.2}$$

Mention should also be made of the case where an operator depends explicitly on position. An example is the operator $\rho(\boldsymbol{x})$ for the charge density at a point $\boldsymbol{x}$ for a particle with charge e. It is given by the operator valued δ-function:

$$\rho(\boldsymbol{x}) = e\delta^3(\vec{\mathbf{X}} - \boldsymbol{x}).$$

The matrix element (expectation value) for this rather unusual operator is given as follows and results in the normal expression for charge density:

$$\begin{aligned} \langle t|\rho(\boldsymbol{x})|t\rangle &= e\int \mathrm{d}^3y\, \psi^*(\boldsymbol{y}, t)\delta^3(\boldsymbol{y} - \boldsymbol{x})\psi(\boldsymbol{y}, t) \\ &= e|\psi(\boldsymbol{x}, t)|^2. \end{aligned} \tag{3.3}$$

This operator $\rho(\boldsymbol{x})$ serves as a prototype of a field operator, i.e., it is an operator whose matrix elements are classical functions of position $\boldsymbol{x}$. We now derive a further relationship which will be generalized later. With the ordinary momentum operator

$$\vec{\mathbf{P}}\psi(\boldsymbol{y}, t) = \frac{1}{i}\nabla_y\psi(\boldsymbol{y}, t) \tag{3.4}$$

we find (Exercise 3.1):

$$\langle t|[\vec{\mathbf{P}}, \rho(\boldsymbol{x})]|t\rangle = \langle t| -\frac{1}{i}\nabla_x\rho(\boldsymbol{x})|t\rangle. \tag{3.5}$$

An analogous relationship holds for all position-dependent operators that are functions only of $(\vec{\mathbf{X}} - \boldsymbol{x})$. For an operator $\mathbf{A}$ in this category

$$\mathbf{A}(\boldsymbol{x}) = \mathrm{F}(\vec{\mathbf{X}} - \boldsymbol{x}),$$

where F is an arbitrary function. We then have

$$[\vec{\mathbf{P}}, \mathbf{A}(x)] = -\frac{1}{i}\nabla_x \mathbf{A}(x). \tag{3.6}$$

From now onwards it is very important that we put behind us the notion that state vectors are always described by wave functions $\psi(x, t)$. In fact, the state vectors in particle physics are elements of much more general Hilbert spaces as will become apparent. The Dirac notation will be very convenient as it always allows us to write the state vector at time t by the same symbol, $|t\rangle$, independently of the special representation which we choose for the Hilbert space.

We now carry out a simple reformulation in which we transfer part or all of the time dependence of the state vectors to the operators. Let $\mathbf{H}_0$ be an arbitrary Hermitian operator. In practice $\mathbf{H}_0$ will often be the Hamilton operator for free particles. We define states $|\tilde{t}\rangle$ and operators $\tilde{\mathbf{A}}(t)$ as follows:

$$\begin{aligned} |\tilde{t}\rangle &= e^{i\mathbf{H}_0 t}\,|t\rangle, \\ \tilde{\mathbf{A}}(t) &= e^{i\mathbf{H}_0 t}\,\mathbf{A}\,e^{-i\mathbf{H}_0 t}. \end{aligned} \tag{3.7}$$

Thus for every instant in time we carry out a different unitary transformation in the space of the state vectors. The following relationships then hold:

$$\begin{aligned} \frac{\partial}{\partial t}\tilde{\mathbf{A}}(t) &= i[\mathbf{H}_0, \tilde{\mathbf{A}}(t)], \\ i\frac{\partial}{\partial t}|\tilde{t}\rangle &= \mathbf{H}'(t)|\tilde{t}\rangle, \end{aligned} \tag{3.8}$$

where

$$\mathbf{H}'(t) = e^{i\mathbf{H}_0 t}\,(\mathbf{H} - \mathbf{H}_0)\,e^{-i\mathbf{H}_0 t}. \tag{3.9}$$

The measured values, i.e., the expectation values, are not altered by this procedure. Indeed it follows from (3.7) that

$$\langle\tilde{t}|\tilde{\mathbf{A}}(t)|\tilde{t}\rangle = \langle t|\mathbf{A}|t\rangle. \tag{3.10}$$

This general reformulation of the Schrödinger equation was first performed by Dirac (see Dirac 1958) and is thus known as the Dirac picture. At first it may not seem particularly useful to have replaced a single differential equation for the state vector (3.1) by differential equations for both operators and state vectors. However, when we come to discuss the Feynman rules it will become clear how useful this formulation is.

A special case of some importance arises when we choose $\mathbf{H}_0$ to be the Hamilton operator $\mathbf{H}$ of the system. This leads to the Heisenberg picture in which state vectors are time independent and the time dependence is completely shifted to the operators. Here we have

$$\frac{\partial}{\partial t}\mathbf{A}(t) = i[\mathbf{H}, \mathbf{A}(t)]. \tag{3.11}$$

This Heisenberg equation provides the starting point for the Heisenberg formulation of quantum mechanics.

The Heisenberg picture has the advantage that it makes the correspondence to classical physics apparent. The quantum mechanical observables, like the classical observables, obey certain equations of motion. The only difference is that quantum mechanical observables are operators whereas classical observables are merely numbers. In the following we will use the Heisenberg picture. We again consider the example of a Schrödinger particle. Let $\mathsf{A}(\boldsymbol{x}, t)$ be an observable that depends on position and time. By transforming (3.6) into the Heisenberg picture and using (3.11) we find

$$\begin{aligned} \frac{\partial}{\partial t}\mathsf{A}(\boldsymbol{x}, t) &= \mathrm{i}[\mathsf{H}, \mathsf{A}(\boldsymbol{x}, t)], \\ \frac{\partial}{\partial x^j}\mathsf{A}(\boldsymbol{x}, t) &= -\mathrm{i}[\mathsf{P}^j(t), \mathsf{A}(\boldsymbol{x}, t)], \end{aligned} \qquad (j = 1, 2, 3). \tag{3.12}$$

Here the $\mathsf{P}^j(t)$ are the components of the momentum operator in the Heisenberg picture. It is this generalized Heisenberg equation which we shall take over to the quantum field theory.

3.2 Free Fields and the Fock Space

We now wish to make the transition to quantum field theory (QFT). The general framework of quantum theory of course remains unchanged. Observables will be described by operators and states by vectors in Hilbert space. Let us consider the electromagnetic field for which QFT was originally developed. The classically observable quantities are then the electric field strength $\boldsymbol{E}(\boldsymbol{x}, t)$ and the magnetic field strength $\boldsymbol{B}(\boldsymbol{x}, t)$ for which the Maxwell equations are valid. In quantum theory we must expect that these field strengths correspond to operators. Thus we arrive at the concept of the quantum field. The question of whether and how the electromagnetic field can be measured in quantum theory was studied in depth by Bohr and Rosenfeld (1933). The result of these investigations was that the field strengths $\boldsymbol{E}$ and $\boldsymbol{B}$ can indeed be measured individually, but that the corresponding operators do not commute.

The gedanken-experiment which was discussed by Bohr and Rosenfeld runs as follows: Assume that we wish to measure the electric field component E^1 at position $\boldsymbol{x}$ and time t. We can achieve this by introducing a charged test body at the point $\boldsymbol{x}$ and measuring the force exerted on it by the electric field (Fig. 3.1). The force exerted on the probe is determined from the change in the momentum of the latter. An exact measurement of the field component E^1 is thus equivalent to an exact measurement of the momentum component p^1 of the test body. As a result of this, the position of the test body will, according to Heisenberg's uncertainty principle, change in an uncontrollable way. Since the test body must be charged, the electromagnetic field will be likewise influenced in an uncontrollable way. In the formalism of quantum mechanics,

Figure 3.1 Sketch relating to the measurement of the field strength component E^1 in quantum theory.

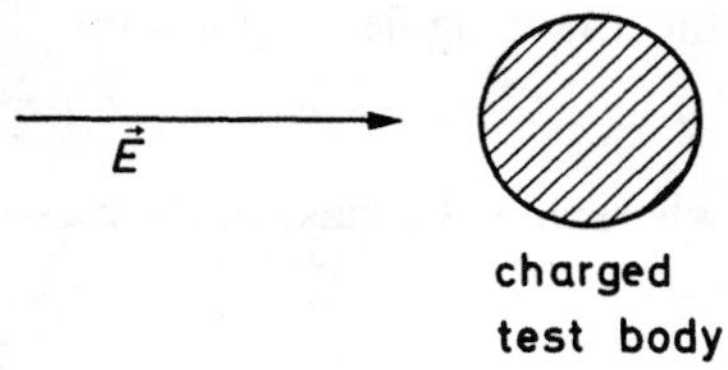

this means that the corresponding operators do not commute. It is also possible to show that commutivity of all field components would be incompatible with the uncertainty principle for the test body.

In actual experiments, one is usually not interested in the electromagnetic field at a particular position. In high energy physics one would like, for example, to know the number and energy of the photons which are produced in a reaction. An experiment of this type is sketched in Fig. 3.2. The spectral components of an electromagnetic wave are separated by a prism. Behind the prism the photons are detected by an array of counters which record the photons by means of the photoelectric effect.

We now wish to investigate the question of the relationship between particles like the photon and the corresponding quantized field operator. In fact, the case of the electromagnetic field suffers from certain complications and will be dealt with in the second part of this book. Here we will consider in more detail a meson field, for example, the field of a neutral pi-meson or that of the neutral K-mesons (K_1^0 or K_2^0). For the time being we will neglect the instability of these particles. Incidentally, with neutral K-mesons it is possible to observe refraction and diffraction effects exactly analogous to those found in experiments with light (see Kleinknecht 1976).

The classical observable of a neutral meson wave is a real scalar field

$$\varphi(x) = \varphi^*(x). \tag{3.13}$$

If Maxwell had developed a theory of the neutral meson he would have found

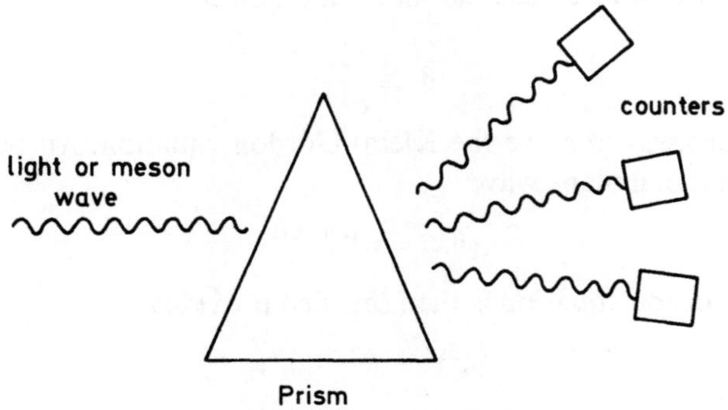

Figure 3.2 Sketch relating to the measurement of the particle content of a light wave or a meson wave.

the following field equation:

$$(\square + m^2)\varphi(x) = 0, \tag{3.14}$$

where m is the mass of the meson and $\square$ is the d'Alembert operator:

$$\begin{aligned}\square &= \frac{\partial^2}{\partial t^2} - \Delta \\ &= g^{\mu\nu}\frac{\partial^2}{\partial x^\mu\, \partial x^\nu}.\end{aligned} \tag{3.15}$$

The equation (3.14) is known as the Klein–Gordon equation (Schrödinger 1926; Klein 1927; Gordon 1927). Apart from the mass term it is identical to the wave equation obeyed by the retarded potentials V and $\boldsymbol{A}$ in electrodynamics:

$$\begin{aligned}\square\, V(x) &= 0, \\ \square\, \boldsymbol{A}(x) &= 0.\end{aligned} \tag{3.16}$$

We also want to consider how the field $\varphi(x)$ behaves in a transition to a different reference system. In a primed system with coordinates

$$x'^\mu = \Lambda^\mu{}_\nu x^\nu + a^\mu \tag{3.17}$$

the transformed function φ' is given by

$$\varphi'(x') = \varphi(x). \tag{3.18}$$

The Klein–Gordon equation is relativistically invariant. Indeed, we find that

$$\begin{aligned}\frac{\partial}{\partial x^\nu} = \frac{\partial x'^\mu}{\partial x^\nu}\frac{\partial}{\partial x'^\mu} &= \Lambda^\mu{}_\nu \frac{\partial}{\partial x'^\mu}, \\ \square &= \square', \\ (\square' + m^2)\varphi'(x') &= 0.\end{aligned} \tag{3.19}$$

From the above we also see that partial derivatives, with respect to coordinates, transform in the same way as a four-vector component with lower indices. Thus we will also use the shorthand notation

$$\partial_\nu \equiv \frac{\partial}{\partial x^\nu}. \tag{3.20}$$

We now proceed to solve the Klein–Gordon equation. An obvious trial solution is that of a plane wave

$$\varphi(x) = \mathrm{e}^{\mathrm{i}(k^0 t - \boldsymbol{k}\boldsymbol{x})}. \tag{3.21}$$

The Klein–Gordon equation is then satisfied provided

$$\begin{aligned}(k^0)^2 - \boldsymbol{k}^2 &= m^2, \\ k^0 &= \pm\sqrt{m^2 + \boldsymbol{k}^2}.\end{aligned} \tag{3.22}$$

Thus we obtain two types of solution, those with positive and those with negative angular frequency k^0. With the definition

$$\omega = +\sqrt{m^2 + \boldsymbol{k}^2} \tag{3.23}$$

we can write the two types of solution as follows[1]

$$\begin{aligned} \varphi_+(x) &= \mathrm{e}^{\mathrm{i}(\omega t - \boldsymbol{k}\boldsymbol{x})}, \\ \varphi_-(x) &= \mathrm{e}^{-\mathrm{i}(\omega t - \boldsymbol{k}\boldsymbol{x})}. \end{aligned} \tag{3.24}$$

Here $\boldsymbol{k}$ is an arbitrary three-dimensional wave vector. Due to the fact that the Klein–Gordon equation is linear, any superposition of solutions is itself also a solution. A general superposition of the solutions (3.24) which gives the desired real solution (3.13) can be written as follows:

$$\varphi(x) = \int \frac{\mathrm{d}^3k}{(2\pi)^3} \frac{1}{2\omega} (\mathrm{e}^{\mathrm{i}kx}\, \alpha^*(\boldsymbol{k}) + \mathrm{e}^{-\mathrm{i}kx}\, \alpha(\boldsymbol{k})), \tag{3.25}$$

where

$$k = \begin{pmatrix} \omega \\ \boldsymbol{k} \end{pmatrix}; \qquad kx = k^\mu x_\mu = \omega t - \boldsymbol{k}\boldsymbol{x}.$$

The function $\alpha(\boldsymbol{k})$ is an arbitrary complex function and we have included in (3.25) certain conventional factors. It can be shown that *every* (well-behaved) solution to the Klein–Gordon equation can be written in this form with a suitable choice of the function $\alpha(\boldsymbol{k})$. The plane waves thus provide a complete system of solutions.

Now let us go over to quantum field theory. We could rely for this step on the Lagrange formalism with canonical quantization rules and derive everything from there. However, we prefer in this section to avoid this formal approach and to begin instead from rather more physically motivated assumptions.

From our discussion of the electromagnetic field strengths, it is expected that a field operator $\Phi(x)$ can also be assigned to a meson field. According to Bohr's correspondence principle, the classical meson field can be interpreted as the expectation value of the quantum mechanical operator for an appropriate (quantum mechanical) state

$$\varphi(x) = \langle |\Phi(x)| \rangle. \tag{3.26}$$

This classical field is real and obeys the Klein–Gordon equation provided that the field operator $\Phi(x)$ is Hermitian and also satisfies the Klein–Gordon equation. We shall adopt this as our first assumption:

Assumption I.

$$\begin{aligned} \Phi(x) &= \Phi^\dagger(x), \\ (\Box + m^2)\Phi(x) &= 0. \end{aligned} \tag{3.27}$$

The superscript † denotes the Hermitian conjugate operator.

[1] Our nomenclature for positive and negative frequency solutions follows Bogoliubov (1959). In the literature one frequently encounters the reverse nomenclature.

The momentum and energy of particles are clearly observable quantities. They must thus correspond to Hermitian operators which, due to relativistic invariance, necessarily combine to a four-vector operator:

$$\mathbf{P} = (\mathbf{P}^\mu) = \begin{pmatrix} \mathbf{H} \\ \vec{\mathbf{P}} \end{pmatrix}. \tag{3.28}$$

The energy operator $\mathbf{P}^0$ is of course one and the same as the Hamilton operator $\mathbf{H}$. As a consequence of energy-momentum conservation we can conclude that the operators $\mathbf{P}^\mu$ are time independent for an isolated system. In particular, this will be true for our free meson field.

As a second basic assumption we introduce the generalized Heisenberg equation as already given in (3.12) within the framework of ordinary quantum mechanics.

Assumption II.

$$\frac{\partial}{\partial x^\mu}\Phi(x) = \mathrm{i}[\mathbf{P}_\mu, \Phi(x)]. \tag{3.29}$$

The requirement of compatibility of assumptions I and II will lead us, in essence, to the particle interpretation for the meson field. We begin with the Klein–Gordon equation for the field operator (3.27). Since the set of plane waves forms a complete system of solutions, it should be possible to write a Fourier expansion for the field operator, analogous to that for the classical field (3.25). The difference in this case is that the coefficients $\mathbf{a}(\boldsymbol{k})$ and $\mathbf{a}^\dagger(\boldsymbol{k})$ are now operators:

$$\Phi(x) = \int \frac{\mathrm{d}^3k}{(2\pi)^3}\frac{1}{2\omega}(\mathrm{e}^{\mathrm{i}kx}\,\mathbf{a}^\dagger(\boldsymbol{k}) + \mathrm{e}^{-\mathrm{i}kx}\,\mathbf{a}(\boldsymbol{k})). \tag{3.30}$$

Choosing $\mathbf{a}^\dagger(\boldsymbol{k})$ to be the Hermitian conjugate of the operator $\mathbf{a}(\boldsymbol{k})$ allows us to satisfy the condition that $\Phi(x)$ is Hermitian. If we substitute this expansion of the field operator into the Heisenberg equation (3.29) we obtain

$$\begin{aligned}\int \frac{\mathrm{d}^3k}{(2\pi)^3}\frac{1}{2\omega}&(\mathrm{e}^{\mathrm{i}kx}\,\mathrm{i}k_\mu\mathbf{a}^\dagger(\boldsymbol{k}) + \mathrm{e}^{-\mathrm{i}kx}\,(-\mathrm{i}k_\mu)\mathbf{a}(\boldsymbol{k}))\\ &= \mathrm{i}\int \frac{\mathrm{d}^3k}{(2\pi)^3}\frac{1}{2\omega}(\mathrm{e}^{\mathrm{i}kx}\,[\mathbf{P}_\mu, \mathbf{a}^\dagger(\boldsymbol{k})] + \mathrm{e}^{-\mathrm{i}kx}\,[\mathbf{P}_\mu, \mathbf{a}(\boldsymbol{k})]).\end{aligned} \tag{3.31}$$

It then follows that

$$[\mathbf{P}_\mu, \mathbf{a}^\dagger(\boldsymbol{k})] = k_\mu\mathbf{a}^\dagger(\boldsymbol{k}), \tag{3.32}$$

$$[\mathbf{P}_\mu, \mathbf{a}(\boldsymbol{k})] = -k_\mu\mathbf{a}(\boldsymbol{k}). \tag{3.33}$$

We must now look at the states which are to be described by vectors in a Hilbert space. The simplest state in particle physics is the one in which no particles are present. This is the vacuum state

$$|0\rangle : \text{vacuum state.} \tag{3.34}$$

Note that this state should not be confused with the null vector. As usual we will normalize the vacuum state to 1:

$$\langle 0|0\rangle = 1. \tag{3.35}$$

If we consider the experiment depicted in Fig. 3.2, then the vacuum state would correspond to the situation in which none of the detectors registers a signal even if the experiment is repeated arbitrarily often.

The vacuum state of course has zero energy and zero momentum:

$$\mathbf{P}_\mu|0\rangle = 0. \tag{3.36}$$

The next step is to apply the operator relation (3.32) to the vacuum state. Using (3.36) we obtain

$$\begin{aligned} [\mathbf{P}^\mu, \mathbf{a}^\dagger(\boldsymbol{k})]|0\rangle &= k^\mu \mathbf{a}^\dagger(\boldsymbol{k})|0\rangle, \\ \mathbf{P}^\mu \mathbf{a}^\dagger(\boldsymbol{k})|0\rangle &= k^\mu \mathbf{a}^\dagger(\boldsymbol{k})|0\rangle. \end{aligned} \tag{3.37}$$

In the case where the state

$$|\boldsymbol{k}\rangle = \mathbf{a}^\dagger(\boldsymbol{k})|0\rangle \tag{3.38}$$

is not equal to the null vector, then $|\boldsymbol{k}\rangle$ is an eigenstate of the energy and momentum operator with eigenvalues k^μ where

$$k = (k^\mu) = \begin{pmatrix} +\sqrt{m^2 + \boldsymbol{k}^2} \\ \boldsymbol{k} \end{pmatrix}. \tag{3.39}$$

However this k is precisely the four-momentum of a relativistic particle of mass m. Thus we can interpret this state as a one-meson state with sharply defined energy and momentum. We note also that if the mass $m = 0$, we obtain the relations between angular frequency and energy and between wave vector and momentum as required by Einstein's quantum hypothesis of light.

If we apply (3.33) to the vacuum state then we obtain

$$\mathbf{P}^\mu \mathbf{a}(\boldsymbol{k})|0\rangle = -k^\mu \mathbf{a}(\boldsymbol{k})|0\rangle. \tag{3.40}$$

Thus the state $\mathbf{a}(\boldsymbol{k})|0\rangle$ would be a state with *negative* energy eigenvalue

$$-k^0 = -\sqrt{m^2 + \boldsymbol{k}^2}. \tag{3.41}$$

Such states are not observed in Nature. All particles have nonnegative energies and thus we must demand

$$\mathbf{a}(\boldsymbol{k})|0\rangle = 0 \qquad \text{for all } \boldsymbol{k}. \tag{3.42}$$

If we now take a general eigenstate of the energy-momentum four-vector

$$\mathbf{P}^\mu|p\rangle = p^\mu|p\rangle, \tag{3.43}$$

then we see from (3.32) and (3.33) that the states $\mathbf{a}^\dagger(\boldsymbol{k})|p\rangle$ and $\mathbf{a}(\boldsymbol{k})|p\rangle$ are eigenstates of $\mathbf{P}^\mu$ with eigenvalues $p^\mu \pm k^\mu$ provided they are not equal to the null vector

$$\begin{aligned} \mathbf{P}^\mu \mathbf{a}^\dagger(\boldsymbol{k})|p\rangle &= (p^\mu + k^\mu)\mathbf{a}^\dagger(\boldsymbol{k})|p\rangle, \\ \mathbf{P}^\mu \mathbf{a}(\boldsymbol{k})|p\rangle &= (p^\mu - k^\mu)\mathbf{a}(\boldsymbol{k})|p\rangle. \end{aligned} \tag{3.44}$$

If we choose $|p\rangle = \mathbf{a}^\dagger(\boldsymbol{k}_2)|0\rangle$, then with $\boldsymbol{k}_1$ in the place of $\boldsymbol{k}$ we obtain

$$\mathbf{P}^\mu \mathbf{a}^\dagger(\boldsymbol{k}_1)\mathbf{a}^\dagger(\boldsymbol{k}_2)|0\rangle = (k_1^\mu + k_2^\mu)\mathbf{a}^\dagger(\boldsymbol{k}_1)\mathbf{a}^\dagger(\boldsymbol{k}_2)|0\rangle. \tag{3.45}$$

This suggests that a state which is produced from the vacuum state by applying two operators $\mathbf{a}^\dagger$, can be interpreted as a state with two mesons. Analogously, by applying the operator $\mathbf{a}^\dagger$ n times to the vacuum state, we obtain an n-meson state. The operators $\mathbf{a}^\dagger$ are therefore known as *creation* operators. The operators $\mathbf{a}$ are called *annihilation* operators because, as we shall see shortly, they remove a meson with the appropriate four-momentum from a state.

Our discussions are not yet complete since we remain to learn about the norm of states with one or more mesons. For this we need to introduce a further physical assumption. We consider the measurement of the meson field at two different space–time points x and y (Fig. 3.3). If these are points with a spacelike separation, i.e.,

$$(x - y)^2 < 0, \tag{3.46}$$

then x lies outside the forward light cone of y and vice versa. If our theory is to satisfy the postulates of special relativity then no signal from the measurement at point x can reach the point y or vice versa. The measurements at x and y cannot influence one another and the corresponding operators thus commute. We will introduce this as our third assumption.

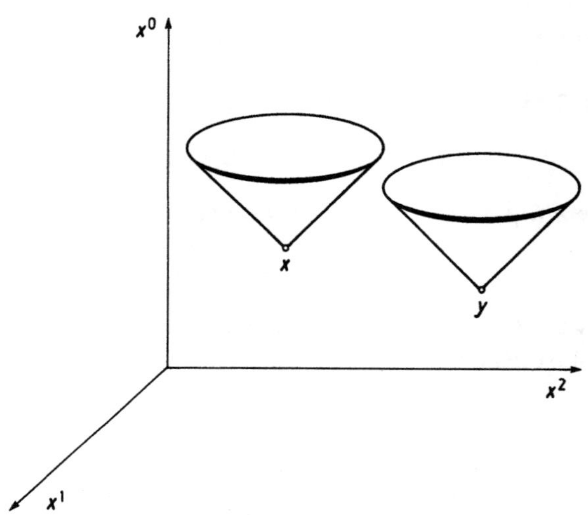

Figure 3.3 Illustration for the measurement of a field operator Φ at two points x and y in Minkowski space with space-like separation. The forwards light cones of x and y are indicated by thick lines.

Assumption III.

$$[\Phi(x), \Phi(y)] = 0 \qquad \text{for} \quad (x-y)^2 < 0. \tag{3.47}$$

This is referred to as the requirement of *microcausality*.

We now take a closer look at (3.47). Let the two points x and y be the following:

$$x = \begin{pmatrix} t \\ \boldsymbol{x} \end{pmatrix}, \qquad y = \begin{pmatrix} t' \\ \boldsymbol{y} \end{pmatrix}, \qquad \boldsymbol{x} \neq \boldsymbol{y}. \tag{3.48}$$

Equation (3.47) can then be written

$$[\Phi(\boldsymbol{x}, t), \Phi(\boldsymbol{y}, t')] = 0 \qquad \text{for} \quad |t'-t| < |\boldsymbol{x}-\boldsymbol{y}| \neq 0. \tag{3.49}$$

The above commutator thus vanishes for $\boldsymbol{x} \neq \boldsymbol{y}$ in a whole interval of t' values around t. In particular, when the two times are equal, $t = t'$, we find

$$\begin{aligned} &[\Phi(\boldsymbol{x}, t), \Phi(\boldsymbol{y}, t)] = 0, \\ &\left[\Phi(\boldsymbol{x}, t), \frac{\partial}{\partial t}\Phi(\boldsymbol{y}, t)\right] = 0 \qquad \text{for} \quad \boldsymbol{x} \neq \boldsymbol{y}. \end{aligned} \tag{3.50}$$

Now we consider the consequences of (3.47) and (3.50) for the creation and annihilation operators and for meson states. From the principle of microcausality it will be possible to derive the Bose statistics of mesons. To do this we consider the plane wave expansion of the field operator and its first derivative with respect to time. From (3.30) we have:

$$\begin{aligned} \Phi(\boldsymbol{x}, t) &= \int \frac{d^3k}{(2\pi)^3}\frac{1}{2\omega} e^{-i\boldsymbol{k}\boldsymbol{x}} \left(e^{i\omega t}\mathbf{a}^\dagger(\boldsymbol{k}) + e^{-i\omega t}\,\mathbf{a}(-\boldsymbol{k})\right), \\ \dot{\Phi}(\boldsymbol{x}, t) &\equiv \frac{\partial}{\partial t}\Phi(\boldsymbol{x}, t) \\ &= \int \frac{d^3k}{(2\pi)^3}\frac{i}{2} e^{-i\boldsymbol{k}\boldsymbol{x}} \left(e^{i\omega t}\mathbf{a}^\dagger(\boldsymbol{k}) - e^{-i\omega t}\,\mathbf{a}(-\boldsymbol{k})\right). \end{aligned} \tag{3.51}$$

Inverting the Fourier decomposition gives

$$\begin{aligned} e^{i\omega t}\,\mathbf{a}^\dagger(\boldsymbol{k}) + e^{-i\omega t}\,\mathbf{a}(-\boldsymbol{k}) &= 2\omega \int d^3x\, e^{i\boldsymbol{k}\boldsymbol{x}}\,\Phi(\boldsymbol{x}, t), \\ e^{i\omega t}\,\mathbf{a}^\dagger(\boldsymbol{k}) - e^{-i\omega t}\,\mathbf{a}(-\boldsymbol{k}) &= -2i \int d^3x\, e^{i\boldsymbol{k}\boldsymbol{x}}\,\dot{\Phi}(\boldsymbol{x}, t). \end{aligned} \tag{3.52}$$

and from this we obtain the commutation relation

$$\begin{aligned} &[e^{i\omega_1 t}\,\mathbf{a}^\dagger(\boldsymbol{k}_1) + e^{-i\omega_1 t}\,\mathbf{a}(-\boldsymbol{k}_1), e^{i\omega_2 t}\,\mathbf{a}^\dagger(\boldsymbol{k}_2) + e^{-i\omega_2 t}\,\mathbf{a}(-\boldsymbol{k}_2)] \\ &\qquad = 2\omega_1 2\omega_2 \int d^3x\, d^3y\, e^{i\boldsymbol{k}_1\boldsymbol{x}}\, e^{i\boldsymbol{k}_2\boldsymbol{y}}\,[\Phi(\boldsymbol{x}, t), \Phi(\boldsymbol{y}, t)]. \end{aligned} \tag{3.53}$$

According to (3.50) the integrand vanishes for $x \neq y$. However, it also vanishes for $x = y$ since it is trivial that the operator $\Phi(x, t)$ commutes with itself. It thus follows that

$$e^{i(\omega_1+\omega_2)t} [a^\dagger(k_1), a^\dagger(k_2)] - e^{i(\omega_1-\omega_2)t} [a(-k_2), a^\dagger(k_1)]$$
$$+ e^{-i(\omega_1-\omega_2)t} [a(-k_1), a^\dagger(k_2)] + e^{-i(\omega_1+\omega_2)t} [a(-k_1), a(-k_2)] = 0 \quad (3.54)$$
$$\text{for all} \quad k_1, k_2, t.$$

The angular frequency $\omega_1 + \omega_2$ is always different from $\omega_1 - \omega_2$, $-(\omega_1 - \omega_2)$ and $-(\omega_1 + \omega_2)$. Since however, (3.54) is valid for all times t, it follows that

$$[a^\dagger(k_1), a^\dagger(k_2)] = 0 \qquad \text{for all} \quad k_1, k_2. \quad (3.55)$$

Analogously we find

$$[a(k_1), a(k_2)] = 0 \qquad \text{for all} \quad k_1, k_2. \quad (3.56)$$

From the requirement of microcausality it therefore follows that the creation operators commute with one another as do the annihilation operators. This already implies that the mesons satisfy Bose statistics. To see this we apply (3.55) to the vacuum state. We find

$$a^\dagger(k_1)a^\dagger(k_2)|0\rangle = a^\dagger(k_2)a^\dagger(k_1)|0\rangle,$$
$$|k_1, k_2\rangle = |k_2, k_1\rangle. \quad (3.57)$$

The state with two mesons of momenta k_1 and k_2 remains unchanged upon exchange of the two momenta. The same is true for states with three or more mesons.

Let us now consider an actual reaction in which two π^0-mesons are produced as, for example, in proton–antiproton annihilation (Fig. 3.4)

$$p + \bar{p} \to \pi^0(k_1) + \pi^0(k_2). \quad (3.58)$$

In the center of mass frame the momenta satisfy $k_2 = -k_1$. According to (3.57) the final state consisting of two π^0-mesons must always be symmetric with respect to the exchange $k_1 \to k_2 = -k_1$, or in other words, with respect to reflection in the origin. From ordinary quantum mechanics we know that the angular momentum must then have an even value ($l = 0, 2, 4, \ldots$) with respect to the center of mass. From (3.57) it also follows that the two π^0-mesons, unlike classical particles, cannot be distinguished from one another.

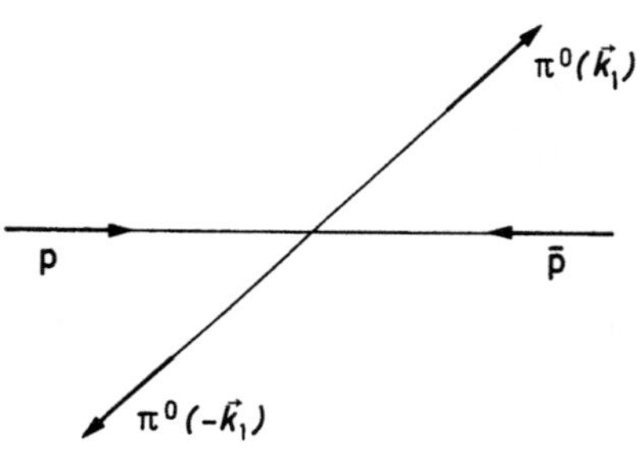

Figure 3.4 Creation of two π^0-mesons in proton–antiproton annihilation. As bosons, the two π^0-mesons must be in a symmetric state.

The relationship between microcausality and the Bose statistics of mesons is not only valid in the case of the free fields which we are considering here. It can be demonstrated quite generally that microcausality imposes Bose statistics upon all particles with integer spin (Pauli 1936, 1940; see also Streater 1964). This is in perfect agreement with observations.

We consider now the commutator of an annihilation operator with a creation operator. If we solve the two equations (3.52) for $\mathbf{a}^\dagger$ and $\mathbf{a}$ we obtain for their commutator

$$[\mathbf{a}(\boldsymbol{k}_1), \mathbf{a}^\dagger(\boldsymbol{k}_2)] = \mathrm{e}^{\mathrm{i}(\omega_1-\omega_2)t} \int \mathrm{d}^3x\, \mathrm{d}^3y\, \mathrm{e}^{-\mathrm{i}\boldsymbol{k}_1\boldsymbol{x}}\, \mathrm{e}^{\mathrm{i}\boldsymbol{k}_2\boldsymbol{y}} \cdot \{\mathrm{i}\omega_2[\dot{\Phi}(\boldsymbol{x}, t), \Phi(\boldsymbol{y}, t)] - \mathrm{i}\omega_1[\Phi(\boldsymbol{x}, t), \dot{\Phi}(\boldsymbol{y}, t)]\}. \quad (3.59)$$

The integrand, in which terms that have already been shown to be zero have been omitted, vanishes for $\boldsymbol{x} \neq \boldsymbol{y}$ from (3.50). If we assume that the commutator of the field operator and its first time derivative also vanishes for $\boldsymbol{x} = \boldsymbol{y}$, then all creation operators would commute with all annihilation operators. The field operator Φ would be a commuting quantity. By generalizing the arguments of Bohr and Rosenfeld already discussed in Sect. 3.1, it is possible to show that this leads to a contradiction of the uncertainty relation. It is also easy to see that in this case all states obtained by applying creation operators to the vacuum state, would be identical to the null vector. Such a theory does not make sense.

As the simplest assumption let us take for the commutator of $\Phi(\boldsymbol{x}, t)$ with $\dot{\Phi}(\boldsymbol{y}, t)$ at equal times, a complex-valued function which, as a consequence of (3.50), can only be nonzero for $\boldsymbol{x} = \boldsymbol{y}$. Such a function is the Dirac δ-function. Thus we will write

$$[\Phi(\boldsymbol{x}, t), \dot{\Phi}(\boldsymbol{y}, t)] = \mathrm{i}Z\delta^3(\boldsymbol{x} - \boldsymbol{y}), \quad (3.60)$$

where Z is a constant. Inserting this into (3.59) gives

$$[\mathbf{a}(\boldsymbol{k}_1), \mathbf{a}^\dagger(\boldsymbol{k}_2)] = Z(2\pi)^3 \cdot 2\omega_1 \delta^3(\boldsymbol{k}_1 - \boldsymbol{k}_2). \quad (3.61)$$

We will now show that the constant Z must be real and positive. For this purpose we consider a general one-meson state which we express as a superposition of momentum eigenstates (3.38)

$$|f\rangle = \int \frac{\mathrm{d}^3k}{(2\pi)^3\sqrt{2\omega}} f(\boldsymbol{k})\mathbf{a}^\dagger(\boldsymbol{k})|0\rangle, \quad (3.62)$$

where $f(\boldsymbol{k})$ is any complex-valued function. We continue by calculating the square of the norm of this state and use the condition that it is nonnegative.

$$\begin{aligned}\langle f|f\rangle &= \int \frac{\mathrm{d}^3k\, \mathrm{d}^3k'}{(2\pi)^6\sqrt{2\omega 2\omega'}} f^*(\boldsymbol{k})f(\boldsymbol{k}')\langle 0|\mathbf{a}(\boldsymbol{k})\mathbf{a}^\dagger(\boldsymbol{k}')|0\rangle \\ &= \int \frac{\mathrm{d}^3k\, \mathrm{d}^3k'}{(2\pi)^6\sqrt{2\omega 2\omega'}} f^*(\boldsymbol{k})f(\boldsymbol{k}')\langle 0|[\mathbf{a}(\boldsymbol{k}), \mathbf{a}^\dagger(\boldsymbol{k}')]|0\rangle \\ &= Z \int \frac{\mathrm{d}^3k}{(2\pi)^3} |f(\boldsymbol{k})|^2 \geq 0. \end{aligned} \quad (3.63)$$

To ensure that not all one-meson states are equal to the null vector we must demand

$$Z > 0. \tag{3.64}$$

By rescaling the field operator

$$\Phi(x) \to \sqrt{Z}\,\Phi(x) \tag{3.65}$$

we can always arrive at $Z = 1$ in (3.60) and thus

$$[\Phi(\boldsymbol{x}, t), \dot{\Phi}(\boldsymbol{y}, t)] = \mathrm{i}\delta^3(\boldsymbol{x} - \boldsymbol{y}). \tag{3.66}$$

This equation constitutes the *canonical commutation rules* which are postulated in the canonical formalism. In quantum field theory they play a role just as fundamental as that of the commutation rules for position and momentum operators in ordinary quantum mechanics.

We have now become acquainted with the most important properties of the quantized field of a free neutral meson. We have expanded the Hermitian field operator in terms of creation and annihilation operators (3.30). For these operators we found the commutation relations

$$\begin{aligned} [\mathsf{a}^\dagger(\boldsymbol{k}_1), \mathsf{a}^\dagger(\boldsymbol{k}_2)] &= 0, \\ [\mathsf{a}(\boldsymbol{k}_1), \mathsf{a}(\boldsymbol{k}_2)] &= 0, \\ [\mathsf{a}(\boldsymbol{k}_1), \mathsf{a}^\dagger(\boldsymbol{k}_2)] &= (2\pi)^3 \cdot 2\omega_1 \delta^3(\boldsymbol{k}_1 - \boldsymbol{k}_2). \end{aligned} \tag{3.67}$$

A basis for the one-meson states is provided by the momentum eigenstates

$$\mathsf{a}^\dagger(\boldsymbol{k})|0\rangle = |\boldsymbol{k}\rangle, \tag{3.68}$$

which satisfy the continuum normalization

$$\langle \boldsymbol{k}|\boldsymbol{k}'\rangle = (2\pi)^3 \cdot 2\omega\delta^3(\boldsymbol{k} - \boldsymbol{k}') \tag{3.69}$$

A basis for n-meson states is obtained by applying n creation operators to the vacuum state:

$$\mathsf{a}^\dagger(\boldsymbol{k}_1)\ldots \mathsf{a}^\dagger(\boldsymbol{k}_n)|0\rangle = |\boldsymbol{k}_1, \ldots, \boldsymbol{k}_n\rangle. \tag{3.70}$$

By generalizing (3.69) it is easy to see that none of these states is equal to the null vector. These states are known as Fock states and they constitute a basis in *Fock space*. (This representation was introduced by Fock in 1932 and is based on results by Dirac (1927) and Jordan (1927, 1927a, 1927b, 1928a).) From the principle of microcausality we have been able to derive the Bose statistics of mesons.

Finally, we wish to discuss the relationship of this theory to the Schrödinger theory. A general one-meson state has been given in (3.62). In the following way it is possible to define a time-dependent state vector in the Schrödinger picture

$$|f, t\rangle = \int \frac{\mathrm{d}^3k}{(2\pi)^3\sqrt{2\omega}} f(\boldsymbol{k}, t)\mathsf{a}^\dagger(\boldsymbol{k})|0\rangle, \tag{3.71}$$

where

$$f(\boldsymbol{k}, t) = f(\boldsymbol{k})\,\mathrm{e}^{-\mathrm{i}\omega t}.$$

We then find

$$\langle f, t|f, t\rangle = \int \frac{d^3k}{(2\pi)^3} |f(\boldsymbol{k}, t)|^2$$

$$= \int \frac{d^3k}{(2\pi)^3} |f(\boldsymbol{k})|^2 = 1. \tag{3.72}$$

The function $f(\boldsymbol{k}, t)$ would be the Schrödinger function in the momentum representation. Its absolute square value $|f(\boldsymbol{k}, t)|^2$ gives the probability of finding the meson at time t with momentum $\boldsymbol{k}$. It is now tempting to introduce the Fourier transform

$$\psi(\boldsymbol{x}, t) = \int \frac{d^3k}{(2\pi)^3} e^{i\boldsymbol{k}\boldsymbol{x}} f(\boldsymbol{k}, t) \tag{3.73}$$

and to interpret its squared modulus as the probability distribution for the position of the particle. However, it is possible to show that this interpretation is only consistent for a *nonrelativistic* meson, i.e., when the momenta present are small compared to the mass of the particle. In the general case this interpretation would mean that particles could travel with speeds in excess of the speed of light, a situation which is clearly unacceptable (see Hegerfeld 1974).

3.3 The Lagrange Formalism and the Noether Theorems

In classical mechanics of point particles it is possible in many cases to derive the equations of motion from Hamilton's principle of least action (see Goldstein 1980; Sommerfeld 1952). In this section we will discuss the analogous formalism for field theory.

We again begin by considering a classical real scalar field $\varphi(x)$. The Lagrange density $\mathcal{L}_0$ of the free field and the action S are defined as

$$\mathcal{L}_0(\varphi, \partial_\mu \varphi) = \tfrac{1}{2}\{\partial_\mu \varphi(x) \partial^\mu \varphi(x) - m^2 \varphi^2(x)\}, \tag{3.74}$$

$$S[\varphi] = \int dx\, \mathcal{L}_0(\varphi, \partial_\mu \varphi). \tag{3.75}$$

We assert the equivalence of the equation of motion (3.14) and the following *principle of least action*: Fields which actually occur in Nature correspond to a stationary value of the action:

$$\delta S[\varphi] = 0. \tag{3.76}$$

In dealing with (3.76) we shall assume that variations in the fields vanish sufficiently fast at infinity (in space and time).

We show first of all that the equation of motion (3.14) follows from (3.76). The variation of the action integral gives

$$\delta S[\varphi] = \int dx \left\{ \frac{\partial \mathcal{L}_0(\varphi, \partial_\mu \varphi)}{\partial(\partial_\lambda \varphi(x))} \partial_\lambda \delta\varphi(x) + \frac{\partial \mathcal{L}_0(\varphi, \partial_\mu \varphi)}{\partial \varphi(x)} \delta\varphi(x) \right\}. \tag{3.77}$$

By means of partial integration of the first term of the integral, in the course of which boundary terms can be neglected due to the condition that $\delta\varphi(x)$ vanishes at infinity, we obtain

$$\delta S[\varphi] = \int \mathrm{d}x \left\{ -\partial_\lambda \frac{\partial \mathscr{L}_0(\varphi, \partial_\mu \varphi)}{\partial(\partial_\lambda \varphi(x))} + \frac{\partial \mathscr{L}_0(\varphi, \partial_\mu \varphi)}{\partial \varphi(x)} \right\} \delta\varphi(x) = 0. \quad (3.78)$$

Since for finite x the variation $\delta\varphi(x)$ of the field is quite arbitrary, it follows from the vanishing of the integral in (3.78) that the integrand must also vanish. Thus we find

$$\begin{aligned} \partial_\lambda \frac{\partial \mathscr{L}_0(\varphi, \partial_\mu \varphi)}{\partial(\partial_\lambda \varphi(x))} - \frac{\partial \mathscr{L}_0(\varphi, \partial_\mu \varphi)}{\partial \varphi(x)} &= 0, \\ \partial_\lambda \partial^\lambda \varphi(x) + m^2 \varphi(x) &= 0. \end{aligned} \quad (3.79)$$

This however is exactly the equation of motion (3.14). By reversing all steps taken, it is seen that the principle of least action can in fact be derived from (3.14).

The principle of least action is of course not limited to free fields. An example of a Lagrange density with interaction is

$$\mathscr{L}(\varphi, \partial_\mu \varphi) = \tfrac{1}{2}\{\partial_\mu \varphi(x) \partial^\mu \varphi(x) - m^2 \varphi^2(x)\} - \tfrac{1}{4}\lambda \varphi^4(x). \quad (3.80)$$

The principle of least action imples the equation of motion

$$\begin{aligned} \partial_\lambda \frac{\partial \mathscr{L}}{\partial(\partial_\lambda \varphi(x))} - \frac{\partial \mathscr{L}}{\partial \varphi(x)} &= 0, \\ \partial_\lambda \partial^\lambda \varphi(x) + m^2 \varphi(x) + \lambda \varphi^3(x) &= 0. \end{aligned} \quad (3.81)$$

The Lagrange formalism turns out to be particularly useful for discussing the symmetries of a theory. We consider a set of real scalar fields $\varphi_j(x)$ (with $j = 1, \ldots, N$). The Lagrange density, which may also be explicitly dependent on x is written as $\mathscr{L}(x, \varphi_j, \partial_\mu \varphi_j)$. The equations of motion are to be derived from the principle of least action

$$\delta \int \mathrm{d}x \, \mathscr{L}(x, \varphi_j, \partial_\mu \varphi_j) = 0. \quad (3.82)$$

We now assume the existence of infinitesimal transformations

$$\begin{aligned} x^\mu \to x'^\mu &= x^\mu + \delta x^\mu, \\ \varphi_j(x) \to \varphi_j'(x') &= \varphi_j(x) + \delta\varphi_j(x), \end{aligned} \quad (3.83)$$

under which $\mathscr{L}\,\mathrm{d}x$ is invariant up to the addition of a total divergence. We thus impose the condition that there exist functions $\delta Q^\lambda(x)$ such that the equation

$$\mathscr{L}(x', \varphi_j'(x'), \partial_\lambda' \varphi_j'(x'))\,\mathrm{d}x' = \{\mathscr{L}(x, \varphi_j(x), \partial_\lambda \varphi_j(x)) + \partial_\lambda \delta Q^\lambda(x)\}\,\mathrm{d}x \quad (3.84)$$

is satisfied. We now assert the validity of a conservation law under these conditions. It reads: For every region B of Minkowski space (Fig. 3.5) with

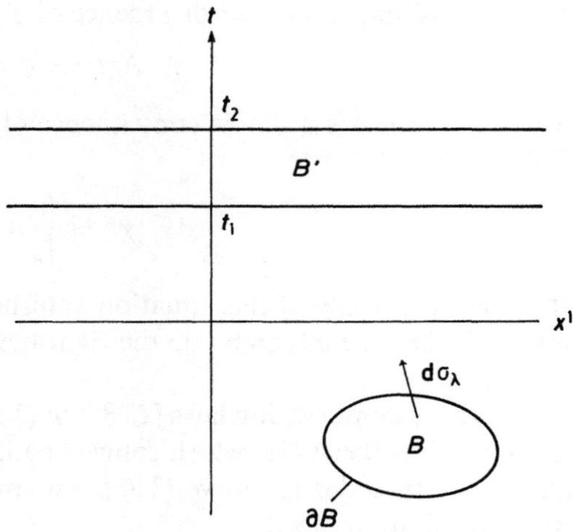

Figure 3.5 A region B in Minkowski space with boundary ∂B and a region B' whose boundary is formed by the hyperplanes $t = t_1$ and $t = t_2$.

boundary ∂B, the three-dimensional surface integral

$$\int_{\partial B} \left(-\frac{\partial \mathcal{L}}{\partial(\partial_\lambda \varphi_j)} \delta\varphi_j + T^\lambda{}_\mu \delta x^\mu + \delta Q^\lambda \right) d\sigma_\lambda = 0 \tag{3.85}$$

vanishes. Here $d\sigma_\lambda$ is a three-dimensional surface element which is defined with the help of the totally antisymmetric tensor $\varepsilon_{\lambda\mu\nu\rho}$, $\varepsilon_{0123} = 1$:

$$d\sigma_\lambda = \frac{1}{3!} \varepsilon_{\lambda\mu\nu\rho}\, dx^\mu\, dx^\nu\, dx^\rho. \tag{3.86}$$

The quantity $T^\lambda{}_\mu$ is the canonical energy-momentum tensor:

$$T^\lambda{}_\mu = \frac{\partial \mathcal{L}}{\partial(\partial_\lambda \varphi_j)} \partial_\mu \varphi_j - \delta^\lambda_\mu \mathcal{L}. \tag{3.87}$$

In order to simplify the notation, we will henceforth omit the arguments of functions provided there is no danger of confusion.

An important special case of (3.85) occurs for a region B' which is bounded by two spatial hyperplanes, $t = t_1$ and $t = t_2$ (Fig. 3.5). For the integral

$$E(t) = \int_{t=\text{const.}} \left(-\frac{\partial \mathcal{L}}{\partial(\partial_\lambda \varphi_j)} \delta\varphi_j + T^\lambda{}_\mu \delta x^\mu + \delta Q^\lambda \right) d\sigma_\lambda \tag{3.88}$$

the following relation then holds

$$E(t_1) - E(t_2) = 0. \tag{3.89}$$

The integral $E(t)$ is thus not time dependent—it is a conserved quantity.

The integrand in (3.85) represents a four-current density $\mathcal{J}^\lambda$:

$$\mathcal{J}^\lambda = -\frac{\partial \mathcal{L}}{\partial(\partial_\lambda \varphi_j)} \delta\varphi_j + T^\lambda{}_\mu \delta x^\mu + \delta Q^\lambda. \tag{3.90}$$

Furthermore, it is found that the validity of (3.85) for all regions B is equivalent to the vanishing of the four-divergence of $\mathscr{J}^\lambda$, i.e., to current conservation:

$$\partial_\lambda \mathscr{J}^\lambda = 0. \tag{3.91}$$

This equivalence is a simple consequence of Gauss's law in four dimensions. This law asserts

$$\int_{\partial B} \mathrm{d}\sigma_\lambda\, \mathscr{J}^\lambda = \int_B \mathrm{d}x\, \partial^\lambda \mathscr{J}_\lambda. \tag{3.92}$$

If the left-hand side of this equation vanishes for the boundary ∂B of every region B, then the integrand on the right-hand side must likewise vanish and vice versa.

The above conservation laws [(3.85) or (3.91)] are examples of the Noether theorems (Noether 1918) which connect continuous symmetries to conserved quantities. In order to prove (3.85), we consider the following expansions which result from (3.83)

$$\begin{aligned} \mathrm{d}x' &= (1 + \partial_\lambda(\delta x^\lambda))\, \mathrm{d}x, \\ \partial'_\lambda \varphi'_j(x') &= \partial_\lambda \varphi_j(x) + \partial_\lambda \delta\varphi_j(x) - \partial_\mu \varphi_j(x) \partial_\lambda \delta x^\mu, \\ \mathscr{L}(x', \varphi'_j(x'), \partial'_\lambda \varphi'_j(x'))\, \mathrm{d}x' &= \{\mathscr{L}(x, \varphi_j(x) + \bar{\delta}\varphi_j(x), \partial_\lambda(\varphi_j(x) + \bar{\delta}\varphi_j(x))) \\ &\quad + \partial_\mu(\delta x^\mu \mathscr{L}(x, \varphi_j(x), \partial_\lambda \varphi_j(x)))\}\, \mathrm{d}x, \end{aligned} \tag{3.93}$$

where

$$\bar{\delta}\varphi_j(x) = \varphi'_j(x) - \varphi_j(x) = \delta\varphi_j(x) - \delta x^\mu \partial_\mu \varphi_j(x). \tag{3.94}$$

Integration of (3.84) over the region B then gives

$$\begin{aligned} &\int_B \{\mathscr{L}(x, \varphi_j(x) + \bar{\delta}\varphi_j(x), \partial_\lambda(\varphi_j(x) + \bar{\delta}\varphi_j(x))) - \mathscr{L}(x, \varphi_j(x), \partial_\lambda \varphi_j(x))\}\, \mathrm{d}x \\ &\qquad + \int_B \partial_\lambda(\mathscr{L}\delta x^\lambda - \delta Q^\lambda)\, \mathrm{d}x = 0. \end{aligned} \tag{3.95}$$

Due to the equation of motion (3.82), the first integral in (3.95) reduces to a boundary term arising in the course of partial integration. The second integral in (3.95) can be re-expressed as a surface integral with the help of Gauss's theorem in four dimensions. It then follows that

$$\int_{\partial B} \mathrm{d}\sigma_\lambda \left\{ \frac{\partial \mathscr{L}}{\partial(\partial_\lambda \varphi_j)} \bar{\delta}\varphi_j + \mathscr{L}\delta x^\lambda - \delta Q^\lambda \right\} = 0. \tag{3.96}$$

Using (3.94) we then obtain from this (3.85). By reversing each of these steps we also see that from a conservation law of the form (3.85) a symmetry of the Lagrange density follows.

As an example we consider a Lagrange density $\mathscr{L}(\varphi_j(x), \partial_\lambda \varphi_j(x))$, which does not depend explicitly on x. The theory is then invariant under translations in time and space:

$$x^\mu \to x'^\mu = x^\mu + \delta a^\mu, \tag{3.97}$$

where δa^μ are arbitrary real quantities. For a transformation of the fields

$$\varphi_j(x) \to \varphi_j'(x') = \varphi_j(x) \tag{3.98}$$

it holds that

$$\mathscr{L}(\varphi_j'(x'), \partial_\lambda' \varphi_j'(x'))\, \mathrm{d}x' = \mathscr{L}(\varphi_j(x), \partial_\lambda \varphi_j(x))\, \mathrm{d}x.$$

Equation (3.88) becomes in this case the time-independent integral

$$E(t) = \int_{t=\text{const.}} T^\lambda{}_\mu \, \mathrm{d}\sigma_\lambda \, \delta a^\mu = \text{const.} \tag{3.99}$$

Since the δa^μ were arbitrary, we obtain four linearly independent conserved quantities

$$P_\mu = \int_{t=\text{const.}} T^\lambda{}_\mu \, \mathrm{d}\sigma_\lambda. \tag{3.100}$$

This represents the general definition of the four-momentum of a classical field configuration.

From the form of the Lagrange density one can usually obtain by inspection the symmetries of the equations of motion. Conversely, the postulation of symmetries gives useful conditions which the Lagrange density has to fulfill. Translational invariance in time and space demands a Lagrange density $\mathscr{L}$ that is not explicitly dependent on x. Poincaré invariance requires in addition that $\mathscr{L}$ transforms as a Lorentz scalar. We have already met some examples of Poincaré-invariant Lagrange densities (3.74), (3.80).

Extremely important for particle physics are *internal symmetries*. These are transformations in which only the fields are changed and not the space–time coordinates:

$$\begin{aligned} x'^\mu &= x^\mu, \\ \varphi_j'(x') &= \varphi_j(x) + \delta\varphi_j(x). \end{aligned} \tag{3.101}$$

If the Lagrange density $\mathscr{L}$ is invariant in the sense of (3.84) under such a transformation, then according to (3.90) and (3.88) there exists a corresponding current $\mathscr{J}^\lambda$ and a charge E:

$$\begin{aligned} \mathscr{J}^\lambda(x) &= -\frac{\partial \mathscr{L}}{\partial(\partial_\lambda \varphi_j)} \delta\varphi_j + \delta Q^\lambda, \\ E(t) &= \int_{t=\text{const.}} \mathscr{J}^\lambda(x)\, \mathrm{d}\sigma_\lambda, \end{aligned} \tag{3.102}$$

which are conserved, i.e., which satisfy

$$\begin{aligned} \partial_\lambda \mathscr{J}^\lambda(x) &= 0, \\ \frac{\mathrm{d}}{\mathrm{d}t} E(t) &= 0. \end{aligned} \tag{3.103}$$

We will meet some examples of such symmetries later on.

3.4 The Canonical Quantization Rules

We consider a theory with scalar fields φ_j ($j = 1, \ldots, N$) and Lagrange density

$$\mathscr{L}(x, \varphi_j, \partial_\mu \varphi_j).$$

In order to quantize this theory, we choose a fixed frame of reference and define the Lagrangian as the integral of the Lagrange density over the space-like slice $t = \text{const.}$:

$$L[t, \varphi_j, \dot{\varphi}_j] = \int_{t=\text{const.}} \mathrm{d}^3x \, \mathscr{L}(x, \varphi_j, \partial_\mu \varphi_j). \tag{3.104}$$

The Lagrangian we regard as a function, or rather a functional of t, $\varphi_j(\boldsymbol{x}, t)$ and $\dot{\varphi}_j(\boldsymbol{x}, t)$. The canonically conjugated momenta are defined via the functional derivatives

$$\Pi_k(\boldsymbol{x}, t) = \frac{\delta L[t, \varphi_j, \dot{\varphi}_j]}{\delta \dot{\varphi}_k(\boldsymbol{x}, t)} \qquad (k = 1, \ldots, N). \tag{3.105}$$

Now, as in mechanics, it is necessary to assume that these relations can be solved for $\dot{\varphi}_j$, i.e., that there exist inverse functionals F:

$$\dot{\varphi}_j = F_j[t, \varphi, \Pi]. \tag{3.106}$$

Then, at every time t, we have a complete system of dynamical variables formed by the $\varphi_j(\boldsymbol{x}, t)$ and the $\Pi_k(\boldsymbol{x}, t)$. The Hamiltonian is also defined as in mechanics:

$$H[t, \varphi, \Pi] = \int \mathrm{d}^3x \, \Pi_j \dot{\varphi}_j - L. \tag{3.107}$$

We move over now to quantum theory and regard φ_j and Π_k as field operators for which we postulate the validity of the canonical commutation rules at all times t:

$$\begin{aligned}
&[\varphi_j(\boldsymbol{x}, t), \varphi_k(\boldsymbol{y}, t)] = 0, \\
&[\Pi_j(\boldsymbol{x}, t), \Pi_k(\boldsymbol{y}, t)] = 0, \\
&[\varphi_j(\boldsymbol{x}, t), \Pi_k(\boldsymbol{y}, t)] = \mathrm{i}\delta_{jk}\delta^3(\boldsymbol{x} - \boldsymbol{y}) \qquad (j, k = 1, \ldots, N).
\end{aligned} \tag{3.108}$$

For the free scalar field with Lagrange density as in (3.74), the canonical quantization produces precisely the quantized field which we discussed in Sect. 3.2.

The canonical quantization enables the Noether theorems to be taken over in a straightforward manner into the quantum theory. We again consider infinitesimal transformations (3.83) and assume the validity of (3.84). The integral (3.88), which now represents an operator $\mathbf{E}$, is again a conserved quantity and generates a unitary representation of the symmetry transformation

$$\mathbf{U} = \mathbb{1} + \mathrm{i}\mathbf{E}. \tag{3.109}$$

This however, is conditional upon $\delta\varphi_j(x)$ and $\delta Q^0(x)$ being independent of the time derivative $\dot{\varphi}_k(x)$. We then have

$$\mathbf{U}^\dagger \mathbf{U} = \mathbb{1},$$

$$\mathbf{U}^{-1} \varphi_j(\boldsymbol{x}, t) \mathbf{U} = \varphi_j(\boldsymbol{x}, t) - \mathrm{i}[\mathbf{E}, \varphi_j(\boldsymbol{x}, t)] \qquad (3.110)$$

$$= \varphi_j(\boldsymbol{x}, t) + \bar{\delta}\varphi_j(\boldsymbol{x}, t).$$

The proof—albeit only formal—of these relations follows from the explicit form of $\mathbf{E}$ (3.88) by using the canonical commutation rules.

As an example we consider the translations (3.97). From (3.99), (3.110) we then obtain the generalized Heisenberg equation (3.29) which we postulated previously.

For the gauge theories that dominate particle physics today, it is only possible to carry out the canonical quantization in special gauges. In Part III this will be treated in detail.

Exercises

3.1 Verify equations (3.5) and (3.6).

3.2 Show that every real solution of the Klein–Gordon equation (3.14) can be written in the form of (3.25).

3.3 Consider a one-meson state with "wave function" (3.73) subject at $t = 0$ to the condition

$$\psi(\boldsymbol{x}, t = 0) = \delta^3(\boldsymbol{x}).$$

Show that $\psi(\boldsymbol{x}, t) \neq 0$ for $|\boldsymbol{x}| > t > 0$. The interpretation of $\psi(\boldsymbol{x}, t)$ as wave function in the sense of the Schrödinger theory, would mean that the meson can travel faster than light, in obvious contradiction to experiment. This is a generally valid conclusion (Hegerfeld 1974).

3.4 Show that $\mathrm{d}^3k/2\omega$ is a Lorentz-invariant measure and that $2\omega\delta^3(\boldsymbol{k} - \boldsymbol{k}')$ is a Lorentz-invariant distribution.

3.5 Consider a Poincaré transformation of the field operator $\Phi(x)$ according to the scheme of equations (3.17), (3.18) where Λ is from the identity component. How do the annihilation and creation operators (3.30), and the one-meson states (3.38) transform?

3.6 Construct the Lagrangian, the canonical momentum, and the Hamiltonian for the free scalar field whose Lagrange density is given in (3.74). Demonstrate that the equations of motion can be written in Hamilton's canonical form:

$$\dot{\varphi}(\boldsymbol{x}, t) = \frac{\delta H}{\delta \Pi(\boldsymbol{x}, t)},$$

$$\dot{\Pi}(\boldsymbol{x}, t) = -\frac{\delta H}{\delta \varphi(\boldsymbol{x}, t)}.$$

Show that the canonical commutation rules (3.108) lead to (3.66).

3.7 Verify equation (3.110). To do so use the relation

$$[\mathsf{F}(\varphi, \Pi), \varphi(\boldsymbol{x}, t)] = -\mathrm{i}\frac{\delta \mathsf{F}(\varphi, \Pi)}{\delta \Pi(\boldsymbol{x}, t)},$$

which, for a polynomial dependence of F on φ, Π, is easily derived by multiplying out the commutator.

3.8 Derive the conserved quantities which follow from the rotational invariance of a theory.

CHAPTER 4

The Dirac Equation and the Dirac Field

4.1 The Dirac Equation

In Chap. 3 we discussed the theory of neutral mesons. Two elementary particles that frequently occur in Nature, the electron and the proton, are however particles which possess intrinsic angular momentum: spin $= \frac{1}{2}$. Since a scalar field offers no opportunity for describing spin we must seek another description for these particles.

For nonrelativistic electrons, Pauli found an appropriate description (Pauli 1927). Within the framework of the Schrödinger picture, a nonrelativistic electron is represented by a wave function with two components:

$$\psi(\boldsymbol{x}, t) = \begin{pmatrix} \psi_1(\boldsymbol{x}, t) \\ \psi_2(\boldsymbol{x}, t) \end{pmatrix}. \tag{4.1}$$

The quantities

$$|\psi_i(\boldsymbol{x}, t)|^2 \, \mathrm{d}^3x; \qquad i = 1, 2, \tag{4.2}$$

are the probabilities of finding the electron with spin in the positive ($i = 1$) or negative ($i = 2$) z-direction in the volume element d^3x around the point $\boldsymbol{x}$. The total angular momentum operator $\vec{\mathbf{J}}$ is the sum of the orbital angular momentum operator $\vec{\mathbf{L}}$ and the spin angular momentum operator $\boldsymbol{\sigma}/2$:

$$\vec{\mathbf{J}} = \vec{\mathbf{L}} + \tfrac{1}{2}\boldsymbol{\sigma},$$

$$\vec{\mathbf{L}} = \boldsymbol{x} \times \frac{1}{\mathrm{i}} \boldsymbol{\nabla}, \tag{4.3}$$

$$\sigma^1 = \begin{pmatrix} 0 & 1 \\ 1 & 0 \end{pmatrix}, \quad \sigma^2 = \begin{pmatrix} 0 & -\mathrm{i} \\ \mathrm{i} & 0 \end{pmatrix}, \quad \sigma^3 = \begin{pmatrix} 1 & 0 \\ 0 & -1 \end{pmatrix}.$$

Here the σ^j (with $j = 1, 2, 3$) are the Pauli spin matrices. According to Pauli's theory, the wave function of a free electron must obey a Schrödinger equation:

$$\mathrm{i}\frac{\partial}{\partial t}\psi(\boldsymbol{x}, t) = -\frac{\Delta}{2m}\psi(\boldsymbol{x}, t). \tag{4.4}$$

This equation is certainly not relativistically invariant since it contains only a first derivative with respect to time but second derivatives with respect to spatial coordinates.

Dirac attempted to find a relativistically invariant equation for a field $\psi(x)$ which could describe free electrons. For reasons which appear to us today to be less than convincing, he looked for a field equation which is linear in time and spatial derivatives. We will regard this as an heuristic principle and adopt the general linear form:

$$\left\{\mathrm{i}\gamma^\mu\frac{\partial}{\partial x^\mu} - a\right\}\psi(x) = 0, \tag{4.5}$$

where the nature of the coefficients γ^μ and the constant a remains to be determined.

According to special relativity, the energy and momentum of a free electron obey the relation

$$(p^0)^2 - \boldsymbol{p}^2 = m^2 \tag{4.6}$$

where m is the mass of the electron. From our earlier discussion of mesons, we expect that the field $\psi(x)$ for an electron of definite momentum will contain a factor which corresponds to a plane wave

$$\psi(x) \propto \mathrm{e}^{-\mathrm{i}px} = \mathrm{e}^{-\mathrm{i}(p^0t - \boldsymbol{p}\boldsymbol{x})}. \tag{4.7}$$

A field of this type will then also satisfy the Klein–Gordon equation

$$(\Box + m^2)\psi(x) = 0. \tag{4.8}$$

How can we make the form (4.5) compatible with (4.8)?

We recall the simple relationship for two complex numbers u and v:

$$u^2 + v^2 = (-\mathrm{i}u - v)(\mathrm{i}u - v). \tag{4.9}$$

Using this analogy we now attempt to linearize the Klein–Gordon equation:

$$\left(-\mathrm{i}\gamma^\mu\frac{\partial}{\partial x^\mu} - m\right)\left(\mathrm{i}\gamma^\nu\frac{\partial}{\partial x^\nu} - m\right)\psi(x) = 0. \tag{4.10}$$

By multiplying this out, the following equation is obtained by making use of the commutability of the partial derivatives, but respecting the order of the coefficients γ^μ, γ^ν since we do not want to exclude the possibility that they are matrices:

$$\left\{\tfrac{1}{2}(\gamma^\mu\gamma^\nu + \gamma^\nu\gamma^\mu)\frac{\partial^2}{\partial x^\mu \partial x^\nu} + m^2\right\}\psi(x) = 0. \tag{4.11}$$

This equation is identical to the Klein–Gordon equation provided we insist that

$$\{\gamma^\mu, \gamma^\nu\} \equiv \gamma^\mu\gamma^\nu + \gamma^\nu\gamma^\mu = 2g^{\mu\nu} \qquad (\mu, \nu = 0, 1, 2, 3). \tag{4.12}$$

How can we ensure that these algebraic relations are satisfied? It is easy to see that the γ^μ cannot be ordinary complex numbers. Indeed it follows from (4.12) that

$$\begin{aligned} \gamma^0\gamma^0 &= 1, \\ \gamma^0\gamma^1 + \gamma^1\gamma^0 &= 0, \\ \gamma^1\gamma^1 &= -1. \end{aligned} \tag{4.13}$$

With complex numbers, which commute with one another, it is not possible to satisfy these equations.

The next simplest approach would be to write the γ^μ as 2×2 matrices. Every 2×2 matrix $\mathbf{C}$, can be written as a linear combination of the Pauli matrices (4.3) and the unit matrix

$$\mathbf{C} = c_0 \mathbb{1} + c^j\sigma^j. \tag{4.14}$$

The Pauli matrices themselves satisfy the relations

$$\{\sigma^j, \sigma^k\} = 2\delta^{jk} \qquad (j, k = 1, 2, 3). \tag{4.15}$$

It would be tempting to set

$$\gamma^j = \mathrm{i}\sigma^j \qquad (j = 1, 2, 3), \tag{4.16}$$

since then, the spatial part of (4.12) would be satisfied. However, we also need a further matrix γ^0 that anticommutes with all the γ^j:

$$\gamma^0\gamma^j + \gamma^j\gamma^0 = 0 \qquad (j = 1, 2, 3). \tag{4.17}$$

The only linearly independent 2×2 matrix still available is, however, the unit matrix and we have

$$\mathbb{1}\cdot\sigma^j + \sigma^j\cdot\mathbb{1} = 2\sigma^j \neq 0. \tag{4.18}$$

Thus we see that 2×2 matrices are also incapable of satisfying (4.12). But with 4×4 matrices it becomes possible, as was shown by Dirac. A standard choice for the Dirac matrices is as follows:

$$\gamma^0 = \left(\begin{array}{c|c} 1 & 0 \\ \hline 0 & -1 \end{array}\right), \qquad \gamma^j = \left(\begin{array}{c|c} 0 & \sigma^j \\ \hline -\sigma^j & 0 \end{array}\right) \qquad (j = 1, 2, 3), \tag{4.19}$$

where each of the entries indicated represents a 2×2 matrix. Note that γ^0 is a Hermitian matrix and that the γ^j are anti-Hermitian ($j = 1, 2, 3$). It is easy to demonstrate that

$$\gamma^\mu\gamma^\nu + \gamma^\nu\gamma^\mu = 2g^{\mu\nu}\cdot\mathbb{1}, \tag{4.20}$$

where $\mathbb{1}$ is the unit matrix in four dimensions. In the following, as in (4.12), the unit matrix will not be written explicitly provided there is no risk of confusion.

The choice of Dirac matrices given in (4.19) is, of course, not the only possibility. We obtain an equivalent representation of the Dirac matrices if we transform the γ^μ with an arbitrary nonsingular matrix $\mathbf{S}$. In doing so, the relation (4.20) remains unchanged. For

$$\gamma'^\mu = \mathbf{S}^{-1}\gamma^\mu\mathbf{S} \tag{4.21}$$

we likewise have

$$\gamma'^\mu\gamma'^\nu + \gamma'^\nu\gamma'^\mu = 2g^{\mu\nu}.$$

The equation thus postulated by Dirac (1928) was

$$\left\{\mathrm{i}\gamma^\mu\frac{\partial}{\partial x^\mu} - m\right\}\psi(x) = 0. \tag{4.22}$$

In this, $\psi(x)$ is a four-component field known as a Dirac spinor:

$$\psi(x) = \begin{pmatrix} \psi_1(x) \\ \psi_2(x) \\ \psi_3(x) \\ \psi_4(x) \end{pmatrix}. \tag{4.23}$$

An equivalent form, which resembles the Schrödinger equation (4.4), is obtained by multiplying (4.22) by γ^0:

$$\mathrm{i}\frac{\partial}{\partial t}\psi(x) = \left(\boldsymbol{\alpha}\frac{1}{\mathrm{i}}\boldsymbol{\nabla} + \beta m\right)\psi(x). \tag{4.24}$$

Here $\boldsymbol{\alpha}$ and β are defined as

$$\boldsymbol{\alpha} = \gamma^0\boldsymbol{\gamma}; \qquad \beta = \gamma^0.$$

In our standard representation (4.19) β and the components of $\boldsymbol{\alpha}$ are Hermitian matrices with

$$\boldsymbol{\alpha} = \left(\begin{array}{c|c} 0 & \boldsymbol{\sigma} \\ \hline \boldsymbol{\sigma} & 0 \end{array}\right); \qquad \beta = \left(\begin{array}{c|c} 1 & 0 \\ \hline 0 & -1 \end{array}\right). \tag{4.25}$$

Before moving on to discuss the interpretation of the Dirac spinor, we will look for solutions to the Dirac equation and familiarize ourselves with their properties. In this we consider $\psi_1(x), \ldots, \psi_4(x)$ as classical complex-valued functions of x.

4.2 Solutions of the Dirac Equation

An obvious approach towards solving the Dirac equation is to make an ansatz in terms of the plane waves (4.7). Since every solution of the Dirac equation also satisfies the Klein–Gordon equation, we already know that the relativistic relationship between angular frequency and wave vector (4.6) must be valid. We therefore expect, as in the case of the solutions for the meson field, both positive and negative frequencies (3.24).

To begin with we consider the solutions of negative frequency and write

$$\psi(x) = \mathrm{e}^{-\mathrm{i}px}\,u(p). \tag{4.26}$$

Here, $u(p)$ is a yet to be determined spinor and

$$px = p^0 t - \boldsymbol{p}\boldsymbol{x},$$
$$p^0 = +\sqrt{\boldsymbol{p}^2 + m^2}.$$

Inserting this in the Dirac equation (4.22) gives

$$(\not{p} - m)u(p) = 0. \tag{4.27}$$

We have introduced here the symbol $\not{p}$ to represent the frequently occurring 4×4 matrix $p^\mu \gamma_\mu$:

$$\not{p} \equiv p^\mu \gamma_\mu.$$

We assume initially $\boldsymbol{p} = \boldsymbol{0}$, or for the four-vector

$$p = p_{\mathrm{R}} \equiv \begin{pmatrix} m \\ 0 \\ 0 \\ 0 \end{pmatrix}. \tag{4.28}$$

As we will show, this corresponds to an electron at rest (subscript R). Making use in (4.27) of the explicit form (4.19) of the γ-matrices gives

$$\left(\begin{array}{c|c} 0 & 0 \\ \hline 0 & -2m \end{array}\right) u(p_{\mathrm{R}}) = 0. \tag{4.29}$$

This equation has two linearly independent solutions given below in a notation where the two upper and two lower components of the Dirac spinor u are each combined to form a two-component spinor:

$$u_s(p_{\mathrm{R}}) = \sqrt{2m} \begin{pmatrix} \chi_s \\ 0 \end{pmatrix} \qquad (s = \pm\tfrac{1}{2}). \tag{4.30}$$

The two-component spinors χ_s are defined as

$$\chi_{1/2} = \begin{pmatrix} 1 \\ 0 \end{pmatrix}, \qquad \chi_{-1/2} = \begin{pmatrix} 0 \\ 1 \end{pmatrix}. \tag{4.31}$$

For a general four-vector p we write

$$u(p) = \begin{pmatrix} \xi \\ \eta \end{pmatrix}, \tag{4.32}$$

where ξ and η are two-component spinors. From (4.27) we obtain

$$(p^0\gamma^0 - \boldsymbol{p}\boldsymbol{\gamma} - m)u(p) = 0, \tag{4.33}$$

$$\begin{pmatrix} p^0 - m & -\boldsymbol{p}\boldsymbol{\sigma} \\ \boldsymbol{p}\boldsymbol{\sigma} & -p^0 - m \end{pmatrix} \begin{pmatrix} \xi \\ \eta \end{pmatrix} = 0, \tag{4.34}$$

$$(p^0 - m)\xi - (\boldsymbol{p}\boldsymbol{\sigma})\eta = 0, \tag{4.35a}$$

$$(\boldsymbol{p}\boldsymbol{\sigma})\xi - (p^0 + m)\eta = 0. \tag{4.35b}$$

As a consequence of (4.35b)

$$\eta = \frac{\boldsymbol{p\sigma}}{p^0 + m}\xi. \tag{4.36}$$

With this (4.35a) is also satisfied since

$$\left(p^0 - m - \frac{(\boldsymbol{p\sigma})^2}{p^0 + m}\right)\xi = \left(p^0 - m - \frac{\boldsymbol{p}^2}{p^0 + m}\right)\xi$$
$$= \frac{(p^0)^2 - m^2 - \boldsymbol{p}^2}{p^0 + m}\xi = 0. \tag{4.37}$$

The two-component spinor ξ is therefore completely arbitrary. We obtain, also for general four-momentum p, two linearly independent solutions of negative frequency, which we will normalize in the following way:

$$u_s(p) = \sqrt{p^0 + m}\cdot\begin{pmatrix} \chi_s \\ \dfrac{\boldsymbol{\sigma p}}{p^0 + m}\chi_s \end{pmatrix} \qquad (s = \pm\tfrac{1}{2}). \tag{4.38}$$

In complete analogy we may look for solutions with positive frequency by adopting the form

$$\psi(x) = \mathrm{e}^{\mathrm{i}px}\, v(p), \tag{4.39}$$

where $v(p)$ represents the Dirac spinor to be determined. Substitution of this form in the Dirac equation gives

$$(\not{p} + m)v(p) = 0, \tag{4.40}$$

and again, for every momentum, there are two linearly independent solutions. These will be chosen, for reasons which will later become clear, as follows:

$$v_s(p) = -\sqrt{p^0 + m}\begin{pmatrix} \dfrac{(\boldsymbol{\sigma p})}{p^0 + m}\varepsilon\chi_s \\ \varepsilon\chi_s \end{pmatrix} \qquad (s = \pm\tfrac{1}{2}), \tag{4.41}$$

where

$$\varepsilon = \begin{pmatrix} 0 & 1 \\ -1 & 0 \end{pmatrix}.$$

The Dirac equation is a linear differential equation for the spinor $\psi(x)$. Thus a superposition of solutions to the equation is again a solution. A general superposition of our plane waves of positive and negative frequency can be written as a Fourier integral:

$$\psi(x) = \int \frac{\mathrm{d}^3 p}{(2\pi)^3}\frac{1}{2p^0}\cdot \sum_{s=\pm 1/2} \{\mathrm{e}^{\mathrm{i}px}\, v_s(p)\beta_s^*(\boldsymbol{p}) + \mathrm{e}^{-\mathrm{i}px}\, u_s(p)\alpha_s(\boldsymbol{p})\}. \tag{4.42}$$

Here, $\alpha_s(\boldsymbol{p})$ and $\beta_s^*(\boldsymbol{p})$ are arbitrary complex-valued functions. They are independent because, in contrast to the case of the meson field, we do not demand that the Dirac field should be real. It is possible to show that all sufficiently well-behaved solutions to the Dirac equation may be represented in the above form. The plane wave solutions which we have considered form a complete system of solutions.

4.3 Transformation Rules for the Dirac Field

We will now examine whether, and in what sense, the Dirac equation is relativistically invariant. We consider the transition to a new inertial system (2.28) where Λ belongs to the identity component:

$$x'^\mu = \Lambda^\mu{}_\nu x^\nu + a^\mu. \tag{4.43}$$

We begin, in analogy to the scalar field (3.18), by writing the Dirac field in the primed frame as

$$\tilde{\psi}(x') = \psi(x) = \psi(\Lambda^{-1}(x' - a)). \tag{4.44}$$

From the Dirac equation for $\psi(x)$ it follows, however, for $\tilde{\psi}(x)$ that

$$\left\{ \mathrm{i}(\Lambda^\mu{}_\rho \gamma^\rho) \frac{\partial}{\partial x'^\mu} - m \right\} \tilde{\psi}(x') = 0. \tag{4.45}$$

This is not identical to the original Dirac equation but, on the other hand, it differs only slightly. If we set

$$\gamma'^\mu = \Lambda^\mu{}_\rho \gamma^\rho, \tag{4.46}$$

then the γ'^μ likewise satisfy the fundamental anticommutation rules (4.12)

$$\{\gamma'^\mu, \gamma'^\nu\} = 2g^{\mu\nu}. \tag{4.47}$$

Such matrices γ'^μ have already been considered in (4.21). An obvious question is whether there exists a nonsingular matrix $\mathbf{S}(\Lambda)$, such that

$$\gamma'^\mu = \Lambda^\mu{}_\rho \gamma^\rho \stackrel{?}{=} \mathbf{S}^{-1}(\Lambda) \gamma^\mu \mathbf{S}(\Lambda). \tag{4.48}$$

As we will show, it is in fact possible to satisfy (4.48). Thence follows the relativistic invariance of the Dirac equation. From (4.45) we obtain

$$\mathbf{S}^{-1}(\Lambda) \left\{ \mathrm{i}\gamma^\mu \frac{\partial}{\partial x'^\mu} - m \right\} \mathbf{S}(\Lambda) \tilde{\psi}(x') = 0. \tag{4.49}$$

Therefore, if we write the Dirac field in the primed frame as

$$\psi'(x') = \mathbf{S}(\Lambda)\psi(x) = \mathbf{S}(\Lambda)\psi(\Lambda^{-1}(x' - a)), \tag{4.50}$$

then $\psi'(x')$ again satisfies the Dirac equation

$$\left\{ \mathrm{i}\gamma^\mu \frac{\partial}{\partial x'^\mu} - m \right\} \psi'(x') = 0. \tag{4.51}$$

In conclusion, not only the argument, but also the components of the Dirac field, must be linearly transformed.

We now wish to construct the matrix $\mathbf{S}(\Lambda)$ of (4.48). We consider first an infinitesimal transformation Λ with

$$\Lambda^{\mu}{}_{\nu} = \delta^{\mu}_{\nu} + \delta h^{\mu}{}_{\nu}, \tag{4.52}$$

where the $\delta h^{\mu}{}_{\nu}$ are to be treated as first-order quantities.

From equation (2.26) for $\Lambda^{\mu}{}_{\nu}$ it follows that $\delta h^{\mu\nu}$ must be antisymmetric

$$\delta h^{\mu\nu} + \delta h^{\nu\mu} = 0. \tag{4.53}$$

The matrix $\mathbf{S}(\Lambda)$ that we are seeking for the infinitesimal transformation (4.52) is given by

$$\mathbf{S}(\Lambda) = \mathbb{1} - \frac{\mathrm{i}}{4}\delta h^{\rho\sigma}\sigma_{\rho\sigma}, \tag{4.54}$$

where

$$\sigma_{\rho\sigma} = \frac{\mathrm{i}}{2}(\gamma_\rho\gamma_\sigma - \gamma_\sigma\gamma_\rho) = \frac{\mathrm{i}}{2}[\gamma_\rho, \gamma_\sigma].$$

With the help of the identity

$$[[\mathbf{A}, \mathbf{B}], \mathbf{C}] = \{\mathbf{A}, \{\mathbf{B}, \mathbf{C}\}\} - \{\mathbf{B}, \{\mathbf{A}, \mathbf{C}\}\}, \tag{4.55}$$

valid for arbitrary matrices $\mathbf{A}$, $\mathbf{B}$, and $\mathbf{C}$, we find

$$\begin{aligned}
\mathbf{S}^{-1}(\Lambda)\gamma^\mu\mathbf{S}(\Lambda) &= \gamma^\mu + \frac{\mathrm{i}}{4}\delta h_{\rho\sigma}[\sigma^{\rho\sigma}, \gamma^\mu] \\
&= \gamma^\mu - \tfrac{1}{8}\delta h_{\rho\sigma}[[\gamma^\rho, \gamma^\sigma], \gamma^\mu] \\
&= \gamma^\mu - \tfrac{1}{8}\delta h_{\rho\sigma}(\{\gamma^\rho, \{\gamma^\sigma, \gamma^\mu\}\} - \{\gamma^\sigma, \{\gamma^\rho, \gamma^\mu\}\}) \\
&= \gamma^\mu - \tfrac{1}{8}\delta h_{\rho\sigma}(4g^{\sigma\mu}\gamma^\rho - 4g^{\rho\mu}\gamma^\sigma) \\
&= \gamma^\mu + \delta h^{\mu}{}_{\sigma}\gamma^\sigma = \Lambda^{\mu}{}_{\sigma}\gamma^\sigma.
\end{aligned} \tag{4.56}$$

Thus we have demonstrated the existence of a matrix $\mathbf{S}(\Lambda)$ which, for infinitesimal Λ, satisfies (4.48). As can be shown using group-theoretical methods, it follows from this that for all finite transformations Λ belonging to the identity component, a matrix $\mathbf{S}(\Lambda)$ exists. In essence, this stems from the fact that all such finite transformations can be constructed from infinitesimal transformations.

We will examine in more detail the special case of the three-dimensional rotations. For an infinitesimal rotation we have

$$\begin{aligned}
t' &= t, \\
x'^j &= R^{jk}x^k = (\delta^{jk} + \varepsilon^{jkl}\delta\varphi^l)x^k.
\end{aligned} \tag{4.57}$$

In this, the magnitude and direction of $\delta\boldsymbol{\varphi}$ give the angle and axis of rotation, respectively. The following further relations are valid

$$\delta h^{jk} = -\varepsilon^{jkl}\delta\varphi^l, \tag{4.58}$$

$$\mathsf{S}(\mathsf{R}) = \mathbb{1} - \tfrac{1}{8}[\gamma_j, \gamma_k]\varepsilon^{jkl}\delta\varphi^l = \mathbb{1} + \mathrm{i}\delta\boldsymbol{\varphi}\left(\begin{array}{c|c} \frac{1}{2}\boldsymbol{\sigma} & 0 \\ \hline 0 & \frac{1}{2}\boldsymbol{\sigma} \end{array}\right), \tag{4.59}$$

$$\psi'(\boldsymbol{x}', t) = \mathsf{S}(\mathsf{R})\psi(\mathsf{R}^{-1}\boldsymbol{x}', t),$$

$$\psi'(\boldsymbol{x}', t) = \left[\mathbb{1} + \mathrm{i}\delta\varphi^l \cdot \left\{\varepsilon^{ljk}x'^j \frac{\partial}{\mathrm{i}\partial x'^k} + \left(\begin{array}{c|c} \frac{1}{2}\sigma^l & 0 \\ \hline 0 & \frac{1}{2}\sigma^l \end{array}\right)\right\}\right] \cdot \psi(\boldsymbol{x}', t). \tag{4.60}$$

The first term in curly brackets (4.60) corresponds to the orbital angular momentum. For a scalar meson field, only this term is present. What is new here, is the additional term with the Pauli matrices. As we will show in more detail in the next section, this term describes the spin angular momentum of the electron.

We also want to study the behavior of the Dirac field at the point $\boldsymbol{x} = 0$ under a rotation. From (4.60) we obtain for an infinitesimal rotation about the third coordinate axis:

$$\psi'(0, t) = \left\{\mathbb{1} + \mathrm{i}\delta\varphi\left(\begin{array}{c|c} \frac{1}{2}\sigma^3 & 0 \\ \hline 0 & \frac{1}{2}\sigma^3 \end{array}\right)\right\}\psi(0, t). \tag{4.61}$$

For a finite rotation about the third axis it is then easy to deduce that

$$\psi'(0, t) = \left\{\exp\left[\mathrm{i}\varphi\left(\begin{array}{c|c} \frac{1}{2}\sigma^3 & 0 \\ \hline 0 & \frac{1}{2}\sigma^3 \end{array}\right)\right]\right\} \cdot \psi(0, t). \tag{4.62}$$

If we take a rotation through an angle $\varphi = 2\pi$, then we obtain for the matrix in (4.62)

$$\exp\left[2\pi\mathrm{i}\left(\begin{array}{c|c} \frac{1}{2}\sigma^3 & 0 \\ \hline 0 & \frac{1}{2}\sigma^3 \end{array}\right)\right]$$

$$= \exp\begin{pmatrix} \mathrm{i}\pi & 0 & 0 & 0 \\ 0 & -\mathrm{i}\pi & 0 & 0 \\ 0 & 0 & \mathrm{i}\pi & 0 \\ 0 & 0 & 0 & -\mathrm{i}\pi \end{pmatrix} = \begin{pmatrix} -1 & 0 & 0 & 0 \\ 0 & -1 & 0 & 0 \\ 0 & 0 & -1 & 0 \\ 0 & 0 & 0 & -1 \end{pmatrix}. \tag{4.63}$$

A rotation through an angle of 2π, which leaves everything unchanged in space, produces a change in sign of the Dirac spinor! We must conclude from this that the Dirac spinor itself is not directly observable. For neutrons, which, as spin-$\frac{1}{2}$ particles, are also described by the Dirac equation, it has been possible to directly demonstrate experimentally this change in sign of the Dirac field under a rotation of 2π (Rauch 1975; Werner 1975). What is actually done in these experiments is a phase comparison between a reference field and a rotated field in an interference device. Thus *two* fields are involved and there is no contradiction of our above statement that a *single* Dirac field is not directly observable.

Observable quantities should be real and may not change their sign under a rotation of 2π. Candidates for such quantities in the framework of the Dirac theory can be readily written down; for example,

$$\tilde{A}(x) = \psi^\dagger(x)\tilde{\mathbf{M}}\psi(x), \tag{4.64}$$

where $\tilde{\mathbf{M}} = \tilde{\mathbf{M}}^\dagger$ is an arbitrary Hermitian 4×4 matrix and $\psi^\dagger$ is the Hermitian conjugate of the spinor ψ. How do these quantities behave under Poincaré transformations? From (4.50) we find for the Hermitian conjugate spinor field

$$\psi'^\dagger(x') = \psi^\dagger(\Lambda^{-1}(x' - a))\mathbf{S}^\dagger(\Lambda). \tag{4.65}$$

The matrix $\mathbf{S}(\Lambda)$, as is easily seen from (4.54), is, in general, not a unitary matrix. The bilinear expression (4.64) with $\mathbf{M} = \mathbb{1}$ is therefore not a relativistic scalar field. The transformation properties of expressions of the form (4.64) becomes more apparent when we use, instead of the Hermitian conjugate spinor, the so-called adjoint spinor.

We define quite generally the *Dirac adjoint* for spinors and 4×4 matrices $\mathbf{M}$ via

$$\begin{aligned} \bar{\psi}(x) &= \psi^\dagger(x)\gamma^0, \\ \bar{\mathbf{M}} &= \gamma^0\mathbf{M}^\dagger\gamma^0. \end{aligned} \tag{4.66}$$

According to our convention (4.19), γ^0 is a Hermitian matrix with $\gamma^0\gamma^0 = \mathbb{1}$. The γ^j (with $j = 1, 2, 3$) are anti-Hermitian. From this we readily arrive at the following relations:

$$\begin{aligned} \overline{\mathbf{M}_1 \cdot \mathbf{M}_2} &= \bar{\mathbf{M}}_2 \cdot \bar{\mathbf{M}}_1, \\ (\bar{\psi}_1\mathbf{M}\psi_2)^* &= \bar{\psi}_2\bar{\mathbf{M}}\psi_1, \\ \bar{\mathbb{1}} &= \mathbb{1}, \\ \bar{\gamma}^\mu &= \gamma^\mu. \end{aligned} \tag{4.67}$$

Furthermore, we assert that

$$\bar{\mathbf{S}}(\Lambda) = \gamma^0\mathbf{S}^\dagger(\Lambda)\gamma^0 = \mathbf{S}^{-1}(\Lambda). \tag{4.68}$$

For the time being we demonstrate the validity of this relation for infinitesimal transformations Λ. According to (4.54) we have

$$\begin{aligned} \bar{\mathbf{S}}(\Lambda) &= \mathbb{1} + \tfrac{1}{8}\delta h^{\rho\sigma}\overline{[\gamma_\rho, \gamma_\sigma]} \\ &= \mathbb{1} + \tfrac{1}{8}\delta h^{\rho\sigma}[\gamma_\sigma, \gamma_\rho] \\ &= \mathbb{1} + \frac{\mathrm{i}}{4}\delta h^{\rho\sigma}\sigma_{\rho\sigma} \\ &= \mathbf{S}^{-1}(\Lambda). \end{aligned} \tag{4.69}$$

The extension of this proof to finite transformations is the subject of Exercise 4.3. From (4.68) and (4.65) we find the following transformation behavior for the adjoint Dirac field:

$$\bar{\psi}'(x') = \bar{\psi}(\Lambda^{-1}(x' - a))\mathbf{S}^{-1}(\Lambda). \tag{4.70}$$

We can now write the bilinear expression (4.64) in an equivalent form where $\mathbf{M}$ is a 4×4 matrix:

$$A(x) = \bar{\psi}(x)\mathbf{M}\psi(x). \tag{4.71}$$

This field is real if

$$\mathbf{M} = \overline{\mathbf{M}}. \tag{4.72}$$

For Poincaré transformations we have, according to (4.50), (4.70)

$$\begin{aligned} A(x) \to A'(x') &= \bar{\psi}'(x')\mathbf{M}\psi'(x') \\ &= \bar{\psi}(\Lambda^{-1}(x' - a))\mathbf{M}'\psi(\Lambda^{-1}(x' - a)), \end{aligned} \tag{4.73}$$

where

$$\mathbf{M}' = \mathbf{S}^{-1}(\Lambda)\mathbf{M}\mathbf{S}(\Lambda).$$

There are clearly 16 linearly independent matrices $\mathbf{M}$. We wish to choose a basis that has simple transformation properties under the Lorentz transformation $\mathbf{M} \to \mathbf{M}'$. We begin by seeking invariant matrices. The unit matrix $\mathbb{1}$ is certainly invariant:

$$\mathbb{1} \to \mathbf{S}^{-1}(\Lambda)\mathbb{1}\mathbf{S}(\Lambda) = \mathbb{1}. \tag{4.74}$$

There is, however, a further 4×4 matrix which, for proper Lorentz transformations, is invariant. This is the matrix

$$\gamma_5 = \mathrm{i}\gamma^0\gamma^1\gamma^2\gamma^3. \tag{4.75}$$

Since γ-matrices with differing indices anticommute, it is also possible to write γ_5 with the help of the totally antisymmetric tensor $\varepsilon_{\mu\nu\rho\sigma}$ as

$$\gamma_5 = \frac{\mathrm{i}}{4!}\varepsilon_{\mu\nu\rho\sigma}\gamma^\mu\gamma^\nu\gamma^\rho\gamma^\sigma. \tag{4.76}$$

If we transform γ_5 then we find with the help of (4.48)

$$\begin{aligned} \mathbf{S}^{-1}(\Lambda)\gamma_5\mathbf{S}(\Lambda) &= \frac{\mathrm{i}}{4!}\varepsilon_{\mu\nu\rho\sigma}\Lambda^\mu{}_{\mu'}\Lambda^\nu{}_{\nu'}\Lambda^\rho{}_{\rho'}\Lambda^\sigma{}_{\sigma'}\gamma^{\mu'}\gamma^{\nu'}\gamma^{\rho'}\gamma^{\sigma'} \\ &= (\det\Lambda)\gamma_5. \end{aligned} \tag{4.77}$$

Thus for proper Lorentz transformations ($\det\Lambda = 1$) the matrix γ_5 is invariant. As we will see it reverses its sign upon a reflection in space. The matrix γ_5 thus behaves as a pseudoscalar quantity. It is easily shown that it anticommutes with all matrices γ^μ:

$$\gamma^\mu\gamma_5 + \gamma_5\gamma^\mu = 0. \tag{4.78}$$

In our standard representation for the matrices γ^μ (4.19), γ_5 has the form

$$\gamma_5 = \left(\begin{array}{c|c} 0 & 1 \\ \hline 1 & 0 \end{array}\right). \tag{4.79}$$

It now becomes a simple matter to list 16 suitable matrices forming a basis. We choose and denote them as follows:

$$\mathbf{M} = \begin{cases} \mathbb{1} & \text{scalar,} \\ \gamma^\mu & \text{vector,} \\ \sigma^{\mu\nu} = \frac{i}{2}[\gamma^\mu, \gamma^\nu] & \text{antisymmetric tensor,} \\ \gamma^\mu \gamma_5 & \text{pseudo-vector,} \\ i\gamma_5 & \text{pseudo-scalar.} \end{cases} \tag{4.80}$$

It can easily be checked that all these matrices satisfy the reality condition (4.72). We will often be meeting bilinear expressions of the form $\bar{\psi}(x)\mathbf{M}\psi(x)$ that are constructed with these matrices. For example, the four-vector $\bar{\psi}(x)\gamma^\mu\psi(x)$ is proportional to the electromagnetic current density of the electron field.

4.4 Quantization and Interpretation of the Dirac Field

Dirac's original motivation was to find a relativistic analogue of the Schrödinger equation. Consequently, the first attempts to interpret the Dirac spinor were in terms of a Schrödinger-like wave function. However, it very soon became clear that such an interpretation leads to insurmountable difficulties. The reasons for this are the following physical facts:

The probability distribution for finding a particle at a particular point in space—a concept which plays such an important role in nonrelativistic quantum mechanics—is a notion of only limited significance within a *relativistic* theory. To see this, let us make a measurement of the position of a particle with an uncertainty Δx. According to Heisenberg's uncertainty principle, the particle then acquires an uncertainty in its momentum of

$$\Delta p \gtrsim \frac{1}{\Delta x}. \tag{4.81}$$

If we choose Δx to be sufficiently small, then the energy uncertainty ΔE, becomes sufficiently large that further particles can be produced. This will occur for

$$\Delta x \lesssim \frac{1}{m}, \tag{4.82}$$

when we demand a spatial resolution smaller than the *Compton wavelength* $1/m$ of the observed particle. Then we have

$$\Delta E \sim \Delta p \gtrsim \frac{1}{\Delta x} \gtrsim m, \tag{4.83}$$

so that the energy uncertainty is greater than the mass of the particle. In this event, the measurement of the position of *one* particle, will generally lead to a state with *several* particles. Thus we are forced to give up the one-particle theory and must look instead to the many-particle theory, which means to the quantum field theory.

We therefore regard the Dirac field as a *field operator*. As in the case of the meson field, we expand the field operator in terms of plane waves where the expansion coefficients will themselves be operators. From the Fourier expansion (4.42), with the substitutions $\alpha_s(\boldsymbol{p}) \to \mathbf{a}_s(\boldsymbol{p})$, $\beta_s^*(\boldsymbol{p}) \to \mathbf{b}_s^\dagger(\boldsymbol{p})$, we obtain

$$\psi(x) = \int \frac{d^3p}{(2\pi)^3} \frac{1}{2p^0} \sum_{s=\pm 1/2} \{e^{ipx} v_s(p)\mathbf{b}_s^\dagger(\boldsymbol{p}) + e^{-ipx} u_s(p)\mathbf{a}_s(\boldsymbol{p})\}. \quad (4.84)$$

The nature of the operators $\mathbf{a}$ and $\mathbf{b}^\dagger$ remains to be clarified. We also demand that the Heisenberg equation retains its validity

$$\frac{\partial \psi(x)}{\partial x^\mu} = i[\mathbf{P}_\mu, \psi(x)]. \quad (4.85)$$

As in the case of the meson field, we derive from this

$$\begin{aligned} &[\mathbf{P}_\mu, \mathbf{a}_s^\dagger(\boldsymbol{p})] = p_\mu \mathbf{a}_s^\dagger(\boldsymbol{p}), \\ &[\mathbf{P}_\mu, \mathbf{b}_s^\dagger(\boldsymbol{p})] = p_\mu \mathbf{b}_s^\dagger(\boldsymbol{p}), \\ &[\mathbf{P}_\mu, \mathbf{a}_s(\boldsymbol{p})] = -p_\mu \mathbf{a}_s(\boldsymbol{p}), \\ &[\mathbf{P}_\mu, \mathbf{b}_s(\boldsymbol{p})] = -p_\mu \mathbf{b}_s(\boldsymbol{p}), \\ &(s = \pm\tfrac{1}{2}). \end{aligned} \quad (4.86)$$

Consistent with the signs in (4.86) we require that the annihilation operators $\mathbf{a}$ and $\mathbf{b}$ give zero when applied to the vacuum state

$$\begin{aligned} \mathbf{a}_s(\boldsymbol{p})|0\rangle &= 0, \\ \mathbf{b}_s(\boldsymbol{p})|0\rangle &= 0. \end{aligned} \quad (4.87)$$

Instead of a single set of creation operators, we here have four of them. Correspondingly, for every fixed momentum $\boldsymbol{p}$ we can construct four one-particle states

$$\mathbf{a}_s^\dagger(\boldsymbol{p})|0\rangle, \quad (4.88a)$$

$$\mathbf{b}_s^\dagger(\boldsymbol{p})|0\rangle, \quad (4.88b)$$

$$(s = \pm\tfrac{1}{2}).$$

As we shall see, none of these four states is equal to the null vector. An electron of definite momentum has two linearly independent spin states and we will attempt to identify these with the states (4.88a). If we are to take the theory seriously, we must also consider the states (4.88b) and demand that a further particle exists with exactly the same mass as the electron. This consequence of Dirac's theory (Dirac 1930; Oppenheimer 1930) was brilliantly confirmed through the discovery of the positron (Anderson 1932, 1933). The states (4.88b) will thus be identified with positron states. We will also see that the Dirac theory automatically endows electrons and positrons with opposite charges.

We now have to give some thought to the question of what algebra is required for the creation and annihilation operators of the Dirac field. We could attempt to use the same commutation rules as for the meson field:

$$\begin{aligned}
[\mathbf{a}_r(p), \mathbf{a}_s^\dagger(p')] &= \delta_{rs}(2\pi)^3 2p^0 \delta^3(p - p'), \\
[\mathbf{b}_r(p), \mathbf{b}_s^\dagger(p')] &= \delta_{rs}(2\pi)^3 2p^0 \delta^3(p - p'),
\end{aligned} \tag{4.89}$$

with all other commutators being equal to zero. However, with this assumption we are led to a nonzero commutator for space-like separations. For equal times we would find

$$[\psi(x, t), \bar{\psi}(y, t)] \neq 0 \qquad \text{for} \quad x \neq y, \tag{4.90}$$

in contradiction to the demands of microcausality [see (3.50)]. It could be argued that this is not important since the Dirac spinor is not directly observable. However, (4.90) also leads to a breaking of microcausality for bilinear expressions in the Dirac field operator (4.71) which we want to identify with observable fields. Thus the requirement of microcausality allows us to conclude that electrons cannot be bosons. In actual fact, we know from experiment that electrons obey the Pauli exclusion principle and Fermi statistics.

The correct algebraic rules for the creation and annihilation operators were put forward by Jordan and Wigner (Jordan 1927, 1928a). Instead of (4.89), what is required are anticommutation rules:

$$\begin{aligned}
\{\mathbf{a}_r(p), \mathbf{a}_s^\dagger(p')\} &= \delta_{rs}(2\pi)^3 2p^0 \delta^3(p - p'), \\
\{b_r(p), b_s^\dagger(p')\} &= \delta_{rs}(2\pi)^3 2p^0 \delta^3(p - p'), \\
\{\mathbf{a}_r^\dagger(p), \mathbf{a}_s^\dagger(p')\} &= \{\mathbf{a}_r(p), \mathbf{a}_s(p')\} \\
&= \{\mathbf{b}_r^\dagger(p), \mathbf{b}_s^\dagger(p')\} = \{\mathbf{b}_r(p), \mathbf{b}_s(p')\} \\
&= \{\mathbf{a}_r^\dagger(p), \mathbf{b}_s(p')\} = \{\mathbf{a}_r^\dagger(p), \mathbf{b}_s^\dagger(p')\} = 0.
\end{aligned} \tag{4.91}$$

With these anticommutation rules, a simple calculation leads to

$$\{\psi(x, t), \psi(y, t)\} = \{\bar{\psi}(x, t), \bar{\psi}(y, t)\} = 0, \tag{4.92a}$$

$$\{\psi(x, t), \bar{\psi}(y, t)\} = \gamma^0 \delta^3(x - y). \tag{4.92b}$$

Let us re-examine (4.92b). If we substitute into this the expansion of the field operator (4.84) then we obtain:

$$\begin{aligned}
\{\psi(x, t), \bar{\psi}(y, t)\} = &\int \frac{\mathrm{d}^3p\, \mathrm{d}^3p'}{(2\pi)^6} \frac{1}{2p_0 2p_0'} \\
&\cdot \sum_{r,s} [\mathrm{e}^{\mathrm{i}px} \mathrm{e}^{-\mathrm{i}p'y} v_r(p)\bar{v}_s(p') \{\mathbf{b}_r^\dagger(p), \mathbf{b}_s(p')\} \\
&+ \mathrm{e}^{-\mathrm{i}px} \mathrm{e}^{\mathrm{i}p'y} u_r(p)\bar{u}_s(p') \{\mathbf{a}_r(p), \mathbf{a}_s^\dagger(p')\}] \\
= &\int \frac{\mathrm{d}^3p}{(2\pi)^3} \frac{1}{2p^0} \cdot \Bigg(\mathrm{e}^{-\mathrm{i}p(x-y)} \sum_s v_s(p)\bar{v}_s(p) \\
&+ \mathrm{e}^{\mathrm{i}p(x-y)} \sum_s u_s(p)\bar{u}_s(p) \Bigg).
\end{aligned} \tag{4.93}$$

We meet here, for the first time, the spin sum of the spinors u and v which we will later be requiring frequently. From the explicit expressions (4.38), (4.41) we find

$$\sum_{s=\pm 1/2} u_s(p)\bar{u}_s(p) = \not{p} + m, \qquad \sum_{s=\pm 1/2} v_s(p)\bar{v}_s(p) = \not{p} - m. \tag{4.94}$$

Substituting this into (4.93) gives

$$\begin{aligned}\{\psi(\boldsymbol{x}, t), \bar{\psi}(\boldsymbol{y}, t)\} &= \int \frac{d^3p}{(2\pi)^3} \frac{1}{2p^0} (e^{-i\boldsymbol{p}(\boldsymbol{x}-\boldsymbol{y})} (p^0\gamma^0 - \boldsymbol{p\gamma} - m) \\ &\quad + e^{i\boldsymbol{p}(\boldsymbol{x}-\boldsymbol{y})} (p^0\gamma^0 - \boldsymbol{p\gamma} + m)) \\ &= \gamma_0 \int \frac{d^3p}{(2\pi)^3} e^{i\boldsymbol{p}(\boldsymbol{x}-\boldsymbol{y})} = \gamma_0 \delta^3(\boldsymbol{x} - \boldsymbol{y}).\end{aligned} \tag{4.95}$$

For observable fields, such as those given in (4.71), this now leads to commutation rules which are compatible with the principle of *microcausality* (3.47). For example, with any 4×4 matrices $\mathbf{M}_1$ and $\mathbf{M}_2$, we find

$$[\bar{\psi}(\boldsymbol{x}, t)\mathbf{M}_1\psi(\boldsymbol{x}, t), \bar{\psi}(\boldsymbol{y}, t)\mathbf{M}_2\psi(\boldsymbol{y}, t)] = 0 \qquad \text{for} \quad \boldsymbol{x} \neq \boldsymbol{y}. \tag{4.96}$$

This equation follows immediately from (4.92) and from the identity

$$[\mathbf{AB}, \mathbf{CD}] = \mathbf{A}\{\mathbf{B}, \mathbf{C}\}\mathbf{D} - \mathbf{AC}\{\mathbf{B}, \mathbf{D}\} - \mathbf{C}\{\mathbf{A}, \mathbf{D}\}\mathbf{B} + \{\mathbf{C}, \mathbf{A}\}\mathbf{DB}, \tag{4.97}$$

which is valid for arbitrary operators **A**, **B**, **C**, and **D**.

We next move on to study the consequences of the anticommutation rules (4.91) for the states. For one-electron and one-positron states of definite momentum

$$|e^-(\boldsymbol{p}, s)\rangle = \mathbf{a}_s^\dagger(\boldsymbol{p})|0\rangle,$$
$$|e^+(\boldsymbol{p}, s)\rangle = \mathbf{b}_s^\dagger(\boldsymbol{p})|0\rangle,$$

we find from (4.91) the expected continuum normalization:

$$\begin{aligned}\langle e^-(\boldsymbol{p}', r)|e^-(\boldsymbol{p}, s)\rangle &= \langle 0|\{\mathbf{a}_r(\boldsymbol{p}'), \mathbf{a}_s^\dagger(\boldsymbol{p})\}|0\rangle \\ &= \delta_{rs}(2\pi)^3 2p^0 \delta^3(\boldsymbol{p}' - \boldsymbol{p}), \\ \langle e^+(\boldsymbol{p}', r)|e^+(\boldsymbol{p}, s)\rangle &= \delta_{rs}(2\pi)^3 2p^0 \delta^3(\boldsymbol{p}' - \boldsymbol{p}).\end{aligned} \tag{4.98}$$

For two-electron states (4.91) implies that the state vector must be *antisymmetric* upon exchange of the two particles:

$$\begin{aligned}\mathbf{a}_r^\dagger(\boldsymbol{p}_1)\mathbf{a}_s^\dagger(\boldsymbol{p}_2)|0\rangle &= -\mathbf{a}_s^\dagger(\boldsymbol{p}_2)\mathbf{a}_r^\dagger(\boldsymbol{p}_1)|0\rangle, \\ |e^-(\boldsymbol{p}_1, r)e^-(\boldsymbol{p}_2, s)\rangle &= -|e^-(\boldsymbol{p}_2, s)e^-(\boldsymbol{p}_1, r)\rangle.\end{aligned} \tag{4.99}$$

Compare this behavior with that of the two-meson state (3.57).

We now consider a general one-electron state which we write as a superposition of plane waves

$$|f\rangle = \int \frac{d^3p}{(2\pi)^3\sqrt{2p^0}} \sum_s f_s(\boldsymbol{p})\mathbf{a}_s^\dagger(\boldsymbol{p})|0\rangle, \tag{4.100}$$

where $f_s(\boldsymbol{p})$ (with $s = \pm\frac{1}{2}$) is an arbitrary complex-valued function. A state vector with two electrons in the same quantum state would have the form

$$|2f\rangle = \int \frac{\mathrm{d}^3 p_1\, \mathrm{d}^3 p_2}{(2\pi)^6 \sqrt{2p_1^0 2p_2^0}} \sum_{r,s} f_r(\boldsymbol{p}_1) f_s(\boldsymbol{p}_2) \mathbf{a}_r^\dagger(\boldsymbol{p}_1) \mathbf{a}_s^\dagger(\boldsymbol{p}_2)|0\rangle. \tag{4.101}$$

However, as a result of (4.99), this state is identically equal to zero

$$|2f\rangle \equiv 0. \tag{4.102}$$

We have thus demonstrated the *Pauli exclusion principle*: Two electrons cannot occupy the same quantum state.

The requirement of microcausality for the free Dirac field has led us to the Pauli principle for the accompanying particles. As can be shown, this also holds when interactions are present (Pauli 1936, 1940). This statement, together with the Bose principle for particles of integer spin (Chap. 3), forms the content of the *spin statistics theorem* which can be derived using only a few very basic axioms (see Streater 1964).

We next wish to study the angular momentum carried by the field quanta. In Sect. 4.3 we have already got to know the transformation law of the Dirac field under rotations. As follows from the results of Chap. 3 (see Exercise 3.8), the connection between the angular momentum operator and infinitesimal rotations of the Dirac field, is completely analogous to that between the four-momentum operator and infinitesimal translations in time and space. According to the Heisenberg equation (4.85), for an infinitesimal translation parametrized by the four-vector δa, we obtain

$$\begin{aligned} \psi(x - \delta a) &= \psi(x) - \delta a^\mu \frac{\partial}{\partial x^\mu} \psi(x) \\ &= \psi(x) - \delta a^\mu \mathrm{i}[\mathbf{P}_\mu, \psi(x)]. \end{aligned} \tag{4.103}$$

The analogous relationship for the infinitesimal rotations defined in (4.57) is

$$\mathsf{S}(\mathsf{R})\psi(\mathsf{R}^{-1}\boldsymbol{x}, t) = \psi(\boldsymbol{x}, t) - \delta\varphi^j \mathrm{i}[\mathbf{J}^j, \psi(\boldsymbol{x}, t)], \tag{4.104}$$

where $\vec{\mathbf{J}}$ is the angular momentum operator. Making use of (4.60) we arrive at the "Heisenberg equation" for the angular momentum:

$$-\left\{\boldsymbol{x} \times \frac{1}{\mathrm{i}}\nabla + \begin{pmatrix} \frac{1}{2}\boldsymbol{\sigma} & 0 \\ 0 & \frac{1}{2}\boldsymbol{\sigma} \end{pmatrix}\right\} \psi(\boldsymbol{x}, t) = [\vec{\mathbf{J}}, \psi(\boldsymbol{x}, t)]. \tag{4.105}$$

If we insert into this equation the Fourier decomposition of the Dirac field (4.84), then after some calculation we obtain

$$[\mathbf{J}^j, \mathbf{a}_s^\dagger(\boldsymbol{p})] = \varepsilon^{jkl} p^k \mathrm{i} \frac{\partial}{\partial p^l} \mathbf{a}_s^\dagger(\boldsymbol{p}) + \sum_r \mathbf{a}_r^\dagger(\boldsymbol{p}) \tfrac{1}{2}\sigma_{rs}^j, \tag{4.106a}$$

$$[\mathbf{J}^j, \mathbf{b}_s^\dagger(\boldsymbol{p})] = \varepsilon^{jkl} p^k \mathrm{i} \frac{\partial}{\partial p^l} \mathbf{b}_s^\dagger(\boldsymbol{p}) + \sum_r \mathbf{b}_r^\dagger(\boldsymbol{p}) \tfrac{1}{2}\sigma_{rs}^j. \tag{4.106b}$$

The first term on the right-hand side of (4.106) describes the orbital angular momentum; this term would be identical for a scalar meson field as was treated in Chap. 3. The second term describes the spin of the electron or positron. In order to see this clearly, we consider an electron at rest. We then have

$$\begin{aligned} |e^-(\boldsymbol{p}=0,s)\rangle &= \mathbf{a}_s^\dagger(\boldsymbol{p}=0)|0\rangle, \\ \vec{\mathbf{J}}|e^-(\boldsymbol{p}=0,s)\rangle &= \sum_r |e^-(\boldsymbol{p}=0,r)\rangle \tfrac{1}{2}\boldsymbol{\sigma}_{rs}, \\ \mathbf{J}^3|e^-(\boldsymbol{p}=0,\pm\tfrac{1}{2})\rangle &= \pm\tfrac{1}{2}|e^-(\boldsymbol{p}=0,\pm\tfrac{1}{2})\rangle. \end{aligned} \tag{4.107}$$

Thus the operator $\mathbf{a}_s^\dagger(\boldsymbol{p}=0)$ for $s=+\frac{1}{2}$ $(s=-\frac{1}{2})$ creates an electron with spin in the positive (negative) z-direction. Analogously, one sees that $\mathbf{b}_s^\dagger(\boldsymbol{p}=0)$, $s=\pm\frac{1}{2}$ creates a positron with z-component of spin $\pm\frac{1}{2}$. This demonstrates the usefulness of the special choice of the solutions $v_s(p)$ in (4.41) since with this choice our positron states with $s=\pm\frac{1}{2}$ also form a canonical basis for spin $\frac{1}{2}$, i.e., they satisfy the same relations as the electron states in (4.107).

4.5 Parity, Charge-Conjugation, and Time-Reversal Invariance of the Free Dirac Field

4.5.1 The Parity Transformation P

In this section we will get to know a few further invariance properties of the free Dirac field. To begin with, we consider the parity operation, P, representing the transition from a spatial Cartesian right-handed system to the corresponding left-handed system (2.40):

$$\begin{aligned} \mathrm{P}\colon t &\to t' = t, \\ \boldsymbol{x} &\to \boldsymbol{x}' = -\boldsymbol{x}. \end{aligned} \tag{4.108}$$

We will investigate whether, and in which sense, the Dirac equation remains invariant under this transformation. From our experience with Lorentz transformations, we write the Dirac spinor in the primed system as

$$\psi'(\boldsymbol{x}',t') = \mathbf{S}(\mathrm{P})\psi(\boldsymbol{x},t). \tag{4.109}$$

We then ask the question of whether it is possible to choose $\mathbf{S}(\mathrm{P})$ such that the Dirac equation is valid for ψ':

$$\left\{\mathrm{i}\gamma^\mu\frac{\partial}{\partial x'^\mu}-m\right\}\psi'(\boldsymbol{x}',t') \overset{?}{=} 0,$$

$$\mathbf{S}^{-1}(\mathrm{P})\left\{\mathrm{i}\gamma^0\frac{\partial}{\partial t}+\mathrm{i}\gamma^j\frac{\partial}{\partial x'^j}-m\right\}\mathbf{S}(\mathrm{P})\psi(\boldsymbol{x},t) \overset{?}{=} 0, \tag{4.110}$$

$$\mathbf{S}^{-1}(\mathrm{P})\left\{\mathrm{i}\gamma^0\frac{\partial}{\partial t}-\mathrm{i}\gamma^j\frac{\partial}{\partial x^j}-m\right\}\mathbf{S}(\mathrm{P})\psi(\boldsymbol{x},t) \overset{?}{=} 0. \tag{4.111}$$

It is possible to reduce (4.111) to the Dirac equation for $\psi(\boldsymbol{x}, t)$ provided

$$\begin{aligned} \mathbf{S}^{-1}(\mathrm{P})\gamma^0\mathbf{S}(\mathrm{P}) &= \gamma^0, \\ \mathbf{S}^{-1}(\mathrm{P})\gamma^j\mathbf{S}(\mathrm{P}) &= -\gamma^j. \end{aligned} \tag{4.112}$$

This can easily be achieved by setting

$$\mathbf{S}(\mathrm{P}) = \gamma^0 \tag{4.113}$$

Thus we choose the parity-transformed Dirac field as follows

$$\begin{aligned} \psi'(\boldsymbol{x}', t') &= \gamma^0\psi(\boldsymbol{x}, t) \\ &= \gamma^0\psi(-\boldsymbol{x}', t'). \end{aligned} \tag{4.114}$$

If we interpret the Dirac spinor as a quantum field, then we see from (4.114) that $\psi'(\boldsymbol{x}, t)$ satisfies the same equation of motion as $\psi(\boldsymbol{x}, t)$ and, furthermore, it is readily shown that it obeys the same anticommutation rules (4.92). Thus it seems plausible to expect that the operators $\psi(\boldsymbol{x}, t)$ and $\psi'(\boldsymbol{x}, t)$ are equivalent, i.e., that they can be obtained from one another by means of a unitary transformation in the space of state vectors. We therefore seek a unitary operator, $\mathbf{U}(\mathrm{P})$ which satisfies the equation

$$\mathbf{U}(\mathrm{P})\psi(\boldsymbol{x}, t)\mathbf{U}^{-1}(\mathrm{P}) \overset{?}{=} \psi'(\boldsymbol{x}, t) = \gamma_0\psi(-\boldsymbol{x}, t). \tag{4.115}$$

Replacing the field operator by its Fourier expansion (4.84) gives

$$\begin{aligned} &\int \frac{\mathrm{d}^3p}{(2\pi)^3}\frac{1}{2p^0}\sum_s \{\mathrm{e}^{\mathrm{i}px}\, v_s(p)\mathbf{U}(\mathrm{P})\mathbf{b}_s^\dagger(\boldsymbol{p})\mathbf{U}^{-1}(\mathrm{P}) \\ &\qquad + \mathrm{e}^{-\mathrm{i}px}\, u_s(p)\mathbf{U}(\mathrm{P})\mathbf{a}_s(\boldsymbol{p})\mathbf{U}^{-1}(\mathrm{P})\} \\ &\overset{?}{=} \int \frac{\mathrm{d}^3p}{(2\pi)^3}\frac{1}{2p_0}\sum_s \{\mathrm{e}^{\mathrm{i}px}\,\gamma_0 v_s(p')\mathbf{b}_s^\dagger(-\boldsymbol{p}) + \mathrm{e}^{-\mathrm{i}px}\,\gamma_0 u_s(p')\mathbf{a}_s(-\boldsymbol{p})\}, \end{aligned} \tag{4.116}$$

where

$$p' = \begin{pmatrix} p^0 \\ -\boldsymbol{p} \end{pmatrix}.$$

We find readily from the explicit form of the spinors u and v (4.38), (4.41):

$$\begin{aligned} \gamma_0 u_s(p') &= u_s(p), \\ \gamma_0 v_s(p') &= -v_s(p). \end{aligned} \tag{4.117}$$

With this, we obtain from (4.116) as a condition on the unitary operator $\mathbf{U}(\mathrm{P})$, the relations

$$\begin{aligned} \mathbf{U}(\mathrm{P})\mathbf{a}_s^\dagger(\boldsymbol{p})\mathbf{U}^{-1}(\mathrm{P}) &= \mathbf{a}_s^\dagger(-\boldsymbol{p}), \\ \mathbf{U}(\mathrm{P})\mathbf{b}_s^\dagger(\boldsymbol{p})\mathbf{U}^{-1}(\mathrm{P}) &= -\mathbf{b}_s^\dagger(-\boldsymbol{p}). \end{aligned} \tag{4.118}$$

If we further demand that the vacuum remains invariant under parity transformations

$$\mathbf{U}(\mathrm{P})|0\rangle = |0\rangle, \tag{4.119}$$

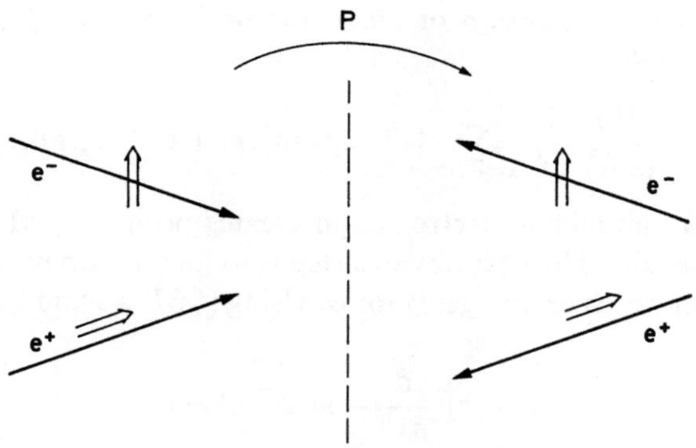

Figure 4.1 Parity transformation of an electron–positron state. The single arrows denote the momentum directions and the double arrows give the spin orientations.

then we see that (4.118) does indeed define a unitary operator in the space of electron and positron states that satisfies (4.115).

For one-particle states we have, according to (4.118),

$$\begin{aligned} \mathsf{U}(\mathrm{P})|\mathrm{e}^-(\boldsymbol{p}, s)\rangle &= |\mathrm{e}^-(-\boldsymbol{p}, s)\rangle, \\ \mathsf{U}(\mathrm{P})|\mathrm{e}^+(\boldsymbol{p}, s)\rangle &= -|\mathrm{e}^+(-\boldsymbol{p}, s)\rangle. \end{aligned} \tag{4.120}$$

It should be noted that, according to the Dirac theory, electrons and positrons have opposite (*intrinsic*) parity. This can be directly verified experimentally, for example, in the decay of the hydrogen-like bound state of e^+ and e^- (positronium) into two photons.

The physical predictions to which the above result leads are the following: For every state of free electrons and positrons there exists a parity transformed state in which all momenta are reversed, all spins are unchanged, and a factor (-1) appears for every positron (Fig. 4.1). The Dirac operator ψ bears the same relation to the original state as the Dirac operator ψ' does to the transformed state. Both ψ and ψ' satisfy the Dirac equation. By means of observations on a system of free electrons and positrons it is thus impossible to distinguish a spatial Cartesian right-handed system from a left-handed system. As we will see later, in Nature, the equivalence of right- and left-handed systems is violated by the weak interaction.

4.5.2 The Charge-Conjugation Transformation C

We next pose the question of whether an observer who exchanges the roles of electrons and positrons, would find different laws of Nature for the free particles. We claim that within the framework of the Dirac theory, it is purely a matter of convention which particle we call an electron and which a positron. In order to show this we shall construct an operator for *charge-conjugation*.

We consider the expansion of the Dirac field operator ψ (4.84) and its adjoint operator $\bar{\psi}$:

$$\bar{\psi}(x) = \int \frac{d^3p}{(2\pi)^3} \frac{1}{2p^0} \sum_{s=\pm 1/2} \{e^{ipx}\, \bar{u}_s(p)\mathbf{a}_s^\dagger(\boldsymbol{p}) + e^{-ipx}\, \bar{v}_s(p)\mathbf{b}_s(\boldsymbol{p})\}. \quad (4.121)$$

We see that ψ annihilates electrons and creates positrons, whereas $\bar{\psi}$ does exactly the opposite. The next obvious step is to look for an invariance of the Dirac equation which exchanges ψ for $\bar{\psi}$. Using (4.67) we find from the Dirac equation

$$\{i(-\gamma^\mu)^T \frac{\partial}{\partial x^\mu} - m\}\bar{\psi}^T(x) = 0. \quad (4.122)$$

It can easily be shown that the matrices $\gamma'^\mu = (-\gamma^\mu)^T$ again satisfy the fundamental anticommutation rules (4.12). From our previous experience we expect to find an equivalence transformation of the form

$$\mathbf{S}^{-1}(C)\gamma^\mu \mathbf{S}(C) = (-\gamma^\mu)^T. \quad (4.123)$$

This is indeed satisfied for

$$\mathbf{S}(C) = i\gamma^2\gamma^0 = \left(\begin{array}{c|c} 0 & -\varepsilon \\ \hline -\varepsilon & 0 \end{array}\right). \quad (4.124)$$

With this we obtain from (4.122)

$$\mathbf{S}^{-1}(C)\left\{i\gamma^\mu \frac{\partial}{\partial x^\mu} - m\right\}\mathbf{S}(C)\bar{\psi}^T(x) = 0. \quad (4.125)$$

Therefore the so-called charge-conjugated field

$$\psi^c(x) = \mathbf{S}(C)\bar{\psi}^T(x) \quad (4.126)$$

likewise satisfies the Dirac equation and, as is readily shown, also the same anticommutation rules (4.92) as $\psi(x)$.

In order to demonstrate the equivalence of electrons and positrons in the free Dirac theory we search for a unitary operator, $\mathbf{U}(C)$, which transforms $\psi(x)$ into $\psi^c(x)$. Thus we are looking for an operator $\mathbf{U}(C)$ which performs the following:

$$\mathbf{U}(C)\psi(x)\mathbf{U}^{-1}(C) \stackrel{?}{=} \psi^c(x) = \mathbf{S}(C)\bar{\psi}^T(x). \quad (4.127)$$

If we replace the field operator by its Fourier expansion, and use the relations

$$\begin{aligned} \mathbf{S}(C)\bar{u}_s^T(p) &= v_s(p), \\ \mathbf{S}(C)\bar{v}_s^T(p) &= u_s(p) \qquad (s = \pm\tfrac{1}{2}), \end{aligned} \quad (4.128)$$

which can be confirmed using the explicit expressions (4.38), (4.41), we arrive at the requirement

$$\begin{aligned} \mathbf{U}(C)\mathbf{a}_s^\dagger(\boldsymbol{p})\mathbf{U}^{-1}(C) &= \mathbf{b}_s^\dagger(\boldsymbol{p}), \\ \mathbf{U}(C)\mathbf{b}_s^\dagger(\boldsymbol{p})\mathbf{U}^{-1}(C) &= \mathbf{a}_s^\dagger(\boldsymbol{p}). \end{aligned} \quad (4.129)$$

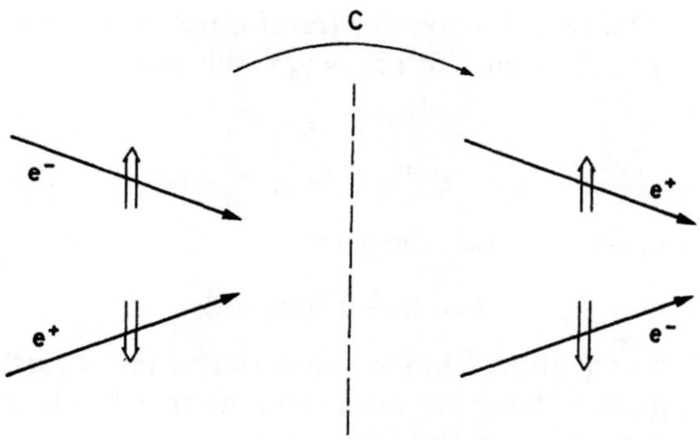

Figure 4.2 Charge-conjugation of an electron–positron state.

By means of these equations and the requirement that the vacuum state is invariant

$$\mathsf{U}(C)|0\rangle = |0\rangle \qquad (4.130)$$

we have in fact defined a unitary transformation in Fock space which satisfies (4.127). Under this charge-conjugation electron and positron labels are interchanged, with momentum and spin variables remaining unchanged (Fig. 4.2).

The charge-conjugation invariance, which we have demonstrated here for the *free* Dirac field, again is not strictly valid in Nature: it too is violated by the weak interaction.

4.5.3 The Time-Reversal Transformation T

Here again, we wish to pose a physical question: Is it possible, by observing a free Dirac field, to distinguish the positive from the negative time direction? Or, in other words: If we make a film of free Dirac particles and allow this film to run backwards, will we observe a physically possible, or impossible, course of events? We will show here that the free Dirac field does not possess a unique time direction.

We begin by recalling the time-reversal transformation of the coordinates (2.41)

$$\begin{aligned} \mathrm{T}\colon t &\to t' = -t, \\ \boldsymbol{x} &\to \boldsymbol{x}' = \boldsymbol{x}. \end{aligned} \qquad (4.131)$$

The Dirac field with reversed time argument, $\psi(\boldsymbol{x}, -t)$, satisfies the differential equation

$$\left\{\mathrm{i}(-\gamma^0)\frac{\partial}{\partial t} + \mathrm{i}\gamma^j\frac{\partial}{\partial x^j} - m\right\}\psi(\boldsymbol{x}, -t) = 0. \qquad (4.132)$$

Using (4.78), (4.79) we easily arrive at a transformation which converts γ^0 into $-\gamma^0$ but leaves $\boldsymbol{\gamma}$ unchanged. With $\mathbf{S}' = \gamma_5\gamma^0$ this reads

$$\begin{aligned} \mathbf{S}'^{-1}\gamma^0\mathbf{S}' &= -\gamma^0, \\ \mathbf{S}'^{-1}\boldsymbol{\gamma}\mathbf{S}' &= \boldsymbol{\gamma}. \end{aligned} \tag{4.133}$$

As before, we can conclude that the field

$$\tilde{\psi}(\boldsymbol{x}, t) = \mathbf{S}'\psi(\boldsymbol{x}, -t) \tag{4.134}$$

satisfies the Dirac equation. Thus we have established a further invariance of the Dirac equation. Since we already know that the Dirac equation is charge-conjugation invariant, the field

$$\begin{aligned} \psi'(\boldsymbol{x}, t) &= \mathbf{S}'\psi^c(\boldsymbol{x}, -t) \\ &= \mathbf{S}(\mathrm{T})\bar{\psi}^{\mathrm{T}}(\boldsymbol{x}, -t), \end{aligned} \tag{4.135}$$

with

$$\mathbf{S}(\mathrm{T}) = \mathbf{S}' \cdot \mathbf{S}(\mathrm{C}) = \mathrm{i}\gamma^2\gamma_5,$$

also satisfies the Dirac equation. In the following we will denote the field ψ' (4.135) as time-reversed field. It can easily be shown that it satisfies the anticommutation rules (4.92).

The question now is whether we can find a unitary transformation $\mathbf{U}(\mathrm{T})$ in Fock space which transforms ψ into ψ':

$$\mathbf{U}(\mathrm{T})\psi(\boldsymbol{x}, t)\mathbf{U}^{-1}(\mathrm{T}) \stackrel{?}{=} \psi'(\boldsymbol{x}, t) = \mathbf{S}(\mathrm{T})\bar{\psi}^{\mathrm{T}}(\boldsymbol{x}, -t). \tag{4.136}$$

By substituting the expansion of the field operators, we see readily that a unitary operator $\mathbf{U}(\mathrm{T})$ would have to transform creation operators into annihilation operators. This cannot be an equivalence transformation as far as the physics is concerned and, as can be shown, it leads to a mathematical contradiction when we further demand the invariance of the vacuum state under $\mathbf{U}(\mathrm{T})$.

However, there exists an *antiunitary* transformation in Fock space which transforms ψ into ψ'. In general we shall call an operator $\mathbf{V}$ *antilinear* if, for all states $|\mathrm{a}\rangle$ and $|\mathrm{b}\rangle$, and for all complex numbers c_1 and c_2, it satisfies the condition

$$\mathbf{V}(c_1|\mathrm{a}\rangle + c_2|\mathrm{b}\rangle) = c_1^*\mathbf{V}|\mathrm{a}\rangle + c_2^*\mathbf{V}|\mathrm{b}\rangle \tag{4.137}$$

The Hermitian conjugate operator $\mathbf{V}^\dagger$ of an antilinear operator $\mathbf{V}$ is defined by

$$\langle\mathrm{a}|\mathbf{V}^\dagger|\mathrm{b}\rangle = \langle\mathrm{b}|\mathbf{V}|\mathrm{a}\rangle. \tag{4.138}$$

This operator $\mathbf{V}^\dagger$ is also antilinear. It can easily be seen that, with these definitions, the usual rules of Hermitian conjugation remain valid for all products of linear and antilinear operators. An operator $\mathbf{V}$ will be denoted *antiunitary* if it is antilinear and satisfies the further condition

$$\mathbf{V}^\dagger\mathbf{V} = \mathbf{V}\mathbf{V}^\dagger = 1. \tag{4.139}$$

We then have, for states $|a\rangle$ and $|b\rangle$, and for states

$$|a'\rangle = \mathbf{V}|a\rangle, \qquad |b'\rangle = \mathbf{V}|b\rangle, \tag{4.140}$$

the following relation

$$\langle a'|b'\rangle = \langle b|a\rangle = \langle a|b\rangle^*. \tag{4.141}$$

Thus in comparison to a unitary transformation, an antinuitary transformation has the additional effect of complex conjugation or, in other words, of an exchange of the "bra" and "ket" vectors. From a physical standpoint this represents the exchange of the initial and the final states in a reaction.

Let $\mathbf{A}$ be any linear operator and $\mathbf{V}$ an antiunitary operator. We will define $\mathbf{A}'$, the operator obtained by an antiunitary transformation of $\mathbf{A}$ by

$$\mathbf{A}' := (\mathbf{V}\mathbf{A}\mathbf{V}^{-1})^\dagger. \tag{4.142}$$

This differs from the unitary transformation in the appearance here of the Hermitian conjugate. Its effect is to yield a linear relationship between $\mathbf{A}$ and $\mathbf{A}'$, i.e., for any complex number c we have

$$(c\cdot\mathbf{A})' = c\mathbf{A}'. \tag{4.143}$$

Furthermore, the matrix elements of the states defined in (4.140) obey the relation

$$\langle a|\mathbf{A}|b\rangle = \langle b'|\mathbf{A}'|a'\rangle. \tag{4.144}$$

We now claim the existence of an antiunitary operator $\mathbf{V}(\mathrm{T})$ such that

$$(\mathbf{V}(\mathrm{T})\psi(\boldsymbol{x}, t)\mathbf{V}^{-1}(\mathrm{T}))^\dagger = \psi'(\boldsymbol{x}, t) = \mathbf{S}(\mathrm{T})\bar{\psi}^{\mathrm{T}}(\boldsymbol{x}, -t). \tag{4.145}$$

This antiunitary operator is again found by replacing the field operators by their Fourier expansion and by making use of the following relation, which is easily derived from the explicit expressions (4.38), (4.41)

$$\begin{aligned}\mathbf{S}(\mathrm{T})\bar{u}_s^{\mathrm{T}}(p) &= (-1)^{s+1/2}u_{-s}(p'),\\ \mathbf{S}(\mathrm{T})\bar{v}_s^{\mathrm{T}}(p) &= (-1)^{s+1/2}v_{-s}(p'),\end{aligned} \tag{4.146}$$

where $s = \pm\frac{1}{2}$ and

$$p' = \begin{pmatrix} p^0 \\ -\boldsymbol{p}\end{pmatrix}.$$

The antiunitary operator $\mathbf{V}(\mathrm{T})$ that satisfies (4.144) is then defined by the following relations:

$$\begin{aligned}\mathbf{V}(\mathrm{T})\mathbf{a}_s^\dagger(\boldsymbol{p})\mathbf{V}^{-1}(\mathrm{T}) &= (-1)^{s-1/2}\mathbf{a}_{-s}^\dagger(-\boldsymbol{p}),\\ \mathbf{V}(\mathrm{T})\mathbf{b}_s^\dagger(\boldsymbol{p})\mathbf{V}^{-1}(\mathrm{T}) &= (-1)^{s-1/2}\mathbf{b}_{-s}^\dagger(-\boldsymbol{p}),\\ \mathbf{V}(\mathrm{T})|0\rangle &= |0\rangle.\end{aligned} \tag{4.147}$$

The time-reversal transformation thus reverses momenta and spins (Fig. 4.3).

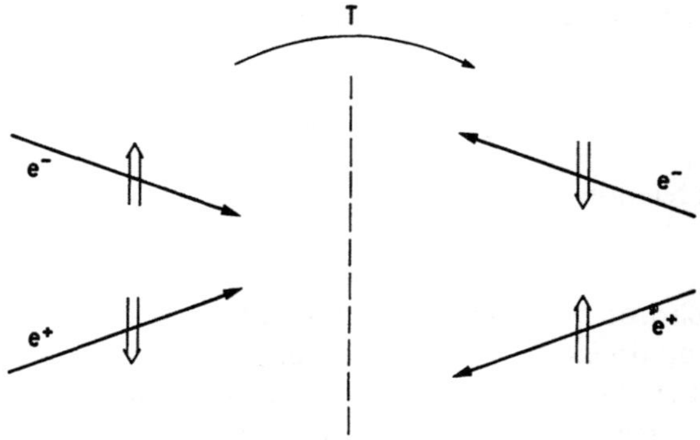

Figure 4.3 Time-reversal transformation of an electron–positron state.

Can the antiunitary transformation constructed on the basis of (4.147) be used to represent time reversal as a physical equivalence transformation? The answer to this question is that indeed it can. To each state $|a\rangle$ of the first observer, the second observer, for whom time is reversed, can assign a state $|a'\rangle = \mathbf{V}(T)|a\rangle$. To every observable $\mathbf{A}$ which the first observer constructs from the field operator, there corresponds an observable $\mathbf{A}'$, given by (4.142), which the second observer constructs in the same way from ψ' (4.145). According to (4.144) the expectation values satisfy

$$\langle a|\mathbf{A}|a\rangle = \langle a'|\mathbf{A}'|a'\rangle, \tag{4.148}$$

This means that the results of measurements, in the appropriately corresponding states, are identical!

For a free Dirac field we have demonstrated time-reversal invariance. In Nature, time-reversal invariance is valid for nearly all processes which have so far been observed. It was not until the experimental investigation of the decay of the neutral K-meson that T-violating effects were discovered.

Thus in Nature each of the invariances C, P, and T, is individually violated. However, in a relativistic field theory, the product $\Theta = CPT$ must always be an invariance transformation. This theorem is derived from general axioms (Lüders 1954, 1957; Pauli 1955; see also Streater 1964) and up to now it is fully confirmed by experiment.

Exercises

4.1 Prove (4.55) and also the following identity of Jacobi

$$[[\mathbf{A}, \mathbf{B}], \mathbf{C}] + [[\mathbf{B}, \mathbf{C}], \mathbf{A}] + [[\mathbf{C}, \mathbf{A}], \mathbf{B}] = 0, \tag{4.149}$$

which is valid for all operators $\mathbf{A}, \mathbf{B}, \mathbf{C}$. Prove the validity of the following series expansion

$$e^{\mathbf{A}}\mathbf{B}\,e^{-\mathbf{A}} = \mathbf{B} + \frac{1}{1!}[\mathbf{A}, \mathbf{B}] + \frac{1}{2!}[\mathbf{A}, [\mathbf{A}, \mathbf{B}]] + \cdots. \tag{4.150}$$

4.2 Demonstrate the existence of a matrix $\mathbf{S}(\Lambda)$ that satisfies (4.48) for all Lorentz transformations Λ which belong to the identity component.

Hint: Demonstrate first the validity of (4.48) for all Lorentz transformations of the form

$$\Lambda(\tau) = \exp(\mathbf{h}\tau),$$

where τ is any real number and $\mathbf{h} = (h^\mu{}_\nu)$ is any matrix whose elements satisfy $h_{\mu\nu} = -h_{\nu\mu}$. Then apply the result of Exercise 2.1.

4.3 Prove (4.68) for all Lorentz transformations Λ belonging to the identity component.

4.4 Verify the completeness relations (4.94) and evaluate the expressions

$$\bar{u}_r(p)\gamma^\mu u_s(p),$$

$$\bar{v}_r(p)\gamma^\mu v_s(p).$$

4.5 Derive the explicit form of the CPT transformation of the Dirac field and discuss CPT in analogy to our discussion of the time-reversal transformation. What do the analogous relations to (4.146), (4.147) look like for $\Theta = \mathrm{CPT}$?

CHAPTER 5

The Scattering Matrix and the Scattering Cross-Section

In this chapter we will get to know the formalism that is used to describe reactions in high energy physics. These reactions might involve scattering processes or decay and production of particles. The central role here is played by the scattering matrix **S**, which was introduced by Wheeler and Heisenberg (Wheeler 1937; Heisenberg 1943, 1944). Before presenting a general formulation we will look at a simple example.

5.1 The Scattering of Electrons by a Heavy Nucleus

We will investigate here the scattering of electrons by a heavy nucleus which will be considered to be a fixed scattering center. The charge of the nucleus is Ze and the distribution of this charge will be described by the density function

$$Ze\rho(\boldsymbol{x}),$$

which thus satisfies the normalization condition

$$\int d^3x \, \rho(\boldsymbol{x}) = 1. \tag{5.1}$$

According to the laws of electrostatics, the charge distribution produces a potential $V(\boldsymbol{x})$ which obeys the Poisson equation:

$$\Delta V(\boldsymbol{x}) = -Ze\rho(\boldsymbol{x}). \tag{5.2}$$

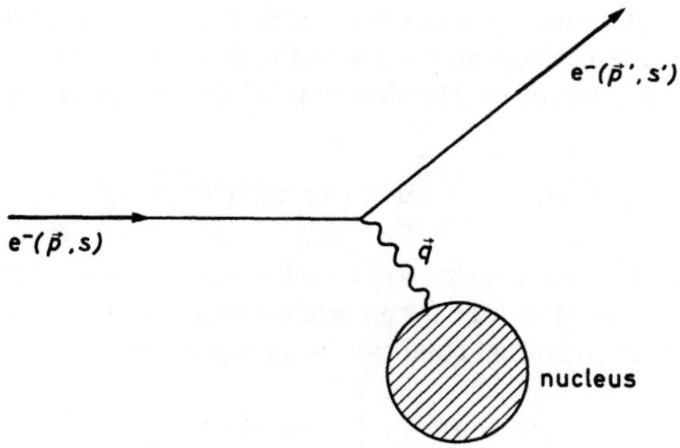

Figure 5.1 Scattering of an electron by a heavy nucleus with charge *Ze*.

Our aim is to calculate the scattering of electrons from this potential (Fig. 5.1). This problem is the relativistic generalization of Rutherford scattering.

In order to describe the scattering within the framework of the quantized Dirac theory, we will make use of the interaction, or Dirac picture which was introduced in Chap. 3. In this formulation the operators satisfy the free equations of motion, whilst the time development of the states is given by the interaction (3.8)

$$\mathrm{i}\frac{\partial}{\partial t}|t\rangle = \mathbf{H}'(t)|t\rangle. \tag{5.3}$$

For times $t \to -\infty$ we must obtain an incoming state which, in our case, consists of an electron with momentum $\boldsymbol{p}$ and spin s

$$\lim_{t\to-\infty} |t\rangle = |e^-(\boldsymbol{p}, s)\rangle. \tag{5.4}$$

This is the initial boundary condition for the differential equation (5.3). For times $t \to +\infty$ the state will look different and in our example it will contain a scattered electron wave. Since (5.3) is a linear differential equation for the state vector, the relationship between the states at times $t = -\infty$ and $t = +\infty$ must likewise be linear. We shall write the initial state as i

$$\lim_{t\to-\infty} |t\rangle = |i\rangle, \tag{5.5}$$

and then we may put

$$\lim_{t\to+\infty} |t\rangle = \mathbf{S}|i\rangle, \tag{5.6}$$

where $\mathbf{S}$ is a linear operator. If we want to know the probability amplitude for finding a particular final state after the scattering, then we simply need to form the matrix element

$$\langle f|t = +\infty\rangle = \langle f|\mathbf{S}|i\rangle \equiv S_{fi}. \tag{5.7}$$

With this we have introduced the scattering matrix $\mathbf{S}$.

In order to determine the scattering matrix for our particular problem, we need to know how electrons interact with the external potential. We give here the appropriate interaction Hamiltonian $\mathbf{H}'$ but defer its derivation until Part II:

$$\mathbf{H}'(t) = -e \int d^3x\, \bar{\psi}(\boldsymbol{x}, t)\gamma^0\psi(\boldsymbol{x}, t)V(\boldsymbol{x}). \tag{5.8}$$

With this interaction it is not easy to solve the differential equation (5.3) in an exact manner. Thus we look for an expansion; we rewrite (5.3) as an integral equation in which the initial condition is incorporated

$$|t\rangle = |i\rangle - \mathrm{i} \int_{-\infty}^{t} dt'\, \mathbf{H}'(t')|t'\rangle. \tag{5.9}$$

If we assume that $\mathbf{H}'$ is small, then we can regard the second term on the right-hand side as a correction term in which the state vector can be replaced by its initial value. This is the *Born approximation.* We then obtain

$$\begin{aligned} |t = +\infty\rangle &\cong |i\rangle - \mathrm{i} \int_{-\infty}^{\infty} dt'\, \mathbf{H}'(t')|i\rangle, \\ \langle f|t = +\infty\rangle &= \langle f|\mathbf{S}|i\rangle \equiv S_{fi} \\ &\cong \delta_{fi} - \mathrm{i} \int_{-\infty}^{\infty} dt' \langle f|\mathbf{H}'(t')|i\rangle. \end{aligned} \tag{5.10}$$

We now insert the $\mathbf{H}'$ appropriate to our example (5.8) together with

$$\begin{aligned} |i\rangle &= |e^-(\boldsymbol{p}, s)\rangle = \mathbf{a}_s^\dagger(\boldsymbol{p})|0\rangle, \\ |f\rangle &= |e^-(\boldsymbol{p}', s')\rangle = \mathbf{a}_{s'}^\dagger(\boldsymbol{p}')|0\rangle. \end{aligned} \tag{5.11}$$

We are interested in the deflected electrons, i.e., those for which $\boldsymbol{p}' \neq \boldsymbol{p}$. We then obtain

$$S_{fi} = -\mathrm{i} \int dt\, \langle e^-(\boldsymbol{p}', s')|(-e) \int d^3x\, \bar{\psi}(\boldsymbol{x}, t)\gamma^0\psi(\boldsymbol{x}, t)V(\boldsymbol{x})|e^-(\boldsymbol{p}, s)\rangle. \tag{5.12}$$

By substituting the expansion of the field operators (4.84), (4.121) we have

$$\begin{aligned} S_{fi} = \mathrm{i}e \int dt\, d^3x\, V(\boldsymbol{x}) &\int \frac{d^3p_1\, d^3p_2}{(2\pi)^6} \frac{1}{2p_1^0 2p_2^0} \\ &\cdot \sum_{r,r'} \langle 0|\mathbf{a}_{s'}(\boldsymbol{p}')\{e^{\mathrm{i}p_2x}\,\bar{u}_{r'}(p_2)\mathbf{a}_{r'}^\dagger(\boldsymbol{p}_2) + e^{-\mathrm{i}p_2x}\,\bar{v}_{r'}(p_2)\mathbf{b}_{r'}(\boldsymbol{p}_2)\}\gamma^0 \\ &\cdot \{e^{\mathrm{i}p_1x}v_r(p_1)\mathbf{b}_r^\dagger(\boldsymbol{p}_1) + e^{-\mathrm{i}p_1x}\,u_r(p_1)\mathbf{a}_r(\boldsymbol{p}_1)\}\mathbf{a}_s^\dagger(\boldsymbol{p})|0\rangle. \end{aligned} \tag{5.13}$$

Making use of (4.87) and the anticommutation rules (4.91) we readily find

$$\begin{aligned} S_{fi} &= \mathrm{i}e \int dt\, d^3x\, V(\boldsymbol{x})\, e^{\mathrm{i}((p'^0-p^0)t-(\boldsymbol{p}'-\boldsymbol{p})\boldsymbol{x})}\,\bar{u}_{s'}(p')\gamma^0u_s(p) \\ &= \mathrm{i}e2\pi\delta(p'^0 - p^0) \int d^3x\, V(\boldsymbol{x})\, e^{-\mathrm{i}(\boldsymbol{p}'-\boldsymbol{p})\boldsymbol{x}}\,\bar{u}_{s'}(p')\gamma^0u_s(p). \end{aligned} \tag{5.14}$$

For the incoming and outgoing electron, the **S**-matrix element contains the corresponding Dirac spinors u and $\bar{u}$. The δ-function ensures that the scattered electron has the same energy as the incident electron as required in a static potential which can neither deliver nor absorb energy. The Fourier integral of the potential can be rewritten with the help of the Poisson equation. Denoting the momentum transfer by $\boldsymbol{q}$

$$\boldsymbol{q} = \boldsymbol{p} - \boldsymbol{p}', \tag{5.15}$$

we have

$$\begin{aligned}\int d^3x\, V(\boldsymbol{x}) e^{i\boldsymbol{q}\boldsymbol{x}} &= -\frac{1}{\boldsymbol{q}^2}\int d^3x\, V(\boldsymbol{x}) \Delta\, e^{i\boldsymbol{q}\boldsymbol{x}} \\ &= -\frac{1}{\boldsymbol{q}^2}\int d^3x\, e^{i\boldsymbol{q}\boldsymbol{x}} \Delta\, V(\boldsymbol{x}) \\ &= \frac{1}{\boldsymbol{q}^2}\int d^3x\, e^{i\boldsymbol{q}\boldsymbol{x}}\, Ze\rho(\boldsymbol{x}).\end{aligned} \tag{5.16}$$

The Fourier transform of the normalized charge distribution is known as its *form factor*

$$F(\boldsymbol{q}) = \int d^3x\, e^{i\boldsymbol{q}\boldsymbol{x}} \rho(\boldsymbol{x}). \tag{5.17}$$

Inserting all of this we obtain

$$S_{fi} = ie2\pi\delta(p'^0 - p^0)\frac{Ze}{\boldsymbol{q}^2} F(\boldsymbol{q}) \bar{u}_{s'}(p')\gamma^0 u_s(p). \tag{5.18}$$

From this we now wish to determine the transition probability and the scattering cross-section. We assume that the incident electron is unpolarized and that we do not observe the polarization of the scattered electron. We then find for the transition probability dw to final states in a momentum interval d^3p':

$$\begin{aligned} dw &= \frac{d^3p'}{(2\pi)^3 2p'^0} \sum_{s,s'}{}' |\langle e^-(\boldsymbol{p}', s')|\mathbf{S}|e^-(\boldsymbol{p}, s)\rangle|^2 \\ &= \frac{d^3p'}{(2\pi)^3 2p'^0} e^2 (Ze)^2 [2\pi\delta(p'^0 - p^0)]^2 \\ &\quad \cdot \frac{1}{|\boldsymbol{q}|^4} |F(\boldsymbol{q})|^2 \sum_{s,s'}{}' |\bar{u}_{s'}(p')\gamma^0 u_s(p)|^2 . \end{aligned} \tag{5.19}$$

Here $\sum'$ means an average over the polarization directions in the initial state and a summation over the polarization directions in the final state. The phase-space factor $(2\pi)^3 2p'^0$ derives from the condition that the sum over the final states must yield a projection operator (Exercise 5.1).

It would seem as though (5.19), since it contains the square of a δ-function, represents a mathematical catastrophe. But in fact the situation can be rescued by means of a trick due to Fermi. We write one δ-function in the form of an integral and obtain

$$(2\pi\delta(p'^0 - p^0))^2 = \int_{-\infty}^{\infty} \mathrm{d}t\, \mathrm{e}^{\mathrm{i}(p'^0 - p^0)t} \cdot 2\pi\delta(p'^0 - p^0)$$
$$= \left(\int_{-\infty}^{\infty} \mathrm{d}t\right) 2\pi\delta(p'^0 - p^0).$$

We now imagine that we switch the potential on and off very slowly, adiabatically, so that the integral over all time can be replaced by the total time T during which the potential is switched on. This procedure makes perfect physical sense: clearly no experiment can last infinitely long. With the substitutions

$$\int_{-\infty}^{\infty} \mathrm{d}t \to T, \qquad (2\pi\delta(p'^0 - p^0))^2 \to T 2\pi\delta(p'^0 - p^0), \tag{5.20}$$

we then obtain from (5.19) a mathematically well-defined result for the transition *rate* $\mathrm{d}w/T$:

$$\frac{\mathrm{d}w}{T} = \frac{\mathrm{d}^3 p'}{(2\pi)^3 2p'_0} e^2 (Ze)^2 2\pi\delta(p'^0 - p^0)$$
$$\cdot \frac{1}{|\boldsymbol{q}|^4} |F(\boldsymbol{q})|^2 \sum_{s,s'}{}' |\bar{u}_{s'}(p')\gamma^0 u_s(p)|^2. \tag{5.21}$$

(It is possible to do without Fermi's trick if, instead of states with fixed momenta, one considers wave packets; see Exercise 5.3.)

The scattering cross-section σ is obtained quite generally as the ratio of the transition rate to the flux Φ of incoming particles. In differential form this reads

$$\mathrm{d}\sigma = \frac{\mathrm{d}w/T}{\Phi}. \tag{5.22}$$

It remains to determine the flux. Since our states are normalized to δ-functions in momentum, we use a similar trick to that of (5.20). We have

$$\langle \mathrm{e}^-(\boldsymbol{p}_1, s_1) | \mathrm{e}^-(\boldsymbol{p}_2, s_2) \rangle = \delta_{s_1 s_2} (2\pi)^3 2p_1^0 \delta^3(\boldsymbol{p}_1 - \boldsymbol{p}_2)$$
$$= \delta_{s_1 s_2} 2p_1^0 \int \mathrm{d}^3 x\, \mathrm{e}^{\mathrm{i}(\boldsymbol{p}_1 - \boldsymbol{p}_2)\boldsymbol{x}}. \tag{5.23}$$

The integral in (5.23) extends over the whole of space and thus diverges for $\boldsymbol{p}_1 = \boldsymbol{p}_2$. In reality however, an infinitely large volume is never available. Let V be the available volume. Then, for the normalization of a state of definite momentum and fixed spin direction, we can write:

$$\langle \mathrm{e}^-(\boldsymbol{p}, s_1) | \mathrm{e}^-(\boldsymbol{p}, s_2) \rangle |_{s_1 = s_2} = 2p^0 \cdot \int_V \mathrm{d}^3 x = 2p^0 V. \tag{5.24}$$

The probability density (probability per unit volume) for this state is $2p^0$ and if $\mathbf{v}$ is the velocity of the incident electrons, we obtain for the flux

$$\Phi = |\mathbf{v}| \cdot 2p^0 = \frac{|\boldsymbol{p}|}{p^0} 2p^0 = 2|\boldsymbol{p}|. \tag{5.25}$$

For the differential of the cross-section we thus have, according to (5.21), (5.22)

$$\mathrm{d}\sigma = \frac{1}{2|\boldsymbol{p}|} \frac{\mathrm{d}^3 p'}{(2\pi)^3 2p_0'} e^2 (Ze)^2 2\pi\delta(p'^0 - p^0) \frac{1}{|\boldsymbol{q}|^4} |F(\boldsymbol{q})|^2 \cdot \sum_{s,s'}' |\bar{u}_{s'}(p')\gamma^0 u_s(p)|^2. \tag{5.26}$$

The differential cross-section with respect to the solid angle Ω of the scattered electrons is obtained by integrating over $|\boldsymbol{p}'|$:

$$\begin{aligned}\frac{\mathrm{d}\sigma}{\mathrm{d}\Omega} &= \frac{1}{2|\boldsymbol{p}|} e^2 (Ze)^2 \int_0^\infty \frac{\mathrm{d}|\boldsymbol{p}'| |\boldsymbol{p}'|^2}{2p'^0 (2\pi)^3} \\ &\quad \cdot 2\pi\delta(p'^0 - p^0) \frac{1}{|\boldsymbol{q}|^4} |F(\boldsymbol{q})|^2 \sum_{s,s'}' |\bar{u}_{s'}(p')\gamma^0 u_s(p)|^2 \\ &= \frac{e^2 (Ze)^2}{(4\pi)^2} \frac{1}{|\boldsymbol{q}|^4} |F(\boldsymbol{q})|^2 \sum_{s,s'}' |\bar{u}_{s'}(p')\gamma^0 u_s(p)|^2. \end{aligned} \tag{5.27}$$

We now look at the averaging and summation over the spin degrees of freedom. Here we meet again with the spin sums (4.94). We have

$$\begin{aligned}\sum_{s,s'}' |\bar{u}_{s'}(p')\gamma^0 u_s(p)|^2 &= \frac{1}{2} \sum_{s,s'} \bar{u}_{s'}(p')\gamma^0 u_s(p) \bar{u}_s(p')\gamma^0 u_{s'}(p') \\ &= \tfrac{1}{2} \mathrm{Tr}[(\not{p}' + m)\gamma^0 (\not{p} + m)\gamma^0]. \end{aligned} \tag{5.28}$$

A possible means of calculating the trace (Tr) of the γ-matrices which appears in (5.28) is to substitute the explicit expressions from (4.19). It is much more elegant however, to use the formulas for the trace of a product of γ-matrices which are given in Appendix A. This leads to

$$\begin{aligned}\sum_{s,s'}' |\bar{u}_{s'}(p')\gamma^0 u_s(p)|^2 &= 2(p_0^2 + \boldsymbol{p}'\boldsymbol{p} + m^2) \\ &= 4\left(\boldsymbol{p}^2 \cos^2\frac{\vartheta}{2} + m^2\right), \end{aligned} \tag{5.29}$$

where ϑ is the scattering angle:

$$\cos\vartheta = \frac{\boldsymbol{p}' \cdot \boldsymbol{p}}{|\boldsymbol{p}'||\boldsymbol{p}|} \qquad (0 \le \vartheta \le \pi).$$

Substituting these relations into (5.27) and noting that

$$|\boldsymbol{q}|^2 = (\boldsymbol{p} - \boldsymbol{p}')^2 = 4|\boldsymbol{p}|^2 \sin^2\frac{\vartheta}{2}, \tag{5.30}$$

we finally obtain

$$\frac{\mathrm{d}\sigma}{\mathrm{d}\Omega} = Z^2 \left(\frac{e^2}{4\pi}\right)^2 \frac{1}{4|\boldsymbol{p}|^2} \frac{\cos^2\frac{\vartheta}{2} + \frac{m^2}{|\boldsymbol{p}|^2}}{\sin^4\frac{\vartheta}{2}} |F(\boldsymbol{q})|^2. \tag{5.31}$$

For a scattering center which is a point charge we have

$$\begin{aligned} \rho(\boldsymbol{x}) &= \delta^3(\boldsymbol{x}), \\ F(\boldsymbol{q}) &= \int \mathrm{d}^3x\, \mathrm{e}^{\mathrm{i}\boldsymbol{q}\boldsymbol{x}} \delta^3(\boldsymbol{x}) = 1. \end{aligned} \tag{5.32}$$

From (5.31) we then obtain

$$\frac{\mathrm{d}\sigma}{\mathrm{d}\Omega} = Z^2 \left(\frac{e^2}{4\pi}\right)^2 \frac{\cos^2\frac{\vartheta}{2} + \frac{m^2}{|\boldsymbol{p}|^2}}{4|\boldsymbol{p}|^2 \sin^4\frac{\vartheta}{2}}. \tag{5.33}$$

This is referred to as the formula for *Mott scattering* and is the relativistic generalization of the Rutherford formula. Indeed, in the nonrelativistic limit, i.e., for momenta which are small compared to the electron's rest mass, we find

$$\frac{\mathrm{d}\sigma}{\mathrm{d}\Omega} = Z^2 \left(\frac{e^2}{4\pi}\right)^2 \frac{1}{(E_{\mathrm{kin}})^2} \frac{1}{16 \sin^4\frac{\vartheta}{2}} \qquad (|\boldsymbol{p}| \ll m). \tag{5.34}$$

Here $E_{\mathrm{kin}} = \boldsymbol{p}^2/(2m)$ is the nonrelativistic kinetic energy.

In contrast to this, in the extreme relativistic limit, we have

$$\frac{\mathrm{d}\sigma}{\mathrm{d}\Omega} = Z^2 \left(\frac{e^2}{4\pi}\right)^2 \frac{\cos^2\frac{\vartheta}{2}}{4|\boldsymbol{p}|^2 \sin^4\frac{\vartheta}{2}} \qquad (|\boldsymbol{p}| \gg m). \tag{5.35}$$

In this case, the cross-section for a point charge, which has no intrinsic length scale, is proportional to the square of the "external" kinematic length scale $1/|\boldsymbol{p}|$. Multiplying the scattering cross-section by $|\boldsymbol{p}|^2$ we obtain a dimensionless quantity:

$$|\boldsymbol{p}|^2 \frac{\mathrm{d}\sigma}{\mathrm{d}\Omega} = Z^2 \left(\frac{e^2}{4\pi}\right)^2 \frac{\cos^2\frac{\vartheta}{2}}{4 \sin^4\frac{\vartheta}{2}} \tag{5.36}$$

which, in the high energy limit, is completely independent of the momentum of the incoming particle; it shows a *scaling behavior*. Observations of such scaling behavior were of great significance in investigations of nucleon structure by deep inelastic electron–nucleon scattering and led to the current picture of the nucleon as a bound state of point-like quarks and gluons. Reversing the above line of reasoning, one can conclude from an observed scaling behavior of a scattering cross-section that the scattering centers must be point-like.

5.2 General Definition of the S- and T-Matrix

We will now consider a general process in which n initial particles a_i with four-momenta p_i scatter to give m particles b_j with four-momenta p'_j,

$$\mathrm{a}_1(p_1) + \cdots + \mathrm{a}_n(p_n) \to \mathrm{b}_1(p'_1) + \cdots + \mathrm{b}_m(p'_m). \tag{5.37}$$

We normalize the one-particle states as in (3.69) and again make use of the interaction picture. The time development of the state vector is then given by a differential equation with a Hamilton operator given by the interaction:

$$\mathrm{i}\frac{\partial}{\partial t}|t\rangle = \mathbf{H}'(t)|t\rangle. \tag{5.38}$$

The initial condition is

$$\lim_{t\to-\infty} |t\rangle = |i\rangle = |\mathrm{a}_1(p_1)\ldots\mathrm{a}_n(p_n)\rangle. \tag{5.39}$$

The S-operator is defined as previously, by

$$\lim_{t\to+\infty} |t\rangle = \mathbf{S}|i\rangle. \tag{5.40}$$

The corresponding S-matrix element for the final state

$$|\mathrm{f}\rangle = |\mathrm{b}_1(p'_1)\ldots\mathrm{b}_m(p'_m)\rangle \tag{5.41}$$

is then

$$S_{fi} = \langle\mathrm{b}_1(\mathrm{p}'_1)\ldots\mathrm{b}_m(p'_m)|\mathbf{S}|\mathrm{a}_1(p_1)\ldots\mathrm{a}_n(p_n)\rangle. \tag{5.42}$$

As in (5.10), we will separate out the part of the S-matrix corresponding to a situation where the particles do not interact:

$$S_{fi} = \delta_{fi} + \text{rest}.$$

Since we are excluding the possibility of external potentials, the four-momentum is conserved. It is then plausible, and can in fact be shown to be generally true, that the "rest" term contains a δ-function for the four-momenta. Thus we write

$$S_{fi} = \delta_{fi} + \mathrm{i}(2\pi)^4\delta(P_f - P_i)\langle f|\mathbf{T}|i\rangle, \tag{5.43}$$

where

$$\begin{aligned} P_f &= p'_1 + \cdots + p'_m, \\ P_i &= p_1 + \cdots + p_n, \\ \langle f|\mathbf{T}|i\rangle &\equiv \langle\mathrm{b}_1(p'_1)\ldots\mathrm{b}_m(p'_m)|\mathbf{T}|\mathrm{a}_1(p_1)\ldots\mathrm{a}_n(p_n)\rangle. \end{aligned} \tag{5.44}$$

Equation (5.43) provides the definition of the T-matrix. According to our normalization conventions, the S- and T-matrix elements are Lorentz-invariant functions. This will be seen explicitly in Part II when we consider some examples.

We now look at a reaction in which the initial state, as is usual in scattering processes, consists of two particles

$$\mathrm{a}_1(p_1) + \mathrm{a}_2(p_2) \rightarrow \mathrm{b}_1(p_1') + \cdots + \mathrm{b}_m(p_m'). \tag{5.45}$$

The definition of the scattering cross-section for this reaction is

$$\sigma = \frac{\text{Transition rate}}{\text{Flux of incident particles}}. \tag{5.46}$$

It should be noted here, that the transition rate and the flux depend on the particular Lotentz system chosen to describe the reaction. It is conventional to choose the rest frame of one of the particles in the initial state. As we shall see, the result for σ is independent of which particle we select from the initial state.

We will next calculate, for a general case, the transition probability $\mathrm{d}w$ for the process (5.45) in which we assume a normalization volume V for each of the particles a_1 and a_2. Making use of the normalization of plane-wave states (5.24), we obtain from (5.43):

$$\mathrm{d}w_{fi} = [(2\pi)^4 \delta(p_1' + \cdots + p_m' - p_1 - p_2)]^2 \frac{1}{2p_1^2 p_2^0}\left(\frac{1}{V}\right)^2$$
$$\cdot \prod_{i=1}^{m} \frac{\mathrm{d}^3 p_i'}{(2\pi)^3 2p_i'^0} |\langle \mathrm{b}_1(p_1') \ldots \mathrm{b}_m(p_m')| \mathbf{T} | \mathrm{a}_1(p_1)\mathrm{a}_2(p_2)\rangle|^2. \tag{5.47}$$

For the square of the δ-function we again apply Fermi's trick. We imagine that the interaction is switched on over the volume V for a time T and write

$$[(2\pi)^4 \delta(p_1' + \cdots + p_m' - p_1 - p_2)]^2$$
$$= \int_{V,T} \mathrm{d}x \, \mathrm{e}^{\mathrm{i}(p_1' + \cdots + p_m' - p_1 - p_2)x} \cdot (2\pi)^4 \delta(p_1' + \cdots + p_m' - p_1 - p_2)$$
$$= VT(2\pi)^4 \delta(p_1' + \cdots + p_m' - p_1 - p_2). \tag{5.48}$$

Then, for the transition rate we have

$$\frac{\mathrm{d}w_{fi}}{T} = \frac{1}{V} \frac{1}{2p_1^0 p_2^0} (2\pi)^4 \delta(p_1' + \cdots + p_m' - p_1 - p_2)$$
$$\cdot \prod_{i=1}^{m} \frac{\mathrm{d}^3 p_i'}{(2\pi)^3 2p_i'^0} |\langle \mathrm{b}_1(p_1') \ldots \mathrm{b}_m(p_m')| \mathbf{T} | \mathrm{a}_1(p_1)\mathrm{a}_2(p_2)\rangle|^2. \tag{5.49}$$

This formula, in its present form, is valid quite generally for any system of reference. We now take the special case of the rest frame of particle a_2. We then find the relations

$$p_2^0 = m_2.$$

$$\Phi = \frac{1}{V}\frac{|\boldsymbol{p}_1|}{p_1^0} = \frac{1}{V}\frac{1}{p_1^0}\frac{w(s, m_1^2, m_2^2)}{2m_2}, \tag{5.50}$$

$$s = (p_1 + p_2)^2.$$

where m_1 and m_2 are the masses of a_1 and a_2 and where Φ represents the flux of a_1 particles. We have made use of (2.71) for the momentum of a_1 in the rest frame of a_2. For the scattering cross-section we then obtain

$$\begin{aligned} \mathrm{d}\sigma &= \frac{\mathrm{d}w_{fi}/T}{\Phi} \\ &= \frac{1}{2w(s, m_1^2, m_2^2)} \prod_{i=1}^{m} \frac{\mathrm{d}^3 p_i'}{(2\pi)^3 2p_i'^0} (2\pi)^4 \delta(p_1' + \cdots + p_m' - p_1 - p_2) \\ &\quad \cdot |\langle \mathrm{b}_1(p_1') \ldots \mathrm{b}_m(p_m')|\mathbf{T}|\mathrm{a}_1(p_1)\mathrm{a}_2(p_2)\rangle|^2. \end{aligned} \tag{5.51}$$

We see from this equation that the particles a_1 and a_2 play completely symmetric roles as was mentioned above. For particles with spin, an average over the spin states is taken for every unpolarized particle in the initial state and, for every unobserved spin of a particle in the final state, a summation is made over the corresponding spin states.

From (5.51) we can readily calculate the total scattering cross-section σ_{tot} for our reaction. In doing so we note that $\mathrm{d}^3 p_i'/2p_i'^0$ is a Lorentz-invariant measure. This can easily be seen from the equation

$$\frac{\mathrm{d}^3 p_i'}{2p_i'^0} = \mathrm{d}p_i' \, \delta_+(p_i'^2 - m_i'^2), \tag{5.52}$$

where

$$\delta_+(p_i'^2 - m_i'^2) = \delta(p_i'^2 - m_i'^2)\theta(p_i'^0),$$

$$\mathrm{d}p_i' = \mathrm{d}p_i'^0 \, \mathrm{d}p_i'^1 \, \mathrm{d}p_i'^2 \, \mathrm{d}p_i'^3.$$

We then obtain

$$\begin{aligned} \sigma_{\mathrm{tot}}(\mathrm{a}_1 + \mathrm{a}_2 \to \mathrm{b}_1' + \cdots + \mathrm{b}_m') &= \frac{1}{2w(s, m_1^2, m_2^2)} \\ &\quad \cdot \int \prod_{i=1}^{m} \left(\frac{\mathrm{d}p_i'}{(2\pi)^3} \delta_+(p_i'^2 - m_i'^2) \right) (2\pi)^4 \delta(p_1' + \cdots + p_m' - p_1 - p_2) \\ &\quad \cdot |\langle \mathrm{b}_1(p_1') \ldots \mathrm{b}_m(p_m')|\mathbf{T}|\mathrm{a}_1(p_1)\mathrm{a}_2(p_2)\rangle|^2. \end{aligned} \tag{5.53}$$

As we have pointed out, the T-matrix element is a Lorentz-invariant function. By means of (5.53) we have thus written the total scattering cross-section in Lorentz-invariant form. However one should not be tempted to think that (5.46) should now be used to calculate the transition rate for a general reference system. The correct expression for the transition rate is always obtained from

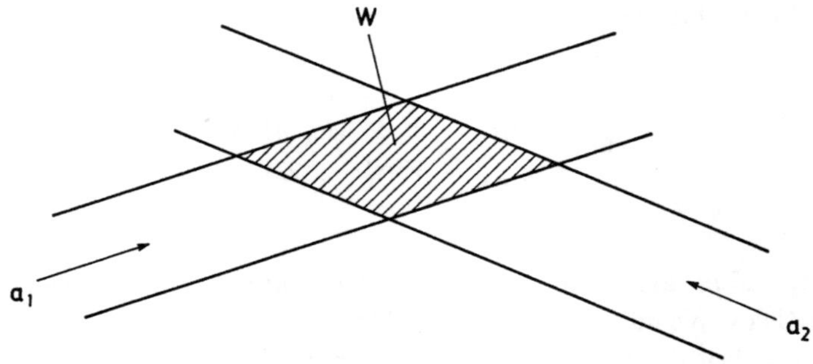

Figure 5.2 Two colliding particle beams with interaction volume W (shaded).

(5.49). Making use of (5.51) we can now write this as

$$\frac{\mathrm{d}w_{fi}}{T} = \frac{1}{V} \frac{w(s, m_1^2, m_2^2)}{2p_1^0 p_2^0} \mathrm{d}\sigma. \tag{5.54}$$

We will now calculate this for an example of some practical importance. We consider the case of two crossed particle beams as has been realized, for example, in the proton–proton storage ring ISR at CERN (Fig. 5.2). For the sake of simplicity, we consider homogeneous continuous beams. The particle densities in the two beams are denoted by n_1 and n_2. The transition rate (5.54) refers to a particle density of $1/V$. In our case we obtain for the transition rate per unit volume

$$\frac{\mathrm{d}w_{fi}}{VT} = n_1 n_2 \frac{w(s, m_1^2, m_2^2)}{2p_1^0 p_2^0} \mathrm{d}\sigma. \tag{5.55}$$

Multiplying by the volume W of the interaction region, the reaction rate becomes

$$\mathrm{d}N = L \cdot \mathrm{d}\sigma, \tag{5.56}$$

where

$$L = W \cdot n_1 n_2 \frac{w(s, m_1^2, m_2^2)}{2p_1^0 p_2^0}. \tag{5.57}$$

The quantity L is known as the luminosity. For equal beam energies we have, in the high energy limit, i.e., for $s \gg m_1^2, m_2^2$:

$$\begin{gathered} p_1^0 \cong p_2^0 \cong \frac{\sqrt{s}}{2}, \\ L \cong 2W \cdot n_1 \cdot n_2. \end{gathered} \tag{5.58}$$

The dimensions of luminosity are $\mathrm{s}^{-1}\,\mathrm{cm}^{-2}$. If F_1 and F_2 are the cross-sectional areas of the beams, then the incident particle currents, I, for velocities $\simeq 1$ are

$$\begin{aligned} I_1 &= n_1 F_1, \\ I_2 &= n_2 F_2. \end{aligned} \tag{5.59}$$

For the luminosity this then gives

$$L \cong 2\frac{W}{F_1 F_2} I_1 I_2. \tag{5.60}$$

For given incident particle currents one thus achieves the highest luminosity by focusing the beams as much as possible; this is due to the fact that W decreases with the third power, whereas $F_1 \cdot F_2$ decreases with the fourth power of the linear size of the beam.

5.3 The Unitarity Relation and the Optical Theorem

Let us start from a normalized initial state $|i\rangle$

$$\langle i|i\rangle = 1. \tag{5.61}$$

The probability of finding the final state $|f\rangle$ as $t \to +\infty$ is given by the absolute square of the S-matrix element

$$|\langle f|\mathbf{S}|i\rangle|^2$$

The probability of finding any final state is clearly equal to one and therefore it must hold that

$$\begin{aligned}\sum_f |\langle f|\mathbf{S}|i\rangle|^2 &= \sum_f \langle i|\mathbf{S}^\dagger|f\rangle\langle f|\mathbf{S}|i\rangle \\ &= \langle i|\mathbf{S}^\dagger\mathbf{S}|i\rangle \\ &= 1.\end{aligned} \tag{5.62}$$

Since this relation must be true for all normalized states $|i\rangle$, there follows the operator relationship

$$\mathbf{S}^\dagger\mathbf{S} = 1. \tag{5.63}$$

This is the unitarity relation for the S-matrix. If we form the matrix element of this with arbitrary states $|i\rangle$ and $|i'\rangle$, then (5.63) reads

$$\langle i'|\mathbf{S}^\dagger\mathbf{S}|i\rangle = \sum_f \langle f|\mathbf{S}|i'\rangle^* \langle f|\mathbf{S}|i\rangle = \langle i'|i\rangle. \tag{5.64}$$

For states $|i\rangle$ and $|i'\rangle$ of the type given in (5.39) with *equal* total momentum $P_i = P_{i'}$ we can now express the S-matrix in (5.64) in terms of the T-matrix (5.43) and obtain:

$$\frac{1}{2\mathrm{i}}\{\langle i'|\mathbf{T}|i\rangle - \langle i|\mathbf{T}|i'\rangle^*\} = \frac{1}{2}\sum_f (2\pi)^4\delta(P_f - P_i)\langle f|\mathbf{T}|i'\rangle^* \langle f|\mathbf{T}|i\rangle. \tag{5.65}$$

An important special case occurs for

$$|i\rangle = |i'\rangle = |\mathrm{a}_1(p_1)\mathrm{a}_2(p_2)\rangle. \tag{5.66}$$

By using (5.53) for the total scattering cross-section we get

$$\mathrm{Im}\langle i|\mathbf{T}|i\rangle = w(s, m_1^2, m_2^2)\cdot\sigma_{\mathrm{tot}}(\mathrm{a}_1 + \mathrm{a}_2 \to \mathrm{X}), \tag{5.67}$$

where X stands for the sum over all possible final states. This is the *optical theorem* which relates the imaginary part of the forward scattering amplitude to the total scattering cross-section.

5.4 The Decay Rate of an Unstable Particle

In particle physics one is often interested in the decay of an unstable particle, i.e., in a reaction of the type

$$\mathrm{a}(p) \rightarrow \mathrm{b}_1(p'_1) + \cdots + \mathrm{b}_m(p'_m). \tag{5.68}$$

We will define the S- and T-matrix elements for this decay according to the general rule (5.43). The differential decay rate, i.e., the rate of transition of the particle a into the above final states, can be obtained by a procedure analogous to that used in deriving the scattering cross-section (5.49). The notion of a decay rate implicitly refers to the rest frame of the decaying particle. Written in Lorentz-invariant form we have

$$\mathrm{d}\Gamma(\mathrm{a}(p) \rightarrow \mathrm{b}_1(p'_1) + \cdots + \mathrm{b}_m(p'_m)) = \frac{1}{2m_\mathrm{a}}(2\pi)^4\delta(p'_1 + \cdots + p'_m - p)$$

$$\cdot \prod_{i=1}^{m} \frac{\mathrm{d}^3 p'_i}{(2\pi)^3 2p'^0_i} |\langle \mathrm{b}_1(p'_1)\ldots \mathrm{b}_m(p'_m)|\mathbf{T}|\mathrm{a}(p)\rangle|^2. \tag{5.69}$$

The total decay rate Γ_a is obtained from this by integrating over the whole of phase space and over all decay channels.

For additional insight the reader is referred to the discussion of particle decay in the framework of the Wigner–Weisskopf method in Appendix I.

Exercises

5.1 Show that the following expression represents a projection operator

$$\mathbf{P}(B) = \int_B \frac{\mathrm{d}^3 p}{(2\pi)^3 2p^0} \sum_s |\mathrm{e}^-(\boldsymbol{p}, s)\rangle\langle \mathrm{e}^-(\boldsymbol{p}, s)|, \tag{5.70}$$

where B is an arbitrary region of three-dimensional momentum space.

5.2 Consider the decay of a scalar Higgs particle h with mass $m_\mathrm{h} > 2m_\mathrm{e}$ into an electron–positron pair. The Hamiltonian for the interaction is

$$\mathbf{H}'(t) = \int \mathrm{d}^3 x \frac{m_\mathrm{e}}{\rho_0} \rho'(\boldsymbol{x}, t)\bar{\psi}(\boldsymbol{x}, t)\psi(\boldsymbol{x}, t), \tag{5.71}$$

where $\rho'(\boldsymbol{x}, t)$ is the Higgs field operator, ρ_0 is a constant, and $\psi(\boldsymbol{x}, t)$ is the Dirac field operator. Calculate the decay rate $\Gamma(\mathrm{h} \rightarrow \mathrm{e}^+\mathrm{e}^-)$ to lowest order in the Born approximation.

Hint: Expand the Higgs field in terms of creation and annihilation operators as in (3.30).

5.3 Consider instead of the momentum eigenstates (5.11), normalized states $|i\rangle$ and $|f\rangle$ corresponding to wave packets with average momentum $\boldsymbol{p}$ and $\boldsymbol{p}'$. Determine the scattering cross-section for these states and show that for sufficiently narrow wave packets one re-obtains the result (5.26) derived in the text using Fermi's trick.

PART II

QUANTUM ELECTRODYNAMICS

CHAPTER 6

Introductory Remarks

In this part we will discuss quantum electrodynamics (QED), the theory of photons, and electrons and their interaction. The emphasis here will lie in the practical treatment of physical problems. For formal aspects we will often simply refer to the relevant literature.

The basic equations which govern the electromagnetic field were discovered by Maxwell in the last century. In rational units (Heavyside units) these read as follows:

$$\begin{aligned} \operatorname{rot} \boldsymbol{B} &= \boldsymbol{j} + \frac{\partial}{\partial t}\boldsymbol{E}, \\ \operatorname{rot} \boldsymbol{E} &= -\frac{\partial}{\partial t}\boldsymbol{B}, \\ \operatorname{div} \boldsymbol{B} &= 0, \\ \operatorname{div} \boldsymbol{E} &= \rho. \end{aligned} \tag{6.1}$$

Here ρ and $\boldsymbol{j}$ are the charge density and current density; $\boldsymbol{E}$ and $\boldsymbol{B}$ are the electric and magnetic field vectors. The units are so chosen that no inconvenient factors of 4π appear in the basic equations. The Coulomb interaction of two electrons at a separation r is then

$$E_{\text{Coulomb}} = \frac{e^2}{4\pi r}. \tag{6.2}$$

Here, and in the following, the symbol e will represent the magnitude of the electron charge. Thus the charge of an electron itself will be written $-e$. In our system of units, which was defined in Chap. 1, energies and reciprocal lengths have the same dimensions. The factor $e^2/4\pi$ is thus dimensionless and characterizes the strength of the interaction between two electrons. This factor is one of the fundamental constants of Nature, the Sommerfeld fine-structure constant α. Its currently accepted experimental value is

$$\frac{e^2}{4\pi} \equiv \alpha = 1/137.036\,04\,(11). \tag{6.3}$$

The smallness of this number has turned out to be a great blessing for the theory. It has the consequence that the Maxwell and Dirac equations, together with the quantum postulates, allow the development of a splendid theory of electromagnetism, namely, quantum electrodynamics. An essential part of this theory involves series expansions in α and obviously one can only expect to achieve good results if α has a small value.

As far as we know, QED describes all electromagnetic phenomena—from the longest radio waves to the shortest wavelengths of 10^{-16} cm, corresponding to the highest presently obtainable energies. (For an overview of high energy tests of QED see Hepp 1981.) The predictions of the theory have been tested and confirmed to an extent which is matched in hardly any other area of physics. An example of a highly successful test is the anomalous magnetic moment of the electron.

As we shall see in more detail in Chap. 8, a free electron behaves like a small magnetic dipole. The magnitude μ_e of the magnetic dipole moment of the electron is quoted as a multiple of the Bohr magneton $e/2m_e$ where m_e is the electron mass. We write:

$$\mu_e = \tfrac{1}{2} g_e \frac{e}{2m_e}, \tag{6.4}$$

where g_e is the Landé g-factor. According to the Dirac theory, g_e should be exactly equal to 2. Experimentally one finds that g_e deviates from this value. The most exact measurements to date (van Dyck 1979) have given the result

$$a_e^{\text{exp.}} = \tfrac{1}{2}(g_e - 2)^{\text{exp.}} = 1\,159\,652\,200(40) \cdot 10^{-12}. \tag{6.5}$$

In the framework of quantum electrodynamics a value different from 2 is indeed predicted for the electron g-factor. For the anomaly a_e this theory gives a series expansion in α with expansion coefficients C_j:

$$a_e = \frac{1}{2}\frac{\alpha}{\pi} + C_2\left(\frac{\alpha}{\pi}\right)^2 + C_3\left(\frac{\alpha}{\pi}\right)^3 + C_4\left(\frac{\alpha}{\pi}\right)^4 + \cdots. \tag{6.6}$$

The calculation of the first-order term by Schwinger (1949) was one of the first great triumphs for QED. In the meantime, all the coefficients up to and including C_4 are known (Kinoshita 1981). The resulting theoretical prediction for a_e is

$$a_e^{\text{theor.}} = 1\,159\,652\,460\,(127)\,(75) \cdot 10^{-12}. \tag{6.7}$$

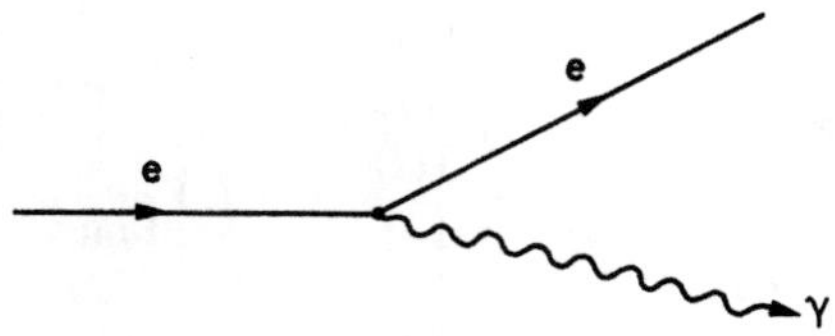

Figure 6.1 Emission of a photon by an accelerating electron.

The first error quoted in this value is due to the uncertainty in the experimental determination of the fine-structure constant α, and the second stems from the uncertainty in the numerical calculation of the coefficients C_3 and C_4 in (6.6). As can be seen, theory and experiment agree to within two standard deviations, i.e., the theory is confirmed to a precision of 2 in 10^7.

By means of this example, we will describe the physical picture underlying QED. We already know from classical electrodynamics that accelerated electrons emit radiation (Fig. 6.1). An electron at rest or in uniform motion in a straight line cannot emit a photon because classical energy-momentum conservation precludes this.

However, in the quantum theory we have the uncertainty relation between energy and time:

$$\Delta E \cdot \Delta t \gtrsim 1. \tag{6.8}$$

As a result of this, even an electron at rest is able to emit a photon. This process does not conserve energy but, due to the uncertainty relation, it is allowed provided the photon is reabsorbed within a time $\Delta t \sim 1/\Delta E$ (Fig. 6.2). Photons with a "transitory" existence such as this are termed *virtual photons.*[1]

In QED one imagines that an electron, or any other charged particle, is surrounded by a cloud of such virtual photons. The amplitude for the emission of a photon is proportional to the charge e and the probability is proportional to the fine-structure constant α.

These virtual photons enable us to understand why the magnetic moment

[1] A "true" physical particle on the other hand, i.e., one whose four-momentum satisfies the mass shell condition (2.54) for $m > 0$ (respectively (2.57) for $m = 0$), is often called a *real* particle. One speaks thus of real and virtual photons, electrons, etc.

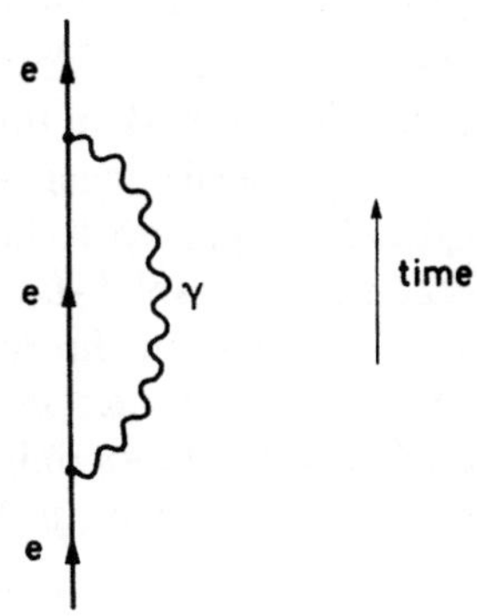

Figure 6.2 Emission and reabsorption of a virtual photon by an electron.

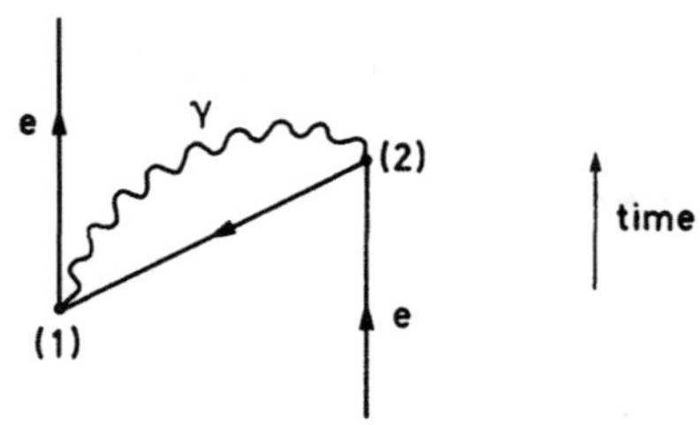

Figure 6.3 Emission and reabsorption of a virtual photon with an electron traveling backwards in time.

of the electron is not exactly equal to the Dirac value ($g_e = 2$). The Dirac value would correspond to a "naked" electron. Upon emission of a virtual photon the electron recoils and, since it is charged, thereby produces an electrical current distribution. If we calculate the magnetic moment for a physical electron then we find in addition to the Dirac term a term due to the convective current associated with the emission of a virtual photon. To leading order, the anomaly a_e should be proportional to α since the probability of virtual photon emission is itself proportional to α. This is confirmed by the explicit result given in (6.6).

It is important to note that the recoil may also cause the electron to travel backwards in time (Fig. 6.3). A more conventional interpretation of this diagram would be as follows: At point (1) an electron–positron pair is created with the emission of a virtual photon. The electron travels off to infinity while the position annihilates with the incoming electron at point (2) accompanied by the reabsorption of the virtual photon.

That electron lines pointing backwards in time may be associated with positrons was first suggested by Stueckelberg in 1941. In the framework of the Feynman rules we will find a consistent relativistically covariant formalism that describes all aspects of propagation and interaction. Many other famous effects such as the Lamb shift and the Casimir effect, can be understood physically in terms of the influence of virtual photons. However, it turned out to be rather difficult to find a procedure to enable the quantitative calculation of these effects. The problem lay in the infinite results that were obtained when attempts were made to use in a naive manner a perturbation theory of the type so successful in nonrelativistic quantum mechanics. The breakthrough came with the development of the renormalization theory at the end of the 1940s by Feynman, Tomonaga, Schwinger, and others. A collection of important original work in this field can be found in Schwinger (1958).

The physical concepts and calculational methods which were developed for QED have proven to be extremely fruitful, not only in this area but also in others such as solid state physics. All modern theories of elementary particles such as quantum chromodynamics or the unified theory for the electromagnetic and weak interactions, are built up in a way similar to QED. Thus the study of QED is not only interesting in itself but also serves as an introduction to modern concepts in the theory of elementary particles.

CHAPTER 7

The Quantization of the Free Electromagnetic Field

7.1 Commutation Rules and Indefinite Metric

In this section we will discuss the quantization of the free electromagnetic field. As we shall see, there are some complications in comparison to the free meson field which we treated in Part I.

To begin with we recall the relativistic form of the Maxwell equations. Electric and magnetic fields are combined into the field-strength tensor and charge and current density into the four-vector current density.

Field-strength tensor:

$$(F^{\mu\nu}) = \begin{pmatrix} 0 & -E^1 & -E^2 & -E^3 \\ E^1 & 0 & -B^3 & B^2 \\ E^2 & B^3 & 0 & -B^1 \\ E^3 & -B^2 & B^1 & 0 \end{pmatrix}. \tag{7.1}$$

Current density four-vector:

$$(j^\mu) = \begin{pmatrix} \rho \\ \boldsymbol{j} \end{pmatrix}. \tag{7.2}$$

The Maxwell equations then read:

$$\partial_\mu F^{\mu\nu} = j^\nu, \tag{7.3a}$$

$$\varepsilon^{\mu\nu\rho\sigma} \partial_\nu F_{\rho\sigma} = 0. \tag{7.3b}$$

The retarded potentials V and $\boldsymbol{A}$ are combined to the four-vector potential A^μ:

$$(A^\mu) = \begin{pmatrix} V \\ A \end{pmatrix}. \tag{7.4}$$

By means of the ansatz

$$F^{\mu\nu} = \partial^\mu A^\nu - \partial^\nu A^\mu, \tag{7.5}$$

which for the field strengths reads

$$\begin{aligned} \boldsymbol{E} &= -\nabla V - \dot{\boldsymbol{A}}, \\ \boldsymbol{B} &= \operatorname{curl} \boldsymbol{A} \end{aligned} \tag{7.6}$$

the second Maxwell equation (7.3b) is identically satisfied.

Equation (7.5) does not completely determine A^μ. With the substitution

$$A^\mu(x) \to A^\mu(x) + \partial^\mu \Lambda(x), \tag{7.7}$$

where $\Lambda(x)$ is an arbitrary function of x, $F^{\mu\nu}$ remains unchanged. Such a transformation is known as a *gauge transformation*. We can exploit the gauge freedom to subject the four-vector potential to a subsidiary condition, for example, the Lorentz condition

$$\partial_\mu A^\mu = 0. \tag{7.8}$$

With this the first Maxwell equation (7.3a) becomes

$$\Box A_\mu = j_\mu. \tag{7.9}$$

In this chapter we will study the free photon field, i.e., the photon field in the absence of electromagnetic currents. In this case the four-vector potential satisfies the d'Alembert equation

$$\Box A_\mu = 0. \tag{7.10}$$

Being observables, the components of the electromagnetic field-strength tensor must be describable in the quantum theory by means of Hermitian operators. For the vector potential A^μ this is not so obvious since A^μ is not directly observable. Nonetheless, it is convenient to start off with A^μ as operator field and pay the price, which is the associated indefinite metric in Hilbert space.

We take over the d'Alembert equation (7.10) to the quantum theory and treat it as an operator equation. In analogy to the procedure used for the meson field in Chap. 3, we will write a Fourier expansion for the field operator $\mathbf{A}_\mu$ with expansion coefficients that are also operators:

$$\mathbf{A}_\mu(x) = \int \frac{\mathrm{d}^3 k}{(2\pi)^3 2\omega} \{ \mathrm{e}^{\mathrm{i}kx}\, \mathbf{a}_\mu^\dagger(\boldsymbol{k}) + \mathrm{e}^{-\mathrm{i}kx}\, \mathbf{a}_\mu(\boldsymbol{k}) \}, \tag{7.11}$$

where

$$k = \begin{pmatrix} \omega \\ \boldsymbol{k} \end{pmatrix}, \qquad \omega = |\boldsymbol{k}|,$$

$$kx = k^\mu x_\mu = \omega t - \boldsymbol{k}\boldsymbol{x}.$$

The generalization of the arguments used in Chap. 3 (3.55), (3.56), (3.61) leads us here to the following commutation rules:

$$[\mathbf{a}_\mu^\dagger(\boldsymbol{k}), \mathbf{a}_\nu^\dagger(\boldsymbol{k}')] = 0, \tag{7.12}$$

$$[\mathbf{a}_\mu(\boldsymbol{k}), \mathbf{a}_\nu(\boldsymbol{k}')] = 0, \tag{7.13}$$

$$[\mathbf{a}_\mu(\boldsymbol{k}), \mathbf{a}_\nu^\dagger(\boldsymbol{k}')] = Z_{\mu\nu}(2\pi)^3 2\omega\delta^3(\boldsymbol{k} - \boldsymbol{k}'), \tag{7.14}$$

with as yet unknown constants $Z_{\mu\nu}$.

The operators $\mathbf{a}$ and $\mathbf{a}^\dagger$ will be realized in Fock space. For the vacuum state we demand in analogy to (3.42):

$$\mathbf{a}_\mu(\boldsymbol{k})|0\rangle = 0 \qquad \text{for all } \mu, \boldsymbol{k}. \tag{7.15}$$

The indices μ and ν in (7.14) are four-vector indices, thus $Z_{\mu\nu}$ must form a constant Lorentz tensor of second rank if we wish the theory to have explicit Lorentz invariance. But the only tensors with this property are the constant multiples of the metric tensor $g_{\mu\nu}$. Thus we obtain, possibly after a rescaling analogous to (3.65):

$$(Z_{\mu\nu}) = \pm(g_{\mu\nu}) = \pm\begin{pmatrix} 1 & 0 & 0 & 0 \\ 0 & -1 & 0 & 0 \\ 0 & 0 & -1 & 0 \\ 0 & 0 & 0 & -1 \end{pmatrix}. \tag{7.16}$$

In Chap. 3 we were able to deduce the sign of the constant Z from the requirement that the states have positive norm. Here it is not so simple since $g_{\mu\nu}$ has both positive and negative eigenvalues. Let us put, as in fact turns out to be correct[1]

$$[\mathbf{a}_\mu(\boldsymbol{k}), \mathbf{a}_\nu^\dagger(\boldsymbol{k}')] = -g_{\mu\nu}(2\pi)^3 2\omega\delta^3(\boldsymbol{k} - \boldsymbol{k}'), \tag{7.17}$$

then we find, in analogy to (3.62), (3.63), that the operators $\mathbf{a}_\mu^\dagger$, with $\mu = 1, 2, 3$ when applied to the vacuum state, lead to states with positive norm, and that $\mathbf{a}_0^\dagger$ leads in a similar way to states with negative norm. If we had chosen the opposite sign in (7.16), we would have obtained just the reverse of this. Thus in both cases, the space of state vectors has indefinite metric. At first this appears to be something of a disaster since for states with negative norm the probability interpretation of quantum mechanics breaks down.

Our formalism has a further unphysical feature. To see this, we consider a general state with a photon of well-defined wave vector $\boldsymbol{k}$. Such a state has the form

$$|\boldsymbol{k}, \varepsilon\rangle = -\varepsilon^\mu \mathbf{a}_\mu^\dagger(\boldsymbol{k})|0\rangle, \tag{7.18}$$

where ε is any complex four-vector and represents the four-polarization of the photon. There are clearly four linearly independent vectors ε and thus we obtain *four* linearly independent one-photon states for a fixed $\boldsymbol{k}$-value. From experiment, however, we know that a real photon has only two polarization

[1] As will be seen, this choice of sign leads to a positive norm for physical states.

degrees of freedom. We will now show that it is possible to dispose of the two unwanted polarization degrees of freedom and, at the same time, to guarantee a positive norm for physical states. The procedure for this was first developed by Gupta and Bleuler (Gupta 1950; Bleuler 1950).

We recall that by means of (7.11) we have satisfied the d'Alembert equation (7.10) in operator form. The d'Alembert equation, however, is only equivalent to the Maxwell equations (7.3) upon imposing the Lorentz condition. Thus we must also bring this Lorentz condition into the quantum theory.

It has not been possible to formulate this condition in operator form and simultaneously to keep manifest Lorentz invariance in the formalism. The way out of these difficulties which was suggested by Gupta and Bleuler, is to demand that the *states* satisfy a subsidiary condition. Only those state vectors in Fock space are assigned a physical meaning which in some way satisfy the Lorentz condition. More precisely, we take the part of the vector potential (7.11) that only contains annihilation operators

$$\mathbf{A}_\mu^{(-)}(x) = \int \frac{d^3k}{(2\pi)^3 2\omega} e^{-ikx}\, \mathbf{a}_\mu(\boldsymbol{k}), \tag{7.19}$$

and demand for physical state vectors

$$\partial^\mu \mathbf{A}_\mu^{(-)}(x)|\text{phys. state}\rangle = 0, \tag{7.20}$$

or, expressed in terms of Fourier components

$$k^\mu \mathbf{a}_\mu(\boldsymbol{k})|\text{phys. state}\rangle = 0 \tag{7.21}$$

for all $\boldsymbol{k}$. By means of this subsidiary condition it is guaranteed that the expectation value of the divergence $\partial^\mu \mathbf{A}_\mu(x)$ of the entire field vanishes for all physical states

$$\langle\text{phys. state}|\partial^\mu \mathbf{A}_\mu(x)|\text{phys. state}\rangle = 0. \tag{7.22}$$

We simply need to let the part of $\partial^\mu \mathbf{A}_\mu(x)$ containing creation operators act on the left side of the matrix element.

The subspace of physical states is clearly a linear space. We now show that, although its metric is not positive definite, it does at least have a positive semidefinite metric. We assert:

$$\langle\text{phys. state}|\text{phys. state}\rangle \geq 0. \tag{7.23}$$

We will also find state vectors that are not equal to the null vector which nonetheless satisfy (7.21) and have zero norm.

In order to prove (7.23) we choose a new basis for the creation and annihilation operators. We consider the operators $\mathbf{a}_\mu^\dagger(\boldsymbol{k})$ for a fixed wave vector $\boldsymbol{k}$ and choose two unit vectors $\boldsymbol{e}_1$ and $\boldsymbol{e}_2$ perpendicular to $\boldsymbol{k}$ such that $\boldsymbol{e}_1, \boldsymbol{e}_2$ and $\boldsymbol{e}_3 = \hat{\boldsymbol{k}} = \boldsymbol{k}/|\boldsymbol{k}|$ form an orthogonal set of basis vectors

$$\boldsymbol{e}_i \boldsymbol{e}_j = \delta_{ij}. \tag{7.24}$$

We define operators α as follows

$$\begin{aligned}
\alpha_0^\dagger(\boldsymbol{k}) &= \frac{1}{\sqrt{2}}(\mathrm{a}_0^\dagger(\boldsymbol{k}) - \hat{\boldsymbol{k}}\,\vec{\mathrm{a}}^\dagger(\boldsymbol{k})),\\
\alpha_1^\dagger(\boldsymbol{k}) &= \boldsymbol{e}_1\,\vec{\mathrm{a}}^\dagger(\boldsymbol{k}),\\
\alpha_2^\dagger(\boldsymbol{k}) &= \boldsymbol{e}_2\,\vec{\mathrm{a}}^\dagger(\boldsymbol{k}),\\
\alpha_3^\dagger(\boldsymbol{k}) &= \frac{1}{\sqrt{2}}(\mathrm{a}_0^\dagger(\boldsymbol{k}) + \hat{\boldsymbol{k}}\,\vec{\mathrm{a}}^\dagger(\boldsymbol{k})).
\end{aligned} \tag{7.25}$$

The commutation rules for the operators α follow from (7.12), (7.13), (7.17):

$$\begin{aligned}
[\alpha_0(\boldsymbol{k}), \alpha_0^\dagger(\boldsymbol{k}')] &= [\alpha_3(\boldsymbol{k}), \alpha_3^\dagger(\boldsymbol{k}')] = 0,\\
[\alpha_0(\boldsymbol{k}), \alpha_3^\dagger(\boldsymbol{k}')] &= [\alpha_3(\boldsymbol{k}), \alpha_0^\dagger(\boldsymbol{k}')] = -(2\pi)^3 2\omega\delta^3(\boldsymbol{k}-\boldsymbol{k}'),\\
[\alpha_1(\boldsymbol{k}), \alpha_1^\dagger(\boldsymbol{k}')] &= [\alpha_2(\boldsymbol{k}), \alpha_2^\dagger(\boldsymbol{k}')] = (2\pi)^3 2\omega\delta^3(\boldsymbol{k}-\boldsymbol{k}').
\end{aligned} \tag{7.26}$$

All other commutators vanish. Our subsidiary condition (7.21) reads:

$$\alpha_0(\boldsymbol{k})|\text{phys. state}\rangle = 0. \tag{7.27}$$

We now consider a state vector in Hilbert space of the form

$$\alpha_1^\dagger(\boldsymbol{k}_1)\alpha_1^\dagger(\boldsymbol{k}_2)\ldots\alpha_2^\dagger\ldots\alpha_0^\dagger\ldots\alpha_3^\dagger\ldots|0\rangle, \tag{7.28}$$

in other words an arbitrary product of creation operators acting on the vacuum state. It is easy to see that, as a result of the commutation rules (7.26), the subsidiary condition (7.27) is satisfied when, and only when, no operator $\alpha_3^\dagger$ appears in (7.28). For example, of the possible states obtained by applying an operator $\alpha^\dagger$ to the vacuum state, only the following are physical:

$$\alpha_1^\dagger(\boldsymbol{k})|0\rangle, \qquad \alpha_2^\dagger(\boldsymbol{k})|0\rangle, \qquad \alpha_0^\dagger(\boldsymbol{k})|0\rangle. \tag{7.29}$$

These states are orthogonal to one another and their norms are greater than or equal to zero

$$\begin{aligned}
\langle 0|\alpha_1(\boldsymbol{k})\alpha_1^\dagger(\boldsymbol{k}')|0\rangle &= (2\pi)^3 2\omega\delta^3(\boldsymbol{k}-\boldsymbol{k}'),\\
\langle 0|\alpha_2(\boldsymbol{k})\alpha_2^\dagger(\boldsymbol{k}')|0\rangle &= (2\pi)^3 2\omega\delta^3(\boldsymbol{k}-\boldsymbol{k}'),\\
\langle 0|\alpha_0(\boldsymbol{k})\alpha_0^\dagger(\boldsymbol{k}')|0\rangle &= 0.
\end{aligned} \tag{7.30}$$

Any physical state vector can easily be seen to be a combination of vectors of the form (7.28) which *do not* contain $\alpha_3^\dagger$. By generalizing (7.30) one can readily convince oneself that the norm of a physical state is greater than or equal to zero. Thus the original assertion (7.23) has been proved.

Now it is quite easy to arrive at a Hilbert space with positive definite metric that allows a probability interpretation in the quantum mechanical sense. We will declare two physical state vectors $|1\rangle$ and $|2\rangle$ to be equivalent if their difference has norm zero, i.e.,

$$|1\rangle \sim |2\rangle,$$

if

$$(\langle 1| - \langle 2|)(|1\rangle - |2\rangle) = 0. \tag{7.31}$$

As is easily seen, the equivalence classes then form a *bona fide* Hilbert space in the mathematical sense.

The physical interpretation of our formalism is the following: The state of a system of photons is described by a whole class of equivalent state vectors in Fock space. This can only make sense physically when the expectation values of *observable* quantities such as the field-strength tensor, the energy, etc., are identical for equivalent state vectors. It is possible to give a general proof of this, and we will demonstrate the assertion by means of some examples. For practical purposes, physical matrix elements can consequently be calculated by selecting any representative of an equivalence class as state vector.

Let us look more closely at the one-photon state vectors of (7.29). As a result of (7.30) the state $\alpha_0^\dagger(\boldsymbol{k})|0\rangle$ is equivalent to the null vector. Only the state vectors

$$\begin{aligned} |\boldsymbol{k}, \varepsilon_1\rangle &= \alpha_1^\dagger(\boldsymbol{k})|0\rangle = \boldsymbol{e}_1 \vec{\mathbf{a}}^\dagger(\boldsymbol{k})|0\rangle, \\ |\boldsymbol{k}, \varepsilon_2\rangle &= \alpha_2^\dagger(\boldsymbol{k})|0\rangle = \boldsymbol{e}_2 \vec{\mathbf{a}}^\dagger(\boldsymbol{k})|0\rangle, \end{aligned} \tag{7.32}$$

where

$$\varepsilon_{1,2} = \begin{pmatrix} 0 \\ \boldsymbol{e}_{1,2} \end{pmatrix},$$

correspond to nonzero physical states. In agreement with experiment, our formalism thus gives *two* linearly independent one-photon states for each value of $\boldsymbol{k}$. A general physical one-photon state with fixed $\boldsymbol{k}$-value can be written in the form

$$|\boldsymbol{k}, \varepsilon\rangle = -\varepsilon^\mu \mathbf{a}_\mu^\dagger(\boldsymbol{k})|0\rangle = \boldsymbol{\varepsilon}\vec{\mathbf{a}}^\dagger(\boldsymbol{k})|0\rangle. \tag{7.33}$$

Here ε, or rather its spatial part $\boldsymbol{\varepsilon}$ obeys the transversality and normalization conditions

$$\begin{aligned} \varepsilon^0 &= 0, \\ \boldsymbol{\varepsilon k} &= 0, \\ |\boldsymbol{\varepsilon}| &= 1. \end{aligned} \tag{7.34}$$

From this, it follows in particular that

$$\varepsilon^\mu k_\mu = 0. \tag{7.35}$$

The states of (7.33) satisfy the continuum normalization

$$\langle \boldsymbol{k}', \varepsilon' | \boldsymbol{k}, \varepsilon\rangle = (\boldsymbol{\varepsilon}'^* \cdot \boldsymbol{\varepsilon})(2\pi)^3 2\omega \delta^3(\boldsymbol{k}' - \boldsymbol{k}). \tag{7.36}$$

We further recall that a linearly polarized photon is described by a real polarization vector $\boldsymbol{\varepsilon} = \boldsymbol{\varepsilon}^*$, whereas right- and left-hand circularly polarized photons have the polarization vectors

$$\varepsilon_{\pm} = \mp \frac{1}{\sqrt{2}}(\boldsymbol{e}_1 \pm i\boldsymbol{e}_2). \tag{7.37}$$

One might inquire why we have not immediately confined ourselves to the state vectors of (7.33). The reason is that the transversality condition on the polarization vector (7.34) is not Lorentz invariant. If we consider a Lorentz transformation Λ, under which the four-momentum k and polarization vector ε of a photon transform as follows:

$$k \to k' = \Lambda k,$$

$$\varepsilon = \begin{pmatrix} 0 \\ \boldsymbol{\varepsilon} \end{pmatrix} \to \varepsilon' = \Lambda \varepsilon,$$

then it follows from (7.35) that

$$\varepsilon'^{\mu} k'_{\mu} = 0.$$

From this it is easy to see that ε' can be written in the form

$$\varepsilon' = \begin{pmatrix} \varepsilon'^0 \\ \boldsymbol{e}' + \varepsilon'^0 \hat{\boldsymbol{k}}' \end{pmatrix}$$

where $\boldsymbol{e}'\boldsymbol{k}' = 0$ and $\hat{\boldsymbol{k}}' = \boldsymbol{k}'/|\boldsymbol{k}'|$. But in general, we have $\varepsilon'^0 \neq 0$, that is, ε' no longer satisfies (7.34). The corresponding state vector, however, is equivalent to that of a transversely polarized photon

$$-\varepsilon'^{\mu} \mathrm{a}^{\dagger}_{\mu}(\boldsymbol{k}')|0\rangle = \{\boldsymbol{e}'\vec{\mathrm{a}}^{\dagger}(\boldsymbol{k}') - \varepsilon'^0[\mathrm{a}^{\dagger}_0(\boldsymbol{k}') - \hat{\boldsymbol{k}}'\vec{\mathrm{a}}^{\dagger}(\boldsymbol{k}')]\}|0\rangle \sim \boldsymbol{e}'\vec{\mathrm{a}}^{\dagger}(\boldsymbol{k}')|0\rangle.$$

7.2 Normal-Ordered and Time-Ordered Products

We will now consider the expectation values of a few physically observable quantities. In the following, $\langle\ \rangle$ indicates the expectation value in an arbitrary physical state; $\boldsymbol{e}_1$ and $\boldsymbol{e}_2$ are as in (7.25) and depend of course on $\boldsymbol{k}$. According to (7.6) we find for the field strengths:

$$\begin{aligned}\langle\vec{\mathbf{B}}(x)\rangle = \int \frac{d^3k}{(2\pi)^3 2\omega} &\cdot \{e^{ikx}\langle -i\boldsymbol{k} \times [\boldsymbol{e}_1\alpha^{\dagger}_1(\boldsymbol{k}) + \boldsymbol{e}_2\alpha^{\dagger}_2(\boldsymbol{k})]\rangle \\ &+ e^{-ikx}\langle i\boldsymbol{k} \times [\boldsymbol{e}_1\alpha_1(\boldsymbol{k}) + \boldsymbol{e}_2\alpha_2(\boldsymbol{k})]\rangle\},\end{aligned} \tag{7.38}$$

$$\begin{aligned}\langle\vec{\mathbf{E}}(x)\rangle = \int \frac{d^3k}{(2\pi)^3 2\omega} &\cdot \{-i\omega\, e^{ikx}\langle \boldsymbol{e}_1\alpha^{\dagger}_1(\boldsymbol{k}) + \boldsymbol{e}_2\alpha^{\dagger}_2(\boldsymbol{k})\rangle + i\omega\, e^{-ikx}\langle \boldsymbol{e}_1\alpha_1(\boldsymbol{k}) + \boldsymbol{e}_2\alpha_2(\boldsymbol{k})\rangle\}.\end{aligned} \tag{7.39}$$

We see that only the transverse degrees of freedom of the photons contribute here, in agreement with experiment.

The energy and momentum of photons are certainly observable quantities. What do the corresponding operators look like? In our system of units, the classical expressions for energy P^0 and momentum $\boldsymbol{P}$ of an electromagnetic field are

$$\begin{aligned} P^0 &= \int_{t=\text{const.}} \mathrm{d}^3x\,\tfrac{1}{2}[\boldsymbol{E}^2(x) + \boldsymbol{B}^2(x)], \\ \boldsymbol{P} &= \int_{t=\text{const.}} \mathrm{d}^3x\,\boldsymbol{E}(x) \times \boldsymbol{B}(x). \end{aligned} \tag{7.40}$$

If we regard $\boldsymbol{E}$ and $\boldsymbol{B}$ here as field operators, then we find for the expectation values of the integrals in (7.40), which now represent operators P'^0 and $\vec{\mathsf{P}}'$, in an arbitrary physical state

$$\langle \mathsf{P}'^0 \rangle = \frac{1}{2}\int \frac{\mathrm{d}^3k}{(2\pi)^3 2\omega}\,\omega\cdot\left\langle \sum_{i=1}^{2} \{\alpha_i^\dagger(\boldsymbol{k})\alpha_i(\boldsymbol{k}) + \alpha_i(\boldsymbol{k})\alpha_i^\dagger(\boldsymbol{k})\}\right\rangle, \tag{7.41}$$

$$\langle \vec{\mathsf{P}}' \rangle = \frac{1}{2}\int \frac{\mathrm{d}^3k}{(2\pi)^3 2\omega}\,\boldsymbol{k}\cdot\left\langle \sum_{i=1}^{2} \{\alpha_i^\dagger(\boldsymbol{k})\alpha_i(\boldsymbol{k}) + \alpha_i(\boldsymbol{k})\alpha_i^\dagger(\boldsymbol{k})\}\right\rangle. \tag{7.42}$$

Again, as must be the case, only the physical degrees of freedom of the photons contribute. However, we now face a new problem. If we look at the vacuum expectation values of the energy and momentum, then with (7.41), (7.42) we find divergent results!

$$\begin{aligned} \langle 0|\mathsf{P}'^0|0\rangle &= \frac{1}{2}\int \frac{\mathrm{d}^3k}{(2\pi)^3 2\omega}\,\omega\cdot\sum_{i=1}^{2}\langle 0|\alpha_i(\boldsymbol{k})\alpha_i^\dagger(\boldsymbol{k})|0\rangle \\ &= \frac{1}{2}\int \mathrm{d}^3k\,\omega\cdot 2\delta^3(0) \\ &= \frac{1}{2}\int \mathrm{d}^3k\,\omega\cdot 2\frac{V}{(2\pi)^3}, \end{aligned} \tag{7.43}$$

$$\begin{aligned} \langle 0|\vec{\mathsf{P}}'|0\rangle &= \frac{1}{2}\int \mathrm{d}^3k\cdot\boldsymbol{k}\cdot 2\delta^3(0) \\ &= \frac{1}{2}\int \mathrm{d}^3k\cdot\boldsymbol{k}\cdot 2\frac{V}{(2\pi)^3}. \end{aligned} \tag{7.44}$$

Here we have followed Fermi's procedure (5.24) and replaced $(2\pi)^3\delta^3(0)$ by a normalization volume V, which we can make as large as we wish. The integral in (7.44) is relatively harmless: if we integrate in a symmetric manner over all momenta it vanishes. But the integral (7.43) is not so easy to get rid of. It represents the zero-point energy of the electromagnetic field in a volume

V. We argue here that it is only ever possible to measure energy differences.[2] The zero value on the energy scale can then be chosen at will, and we shall assign it to the vacuum energy. The energy operator which we have used up to now must thus be replaced by

$$\mathbf{P}^0 = \mathbf{P}'^0 - \langle 0|\,\mathbf{P}'^0\,|0\rangle . \tag{7.45}$$

For this new energy operator it is then apparent that

$$\begin{aligned}
\langle \mathbf{P}^0 \rangle &= \int \frac{d^3k}{(2\pi)^3 2\omega}\,\omega \left\langle \frac{1}{2}\sum_{i=1}^{2} \{\alpha_i^\dagger(\boldsymbol{k})\alpha_i(\boldsymbol{k}) + \alpha_i(\boldsymbol{k})\alpha_i^\dagger(\boldsymbol{k}) - \langle 0|\,\alpha_i(\boldsymbol{k})\alpha_i^\dagger(\boldsymbol{k})\,|0\rangle\} \right\rangle \\
&= \int \frac{d^3k}{(2\pi)^3 2\omega}\,\omega \left\langle \frac{1}{2}\sum_{i=1}^{2} \{\alpha_i^\dagger(\boldsymbol{k})\alpha_i(\boldsymbol{k}) + \alpha_i(\boldsymbol{k})\alpha_i^\dagger(\boldsymbol{k}) - [\alpha_i(\boldsymbol{k}), \alpha_i^\dagger(\boldsymbol{k})]\} \right\rangle \\
&= \int \frac{d^3k}{(2\pi)^3 2\omega}\,\omega \left\langle \sum_{i=1}^{2} \alpha_i^\dagger(\boldsymbol{k})\alpha_i(\boldsymbol{k}) \right\rangle .
\end{aligned} \tag{7.46}$$

All creation operators appear here to the left of annihilation operators and this leads to a well-defined operator without divergences.

From a mathematical point of view the divergence problems arose since products of field operators taken at the same point, e.g., $[\vec{\mathbf{E}}(x)]^2$ are too singular (7.40). It is possible to carry out the subtraction of the vacuum energy automatically by introducing a new type of product for field operators. This is the *normal-ordered product* or *normal product*. In the normal product, all creation operators are made to operate as if they stood to the left of annihilation operators. This rule will be expressed by writing the operators between a pair of colons. Thus we make the definitions

$$\begin{aligned}
:\mathbf{a}^\dagger \mathbf{a}'^\dagger: &= \mathbf{a}^\dagger \mathbf{a}'^\dagger , \\
:\mathbf{a}^\dagger \mathbf{a}': &= \mathbf{a}^\dagger \mathbf{a}' , \\
:\mathbf{a}' \mathbf{a}^\dagger: &= \mathbf{a}^\dagger \mathbf{a}' , \\
:\mathbf{a}\mathbf{a}': &= \mathbf{a}\mathbf{a}' .
\end{aligned} \tag{7.47}$$

Here $\mathbf{a}$ and $\mathbf{a}'$ are any creation operators for a Bose field. For Fermi fields an additional minus sign appears on commutation as will be discussed in the next chapter. The extension of the above definition to products of linear combinations of creation and annihilation operators is obvious. If, for example, we write the electric field strength schematically as

$$\vec{\mathbf{E}}(x) \sim \mathbf{a}^\dagger + \mathbf{a} , \tag{7.48}$$

[2] It is an open question whether and how one must take account of the zero-point energy of the fields when considering gravitational phenomena. Since the zero-point energy of Fermi fields turns out to be negative, the possibility exists that the contributions of all fields taken together exactly cancel. This is the case in the so-called supersymmetric theories.

then we have

$$
\begin{aligned}
:\vec{\mathbf{E}}^2(x): &\sim :(\mathbf{a}^\dagger + \mathbf{a})(\mathbf{a}^\dagger + \mathbf{a}): \\
&= :\mathbf{a}^\dagger\mathbf{a}^\dagger + \mathbf{a}^\dagger\mathbf{a} + \mathbf{a}\mathbf{a}^\dagger + \mathbf{a}\mathbf{a}: \\
&= \mathbf{a}^\dagger\mathbf{a}^\dagger + 2\mathbf{a}^\dagger\mathbf{a} + \mathbf{a}\mathbf{a}.
\end{aligned}
\tag{7.49}
$$

We then obtain the correct expressions for energy and momentum by replacing the products of classical fields (7.40) by the normal products of the field operators.

$$
\begin{aligned}
\mathbf{P}^0 &= \int_{t=\mathrm{const.}} \mathrm{d}^3x\, \tfrac{1}{2}:[\vec{\mathbf{E}}^2(x) + \vec{\mathbf{B}}^2(x)]:, \\
\vec{\mathbf{P}} &= \int_{t=\mathrm{const.}} \mathrm{d}^3x\, :[\vec{\mathbf{E}}^2(x) \times \vec{\mathbf{B}}(x)]:.
\end{aligned}
\tag{7.50}
$$

It still remains to calculate expectation values of the vector potential which will be needed for the Feynman rules. These quantities are not directly observable and thus the expectation values are *not* identical for all members of an equivalence class of state vectors. Using (7.11)–(7.17) we find for the product involving the vector potential at x and y:

$$
\langle 0|\mathbf{A}_\mu(x)\mathbf{A}_\nu(y)|0\rangle = -g_{\mu\nu}\int \frac{\mathrm{d}^3k}{(2\pi)^3}\frac{1}{2\omega}\mathrm{e}^{-\mathrm{i}k(x-y)}, \tag{7.51}
$$

where

$$
k = \begin{pmatrix} |\boldsymbol{k}| \\ \boldsymbol{k} \end{pmatrix}.
$$

We will now introduce the time-ordered product, or T-product, of two operators $\mathbf{A}_\mu(x)$ through the definition

$$
\mathrm{T}(\mathbf{A}_\mu(x)\mathbf{A}_\nu(y)) = \theta(x^0 - y^0)\mathbf{A}_\mu(x)\mathbf{A}_\nu(y) + \theta(y^0 - x^0)\mathbf{A}_\nu(y)\mathbf{A}_\mu(x). \tag{7.52}
$$

In an analogous manner, we define the T-product for arbitrary Bose operators, i.e., for field operators which behave as a scalar, a four-vector, or a tensor of higher rank under Lorentz transformations. The T-product in (7.52) may at first give the impression of being dependent on which frame of reference is chosen since the time coordinate plays a special role. However, this only appears to be so. From the commutation rules (7.12)–(7.17) it is easy to derive that for spatial separations $\mathbf{A}_\mu(x)$ and $\mathbf{A}_\nu(y)$ commute (Microcausality!):

$$
[\mathbf{A}_\mu(x), \mathbf{A}_\nu(y)] = 0 \qquad \text{for} \quad (x-y)^2 < 0. \tag{7.53}
$$

Thus the T-product only determines the order of the operators for values of x within the forward and backward light cones of y, and this represents a Lorentz-invariant property (Fig. 7.1).

Particularly important is the vacuum expectation value of the T-product of two vector potentials, for which we will derive an integral representation. We assert the validity of the following relation

$$
\langle 0|\mathrm{T}(\mathbf{A}_\mu(x)\mathbf{A}_\nu(y))|0\rangle = \mathrm{i}g_{\mu\nu}D_\mathrm{F}(x-y), \tag{7.54}
$$

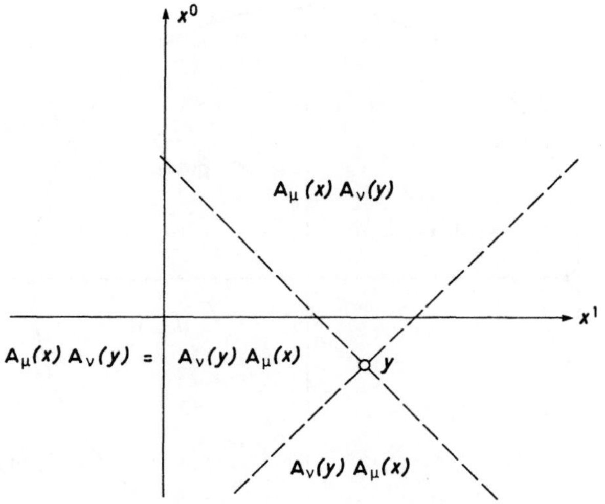

Figure 7.1 Diagram to illustrate the definition of the time-ordered product (7.52).

where the function D_F is defined by

$$\mathrm{i}g_{\mu\nu}D_F(x-y) = \lim_{\varepsilon\to+0}\int\frac{\mathrm{d}k}{(2\pi)^4}\,\mathrm{e}^{-\mathrm{i}k(x-y)}\frac{-\mathrm{i}g_{\mu\nu}}{k^2+\mathrm{i}\varepsilon}. \tag{7.55}$$

The integral in (7.55) extends over the whole of four-dimensional momentum space with the measure $\mathrm{d}k = \mathrm{d}k^0\,\mathrm{d}k^1\,\mathrm{d}k^2\,\mathrm{d}k^3$. In the following we will cease to write the limit sign explicitly for expressions of the above form, and will take ε to be a positive number which tends to zero. In order to prove (7.54) we first examine the integral over k^0 in (7.55) and consider this as an integral in the complex k^0-plane:

$$\mathrm{i}g_{\mu\nu}D_F(x-y) = g_{\mu\nu}\int\frac{\mathrm{d}^3k}{(2\pi)^3}\,\mathrm{e}^{\mathrm{i}\boldsymbol{k}(\boldsymbol{x}-\boldsymbol{y})}\int_{-\infty}^{\infty}\frac{\mathrm{d}k^0}{2\pi\mathrm{i}}\,\mathrm{e}^{-\mathrm{i}k^0(x^0-y^0)}\cdot\frac{1}{(k^0)^2-\boldsymbol{k}^2+\mathrm{i}\varepsilon}. \tag{7.56}$$

The poles of the integrand lie at

$$k^0 = \pm\sqrt{\boldsymbol{k}^2-\mathrm{i}\varepsilon}\to\pm|\boldsymbol{k}|,$$

as is illustrated in Fig. 7.2. For $x^0 > y^0$ we have

$$\mathrm{e}^{-\mathrm{i}k^0(x^0-y^0)}\to 0 \qquad \text{for} \quad \mathrm{Im}\,k^0\to-\infty. \tag{7.57}$$

We can then add to the integration path a very large semicircle in the lower half of the k^0-plane and apply the residue theorem. We find

$$\mathrm{i}g_{\mu\nu}D_F(x-y) = -g_{\mu\nu}\int\frac{\mathrm{d}^3k}{(2\pi)^3}\frac{\mathrm{e}^{-\mathrm{i}[|\boldsymbol{k}|(x^0-y^0)-\boldsymbol{k}(\boldsymbol{x}-\boldsymbol{y})]}}{2|\boldsymbol{k}|} \tag{7.58}$$

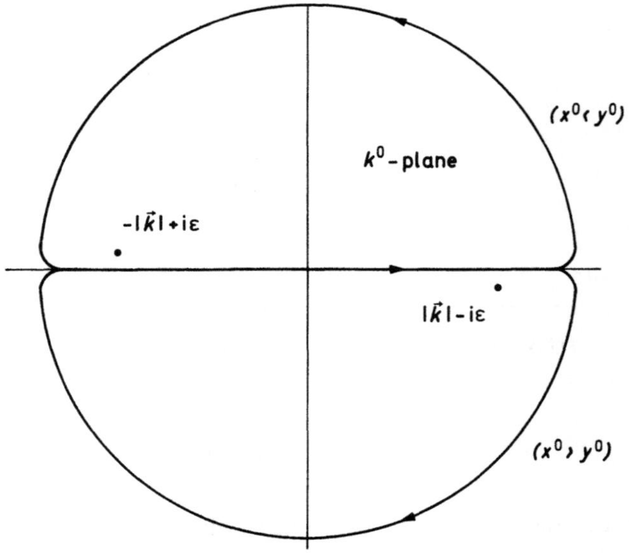

Figure 7.2 Position in the k^0-plane of the poles and integration paths for the integral in (7.56).

and by comparison with (7.51)

$$ig_{\mu\nu}D_F(x - y) = \langle 0|\mathbf{A}_\mu(x)\mathbf{A}_\nu(y)|0\rangle \qquad \text{for} \quad x^0 > y^0. \tag{7.59}$$

For $x^0 < y^0$ we can close the integration path in the upper half plane to give

$$\begin{aligned} ig_{\mu\nu}D_F(x - y) &= -g_{\mu\nu}\int \frac{d^3k}{(2\pi)^3}\frac{e^{-i[|\boldsymbol{k}|(y^0-x^0)-\boldsymbol{k}(\boldsymbol{x}-\boldsymbol{y})]}}{2|\boldsymbol{k}|} \\ &= \langle 0|\mathbf{A}_\nu(y)\mathbf{A}_\mu(x)|0\rangle \qquad \text{for} \quad x^0 < y^0. \end{aligned} \tag{7.60}$$

This completes the proof of (7.54). The function $ig_{\mu\nu}D_F(x - y)$ and its Fourier transform $-ig_{\mu\nu}/(k^2 + i\varepsilon)$ are known as the Feynman propagator in Minkowski space and in momentum space for the free photon field. They play a decisive role in the Feynman rules.

Exercises

7.1 A physical quantity must be gauge invariant, i.e., it must be a functional F of the potentials A_μ that satisfies the condition [see (7.7)]

$$F[A_\mu] = F[A_\mu + \partial_\mu\Lambda]. \tag{7.61}$$

In the quantum theory $\mathbf{F}[\mathbf{A}_\mu]$ takes the form of an operator. Show that its expectation value is equal for all state vectors within an equivalence class.

7.2 Verify (7.38) and (7.39).

7.3 **Under a Poincaré transformation (2.28), with $\Lambda^0{}_0 > 0$, the operator of the vector potential transforms as follows:**

$$(\Lambda, a)\colon \mathbf{A}^\mu(x) \to \Lambda^\mu{}_\nu \mathbf{A}^\nu(\Lambda^{-1}(x - a)). \tag{7.62}$$

Show that this leads to a unitary transformation in the space of state vectors if we assume that the vacuum state is invariant. How do the one-photon states transform?

CHAPTER 8

Further Aspects of the Theory of the Free Dirac Field

In this chapter we will discuss the Dirac current and the magnetic moment of the electron according to the Dirac Theory. We shall be making use of some of the results which were developed in Chap. 4.

8.1 The Dirac Current

As was shown in Sect. 4.3, the Dirac current

$$\bar{\psi}(x)\gamma^{\mu}\psi(x) \tag{8.1}$$

transforms like a four-vector density. The original interpretation of the Dirac field was as a relativistic probability amplitude. The zeroth component of the Dirac current was regarded as a probability density for the location of the electron in space. Classically, it is indeed true that this zeroth component is always positive

$$\bar{\psi}(x)\gamma^{0}\psi(x) = \psi^{\dagger}(x)\psi(x) > 0 \qquad \text{for} \quad \psi(x) \neq 0. \tag{8.2}$$

However, we have seen in Chap. 4 that it is impossible to maintain a one-particle interpretation of the Dirac spinor. What role is played by the Dirac current in the theory of the free quantized Dirac field?

The charge and current distribution, i.e., the electromagnetic current density four-vector $j^{\mu}(x)$ of a system of electrons and positrons is clearly an observable quantity. It is reasonable to suppose that the corresponding operator can be written as

$$\mathbf{j}^\mu(x) = -e\bar{\psi}(x)\gamma^\mu\psi(x). \tag{8.3}$$

We recall here that the charge of the electron is denoted by $-e$. If the assignment (8.3) is correct, then the operator for the total charge would be

$$\mathbf{Q}' = \int d^3x\, \mathbf{j}^0(\boldsymbol{x}, t) = -e \int d^3x\, \bar{\psi}(\boldsymbol{x}, t)\gamma^0\psi(\boldsymbol{x}, t). \tag{8.4}$$

By substituting the expansion for the Dirac field operator (4.84) we obtain

$$\mathbf{Q}' = -e \int \frac{d^3p}{(2\pi)^3 2p_0} \sum_{s=\pm 1/2} [\mathbf{a}_s^\dagger(\boldsymbol{p})\mathbf{a}_s(\boldsymbol{p}) + \mathbf{b}_s(\boldsymbol{p})\mathbf{b}_s^\dagger(\boldsymbol{p})]. \tag{8.5}$$

Now we encounter a similar problem to that of assigning a zero point to the energy scale. If we look at the expectation value of the charge in the vacuum state we obtain infinity. According to the anticommutation rules (4.91) we find

$$\begin{aligned}
\langle 0|\mathbf{Q}'|0\rangle &= -e \int \frac{d^3p}{(2\pi)^3 2p_0} \sum_{s=\pm 1/2} \langle 0|\mathbf{b}_s(\boldsymbol{p})\mathbf{b}_s^\dagger(\boldsymbol{p})|0\rangle \\
&= -e \int \frac{d^3p}{(2\pi)^3 2p_0} \sum_{s=\pm 1/2} \{\mathbf{b}_s(\boldsymbol{p}), \mathbf{b}_s^\dagger(\boldsymbol{p})\} \\
&= -e \int d^3p\, 2\cdot\delta^3(0) = \infty.
\end{aligned} \tag{8.6}$$

Again this simply means that we have not made an appropriate choice of the zero value from which charge is counted. We obtain a "good" charge operator $\mathbf{Q}$ if we select the total charge of the vacuum as the zero of our charge scale. Then we have

$$\begin{aligned}
\mathbf{Q} &= \mathbf{Q}' - \langle 0|\mathbf{Q}'|0\rangle \\
&= -e \int \frac{d^3p}{(2\pi)^3 2p_0} \sum_{s=\pm 1/2} (\mathbf{a}_s^\dagger(\boldsymbol{p})\mathbf{a}_s(\boldsymbol{p}) + \mathbf{b}_s(\boldsymbol{p})\mathbf{b}_s^\dagger(\boldsymbol{p}) - \{\mathbf{b}_s(\boldsymbol{p}), \mathbf{b}_s^\dagger(\boldsymbol{p})\}) \\
&= -e \int \frac{d^3p}{(2\pi)^3 2p_0} \sum_{s=\pm 1/2} (\mathbf{a}_s^\dagger(\boldsymbol{p})\mathbf{a}_s(\boldsymbol{p}) - \mathbf{b}_s^\dagger(\boldsymbol{p})\mathbf{b}_s(\boldsymbol{p})).
\end{aligned} \tag{8.7}$$

The operator $\mathbf{Q}$ clearly has positive and negative eigenvalues with electrons having charge $-e$ and positrons $+e$ in agreement with experiment.

The infinite self-charge of the vacuum is, mathematically seen, a consequence of forming in (8.3) the product of two field operators at the same point. It is possible to avoid infinities by defining a normal ordering for products of fermion field operators. For every exchange of Fermi operators we introduce a factor -1 corresponding to their anticommutation rules. From this definition we thus have:

$$:\mathbf{a}_r(\boldsymbol{p})\mathbf{a}_s^\dagger(\boldsymbol{p}'): = -\mathbf{a}_s^\dagger(\boldsymbol{p}')\mathbf{a}_r(\boldsymbol{p}). \tag{8.8}$$

We obtain a current that automatically gives zero total charge for the vacuum if we replace (8.3) by

$$\mathbf{j}^\mu(x) = -e:\bar{\psi}(x)\gamma^\mu\psi(x):. \tag{8.9}$$

It should be noted that the normal product $:\bar{\psi}\gamma^0\psi:$ does not satisfy a positivity condition analogous to (8.2).

8.2 The Magnetic Moment of the Electron in the Dirac Theory

We now wish to calculate the magnetic moment of an electron according to the free Dirac theory. In general, the operator for the magnetic moment at time t is given by

$$\vec{\mu}(t) = \frac{1}{2}\int d^3x\, \boldsymbol{x} \times \vec{\mathbf{j}}(\boldsymbol{x}, t). \tag{8.10}$$

We consider the matrix element of this operator between one-electron states of definite momentum at time $t = 0$. Writing $\vec{\mu}(0) \equiv \vec{\mu}$ and

$$|\mathrm{e}(\boldsymbol{p}, s)\rangle = \mathbf{a}_s^\dagger(\boldsymbol{p})|0\rangle, \tag{8.11}$$

we find

$$\langle \mathrm{e}(\boldsymbol{p}', r)|\vec{\mu}|\mathrm{e}(\boldsymbol{p}, s)\rangle = -e\frac{1}{2}\int d^3x\, \boldsymbol{x} \times \langle \mathrm{e}(\boldsymbol{p}', r)|:\bar{\psi}(\boldsymbol{x}, 0)\boldsymbol{\gamma}\psi(\boldsymbol{x}, 0):|\mathrm{e}(\boldsymbol{p}, s)\rangle. \tag{8.12}$$

It is clear that substitution of the expansion (4.84) of the field operators will lead us to an expression whose schematic form is

$$\langle 0|\mathbf{a}:(\mathbf{b} + \mathbf{a}^\dagger)(\mathbf{b}^\dagger + \mathbf{a}):\mathbf{a}^\dagger|0\rangle = \langle 0|\mathbf{a}(-\mathbf{b}^\dagger\mathbf{b} + \mathbf{a}^\dagger\mathbf{b}^\dagger + \mathbf{b}\mathbf{a} + \mathbf{a}^\dagger\mathbf{a})\mathbf{a}^\dagger|0\rangle. \tag{8.13}$$

Such an expression is best calculated by using the anticommutation rules to shift all annihilation operators to the right until they act on the vacuum state thereby producing zero. As an example we consider annihilation operators $\mathbf{a}_i$ (with $i = 1, \ldots, 4$), where the index i momentarily replaces momentum and spin values. Making use of (4.91) we then find

$$\langle 0|\mathbf{a}_i\mathbf{a}_j^\dagger|0\rangle = \{\mathbf{a}_i, \mathbf{a}_j^\dagger\}, \tag{8.14}$$

$$\begin{aligned}
\langle 0|\mathbf{a}_1\mathbf{a}_2^\dagger\mathbf{a}_3\mathbf{a}_4^\dagger|0\rangle &= \langle 0|(\{\mathbf{a}_1, \mathbf{a}_2^\dagger\} - \mathbf{a}_2^\dagger\mathbf{a}_1)\mathbf{a}_3\mathbf{a}_4^\dagger|0\rangle \\
&= \langle 0|\mathbf{a}_1\mathbf{a}_2^\dagger|0\rangle\cdot\langle 0|\mathbf{a}_3\mathbf{a}_4^\dagger|0\rangle - \langle 0|\mathbf{a}_2^\dagger\mathbf{a}_1\mathbf{a}_3\mathbf{a}_4^\dagger|0\rangle \\
&= \langle 0|\mathbf{a}_1\mathbf{a}_2^\dagger|0\rangle\cdot\langle 0|\mathbf{a}_3\mathbf{a}_4^\dagger|0\rangle - \langle 0|\mathbf{a}_2^\dagger\mathbf{a}_1\{\mathbf{a}_3, \mathbf{a}_4^\dagger\}|0\rangle \\
&= \langle 0|\mathbf{a}_1\mathbf{a}_2^\dagger|0\rangle\cdot\langle 0|\mathbf{a}_3\mathbf{a}_4^\dagger|0\rangle.
\end{aligned} \tag{8.15}$$

The vacuum expectation value of the product of two operators **A**, **B** is known as the *contraction* of **A** and **B** and is denoted as follows:

$$\underbrace{\mathbf{A}\mathbf{B}} \equiv \langle 0|\mathbf{A}\mathbf{B}|0\rangle. \tag{8.16}$$

The contraction of two creation or two annihilation operators clearly gives zero. The contraction of an annihilation with a creation operator is only nonzero when the creation operator appears to the right of the annihilation operator. In terms of contractions our result (8.15) means

$$\langle 0|\mathbf{a}_1\mathbf{a}_2^\dagger\mathbf{a}_3\mathbf{a}_4^\dagger|0\rangle = \underbracket{\mathbf{a}_1\mathbf{a}_2^\dagger}\,\underbracket{\mathbf{a}_3\mathbf{a}_4^\dagger}. \tag{8.17}$$

Since all other possible contractions of the operators on the left-hand side yield a vanishing result, we can also write

$$\langle 0|\mathbf{a}_1\mathbf{a}_2^\dagger\mathbf{a}_3\mathbf{a}_4^\dagger|0\rangle = \text{sum of all contractions.} \tag{8.18}$$

It is easy to see that this result is generally valid for the vacuum expectation value of any product of sums of creation and annihilation operators (Wick's theorem; Wick 1950). One should not forget that in order to contract two Fermi operators they have to be brought together by means of anticommutation which gives additional factors of -1. For Fermi operators we have, for example,

$$\begin{aligned}\langle 0|\mathbf{a}_1\mathbf{a}_2\mathbf{a}_3^\dagger\mathbf{a}_4^\dagger|0\rangle &= \langle 0|\mathbf{a}_1\mathbf{a}_2\mathbf{a}_3^\dagger\mathbf{a}_4^\dagger|0\rangle + \langle 0|\underbracket{\mathbf{a}_1\underbracket{\mathbf{a}_2\mathbf{a}_3^\dagger}\mathbf{a}_4^\dagger}|0\rangle\\ &= -\underbracket{\mathbf{a}_1\mathbf{a}_3^\dagger}\,\underbracket{\mathbf{a}_2\mathbf{a}_4^\dagger} + \underbracket{\mathbf{a}_1\mathbf{a}_4^\dagger}\,\underbracket{\mathbf{a}_2\mathbf{a}_3^\dagger}\\ &= -\langle 0|\mathbf{a}_1\mathbf{a}_3^\dagger|0\rangle\langle 0|\mathbf{a}_2\mathbf{a}_4^\dagger|0\rangle + \langle 0|\mathbf{a}_1\mathbf{a}_4^\dagger|0\rangle\langle 0|\mathbf{a}_2\mathbf{a}_3^\dagger|0\rangle.\end{aligned} \tag{8.19}$$

After these introductory remarks we can now proceed to calculate the expression (8.12) for the magnetic moment. We find:

$$\langle e(\boldsymbol{p}',r)|\vec{\mu}|e(\boldsymbol{p},s)\rangle = -\frac{e}{2}\int d^3x\, \boldsymbol{x}\times \underbracket{\mathbf{a}_r(\boldsymbol{p}')\bar{\psi}}(\boldsymbol{x},0)\boldsymbol{\gamma}\underbracket{\psi(\boldsymbol{x},0)\mathbf{a}_s^\dagger(\boldsymbol{p})}, \tag{8.20}$$

$$\begin{aligned}\underbracket{\psi(\boldsymbol{x},t)\mathbf{a}_s^\dagger(\boldsymbol{p})} &= \langle 0|\psi(\boldsymbol{x},t)\mathbf{a}_s^\dagger(\boldsymbol{p})|0\rangle\\ &= \int\frac{d^3p_1}{(2\pi)^3 2p_1^0}\sum_{s'=\pm 1/2}\langle 0|[e^{ip_1x}\,\mathbf{b}_{s'}^\dagger(p_1)v_{s'}(p_1)\\ &\quad + e^{-ip_1x}\,\mathbf{a}_{s'}(p_1)u_{s'}(p_1)]\mathbf{a}_s^\dagger(\boldsymbol{p})|0\rangle\\ &= u_s(p)e^{-ipx},\end{aligned} \tag{8.21}$$

$$\langle e(\boldsymbol{p}',r)|\vec{\mu}|e(\boldsymbol{p},s)\rangle = -\frac{e}{2}\int d^3x\, e^{i(\boldsymbol{p}-\boldsymbol{p}')\boldsymbol{x}}\,\boldsymbol{x}\times\bar{u}_r(p')\boldsymbol{\gamma}u_s(p). \tag{8.22}$$

In the nonrelativistic limit $|\boldsymbol{p}| \ll m$, $|\boldsymbol{p}'| \ll m$ we have

$$u_s(p) \cong \sqrt{2m}\begin{pmatrix}\chi_s\\ \dfrac{\boldsymbol{\sigma p}}{2m}\chi_s\end{pmatrix}. \tag{8.23}$$

With the analogous expression for $u_r(p')$ we find

$$\bar{u}_r(p')\gamma u_s(p) \cong \chi_r^\dagger\{\boldsymbol{p} + \boldsymbol{p}' + \mathrm{i}\boldsymbol{\sigma} \times \boldsymbol{q}\}\chi_s, \tag{8.24}$$

where $\boldsymbol{q} = \boldsymbol{p}' - \boldsymbol{p}$. From (8.22) we then obtain

$$\langle \mathrm{e}(\boldsymbol{p}', r)|\vec{\mu}|\mathrm{e}(\boldsymbol{p}, s)\rangle \cong -\frac{e}{2}\int \mathrm{d}^3x\, \mathrm{e}^{\mathrm{i}(\boldsymbol{p}-\boldsymbol{p}')\boldsymbol{x}}\, \chi_r^\dagger \boldsymbol{x} \times (\boldsymbol{p} + \boldsymbol{p}' + \mathrm{i}\boldsymbol{\sigma} \times \boldsymbol{q})\chi_s. \tag{8.25}$$

The form of this expression does not make its interpretation self-evident. In order to make contact with the familiar formula for the magnetic moment in nonrelativistic quantum mechanics we consider wave packets, i.e., normalized one-electron states of the form

$$|f\rangle = \int \frac{\mathrm{d}^3p}{(2\pi)^3\sqrt{2p^0}} \sum_{s=\pm 1/2} f_s(\boldsymbol{p})\mathbf{a}_s^\dagger(\boldsymbol{p})|0\rangle, \tag{8.26}$$

where according to the nonrelativistic regime the functions $f_s(\boldsymbol{p})$ are supposed to satisfy

$$f_s(\boldsymbol{p}) \neq 0 \quad \text{only for} \quad |\boldsymbol{p}| \ll m,$$

$$\int \frac{\mathrm{d}^3p}{(2\pi)^3} \sum_{s=\pm 1/2} |f_s(\boldsymbol{p})|^2 = 1. \tag{8.27}$$

Under these restrictions, we can regard the functions $f_s(\boldsymbol{p})$ as Schrödinger functions in momentum space. The Fourier transforms

$$\psi_s(\boldsymbol{x}) = \int \frac{\mathrm{d}^3p}{(2\pi)^3}\, \mathrm{e}^{\mathrm{i}\boldsymbol{p}\boldsymbol{x}} f_s(\boldsymbol{p}) \tag{8.28}$$

are then, according to (3.73), (4.1), the Schrödinger functions of a "Pauli electron" in position space at time $t = 0$. From (8.27) we indeed obtain the correct normalization for ψ_s:

$$\int \mathrm{d}^3x \sum_{s=\pm 1/2} |\psi_s(\boldsymbol{x})|^2 = 1. \tag{8.29}$$

A simple calculation then gives

$$\langle f|\vec{\mu}|f\rangle = \int \frac{\mathrm{d}^3p'\,\mathrm{d}^3p}{(2\pi)^6 2m} \sum_{r,s} f_r^*(\boldsymbol{p}')f_s(\boldsymbol{p})\langle \mathrm{e}(\boldsymbol{p}', r)|\vec{\mu}|\mathrm{e}(\boldsymbol{p}, s)\rangle$$

$$= -\frac{e}{2m}\int \mathrm{d}^3x \sum_{r,s} \psi_r^*(\boldsymbol{x})\left(\delta_{rs}\boldsymbol{x} \times \frac{1}{i}\boldsymbol{\nabla} + \boldsymbol{\sigma}_{rs}\right)\psi_s(\boldsymbol{x}),$$

that is

$$\langle f|\vec{\mu}|f\rangle = -\frac{e}{2m}\langle f|\vec{\mathbf{L}} + \boldsymbol{\sigma}|f\rangle, \tag{8.30}$$

where $\vec{\mathbf{L}}$ is the angular momentum operator (4.3). Recalling that the operator for total angular momentum is

$$\vec{\mathbf{J}} = \vec{\mathbf{L}} + \tfrac{1}{2}\boldsymbol{\sigma},$$

we see that, in general, the magnetic moment and the angular momentum of an electron do not point in the same direction. The spin angular momentum contributes with an additional factor of 2 to the magnetic moment. A great triumph of the Dirac theory is its ability to accommodate this fact which had previously been established empirically.

For an electron *at rest* we have for the expectation values (see Exercise 8.4)

$$\langle\vec{\mathbf{L}}\rangle = 0,$$
$$\langle\vec{\mu}\rangle_{e^-} = -\frac{e}{2m}\langle\boldsymbol{\sigma}\rangle. \tag{8.31}$$

Here the nonrelativistic expression of course represents the exact answer in the framework of the Dirac theory. Analogously, one finds for a positron at rest

$$\langle\vec{\mu}\rangle_{e^+} = \frac{e}{2m}\langle\boldsymbol{\sigma}\rangle. \tag{8.32}$$

However, these relationships (8.31), (8.32) are not in precise agreement with experiment. Experimentally one finds

$$\langle\vec{\mu}\rangle_{e^-} = -\frac{e}{2m}\frac{1}{2}g_e\langle\boldsymbol{\sigma}\rangle \tag{8.33}$$

with $g_e \neq 2$ as has already been discussed in Chap. 6.

8.3 The Free Electron Propagator

To end this chapter we will define the time-ordered product of two Dirac operators and give its vacuum expectation value. We use indices α, β to enumerate the four components of the Dirac spinor and define

$$\mathrm{T}(\psi_\alpha(x)\bar{\psi}_\beta(y)) = \theta(x^0 - y^0)\psi_\alpha(x)\bar{\psi}_\beta(y) - \theta(y^0 - x^0)\bar{\psi}_\beta(y)\psi_\alpha(x). \tag{8.34}$$

Note, in particular, the appearance of the minus sign in contrast to the T-product of Bose operators (7.52). In order to express the vacuum expectation value of this T-product we define a function $\mathbf{S}_\mathrm{F}(z)$ which, for a fixed four-vector z, represents a 4×4 matrix:

$$\mathbf{S}_\mathrm{F}(z) = -\int \frac{\mathrm{d}p}{(2\pi)^4}\,\mathrm{e}^{-ipz}\,\frac{1}{\not{p} - m + \mathrm{i}\varepsilon}. \tag{8.35}$$

We now assert the validity of the following relations:

$$\begin{aligned}\langle 0|\mathrm{T}(\psi(x)\bar{\psi}(y))|0\rangle &= \frac{1}{i}\mathbf{S}_\mathrm{F}(x-y)\\ &= \int \frac{\mathrm{d}p}{(2\pi)^4}\,\mathrm{e}^{-ip(x-y)}\,\frac{\mathrm{i}}{\not{p} - m + \mathrm{i}\varepsilon}\\ &= \int \frac{\mathrm{d}p}{(2\pi)^4}\,\mathrm{e}^{-ip(x-y)}\,\frac{\mathrm{i}(\not{p} + m)}{p^2 - m^2 + \mathrm{i}\varepsilon}.\end{aligned} \tag{8.36}$$

The function $-\mathrm{i}\mathbf{S}_{\mathrm{F}}(x-y)$ or its Fourier transform $\mathrm{i}/(\not{p}-m+\mathrm{i}\varepsilon)$ are known as the propagator of the free electron. The proof of (8.36) follows the same lines as the proof of (7.54) for the photon field.

Exercises

8.1 Verify (8.5) using the relations given in Appendix A.

8.2 Determine the following expectation value by means of Wick's theorem:

$$\langle 0|\mathbf{a}_1\mathbf{a}_2^\dagger\mathbf{a}_3\mathbf{a}_4\mathbf{a}_5^\dagger\mathbf{a}_6^\dagger|0\rangle.$$

Regard the $\mathbf{a}_i$ first as Fermi operators and then as Bose operators.

8.3 Prove (8.30)

8.4 Consider an electron state as in (8.26) with functions

$$f_s(\boldsymbol{p}) = c_s \cdot \sigma^{-3/2}(2\pi)^{3/4} \exp\left(-\frac{\boldsymbol{p}^2}{4\sigma^2}\right).$$

where $\sigma > 0$ and c_s are arbitrary complex numbers satisfying

$$\sum_{s=\pm 1/2} |c_s|^2 = 1.$$

An electron at rest is defined as the limit $\sigma \to 0$ in the above states. Demonstrate the validity of (8.31) with

$$\langle\boldsymbol{\sigma}\rangle \equiv \sum_{r,s} c_r^* \boldsymbol{\sigma}_{rs} c_s.$$

8.5 Prove (8.36).

CHAPTER 9

Electromagnetic Coupling and the Perturbation Expansion

9.1 The Electromagnetic Coupling of the Dirac Field

When an electrically charged particle traverses an electric or magnetic field it suffers a deflection. If it is a classical, nonrelativistic point particle with mass m and charge q, then the force $\boldsymbol{K}$ has the form given by Lorentz. The equation of motion reads (see Jackson 1975)

$$m\ddot{\boldsymbol{z}}(t) = \boldsymbol{K} = q\boldsymbol{E}(\boldsymbol{z}, t) + q\dot{\boldsymbol{z}}(t) \times \boldsymbol{B}(\boldsymbol{z}, t), \tag{9.1}$$

where $\boldsymbol{z}(t)$ and $\dot{\boldsymbol{z}}(t)$ are the position and velocity of the particle at time t. The Lagrangian which leads to this equation of motion is

$$L = L_0 + L', \tag{9.2}$$

where

$$L_0 = \frac{m\dot{\boldsymbol{z}}(t)^2}{2},$$

$$L' = -q(A^0(\boldsymbol{z}(t), t) - \dot{\boldsymbol{z}}(t)\boldsymbol{A}(\boldsymbol{z}(t), t)).$$

With the help of the current density four-vector j^μ

$$(j^\mu(\boldsymbol{x}, t)) = q\begin{pmatrix} 1 \\ \dot{\boldsymbol{z}}(t) \end{pmatrix} \delta^3(\boldsymbol{x} - \boldsymbol{z}(t)) \tag{9.3}$$

we can write the interaction term L' as follows:

$$L' = -\int \mathrm{d}^3x\, j^\mu(\boldsymbol{x}, t) A_\mu(\boldsymbol{x}, t). \tag{9.4}$$

It turns out that in this form the coupling can be simply taken over to the quantized field theory if for j^μ we write the current density of the Dirac field (8.9).

Within the framework of a deductive construction of QED we would rely on the Lagrangian of the coupled Maxwell–Dirac system, beginning with that of the classical fields. This Lagrangian L is an integral over a Lagrange density $\mathscr{L}$ (see Sect. 3.3):

$$L = \int \mathrm{d}^3x\, \mathscr{L}(\boldsymbol{x}, t),$$

where

$$\begin{aligned} \mathscr{L}(x) &= \mathscr{L}_0(x) + \mathscr{L}'(x), \\ \mathscr{L}_0(x) &= -\tfrac{1}{4}F_{\mu\nu}(x)F^{\mu\nu}(x) + \bar{\psi}(x)(\mathrm{i}\gamma^\mu\partial_\mu - m)\psi(x), \\ \mathscr{L}'(x) &= -j^\mu(x)A_\mu(x) = e\bar{\psi}(x)\gamma^\mu A_\mu(x)\psi(x). \end{aligned} \tag{9.5}$$

The term $\mathscr{L}'$ describes the interaction. It should be noted that the total Lagrange density $\mathscr{L}$ is obtained from that of the free fields ($\mathscr{L}_0$) by way of the following substitution:

$$\mathrm{i}\partial_\mu \to \mathrm{i}\partial_\mu + \mathrm{e}A_\mu(x). \tag{9.6}$$

This is called a "minimal substitution". It guarantees the *gauge invariance* of the full theory. Indeed, it is possible to supplement the transformation of the vector potential (7.7) by a phase transformation of the Dirac spinor such that the Lagrange density remains invariant. With the replacements

$$\begin{aligned} A_\mu(x) &\to A_\mu(x) + \partial_\mu\Lambda(x), \\ \psi(x) &\to \exp(\mathrm{i}e\Lambda(x))\cdot\psi(x), \end{aligned} \tag{9.7}$$

we in fact find

$$\mathscr{L}(x) \to \mathscr{L}(x).$$

Here again $\Lambda(x)$ is any real function of x. This invariance of the coupled Maxwell–Dirac system, which was discovered by H. Weyl (1929) is known in modern terminology as U(1) gauge symmetry. Gauge symmetries form the cornerstone of modern theories of elementary particles. We will see later that both the strong and the weak interactions are governed by gauge symmetries that represent a generalization of the gauge symmetry of quantum electrodynamics.

If we neglect the coupling term in the Lagrange density $\mathscr{L}$ (9.5), i.e., we set the electric charge $e = 0$, then by the principle of least action (Sect. 3.3) we obtain as equations of motion the free Maxwell and Dirac equations. The quantization of the corresponding fields has been discussed in Chaps. 4, 7, and 8. The idea now is to make a series expansion in e in order to take the coupling into account. For this we will rely largely on the methods of perturbation theory which we know from nonrelativistic quantum mechanics. We shall be making use of the Dirac or interaction picture which was discussed in

Chap. 3. The field operators then satisfy the free equations and the state vector satisfies a Schrödinger equation

$$\mathrm{i}\frac{\partial}{\partial t}|t\rangle = \mathbf{H}'(t)|t\rangle, \tag{9.8}$$

where instead of the total Hamiltonian, only the interaction term appears. It turns out that the interaction energy $\mathbf{H}'(t)$ differs from the coupling term in the Lagrangian only in its sign:

$$\begin{aligned}\mathbf{H}'(t) &= -\int d^3x\, \mathscr{L}'(\boldsymbol{x}, t) = \int d^3x\, \mathbf{j}^\mu(\boldsymbol{x}, t)\mathbf{A}_\mu(\boldsymbol{x}, t) \\ &= -e\int d^3x :\bar{\psi}(\boldsymbol{x}, t)\gamma^\mu\psi(\boldsymbol{x}, t): \mathbf{A}_\mu(\boldsymbol{x}, t).\end{aligned} \tag{9.9}$$

This is plausible since the coupling term in (9.5) contains no time derivatives. A formal derivation of (9.9) is considered in Exercise 9.3. Here we will not deepen this any further but will regard (9.9) as a basic postulate about the form of the electromagnetic interaction between electrons and positrons.

Equations (9.8), (9.9) form the basis for the construction of the Feynman rules of QED.

9.2 The Feynman Rules

In this section we will treat the following physical situation: At time $t \to -\infty$ let a certain number of electrons, positrons, and photons be present, all of which are well separated from one another. In the course of time these particles may meet, scatter from one another, and annihilate or produce new particles. We would like to know the state at time $t \to +\infty$ and, in particular, the amplitude for the transition to a given state with a certain number of electrons, positrons, and photons:

$$\begin{aligned}&\mathrm{e}^-(p_1) + \cdots + \mathrm{e}^+(q_1) + \cdots + \gamma(k_1) + \cdots \\ &\quad \to \mathrm{e}^-(p_1') + \cdots + \mathrm{e}^+(q_1') + \cdots + \gamma(k_1') + \cdots .\end{aligned} \tag{9.10}$$

The starting point is to consider (9.8) which gives the time evolution of the states in the Dirac picture. If we expand $\bar{\psi}$, ψ and $\mathbf{A}_\mu$ in (9.9) in terms of creation and annihilation operators, indicated schematically by $\bar{\psi} \sim \mathbf{b} + \mathbf{a}^\dagger$, $\psi \sim \mathbf{b}^\dagger + \mathbf{a}$, $\mathbf{A}_\mu \sim \alpha^\dagger + \alpha$ then for $\mathbf{H}'$ we obtain the following structure:

$$\begin{aligned}\mathbf{H}' &\sim :(\mathbf{b} + \mathbf{a}^\dagger)(\mathbf{b}^\dagger + \mathbf{a}):(\alpha^\dagger + \alpha) \\ &\sim \underset{(1)}{-\mathbf{b}^\dagger\mathbf{b}(\alpha^\dagger + \alpha)} + \underset{(2)}{\mathbf{a}^\dagger\mathbf{b}^\dagger(\alpha^\dagger + \alpha)} + \underset{(3)}{\mathbf{b}\mathbf{a}(\alpha^\dagger + \alpha)} + \underset{(4)}{\mathbf{a}^\dagger\mathbf{a}(\alpha^\dagger + \alpha)}.\end{aligned} \tag{9.11}$$

If we apply $\mathbf{H}'$ to an arbitrary state, then term (4), for example, has the following effect: The operator a destroys an electron and $\mathbf{a}^\dagger$ creates another electron, in general with a different momentum. In this process a photon is

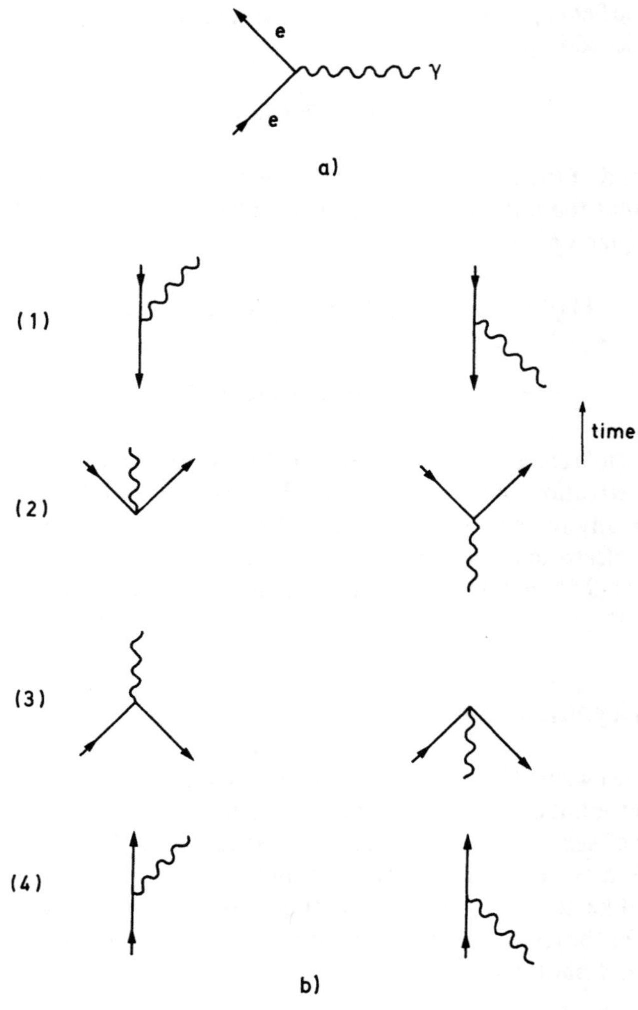

Figure 9.1 Feynman diagram representing **H**′ of (9.9), (9.11) (a). Interpretation of this diagram according to the four terms of (9.11) (b).

either emitted ($\alpha^\dagger$) or absorbed (α). An "elementary" process of this sort can be represented pictorially by a diagram in which electrons are symbolized by straight lines with arrows and photons by wavy lines (Fig. 9.1). An analogous interpretation may be given to the terms (1) to (3).

(1) Emission or absorption of a photon by a positron.
(2) Creation of an electron–positron pair with the emission or absorption of a photon.
(3) Annihilation of an electron–positron pair with the emission or absorption of a photon.
(4) Emission or absorption of a photon by an electron.

It is possible to represent all these processes by a single diagram provided we symbolize positrons as electrons traveling backwards in time (Fig. 9.1).

We now seek a way to visualize the time evolution of an arbitrary state. At time $t \to -\infty$ we have a certain number of electrons, positrons, and photons, which are represented by their corresponding lines. Equation (9.8) tells us that there exists a probability per unit time for the transition into another state due to the occurrence of one of the processes (1)–(4). As the time proceeds this may happen repeatedly. According to the rules of quantum mechanics, the transition amplitudes for a particular final state must be added coherently, independently of the intermediate steps which led to this state. The correct superposition of all these amplitudes is obtained by solving the differential equation (9.8) with the appropriate initial condition at $t \to -\infty$. It is straght-forward to verify that at least formally the solution is given by

$$|t\rangle = \left\{1 + (-\mathrm{i}) \int_{-\infty}^{t} \mathrm{d}t'\, \mathbf{H}'(t') + (-\mathrm{i})^2 \int_{-\infty}^{t} \mathrm{d}t' \cdot \int_{-\infty}^{t'} \mathrm{d}t''\, \mathbf{H}'(t')\mathbf{H}'(t'') + \cdots \right\} |t = -\infty\rangle. \quad (9.12)$$

From this we obtain the S-operator which describes the time evolution of the states from $t \to -\infty$ to $t \to +\infty$:

$$\begin{aligned} |t = +\infty\rangle &= \mathbf{S}|t = -\infty\rangle \\ &= \left\{1 + (-\mathrm{i}) \int_{-\infty}^{\infty} \mathrm{d}t'\, \mathbf{H}'(t') \right. \\ &\quad \left. + (-\mathrm{i})^2 \int_{-\infty}^{\infty} \mathrm{d}t' \int_{-\infty}^{t'} \mathrm{d}t''\, \mathbf{H}'(t')\mathbf{H}'(t'') + \cdots \right\} |t = -\infty\rangle. \end{aligned} \quad (9.13)$$

The S-operator thus constructed, can be written rather more elegantly with the help of the time-ordered product:

$$\begin{aligned} \mathbf{S} &= \left\{1 + (-\mathrm{i}) \int_{-\infty}^{\infty} \mathrm{d}t'\, \mathbf{H}'(t') \right. \\ &\quad \left. + \frac{(-\mathrm{i})^2}{2!} \int_{-\infty}^{\infty} \mathrm{d}t' \int_{-\infty}^{\infty} \mathrm{d}t''\, \mathrm{T}(\mathbf{H}'(t')\mathbf{H}'(t'')) + \cdots \right\} \\ &= \sum_{n=0}^{\infty} \frac{(-\mathrm{i})^n}{n!} \int_{-\infty}^{\infty} \mathrm{d}t_1 \ldots \int_{-\infty}^{\infty} \mathrm{d}t_n\, \mathrm{T}(\mathbf{H}'(t_1) \ldots \mathbf{H}'(t_n)) \\ &= \mathrm{T}\left\{\exp\left[-\mathrm{i} \int_{-\infty}^{\infty} \mathrm{d}t\, \mathbf{H}'(t)\right]\right\}. \end{aligned} \quad (9.14)$$

Since $\mathbf{H}'$ is proportional to e, we have in (9.14) an expansion of the S-operator in powers of e. The successive terms in this expansion clearly represent the repetition of the elementary processes (1)–(4) once, twice, etc. The integrations over time represent the coherent superposition of processes that occur at different times. The individual contributions to a transition amplitude can

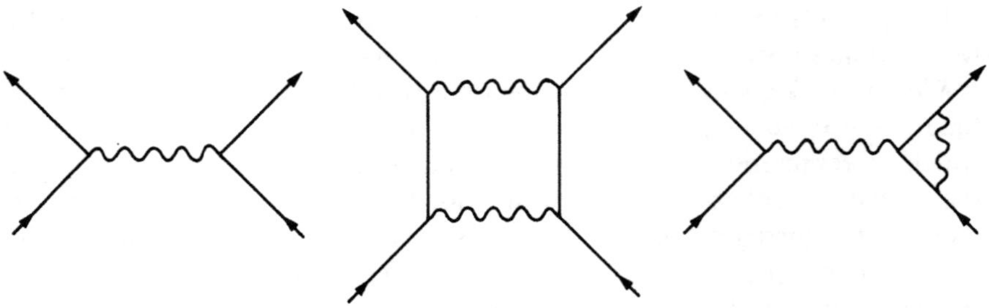

Figure 9.2 A few diagrams for the reaction $e^- + e^- \rightarrow e^- + e^-$.

again be visualized using diagrams in which we connect the appropriate number of elementary processes. As an example, Fig. 9.2 shows a few possible diagrams for the scattering of two electrons from one another. One must be on one's guard however, not to ask when, or how many, elementary processes took place in a particular experiment. This would be a completely meaningless question, since the decomposition of the reaction into elementary processes is purely a theoretical aid for calculating the transition amplitudes.

It is now a question of mathematics to analyze the series for the S-operator (9.14) and to assign an analytical expression to each of the diagrams occurring in a transition amplitude. The results are the general Feynman rules which are summarized in Appendix B. The methods for deriving these rules will be explained in more detail in the next chapter with reference to the example of electron–electron scattering.

In order to calculate the transition amplitude for a particular process, one begins by drawing all the diagrams whose external lines correspond to the incoming and outgoing particles. Inside a diagram, any number of vertices are allowed. However, every vertex introduces a factor of $e = \sqrt{4\pi\alpha}$. If we arrange the diagrams according to increasing number of vertices, then we obtain an expansion of the amplitude in powers of e. In practical cases, one must usually make do with the lowest order terms. This is actually a good approximation since the fine-structure constant α is much smaller than 1. Those diagrams, contributing to the order considered, are then translated into analytical expressions according to the rules given in Appendix B. Since one is only interested in effects which truly involve scattering, i.e., processes in which all particles interact with one another, it is sufficient to consider the connected diagrams.

Exercises

9.1 Derive the equation of motion (9.1) from the Lagrangian L (9.2).

9.2 Derive the equations of motion that follow according to the principle of least action from the Lagrange density (9.5). Show that for $e = 0$ these lead to the free Maxwell and Dirac equations.

9.3 Consider the Lagrange density $\mathscr{L}$ (9.5) of the classical Maxwell–Dirac system. We write the corresponding Lagrangian L as a functional of the fields

$$L(A_\mu(\boldsymbol{x}, t), \dot{A}_\mu(\boldsymbol{x}, t), \psi(\boldsymbol{x}, t), \dot{\psi}(\boldsymbol{x}, t), \bar{\psi}(\boldsymbol{x}, t), \dot{\bar{\psi}}(\boldsymbol{x}, t)] = \int \mathrm{d}^3 x\, \mathscr{L}(\boldsymbol{x}, t), \quad (9.15)$$

where for $F^{\mu\nu}$ we have inserted the expression from (7.5). Show that

$$\frac{\delta L}{\delta \dot{A}_0(\boldsymbol{x}, t)} \equiv 0, \quad (9.16)$$

and that it is thus not possible to apply the canonical quantization rules of Sect. 3.4 with this Lagrangian. A way out of this difficulty, and one which is practicable for QED, is as follows: To the Lagrange density $\mathscr{L}$ of (9.5) we add another two terms $\mathscr{L}_1$ and $\mathscr{L}_2$,

$$\mathscr{L}_1 = -\tfrac{1}{2}\partial_\mu\{A^\nu\partial_\nu A^\mu - A^\mu\partial_\nu A^\nu\} - \tfrac{1}{2}\partial_\mu(\bar{\psi}\mathrm{i}\gamma^\mu\psi), \quad (9.17)$$

$$\mathscr{L}_2 = -\tfrac{1}{2}(\partial_\mu A^\mu)^2, \quad (9.18)$$

and consider the density $\tilde{\mathscr{L}}$

$$\tilde{\mathscr{L}} = \mathscr{L} + \mathscr{L}_1 + \mathscr{L}_2 \quad (9.19)$$

as a starting point for QED. This approach was taken, for example, by Bogoliubov (1959). The addition of $\mathscr{L}_1$ is completely harmless since it represents a total divergence and thus leaves the equation of motion unchanged. The addition of $\mathscr{L}_2$ is motivated by the results of Chap. 7. Show that for all physical states

$$\langle \text{phys. state} \,|\!:\!(\partial_\mu \mathbf{A}^\mu(x))^2\!:\!|\, \text{phys. state}\rangle = 0. \quad (9.20)$$

At the quantized level, $\mathscr{L}_2$ thus corresponds to an operator that is equivalent to zero in the space of physical states. Using the principle of least action derive the equations of motion that follow from the Lagrange density $\tilde{\mathscr{L}}$. Apply the quantization rules from Sect. 3.4 to the Lagrangian $\tilde{L}$

$$\tilde{L} = \int \mathrm{d}^3 x\, \tilde{\mathscr{L}}, \quad (9.21)$$

and show that the Hamilton operator in the interaction picture is given by (9.9).

CHAPTER 10

Simple Reactions in Quantum Electrodynamics

In this chapter we will discuss a series of simple reactions in QED. To begin with we will take electron–electron scattering and will derive the Feynman rules given in Appendix B from the general expression for the S-operator (9.14).

10.1 Electron–Electron Scattering (Møller Scattering)

When an electron approaches another electron, it can be scattered. We shall consider the scattering in the center of mass frame. The momenta are as given in Fig. 10.1 and ϑ is the scattering angle in the center of mass frame. It naturally follows that

$$|\boldsymbol{p}| = |\boldsymbol{p}'|. \tag{10.1}$$

For classical, nonrelativistic electrons we can calculate the cross-section using Rutherford's method. The motion of two particles around one another is, of course, mathematically equivalent to the motion of a fictitious particle with the reduced mass around a fixed center. Taking account of this, we obtain from the classical Rutherford scattering formula (see Goldstein 1980)

$$\left.\frac{d\sigma}{d\Omega}\right|_{\text{classical}} = \frac{\alpha^2 m^2}{16|\boldsymbol{p}|^4}\left\{\frac{1}{\sin^4 \vartheta/2} + \frac{1}{\cos^4 \vartheta/2}\right\}, \tag{10.2}$$

Figure 10.1 Electron–electron scattering in the center of mass system. The momenta of the particles are given in brackets.

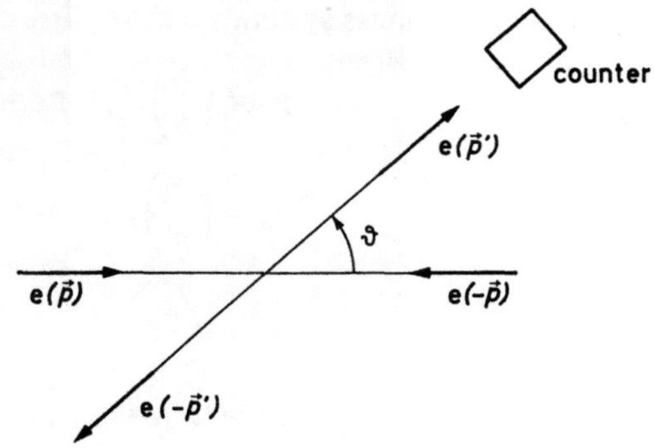

where $d\Omega = \sin\vartheta \, d\vartheta \, d\varphi$ and the range of ϑ and φ is given classically by

$$\begin{aligned} 0 &\le \vartheta \le \pi, \\ 0 &\le \varphi \le 2\pi. \end{aligned} \tag{10.3}$$

The term proportional to $\sin^{-4} \vartheta/2$ is familiar. The term proportional to $\cos^{-4} \vartheta/2$ arises because both particles are electrons. If we observe the scattered electrons at a particular angle ϑ, then there is a probability proportional to $\sin^{-4} \vartheta/2$ of observing the electron incident from the left. For symmetry reasons, the probability of encountering here the electron incident from the right is proportional to

$$\sin^{-4}\left(\frac{\pi - \vartheta}{2}\right) = \cos^{-4}\left(\frac{\vartheta}{2}\right). \tag{10.4}$$

Classically, we simply need to add the two probabilities.

We will in fact see, that even for extremely slow electrons, the quantum mechanical scattering formula does not agree with the classical formula (10.2) since quantum mechanically, it is not possible to determine whether the scattered electron came from the right or from the left. An additional term then appears which describes the interference of the two electrons.

The quantum mechanical problem can be stated as follows: We have two incoming electrons with four-momenta p_1 and p_2, and spin orientations r_1 and r_2. We wish to know the probability of finding, as a result of the scattering, two outgoing electrons with momenta and spin orientations p_3, p_4, r_3, r_4:

$$e^-(p_1, r_1) + e^-(p_2, r_2) \to e^-(p_3, r_3) + e^-(p_4, r_4). \tag{10.5}$$

The kinematics of such a two-particle reaction was discussed in detail in Chap. 2 and we will again use the variables s, t, and u that were introduced there:

$$\begin{aligned} s &= (p_1 + p_2)^2, \\ t &= (p_1 - p_3)^2, \\ u &= (p_1 - p_4)^2. \end{aligned} \tag{10.6}$$

In the center of mass system we have

$$\begin{aligned} p_1 &= \begin{pmatrix} E \\ \boldsymbol{p} \end{pmatrix}, \qquad p_2 = \begin{pmatrix} E \\ -\boldsymbol{p} \end{pmatrix}, \\ p_3 &= \begin{pmatrix} E \\ \boldsymbol{p}' \end{pmatrix}, \qquad p_4 = \begin{pmatrix} E \\ -\boldsymbol{p}' \end{pmatrix}, \\ E &= \frac{\sqrt{s}}{2}, \\ |\boldsymbol{p}| &= |\boldsymbol{p}'| = \tfrac{1}{2}\sqrt{s - 4m^2}, \\ t &= -(\boldsymbol{p} - \boldsymbol{p}')^2 = -4|\boldsymbol{p}|^2 \sin^2 \frac{\vartheta}{2}, \\ u &= -(\boldsymbol{p} + \boldsymbol{p}')^2 = -4|\boldsymbol{p}|^2 \cos^2 \frac{\vartheta}{2}. \end{aligned} \tag{10.7}$$

Before the scattering (at time $t = -\infty$) we thus have the following state:

$$|t = -\infty\rangle = \mathbf{a}^\dagger_{r_1}(\boldsymbol{p}_1)\mathbf{a}^\dagger_{r_2}(\boldsymbol{p}_2)|0\rangle. \tag{10.8}$$

We are interested in the amplitude for finding in the final state $|t = +\infty\rangle$ two electrons of momenta p_3, p_4 and spin orientations r_3, r_4. The corresponding transition amplitude is

$$\begin{aligned} S_{fi} &= \langle \mathrm{e}(p_3, r_3)\mathrm{e}(p_4, r_4)|\mathbf{S}|\mathrm{e}(p_1, r_1)\mathrm{e}(p_2, r_2)\rangle \\ &= \langle 0|\mathbf{a}_{r_3}(\boldsymbol{p}_3)\mathbf{a}_{r_4}(\boldsymbol{p}_4)\mathbf{S}\mathbf{a}^\dagger_{r_1}(\boldsymbol{p}_1)\mathbf{a}^\dagger_{r_2}(\boldsymbol{p}_2)|0\rangle. \end{aligned} \tag{10.9}$$

We will consider only the case in which a scattering has actually taken place. Thus we have

$$(p_1, r_1), (p_2, r_2) \neq (p_3, r_3), (p_4, r_4). \tag{10.10}$$

For this case the unit operator in the expansion of the S-operator (9.14) does not contribute. The first correction term also does not contribute for the following reason: if we express $\mathbf{H}'$ in terms of creation and annihilation operators as in (9.11), then the photon creation and annihilation operators give zero when acting to the left and to the right respectively. Thus we only obtain scattering upon expansion to order e^2. We shall also truncate the expansion at this term. Then we have

$$\begin{aligned} S_{fi} = \frac{(-\mathrm{i})^2}{2!} \int_{-\infty}^{\infty} \mathrm{d}t' \int_{-\infty}^{\infty} \mathrm{d}t'' \\ \cdot \langle \mathrm{e}(p_3, r_3)\mathrm{e}(p_4, r_4)|\mathrm{T}(\mathbf{H}'(t')\mathbf{H}'(t''))|\mathrm{e}(p_1, r_1)\mathrm{e}(p_2, r_2)\rangle. \end{aligned} \tag{10.11}$$

On substitution of the explicit form of $\mathbf{H}'$ (9.9) this becomes

$$\begin{aligned} S_{fi} = \frac{(\mathrm{i})^2}{2!} e^2 \int \mathrm{d}x'\, \mathrm{d}x''\, \langle 0|\mathbf{a}_{r_3}(\boldsymbol{p}_3)\mathbf{a}_{r_4}(\boldsymbol{p}_4) \\ \cdot \mathrm{T}\{:\bar{\psi}(x')\gamma^\mu\psi(x'): \mathbf{A}_\mu(x') :\bar{\psi}(x'')\gamma^\nu\psi(x''): \mathbf{A}_\nu(x'')\} \\ \cdot \mathbf{a}^\dagger_{r_1}(\boldsymbol{p}_1)\mathbf{a}^\dagger_{r_2}(\boldsymbol{p}_2)|0\rangle. \end{aligned} \tag{10.12}$$

This expression may at first seem rather complicated. However, we only need to insert the expansions of the fields in terms of creation and annihilation operators, and to use Wick's theorem in order to achieve a much simpler form for this expression. Our matrix element has the following structure:

$$\begin{aligned}&\langle 0|\mathbf{a}\mathbf{a}{:}(\mathbf{b}+\mathbf{a}^\dagger)(\mathbf{b}^\dagger+\mathbf{a}){:}(\alpha+\alpha^\dagger)\\ &\quad{:}(\mathbf{b}+\mathbf{a}^\dagger)(\mathbf{b}^\dagger+\mathbf{a}){:}(\alpha+\alpha^\dagger)\mathbf{a}^\dagger\mathbf{a}^\dagger|0\rangle .\end{aligned} \tag{10.13}$$

We see that the operators of $\mathbf{A}_\mu(x')$ can only be contracted with those from $\mathbf{A}_\nu(x'')$. If we further contract all Fermi operators then we find no contribution when operators from $\psi(x')$ and $\bar{\psi}(x'')$ are paired. This is because, here, at least one contraction of an incoming with an outgoing electron operator occurs and, due to the condition (10.10), we get zero. The only nonzero contribution is obtained by contracting all operators of incoming and outgoing electrons with the field operators ψ and $\bar{\psi}$. For example, we must contract $\mathbf{a}^\dagger_{r_1}(\boldsymbol{p}_1)$ in (10.12) with $\psi(x')$ and $\psi(x'')$ in turn. Thus we have

$$\begin{aligned}&\langle 0|\mathbf{a}(\boldsymbol{p}_3)\mathbf{a}(\boldsymbol{p}_4)\,{:}\bar{\psi}(x')\gamma^\mu\psi(x'){:}\,{:}\bar{\psi}(x'')\gamma^\nu\psi(x''){:}\,\mathbf{a}^\dagger(\boldsymbol{p}_1)\mathbf{a}^\dagger(\boldsymbol{p}_2)|0\rangle\\
&\quad=\mathbf{a}(\boldsymbol{p}_3)\mathbf{a}(\boldsymbol{p}_4)\,{:}\bar{\psi}(x')\gamma^\mu\psi(x'){:}\,{:}\bar{\psi}(x'')\gamma^\nu\psi(x''){:}\,\mathbf{a}^\dagger(\boldsymbol{p}_1)\mathbf{a}^\dagger(\boldsymbol{p}_2)\\
&\qquad+\mathbf{a}(\boldsymbol{p}_3)\mathbf{a}(\boldsymbol{p}_4)\,{:}\bar{\psi}(x')\gamma^\mu\psi(x'){:}\,{:}\bar{\psi}(x'')\gamma^\nu\psi(x''){:}\,\mathbf{a}^\dagger(\boldsymbol{p}_1)\mathbf{a}^\dagger(\boldsymbol{p}_2)\\
&\qquad+\mathbf{a}(\boldsymbol{p}_3)\mathbf{a}(\boldsymbol{p}_4)\,{:}\bar{\psi}(x')\gamma^\mu\psi(x'){:}\,{:}\bar{\psi}(x'')\gamma^\nu\psi(x''){:}\,\mathbf{a}^\dagger(\boldsymbol{p}_1)\mathbf{a}^\dagger(\boldsymbol{p}_2)\\
&\qquad+\mathbf{a}(\boldsymbol{p}_3)\mathbf{a}(\boldsymbol{p}_4)\,{:}\bar{\psi}(x')\gamma^\mu\psi(x'){:}\,{:}\bar{\psi}(x'')\gamma^\nu\psi(x''){:}\,\mathbf{a}^\dagger(\boldsymbol{p}_1)\mathbf{a}^\dagger(\boldsymbol{p}_2)\\
&\quad=\{\bar{u}(p_4)\,\mathrm{e}^{\mathrm{i}p_4x'}\,\gamma^\mu u(p_1)\,\mathrm{e}^{-\mathrm{i}p_1x'}\cdot\bar{u}(p_3)\,\mathrm{e}^{\mathrm{i}p_3x''}\,\gamma^\nu u(p_2)\,\mathrm{e}^{-\mathrm{i}p_2x''}\\
&\qquad-(1\leftrightarrow 2)-(3\leftrightarrow 4)+(1\leftrightarrow 2,\,3\leftrightarrow 4)\}.\end{aligned} \tag{10.14}$$

Here we have made use of (8.21) and noted that the contraction of Fermi operators involves bringing them together by anticommutation. This leads to a factor -1 in the second and third terms of (10.14). The spin indices have been supressed. If this result is inserted into (10.12) then, since the contraction of Fermi operators for $x_0' > x_0''$ and for $x_0' < x_0''$ is the same, we obtain

$$\begin{aligned}S_{fi}&=\frac{(\mathrm{i}e)^2}{2}\int \mathrm{d}x'\,\mathrm{d}x''\\
&\quad\cdot\{\theta(x_0'-x_0'')\langle 0|\mathbf{A}_\mu(x')\mathbf{A}_\nu(x'')|0\rangle+\theta(x_0''-x_0')\langle 0|\mathbf{A}_\nu(x'')\mathbf{A}_\mu(x')|0\rangle\}\\
&\quad\cdot[\bar{u}(p_4)\gamma^\mu u(p_1)\,\mathrm{e}^{\mathrm{i}(p_4-p_1)x'}\cdot\bar{u}(p_3)\gamma^\nu u(p_2)\,\mathrm{e}^{\mathrm{i}(p_3-p_2)x''}\\
&\quad-(1\leftrightarrow 2)-(3\leftrightarrow 4)+(1\leftrightarrow 2,\,3\leftrightarrow 4)].\end{aligned} \tag{10.15}$$

Collecting together terms which differ only in the symbols used for integration or summation variables, finally leads to

$$S_{fi} = (\mathrm{i}e)^2 \int \mathrm{d}x'\,\mathrm{d}x''\,\langle 0|\mathrm{T}(\mathbf{A}_\mu(x')\mathbf{A}_\nu(x''))|0\rangle$$

$$\cdot\{\bar{u}(p_4)\gamma^\mu u(p_1)\bar{u}(p_3)\gamma^\nu u(p_2)\,\mathrm{e}^{\mathrm{i}(p_4-p_1)x'}\,\mathrm{e}^{\mathrm{i}(p_3-p_2)x''}$$

$$-\,\bar{u}(p_3)\gamma^\mu u(p_1)\bar{u}(p_4)\gamma^\nu u(p_2)\,\mathrm{e}^{\mathrm{i}(p_3-p_1)x'}\,\mathrm{e}^{\mathrm{i}(p_4-p_2)x''}\}. \tag{10.16}$$

We see that the propagator function of the photon field appears in a natural way. Replacing this by its Fourier representation (7.55) enables the integrations over x' and x'' to be easily carried out; they give δ-functions of four-momenta. The final result for the scattering amplitude to order e^2 is then

$$S_{fi} = \mathrm{i}(2\pi)^4\delta(p_4 + p_3 - p_2 - p_1)T_{fi}$$

with a T-matrix element T_{fi} (cf. Sect. 5.2) given by

$$T_{fi} = \frac{1}{i}\left\{\bar{u}(p_4)(\mathrm{i}e\gamma^\mu)u(p_1)\frac{-\mathrm{i}g_{\mu\nu}}{(p_4-p_1)^2}\bar{u}(p_3)(\mathrm{i}e\gamma^\nu)u(p_2)\right.$$

$$\left.-\,\bar{u}(p_3)(\mathrm{i}e\gamma^\mu)u(p_1)\frac{-\mathrm{i}g_{\mu\nu}}{(p_3-p_1)^2}\bar{u}(p_4)(\mathrm{i}e\gamma^\nu)u(p_2)\right\}. \tag{10.17}$$

We now compare our result with the Feynman rules given in Appendix B: According to these, there are two diagrams for electron–electron scattering to order e^2 which are drawn in Fig. 10.2. Translating these into mathematical form, we obtain exactly (10.17). Thus, for this case, we have explicitly checked the validity of the Feynman rules. To derive the rules for the general case one starts from (9.14) and applies similar considerations as above.

We move on now to calculate the scattering cross section from the transition amplitude. We consider an initial state consisting of unpolarized electrons and are not interested in the polarization of the electrons in the final state. According to the general formulas of Chap. 5 we then obtain:

$$d\sigma = \frac{1}{2w(s, m_\mathrm{e}^2, m_\mathrm{e}^2)}\frac{\mathrm{d}^3p_3\,\mathrm{d}^3p_4}{(2\pi)^6 2p_3^0 2p_4^0}$$

$$\cdot(2\pi)^4\delta(p_1 + p_2 - p_3 - p_4)\sum_{\text{Spins}}{}' |T_{fi}|^2. \tag{10.18}$$

Here $\sum'$ denotes an average (summation) over the spin directions of the

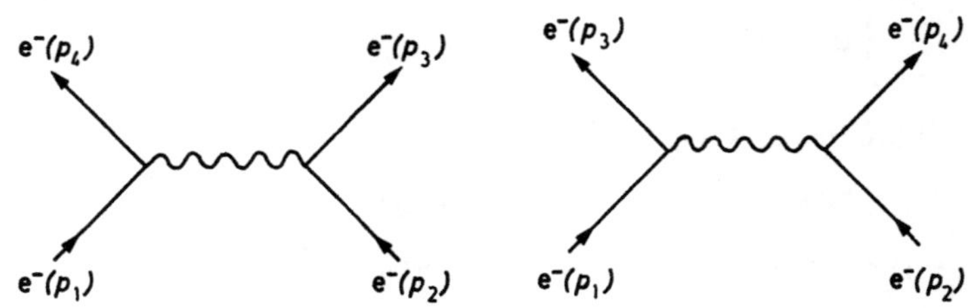

Figure 10.2 Feynman diagrams to order e^2 for electron–electron scattering. These are examples of so-called "tree diagrams", i.e., diagrams without loops.

electrons in the initial (final) state. Using the expressions for spin sums (4.94) we find

$$\sum_{\text{Spins}}' |T_{fi}|^2 = e^4 \sum_{\text{Spins}}' \left\{ \frac{1}{(p_4 - p_1)^2} \bar{u}(p_2)\gamma_\mu u(p_3)\bar{u}(p_1)\gamma^\mu u(p_4) - \frac{1}{(p_3 - p_1)^2} \bar{u}(p_1)\gamma_\mu u(p_3)\bar{u}(p_2)\gamma^\mu u(p_4) \right\}$$

$$\cdot \left\{ \frac{1}{(p_4 - p_1)^2} \bar{u}(p_4)\gamma_\nu u(p_1)\bar{u}(p_3)\gamma^\nu u(p_2) - \frac{1}{(p_3 - p_1)^2} \bar{u}(p_4)\gamma_\nu u(p_2)\bar{u}(p_3)\gamma^\nu u(p_1) \right\}$$

$$= \frac{e^4}{4} \left\{ \frac{1}{u^2} \mathrm{Tr}[(\not{p}_2 + m)\gamma_\mu(\not{p}_3 + m)\gamma_\nu] \cdot \mathrm{Tr}[(\not{p}_1 + m)\gamma^\mu(\not{p}_4 + m)\gamma^\nu] - \frac{1}{tu} \mathrm{Tr}[(\not{p}_2 + m)\gamma_\mu(\not{p}_3 + m)\gamma_\nu(\not{p}_1 + m)\gamma^\mu(\not{p}_4 + m)\gamma^\nu] + (3 \leftrightarrow 4) \right\}. \tag{10.19}$$

The traces can be evaluated with the help of the formulas given in Appendix A. After some calculation we find

$$\sum' |T_{fi}|^2 = \frac{64\pi^2\alpha^2}{t^2u^2} \{(s - 2m^2)^2(t^2 + u^2) + ut(-4m^2s + 12m^4 + ut)\}. \tag{10.20}$$

Putting this result into (10.18) we obtain the differential cross-section in the center of mass frame [see Fig. 10.1 and eq. (10.7)]:

$$d\sigma = d\Omega \int_0^\infty d|\boldsymbol{p}_3||\boldsymbol{p}_3|^2 \int d^3p_4 \frac{1}{2[s(s - 4m^2)]^{1/2}} \cdot \frac{1}{(2\pi)^2} \frac{1}{2p_3^0 2p_4^0} \delta(\sqrt{s} - p_3^0 - p_4^0)\delta^3(\boldsymbol{p}_3 + \boldsymbol{p}_4) \sum_{\text{Spins}}' |T_{fi}|^2,$$

$$\frac{d\sigma}{d\Omega} = \frac{\alpha^2}{st^2u^2} \{(s - 2m^2)^2(t^2 + u^2) + ut(-4m^2s + 12m^4 + ut)\}. \tag{10.21}$$

Here t and u are functions of the scattering angle ϑ as given in (10.7). A simple calculation gives the differential cross-section in terms of the invariant variable t

$$\frac{d\sigma}{dt} = \frac{4\pi\alpha^2}{s(s - 4m^2)} \frac{1}{t^2u^2} \cdot \{(s - 2m^2)^2(t^2 + u^2) + ut(-4m^2s + 12m^4 + ut)\}. \tag{10.22}$$

Formulas equivalent to (10.21), (10.22) were first derived by Møller (1932).

We must now turn our attention to the kinematical bounds on the values of ϑ and t. For this purpose we consider Fig. 10.1. Quantum mechanically, the two electrons in the initial and final states must be regarded as identical particles. The question as to which of the two incoming electrons enters the counter at scattering angle ϑ is a meaningless one. We can only ascertain that from the point of interaction, one electron flies off at an angle ϑ and the other at an angle $\pi - \vartheta$. Therefore, *all possible* final states are included when ϑ is allowed to take values in the interval

$$0 \leq \vartheta \leq \frac{\pi}{2}. \tag{10.23}$$

For the variable t this implies the following range of values:

$$0 \geq t \geq -\tfrac{1}{2}(s - 4m^2). \tag{10.24}$$

We will now consider the scattering cross-section in the nonrelativistic and in the extreme relativistic limits. In the nonrelativistic limit we obtain

$$\frac{d\sigma}{d\Omega} = \frac{\alpha^2 m^2}{16|\boldsymbol{p}|^4}\left\{\frac{1}{\sin^4\vartheta/2} + \frac{1}{\cos^4\vartheta/2} - \frac{1}{\sin^2\vartheta/2\cos^2\vartheta/2}\right\} \quad \text{for } |\boldsymbol{p}| \ll m. \tag{10.25}$$

Comparing this with the classical result (10.2), we note the appearance of an additional term. This term stems from the quantum mechanical interference of the two electrons. Quantum mechanically, we must add the *amplitudes* that correspond to the two Feynman diagrams (Fig. 10.2) and then take the absolute square in order to calculate the scattering cross-section. The sign of the interference term is due to the Fermi statistics obeyed by the electrons. For Bose particles the sign would be reversed. This formula (10.25) was first derived by Mott in 1930.

In the extreme relativistic limit the cross-section is given by

$$\begin{aligned}\frac{d\sigma}{d\Omega} &= \frac{\alpha^2}{s}\left\{\frac{1}{\sin^4\vartheta/2} + \frac{1}{\cos^4\vartheta/2} + 1\right\}\\ &= \frac{\alpha^2}{s}\frac{(3+\cos^2\vartheta)^2}{\sin^4\vartheta} \quad \text{for } |\boldsymbol{p}| \gg m.\end{aligned} \tag{10.26}$$

It is noteworthy that the dimensionless cross-section, given by $s \cdot d\sigma/d\Omega$, is, in this high energy limit, no longer dependent on s. This is referred to as *scaling* behavior and is understood as a consequence of the point-like nature of the electron in QED. If the electron were to have a size $r = 1/\Lambda$, then one would expect $s \cdot d\sigma/d\Omega$ to be a nontrivial function of the dimensionless variable s/Λ^2 or, in other words, $s \cdot d\sigma/d\Omega$ would not be independent of s. We will frequently encounter arguments of this nature when we come to discuss the strong interaction.

The predictions of the theory of electron–electron scattering have been confirmed without exception in experiments. A few references on this are Mott (1965), Ashkin (1954), and Barber (1966, 1971). In recent years, the interest in electron–electron scattering has waned in favor of investigations of electron–positron reactions.

In the following we will apply the Feynman rules in order to evaluate a few simple processes. We will only be concerned with the relevant diagrams and results. The intermediate steps follow exactly the same lines as those for electron–electron scattering.

10.2 Electron–Positron Scattering (Bhabha Scattering)

Electron–positron scattering is described by the reaction formula

$$e^-(p_1) + e^+(p_2) \rightarrow e^-(p_3) + e^+(p_4). \tag{10.27}$$

The kinematics is the same as for electron–electron scattering (10.6), (10.7). To lowest order we must take into account the diagrams shown in Fig. 10.3. In the "scattering diagram" the four-momentum of the virtual photon is space-like or zero, and in the "annihilation diagram" it is time-like:

$$\begin{aligned} (p_1 - p_3)^2 &\leq 0, \\ (p_1 + p_2)^2 &\geq 4m^2. \end{aligned} \tag{10.28}$$

For unpolarized electrons and positrons we find that for ultra relativistic energies the cross-section is given in the center of mass frame by

$$\begin{aligned} \frac{d\sigma}{d\Omega} &= \frac{\alpha^2}{s}\left\{\frac{1}{2}\frac{1+\cos^4 \vartheta/2}{\sin^4 \vartheta/2} + \frac{1}{4}(1+\cos^2\vartheta) - \frac{\cos^4 \vartheta/2}{\sin^2 \vartheta/2}\right\} \\ &= \frac{\alpha^2}{16s}\frac{(3+\cos^2\vartheta)^2}{\sin^4 \vartheta/2} \\ &\text{for} \quad s \gg 4m^2. \end{aligned} \tag{10.29}$$

The difference between this and electron–electron scattering lies in the fact

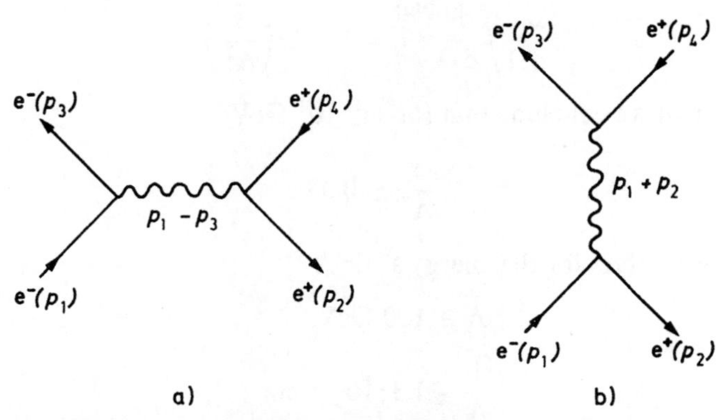

Figure 10.3 Feynman diagrams to order e^2 for electron–positron scattering; scattering diagram (a) and annihilation diagram (b).

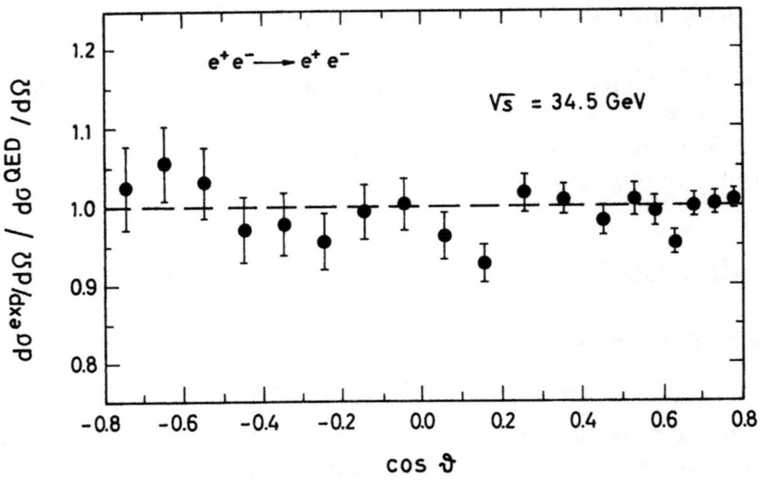

Figure 10.4 Data from the TASSO group for the differential cross-section of the reaction $e^-e^+ \rightarrow e^-e^+$ divided by the prediction (10.29) of QED. The center of mass energy is $\sqrt{s} = 34.5$ GeV. Corrections for radiative effects to order α^3 have been taken into account (after Althoff 1984).

that here the initial and final states contain two distinguishable particles. The kinematic range for the scattering angle ϑ is thus

$$0 \leq \vartheta \leq \pi. \tag{10.30}$$

Bhabha scattering has recently been studied in great detail, for example, at the storage ring PETRA. In these experiments the highest center of mass energy was $\sqrt{s} \cong 45$ GeV. A result typical of those obtained by the groups working at PETRA is shown in Fig. 10.4. As can be seen, the agreement between experiment and theory is very good. In order to derive the range of validity of QED from this, we proceed as follows: Assuming that QED becomes modified at an energy scale Λ, then for $s \ll \Lambda^2$ we expect:

$$\frac{d\sigma}{d\Omega} \bigg/ \frac{d\sigma}{d\Omega}^{\text{QED}} = 1 + O\left(\frac{s}{\Lambda^2}\right). \tag{10.31}$$

From Fig. 10.4 we conclude that for $s \cong 10^3$ GeV^2

$$\frac{s}{\Lambda^2} \lesssim 0.05 \tag{10.32}$$

From this we obtain for the energy scale Λ

$$\begin{aligned} &\Lambda \gtrsim 150 \text{ GeV}, \\ &\frac{1}{\Lambda} \lesssim 1.3 \cdot 10^{-16} \text{ cm}. \end{aligned} \tag{10.33}$$

In this reaction, the validity of QED has thus been established down to length scales as small as $1.3 \cdot 10^{-16}$ cm. As we will see in Part IV, however, even in this energy range QED must in fact be considered as part of the more general theory of the electroweak interaction.

10.3 Compton Scattering

In this section we will discuss the scattering of a photon by a free electron:

$$\gamma(k) + e^-(p) \to \gamma(k') + e^-(p'). \tag{10.34}$$

Historically, this process played an important role in connection with Einstein's hypothesis of light quanta. In papers published in 1905 and 1917, Einstein postulated that an electromagnetic wave can also have particle-like character; he introduced photons into physics. This idea was received with great scepticism by many physicists of the time who were used to the wave nature of light. It was only in 1923, when Compton showed experimentally that the scattering of light from electron follows the rules for the elastic collision between two particles, in contradiction to the predictions of the older theory, that Einstein's hypothesis attained general recognition. (For the history of these developments, see Pais (1979).)

The laws of an elastic collision demand energy and momentum conservation. We consider this for the reaction (10.34) seen in the laboratory system in which the electron is at rest before the collision (Fig. 10.5). The four-momenta are then

$$\begin{aligned} p &= \begin{pmatrix} m \\ 0 \end{pmatrix}, \quad & p' &= \begin{pmatrix} \sqrt{m^2 + \boldsymbol{p}'^2} \\ \boldsymbol{p}' \end{pmatrix}, \\ k &= \begin{pmatrix} \omega \\ \boldsymbol{k} \end{pmatrix}, \quad & k' &= \begin{pmatrix} \omega' \\ \boldsymbol{k}' \end{pmatrix}, \end{aligned} \tag{10.35}$$

where $\omega = |\boldsymbol{k}|$ and $\omega' = |\boldsymbol{k}'|$. The conservation rules for energy and momentum read

$$\begin{aligned} m + \omega &= \sqrt{m^2 + \boldsymbol{p}'^2} + \omega', \\ \boldsymbol{k} &= \boldsymbol{p}' + \boldsymbol{k}'. \end{aligned} \tag{10.36}$$

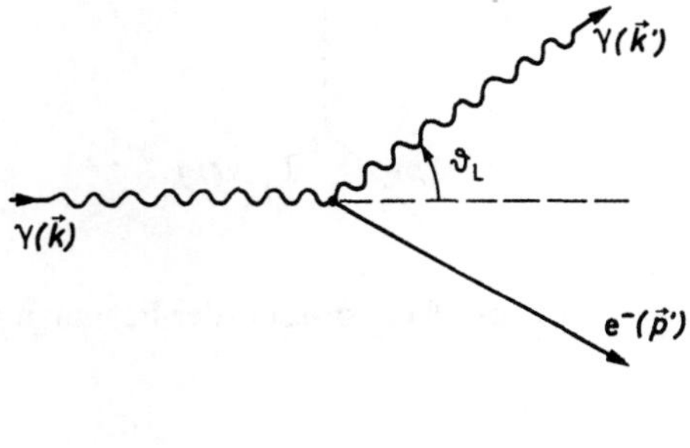

Figure 10.5 Compton scattering in the laboratory system.

For a scattering angle ϑ_L (Fig. 10.5) it follows from this that

$$m + \omega = \sqrt{m^2 + (\boldsymbol{k} - \boldsymbol{k}')^2} + \omega',$$

$$\frac{\omega - \omega'}{\omega\omega'} = \frac{1}{m}(1 - \cos\vartheta_L), \tag{10.37}$$

$$\omega' = \omega\left[1 + \frac{\omega}{m}(1 - \cos\vartheta_L)\right]^{-1}.$$

If we denote the wavelengths of the incident and scattered photon by λ and λ' then with

$$\omega = \frac{2\pi}{\lambda}, \qquad \omega' = \frac{2\pi}{\lambda'}, \tag{10.38}$$

we obtain from (10.37)

$$\lambda' - \lambda = \frac{2\pi}{m}(1 - \cos\vartheta_L). \tag{10.39}$$

This is the equation that was experimentally confirmed by Compton. The $1/m$ on the right-hand side plays the role of a wavelength. Thus $1/m$ is referred to as the *Compton wavelength* of the electron. Its experimental value is

$$m^{-1} = 3.861\,590\,5(64) \cdot 10^{-11}\ \text{cm}. \tag{10.40}$$

To determine the scattering cross-section for our reaction, we draw the corresponding Feynman diagrams of lowest order (Fig. 10.6). On application of the Feynman rules we obtain for the cross-section of unpolarized electrons and photons the following expression:

$$\frac{d\sigma}{d\Omega_L} = \frac{\alpha^2}{2}\frac{1}{[m + \omega(1 - \cos\vartheta_L)]^2}\cdot\left\{\frac{\omega^2(1 - \cos\vartheta_L)^2}{m[m + \omega(1 - \cos\vartheta_L)]} + 1 + \cos^2\vartheta_L\right\}. \tag{10.41}$$

Here ϑ_L and Ω_L are the polar angle and solid angle, respectively, of the scattered photon in the laboratory frame. This result was first derived by Klein and Nishina in 1929.

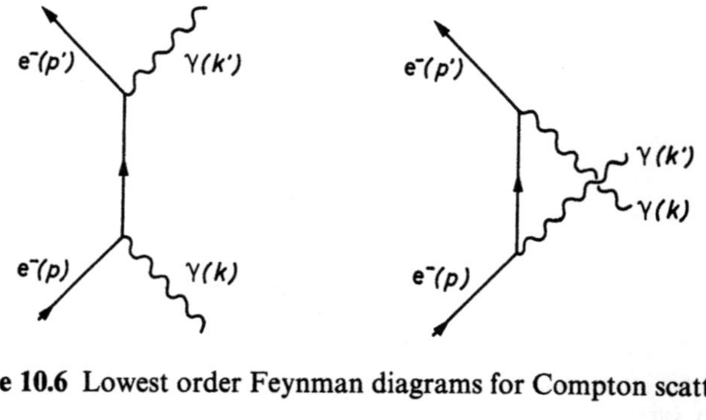

Figure 10.6 Lowest order Feynman diagrams for Compton scattering.

We first consider the nonrelativistic limit:

$$\omega \ll m. \tag{10.42}$$

We then find from (10.37), (10.41) the formulas that were derived classically by Thomson

$$\begin{aligned} \omega' &= \omega, \\ \frac{d\sigma}{d\Omega_L} &= \frac{\alpha^2}{m^2}\frac{1+\cos^2\vartheta_L}{2}, \\ \sigma_{tot} &= \int d\Omega_L \frac{d\sigma}{d\Omega_L} = \frac{8\pi}{3}\frac{\alpha^2}{m^2}. \end{aligned} \tag{10.43}$$

In the ultra relativistic limit, i.e., for

$$\omega \gg m \quad \text{or} \quad \lambda \ll \frac{1}{m}, \tag{10.44}$$

we distinguish two cases. In the extreme forward direction

$$1 - \cos\vartheta_L \ll \left(\frac{m}{\omega}\right) \tag{10.45}$$

we find from (10.39), (10.41) the classical values for the wavelength of the scattered photon and for the cross-section

$$\begin{aligned} \lambda' &\cong \lambda, \\ \frac{d\sigma}{d\Omega_L} &\cong \frac{\alpha^2}{m^2}. \end{aligned} \tag{10.46}$$

For large angles

$$1 - \cos\vartheta_L \gg \frac{m}{\omega} = \frac{m\lambda}{2\pi}, \tag{10.47}$$

we find, on the other hand,

$$\lambda' \cong \frac{2\pi}{m}(1 - \cos\vartheta_L), \tag{10.48}$$

$$\frac{d\sigma}{d\Omega_L} \cong \frac{\alpha^2}{2m\omega(1-\cos\vartheta_L)}. \tag{10.49}$$

In the high energy limit the wavelength of photons scattered through large angles is thus independent of the wavelength of the incoming photon. The corresponding cross-section decreases as $1/\omega$ with the photon energy.

An experimental test of the Klein–Nishina formula (10.41) is shown in Fig. 10.7.

Also of interest is the total scattering cross-section. By integrating (10.41)

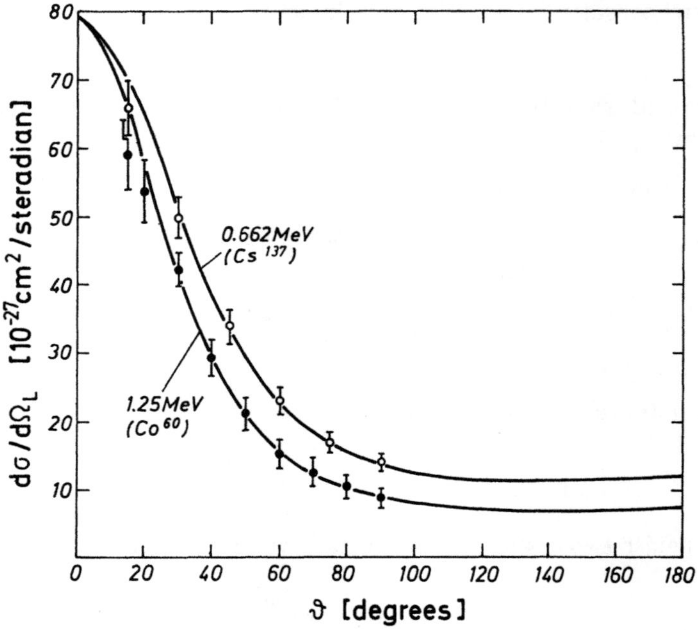

Figure 10.7 Experimental results for Compton scattering. The curves correspond to the Klein–Nishina formula (10.41) for photon energies $\omega = 0.662$ MeV and $\omega = 1.25$ MeV. The experimental data are from Hofstadter (1949) and Bernstein (1956) (after Evans 1958).

over all angles we obtain

$$\sigma_{\text{tot}} = \frac{\pi\alpha^2}{\omega m}\left\{\left[1 - 2\frac{m}{\omega} - 2\left(\frac{m}{\omega}\right)^2\right]\ln\frac{2\omega + m}{m} + \frac{1}{2} + 4\frac{m}{\omega} - \frac{m^2}{2(2\omega + m)^2}\right\}. \tag{10.50}$$

For $\omega \gg m$ this becomes

$$\sigma_{\text{tot}} = \frac{\pi\alpha^2}{\omega m}\left\{\ln\frac{2\omega}{m} + \frac{1}{2} + \mathrm{O}\left(\frac{1}{\omega}\right)\right\}, \tag{10.51}$$

i.e., the total cross-section also decreases for large photon energies as $1/\omega$.

With this we bring our discussion of the Compton effect to a close. For further discussion we refer, for example, to Evans (1958). We also mention that the reaction

$$e^- + e^+ \to \gamma + \gamma \tag{10.52}$$

has been studied in detail in modern electron–positron storage rings. This process is obtained from the Compton process by a *crossing* transformation (see Sect. 2.3). The corresponding Feynman diagrams of lowest order are those of Fig. 10.6, only the interpretation of the external lines as incoming or outgoing particles is different. The validity of QED for electron–positron annihilation into two photons (10.52) has been confirmed to a similar degree to the Bhabha scattering (see Dittmann 1981; Bartel 1983)

Exercises

10.1 Verify equations (10.20)–(10.22)

10.2 Calculate the cross-section for Bhabha scattering according to the Feynman rules. Verify (10.29).

10.3 Calculate the cross-section for Compton scattering (10.34) according to the Feynman rules.

10.4 Calculate the cross-section for e^+e^- annihilation into two photons

$$e^+ + e^- \to \gamma + \gamma$$

according to the Feynman rules in the high energy limit ($s \gg m^2$).

CHAPTER 11

The Muon and Muon Pair Production in Electron–Positron Annihilation

11.1 Properties of the Muon

The muon was discovered in cosmic rays in 1937 (Anderson 1937, 1938; Street 1937). It behaves in essentially the same way as the electron, but its mass is about 200 times larger:

$$\begin{aligned} m_e &= 0.511\,003\,4(14)\ \text{MeV}, \\ m_\mu &= 105.659\,32(29)\ \text{MeV}. \end{aligned} \tag{11.1}$$

The muon decays with a mean lifetime of

$$\tau_\mu = 2.197\,09(5) \cdot 10^{-6}\ \text{s}. \tag{11.2}$$

The most common decay, with a branching ratio of almost 100% is the decay into an electron, an electron antineutrino and a muon neutrino:

$$\mu^- \to e^- + \bar{\nu}_e + \nu_\mu. \tag{11.3}$$

It is noteworthy that the energetically allowed process

$$\mu^- \to e^- + \gamma \tag{11.4}$$

has, to date, never been observed. From this fact one may conclude the existence of a conserved quantum number which makes this a forbidden process, in the same way as the process

$$e^- + e^- \to e^- + e^- + e^- + e^- \tag{11.5}$$

is forbidden by the conservation of charge. This quantum number is known

Table 11.1 The electron, muon, and tau numbers of the elementary particles. The lepton number L is given by $L = L_e + L_\mu + L_\tau$.

	L_e	L_μ	L_τ
e^-, ν_e	+1	0	0
$e^+, \bar{\nu}_e$	−1	0	0
μ^-, ν_μ	0	+1	0
$\mu^+, \bar{\nu}_\mu$	0	−1	0
τ^-, ν_τ	0	0	+1
$\tau^+, \bar{\nu}_\tau$	0	0	−1
hadrons (p, n, π, ...)	0	0	0
γ, W, Z	0	0	0

as the *muon number* L_μ. In an analogous way one defines the *electron number* L_e. The quantum numbers are assigned to the elementary particles as shown in Table 11.1. The statement "the electron number is a conserved quantum number" simply means that in all reactions so far observed, the sum of the electron numbers of the particles in the initial state equals the sum of the electron numbers of the particles in the final state. The same holds for the muon number. Thus one can "explain" the failure to observe, for example, the likewise energetically allowed decay of a proton into a positron and a photon:

$$p \not\to e^+ + \gamma. \tag{11.6}$$

In 1975 a further lepton the tau, τ, was discovered (Perl 1975). The mass of the tau is

$$m_\tau = 1784.2 \pm 3.2 \text{ MeV}. \tag{11.7}$$

It seems highly probable that the tau possesses its own massless neutrino ν_τ and that in analogy with the muon number there exists a new conserved quantum number L_τ, the tau number (see Part IV). The sum of L_e, L_μ, and L_τ is known as the lepton number L.

We return now to the muon. If we consider only the interaction of muons, electrons, and photons then the interaction Hamiltonian is as before (9.9), and we merely have to add the contributions of the electrons and muons in the electromagnetic current. With the field operators ψ_e and ψ_μ for electrons and muons, respectively, we have

$$\mathbf{H}'(t) = -e \int d^3x \, (:\bar{\psi}_e(\boldsymbol{x}, t)\gamma_\lambda \psi_e(\boldsymbol{x}, t): + :\bar{\psi}_\mu(\boldsymbol{x}, t)\gamma_\lambda \psi_\mu(\boldsymbol{x}, t):)\mathbf{A}^\lambda(\boldsymbol{x}, t). \tag{11.8}$$

In the Feynman diagrams we then have two types of fermion lines, the electron and the muon lines. An electron line cannot turn into a muon line and vice versa; this corresponds to the conservation of muon and electron number. The only vertices in the theory are the electron–photon vertex and the muon–photon vertex and these are identical (Fig. 11.1).

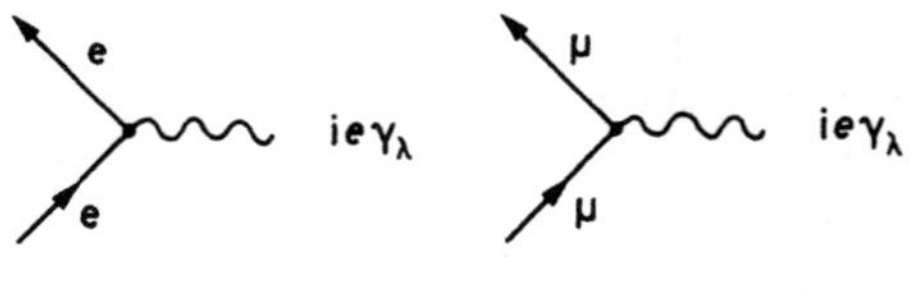

Figure 11.1 The electron–photon and the muon–photon vertex of QED together with the corresponding analytical expressions.

11.2 The Reaction $e^-e^+ \to \mu^-\mu^+$

Of great importance for the physics that is investigated at electron–positron storage rings, is the reaction $e^- + e^+ \to \mu^- + \mu^+$ since this represents the simplest example of the creation of two point-like particles with spin $\frac{1}{2}$. The study of muon pair production also serves as a means of establishing a standard for comparison with other reactions such as the production of hadrons: $e^- + e^+ \to$ hadrons.

We shall study the reaction

$$e^-(p_1) + e^+(p_2) \to \mu^-(p_3) + \mu^+(p_4). \tag{11.9}$$

Again s is the square of the total energy and ϑ the scattering angle in the center of mass frame (Fig. 11.2):

$$\begin{aligned} s &= (p_1 + p_2)^2, \\ \cos\vartheta &= \frac{\boldsymbol{p}_1 \boldsymbol{p}_3}{|\boldsymbol{p}_1||\boldsymbol{p}_3|}. \end{aligned} \tag{11.10}$$

We want to determine the cross-section for reaction (11.9) to lowest order in α. Only one Feynman diagram contributes (Fig. 11.3). According to the Feynman rules we obtain the S-matrix element as follows:

$$\langle \mu^-(p_3)\mu^+(p_4)|\mathbf{S}|e^-(p_1)e^+(p_2)\rangle$$

$$= (2\pi)^4\delta(p_3 + p_4 - p_1 - p_2)\frac{-\mathrm{i}g^{\rho\lambda}}{s}\bar{v}_e(p_2)\mathrm{i}e\gamma_\lambda u_e(p_1)\cdot\bar{u}_\mu(p_3)\mathrm{i}e\gamma_\rho v_\mu(p_4). \tag{11.11}$$

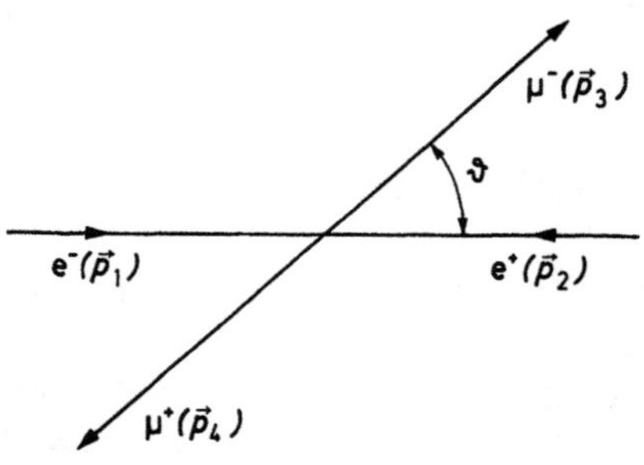

Figure 11.2 The reaction $e^-e^+ \to \mu^-\mu^+$ in the center of mass frame.

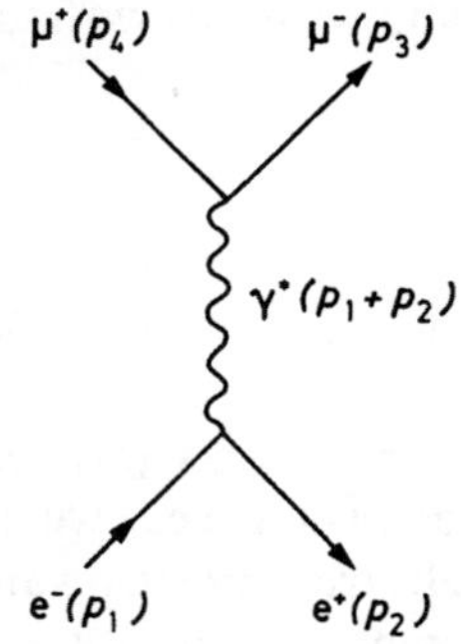

Figure 11.3 The lowest order Feynman diagram for the reaction $e^-e^+ \to \mu^-\mu^+$. The virtual photon is denoted by γ^*.

For the cross-section we obtain, using the formulas from Chap. 5:

$$\mathrm{d}\sigma = \frac{1}{2w(s, m_e^2, m_e^2)}\frac{1}{(2\pi)^6}\frac{\mathrm{d}^3p_3\,\mathrm{d}^3p_4}{2p_3^0 2p_4^0}\cdot(2\pi)^4\delta(p_3+p_4-p_1-p_2)\frac{e^4}{s^2}$$
$$\cdot\sum_{\text{Spins}}{}' |\bar{u}_\mu(p_3)\gamma^\lambda v_\mu(p_4)\bar{v}_e(p_2)\gamma_\lambda u_e(p_1)|^2. \tag{11.12}$$

For unpolarized particles in the initial state and upon summation over the polarizations of the particles in the final state this becomes

$$\mathrm{d}\sigma = \frac{2\alpha^2}{s^3\left(1-\dfrac{4m_e^2}{s}\right)^{1/2}}\frac{\mathrm{d}^3p_3\,\mathrm{d}^3p_4}{2p_3^0 2p_4^0}\delta(p_3+p_4-p_1-p_2)$$
$$\cdot\tfrac{1}{4}\,\mathrm{Tr}[\gamma^\lambda(\not{p}_4 - m_\mu)\gamma^\rho(\not{p}_3+m_\mu)]\cdot\mathrm{Tr}[\gamma_\lambda(\not{p}_1+m_e)\gamma_\rho(\not{p}_2-m_e)]. \tag{11.13}$$

We only consider the extreme relativistic case

$$s \gg m_\mu^2, m_e^2. \tag{11.14}$$

Then, using the expressions given in Appendix A for the traces, we obtain for the differential cross-section with respect to the solid angle $\mathrm{d}\Omega$ of the μ^- particle

$$\frac{\mathrm{d}\sigma}{\mathrm{d}\Omega} = \frac{\alpha^2}{2s^3}\int_0^\infty \mathrm{d}|\boldsymbol{p}_3||\boldsymbol{p}_3|\int\frac{\mathrm{d}^3p_4}{|\boldsymbol{p}_4|}\cdot\delta(\sqrt{s}-|\boldsymbol{p}_3|-|\boldsymbol{p}_4|)\delta^3(\boldsymbol{p}_3+\boldsymbol{p}_4)$$
$$\cdot 4(p_4^\lambda p_3^\rho + p_3^\lambda p_4^\rho - (p_3p_4)g^{\lambda\rho})\cdot(p_{1\lambda}p_{2\rho}+p_{2\lambda}p_{1\rho}-(p_1p_2)g_{\lambda\rho})$$
$$= \frac{2\alpha^2}{s^3}[(p_1p_4)(p_2p_3)+(p_2p_4)(p_1p_3)]. \tag{11.15}$$

In the high energy limit, i.e., for (11.14), we have

$$p_1p_3 = p_2p_4 = \frac{s}{4}(1-\cos\vartheta),$$
$$p_1p_4 = p_2p_3 = \frac{s}{4}(1+\cos\vartheta), \tag{11.16}$$

and with this we find the differential and total cross-sections:

$$\frac{d\sigma}{d\Omega} = \frac{\alpha^2}{4s}(1 + \cos^2 \vartheta), \tag{11.17}$$

$$\sigma_{tot}(e^-e^+ \to \mu^-\mu^+) = \int d\Omega \frac{d\sigma}{d\Omega} = \frac{4\pi\alpha^2}{3s}. \tag{11.18}$$

Equations (11.17), (11.18) are standard results for the production of a point-like, i.e., structureless, zero mass, spin $\frac{1}{2}$ particle and its antiparticle in electron–positron annihilation. We again find a scaling behavior: The dimensionless quantity $s \cdot d\sigma/d\Omega$ is independent of s. The angular distribution, proportional to $1 + \cos^2 \vartheta$ (11.7), is typical for produced particles of spin $\frac{1}{2}$. For comparison we may consider the production of two spin zero particles, for example, π-mesons:

$$e^- + e^+ \to \pi^- + \pi^+, \tag{11.19}$$

in which case one finds a distribution proportional to $\sin^2 \vartheta$:

$$\frac{d\sigma}{d\Omega}(e^- + e^+ \to \pi^- + \pi^+) \propto \sin^2 \vartheta. \tag{11.20}$$

The production of muon pairs has been studied in detail at electron–positron storage rings. The results provide good confirmation of the theoretical formulas. In Figs. 11.4 and 11.5 we show some results obtained by groups working at PETRA. Only at center of mass energies $\sqrt{s} \geq 30$ GeV

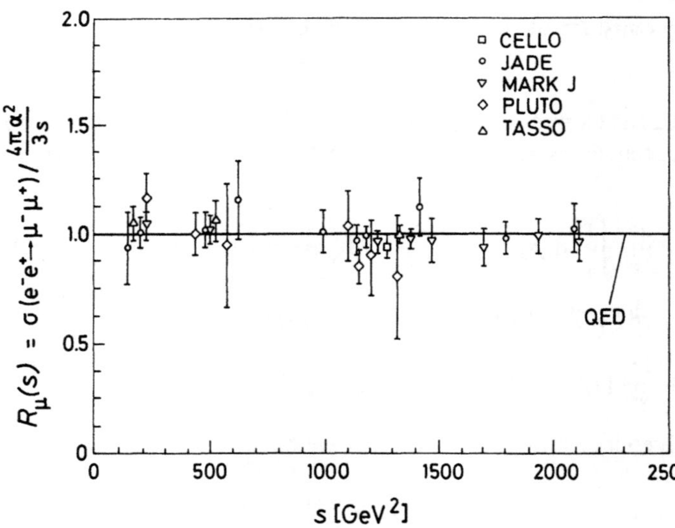

Figure 11.4 Data for the total cross-section of the reaction $e^-e^+ \to \mu^-\mu^+$ as a function of s, divided by the prediction (11.18) of QED. The experimental groups from which the data stem are indicated on the figure (after Haidt 1984).

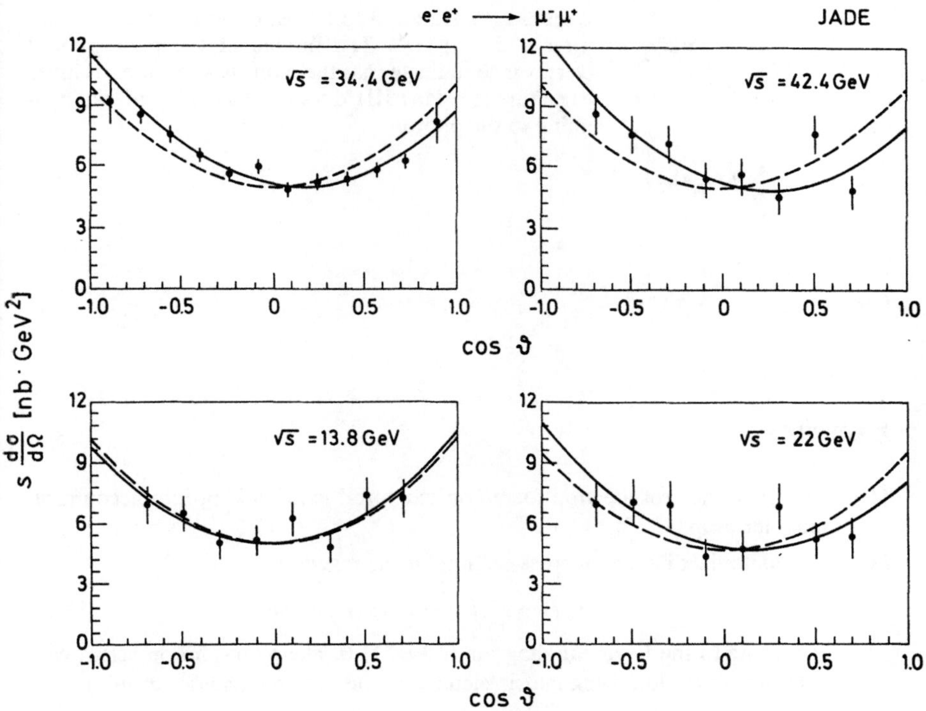

Figure 11.5 The angular distribution of muons in the reaction $e^-e^+ \to \mu^-\mu^+$. Data from the JADE group with center of mass energies $\sqrt{s}$ as indicated. The dashed lines are the predictions of QED (11.17). The solid lines are obtained when the effects of the weak interaction are included (after Bartel 1985).

were discrepancies found between the observed angular distribution of muons and the predictions (11.17) of QED (Bartel 1982; Brandelik 1982; Aveda 1982; Behrend 1982; Berger 1983; Fernandez 1983). Today this is regarded as a manifestation of the heavy Z-boson ($m_Z \sim 90$ GeV), which is one of the carriers of the weak interaction. This phenomenon will be treated in detail in Part IV.

The formulas (11.17), (11.18) have also turned out to be of fundamental importance for the theory of the strong interaction. In experimental studies of the production of hadrons in e^+e^- storage rings, the cross-section at high energies is found to be of the same order of magnitude, and to have the same s dependence, as that for muon pair production. The analysis of these events also provides strong evidence for an angular distribution proportional to $1 + \cos^2 \vartheta$. This and other observations are today interpreted as evidence that the primary process in hadron production is the creation of a pair of point-like spin $\frac{1}{2}$ particles, the so-called quarks. An extensive discussion of this will be given in Part III.

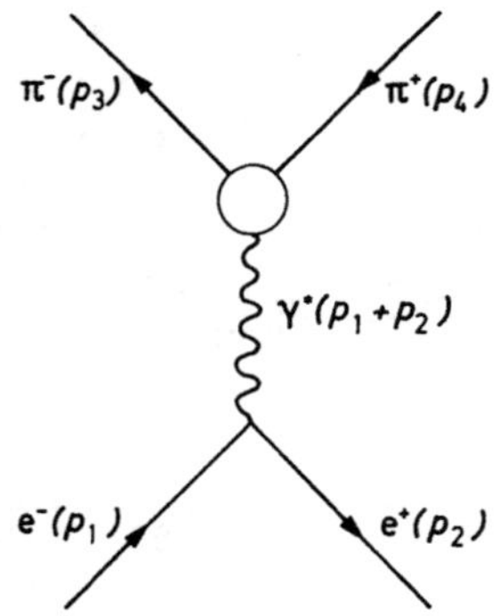

Figure 11.6 Diagram to lowest order in α for the reaction $e^-e^+ \to \pi^-\pi^+$. The bubble at the pion–photon vertex is to indicate that the pion, as a result of its inner structure (see Part III), does not have a point-like coupling to the photon.

Exercises

11.1 Calculate the cross-section for muon pair production taking into account all mass terms in (11.13).

11.2 Calculate the differential cross-section for the reaction

$$e^-(p_1) + e^+(p_2) \to \pi^-(p_3) + \pi^+(p_4) \tag{11.21}$$

according to the Feynman diagram in Fig. 11.6. The pion–photon vertex corresponds to the following matrix element of the electromagnetic current $\mathbf{j}_\mu$:

$$-\mathrm{i}\langle \pi^+(p_4)\pi^-(p_3)|\mathbf{j}_\mu(0)|0\rangle.$$

Show that, as a result of Lorentz invariance and current conservation, the following representation of this matrix element in terms of an initially undetermined function $F_\pi(q^2)$—the form factor of the pion—holds:

$$-\mathrm{i}\langle \pi^+(p_4)\pi^-(p_3)|\mathbf{j}_\mu(0)|0\rangle = ie(p_3 - p_4)_\mu F_\pi(q^2). \tag{11.22}$$

Here $q = p_3 + p_4$. Show that the known total charge of the pion leads to the condition $F_\pi(0) = 1$ (*Hint*: Apply crossing to bring one pion into the initial state.) If the pion were a point-like particle we would have $F_\pi(q^2) \equiv 1$. Compare the theoretical result for the differential cross-section with the experimental results of Quenzer (1978).

CHAPTER 12

External Fields

Until now, we have treated only reactions taking place in the vacuum. In practice, however, one is often interested in processes which occur in the presence of a given electric or magnetic field. The external field may be produced by some macroscopic device such as a capacitor or an electromagnet, but it could equally well be the field due to an atomic nucleus. Some typical problems of this nature are the following:

(i) The scattering of an electron by a given charge distribution, e.g., by a heavy nucleus whose recoil can be neglected.
(ii) The emission of synchrotron radiation by an electron circulating in a storage ring.
(iii) "Bremsstrahlung" radiation emitted by an electron as it is decelerated in the field of a nucleus. This process is used to produce X-rays in cathode-ray tubes.
(iv) The production of an electron–positron pair by a photon in the field of a heavy nucleus. This process led to the discovery of the positron by Anderson in 1932. Today it serves as an experimental means of detecting photons in high energy physics.

In this chapter we will describe the general methods that are used for solving such problems, and discuss a few examples. Our starting point is again the Hamiltonian for the electromagnetic interaction (9.9) which played a prominant role in the discussions in Chap. 9:

$$\mathbf{H}'(t) = \int d^3x\, \mathbf{j}^\mu(\boldsymbol{x}, t)\mathbf{A}_\mu(\boldsymbol{x}, t). \tag{12.1}$$

Here, however, the electromagnetic four-vector potential becomes a sum of the external potential A_μ^{ext} and the quantum field $\mathbf{A}'_\mu$:

$$\mathbf{A}_\mu = \mathbf{A}'_\mu + A_\mu^{\text{ext}}. \tag{12.2}$$

In the Dirac picture to be used here, we will take $\mathbf{A}'_\mu$ to be the free field operator known from Chap. 7. The external potential $A_\mu^{\text{ext}}(x)$ is to be calculated from the external charge and current distribution described by the current density j_μ^{ext}. If we again choose the Lorentz condition (7.8) then we have

$$\begin{aligned} \Box A_\mu^{\text{ext}}(x) &= j_\mu^{\text{ext}}(x), \\ \partial^\mu A_\mu^{\text{ext}}(x) &= 0. \end{aligned} \tag{12.3}$$

We will allow $A_\mu^{\text{ext}}(x)$ to vary with time, although in practice, the external fields are often time independent.

In order to discuss the influence of the external potential we again consider the "Schrödinger equation" in the Dirac picture (9.8). On inserting the expansion of the field operators in terms of annihilation and creation operators, we obtain, in analogy to (9.11), the following structure for the part of $\mathbf{H}'$ containing the external potential

$$\begin{aligned} \mathbf{H}' &\sim :(\mathbf{b} + \mathbf{a}^\dagger)(\mathbf{b}^\dagger + \mathbf{a}): A^{\text{ext}} \\ &\sim \underset{(1')}{-\mathbf{b}^\dagger\mathbf{b}A^{\text{ext}}} + \underset{(2')}{\mathbf{a}^\dagger\mathbf{b}^\dagger A^{\text{ext}}} + \underset{(3')}{\mathbf{b}\mathbf{a}A^{\text{ext}}} + \underset{(4')}{\mathbf{a}^\dagger\mathbf{a}A^{\text{ext}}}. \end{aligned} \tag{12.4}$$

The interpretation of these terms is as in Chap. 9 except that now, in place of the emission or absorption of a photon, we have the interaction with the external field. The individual terms in (12.4) describe:

(1′) Scattering of a positron in the external potential.
(2′) Creation of an electron–positron pair in the external potential.
(3′) Annihilation of an electron–positron pair in the external potential.
(4′) Scattering of an electron in the external potential.

We can again visualize these elementary processes by means of a single diagram (Fig. 12.1).

For the time evolution of states, or for S-matrix elements, we now have to take the elementary processes involving the external potential into account and must make corresponding additions to the Feynman rules. In the next section this will be discussed for a few examples and the rules for the general case will then be presented.

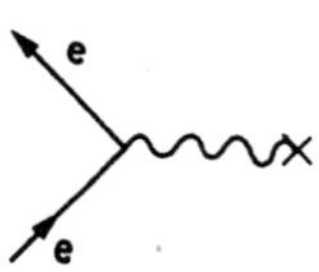

Figure 12.1 The diagram for the elementary interaction of an electron or positron with an external field.

Figure 12.2 Scattering of an electron by a nucleus.

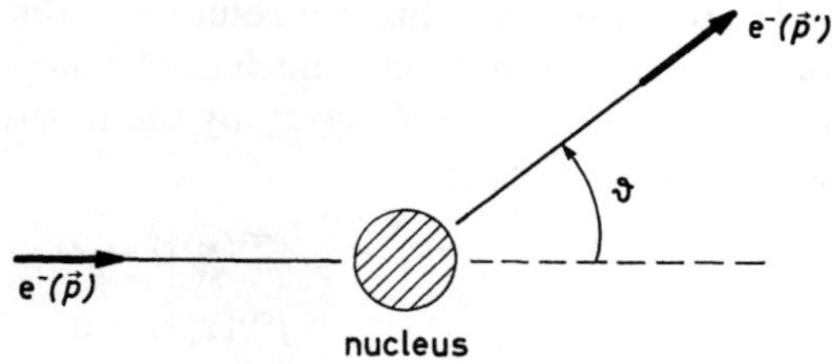

12.1 The Scattering of Electrons in an External Potential

We consider the scattering of an electron in an external potential. The potential might, for example, be that of a nucleus, which we will regard as fixed, i.e., whose recoil can be neglected (Fig. 12.2). The four-momenta of the incoming and outgoing electron will be denoted by p and p'.

The lowest order term that contributes to the S-matrix is obtained according to (9.14) as follows, noting that we are again only interested in actual scattering processes ($p \neq p'$) and that the spin indices have been suppressed:

$$\begin{aligned} \langle \mathrm{e}(p')|\,\mathbf{S}\,|\mathrm{e}(p)\rangle &= \left\langle \mathrm{e}(p') \middle| \left\{ 1 - \mathrm{i} \int_{-\infty}^{\infty} \mathrm{d}t\, \mathbf{H}'(t) \right\} \middle| \mathrm{e}(p) \right\rangle \\ &= \int \mathrm{d}t\, \mathrm{d}^3x\, A_\mu^{\mathrm{ext}}(\boldsymbol{x}, t) \cdot \langle \mathrm{e}(p')|\, :(\mathrm{i}e)\bar{\psi}(\boldsymbol{x}, t)\gamma^\mu \psi(\boldsymbol{x}, t):\, |\mathrm{e}(p)\rangle \\ &= \int \mathrm{d}t\, \mathrm{d}^3x\, A_\mu^{\mathrm{ext}}(\boldsymbol{x}, t)\, \mathrm{e}^{\mathrm{i}(p'^0 - p^0)t - \mathrm{i}(\boldsymbol{p}' - \boldsymbol{p})\boldsymbol{x}} \cdot \bar{u}(p')\mathrm{i}e\gamma^\mu u(p). \end{aligned} \tag{12.5}$$

The amplitude corresponds to the occurrence of a single elementary process in an external field. The relevant Feynman diagram is thus given directly by Fig. 12.1 with the electron lines interpreted as incoming and outgoing electrons. To the electron lines we will again assign analytical expressions as given in Appendix B. We then see from (12.5) that the factor

$$\mathrm{i}e\gamma^\mu \int \mathrm{d}x\, \mathrm{e}^{\mathrm{i}(p'-p)x} A_\mu^{\mathrm{ext}}(x) = -\mathrm{i}e\gamma^\mu \frac{1}{(p'-p)^2} \int \mathrm{d}x\, \mathrm{e}^{\mathrm{i}(p'-p)x} j_\mu^{\mathrm{ext}}(x) \tag{12.6}$$

corresponds to the interaction of the electron with the external field. Here we have used (12.3) in order to express everything in terms of the external current and charge distributions.

A generalization of this procedure readily leads us to the Feynman rules for problems involving external fields. One draws all possible diagrams with given incoming and outgoing lines where both types of vertices, representing interactions with the quantum field and with the external field, have to be taken into account. Every interaction with the external field corresponds to a factor as in (12.6). All other lines and vertices are translated into analytical language as before; only the δ-function (B.2) representing four-momentum conservation is no longer included.

We will now specialize our solution to the particular case of the scattering of an electron from a static nucleus of charge Ze (Fig. 12.2). For this case the external source is simply given by the time-independent charge distribution $Ze\rho(\boldsymbol{x})$ of the nucleus:

$$\begin{aligned} j_0^{\mathrm{ext}}(\boldsymbol{x}, t) &= Ze\rho(\boldsymbol{x}), \\ \boldsymbol{j}^{\mathrm{ext}}(\boldsymbol{x}, t) &= 0. \end{aligned} \tag{12.7}$$

This problem has already been treated in Sect. 5.1. We have now justified the form of the Hamiltonian for the interaction used in that section.

12.2 Bremsstrahlung

In general, when an electron comes near to a nucleus or an atom it is deflected. The laws of electrodynamics imply that even on the classical level this will cause the emission of electromagnetic radiation. We will consider the acceleration of an electron in the field of a nucleus and will treat the latter as an external field. This corresponds to the following reaction:

$$\mathrm{e}(p) + \text{Nucleus} \to \mathrm{e}(p') + \gamma(k) + \text{Nucleus}. \tag{12.8}$$

We will regard the nucleus as a point charge at the coordinate origin. The external potential is then simply the Coulomb potential of the nuclear charge Ze and we have

$$\begin{aligned} (j^{\mu}_{\mathrm{ext}}(\boldsymbol{x}, t)) &= \begin{pmatrix} Ze\delta^3(\boldsymbol{x}) \\ 0 \\ 0 \\ 0 \end{pmatrix}, \\ (A^{\mu}_{\mathrm{ext}}(\boldsymbol{x}, t)) &= \begin{pmatrix} \dfrac{Ze}{4\pi|\boldsymbol{x}|} \\ 0 \\ 0 \\ 0 \end{pmatrix}. \end{aligned} \tag{12.9}$$

According to QED, we obtain to lowest order the diagrams shown in Fig. 12.3; they represent the emission of a single photon by the electron. The evaluation of these diagrams leads to complicated expressions. We shall confine ourselves here to a few qualitative comments and will state a few results. Firstly, some remarks about the kinematics: The four-momenta are denoted as follows:

$$p = \begin{pmatrix} E \\ \boldsymbol{p} \end{pmatrix}, \qquad p' = \begin{pmatrix} E' \\ \boldsymbol{p}' \end{pmatrix}, \qquad k = \begin{pmatrix} \omega \\ \boldsymbol{k} \end{pmatrix}. \tag{12.10}$$

Figure 12.3 Lowest order Feynman diagrams for the emission of bremsstrahlung by an electron in the field of of a nucleus.

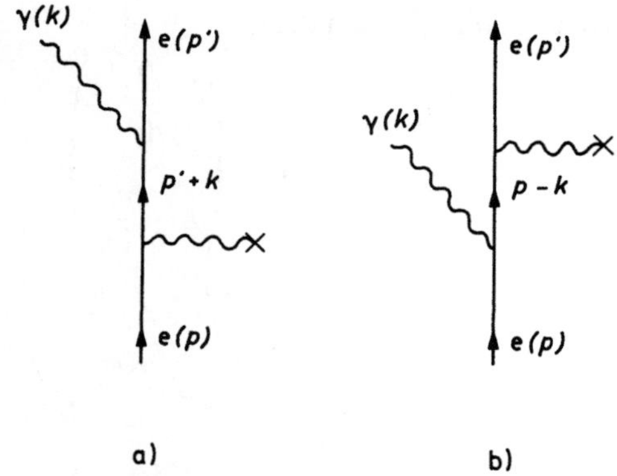

According to the Feynman rules in Appendix B, the interaction with the Coulomb field of the nucleus in the diagrams in Fig. 12.3, corresponds to a factor

$$\mathrm{i}e\gamma^0 \frac{1}{|\boldsymbol{q}|^2} \int \mathrm{d}t\, \mathrm{d}^3x\, \mathrm{e}^{\mathrm{i}(E'-E+\omega)t}\, \mathrm{e}^{\mathrm{i}\boldsymbol{q}\boldsymbol{x}}\, Ze\delta^3(\boldsymbol{x}) = \mathrm{i}e\gamma^0 \frac{Ze}{|\boldsymbol{q}|^2} 2\pi\delta(E' - E + \omega), \tag{12.11}$$

where $\boldsymbol{q} = \boldsymbol{p} - \boldsymbol{p}' - \boldsymbol{k}$. Since the nuclear field is static, energy must be conserved and this is expressed by the δ-function in (12.11). However, momentum is not conserved between incoming and outgoing particles. The stationary nucleus can absorb any amount of recoil momentum $\boldsymbol{q}$. We see from (12.11) that the amplitude is large for small recoil momenta. We are interested in high energies or, more exactly, in the kinematical region

$$\begin{aligned} &E, E', \omega \gg m, \\ &E', \omega = \mathrm{O}(E). \end{aligned} \tag{12.12}$$

We claim that in this case $|\boldsymbol{q}|$ can only be small when the photon and the electron in the final state are moving in almost the same direction as the electron in the initial state. To prove this, we start with energy conservation which implies, under the conditions of (12.12), that

$$|\boldsymbol{p}| \cong |\boldsymbol{p}'| + |\boldsymbol{k}|. \tag{12.13}$$

The condition

$$\boldsymbol{q}^2 = (\boldsymbol{p} - \boldsymbol{p}' - \boldsymbol{k})^2 \cong 0$$

can then only be satisfied for momenta $\boldsymbol{p}, \boldsymbol{p}'$, and $\boldsymbol{k}$ that are almost collinear.

Let us take a closer look at the momenta. We choose the direction of the momentum of the incoming electron as the z-axis. For high energies and for the emission of bremsstrahlung quanta close to the forward direction we have, to a good approximation,

$$\boldsymbol{p} = \begin{pmatrix} 0 \\ 0 \\ E - \dfrac{m^2}{2E} \end{pmatrix},$$

$$\boldsymbol{p}' = \begin{pmatrix} \boldsymbol{p}'_\perp \\ E' - \dfrac{m^2 + \boldsymbol{p}'^2_\perp}{2E'} \end{pmatrix},$$

$$\boldsymbol{k} = \begin{pmatrix} \boldsymbol{k}_\perp \\ \omega - \dfrac{\boldsymbol{k}^2_\perp}{2\omega} \end{pmatrix}, \tag{12.14}$$

$$\boldsymbol{q}^2 = (\boldsymbol{p}'_\perp + \boldsymbol{k}_\perp)^2 + \mathrm{O}\left(\left(\frac{m^2}{E}\right)^2\right). \tag{12.15}$$

Since the scattering cross-section is proportional to $1/(\boldsymbol{q}^2)^2$, the order of magnitude of the correction term in (12.15) gives us a measure of the width of the distribution in $(\boldsymbol{p}'_\perp + \boldsymbol{k}_\perp)^2$. The major contributions to the cross-section are expected for

$$|\boldsymbol{p}'_\perp + \boldsymbol{k}_\perp| \lesssim \mathrm{O}\left(\frac{m^2}{E}\right), \tag{12.16}$$

In other words, the sum of the transverse momenta of the electron and the photon is very small at high energies.

We now want to estimate the order of magnitude of the individual transverse momenta. For this we first consider the diagram of Fig. 12.3(b). The virtual electron line leads, according to Appendix B, to the following factor in the amplitude:

$$\begin{aligned} \mathrm{i}\frac{\not{p} - \not{k} + m}{(p-k)^2 - m^2} &= -\mathrm{i}\frac{\not{p} - \not{k} + m}{2pk} \\ &= -\mathrm{i}\frac{\not{p} - \not{k} + m}{2(E\omega - \boldsymbol{p}\cdot\boldsymbol{k})} \\ &\cong -\mathrm{i}\frac{\not{p} - \not{k} + m}{\dfrac{E}{\omega}\left(\boldsymbol{k}^2_\perp + \left(\dfrac{\omega}{E}\right)^2 m^2\right)}. \end{aligned} \tag{12.17}$$

For ω/E of the order of 1 (eq. (12.12)), we find from this that the major contribution to the reaction stems from

$$|\boldsymbol{k}_\perp| \lesssim \mathrm{O}(m). \tag{12.18}$$

On account of (12.16) we then also have

$$|\boldsymbol{p}'_\perp| \lesssim \mathrm{O}(m). \tag{12.19}$$

An analogous estimate for diagram (a) in Fig. 12.3 leads to the same result. The typical angle for bremsstrahlung emission from high energy electrons is thus of the order of

$$\vartheta \cong \frac{|\boldsymbol{k}_\perp|}{\omega} \lesssim \frac{m}{E}. \tag{12.20}$$

We can use this result to determine the order of magnitude of the scattering cross-section for the reaction (12.8). Generally speaking, a scattering cross-section can be imagined as a product of a geometrical factor and a probability factor, the opacity. The geometrical factor gives the area which the projectile must strike in order that anything should happen at all. Even if the geometrical conditions are satisfied, it is not necessarily true that a reaction will take place. As a classical example, we might think of the absorption of light by a grey glass sphere. Depending on the shade of grey, more or less of the incident light will be absorbed. A black sphere will absorb all the incident light; this corresponds to a probability factor of unity and a absorption cross-section which is thus equal to the geometrical cross-section.

In our case, with transverse momenta of the order of m, we can use the uncertainty principle to estimate a value of $1/m^2$ for the geometrical factor. The probability factor can be estimated from the coupling constants appearing in the diagrams of Fig. 12.3

$$\text{Probability factor} \propto (Ze^3)^2 \propto Z^2\alpha^3. \tag{12.21}$$

This is much less than 1 for all nuclei. With this we obtain the following estimate for the total scattering cross-section

$$\sigma_{\text{tot}} \cong \bar{\sigma} \equiv \frac{Z^2\alpha^3}{m^2}. \tag{12.22}$$

The quantity $\bar{\sigma}$ defined here proves in fact to be a suitable unit for bremsstrahlung and similar processes.

The explicit calculation gives the following (see Heitler 1970): In the limit of high energies for the incoming and outgoing electrons we find, from the diagrams in Fig. 12.3 for the differential cross-section with respect to the photon energy,

$$\mathrm{d}\sigma(\omega) = \tfrac{2}{3}\bar{\sigma}\frac{\mathrm{d}\omega}{\omega}\left[4 - 4\frac{\omega}{E} + 3\left(\frac{\omega}{E}\right)^2\right]\cdot\left[2\ln\frac{2E(E-\omega)}{m\omega} - 1\right]$$

$$(E \gg m,\ E' = E - \omega \gg m). \tag{12.23}$$

The differential cross-section is indeed of order $\bar{\sigma}$, but for $\omega \to 0$ it shows the characteristic bremsstrahlung behavior of being proportional to $\mathrm{d}\omega/\omega$. The total cross-section would appear to diverge! If we integrate over all frequencies above a minimum frequency ω_{min}, then we find:

$$\int_{\omega_{\text{min}}}^{E} \mathrm{d}\sigma(\omega) \propto \bar{\sigma}\cdot\ln\frac{E}{\omega_{\text{min}}} \to \infty \quad \text{for} \quad \omega_{\text{min}} \to 0. \tag{12.24}$$

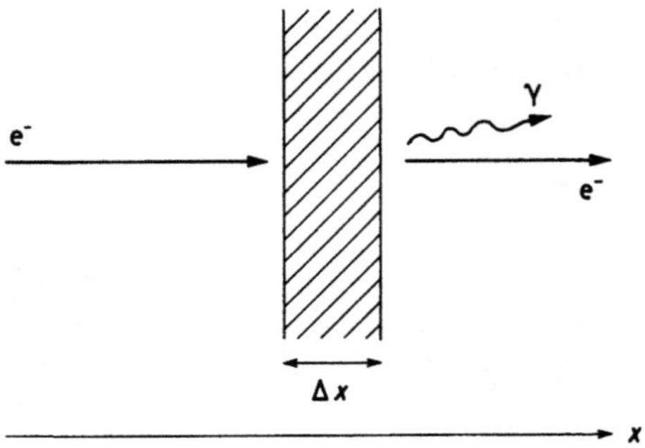

Figure 12.4 Schematic diagram for the measurement of the energy loss by electrons traveling through a piece of matter of thickness Δx.

We will return in detail to this infrared divergence in Chap. 14. Our estimate in (12.22) is, incidentally, not in conflict with (12.24) for which $\omega \gg m$ was assumed in (12.12).

Another important quantity is the cross-section for electron energy loss through bremsstrahlung. If a fast electron travels through matter is loses energy by collisions with bound electrons, which lead to excitation and ionization, and also by bremsstrahlung. For large enough energies the bremsstrahlung dominates (see Heitler 1970). In order to calculate the average energy lost through bremsstrahlung per unit length of path, we imagine the following experiment (Fig. 12.4): An electron beam of flux Φ and cross-sectional area F is incident vertically on a piece of material of thickness Δx. The number of atoms per unit volume in the target is N. According to the definition of scattering cross-section, the rate $\mathrm{d}\Gamma$ for the emission of a bremsstrahlung quantum with energy between ω and $\omega + \mathrm{d}\omega$ is given by

$$\mathrm{d}\Gamma(\omega) = \mathrm{d}\sigma(\omega)\Phi N F \Delta x .$$

In this we have taken into accout that there are $NF\Delta x$ scattering centers. In a time T the number of electrons entering the target is ΦFT. For the average energy radiated by an electron $-\Delta E$, which, of course, represents the energy loss of the electron, we thus find

$$-\Delta E = \frac{T\int_0^E \omega \, \mathrm{d}\Gamma(\omega)}{\Phi F T} = \int_0^E \omega \, \mathrm{d}\sigma(\omega) N \Delta x .$$

The decrease in the energy of a fast electron traveling through matter is thus described by the following differential equation where x is the penetration depth:

$$-\frac{\mathrm{d}E}{\mathrm{d}x} = N \int_0^E \omega \, \mathrm{d}\sigma(\omega). \tag{12.25}$$

As can be inferred from (12.23), the frequency distribution of the radiated photons is essentially constant:

$$\omega \frac{\mathrm{d}\sigma(\omega)}{\mathrm{d}\omega} \sim \text{const.} \qquad \text{for} \quad 0 \leq \omega \leq E. \tag{12.26}$$

From this we conclude that, on average, the electron loses its energy by radiating a few hard (energetic) photons, and that the integral in (12.25) is proportional to the primary energy E. It then seems reasonable to define a cross-section σ_{rad} for the energy loss by

$$\sigma_{\text{rad}}(E) = \frac{1}{E} \int_0^E \omega \, d\sigma(\omega). \tag{12.27}$$

With this we obtain

$$-\frac{dE}{dx} = N\sigma_{\text{rad}}(E) \cdot E. \tag{12.28}$$

The order of magnitude of σ_{rad} is again given by $\bar{\sigma}$ (12.22). The integration of (12.23) gives, in fact, for $E \gg m$,

$$\sigma_{\text{rad}} = 4\left(\ln \frac{2E}{m} - \frac{1}{3}\right)\bar{\sigma}. \tag{12.29}$$

These results are valid for bare nuclei. For real atoms, the influence of the electron clouds must be taken into account. These lead, on the one hand to a screening of the nuclear field and, on the other hand, they produce additional bremsstrahlung due to electron–electron collisions. In this case one finds, in the extreme relativistic regime, that σ_{rad} becomes independent of the energy E (Heitler 1970):

$$\sigma_{\text{rad}} = \{4 \ln(183 Z^{-1/3}) + \tfrac{2}{9}\}\bar{\sigma} \qquad \text{for} \quad E \gg m\alpha^{-1} Z^{-1/3}. \tag{12.30}$$

It is then easy to integrate (12.28) and one finds an exponential decrease of the electron energy as a function of penetration depth x. If the surface of the material is at $x = 0$, and the initial energy of the electron is E_0, then

$$E(x) = E_0 \, e^{-x/L_{\text{rad}}}, \tag{12.31}$$

where

$$L_{\text{rad}} = (N \cdot \sigma_{\text{rad}})^{-1}. \tag{12.32}$$

Here we have introduced the *radiation length* L_{rad} as the characteristic length for the energy loss. The radiation lengths of a few materials are listed in Table 12.1. In their calculation, a few corrections to (12.30) have been included (see Tsai (1974) and also Particle Data Group (1984)).

Table 12.1 The radiation lengths of some materials (after Particle Data Group 1986).

Material	Radiation Length L_{rad} [cm]
Fe	1.76
Pb	0.56
Air (at 20°C, 1 atm.)	30420
H_2O	36.1

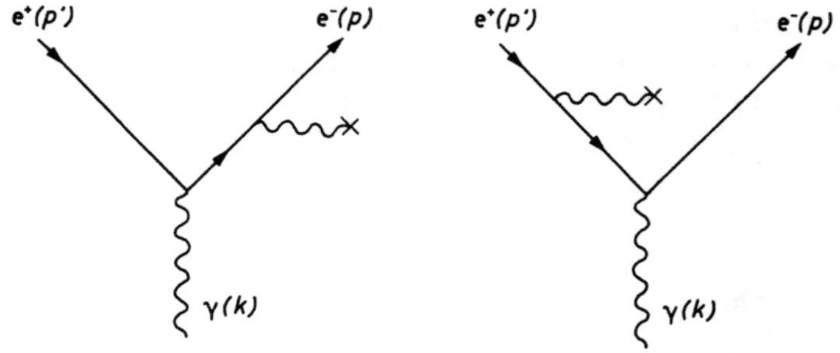

Figure 12.5 Lowest order Feynman diagrams for pair production by a photon in an external field.

12.3 Pair Creation (The Bethe–Heitler Process)

The last topic to be discussed in their chapter is the production of an electron–positron pair by a photon in the field of an atom or a nucleus (Bethe 1934; Racah 1934, 1936; see also Tsai 1974):

$$\gamma(k) + \text{nucleus} \rightarrow \mathrm{e}^-(p) + \mathrm{e}^+(p') + \text{nucleus}. \tag{12.33}$$

Clearly, this reaction can only take place if the photon energy ω is greater than or equal to $2m$:

$$\omega \geq 2m.$$

To lowest order we must again, as for bremsstrahlung, take two diagrams into account (Fig. 12.5). These two diagrams are obtained from the two bremsstrahlung diagrams (Fig. 12.3) by interchanging lines between the initial and final states, i.e., by applying a crossing transformation. The kinematic relations for the two reactions also turn out to be very similar. In particular, the order of magnitude of the scattering cross-section is again given by $\bar{\sigma}$ (12.22). More precisely, in the high energy limit, one finds for bare nuclei (see Heitler 1970):

$$\sigma_{\text{Pair}} = \bar{\sigma}\left(\frac{28}{9}\ln\frac{2\omega}{m} - \frac{218}{27}\right) \quad \text{for} \quad \omega \gg m. \tag{12.34}$$

For atoms with electron clouds we have

$$\sigma_{\text{Pair}} = \bar{\sigma}\left[\frac{28}{9}\ln(183 \cdot Z^{-1/3}) - \frac{2}{27}\right] \quad \text{for} \quad \omega \gg m\alpha^{-1}Z^{-1/3}. \tag{12.35}$$

If we compare this result with the cross-section for Compton scattering (10.51), which falls as $1/\omega$ at high energies, we see that above a certain energy, pair creation with its essentially constant cross-section becomes the dominant process.

The absorption of photons in matter is governed by a total of three effects, namely, the photo-electric effect, the Compton effect, and pair creation. The

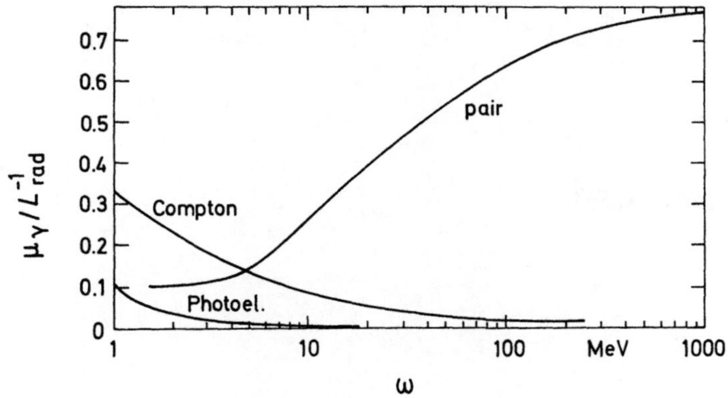

Figure 12.6 The individual contributions to the photon absorption coefficient μ_γ for lead, as a function of the energy ω. As unit the inverse radiation length $L_{\rm rad}^{-1}$ is taken (cf. Table 12.1) (after Particle Data Group 1980).

intensity of a photon beam passing through matter thus decreases with the penetration depth x according to the law

$$-\mathrm{d}I = \mu_\gamma I \,\mathrm{d}x, \tag{12.36}$$

where the absorption coefficient μ_γ is given by

$$\mu_\gamma = N(\sigma_{\rm Photoel.} + \sigma_{\rm Compton} + \sigma_{\rm Pair})$$

and represents the sum of the contributions from all three effects. Here N is again the number of atoms per cm^3 in the material. In Fig. 12.6 we show the individual contributions to the absorption coefficient for lead as a function of photon energy. For high energies, only pair creation is important. Comparing equations (12.35) and (12.30) we find

$$\mu_\gamma \cong N\sigma_{\rm Pair} \cong \frac{7}{9} N\sigma_{\rm rad} = \frac{7}{9}\frac{1}{L_{\rm rad}} \quad \text{for} \quad \omega \gg m\alpha^{-1}Z^{-1/3}. \tag{12.37}$$

Integration of (12.36) then gives an exponential fall-off for the photon intensity in the form

$$I(x) = I_0\,\mathrm{e}^{-(7/9)(x/L_{\rm rad})}. \tag{12.38}$$

Exercises

12.1 Calculate to lowest order the differential cross-section (12.23) for bremsstrahlung according to the diagrams in Fig. 12.3 in the limit of high energies E, $E' \gg m$. Verify (12.29).

12.2 Calculate the total cross-section for the Bethe–Heitler process from the diagrams in Fig. 12.5 for the high energy limit and verify (12.34).

CHAPTER 13

Positronium

13.1 The Spectrum of Positronium States and Their General Properties

An electron and a positron can form bound states in just the same way as an electron and a proton. The bound states of e^- and e^+ are known as *positronium* states. The spectrum of positronium states is similar to that of hydrogen since the Coulomb force is the same in both cases. The proton in hydrogen can, to a good approximation, be taken to be a fixed center around which the electron orbits. In positronium, however, both partners have the same mass. As always, this two-body problem corresponds mathematically to the motion of a single particle with the reduced mass about a fixed center. For positronium the reduced mass μ has the value

$$\mu = \frac{m_e^2}{m_e + m_e} = \tfrac{1}{2} m_e. \tag{13.1}$$

The Bohr radius r_b of positronium is twice as large as that of hydrogen (Fig. 13.1):

$$r_b = \frac{1}{\alpha\mu} = \frac{2}{\alpha m_e}. \tag{13.2}$$

As in hydrogen, the motion of the electron and positron in positronium is nonrelativistic. This can be seen as follows. An electron in the first Bohr orbit has orbital momentum 1. In terms of its velocity v we then have

$$m_e v r_b \cong \frac{v}{\alpha} \cong 1. \tag{13.3}$$

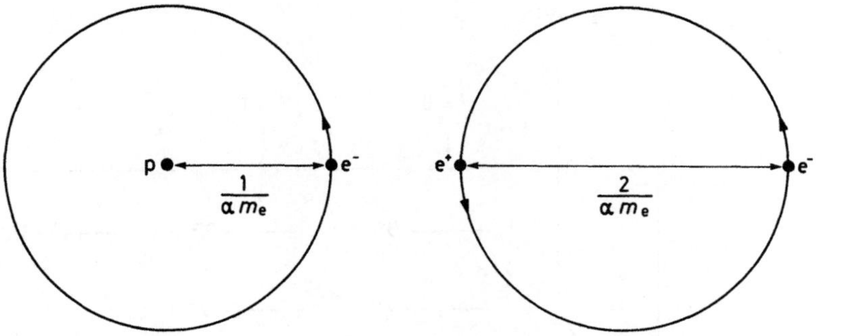

Figure 13.1 The hydrogen and the positronium systems, characterized by their first Bohr orbit.

Typical velocities are thus of the order of α and typical momenta p of the order of αm_e:

$$\begin{aligned} v &\cong \alpha \ll 1, \\ p &\cong \alpha m_e \ll m_e. \end{aligned} \tag{13.4}$$

The binding energy of positronium states can be calculated to a first approximation from the Schrödinger equation with the Coulomb potential. This gives the following Balmer formula;

$$E_n = \frac{\alpha^2 \mu}{2n^2} = -\frac{\alpha^2 m_e}{4n^2} = -\frac{Ry}{2n^2}, \tag{13.5}$$

where $Ry = 13.605\,804(36)$ eV is the Rydberg constant and $n = 1, 2, 3, \ldots$ is the principal quantum number. To each value n of the principal quantum number there correspond a number of degenerate levels with different orbital angular momentum quantum numbers l, l_z, namely $l = 0, 1, \ldots, n-1$ (S, P, D, etc. in spectroscopic notation) and $l_z = -l, \ldots, +l$ (Fig. 13.2).

Electrons and positrons both have spin $\frac{1}{2}$. Every level with a given n, l, l_z is therefore fourfold degenerate corresponding to the two possible values for the z-component of the spin for each of the particles. The two spin angular momenta can be added to give a total spin of $s = 0$ or 1. The corresponding basis vectors labeled by the value of the z-component of the total spin are, according to the addition rules for angular momentum,

$$\begin{aligned} |s=0, s_z=0\rangle &= \frac{1}{\sqrt{2}}(\overset{\uparrow}{e}{}^- \overset{\downarrow}{e}{}^+ - \overset{\downarrow}{e}{}^- \overset{\uparrow}{e}{}^+), \\ |s=1, s_z=1\rangle &= \overset{\uparrow}{e}{}^- \overset{\uparrow}{e}{}^+, \\ |s=1, s_z=0\rangle &= \frac{1}{\sqrt{2}}(\overset{\uparrow}{e}{}^- \overset{\downarrow}{e}{}^+ + \overset{\downarrow}{e}{}^- \overset{\uparrow}{e}{}^+), \\ |s=1, s_z=-1\rangle &= \overset{\downarrow}{e}{}^- \overset{\downarrow}{e}{}^+. \end{aligned} \tag{13.6}$$

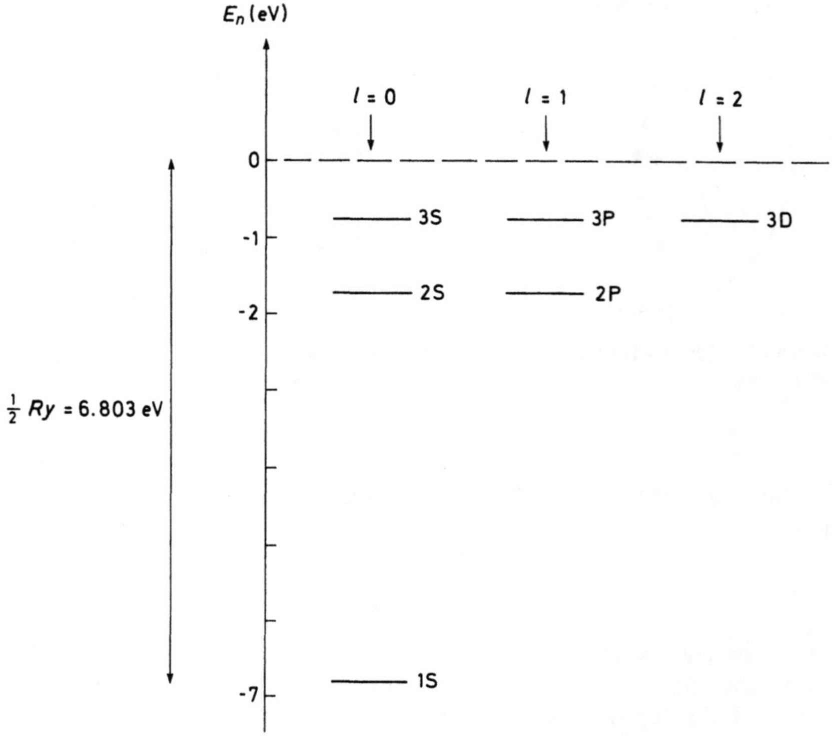

Figure 13.2 Energy levels of positronium for $n = 1, 2, 3$ in the nonrelativistic Coulomb approximation.

Here a notation has been used where the z-component of the electron and positron spins are indicated by arrows.

The degeneracy of the states with various values of the orbital angular momentum and total spin only holds in a nonrelativistic treatment with a pure Coulomb force acting between electron and positron. In reality, there are relativistic and higher order effects such as spin–angular momentum, and spin–spin interactions, and vacuum polarization corrections that lead to an energy splitting of the states. The resulting levels are labeled by the principal quantum number n, the orbital angular momentum l, the total spin s, and the total angular momentum j, according to the scheme

$$n^{2s+1}(l)_j. \tag{13.7}$$

States with $s = 0$ are called singlet states and those with $s = 1$ triplet states. The ground state ($n = 1$, $l = 0$) is split into a singlet state $1\,{}^1S_0$ with $j = 0$ known as parapositronium, and a triplet state $1\,{}^3S_1$ with $j = 1$ known as orthopositronium (Fig. 13.3).

We turn now to the behavior of positronium states under the operation of parity and charge conjugation. Upon reflection in space, a wavefunction with

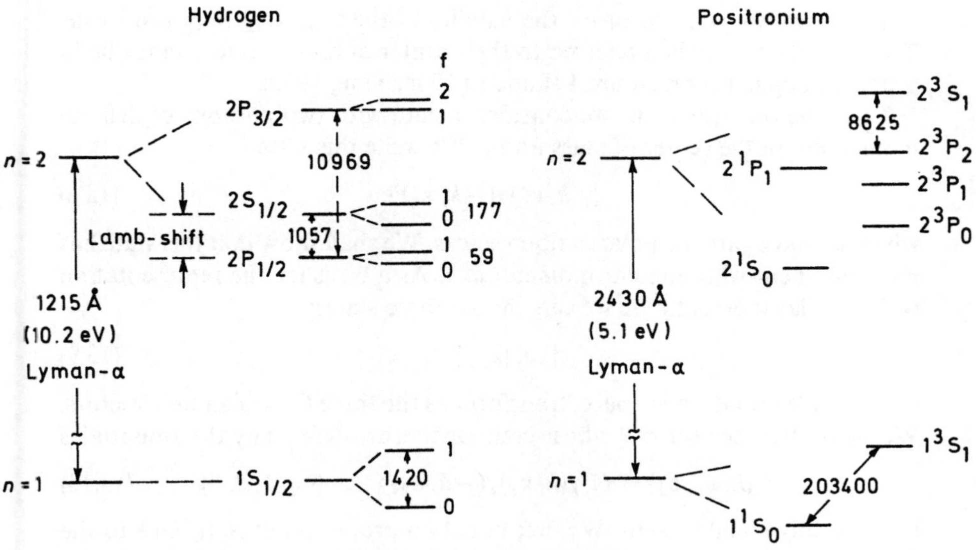

Figure 13.3 Scheme showing the splitting of the ground state ($n = 1$) and the first excited state ($n = 2$) of positronium and hydrogen (after Berko 1980). The energy splitting is given in MHz; 1 MHz $\cong 4.13 \cdot 10^{-9}$ eV. The splitting for the ground state of positronium is thus 203 400 MHz $\cong 8.4 \cdot 10^{-4}$ eV.

Table 13.1 Eigenvalues of parity P, charge-conjugation C, and of CP for the positronium states.

Positroniumsystem	P	C	CP
$n^{2s+1}(l)_j$	$(-1)^{l+1}$	$(-1)^{l+s}$	$(-1)^{s+1}$

orbital angular momentum l acquires the eigenvalue $(-1)^l$ and the spin state remains unchanged. As we saw in Chap. 4, electrons and positrons have negative relative parity. Taking this into account we obtain the parity eigenvalues given in Table 13.1. The eigenvalues of charge conjugation can be determined from (4.129). It is noteworthy that the singlet and triplet states are eigenstates of the product CP with eigenvalues -1 and $+1$, respectively.

13.2 The Decay of Positronium

The two ground states $1\,^1S_0$ and $1\,^3S_1$ are, of course, unstable since the electron and the positron can annihilate. Due to energy and momentum conservation, however, the annihilation into a single photon is impossible. In fact, the $1\,^1S_0$ state decays, almost without exception, into two photons. The $1\,^3S_1$ state cannot decay into two photons as we shall show shortly; it almost always decays into three photons. Before entering into further discussion of positronium, we want to prove the validity of the following important rule: Two real photons, when referred to their center of mass frame, cannot be in a state of angular momentum 1 (Landau 1948; Yang 1950).

To prove this statement we consider a state with two photons of definite momentum in the center of mass frame. We write this state

$$|\gamma(\boldsymbol{k}, \varepsilon_1)\gamma(-\boldsymbol{k}, \varepsilon_2)\rangle, \tag{13.8}$$

where ε_1 and ε_2 are the polarization vectors. We shall show that this state has no component with angular momentum 1. As a basis for the representation with angular momentum 1 we can choose three states:

$$|1\rangle, |2\rangle, |3\rangle, \tag{13.9}$$

which, under rotation in space, transform as the three Cartesian unit vectors. We claim that the vector $\boldsymbol{A}$ whose components are defined by the amplitudes

$$A^i(\boldsymbol{k}, \varepsilon_1, \varepsilon_2) = \langle i|\gamma(\boldsymbol{k}, \varepsilon_1)\gamma(-\boldsymbol{k}, \varepsilon_2)\rangle \qquad (i = 1, 2, 3) \tag{13.10}$$

is identically equal to zero. We first note two properties of $\boldsymbol{A}$. (i) Due to the *superposition principle* of quantum mechanics $\boldsymbol{A}$ must be a homogeneous linear function of ε_1 and ε_2. (ii) The Bose character of the photons demands the symmetry condition

$$\boldsymbol{A}(\boldsymbol{k}, \varepsilon_1, \varepsilon_2) = \boldsymbol{A}(-\boldsymbol{k}, \varepsilon_2, \varepsilon_1). \tag{13.11}$$

The possibilities for $\boldsymbol{A}$ compatible with (i) are easy to enumerate:

$$\begin{aligned}
&\text{(a)} \quad \boldsymbol{A} \propto (\varepsilon_1 \cdot \varepsilon_2)\boldsymbol{k},\\
&\text{(b)} \quad \boldsymbol{A} \propto \varepsilon_1 \times \varepsilon_2,\\
&\text{(c)} \quad \boldsymbol{A} \propto (\varepsilon_1 \times \varepsilon_2) \times \boldsymbol{k} = \varepsilon_2(\varepsilon_1 \cdot \boldsymbol{k}) - \varepsilon_1(\varepsilon_2 \cdot \boldsymbol{k}),\\
&\text{(d)} \quad \boldsymbol{A} \propto [(\varepsilon_1 \times \varepsilon_2)\cdot \boldsymbol{k}]\cdot \boldsymbol{k},\\
&\text{(e)} \quad \boldsymbol{A} \propto (\varepsilon_1 \cdot \boldsymbol{k})\varepsilon_2 + (\varepsilon_2 \cdot \boldsymbol{k})\varepsilon_1,\\
&\text{(f)} \quad \boldsymbol{A} \propto (\varepsilon_1 \times \boldsymbol{k})(\varepsilon_2 \cdot \boldsymbol{k}) + (\varepsilon_2 \times \boldsymbol{k})(\varepsilon_1 \cdot \boldsymbol{k}),\\
&\text{(g)} \quad \boldsymbol{A} \propto (\varepsilon_1 \cdot \boldsymbol{k})(\varepsilon_2 \cdot \boldsymbol{k})\boldsymbol{k}.
\end{aligned} \tag{13.12}$$

No further linearly independent vector $\boldsymbol{A}$ compatible with (i) can be formed as can be shown with the help of the elementary representation theory of the rotation group (cf. Exercise 13.2). We now note that the expressions (a), (b), (d), (e), and (g) are incompatible with Bose symmetry (13.11). Expressions (c), (e), (f), and (g) vanish due to the *transversality* condition (7.34) for real photons[1]

$$\varepsilon_1 \cdot \boldsymbol{k} = \varepsilon_2 \cdot \boldsymbol{k} = 0. \tag{13.13}$$

Thus we have shown that $\boldsymbol{A}$ is equal to zero and therewith the statement is proved. This proof only makes use of the superposition principle, rotational invariance, Bose symmetry, and the transversality of photons.

[1] For virtual photons the condition (13.13) does not hold in general.

From the Landau–Yang theorem it now follows immediately that the decay of orthopositronium, a state with total angular momentum $j = 1$, into two photons, is strictly forbidden.

$$1\,{}^3\mathrm{S}_1 \not\rightarrow 2\gamma . \tag{13.14}$$

Parapositronium, however, can decay into two photons

$$1\,{}^1\mathrm{S}_0 \rightarrow 2\gamma . \tag{13.15}$$

We would like to determine the rate for this process. Since the initial state is a bound state, this problem, at first sight, seems difficult to treat by the Feynman rules used in the preceeding chapters. However, to lowest order, it turns out to be very easy to do so. We regard the positronium as a wave packet composed of electron and positron plane waves. Then, with the help of the Feynman rules, we calculate the amplitude for the reaction

$$\mathrm{e}^-(p_1, r) + \mathrm{e}^+(p_2, s) \rightarrow \gamma(k_1, \varepsilon_1) + \gamma(k_2, \varepsilon_2) \tag{13.16}$$

for particles with well-defined momentum and polarization. The result is then convoluted with the wave packet of the positronium state.

In its rest frame, we can represent the positronium $1\,{}^1\mathrm{S}_0$ state as follows:

$$|1\,{}^1\mathrm{S}_0\rangle = \frac{1}{\sqrt{V}} \int \frac{\mathrm{d}^3 p}{(2\pi)^3 2p^0} f(\boldsymbol{p}) \cdot \frac{1}{\sqrt{2}} (\mathbf{a}^\dagger_{1/2}(\boldsymbol{p})\mathbf{b}^\dagger_{-1/2}(-\boldsymbol{p}) - \mathbf{a}^\dagger_{-1/2}(\boldsymbol{p})\mathbf{b}^\dagger_{1/2}(-\boldsymbol{p}))|0\rangle , \tag{13.17}$$

where V is a large normalization volume as used in Chap. 5. We introduce V in order to be able to normalize the state to 1. Application of Fermi's trick (5.23), (5.24) gives the normalization condition

$$\begin{aligned} \langle 1\,{}^1\mathrm{S}_0 | 1\,{}^1\mathrm{S}_0\rangle &= \frac{1}{V} \int \mathrm{d}^3 p\, \mathrm{d}^3 p'\, f^*(\boldsymbol{p}) f(\boldsymbol{p}') [\delta^3(\boldsymbol{p} - \boldsymbol{p}')]^2 \\ &= \int \frac{\mathrm{d}^3 p}{(2\pi)^3} |f(\boldsymbol{p})|^2 = 1 . \end{aligned} \tag{13.18}$$

The function $f(\boldsymbol{p})$ is the wave function in momentum space, i.e., the Fourier transform of the usual Schrödinger function $\psi(\boldsymbol{x})$ with respect to the relative coordinate $\boldsymbol{x}$ of e^- and e^+ [see (3.71), (3.33)].

These functions obey the following relations:

$$\begin{aligned} \psi(\boldsymbol{x}) &= \int \frac{\mathrm{d}^3 p}{(2\pi)^3} \mathrm{e}^{\mathrm{i}\boldsymbol{p}\boldsymbol{x}} f(\boldsymbol{p}), \\ f(\boldsymbol{p}) &= \int \mathrm{d}3x\, \mathrm{e}^{-\mathrm{i}\boldsymbol{p}\boldsymbol{x}}\, \varphi(\boldsymbol{x}), \\ \int \mathrm{d}^3 x\, |\psi(\boldsymbol{x})|^2 &= \int \frac{\mathrm{d}^3 p}{(2\pi)^3} |f(\boldsymbol{p})|^2 = 1 . \end{aligned} \tag{13.19}$$

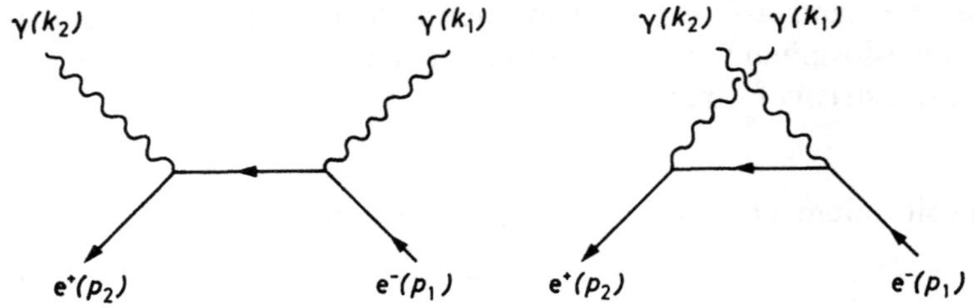

Figure 13.4 Lowest order diagrams for the annihilation of an electron–positron pair into two photons.

In our case, $\psi(\boldsymbol{x})$ is the well-known ground state wave function of a hydrogen-like system

$$\psi(\boldsymbol{x}) = \frac{1}{\sqrt{\pi r_b^{3/2}}} e^{-|\boldsymbol{x}|/r_b}. \tag{13.20}$$

In order to determine the amplitude for the reaction (13.16) we draw the appropriate Feynman diagrams of lowest order (Fig. 13.4). According to the rules in Appendix B, the S-matrix element is

$$\begin{aligned}
&\langle\gamma(k_1, \varepsilon_1)\gamma(k_2, \varepsilon_2)|\,\mathbf{S}\,|e^-(p_1, r)e^+(p_2, s)\rangle \\
&= (2\pi)^4\delta(k_1 + k_2 - p_1 - p_2)\cdot\varepsilon_1^{*\mu}\varepsilon_2^{*\nu}\cdot\bar{v}_s(p_2) \\
&\cdot\left\{(\mathrm{i}e\gamma_\nu)\frac{\mathrm{i}}{\not{p}_1 - \not{k}_1 - m_e}(\mathrm{i}e\gamma_\mu) + (\mathrm{i}e\gamma_\mu)\frac{\mathrm{i}}{\not{p}_1 - \not{k}_2 - m_e}(\mathrm{i}e\gamma_\nu)\right\}u_r(p_1).
\end{aligned} \tag{13.21}$$

The electron and the positron in positronium have very small linear momenta. Thus we will expand the S-matrix element (13.21) in terms of these momenta and only retain the leading term which corresponds to an annihilation at rest. Then, in the S-matrix element, we can substitute

$$p_1 \cong p_2 \cong \begin{pmatrix} m_e \\ 0 \\ 0 \\ 0 \end{pmatrix}, \tag{13.22}$$

$$\begin{aligned}
u_r(p_1) &\cong \sqrt{2m_e}\begin{pmatrix} \chi_r \\ 0 \end{pmatrix}, \\
v_s(p_2) &\cong -\sqrt{2m_e}\begin{pmatrix} 0 \\ \varepsilon\chi_s \end{pmatrix},
\end{aligned} \tag{13.23}$$

where χ_r and χ_s are the Pauli spinors for the electron and positron at rest

(4.30), (4.41). With these approximations we obtain

$$\begin{aligned}
&\langle\gamma(k_1,\varepsilon_1)\gamma(k_2,\varepsilon_2)|\mathbf{S}|e^-(p_1,r)e^+(p_2,s)\rangle\\
&\quad\cong \mathrm{i}e^2(2\pi)^4\delta(k_1^0+k_2^0-2m_e)\delta^3(\boldsymbol{k}_1+\boldsymbol{k}_2)\\
&\quad\cdot\bar{v}_s(p_2)\left\{\not\varepsilon_2^*\frac{\not p_1-\not k_1+m_e}{2p_1k_1}\not\varepsilon_1^*+\not\varepsilon_1^*\frac{\not p_1-\not k_2+m_e}{2p_1k_2}\not\varepsilon_2^*\right\}u_r(p_1)\\
&\quad\cong -\mathrm{i}e^2(2\pi)^4\delta(k_1^0+k_2^0-2m_e)\delta^3(\boldsymbol{k}_1+\boldsymbol{k}_2)\\
&\quad\cdot\frac{1}{2m_e^2}\bar{v}_s(p_2)\left\{\not\varepsilon_2^*\not\mathbf{k}_1\not\varepsilon_1^*+\not\varepsilon_1^*\not\mathbf{k}\not\varepsilon_2^*\right\}u_r(p_1).
\end{aligned}\tag{13.24}$$

We now use the explicit representation of the γ-matrices from (4.19) and carry out the matrix multiplication. In doing so, we note the conditions

$$\begin{aligned}
\boldsymbol{k}_1 &= -\boldsymbol{k}_2,\\
\boldsymbol{\varepsilon}_1\cdot\boldsymbol{k}_1 &= \boldsymbol{\varepsilon}_2\cdot\boldsymbol{k}_2 = 0.
\end{aligned}\tag{13.25}$$

Using the notation with 2×2 submatrices this gives:

$$\begin{aligned}
&\not\varepsilon_2^*\not\mathbf{k}\not\varepsilon_1^*+\not\varepsilon_1^*\not\mathbf{k}\not\varepsilon_2^*\\
&\quad=(\boldsymbol{\varepsilon}_1^*\cdot\boldsymbol{\varepsilon}_2^*)2k_1^0\left(\begin{array}{c|c}1&0\\\hline 0&-1\end{array}\right)+2\mathrm{i}(\boldsymbol{\varepsilon}_1^*\times\boldsymbol{\varepsilon}_2^*)\cdot\boldsymbol{k}_1\left(\begin{array}{c|c}0&1\\\hline -1&0\end{array}\right).
\end{aligned}\tag{13.26}$$

Inserting this result into (13.24) gives for the S-matrix element

$$\begin{aligned}
&\langle\gamma(k_1,\varepsilon_1)\gamma(k_2,\varepsilon_2)|\mathbf{S}|e^-(p_1,r)e^+(p_2,s)\rangle\\
&\quad\cong\frac{4\alpha}{m_e}(2\pi)^5\delta(k_1^0+k_2^0-2m_e)\delta^3(\boldsymbol{k}_1+\boldsymbol{k}_2)(\boldsymbol{\varepsilon}_1^*\times\boldsymbol{\varepsilon}_2^*)\cdot\boldsymbol{k}_1\varepsilon_{sr},
\end{aligned}\tag{13.27}$$

where (ε_{sr}) is the antisymmetric 2×2 matrix as in (4.41).

We now form the convolution with the positronium wavefunction (13.17) and obtain

$$\begin{aligned}
&\langle\gamma(k_1,\varepsilon_1)\gamma(k_2,\varepsilon_2)|\mathbf{S}|1\,{}^1\mathrm{S}_0\rangle\\
&\quad=\frac{1}{\sqrt{V}}\int\frac{d^3p}{(2\pi)^3 2p^0}f(\boldsymbol{p})\frac{1}{\sqrt{2}}\\
&\quad\cdot\{\langle\gamma(k_1,\varepsilon_1)\gamma(k_2,\varepsilon_2)|\mathbf{S}|e^-(\boldsymbol{p},+\tfrac{1}{2})e^+(-\boldsymbol{p},-\tfrac{1}{2})\rangle\\
&\quad-\langle\gamma(k_1,\varepsilon_1)\gamma(k_2,\varepsilon_2)|\mathbf{S}|e^-(\boldsymbol{p},-\tfrac{1}{2})e^+(-\boldsymbol{p},+\tfrac{1}{2})\rangle\}\\
&\quad\cong-\sqrt{\frac{2}{V}}\frac{2\alpha}{m_e^2}(2\pi)^2\delta(k_1^0+k_2^0-2m_e)\delta^3(\boldsymbol{k}_1+\boldsymbol{k}_2)\\
&\quad\cdot(\boldsymbol{\varepsilon}_1^*\times\boldsymbol{\varepsilon}_2^*)\cdot\boldsymbol{k}_1\int d^3p\,f(\boldsymbol{p}).
\end{aligned}\tag{13.28}$$

We have made use of the fact that in our approximation the S-matrix element

(13.27) is not dependent on $\boldsymbol{p}$ and thus can be factored out of the integral. The integral over the wavefunction in momentum space leads, according to (13.19) to the spatial wavefunction at the origin and we find

$$\langle \gamma(k_1, \varepsilon_1)\gamma(k_2, \varepsilon_2)|\mathbf{S}|1\,{}^1\mathrm{S}_0\rangle$$
$$= -\sqrt{\frac{2}{V}}\frac{2\alpha}{m_\mathrm{e}^2}(2\pi)^5\delta(k_1^0 + k_2^0 - 2m_\mathrm{e})\delta^3(\boldsymbol{k}_1 + \boldsymbol{k}_2)(\varepsilon_1^* \times \varepsilon_2^*)\cdot \boldsymbol{k}_1\psi(0). \quad (13.29)$$

From this it is now easy to calculate the decay rate, i.e., the probability of decay per unit time. We again use Fermi's trick from Chap. 5 and find

$$\Gamma(1\,{}^1\mathrm{S}_0 \to 2\gamma) = \frac{1}{2}\int \mathrm{d}^3k_1\,\mathrm{d}^3k_2\,\delta(k_1^0 + k_2^0 - 2m_\mathrm{e})\delta^3(\boldsymbol{k}_1 + \boldsymbol{k}_2)$$
$$\cdot\frac{2\alpha^2}{m_\mathrm{e}^6}\sum_{\mathrm{Spins}}|(\varepsilon_1^* \times \varepsilon_2^*)\cdot \boldsymbol{k}_1|^2|\psi(0)|^2. \quad (13.30)$$

The two photons in the final state are indistinguishable. If in the integral (13.30) we allow $\boldsymbol{k}_1$ and $\boldsymbol{k}_2$ to vary independently of one another over the whole of momentum space, then we have counted every photon state twice in the sum over final states. This is corrected in (13.30) by an explicit factor of $\frac{1}{2}$ in front of the integral. When carrying out the spin sum we must pay attention to the transversality condition for the photons (13.25). If we sum over linearly polarized photon states and take the z-axis to be along $\boldsymbol{k}_1$, then ε_1 and ε_2 can point independently of one another in the x- and y-directions. From these four spin states, however, only two result in a nonzero matrix element. Denoting the unit vectors in the x- and y-directions by $\boldsymbol{e}_x$ and $\boldsymbol{e}_y$, it therefore follows that

$$\sum_{\mathrm{Spins}}|(\varepsilon_1^* \times \varepsilon_2^*)\cdot \boldsymbol{k}_1|^2 = |(\boldsymbol{e}_x \times \boldsymbol{e}_y)\cdot \boldsymbol{k}_1|^2 + |(\boldsymbol{e}_y \times \boldsymbol{e}_x)\cdot \boldsymbol{k}_1|^2$$
$$= 2|\boldsymbol{k}_1|^2. \quad (13.31)$$

Carrying out the integral in (13.30) finally gives

$$\Gamma(1\,{}^1\mathrm{S}_0 \to 2\gamma) = \frac{4\pi\alpha^2}{m_\mathrm{e}^2}|\psi(0)|^2. \quad (13.32)$$

In obtaining this result, we have not used the fact that in positronium the electron and the positron are bound together by the Coulomb force. Equation (13.32) is valid for any form of the potential.

We would like to give our result a physical interpretation. With probability α the electron (or the positron) emits a γ-quantum of energy $\sim m_\mathrm{e}$. In doing so the electron (positron) receives a recoil momentum of order m_e. According to the uncertainty principle, such a fluctuation in the energy can only last for a time $1/m_\mathrm{e}$. The photon can thus become a real photon if, within the time $1/m_\mathrm{e}$, the electron meets the positron. In a time $1/m_\mathrm{e}$ the electron can travel a distance $\lesssim 1/m_\mathrm{e}$ which is much smaller than the "atomic radius" r_b of positronium. The annihilation can thus only proceed when e^- and e^+ are both found in a volume of size $(1/m_e)^3$. The probability of this is

$$|\psi(0)|^2\left(\frac{1}{m_e}\right)^3. \tag{13.33}$$

Finally, we have to estimate the probability that e^- and e^+, when they are sufficiently close together, annihilate into a photon within a time $1/m_e$. This gives another factor of α since a further elementary process must occur. Taking all these factors into account, we obtain for the transition rate (= transition probability/lifetime of fluctuations) the result

$$\Gamma \propto \alpha^2|\psi(0)|^2\left(\frac{1}{m_e}\right)^3 \bigg/ \frac{1}{m_e} \tag{13.34}$$

in agreement with (13.32).

We now insert the Coulomb wavefunction at the origin from (13.20) into (13.32). For the decay rate and the mean lifetime we then obtain

$$\Gamma(1\,{}^1S_0 \to 2\gamma) = \tfrac{1}{2}\alpha^5 m_e \cong 8.03\cdot 10^9\ \mathrm{s}^{-1},$$

$$\tau_{1\,{}^1S_0\to 2\gamma} = \Gamma^{-1}(1\,{}^1S_0 \to 2\gamma) \cong 1.24\cdot 10^{-10}\ \mathrm{s}. \tag{13.35}$$

This agrees very well with the experimentally observed value (Theriot 1970):

$$\Gamma_{\mathrm{exp}}(1\,{}^1S_0 \to 2\gamma) = 7.99(11)\cdot 10^9\ \mathrm{s}^{-1}. \tag{13.36}$$

The $1\,{}^3S_1$ state can only decay into three or more photons. Every additional photon leads to an extra factor of α in the decay rate. In fact, the explicit calculation (Ore 1949) gives

$$\Gamma(1\,{}^3S_1 \to 3\gamma) = \tfrac{16}{9}(\pi^2-9)\frac{\alpha^3}{m_e^2}|\psi(0)|^2. \tag{13.37}$$

Inserting the Coulomb wavefunction at the origin we have

$$\Gamma(1\,{}^3S_1 \to 3\gamma) = \frac{2(\pi^2-9)}{9\pi}\alpha^6 m_e \cong 7.21\cdot 10^6\ \mathrm{s}^{-1},$$

$$\tau_{1\,{}^3S_1\to 3\gamma} \cong 1.39\cdot 10^{-7}\ \mathrm{s}. \tag{13.38}$$

These values are slightly modified when higher order terms are taken into account. One then obtains the following value for the total decay rate of orthopositronium (Rich 1981):

$$\Gamma(1\,{}^3S_1)_{\mathrm{theor.}} = (7.0386 \pm 0.00016)\cdot 10^6\ \mathrm{s}^{-1}, \tag{13.39}$$

with which the average experimental value (Rich 1981),

$$\Gamma(1\,{}^3S_1)_{\mathrm{exp.}} = (7.050 \pm 0.004)\cdot 10^6\ \mathrm{s}^{-1}, \tag{13.40}$$

agrees very well.

We return now to the S-matrix element for the transition $1\,{}^1S_0 \to 2\gamma$ (13.29). From this we deduce that the two decay photons are in a state of negative parity, $P = -1$. Indeed, their wavefunction is proportional to the pseudoscalar quantity

$$(\boldsymbol{\varepsilon}_1 \times \boldsymbol{\varepsilon}_2) \cdot \boldsymbol{k}_1.$$

Even if we were not familiar with the Dirac equation, we could observe the correlation of the polarization vectors of the two photons and thereby determine the parity of the $1\,^1\mathrm{S}_0$ positronium state (cf. Table 13.1). We must, of course, assume that parity is conserved in the decay process. Such considerations are of great importance when it comes to the determination of the intrinsic quantum numbers of new particles.

When we come to discuss the strong interaction we shall meet bound states that behave in a very similar manner to positronium. We will also find processes analogous to the decay into photons. Instead of electrons and positrons we shall be dealing with quarks and antiquarks, and in place of the photons we will have the gluons.

Exercises

13.1 Calculate to lowest order the decay rate of the orthopositronium state $1\,^3\mathrm{S}_1$ into three photons.

13.2 Use the theory of the representation of the rotation group to show that any amplitude $A = A(\boldsymbol{k}, \boldsymbol{\varepsilon}_1, \boldsymbol{\varepsilon}_2)$ which is a homogeneous linear function of $\boldsymbol{\varepsilon}_1$ and $\boldsymbol{\varepsilon}_2$ is a linear combination of the amplitudes (a)–(g) in (13.12) where the coefficients are functions of $|\boldsymbol{k}|$.

CHAPTER 14

Radiative Corrections

In the previous chapters we have calculated a whole series of reactions in QED to lowest order. The results were always finite and agreed well with experiment. We know, however, that, in principle the theory includes higher order terms and that for a precise comparison with experiment, these too must be taken into account. After the development of QED at the end of the 1920s, it quickly became clear that in calculations of higher order terms, the so-called radiative corrections, one arrived at infinite results. It required much effort by many physicists in order to come to grips with these infinities and to find a systematic procedure for calculating higher order contributions with a finite result. This procedure is known as renormalization. Important original contributions on this subject are collected in Schwinger (1958). Here we will not go into the details of renormalization theory but will illustrate the problems and methods of solution by means of a few examples.

14.1 Radiative Corrections to the Scattering in an External Potential

In Sect. 12.1 we discussed the scattering of an electron in an external potential. The lowest order diagram for this is given in Fig. 12.1. To include the leading radiative corrections we must take into account the diagrams shown in Fig. 14.1. In these we note the appearance of loops.

Let us take the diagram of Fig. 14.1(c) as an example. According to the rules in Appendix B, we have to integrate over the loop momentum l. Schematically,

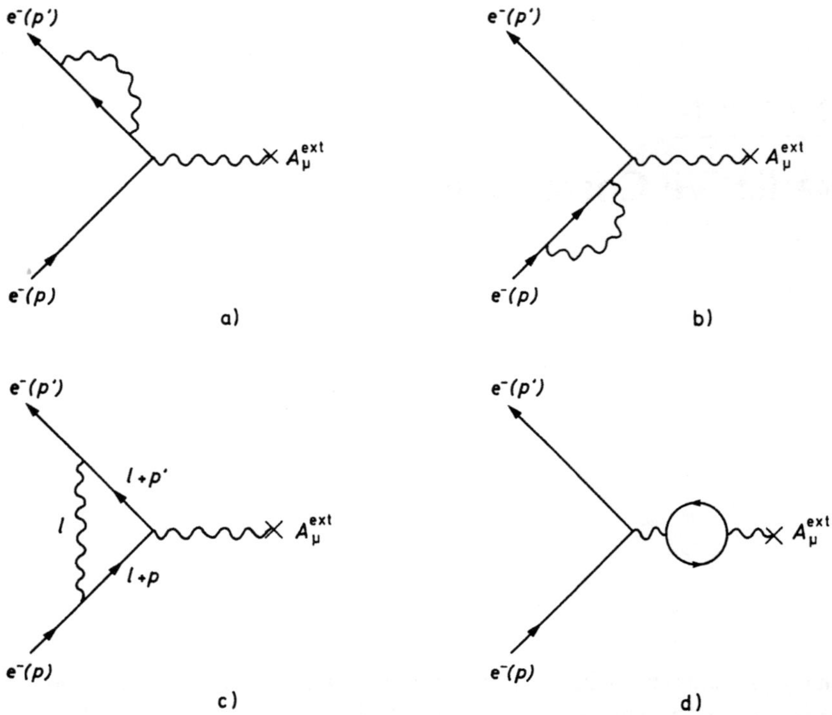

Figure 14.1 Feynman diagrams for the order α radiative corrections to the scattering of an electron in an external potential.

we obtain for this amplitude:

$$A^{(c)} \sim \int d^4 l \frac{1}{l^2} \frac{\not{p} + \not{l} + m_e}{p^2 + 2pl + l^2 - m_e^2} \frac{\not{p}' - \not{l} + m_e}{p'^2 - 2p'l + l^2 - m_e^2}. \tag{14.1}$$

We see immediately that for large momentum l this integral diverges logarithmically. This is the *ultraviolet catastrophe*. The integral also diverges for small loop momentum ($l \to 0$). We have

$$p^2 = p'^2 = m_e^2 \tag{14.2}$$

and as a result of this it follows that

$$A^{(c)} \sim \int d^4 l \frac{1}{l^2} \frac{1}{pl} \frac{1}{p'l} \qquad \text{for} \quad l \to 0. \tag{14.3}$$

This is known as the *infrared catastrophe*.

We first discuss the infrared catastrophe which comes from soft photons. To solve this we take a typically quantum mechanical approach. We ask ourselves what we have really observed when we say that a detector has registered an electron. Consider, for example, a wire chamber in a magnetic field; with the help of this one can determine the momentum, and thus also the energy, of an electron. Even the best detector, however, has a finite energy resolution ΔE. Thus we cannot decide experimentally whether we have observed an isolated electron or whether the electron was accompanied by a very soft photon of energy $\omega \leq \Delta E$. In other words, the experimental result is not the scattering cross-section for an isolated electron in the final state, but the sum of the cross-sections for the following final states:

$$\begin{aligned}\sigma_{\text{exp}} &= \sigma\,(1\text{ electron})\\ &\quad +\sigma\,(1\text{ electron} + 1\text{ photon of energy }\omega \leq \Delta E)\\ &\quad +\sigma\,(1\text{ electron} + 2\text{ photons of total energy} \leq \Delta E) + \ldots . \end{aligned} \tag{14.4}$$

The emission of a photon by an electron in an external field was discussed in Sect. 12.2. In evaluating the diagrams in Fig. 12.3 we also find an infrared divergence. In analogy to (12.24), with an infrared cutoff parameter $\omega_{\min}$ for the photon energy, we find

$$\sigma\,(1\text{ electron} + 1\text{ photon with energy }\omega,\text{ where }\omega_{\min} \leq \omega \leq \Delta E)$$
$$\propto \bar{\sigma} \ln \frac{\Delta E}{\omega_{\min}} \to \infty \qquad \text{for} \quad \omega_{\min} \to 0. \tag{14.5}$$

We must demand that the theory yields finite results for observable quantities. This, however, is already the case when the scattering cross-section, summed over the final states of (14.4) is finite. As can be seen by explicit calculation to the order of α considered here, the infrared divergences in σ(1 electron) which stem from the diagrams in Fig. 14.1, cancel with the infrared divergences in the bremsstrahlung cross-section (14.5). In order to avoid problems from dealing in intermediate steps with infrared divergent integrals, one introduces a small photon mass which can be set to zero in the final result. This final result naturally contains the energy resolution ΔE of the apparatus.

For a further discussion of the infrared problem, we refer the reader to the original literature (Bloch 1937; Yennie 1961).

The ultraviolet catastrophe is considerably more difficult to overcome; it is dealt with by the renormalization theory. The basic ideas are the following: First, with divergent integrals, no progress can be made. Somehow the integrals must be made finite—they must be *regularized*. Various regularization procedures have been suggested. For example, one can truncate all divergent integrals by means of a cutoff parameter Λ:

$$\int dl \frac{1}{l^2} \frac{1}{\not{p} + \not{l} - m} \frac{1}{\not{p}' - \not{l} - m} \to \int_{|l| \leq \Lambda} dl \frac{1}{l^2} \frac{1}{\not{p} + \not{l} - m} \frac{1}{\not{p}' - \not{l} - m}. \tag{14.6}$$

The results which are obtained are then logarithmically dependent on Λ. This simple cutoff procedure does not prove to be particularly useful in QED. Better cutoff procedures are the Pauli–Villars regularization and the dimensional regularization. We will not go into further details of these here.

After the regularization, we have a theory giving finite results for all quantities. The parameters of the theory are

$$e_0, m_0, \Lambda,$$

where e_0 and m_0 represent the charge and mass parameters of the electron as they appear in the Hamiltonian and in the Feynman rules that are derived from this. The starting point of renormalization theory is the observation that these parameters e_0 and m_0 are *not identical to the observed charge* e and mass m of the electron.

How can we measure the charge of an electron? We can measure, for example, the force experienced by a slow electron in an external magnetic field $\boldsymbol{B}$. This is given by the Lorentz formula

$$\boldsymbol{K} = (-e)(\boldsymbol{v} \times \boldsymbol{B}). \tag{14.7}$$

In the framework of QED the problem to be treated is the scattering of an electron in an external potential. If we make a calculation using the regularized theory, then we do indeed find a force of the form (14.7), but with $e \neq e_0$. With constants a_1, a_2, etc., the calculation gives

$$e = e_0\left(1 + a_1 e_0^2 \ln\frac{\Lambda}{m_0} + a_2 e_0^4\left(\ln\frac{\Lambda}{m_0}\right)^2 + \dots\right). \tag{14.8}$$

Analogously, for the observable mass m we have, with constants b_1, b_2, etc.:

$$m = m_0\left(1 + b_1 e_0^2 \ln\frac{\Lambda}{m_0} + b_2 e_0^4\left(\ln\frac{\Lambda}{m_0}\right)^2 \dots\right). \tag{14.9}$$

In the limit $\Lambda \to \infty$, e_0, m_0 fixed the observable quantities e and m would appear to diverge; clearly this is a senseless result.

The way out of this situation is found by noting that e_0 and m_0 are purely mathematical quantities because they have no meaning outside the applied regularization procedure. It is therefore more sensible to take the limit $\Lambda \to \infty$ by holding the observable quantities e and m constant and varying the mathematical parameters e_0 and m_0 with Λ. The original quantities e_0 and m_0 will then diverge as $\Lambda \to \infty$ but this is no serious problem since they are not observable.

In its individual steps the procedure is then the following: We wish to calculate a certain transition amplitude A. Our technique allows us to calculate A in the regularized theory as a function F of the mathematical, or "bare" parameters e_0 and m_0 and of the cutoff parameter Λ,

$$A = F(e_0, m_0, \Lambda, \dots), \tag{14.10}$$

where the dots stand for external momenta and polarizations. By inverting equations (14.8), (14.9) we can express the bare parameters in terms of the physical quantities:

$$\begin{aligned} e_0 &= e_0(e, m, \Lambda), \\ m_0 &= m_0(e, m, \Lambda). \end{aligned} \tag{14.11}$$

If this is inserted in (14.10), we then obtain the amplitude as a function of the parameters e, m, and Λ:

$$A = F(e_0(e, m, \Lambda), m_0(e, m, \Lambda), \Lambda, \ldots).$$

Now we allow Λ to go to infinity and hold e and m constant. The theory of renormalization asserts that this limit exists for all orders of the expansion in terms of e after an appropriate scale factor Z which does not depend on the external momenta[1] has been factored out from A. We define the renormalized amplitude A' as a function F' of the physical parameters e, m and the external variables by

$$\begin{aligned} A' &= \lim_{\Lambda\to\infty} Z(e, m, \Lambda)\cdot F(e_0(e, m, \Lambda), m_0(e, m, \Lambda), \Lambda, \ldots) \\ &= F'(e, m, \ldots). \end{aligned} \tag{14.12}$$

The renormalization theory yields an expansion of the renormalized amplitudes in powers of e:

$$F'(e, m, \ldots) = F'_0(m, \ldots) + eF'_1(m, \ldots) + e^2F'_2(m, \ldots) + \ldots, \tag{14.13}$$

where the expansion coefficients F'_i are finite for all i.

No statement is made concerning the convergence of this infinite series. In fact, all signs indicate that this power series is not convergent but merely asymptotic.

The exact mathematical execution of the ideas sketched above is rather complicated and thus we will not go into further details. The renormalized form of the theory has, until now, received outstanding experimental confirmation. In the introduction we have already discussed the anomalous magnetic moment of the electron. In the next section we will give a short description of yet another precision test of QED, namely the Lamb shift.

14.2 The Lamb Shift

This term refers to a small shift in the energy levels of atoms, which, in the case of hydrogen-like systems, has been calculated and measured to great precision. The bound states of the hydrogen atom can be determined in the nonrelativistic approximation from the Schrödinger equation for a Coulomb potential. The energy levels are given by the Balmer formula:

$$E_n = -\frac{\alpha^2 m_e}{2n^2}, \qquad n = 1, 2, 3, \ldots. \tag{14.14}$$

In this approximation we also have that states with the same principal quantum number n, but different orbital momentum quantum numbers l, are

[1] The factors Z are generally a product of the so-called wavefunction renormalization constants. A consequence of this renormalization is, for example, that diagrams of the type Fig. 14.1(a), (b) need not be taken into account in the calculation of the renormalized amplitude since their contributions are already incorporated into the Z factors.

degenerate in energy. In particular, The 2S-state is degenerate with the 2P-state.

With the discovery of the Dirac equation, it became possible to account exactly for the spin of the electron and all relativistic effects. Indeed, for the Dirac equation of an electron in a Coulomb field with potential energy

$$V(r) = -\frac{e^2}{4\pi r} \tag{14.15}$$

an exact analytical solution was found (Darwin 1928; Gordon 1928). The energy levels for $l \neq 0$ are split due to the spin–orbit interaction into levels with total angular momentum $j = l \pm \frac{1}{2}$. But even in these calculations, states such as $2S_{1/2}$ and $2P_{1/2}$ remain degenerate.

Experimentally however, it was found that the $2S_{1/2}$-level lies very slightly higher than the $2P_{1/2}$-level (Lamb 1947). The first successful theoretical calculation of this effect was that of Bethe (1947). This energy shift is a typical effect of the virtual photons. Here we will not go through all the stages of the argument, but will present an order of magnitude estimate of the effect.

An electron, even when it is left alone, is constantly emitting and reabsorbing virtual photons (Fig. 6.2). If a photon of energy $\omega \lesssim m$ is emitted, then the electron receives a recoil momentum of magnitude ω, corresponding to a velocity

$$v \cong \frac{\omega}{m}. \tag{14.16}$$

According to the uncertainty principle however, such a state can only exist for a time $1/\omega$. In this time the electron can travel a distance

$$|\Delta \boldsymbol{x}| = v\frac{1}{\omega} \sim \frac{1}{m}. \tag{14.17}$$

The photons of course, are emitted quite randomly in all directions. Thus for the average value we have

$$\langle \Delta \boldsymbol{x} \rangle = 0. \tag{14.18}$$

However, the mean square displacement does not vanish. Since the fluctuations occur with a probability α and because the individual coordinate directions are independent, we expect:

$$\begin{aligned} \langle \Delta x^i \Delta x^j \rangle &= \tfrac{1}{3}\delta_{ij}\langle (\Delta \boldsymbol{x})^2 \rangle, \\ \langle (\Delta \boldsymbol{x})^2 \rangle &\cong \alpha (1/m)^2. \end{aligned} \tag{14.19}$$

If the electron is located in a potential, then classically, it will deviate slightly around the position of its expected trajectory due to the emission and reabsorption of virtual photons (Fig. 14.2).

The electron then sees not only the potential on its expected trajectory, but experiences the average value over a small region about this trajectory. Expanding the potential energy about a point x on the trajectory we have

$$V(\boldsymbol{x} + \Delta \boldsymbol{x}) = V(\boldsymbol{x}) + \Delta x^i \nabla_i V(\boldsymbol{x}) + \tfrac{1}{2}\Delta x^i \Delta x^j \nabla_i \nabla_j V(\boldsymbol{x}) + O((\Delta x)^3) \tag{14.20}$$

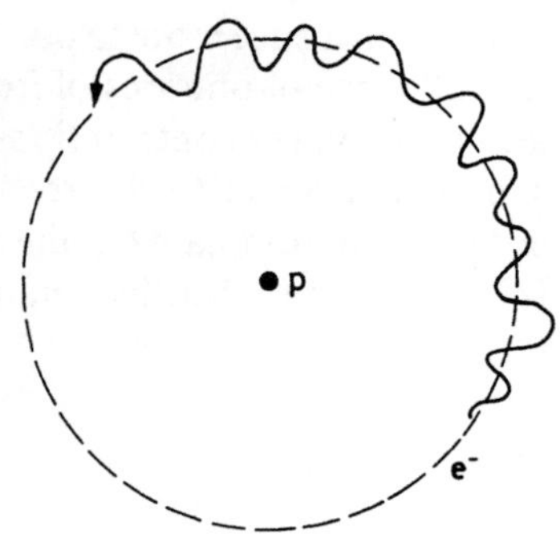

Figure 14.2 Fluctuation in the position of an electron about its expected orbit in a hydrogen atom. The fluctuation is caused by the emission and reabsorption of virtual photons.

and for the average value over the fluctuations

$$\begin{aligned}\langle V(\boldsymbol{x}+\Delta\boldsymbol{x})\rangle &= V(\boldsymbol{x}) + \tfrac{1}{2}\langle \Delta x^i \Delta x^j\rangle \nabla_i \nabla_j V(\boldsymbol{x}) \\ &= V(\boldsymbol{x}) + \tfrac{1}{6}\langle (\Delta\boldsymbol{x})^2\rangle (\boldsymbol{\nabla})^2 V(\boldsymbol{x}). \end{aligned} \tag{14.21}$$

For the energy shift ΔE of a quantum state with wavefunction $\psi(\boldsymbol{x})$ we thus expect

$$\Delta E \cong \tfrac{1}{6}\langle (\Delta\boldsymbol{x})^2\rangle \int \mathrm{d}^3x\, |\psi(\boldsymbol{x})|^2\, \Delta V(\boldsymbol{x}). \tag{14.22}$$

For an electron in a Coulomb field we have from (14.15):

$$\Delta V(\boldsymbol{x}) = e^2 \delta^3(\boldsymbol{x}) = 4\pi\alpha\delta^3(\boldsymbol{x}). \tag{14.23}$$

and for the energy shift this gives

$$\begin{aligned}\Delta E &\cong \tfrac{1}{6}\langle (\Delta\boldsymbol{x})^2\rangle 4\pi\alpha |\psi(0)|^2 \\ &\cong \frac{4\pi\alpha^2}{6m^2}|\psi(0)|^2. \end{aligned} \tag{14.24}$$

In the Coulomb approximation, only the S-states have a wavefunction which is nonzero at the origin. Thus for these states we expect an energy shift, and indeed a shift to higher energies. Inserting the value of the 2S-wavefunction at the origin we obtain

$$\Delta E_{2S} \cong \tfrac{1}{12}\alpha^5 m \cong 2000\ \mathrm{MHz}. \tag{14.25}$$

As was stated previously, according to the Dirac theory, the levels $2S_{1/2}$ and $2P_{1/2}$ are degenerate. The above arguments show that within QED this degeneracy should be lifted since from (14.25) the $2S_{1/2}$-level should be pushed upwards whereas the $2P_{1/2}$-level suffers no shift. The energy difference $2P_{1/2} - 2S_{1/2}$ is a pure QED effect and therefore a sensitive test for the concept of virtual photons.

In experiments one finds, in fact, that the $2S_{1/2}$-level lies approximately 1057 MHz higher than the $2P_{1/2}$-level (Fig. 13.3). This agrees in sign and in order of magnitude with our estimate (14.25). For an extensive comparison of theory and experiment we refer, for example, to Drell (1979) and Kugel (1977).

We return once more to our estimate (14.25). In this we have only considered the influence of photons of frequency $\omega \lesssim m$. The higher frequencies $\omega \gtrsim m$ lead to divergent contributions. It was Bethe who recognized that these terms simply contribute to the renormalization of the mass of the electron. This enabled him to obtain a finite result. Historically, this was a milestone on the road to a comprehensive renormalization theory.

PART III

The Strong Interaction

CHAPTER 15

Historical Overview

The physics of the strong interaction developed in an explosive manner during the seventies. The peak of this development came with the construction of quantum chromodynamics (QCD), which lays claim to being the fundamental theory in this area of physics. In this part of the book we will attempt to trace the historical path which led to QCD. Despite the existence of some impressive comparisons between experimental data and QCD, it should be stressed that this theory is not nearly so well confirmed as, for example, QED.

The strong interaction describes forces which act only between hadrons, i.e., between protons, neutrons, pions, kaons, etc., and which are not directly felt by leptons (e, μ, ...). A list of all the presenty known hadrons has been given in Tables 1.1 and 1.2. A selection of the hadrons from this list, with which we shall be concerned in detail, is presented in Table 15.1.

The strong interaction is responsible for, among other things, the nuclear forces. One "strong" reaction, namely the radioactive α-decay, has been known for a long time. The real history of the strong interaction began, however, with an experiment to measure the magnetic moment of the proton by Stern, Estermann, and Frisch in 1932–3 (Frisch 1933; Estermann 1933). To understand the significance of this experiment we have to recapitulate on the views that were held at that time about the structure of matter.

Up until 1932 the only known elementary particles were the photon γ, the electron e, and the proton p. Following the development of quantum mechanics and the discovery of the Dirac equation, it seemed as if the photon and the electron were understood. Leading physicists such as Max Born were

Table 15.1 List of low lying hadrons with their spin and parity quantum numbers and their masses rounded to the nearest MeV.

Hadrons	(Spin)^{parity}	Mass in MeV
$\pi^\pm$	0^-	140
π^0	0^-	135
$K^\pm$	0^-	494
$K^0, \bar{K}^0$	0^-	498
η	0^-	549
η'	0^-	958
$D^\pm$	0^-	1869
$D^0, \bar{D}^0$	0^-	1865
$F^\pm$	0^-	1971
$B^\pm$	0^-	5271
$B^0, \bar{B}^0$	0^-	5274
$\rho^\pm, \rho^0$	1^-	770
ω	1^-	783
$K^{*\pm}, K^{*0}, \bar{K}^{*0}$	1^-	892
φ	1^-	1020
J/ψ	1^-	3097
γ	1^-	9460
p	$\frac{1}{2}^+$	938
n	$\frac{1}{2}^+$	940
Λ	$\frac{1}{2}^+$	1116
Σ^+	$\frac{1}{2}^+$	1189
Σ^0	$\frac{1}{2}^+$	1192
Σ^-	$\frac{1}{2}^+$	1197
Ξ^0	$\frac{1}{2}^+$	1315
Ξ^-	$\frac{1}{2}^+$	1321
Ω^-	$\frac{3}{2}^+$	1672

of the opinion that the proton too would soon be understood, and that physics would then be complete (Rabi 1975).

One of the triumphs of the Dirac theory was its ability to account for the magnetic moment of the electron. In natural units ($\hbar = c = 1$) the result was (cf. (8.31))

$$\boldsymbol{\mu}_e = -\frac{e}{2m_e}\boldsymbol{\sigma}. \tag{15.1}$$

At that time this agreed with the measured value to within the experimental error.

It was then expected that the proton would also follow the Dirac theory and should thus have a magnetic moment of

$$\boldsymbol{\mu}_p = \frac{e}{2m_p}\boldsymbol{\sigma}, \tag{15.2}$$

where m_p is the proton mass. Since the proton is 1836 times heavier than the electron, its magnetic moment should thus be 1836 times smaller which makes it much more difficult to measure. The quantity $e/2m_{\mathrm{p}}$ is referred to as a nuclear magneton.

The experiments of O. Stern and co-workers were of the same type as the Stern–Gerlach experiment but employing a molecular hydrogen beam. The result was not the expected one: The measured magnetic moment of the proton was not equal to a nuclear magneton but instead approximately 2.5 nuclear magnetons. The current precision value is

$$\boldsymbol{\mu}_{\mathrm{p}} = 2.792\,845\,6(11)\frac{e}{2m_{\mathrm{p}}}\boldsymbol{\sigma}. \tag{15.3}$$

With this observation it had become evident that the proton possesses a structure for which a new interaction had to be responsible. The two forces known at that time, gravitation and electromagnetism, were then supplemented by the strong interaction and slightly later, with Fermi's theory of β-decay, by the weak interaction.

In the following we cite a few major developments in the investigation of the strong interaction:

1932 Discovery of the neutron by Chadwick (Chadwick 1932).
1935 Prediction by Yukawa of the π-meson as carrier of the nuclear force (Yukawa 1935).

Yukawa's argument was approximately as follows: The wave equation of electrodynamics has the form (7.9):

$$\Box A_\mu(x) = j_\mu(x), \tag{15.4}$$

where A_μ is the vector potential and j_μ is the current density four-vector. This equation describes, on the one hand, the free photons as a solution to the homogeneous equation,

$$\Box A_\mu(x) = 0. \tag{15.5}$$

and on the other hand, it yields the Coulomb potential as the solution to the inhomogeneous equation for a static point source. For

$$(j^\mu(\boldsymbol{x}, t)) = \begin{pmatrix} e\delta^3(\boldsymbol{x}) \\ 0 \end{pmatrix}$$

the solution of (15.4) reads

$$\begin{aligned} A^0(\boldsymbol{x}, t) &= \frac{e}{4\pi|\boldsymbol{x}|}, \\ A^j(\boldsymbol{x}, t) &= 0. \qquad (j = 1, 2, 3). \end{aligned} \tag{15.6}$$

Yukawa now assumed that the nuclear force too, is described by a wave field $U(x)$ which should satisfy a wave equation. Practically the only possible generalization of (15.4) is that offered by the Klein–Gordon equation in which the important difference is the additional mass term. This then led to the ansatz

$$(\Box + m^2)U(x) = g\rho(x), \tag{15.7}$$

where $\rho(x)$ is a kind of nucleon density distribution and g is a coupling constant. For a static point source, i.e., for

$$\rho(\boldsymbol{x}, t) = \delta^3(\boldsymbol{x}),$$

this equation has the Yukawa potential as its solution

$$U(\boldsymbol{x}, t) = g\frac{e^{-m|\boldsymbol{x}|}}{4\pi|\boldsymbol{x}|}. \tag{15.8}$$

The homogeneous equation

$$(\Box + m^2)U(x) = 0 \tag{15.9}$$

is solved by plane waves (Chap. 3)

$$U(x) = e^{-ikx}, \tag{15.10}$$

where $k^2 = m^2$. These describe the motion of free particles of mass m. Yukawa formulated the following important relationship:

> The Compton wavelength of the quanta of the field = 1/mass = range of the force field in the static case.

From the binding energy of the deuteron and from neutron–proton scattering data, Yukawa was able to estimate the Compton wavelength and therefore the mass of his quanta

$$\begin{aligned} 1/m &\approx 10^{-12}\text{–}10^{-13}\ \text{cm}, \\ m &\approx 20\text{–}200\ \text{MeV}. \end{aligned} \tag{15.11}$$

In fact, the mass of the Yukawa particles, the π-mesons, is approximately 140 MeV and the radius of the hadrons is of the order of 10^{-13} cm.

1947 Discovery of the charged π-mesons in cosmic rays (Lattes 1947).
1947 Observation of long-lived new particles (V-particles) in cosmic rays (Rochester 1947). This was the first indication of the existence of what we now call "strange" particles (K-mesons, Λ-hyperons, etc.) The earlier name V-particles stemmed from the characteristic topology of the tracks which were produced in cloud chambers and bubble chambers by the decay products of the K^0- and the Λ-particle.
1950 Discovery of the neutral π-meson (Carlson 1950; Bjorkland 1950).

The existence of a neutral partner to the charged meson considered by Yukawa was predicted by Kemmer as early as 1938 (Kemmer 1938) in his generalization of the concept of isospin which Heisenberg had introduced in 1932 (Heisenberg 1932). At this point we will take the opportunity to discuss isospin in more detail.

If we ignore mass differences of the order of 1–5 MeV, then protons and neutrons, as well as the π^+-, π^--, and π^0-mesons, all have the same mass—they are degenerate states. In analogy to the hydrogen atom, in which degeneracy of states that differ only in their magnetic quantum number is attributable to the rotational invariance, it is tempting in this case too, to attribute the almost equal masses of the elementary particles to an invariance group. This invariance group, which, in analogy to spin, is called the isospin, has no intuitive interpretation—instead it is defined as an abstract transformation group in the space of state vectors. A proton and a neutron, both in the same state of motion and the same spin state, can be taken as basis vectors $|\mathrm{p}\rangle$ and $|\mathrm{n}\rangle$ of a two-dimensional vector space. We define an isospin rotation as an SU(2)-transformation in this abstract space:

$$\begin{aligned} |\mathrm{p}\rangle &\to |\mathrm{p}\rangle U_{11} + |\mathrm{n}\rangle U_{21}, \\ |\mathrm{n}\rangle &\to |\mathrm{p}\rangle U_{12} + |\mathrm{n}\rangle U_{22}. \end{aligned} \tag{15.12}$$

Here

$$\mathbf{U} = \begin{pmatrix} U_{11} & U_{12} \\ U_{21} & U_{22} \end{pmatrix} \tag{15.13}$$

is a unitary 2×2 matrix with determinant 1. Thus we have

$$\begin{aligned} \mathbf{U}\mathbf{U}^\dagger &= 1, \\ \det \mathbf{U} &= 1. \end{aligned} \tag{15.14}$$

The transformation of protons and neutrons under the isospin rotation is exactly like the transformation of the states of an electron corresponding to a value $\pm\frac{1}{2}$ for the z-component of the spin under a rotation in space. The proton and the neutron form an isodoublet. To see this in detail it is useful, as in the case of spatial rotations, to consider infinitesimal isospin rotations and to introduce the generators of the isospin group which correspond to the angular momentum operators. For our proton–neutron doublet, an infinitesimal isospin transformation has the form

$$\mathbf{U} = 1 + \mathrm{i} \sum_{a=1}^{3} \delta\varphi_a \frac{\tau_a}{2}. \tag{15.15}$$

Here the $\delta\varphi_a$ are infinitesimal parameters and the τ_a (with $a = 1, 2, 3$) are the Pauli spin matrices (4.3) which, when they refer to isospin, are denoted by τ. Here the generators are

$$\mathbf{I}_a = \tfrac{1}{2}\tau_a \qquad (a = 1, 2, 3). \tag{15.16}$$

From (15.12) follow relations analogous to (4.107), for example,

$$\begin{aligned} \mathbf{I}_3|\mathrm{p}\rangle &= \tfrac{1}{2}|\mathrm{p}\rangle, \\ \mathbf{I}_3|\mathrm{n}\rangle &= -\tfrac{1}{2}|\mathrm{n}\rangle. \end{aligned} \tag{15.17}$$

In the same way as we can combine the spins of several electrons to give states of higher total angular momentum, we can build up states of higher isospin from several nucleons (Table 15.2). We classify these states according to their eigenvalues with respect to the operators $\mathbf{I}_3$ and

$$\vec{\mathbf{I}}^2 = \sum_{a=1}^{3} (\mathbf{I}_a)^2. \tag{15.18}$$

Table 15.2 The (total) isospin I and eigenvalue I_3 of the third component of the isospin for two-nucleon states and for the π-mesons. An example of a two-nucleon state with $I = 0$ is the deuteron.

	I	I_3
pp	1	+1
$\frac{1}{\sqrt{2}}$(pn + np)	1	0
nn	1	−1
$\frac{1}{\sqrt{2}}$(pn − np)	0	0
π^+	1	+1
π^0	1	0
π^-	1	−1

The operator $\vec{I}^2$ corresponds to the square of the angular momentum. The possible eigenvalues of $\vec{I}^2$ in a general representation of the isospin group are thus of the form $I(I + 1)$ with

$$I = 0, \tfrac{1}{2}, 1, \ldots . \qquad (15.19)$$

One refers to I as the (total) isospin of a state.

The concept of isospin was extended to the mesons by Kemmer. He postulated that the Yukawa mesons should form an isotriplet, i.e., a system with $I = 1$ (Table 15.2). This was the only way in which he was able to explain the then-known facts about the nuclear forces. This theory, however, required in addition to the two charged mesons, a further uncharged Yukawa meson.

In order to understand another important advance in our knowledge of the strong interaction, we must discuss the theoretical problems which were presented by the V-particles. Since 1953 it has been possible to produce these V-particles in accelerators. It was found that V-particles are quite frequently produced in reactions such as

$$\pi^- + \mathrm{p} \rightarrow \mathrm{K}^0 + \Lambda. \qquad (15.20)$$

For pion energies, $E_\pi \cong 1.5$ GeV, the following scattering cross-sections were found:

$$\begin{aligned} \sigma(\pi^-\mathrm{p} \rightarrow \mathrm{K}^0\Lambda) &\cong 1\ \mathrm{mb} \equiv 10^{-27}\ \mathrm{cm}^2, \\ \sigma(\pi^-\mathrm{p}, \mathrm{total}) &\cong 40\ \mathrm{mb} \equiv 4 \cdot 10^{-26}\ \mathrm{cm}^2, \\ \frac{\sigma(\pi^-\mathrm{p} \rightarrow \mathrm{K}^0\Lambda)}{\sigma(\pi^-\mathrm{p}, \mathrm{total})} &\cong \frac{1}{40}. \end{aligned} \qquad (15.21)$$

These scattering cross-sections corresponded to the geometrical cross-sections

Table 15.3 The quantum number strangeness for a number of hadrons.

Hadrons	S
p, n, π^+, π^0, π^-	0
$\Lambda, \Sigma^+, \Sigma^0, \Sigma^-$	-1
Ξ^0, Ξ^-	-2
K^0, K^+	$+1$
$\bar{K}^0, K^-$	-1

of the hadrons $\sim(10^{-13}\,\mathrm{cm})^2$. Thus K- and Λ-particles were produced through a strong interaction. The observed decays, for examples, that of the Λ-hyperon

$$\Lambda \to p + \pi^-, n + \pi^0, \tag{15.22}$$

could not, however, be strong interaction processes. The natural lifetime for a strong decay, τ_{strong}, can be estimated using the relationship τ_{strong} = (hadron radius)/(speed of light):

$$\tau_{\mathrm{strong}} \sim 10^{-13}\,\mathrm{cm}/3\cdot 10^{10}\,\mathrm{cm\,s^{-1}} \sim 10^{-23}\,\mathrm{s}. \tag{15.23}$$

In contrast to this, the experimental lifetime of the Λ-particle has a value

$$\tau_\Lambda \cong 2.63 \cdot 10^{-10}\,\mathrm{s}. \tag{15.24}$$

This problem was solved by assigning a new additive quantum number to the V-particles, the *strangeness* S, as given in Table 15.3 (Gell-Mann 1953). The strong and the electromagnetic interactions were supposed to conserve strangeness. With this, it immediately became possible to understand the earlier discovered rules of associated production (Pais 1952). A process such as

$$\pi^- + p \to K^0 + \Lambda, \tag{15.25}$$

in which two strange particles took part was allowed, i.e., it could take place by means of the strong interaction since the sum of the quantum numbers S was equal in the initial and final states. Processes such as

$$\pi^- + p \to K^- + p \tag{15.26}$$

were never observed, and as an "explanation" one could now claim that they could not be mediated by the strong interaction since they violate the conservation of strangeness.

Only the *weak* interaction, which is also responsible for the β-decay of the neutron, was supposed to allow violation of strangeness in processes such as

$$\Lambda \to p + \pi^-. \tag{15.27}$$

This served to explain the small transition rate for this process.

Following these discoveries, one could assign a series of quantum numbers to all the then-known hadrons, namely, the space–time related quantum numbers spin and parity, and the internal quantum numbers charge Q, baryon number B, isospin (I, I_3), strangeness S, and hypercharge Y. Every baryon recived the baryon number $B = 1$, every antibaryon $B = -1$ and every meson $B = 0$. The hypercharge was defined as

$$Y = S + B. \tag{15.28}$$

In addition, one had the relation of Gell-Mann and Nishijima (Nishijima 1955) for the electric charge Q of the hadrons in units of the proton charge

$$Q = I_3 + \tfrac{1}{2}Y. \tag{15.29}$$

Experiment shows that parity and the internal quantum numbers are conserved in strong interactions. Electromagnetic interactions violate total isospin but preserve I_3. The weak interaction violates isospin, strangeness, and, as was discovered following the work of Lee and Yang (Lee 1956), also parity. According to the theoretical ideas of grand unification, it is expected that the baryon number is also weakly violated and that the proton decays, for example, according to the scheme

$$\mathrm{p} \to \mathrm{e}^+ + \pi^0. \tag{15.30}$$

Until now, however, proton decay has not been seen experimentally (Particle Data Group 1986).

We return now to the internal symmetries of the hadrons. With isospin and hypercharge, the generators of an invariance group of the strong interaction had been found. This had, in mathematical language, the structure of a direct product, namely, the group SU(2) × U(1). Many physicists found this unsatisfactory since such an invariance does not, for example, lead to a connection between the properties of the K- and π-mesons or the Λ-hyperon and the nucleons. It is the constant desire of physicists however, to trace varied phenomena back to some unifying principle.

The decisive breakthrough in this respect was made by Gell-Mann and Ne'eman (Gell–Mann 1961, 1964; Ne'eman 1961). They designed a scheme for classifying all the then-known hadrons according to representations of the SU(3)-group. The isospin group SU(2) and the U(1) of hypercharge were understood as subgroups of this SU(3). The best-known baryons and mesons were classified as members of the eight-dimensional representation of SU(3) and because of this one spoke of the "theory of the eight-fold way" (Fig. 15.1). For exact SU(3)-invariance, all particles in an SU(3)-multiplet ought to have the same mass. As shown by a glance at Table 15.1, this is certainly not the case. This SU(3)-group, which we today know as "flavor-SU(3)", was thus itself violated by the strong interaction. The order of magnitude of the symmetry breaking was estimated to be

$$\frac{m_\Lambda - m_\mathrm{p}}{m_\mathrm{p}} \cong 20\,\%. \tag{15.31}$$

This was small enough for the effects of this symmetry breaking to be treated by perturbation theory.

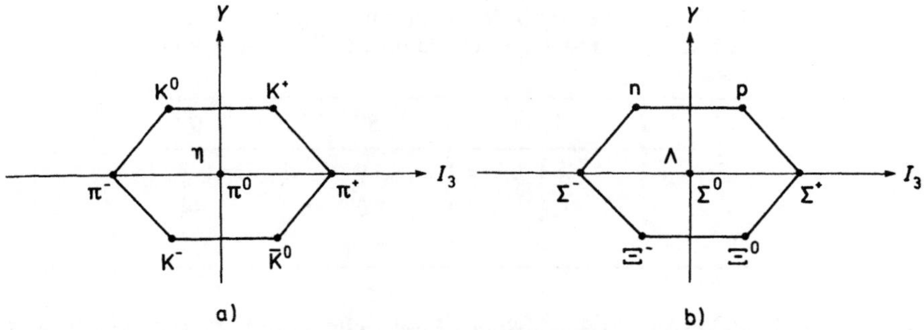

Figure 15.1 The arrangement of the lightest SU(3) multiplets of mesons (a) and baryons (b) in the I_3–Y plane according to the theory of the eightfold way.

The theory of the eight-fold way enabled, for example, the pion–nucleon and the kaon–nucleon coupling strengths to be related to one another. The greatest success, however, was the prediction of the Ω^--particle and of its mass. This particle was actually discovered in 1964 (Barnes 1964) in the reaction

$$K^- + p \rightarrow \Omega^- + K^+ + K^0. \qquad (15.32)$$

The Ω^- decayed weakly according to the scheme

$$\begin{aligned}\Omega^- &\rightarrow \Xi^0 + \pi^- \\ &\quad\;\; \hookrightarrow \Lambda + \pi^0 \\ &\quad\;\;\quad\;\; \hookrightarrow \pi^- + p.\end{aligned} \qquad (15.33)$$

From the production reaction (15.32) in which strangeness was supposed to be conserved, one could immediately conclude that the Ω^- had strangeness $S = -3$.

A notable feature of the SU(3) classification scheme was that none of the then-known hadrons was classified according to the fundamental three-dimensional representation of the SU(3) group. This was quite the opposite for the isospin group SU(2). There, neutrons and protons transformed according to the simplest nontrivial representation with $I = \frac{1}{2}$. All other isospin representations with $I = 0,\ 1,\ \frac{3}{2},\ 2,\ \ldots$ can be constructed from combinations of several nucleons. We find for the deuteron, for example, since it has isospin $I = 0$, the following wavefunction in isospin space (Table 15.2)

$$d \sim pn - np.$$

In the case of the (flavor)-SU(3), the situation was that all hadrons transformed according to the singlet, or according to higher dimensional SU(3) representations, the eight–and the ten-dimensional ones. The analogues of p and n for SU(3) were missing.

When theoreticians require particles that have not yet been observed, they postulate them. In 1964 Gell-Mann and Zweig independently introduced particles that should transform according to the fundamental representation

Table 15.4 The intrinsic quantum numbers of the quarks u, d, and s as postulated by Gell-Mann and Zweig.

	I	I_3	Y	S	B	Q
u	$\frac{1}{2}$	$+\frac{1}{2}$	$+\frac{1}{3}$	0	$\frac{1}{3}$	$\frac{2}{3}$
d	$\frac{1}{2}$	$-\frac{1}{2}$	$+\frac{1}{3}$	0	$\frac{1}{3}$	$-\frac{1}{3}$
s	0	0	$-\frac{2}{3}$	-1	$\frac{1}{3}$	$-\frac{1}{3}$

of SU(3) (Gell-Mann 1964a; Zweig 1964). The name *quarks* which was introduced by Gell-Mann for these particles, has become generally accepted. The quarks were supposed to be spin-$\frac{1}{2}$ particles existing in three varieties u, d, and s for "up", "down", and "strange". The intrinsic quantum numbers were those given in Table 15.4. Under the flavor-SU(3) transformations u, d, and s were supposed to transform according to the fundamental representation as follows:

$$\begin{aligned} |u\rangle &\to |u\rangle U_{11} + |d\rangle U_{21} + |s\rangle U_{31}, \\ |d\rangle &\to |u\rangle U_{12} + |d\rangle U_{22} + |s\rangle U_{32}, \\ |s\rangle &\to |u\rangle U_{13} + |d\rangle U_{23} + |s\rangle U_{33}. \end{aligned} \tag{15.34}$$

In this

$$\mathbf{U} = \begin{pmatrix} U_{11} & U_{12} & U_{13} \\ U_{21} & U_{22} & U_{23} \\ U_{31} & U_{32} & U_{33} \end{pmatrix} \tag{15.35}$$

is a unitary 3×3 matrix with determinant 1:

$$\begin{aligned} \mathbf{U}\mathbf{U}^\dagger &= 1, \\ \det \mathbf{U} &= 1. \end{aligned} \tag{15.36}$$

In this quark theory it was possible to describe mesons and baryons as states which transformed like quark–antiquark and three-quark states respectively. Schematically, one wrote

$$\text{Mesons} \sim q\bar{q},$$

$$\text{Baryons} \sim qqq.$$

For the π- and K-mesons this explicitly meant

$$\begin{aligned} \pi^+ &\sim u\bar{d}, \\ \pi^0 &\sim \frac{1}{\sqrt{2}}(u\bar{u} - d\bar{d}), \\ \pi^- &\sim d\bar{u}, \\ K^+ &\sim u\bar{s}, \\ K^0 &\sim d\bar{s}. \end{aligned} \tag{15.37}$$

The baryons p, n, and Λ were expressed as the following three-quark states:

$$\begin{aligned} p &\sim (\mathrm{uud})_8 , \\ n &\sim (\mathrm{ddu})_8 , \\ \Lambda &\sim (\mathrm{uds})_8 , \end{aligned} \tag{15.38}$$

where the subscript 8 is to indicate that an octet combination of the three quarks should be formed, as will be explained in more detail later. If one ascribed the same mass to the u and d quarks (since they formed an isospin doublet and since isospin was conserved for the strong interaction) and made the s-quark slightly heavier, one could also understand the mass difference within an SU(3) mutiplet. It was then clear that the K-meson should be heavier than the π-meson, and the Λ heavier than the nucleon.

What was revolutionary in the quark theory however, was the noninteger charge of the postulated particles. At that time there were two alternative possibilities:

(i) In a certain sense quarks are just mathematical objects. In nature there are no free quarks and thus no particles with noninteger charge.
(ii) Particles with one-third integral charges do exist in nature.

Up until today it has not been possible to decide experimentally which of these two alternatives is correct. There have been very many experiments searching for particles with noninteger charge. The majority of these experiments have given negative results (Morpurgo 1979). However, a further group has claimed to have found evidence for particles with noninteger charge (La Rue 1979). It is possible that this could represent the discovery of free quarks but this experimental result has so far not been confirmed. Today the common belief is that alternative (i) is correct.

With the quark hypothesis, one had a picture of hadrons as bound states of constituents and soon physicists started to write down wavefunctions for quarks. This picture received more and more empirical support. Earlier constituent models, e.g., that of Sakata (1956), were disproved by experiment. We also mention that, at the beginning, the quark model was regarded as heresy by many physicists. The rule of the day was the so-called bootstrap hypothesis in which the introduction of constituents for the hadrons was carefully avoided!

Very soon it was realized that the quarks, if taken seriously, posed a further puzzle. Let us consider, for example, the Ω^-, a particle with spin $\frac{3}{2}$. The z-component of the spin then has the possible eigenvalues $-\frac{3}{2}$, $-\frac{1}{2}$, $\frac{1}{2}$, $\frac{3}{2}$. We want to take a closer look at the wavefunction of an Ω^- in the quark scheme. As a baryon with strangeness $S = -3$, the Ω^- consists of three s-quarks. Each of the s-quarks has spin $\frac{1}{2}$. The total spin of the Ω^- will in general comprise the spins of the three s-quarks and their orbital angular momenta (Fig. 15.2). The Ω^- is the lightest baryon with $S = -3$; it is the ground state of the three s-quarks. For reasonable potentials, one expects that in the ground state, the orbital angular momenta are zero and that the spatial

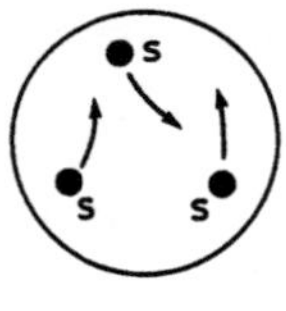

Figure 15.2 Schematic drawing of an Ω^--particle as a bound state of three s-quarks.

wavefunction is symmetric. It is then clear that the spin of the Ω^- must be constructed merely from the spins of the quarks. For an Ω^- with z-component of spin $\frac{3}{2}$, all the quark spins must be aligned. Thus for the total wavefunction we obtain the form

$$\Omega^-(\tfrac{3}{2}, \tfrac{3}{2}) \sim \overset{\uparrow\uparrow\uparrow}{\mathrm{sss}}\psi(\boldsymbol{x}_1, \boldsymbol{x}_2, \boldsymbol{x}_3), \tag{15.39}$$

where $\psi(\boldsymbol{x}_1, \boldsymbol{x}_2, \boldsymbol{x}_3)$ should be totally symmetric under permutations of the $\boldsymbol{x}_i$. With this, however, we have a totally symmetric wavefunction for particles with half integral spin. For all known fermions this contradicts the Pauli principle. To avoid the need to assume some complicated quark dynamics, the solution to this dilemma was either to throw out the quark concept or to ascribe some special properties to the quarks, namely the so-called parastatistics (Greenberg 1964). An equivalent description is to assign to the quarks a new degree of freedom, the color (Gell-Mann 1972; Fritzsch 1973). According to this, one imagines that each of the quarks u, d, and s, i.e., each of the three "flavors", can occur in three different versions or "colors":

$$\mathrm{u}_1, \mathrm{u}_2, \mathrm{u}_3; \quad \mathrm{d}_1, \mathrm{d}_2, \mathrm{d}_3; \quad \mathrm{s}_1, \mathrm{s}_2, \mathrm{s}_3.$$

With the help of the color degree of freedom it is easy to construct a totally antisymmetric wavefunction describing the Ω^--particle:

$$\Omega^-(\tfrac{3}{2}, \tfrac{3}{2}) \sim \overset{\uparrow}{\mathrm{s}}_\alpha \overset{\uparrow}{\mathrm{s}}_\beta \overset{\uparrow}{\mathrm{s}}_\gamma \varepsilon_{\alpha\beta\gamma}\psi(\boldsymbol{x}_1, \boldsymbol{x}_2, \boldsymbol{x}_3), \tag{15.40}$$

in which $\psi(\boldsymbol{x}_1, \boldsymbol{x}_2, \boldsymbol{x}_3)$ is again totally symmetric. This state is invariant under rotations in color space, i.e., under transformations of the form

$$\mathrm{q}_\alpha \to \sum_{\beta=1}^{3} \mathrm{q}_\beta U_{\beta\alpha} \qquad (\mathrm{q} = \mathrm{u}, \mathrm{d}, \mathrm{s}; \alpha, \beta = 1, 2, 3). \tag{15.41}$$

Here $\mathbf{U} = (U_{\beta\alpha})$ is any unitary 3×3 matrix with determinant 1. These rotations in color space again form an SU(3) group, the color-SU(3), which however, should not be confused with the flavor-SU(3) introduced previously. In the modern theory of the strong interaction, quantum chromodynamics, the color-SU(3) group plays the fundamental role.

In order to achieve agreement with experiment it was necessary to assume that all hadrons are described by quark states that are invariant under color-SU(3) transformations. A π^+-meson, for example, is today pictured, in an extension of the simple scheme given in (15.37), as the following quark state:

$$\pi^+ \sim u_1\bar{d}_1 + u_2\bar{d}_2 + u_3\bar{d}_3. \qquad (15.42)$$

Further experimental support for the quark color hypothesis is provided by electron–positron annihilation into hadrons and by the decay of the π^0-meson. We will return to these in more detail later.

The quark constituent scheme has proved to be very efficient in describing the spectrum and static properties of mesons and baryons. A quite different form of evidence for hadron constituents was found in about 1969 in experiments in which electrons were scattered off nucleons. These experiments were basically quite analogous to Rutherford's scattering experiments in which he observed the scattering of α-particles from atoms. The fact that many α-particles were deflected through large angles led Rutherford to conclude the existence of an atomic constituent, namely the nucleus. In the electron–nucleon scattering experiments

$$e + N \rightarrow e + X, \qquad (15.43)$$

which were originally performed at Stanford (USA) with electron energies of 1–20 GeV, it was found that many electrons were likewise deflected through large angles. This suggested the similar conclusion that nucleons too can be considered as bound states of certain constituents; these, following Feynman (1969, 1972), are known as partons. As later became clear, and as we will discuss in depth, some of these partons are nothing other than the previously known quarks. There was evidence however, that there must also be other partons in the nucleon apart from quarks. Even this was not entirely unexpected—indeed, something has to hold the quarks together in the nucleon. As we learnt in our discussion of the Yukawa theory, every force field is associated with a particle, a particular boson. The bosons with are thought to hold the quarks together are known as gluons and it is supposed, in analogy to the photons, that they are vector particles. In the years 1979–80 new and convincing evidence for the existence of gluons was found in electron–positron annihilation (Brandelik 1979; Barber 1979; Berger 1979; Bartel 1980).

Until 1972 it was not possible to claim that a theory of the strong interaction had been found. There were, of course, models, such as the quark model, the parton model, and many others, all of which could describe certain features of the experimental results. However, none could claim to describe everything. Furthermore, most of the models possessed a large number of free parameters (sometimes more than 40), all of which had to be fitted to experiment. This inevitably meant that the predictive power of the models was very small.

The situation changed in a decisive manner with the discovery of asymptotic freedom (t'Hooft 1972; Gross 1973; Politzer 1973). Those theoreticians who tended more towards abstracts concepts had learnt at the beginning of the 1970s to work with the so-called non-Abelian gauge theories. These theories were introduced as early as 1954 by Yang and Mills (Yang 1954). It now turned out that the most natural ansatz for the coupling of quarks and gluons led to exactly such a Yang–Mills theory and, even more notable, that it was even possible to carry out quantitative calculations (Gross 1973a; Fritzsch 1973a; Politzer 1974).

With this was born the modern theory of the strong interaction—quantum chromodynamics (QCD). This theory, like quantum electrodynamics, has only very few fundamental parameters that must be fitted to experiment, namely, a coupling parameter and the mass parameters for the different quark flavors. In quantum electrodynamics the analogous parameters are the fine-structure constant α and the masses of the leptons. The aim of the theoreticians is to use quantum chromodynamics to quantitatively describe all phenomena associated with the strong interaction. It must be said, however, that we still have a long way to go before this goal is attained. With the discovery of asymptotic freedom, it has nevertheless been possible to fulfill the above ambition for a partial domain of the strong interaction. Today we can quantitatively describe the essential features of all reactions in which a large momentum transfer occurs, i.e., reactions in which small distances, much smaller than the hadron radius, play a dominant role. Thus, in our discussions we will be concentrating on such reactions.

An outstanding experimental discovery, that of the J/ψ-particle (Aubert 1974; Augustin 1974), was the decisive factor which persuaded many high energy physicists to take QCD seriously. The J/ψ has the very strange property that with a mass of about 3.1 GeV, it has a decay rate of only about 60 keV. This fact could be explained in a perfectly natural way within the framework of QCD. For this however, it was necessary to introduce a further quark, the charm quark (c). The J/ψ-particle was understood as a c-$\bar{c}$ bound state

$$J/\psi \sim c_1\bar{c}_1 + c_2\bar{c}_2 + c_3\bar{c}_3 \tag{15.44}$$

and could be treated as a quark analogue to orthopositronium (Chap. 13).

With this the number of known quarks rose to four: u, d, s, c. The charm quark, with the correct properties such as charge $Q = \frac{2}{3}$, had been predicted by Glashow (1970). The number of known quarks increased again with the discovery of the Υ-particle (Herb 1977) which was interpreted as a bound state of a quark b and its antiquark $\bar{b}$. Here b stands for "bottom" or "beauty". Thus, as of 1977, five quarks of different flavors were known. The theoretical picture that we will develop in Part IV has, in fact, suggested the existence of a sixth quark, t ("top" or "truth"). Experimental evidence for this t-quark has been keenly sought. A first experimental indication for the existence of the t-quark was presented in 1984 (Rubbia 1984), but this has not been confirmed to date. Indications from more recent experiments are that the t-quark—if it exists at all—must have a mass $m_t > 50$ GeV. Thus we are still waiting for all three families of fundamental fermions, with the members shown in Table 1.3, to be completed.

CHAPTER 16

Phenomenology of Hadronic Reactions

A large part of our knowledge about hadrons stems from various scattering experiments of types differing according to the projectile used. We can study lepton–hadron, photon–hadron, or hadron–hadron scattering. Some examples of these reactions are

$$e^- + N \to e^- + X,$$
$$\nu_\mu + N \to \mu^- + X,$$

$$\gamma + p \to \gamma + p,$$
$$\gamma + p \to \pi^0 + p,$$

$$\pi^- + p \to \pi^- + p,$$
$$\pi^+ + p \to \pi^+ + \pi^0 + p,$$
$$\pi^- + p \to K^0 + \Lambda.$$

Here X stands for an arbitrary final state. In lepton–hadron and photon–hadron scattering we have an interplay of the electroweak and the strong interaction. These processes were extremely useful in the development of our ideas about the structure of the nucleons and we will later discuss them in detail. In this chapter we will get to know a few simple facts about hadron–hadron scattering.

One of the best-investigated reactions is the pion–nucleon scattering. The simplest quantity that we can consider is the total cross-section (Fig. 16.1). As

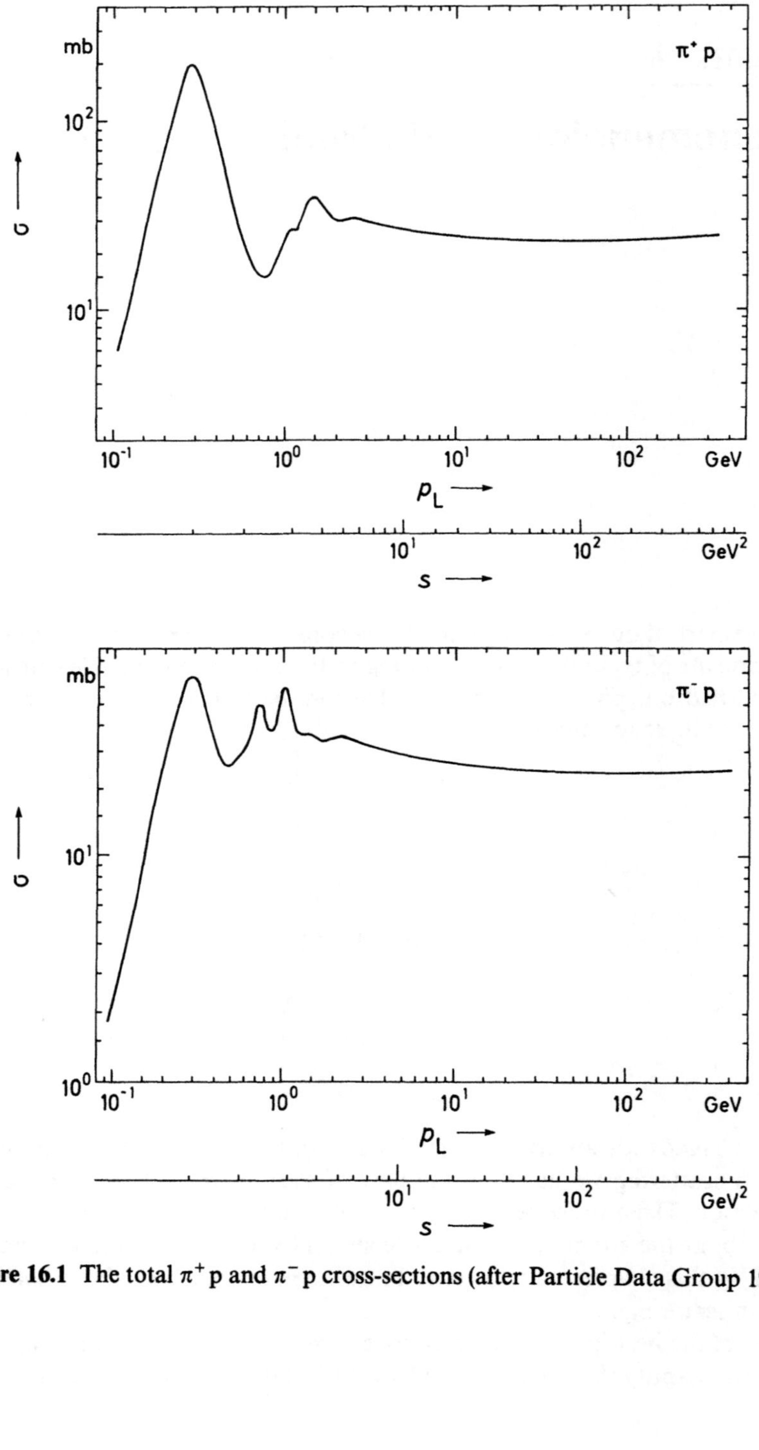

Figure 16.1 The total π^+p and π^-p cross-sections (after Particle Data Group 1984).

we see, this is of the order of 30–100 mb, or $3\text{–}10 \times 10^{-26}$ cm^2. This corresponds to the geometrical cross-section of the hadrons, as was noted earlier. We also see that both the π^+p and the π^-p cross-sections have pronounced maxima and minima for small laboratory momenta $p_L \lesssim 2$ GeV. Above $p_L \sim 2$ GeV both cross-sections first decrease slowly with increasing momentum and then start to rise again without exhibiting any noticeable structure. With increasing energy the total cross-sections for π^+p and π^-p also become more and more similar, i.e., the charge of the projectile obviously ceases to play an important role. A similar pattern is observed for other hadronic total cross-sections.

16.1 Resonance Physics

We will first consider the low energy region. The maxima observed in the total cross-section are interpreted as resonances, i.e., as formation of unstable particles. Analogous processes are well known in atomic and nuclear physics. For example, we might think of the excitation of a higher atomic energy level by an incident beam of photons. The position of a maximum determines (approximately) the mass of the unstable particle. From the width of the resonance, which gives the energy uncertainty ΔE of the corresponding energy level, we can determine the lifetime of the resonance using the uncertainty relation

$$\Delta E \cdot \Delta t \sim 1. \tag{16.1}$$

Let us look at the first large resonance in π^+p scattering. We consider the total cross-section, i.e., the reaction

$$\pi^+(p_1) + \mathrm{p}(p_2) \to \mathrm{X}, \tag{16.2}$$

where p_1 and p_2 are the four-momenta of the pion and the proton, respectively, and X stands for any final state. The relation between the center of mass energy $\sqrt{s}$ and the laboratory momentum of the pion p_L is given by the equation

$$s = (p_1 + p_2)^2 = 2\sqrt{p_L^2 + m_\pi^2}\, m_p + m_p^2 + m_\pi^2. \tag{16.3}$$

The π^+p cross-section has its maximum at $p_L \cong 0.3$ GeV or $\sqrt{s} \cong 1.23$ GeV. This is the mass of the corresponding resonance which is known today as the $\Delta(1232)$-resonance.

A simple diagram for the production of the Δ-resonance is shown in Fig. 16.2. The corresponding matrix element T_{fi} is proportional to the propagator of the Δ-particle. We know what the propagator for a free, stable Δ-particle would look like: apart from its different spin matrices, it would have a form resembling the electron propagator (8.36) with $m_e \to m_\Delta$. Thus we would have

$$T_{fi} \propto \frac{1}{s - m_\Delta^2}. \tag{16.4}$$

According to Appendix I, the propagation of the unstable Δ-particle is governed

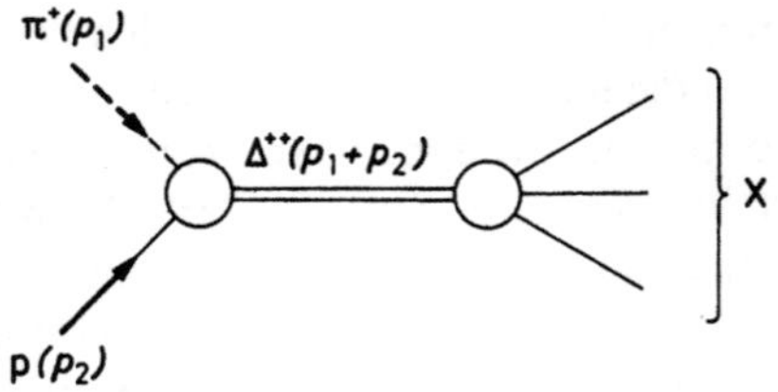

Figure 16.2 Diagram for the production and decay of the Δ^{++} resonance in π^+p scattering. The final state X consists to nearly 100% of π^+p again.

by a complex mass parameter. The real part of this mass parameter is equal to the mass m_Δ introduced before, and the imaginary part is equal to minus one-half of the decay rate Γ_Δ. Thus in (16.4) we replace m_Δ by $m_\Delta - \mathrm{i}\Gamma_\Delta/2$ and with the assumption that $m_\Delta^2 \gg \Gamma_\Delta^2$ we obtain

$$T_{fi} \propto \frac{1}{s - m_\Delta^2 + \mathrm{i} m_\Delta \Gamma_\Delta}. \tag{16.5}$$

The scattering cross-section, which is proportional to the absolute square of the matrix element, then has the form

$$\sigma \propto \frac{1}{(s - m_\Delta^2)^2 + m_\Delta^2 \Gamma_\Delta^2}. \tag{16.6}$$

In elementary particle physics, equations (16.5), (16.6) are known as the "Breit–Wigner formulas". They describe the typical resonance behavior already familiar to us from mechanics. The decay rate Γ_Δ is equal to the halfwidth of the resonance on a scale given by the energy $\sqrt{s}$ (Fig. 16.3). Indeed, for values of s close to m_Δ^2 we have

$$\sigma \propto \frac{1}{4m_\Delta^2[(\sqrt{s} - m_\Delta)^2 + \frac{1}{4}\Gamma_\Delta^2]}. \tag{16.7}$$

Thus the lifetime $\tau_\Delta = 1/\Gamma_\Delta$ is identical to the reciprocal of the energy uncertainty, in agreement with (16.1).

If the parameters m_Δ and Γ_Δ are fitted to the experimental values for the π^+p scattering cross-section shown in Fig. 16.1, then one finds (Particle Data Group 1986):

$$\begin{aligned} m_\Delta &= 1232\ \text{MeV}, \\ \Gamma_\Delta &= \ \ 115\ \text{MeV}. \end{aligned} \tag{16.8}$$

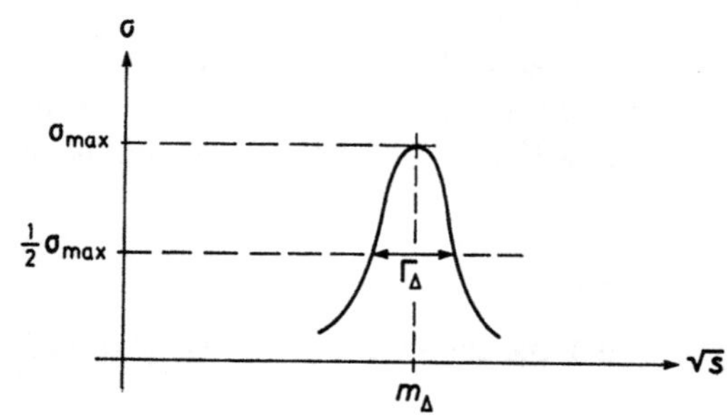

Figure 16.3 Resonance curve and width Γ_Δ corresponding to (16.7) (schematic).

The lifetime of the Δ-resonance is thus

$$\tau_\Delta = \frac{1}{\Gamma_\Delta} = \frac{1}{115}\,\mathrm{MeV}^{-1} = 0.57 \cdot 10^{-23}\ \mathrm{s}. \tag{16.9}$$

Therefore the lifetime of the Δ-particle has indeed the order of magnitude of the lifetime estimated in (15.23) for strong decays.

Once a resonance is discovered, one wants to know its quantum numbers. The analysis of the pion nucleon scattering data showed that the Δ-resonance has spin $\frac{3}{2}$, parity $+1$, and isospin $\frac{3}{2}$. The isospin of the Δ-resonance is established by means of the following argument: A pion ($I = 1$) and a nucleon ($I = \frac{1}{2}$) can form resonant states with total isospin $I = \frac{1}{2}$ and $I = \frac{3}{2}$. For the Δ, however, only $I = \frac{3}{2}$ is possible since it occurs in π^+p scattering where the initial state contains only $I = \frac{3}{2}$. According to the rules for isospin addition, and with the notation $|I, I_3\rangle$, we obtain the following for the states:

$$|\pi^+\rangle|\mathrm{p}\rangle = |1, 1\rangle|\tfrac{1}{2}, \tfrac{1}{2}\rangle = |\tfrac{3}{2}, \tfrac{3}{2}\rangle. \tag{16.10}$$

Since the Δ has strangeness $S = 0$ and hypercharge $Y = S + B = 1$, the Gell-Mann–Nishijima formula (15.29) yields the following charge states:

$$\Delta^-, \Delta^0, \Delta^+, \Delta^{++}.$$

All of these states are actually observed.

In order to determine the spin and the parity of the Δ-resonance, it is not sufficient to consider only the total scattering cross-section. In addition to this one requires information about differential cross-sections such as the π^+p elastic scattering. This will be discussed in more detail in Sect. 16.3.

Since the 1950s, the search for hadron resonances, and the determination of their parameters: mass, halfwidth, quantum numbers, branching ratios, etc., has occupied many physicists. The results of a large number of experiments are reported every two years in the tables of particle properties published by the Particle Data Group. Our Tables 1.1 and 1.2 are taken from this source. They are the analogue to a table of energy levels for an atom or a nucleus. One of the tasks of the theory is to explain the observed hadron spectrum. So far potential models in which the baryons are described as three-quark states and the mesons as quark–antiquark states have been quite successful. The simplest example of such a potential model will be discussed in more detail in Sect. 17.4.

16.2 A Basis for the Hadron States and the Symmetries C, P, and T

In hadron physics, an important role is played by symmetries. As already discussed in Sect. 3.3, symmetries are closely linked with conserved quantities. The symmetry of a theory under Poincaré transformations leads to the laws of conservation of energy, monentum, and angular momentum. In the strong interaction we also have invariance under reflection in space (P), time reversal (T), and charge conjugation (C). Here, we wish to extend these operations, which have already been discussed for free electrons in Chap. 4, to the hadrons. For this purpose, we have to find a suitable basis for the hadron states.

We begin by presupposing only Poincaré invariance for the hadron theory. This implies that no one inertial system is physically singled out from the others and that each leads to an equivalent description of the physical events. This is realized mathematically by representing the Poincaré transformations (Λ, a) by unitary operators $\mathbf{U}(\Lambda, a)$ in the space of state vectors[1] for which we have [see (2.28), (2.29)]:

$$\mathbf{U}(\Lambda, a)\mathbf{U}(\Lambda', a') = \mathbf{U}(\Lambda\Lambda', \Lambda a' + a). \tag{16.11}$$

The infinitesimal generators of these transformations can be constructed for a given Lagrange density by the methods presented in Sects. 3.3 and 3.4.

According to (2.28), (4.52), (4.53), an infinitesimal Poincaré transformation has the form

$$(\Lambda, \delta a)\colon x^\mu \to x'^\mu = x^\mu + \delta h^\mu{}_\nu x^\nu + \delta a^\mu, \tag{16.12}$$

where $\delta h^{\mu\nu} + \delta h^{\nu\mu} = 0$. We expand the corresponding unitary transformation $\mathbf{U}(\Lambda, \delta a)$ to first order in δh and δa,

$$\mathbf{U}(\Lambda, \delta a) = 1 + \frac{\mathrm{i}}{2}\delta h^{\mu\nu}\mathbf{M}_{\mu\nu} + \mathrm{i}\delta a^\mu \mathbf{P}_\mu. \tag{16.13}$$

The expansion coefficients $\mathbf{M}_{\mu\nu}$ and $\mathbf{P}_\mu$ are, of course, operators. On account of the antisymmetry of $\delta h^{\mu\nu}$, $\mathbf{M}_{\mu\nu}$ is likewise antisymmetric; since $\mathbf{U}$ is unitary, $\mathbf{M}_{\mu\nu}$ and $\mathbf{P}_\mu$ must be Hermitian. Thus we have

$$\begin{aligned} &\mathbf{M}_{\mu\nu} + \mathbf{M}_{\nu\mu} = 0, \\ &\mathbf{M}_{\mu\nu} = \mathbf{M}^\dagger_{\mu\nu}, \qquad \mathbf{P}_\mu = \mathbf{P}^\dagger_\mu. \end{aligned} \tag{16.14}$$

From the structure of the Poincaré group follow the commutation rules (Exercise 16.1)

$$\begin{aligned} &[\mathbf{P}_\mu, \mathbf{P}_\nu] = 0, \\ &[\mathbf{P}_\mu, \mathbf{M}_{\rho\sigma}] = \frac{1}{\mathrm{i}}(g_{\mu\rho}\mathbf{P}_\sigma - g_{\mu\sigma}\mathbf{P}_\rho), \\ &[\mathbf{M}_{\mu\nu}, \mathbf{M}_{\rho\sigma}] = \frac{1}{\mathrm{i}}(g_{\mu\sigma}\mathbf{M}_{\nu\rho} + g_{\nu\rho}\mathbf{M}_{\mu\sigma} - g_{\mu\rho}\mathbf{M}_{\nu\sigma} - g_{\nu\sigma}\mathbf{M}_{\mu\rho}). \end{aligned} \tag{16.15}$$

The quantities $\mathbf{P}_\mu$ and $\mathbf{M}_{\mu\nu}$ are of fundamental physical importance since the operators for the four-momentum $\mathbf{P}$ and the angular momentum $\vec{\mathbf{J}}$ are given by [see (3.99), (3.109) and Exercise 3.8]:

$$\begin{aligned} &\mathbf{P} = (\mathbf{P}^\mu), \\ &\mathbf{J}^i = -\tfrac{1}{2}\varepsilon^{ijk}\mathbf{M}^{jk}. \end{aligned} \tag{16.16}$$

[1] The unitary representations of the Poincaré group were first systematically studied and classified by Wigner (1939). We will follow here essentially the same procedure that he used to find basis states and thereby explicit realizations of the representations.

From (16.15) we find the familiar commutation rules for the angular momentum operator:

$$[\mathbf{P}^0, \mathbf{J}^k] = 0,$$
$$[\mathbf{P}^j, \mathbf{J}^k] = \mathrm{i}\varepsilon^{jkl}\mathbf{P}^l, \qquad (16.17)$$
$$[\mathbf{J}^j, \mathbf{J}^k] = \mathrm{i}\varepsilon^{jkl}\mathbf{J}^l.$$

We now want to use these relationships in order to construct a basis for the hadron states. According to (16.15) the operators $\mathbf{P}_\mu$ ($\mu = 0, \ldots, 3$) commute with one another. Thus we can choose the basis states such that they are simultaneous eigenstates of all components $\mathbf{P}_\mu$ of the four-momentum operator. We begin by considering the states of a single hadron a at rest, i.e., with momentum $\boldsymbol{p} = 0$. Let $|\mathrm{a}(\boldsymbol{p} = 0, r)\rangle$ (with $r = 1, \ldots, N$) be a basis for these states and let the space spanned by these basis states be $\mathbb{R}_0$. We have

$$\vec{\mathbf{P}}|\mathrm{a}(\boldsymbol{p} = 0, r)\rangle = 0. \qquad (16.18)$$

If we rotate one of these states in space, then we again obtain a state corresponding to a particle a at rest. Mathematically, this can be inferred from (16.17), according to which the application of an angular momentum operator to $|\mathrm{a}(\boldsymbol{p} = 0, r)\rangle$ again gives a state with $\boldsymbol{p} = 0$:

$$\begin{aligned}\mathbf{P}^j(\mathbf{J}^k|\mathrm{a}(\boldsymbol{p} = 0, r)\rangle) &= [\mathbf{P}^j, \mathbf{J}^k]|\mathrm{a}(\boldsymbol{p} = 0, r)\rangle \\ &= \mathrm{i}\varepsilon^{jkl}\mathbf{P}^l|\mathrm{a}(\boldsymbol{p} = 0, r)\rangle \\ &= 0 \end{aligned} \qquad (16.19)$$

Thus, the application of the angular momentum operators does not lead out of the space $\mathbb{R}_0$ which therefore carries a representation of the rotation group. One speaks of the states of *a* particle if this representation is irreducible[2]. The total angular momentum s in the rest frame is referred to as the spin of the particle. It can take the values $0, \frac{1}{2}, 1, \ldots$. A basis for the states of the particle a at rest is formed by the eigenstates of the component $\mathbf{J}^3$:

$$\begin{aligned}\mathbf{J}^3|\mathrm{a}(\boldsymbol{p} = 0, s_3)\rangle &= s_3|\mathrm{a}(\boldsymbol{p} = 0, s_3)\rangle, \qquad (s_3 = -s, \ldots, s), \\ \vec{\mathbf{J}}^2|\mathrm{a}(\boldsymbol{p} = 0, s_3)\rangle &= s(s + 1)|\mathrm{a}(\boldsymbol{p} = 0, s_3)\rangle.\end{aligned} \qquad (16.20)$$

From (16.17) we can further conclude that all states of a particle at rest have the same eigenvalue of the energy operator $\mathbf{P}^0$, i.e., they have the same mass m_a since $\vec{\mathbf{J}}$ commutes with $\mathbf{P}^0$:

$$\mathbf{P}^0|\mathrm{a}(\boldsymbol{p} = 0, s_3)\rangle = m_a|\mathrm{a}(\boldsymbol{p} = 0, s_3)\rangle. \qquad (16.21)$$

In the following we assume that $m_a > 0$. This is the case for all hadrons. With the canonical choice of phases for the states $|\mathrm{a}(\boldsymbol{p} = 0, s_3)\rangle$ (see, for example, Edmonds 1957), the angular momentum operators are represented in the

[2] If the representation is reducible, it can always be decomposed into its irreducible parts, which then correspond to the states of single particles. The notions of reducible and irreducible representations are explained in Appendix C.

space $\mathbb{R}_0$ by a standard set of matrices $\mathbf{S}^j = (S_{s_3' s_3})$

$$\mathbf{J}^j |a(\boldsymbol{p} = 0, s_3)\rangle = \sum_{s_3' = -s}^{s} |a(\boldsymbol{p} = 0, s_3')\rangle S^j_{s_3', s_3}, \tag{16.22}$$

with $\mathbf{S}^j = 0$ for spin $s = 0$ and $\mathbf{S}^j = \boldsymbol{\sigma}^j/2$ for spin $s = \frac{1}{2}$. The $\mathbf{S}^j$ for larger spin values are given in Appendix C.

Examples of hadrons with spin $s = 0$ are the mesons $\pi^\pm$ and π^0; the ρ-mesons have spin $s = 1$, the proton and neutron have $s = \frac{1}{2}$, and the Δ-particle and the Ω^--particle have $s = \frac{3}{2}$.

We now wish to construct a basis for all states of an isolated particle a. We do this by Lorentz transforming without rotation (boosting) the states of (16.20) to nonzero velocities. We have already become familiar with a Lorentz boost in the x^1-direction, namely $\Lambda(v)$ in (2.33). $\Lambda(v)$ transforms a particle at rest with momentum $\boldsymbol{p} = 0$ into a particle with momentum $\boldsymbol{p}$ in the x^1-direction. The corresponding four-momenta p_R and p are

$$p_R = \begin{pmatrix} m_a \\ 0 \\ 0 \\ 0 \end{pmatrix},$$

$$p = \Lambda(v) p_R = \begin{pmatrix} \dfrac{m_a}{\sqrt{1 - v^2}} \\ \dfrac{m_a v}{\sqrt{1 - v^2}} \\ 0 \\ 0 \end{pmatrix}. \tag{16.23}$$

For general four-momenta p, the Lorentz boost which takes p_R into p has the form

$$(\Lambda_p{}^\mu{}_\nu) = \left(\begin{array}{c|c} p^0/m_a & p^j/m_a \\ \hline p^i/m_a & \delta^{ij} - \dfrac{p^i p^j}{\boldsymbol{p}^2} + \dfrac{p^0}{m_a} \dfrac{p^i p^j}{\boldsymbol{p}^2} \end{array}\right). \tag{16.24}$$

As can easily be seen, we indeed have

$$\Lambda_p p_R = p.$$

We define our basis states $|a(\boldsymbol{p}, s_3)\rangle$ for general momentum $\boldsymbol{p}$ as the states produced from the states at rest by means of the Lorentz transformation Λ_p

$$|a(\boldsymbol{p}, s_3)\rangle = \mathbf{U}(\Lambda_p, 0) |a(\boldsymbol{p} = 0, s_3)\rangle. \tag{16.25}$$

In the general state $|a(\boldsymbol{p}, s_3)\rangle$, s_3 is thus the eigenvalue of the third component of the angular momentum after a boost into the rest frame of the particle. It can easily be shown that

$$\mathbf{P}_\mu |a(\boldsymbol{p}, s_3)\rangle = p_\mu |a(\boldsymbol{p}, s_3)\rangle, \tag{16.26}$$

and that the normalization of the states can be chosen as follows:

$$\langle a(\boldsymbol{p}', s_3')|a(\boldsymbol{p}, s_3)\rangle = (2\pi)^3 2p^0 \delta^3(\boldsymbol{p}' - \boldsymbol{p})\delta_{s_3', s_3}. \tag{16.27}$$

A basis for all hadron states is obtained by the usual procedure of forming the tensor product of the desired number of one-particle states. The basis which we have just constructed is, of course, not the only one possible; for many purposes it is more convenient to choose, for example, the so-called helicity basis (see Jacob 1959).

We now want to consider the effect of the angular momentum operator $\vec{\mathbf{J}}$ on the states $|a(\boldsymbol{p}, s_3)\rangle$. We first look at an arbitrary rotation

$$\begin{aligned} t \to t' &= t, \\ x^j \to x'^j &= R^{jk}x^k, \end{aligned} \tag{16.28a}$$

where $\mathbf{R} = (R^{jk})$ is an orthogonal 3×3 matrix ($\mathbf{R}^{\mathrm{T}}\mathbf{R} = 1$). The matrix of the corresponding Lorentz transformation is

$$\Lambda_R = \left(\begin{array}{c|c} 1 & 0 \\ \hline 0 & \mathbf{R} \end{array}\right) \tag{16.28b}$$

By means of this rotation, a four-momentum p is converted into p':

$$\Lambda_R p = p', \tag{16.29}$$

where $p'^0 = p^0$ and $\boldsymbol{p}' = \mathbf{R}\boldsymbol{p}$. We next consider the boosts Λ_p and $\Lambda_{p'}$ which convert the four-momentum p_R of a particle at rest to p and p', respectively. It evidently follows that

$$(\Lambda_{p'}^{-1}\Lambda_R\Lambda_p)p_R = p_R, \tag{16.30}$$

i.e., $\Lambda_{p'}^{-1}\Lambda_R\Lambda_p$ is a pure rotation. By explicit calculation we further find

$$\Lambda_{p'}^{-1}\Lambda_R\Lambda_p = \Lambda_R. \tag{16.31}$$

We now consider an infinitesimal rotation as in (4.57) where

$$R^{jk} = \delta^{jk} + \varepsilon^{jkl}\delta\varphi^l. \tag{16.32}$$

According to (4.58) and (16.13) we have

$$\mathbf{U}(\Lambda_R, 0) = 1 + \mathrm{i}\delta\boldsymbol{\varphi}\vec{\mathbf{J}}. \tag{16.33}$$

If we apply this rotation to the state $|a(\boldsymbol{p}, s_3)\rangle$, then using (16.11), (16.22) we find

$$\begin{aligned} &(1 + \mathrm{i}\delta\boldsymbol{\varphi}\vec{\mathbf{J}})|a(\boldsymbol{p}, s_3)\rangle \\ &\quad = \mathbf{U}(\Lambda_R, 0)\mathbf{U}(\Lambda_p, 0)|a(0, s_3)\rangle \\ &\quad = \mathbf{U}(\Lambda_R\Lambda_p, 0)|a(0, s_3)\rangle \\ &\quad = \mathbf{U}(\Lambda_{p'}\Lambda_R, 0)|a(0, s_3)\rangle \\ &\quad = \mathbf{U}(\Lambda_{p'}, 0)\mathbf{U}(\Lambda_R, 0)|a(0, s_3)\rangle \\ &\quad = \mathbf{U}(\Lambda_{p'}, 0)(1 + \mathrm{i}\delta\varphi^j\mathbf{J}^j)|a(0, s_3)\rangle \end{aligned}$$

$$= \mathbf{U}(\Lambda_{p'}, 0) \sum_{s_3'=-s}^{s} (\delta_{s_3' s_3} + \mathrm{i}\delta\varphi^j S^j_{s_3' s_3}) |\mathrm{a}(0, s_3')\rangle$$

$$= \sum_{s_3'=-s}^{s} (\delta_{s_3' s_3} + \mathrm{i}\delta\varphi^j S^j_{s_3' s_3}) |\mathrm{a}(\boldsymbol{p}', s_3')\rangle$$

$$= |\mathrm{a}(\boldsymbol{p}, s_3)\rangle + \mathrm{i}\delta\varphi^j \sum_{s_3'=-s}^{s} \left(\delta_{s_3' s_3} \varepsilon^{jkl} p^k \mathrm{i} \frac{\partial}{\partial p^l} + S^j_{s_3' s_3}\right) |\mathrm{a}(\boldsymbol{p}, s_3')\rangle. \qquad (16.34)$$

In the final step we have used the relation

$$p'^j = R^{jk} p^k = p^j + \varepsilon^{jkl} \delta\varphi^l p^k \qquad (16.35)$$

and expanded all quantities to first order in $\delta\boldsymbol{\varphi}$. From (16.34) we obtain the result of applying the angular momentum operators to our basis states:

$$\mathbf{J}^j |\mathrm{a}(\boldsymbol{p}, s_3)\rangle = \sum_{s_3'=-s}^{s} \left(\delta_{s_3' s_3} \varepsilon^{jkl} p^k \mathrm{i} \frac{\partial}{\partial p^l} + S^j_{s_3' s_3}\right) |\mathrm{a}(\boldsymbol{p}, s_3')\rangle. \qquad (16.36)$$

We will be needing this result when we come to discuss partial wave analysis.

We now define the parity operator $\mathbf{U}(\mathrm{P})$. The effect of $\mathbf{U}(\mathrm{P})$ is to reverse the momentum; the spin is unchanged. In addition there appears a factor $\eta_a = \pm 1$ which takes the intrinsic parity of the particle a into account:

$$\mathbf{U}(\mathrm{P}) |\mathrm{a}(\boldsymbol{p}, s_3)\rangle = \eta_a |\mathrm{a}(-\boldsymbol{p}, s_3)\rangle. \qquad (16.37)$$

Parity conservation of the strong interaction can be formulated as follows: one can assign an intrinsic parity η_a to each hadron such that for every reaction the scattering amplitude is identical to the amplitude of the corresponding parity transformed reaction obtained by applying (16.37). To be explicit let us consider a reaction and its parity transformed partner reaction:

$$\begin{aligned} &\sum_i \mathrm{a}^{(i)}(\boldsymbol{p}^{(i)}, s_3^{(i)}) \to \sum_f \mathrm{b}^{(f)}(\boldsymbol{q}^{(f)}, r_3^{(f)}), \\ &\sum_i \mathrm{a}^{(i)}(-\boldsymbol{p}^{(i)}, s_3^{(i)}) \to \sum_f \mathrm{b}^{(f)}(-\boldsymbol{q}^{(f)}, r_3^{(f)}). \end{aligned} \qquad (16.38)$$

Parity conservation for strong processes implies for the corresponding S-matrix

$$\mathbf{U}(\mathrm{P})^{-1} \mathbf{S} \mathbf{U}(\mathrm{P}) = \mathbf{S} \qquad (16.39)$$

and thus for the S-matrix elements

$$\begin{aligned} &\langle \mathrm{b}^{(1)}(\boldsymbol{q}^{(1)}, r_3^{(1)}), \ldots |\mathbf{S}| \mathrm{a}^{(1)}(\boldsymbol{p}^{(1)}, s_3^{(1)}), \ldots \rangle \\ &= \langle \mathrm{b}^{(1)}(\boldsymbol{q}^{(1)}, r_3^{(1)}), \ldots |\mathbf{U}(\mathrm{P})^{-1} \mathbf{S} \mathbf{U}(\mathrm{P})| \mathrm{a}^{(1)}(\boldsymbol{p}^{(1)}, s_3^{(1)}), \ldots \rangle \\ &= \left(\prod_{f,i} \eta_b^{(f)} \eta_a^{(i)}\right) \langle \mathrm{b}^{(1)}(-\boldsymbol{q}^{(1)}, r_3^{(1)}), \ldots |\mathbf{S}| \mathrm{a}^{(1)}(-\boldsymbol{p}^{(1)}, s_3^{(1)}), \ldots \rangle. \end{aligned} \qquad (16.40)$$

It should be mentioned here that the intrinsic parity of some of the hadrons is purely a matter of convention. For example, we can use the conservation of the operator **B** for baryon number in order to set the parity of the proton to $+1$. If we had a parity operator $\mathbf{U}(\mathrm{P}')$ for which the proton had intrinsic parity -1, then in place of $\mathbf{U}(\mathrm{P}')$ we could take the likewise conserved operator

$$\mathbf{U}(\mathrm{P}) = \mathbf{U}(\mathrm{P}')\exp(i\pi \mathbf{B}) \tag{16.41}$$

as our parity operator, for which the proton then has intrinsic parity $+1$. In contrast to this, the relative parity of the proton and antiproton is observable and is equal to -1 in agreement with the Dirac equation. In a similar way, one can use the conservation of strangeness by the strong interaction in order to assign an intrinsic parity of $+1$ to the Λ-hyperon.

The discussion of the charge conjugation is also straightforward. The operator $\mathbf{U}(\mathrm{C})$ should exchange particle and antiparticle in every state, and should leave momentum and spin unchanged:

$$\mathbf{U}(\mathrm{C})|\mathrm{a}(\boldsymbol{p}, s_3)\rangle = \eta'_a|\bar{\mathrm{a}}(\boldsymbol{p}, s_3)\rangle. \tag{16.42}$$

Here η'_a is again a phase factor. It has been experimentally demonstrated that the strong interaction is invariant under charge conjugation. A strong reaction and its corresponding charge-conjugated reaction obtained by replacing particles by antiparticles and vice versa have the same scattering amplitude apart from a phase factor. In operator language this reads

$$\mathbf{U}(\mathrm{C})^{-1}\mathbf{S}\mathbf{U}(\mathrm{C}) = \mathbf{S}. \tag{16.43}$$

An example of charge conjugated reactions is

$$\begin{aligned} \mathrm{p}(\boldsymbol{p}, s_3) + \bar{\mathrm{p}}(\boldsymbol{p}', s'_3) &\rightarrow \pi^+(\boldsymbol{k}_1) + \pi^-(\boldsymbol{k}_2), \\ \bar{\mathrm{p}}(\boldsymbol{p}, s_3) + \mathrm{p}(\boldsymbol{p}', s'_3) &\rightarrow \pi^-(\boldsymbol{k}_1) + \pi^+(\boldsymbol{k}_2). \end{aligned} \tag{16.44}$$

The phase factor η'_a in (16.42) is particularly important for particles that are identical to their antiparticles. Such particles are eigenstates of $\mathbf{U}(\mathrm{C})$ with eigenvalues ± 1. The π^0-meson, for example, has charge conjugation $\mathrm{C} = +1$ and the ρ^0-meson and the photon both have $\mathrm{C} = -1$.

Somewhat more difficult to discuss is the time-reversal operator $\mathbf{V}(\mathrm{T})$. According to Sect. 4.5 $\mathbf{V}(\mathrm{T})$ is defined as an *antiunitary* operator that, for one-particle states, reverses both momentum and spin (Wigner 1932):

$$\mathbf{V}(\mathrm{T})|\mathrm{a}(\boldsymbol{p}, s_3)\rangle = \eta''_a|\mathrm{a}(-\boldsymbol{p}, -s_3)\rangle, \tag{16.45}$$

where η''_a is again a phase factor.

Experimentally, it has been found for strong reactions that the time-reversed reactions, obtained by interchanging initial and final states and by reversing the momenta and spins, have the same amplitudes apart from possible phase factors. An example for two reactions related by time reversal is

$$\begin{aligned} \mathrm{p}(\boldsymbol{p}, s_3) + \pi^-(\boldsymbol{k}) &\rightarrow \mathrm{p}(\boldsymbol{p}', s'_3) + \pi^-(\boldsymbol{k}'), \\ \mathrm{p}(-\boldsymbol{p}', -s'_3) + \pi^-(-\boldsymbol{k}') &\rightarrow \mathrm{p}(-\boldsymbol{p}, -s_3) + \pi^-(-\boldsymbol{k}). \end{aligned} \tag{16.46}$$

In general, the S-matrix for strong interaction processes obeys the following relation, where $|\alpha\rangle$ and $|\beta\rangle$ denote arbitrary hadron states [see (4.142)–(4.144)].

$$\begin{aligned} (\mathbf{V}(\mathrm{T})^{-1}\mathbf{S}\mathbf{V}(\mathrm{T}))^\dagger &= \mathbf{S}, \\ \langle\alpha|\mathbf{S}|\beta\rangle &= \langle\mathbf{V}(\mathrm{T})\beta|\mathbf{S}|\mathbf{V}(\mathrm{T})\alpha\rangle. \end{aligned} \tag{16.47}$$

The question of the phase factors in (16.37), (16.42), (16.45) would at first seem to be rather confusing: As already mentioned, some phases are pure convention whereas others are measurable quantities (see Streater 1964). In hadron physics, however, there is a simple solution. We merely need to know the transformation properties of the quark fields under P, C, and T; this should be exactly the same as that of free Dirac fields (Sect. 4.5). The phases, acquired by a hadronic state under P, C, and T transformations, are then determined by its description in the quark model which will be discussed in Sect. 17.4.

The invariance under P, C, and T transformations only holds for strong and electromagnetic processes. Weak processes violate P, C, and T. For example, the K-meson can decay either into two poins or into three pions:

$$
\begin{aligned}
&K^+ + \pi^+\pi^0 , \\
&K^+ + \pi^+\pi^-\pi^+ .
\end{aligned}
\tag{16.48}
$$

It has been established experimentally that in both cases the pions are in S-states, i.e., they carry zero orbital angular momentum. Since pions have intrinsic parity -1, the two-pion state has parity $+1$ and the three-pion state has parity -1. In the framework of weak interactions it is therefore impossible to find a conserved parity operator.

From very general axioms of local field theory, it follows however that all interactions are invariant under the product $\Theta = \mathrm{CPT}$ as already mentioned in Sect. 4.5. This CPT invariance guarantees equality of the masses and total lifetimes of particle and antiparticle. For example, it follows that

$$
\begin{aligned}
m_\mathrm{p} &= m_{\bar{\mathrm{p}}} , \\
\tau_{\mathrm{K}^+} &= \tau_{\mathrm{K}^-} .
\end{aligned}
\tag{16.49}
$$

The CPT invariance has, until now, been fully confirmed by experiment. As an example we quote the following experimental values (Particle Data Group 1986):

$$
\begin{aligned}
\frac{2(m_\mathrm{p} - m_{\bar{\mathrm{p}}})}{m_\mathrm{p} + m_{\bar{\mathrm{p}}}} &= (7 \pm 4)\cdot 10^{-5} , \\
\frac{2(\tau_{\mathrm{K}^+} - \tau_{\mathrm{K}^-})}{\tau_{\mathrm{K}^+} + \tau_{\mathrm{K}^-}} &= (1.1 \pm 0.9)\cdot 10^{-3} .
\end{aligned}
\tag{16.50}
$$

This is compatible with (16.49).

Below we give the C, P, T, and Θ transformations of a simple reaction

$$
\begin{aligned}
&\mathrm{p}(\boldsymbol{p}, s_3) &&+ \pi^-(\boldsymbol{k}) &&\to n(\boldsymbol{p}', s_3') &&+ \pi^0(\boldsymbol{k}') , \\
\mathrm{P}: \; &\mathrm{p}(-\boldsymbol{p}, s_3) &&+ \pi^-(-\boldsymbol{k}) &&\to \mathrm{n}(-\boldsymbol{p}', s_3') &&+ \pi^0(-\boldsymbol{k}') , \\
\mathrm{C}: \; &\bar{\mathrm{p}}(\boldsymbol{p}, s_3) &&+ \pi^+(\boldsymbol{k}) &&\to \bar{\mathrm{n}}(\boldsymbol{p}', s_3') &&+ \pi^0(\boldsymbol{k}') , \\
\mathrm{T}: \; &\mathrm{n}(-\boldsymbol{p}', -s_3') &&+ \pi^0(-\boldsymbol{k}') &&\to \mathrm{p}(-\boldsymbol{p}, -s_3) &&+ \pi^-(-\boldsymbol{k}) , \\
\Theta: \; &\bar{\mathrm{n}}(\boldsymbol{p}', -s_3') &&+ \pi^0(\boldsymbol{k}') &&\to \bar{\mathrm{p}}(\boldsymbol{p}, -s_3) &&+ \pi^+(\boldsymbol{k}) .
\end{aligned}
\tag{16.51}
$$

The time-reversal invariance is useful, among other reasons, since, assuming its validity, one can indirectly measure the scattering cross-sections for certain reactions with unstable particles in the initial state, for example $n + \pi^0 \to p + \pi^-$ as in (16.51).

16.3 Partial Wave Analysis

Our starting point here is the fact that the S-matrix must, as a result of rotational invariance, commute with the angular momentum operator. Thus the scattering matrix can only connect states that have the same angular momentum. In the basis of angular momentum eigenstates the S-matrix decomposes into submatrices corresponding to the individual angular momentum eigenvalues. The representation of S-matrix elements between momentum eigenstates in terms of these submatrices is known as partial wave analysis. For strong reactions, parity is also conserved and the S-matrix then further reduces into submatrices corresponding to the two parity eigenvalues ± 1. The partial wave analysis is the most important method for determining the spin and the parity of resonances.

We will carry out a partial wave analysis for the elastic $\pi^+ p$ scattering, i.e., for the reaction

$$\pi^+(\boldsymbol{p}_1) + \mathrm{p}(\boldsymbol{p}_2, r) \to \pi^+(\boldsymbol{p}_1') + \mathrm{p}(\boldsymbol{p}_2', r'), \tag{16.52}$$

where r and r' are the spin indices of the proton. As an application we will determine the spin and the parity of the Δ-resonance. The effect of the angular momentum operator $\vec{\mathbf{J}}$ on the initial state is, according to (16.36), as follows:

$$\mathbf{J}^j|\pi^+(\boldsymbol{p}_1), \mathrm{p}(\boldsymbol{p}_2, r)\rangle = \sum_s \left\{\delta_{sr}\varepsilon^{jkl}\left(p_1^k\, \mathrm{i}\frac{\partial}{\partial p_1^l} + p_2^k\, \mathrm{i}\frac{\partial}{\partial p_2^l}\right) + \tfrac{1}{2}\sigma_{sr}^j\right\}|\pi^+(\boldsymbol{p}_1)\mathrm{p}(\boldsymbol{p}_2, s)\rangle. \tag{16.53}$$

We now introduce the center of mass momentum $\boldsymbol{p}_s$ and the relative momentum $\boldsymbol{p}$:

$$\begin{aligned} \boldsymbol{p}_s &= \boldsymbol{p}_1 + \boldsymbol{p}_2, \\ \boldsymbol{p} &= \tfrac{1}{2}(\boldsymbol{p}_1 - \boldsymbol{p}_2). \end{aligned} \tag{16.54}$$

We then have

$$\mathbf{J}^j|\pi^+(\tfrac{1}{2}\boldsymbol{p}_s + \boldsymbol{p}), \mathrm{p}(\tfrac{1}{2}\boldsymbol{p}_s - \boldsymbol{p}, r)\rangle = \sum_s \left\{\delta_{sr}\varepsilon^{jkl}\left(p_s^k\, \mathrm{i}\frac{\partial}{\partial p_s^l} + p^k\, \mathrm{i}\frac{\partial}{\partial p^l}\right) + \tfrac{1}{2}\sigma_{sr}^j\right\}|\pi^+(\tfrac{1}{2}\boldsymbol{p}_s + \boldsymbol{p}), \mathrm{p}(\tfrac{1}{2}\boldsymbol{p}_s - \boldsymbol{p}, s)\rangle. \tag{16.55}$$

In the center of mass frame, i.e., for $\boldsymbol{p}_s = 0$ we obtain

$$\mathbf{J}^j|\pi^+(\boldsymbol{p}), \mathrm{p}(-\boldsymbol{p}, r)\rangle = \sum_s \{\delta_{sr}\mathbf{L}^j + \tfrac{1}{2}\sigma^j_{sr}\}|\pi^+(\boldsymbol{p}), \mathrm{p}(-\boldsymbol{p}, s)\rangle, \quad (16.56)$$

where the operator $\vec{\mathbf{L}}$ for the orbital angular momentum in the center of mass frame is defined as follows:

$$\mathbf{L}^j|\pi^+(\boldsymbol{p}), \mathrm{p}(-\boldsymbol{p}, r)\rangle = \varepsilon^{jkl}p^k\,\mathrm{i}\frac{\partial}{\partial p^l}|\pi^+(\boldsymbol{p}), \mathrm{p}(-\boldsymbol{p}, r)\rangle. \quad (16.57)$$

We are looking for eigenstates of the total angular momentum, i.e., of $\vec{\mathbf{J}}^2$ and $\mathbf{J}^3$. As a first step we construct the eigenstates of the orbital angular momentum, i.e., of $\vec{\mathbf{L}}^2$ and $\mathbf{L}^3$, for fixed center of mass energy $\sqrt{s}$ or for fixed $|\boldsymbol{p}|$. These states can be defined with the help of the spherical harmonics $Y_{l,l_3}(\hat{\boldsymbol{p}})$ ($\hat{\boldsymbol{p}} = \boldsymbol{p}/|\boldsymbol{p}|$) from Appendix C, as follows:

$$|s; l, l_3; \tfrac{1}{2}, r\rangle = \int d\Omega_p\, Y_{l,l_3}(\hat{\boldsymbol{p}})|\pi^+(\boldsymbol{p}), \mathrm{p}(-\boldsymbol{p}, r)\rangle$$
$$(l = 0, 1, 2, \ldots; l_3 = -l, \ldots, l). \quad (16.58)$$

In fact, as a simple calculation shows, we have

$$\vec{\mathbf{L}}^2|s; l, l_3; \tfrac{1}{2}, r\rangle = l(l+1)|s; l, l_3; \tfrac{1}{2}, r\rangle,$$
$$\mathbf{L}^3|s; l, l_3; \tfrac{1}{2}, r\rangle = l_3|s; l, l_3; \tfrac{1}{2}, r\rangle. \quad (16.59)$$

In order to obtain the eigenstates of the total angular momentum, we now include the spin of the proton in accordance with the rules of angular momentum addition. The eigenstates are defined with the help of the Clebsch–Gordan coefficients $(l, l_3; \tfrac{1}{2}, r|j, j_3)$ as

$$|s; l; j, j_3\rangle = \sum_{l_3,r} (l, l_3; \tfrac{1}{2}, r|j, j_3)|s; l, l_3; \tfrac{1}{2}, r\rangle$$
$$(|l-\tfrac{1}{2}| \le j \le l+\tfrac{1}{2};\ j_3 = -j, \ldots, j). \quad (16.60)$$

They satisfy

$$\vec{\mathbf{J}}^2|s; l; j, j_3\rangle = j(j+1)|s; l; j, j_3\rangle,$$
$$\mathbf{J}^3|s; l; j, j_3\rangle = j_3|s; l; j, j_3\rangle. \quad (16.61)$$

The parity of the states given in (16.60) is determined as follows: The intrinsic parity of the proton is set by convention to $+1$ (see Sect. 16.2). If we regard the intrinsic parity of the pion as known to be -1, then from (16.37), (16.58) and from (C.37) in Appendix C, we find:

$$\mathbf{U}(\mathrm{P})|s; l; j, j_3\rangle = (-1)^{l+1}|s; l; j, j_3\rangle. \quad (16.62)$$

For the normalization of the states (16.60) we find from (16.27) using Fermi's trick from Chap. 5:

$$\langle s'; l'; j', j_3'|s; l; j, j_3\rangle = (2\pi)^3 V\frac{8s}{w(s, m_\pi^2, m_\mathrm{p}^2)}\delta_{l',l}\delta_{j',j}\delta_{j_3',j_3}\delta(\sqrt{s'} - \sqrt{s}). \quad (16.63)$$

Here V is the normalization volume and $w(s, m_\pi^2, m_p^2) = 2\sqrt{s}\,|\boldsymbol{p}|$ [see (2.72), (2.76)].

Inverting equations (16.58), (16.60) now gives us the expansion of the incoming state in terms of angular momentum eigenstates (Exercise 16.4):

$$|\pi^+(\boldsymbol{p}), \mathrm{p}(-\boldsymbol{p}, r)\rangle = \sum_{l,l_3,j,j_3} Y^*_{l,l_3}(\hat{\boldsymbol{p}})(l, l_3; \tfrac{1}{2}, r|j, j_3)|s; l; j, j_3\rangle. \quad (16.64)$$

The sum runs over the following values of total angular momentum j and orbital angular momentum l:

$$j = \tfrac{1}{2}, \tfrac{3}{2}, \tfrac{5}{2}, \ldots;$$

$$l = j \pm \tfrac{1}{2}.$$

According to (16.62), states with the same j-value but different l-values have different parity (Table 16.1).

We now consider the elastic π^+p scattering in the center of mass system (Fig. 16.4). The momentum of the outgoing pion is $\boldsymbol{p}'$ and the scattering angle is ϑ. For the final state $|\pi^+(\boldsymbol{p}'), p(-\boldsymbol{p}', r')\rangle$ we can write down an expression in complete analogy to (16.64) with the substitutions $\boldsymbol{p} \to \boldsymbol{p}'$ and $\boldsymbol{r} \to \boldsymbol{r}'$. As already mentioned, the S-matrix cannot change the angular momentum or the parity of a state. Thus we have

$$\langle s'; l'; j', j_3'|\mathbf{S}|s; l; j, j_3\rangle = 0, \quad (16.65)$$

which is valid except when $l' = l$, $j' = j$, and $j_3' = j_3$. On account of the rotational invariance, the diagonal elements are not dependent on j_3 and we parametrize them in terms of their magnitudes $\eta_{l,j}(s) \geq 0$ and the real phases $\delta_{l,j}(s)$ as follows:

$$\langle s'; l; j, j_3|\mathbf{S}|s; l; j, j_3\rangle = \eta_{l,j}(s) \cdot e^{2i\delta_{l,j}(s)} \cdot \langle s'; l; j, j_3|s; l; j, j_3\rangle. \quad (16.66)$$

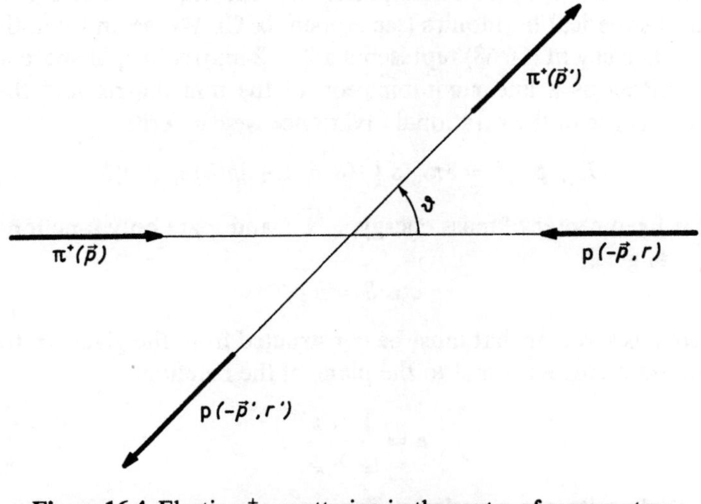

Figure 16.4 Elastic π^+p scattering in the center of mass system.

Here we make use of the fact that, due to energy conservation and according to (16.63), both sides of (16.66) are proportional to $\delta(\sqrt{s'} - \sqrt{s})$. Since **S** is a unitary operator, the magnitude of diagonal matrix elements for normed states must be smaller than or equal to one, i.e.,

$$0 \leq \eta_{l,j}(s) \leq 1. \tag{16.67}$$

The $\delta_{l,j}$ are called phase shifts and $\eta_{l,j}$ the absorption parameters. Pure elastic scattering corresponds to $\eta_{l,j} = 1$, complete absorption to $\eta_{l,j} = 0$.

We now use these relationships to express the T-matrix element for our reaction (16.52). In this we set

$$\langle \pi^+(\boldsymbol{p}'), \mathrm{p}(-\boldsymbol{p}', r') | \mathbf{T} | \pi^+(\boldsymbol{p}), \mathrm{p}(-\boldsymbol{p}, r) \rangle \equiv T_{r',r}(\boldsymbol{p}', \boldsymbol{p}). \tag{16.68}$$

Using Fermi's trick we then obtain from (5.43):

$$\langle \pi^+(\boldsymbol{p}'), \mathrm{p}(-\boldsymbol{p}', r') | (\mathbf{S} - 1) | \pi^+(\boldsymbol{p}), \mathrm{p}(-\boldsymbol{p}, r) \rangle = \mathrm{i} V 2\pi \delta(\sqrt{s'} - \sqrt{s}) T_{r',r}(\boldsymbol{p}', \boldsymbol{p}). \tag{16.69}$$

If we substitute the expansion of the initial and final state from (16.64) on the left-hand side of this equation, then with (16.63)–(16.66) we obtain

$$T_{r',r}(\boldsymbol{p}', \boldsymbol{p}) = \frac{64\pi^2 s}{w(s, m_\pi^2, m_\mathrm{p}^2)} \sum_{l, l_3', l_3, j, j_3} a_{l,j}(s) \cdot Y_{l,l_3'}(\hat{\boldsymbol{p}}') Y^*_{l,l_3}(\hat{\boldsymbol{p}}) (l, l_3'; \tfrac{1}{2}, r' | j, j_3)(l, l_3; \tfrac{1}{2}, r | j, j_3), \tag{16.70}$$

where

$$a_{l,j}(s) = \frac{1}{2\mathrm{i}} (\eta_{l,j}(s) \mathrm{e}^{2\mathrm{i}\delta_{l,j}(s)} - 1). \tag{16.71}$$

Equation (16.70) is the partial wave expansion for $\pi^+\mathrm{p}$ elastic scattering. The methods used in deriving this equation can clearly be easily generalized to the scattering of particles with any spin.

The expansion (16.70) can be simplified with the help of the theorem for the addition of spherical harmonics (see Appendix C). We begin by noting that the T-matrix element (16.68) represents a 2×2 matrix in spin space and can thus be written as a linear combination of the unit matrix and the Pauli matrices. Because of the rotational invariance we can write

$$T_{r',r}(\boldsymbol{p}', \boldsymbol{p}) = 8\pi\sqrt{s}\, \{ f(\vartheta)\delta_{r',r} + \mathrm{i} g(\vartheta) \boldsymbol{\sigma}_{r',r} \cdot \boldsymbol{c} \}. \tag{16.72}$$

where, for fixed center of mass energy, $\sqrt{s}$, f and g are only functions of the scattering angle ϑ,

$$\cos\vartheta = \hat{\boldsymbol{p}}' \hat{\boldsymbol{p}}, \tag{16.73}$$

and where $\boldsymbol{c}$ is a vector that must be constructed from the given vectors $\boldsymbol{p}, \boldsymbol{p}'$ and from the vector $\boldsymbol{n}$ normal to the plane of the reaction:

$$\boldsymbol{n} = \frac{\boldsymbol{p} \times \boldsymbol{p}'}{|\boldsymbol{p} \times \boldsymbol{p}'|}. \tag{16.74}$$

As a result of parity conservation in our reaction, $\boldsymbol{c}$ must be even with respect to the simultaneous reversal of the signs of $\boldsymbol{p}$ and $\boldsymbol{p}'$. Thus $\boldsymbol{c}$ must be proportional to $\boldsymbol{n}$. The proportionality factor can be absorbed into $g(\vartheta)$ enabling us to write

$$T_{r',r}(\boldsymbol{p}',\boldsymbol{p}) = 8\pi\sqrt{s}\,\{f(\vartheta)\delta_{r',r} + \mathrm{i}g(\vartheta)\boldsymbol{\sigma}_{r',r}\cdot\boldsymbol{n}\}. \tag{16.75}$$

We now compare this with (16.70). By forming the trace in the spin indices $f(\vartheta)$ is projected out and using (C.41) from Appendix C for the Clebsch–Gordan coefficients, we obtain the following expansion:

$$\begin{aligned} f(\vartheta) &= \frac{1}{16\pi\sqrt{s}}\sum_r T_{r,r}(\boldsymbol{p}',\boldsymbol{p}) \\ &= \frac{2\pi}{|\boldsymbol{p}|}\sum_{l,j} a_{l,j}\sum_{l'_3,l_3} Y_{l,l'_3}(\hat{\boldsymbol{p}}')Y^*_{l,l_3}(\hat{\boldsymbol{p}})\frac{2j+1}{2l+1}\delta_{l'_3,l_3} \\ &= \frac{1}{|\boldsymbol{p}|}\sum_{l=0}^{\infty}\left(\sum_{j=|l-1/2|}^{l+1/2}(j+\tfrac{1}{2})a_{l,j}\right)P_l(\cos\vartheta). \end{aligned} \tag{16.76}$$

Here P_l are the Legendre polynomials. In a similar way, by multiplying (16.70) and (16.75) by $\sigma_{r,r'}$ and forming the trace we obtain the expansion of $g(\vartheta)$ in the form (Exercise 16.5)

$$g(\vartheta) = \frac{1}{|\boldsymbol{p}|}\sum_{l=1}^{\infty}(a_{l,l+1/2} - a_{l,l-1/2})\sin\vartheta\, P'_l(\cos\vartheta). \tag{16.77}$$

The differential cross-section can be easily calculated from (5.51) and (16.75). For unpolarized protons in the initial state and with summation over the spin directions of the protons in the final state we obtain

$$\frac{\mathrm{d}\sigma}{\mathrm{d}\Omega} = |f(\vartheta)|^2 + |g(\vartheta)|^2, \tag{16.78}$$

where Ω is the solid angle of the outgoing pion in the center of mass system.

With the aid of these preliminaries we now want to determine the spin and the parity of the $\Delta(1232)$-resonance. Since this resonance is so marked in the plot of the total cross-section (Fig. 16.1), we can assume that the scattering in this energy range is dominated by *one* partial wave. Which one this is can be inferred from the observed angular distribution of the outgoing pion. The corresponding values of l and j then give the spin and parity of the resonance.

We will examine two possibilities. Let us first assume that we are dealing with a resonance of spin $\frac{1}{2}$ and negative parity. A state with these quantum numbers can only be produced by pions with orbital angular momentum $l = 0$, as can be seen from Table 16.1. If such a resonance is dominant, i.e., if it is mainly the $l = 0, j = \frac{1}{2}$ term that contributes in the expansion (16.70) then we obtain an isotropic angular distribution for the pion in the final state:

$$\frac{\mathrm{d}\sigma}{\mathrm{d}\Omega} = \text{const.} \qquad \text{for} \quad j^P = \tfrac{1}{2}^-. \tag{16.79}$$

Table 16.1 Angular momentum and parity values occurring in the π^+–p state in the center of mass system (16.64).

Total angular momentum j	Orbital angular momentum l	Parity $(-1)^{l+1}$
$\frac{1}{2}$	0	-1
$\frac{1}{2}$	1	$+1$
$\frac{3}{2}$	1	$+1$
$\frac{3}{2}$	2	-1
$\frac{5}{2}$	2	-1
$\vdots$	$\vdots$	$\vdots$

If, on the other hand, we consider the formation of a resonance with spin $\frac{3}{2}$ and positive parity, then according to Table 16.1, the orbital angular momentum in the initial and final state must be $l = 1$. When the partial wave with $l = 1$ and $j = \frac{3}{2}$ is the dominant term, then from (16.76)–(16.78) we find:

$$f(\vartheta) = \frac{2}{|\boldsymbol{p}|} a_{1,3/2} \cos\vartheta,$$
$$g(\vartheta) = \frac{1}{|\boldsymbol{p}|} a_{1,3/2} \sin\vartheta, \tag{16.80}$$

$$\frac{d\sigma}{d\Omega} = \frac{1}{|\boldsymbol{p}|^2} |a_{1,3/2}|^2 (1 + 3\cos^2\vartheta). \tag{16.81}$$

Figure 16.5 shows data for $d\sigma/d\Omega$ as a function of $\cos\vartheta$ for a center of mass energy $\sqrt{s} = 1235.4$ MeV, i.e., exactly on the Δ-resonance (Bussey 1973). As can be seen, a constant distribution (16.79) is ruled out by this data. However, a $(1 + 3\cos^2\vartheta)$-law (16.81) provides a very reasonable description of the angular distribution. It is clear though, that other partial waves also contribute to some extent. From our simple analysis we are thus able to claim that the Δ-resonance probably has spin $\frac{3}{2}$ and parity $+1$. In fact, this has been confirmed in detailed studies (see Höhler 1983).

We would also like to determine $a_{1,3/2}$ from the data. For $\sqrt{s} = 1235.4$ MeV we find $|\boldsymbol{p}|^{-2} = (230.2\ \text{MeV})^{-2} \cong 7.34$ mb, and, assuming that the $l = 1, j = \frac{3}{2}$ amplitude is dominant, we have

$$\frac{d\sigma}{d\Omega} = 7.34 \cdot |a_{1,3/2}|^2 (1 + 3\cos^2\vartheta)\ \text{mb/sr}. \tag{16.82}$$

On the other hand, (16.71) and (16.67) imply

$$|a_{1,3/2}|^2 = \left(\frac{1 - \eta_{1,3/2}}{2}\right)^2 + \eta_{1,3/2} \sin^2\delta_{1,3/2} \leq 1, \tag{16.83}$$

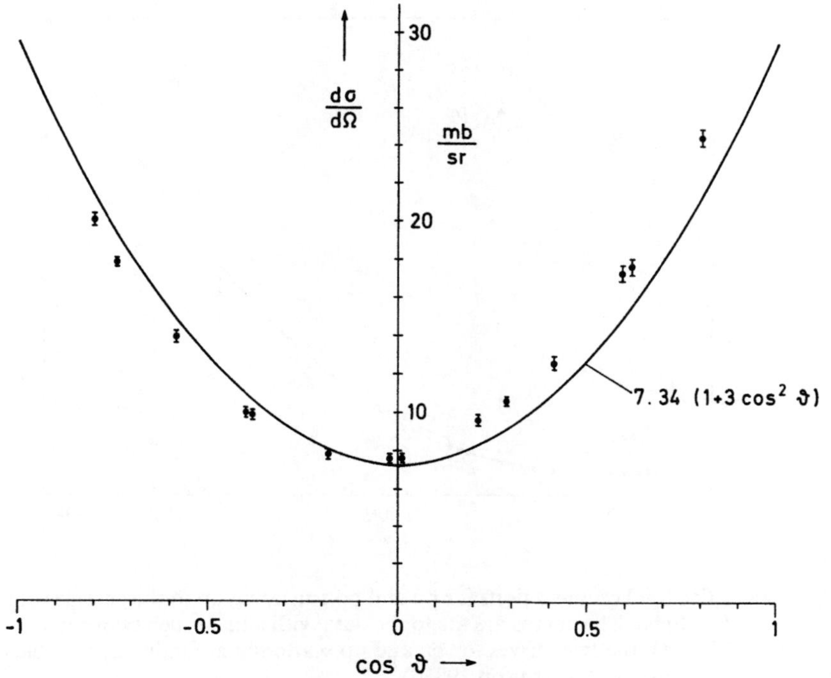

Figure 16.5 Experimental data for the angular distribution of elastic π^+p scattering at a center of mass energy $\sqrt{s} = 1235.4$ MeV (Bussey 1973). The curve corresponds to the maximum of $d\sigma/d\Omega$ under the assumption that only the partial wave with $l = 1$, $j = 3/2$ contributes (16.82), (16.83).

and Fig. 16.5 tells us that

$$|a_{1,3/2}|^2 \cong 1$$
$$\eta_{1,3/2} \cong 1, \qquad \delta_{1,3/2} \cong \frac{\pi}{2}. \tag{16.84}$$

The phase shift $\delta_{1,3/2}$ thus passes through $\pi/2$ at $\sqrt{s} = 1235.4$ MeV $\cong m_\Delta$, as is expected for a resonance. In addition, the result (16.84) means that in the region of the Δ-resonance, π^+p scattering is almost completely elastic.

16.4 Total Cross-Sections at High Energies

The total cross-section for π^+p and π^-p scattering shows a very smooth behavior at high energies (Fig. 16.1). Similar behavior is found for the total cross-section for pp and $\bar{\text{p}}$p scattering as shown in Fig. 16.6. The theoretical calculation of these cross-sections in the framework of quantum chromodynamics (to be discussed later), is an unsolved problem. Thus, in this

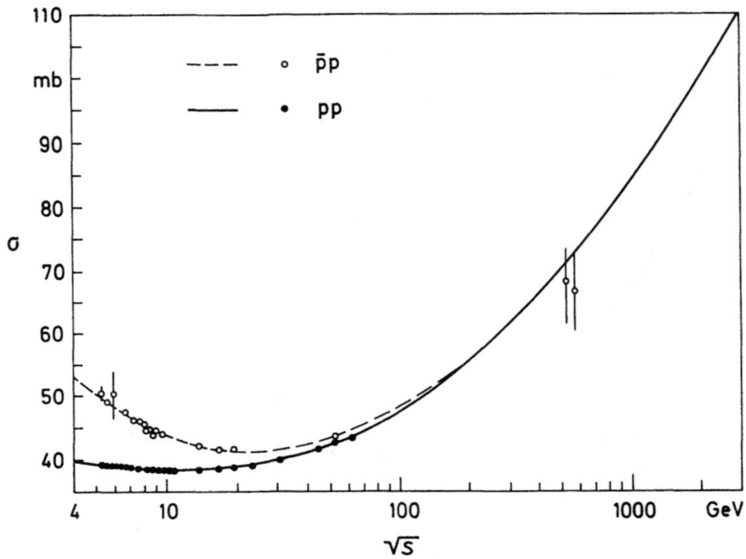

Figure 16.6 The total cross-sections for p̄p and pp scattering up to the energies of the CERN Sp̄pS collider. The curves are fits to the data with a $(\ln s)^2$ behavior for $s \to \infty$. For $\sqrt{s} \geq 150$ GeV the two curves for p̄p and pp scattering are indistinguishable on the scale of this figure (after Arnison 1983d).

section, we will merely discuss two theorems that are derived from general assumptions: the Froissart–Martin bound (Froissart 1961; Martin 1963) and the Pomeranchuk theorem (Pomeranchuk 1958).

We know that hadrons are extended objects. The radius of a hadron is given by the range of the Yukawa potential and is thus of the order of $1/m_\pi$. The Yukawa potential can be thought of as the effect of a cloud of virtual π-mesons which surrounds each hadron. The probability of encountering such a virtual pion decreases in proportion to $\exp(-m_\pi r)$ at the edge of the hadron. If we consider hadron–hadron scattering at very high energies, we can assume that scattering always takes place when the hadrons collide such that they overlap geometrically. Naïvely we would then expect the total scattering cross section to be

$$\sigma_{\text{tot}}(s) \xrightarrow[s\to\infty]{} \frac{1}{m_\pi^2} \cdot c \tag{16.85}$$

with a numerical factor c of the order of one.

It is a great triumph of the axiomatic methods that from very few assumptions of a general nature, it has been possible to derive an upper bound for the total cross-section. The Froissart–Martin bound reads

$$\sigma_{\text{tot}}(s) \underset{s\to\infty}{\leq} \frac{\pi}{m_\pi^2}\left(\ln \frac{s}{s_0}\right)^2 . \tag{16.86}$$

This result corresponds quite closely to the naïve expectation (16.85). Unfortunately, (16.86) contains an unknown scale factor s_0 and no statement is made as to when the asymptotic behavior should set in. It is possible that the increase of the total cross section seen in Fig. 16.6 for $\sqrt{s} \gtrsim 40$ GeV, is already the asymptotic behavior corresponding to the Froissart–Martin bound.

With the help of axiomatic methods, many other bounds for cross-sections have been derived (see Roy 1972 for an overview). Of these, we will only mention the Pomeranchuk theorem. In this, one considers the scattering of a particle a and its antiparticle $\bar{\mathrm{a}}$ by the same target b:

$$\begin{aligned} \mathrm{a} + \mathrm{b} &\to X, \\ \bar{\mathrm{a}} + \mathrm{b} &\to X. \end{aligned} \tag{16.87}$$

From some fairly general assumptions it is possible to show that the difference in the total cross-sections for these two reactions should tend to zero at high energies:

$$\sigma_{\mathrm{tot}}^{\mathrm{ab}}(s) - \sigma_{\mathrm{tot}}^{\bar{\mathrm{a}}\mathrm{b}}(s) \xrightarrow[s \to \infty]{} 0. \tag{16.88}$$

This theorem, too, would seem to be compatible with the experimental data (Figs. 16.1 and 16.6).

We end this section with a brief mention of the so-called Regge pole model. In the 1960s and 1970s this model was extensively used to describe the high energy behavior of amplitudes and cross-sections. The model provided a set of semi-empirical rules and formulas which proved to be qualitatively useful (see Collins 1977). However, to date, it has not been possible to derive these rules and expressions from quantum chromodynamics.

16.5 Multi-Particle Production at High Energies

In this last section of the chapter we shall discuss a few of the features of particle production at high energies. If high energy hadrons collide with one another then, energetically, it is possible to produce many new hadrons. Reactions such as the proton–proton collision have been studied in great detail at the storage ring ISR at CERN (Giacomelli 1979). In reactions with center of mass energy $\sqrt{s} \cong 50$ GeV an average of 12 charged and 6 neutral particles are produced, and in proton–antiproton collisions at $\sqrt{s} =$ 540 GeV, an average of about 30 charged particles (Alpgård 1983). The mean multiplicity $\langle n_{\mathrm{ch}} \rangle$ of charged particles in the final state is shown in Fig. 16.7. The multiplicity grows at large energies presumably with a power of ln s. A fit to the data for so-called "non-single-diffractive" events gives an empirical formula of the form (Breakstone 1984)

$$\langle n_{\mathrm{ch}} \rangle = 0.61 + 0.56 \ln s + 0.129 \ln^2 s \qquad (s \text{ in GeV}^2). \tag{16.89}$$

An inelastic event is called single diffractive if one of the incoming particles is only slightly excited and thus produces correspondingly few secondary particles.

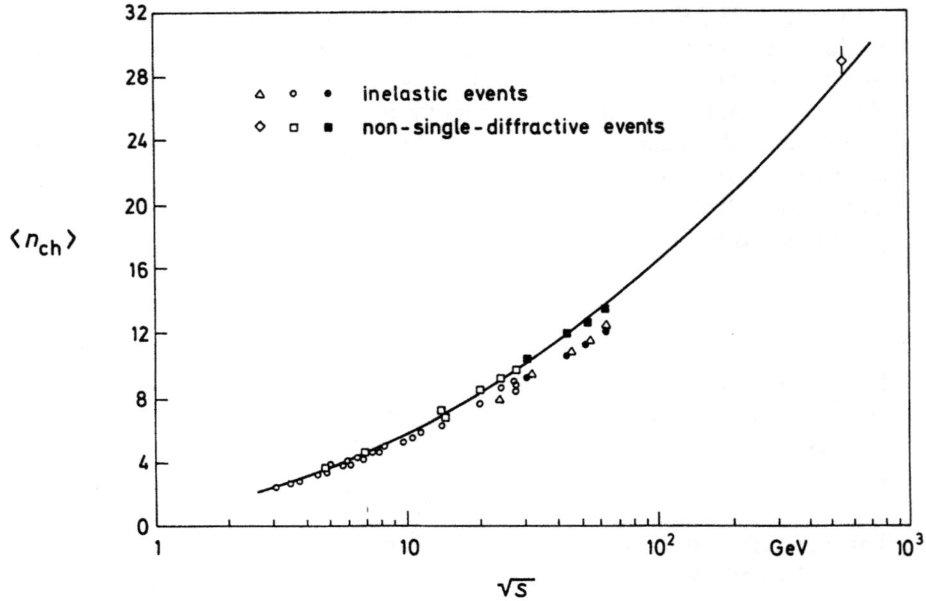

Figure 16.7 The mean multiplicity $\langle n_{ch} \rangle$ of charged particles produced in pp collisions as a function of the center of mass energy $\sqrt{s}$. The point at $\sqrt{s} = 540$ GeV refers to $\bar{p}p$ collisions. As indicated, the dataset consists of inelastic events and of so-called "non-single-diffractive" events. The curve represents a fit to the latter class of events (16.89) (after Breakstone 1984).

Figure 16.8 shows experimental multiplicities for the production of various types of particles in pp scattering. As can be seen, mainly pions are produced. This is understandable since they are the lightest hadrons. The intermediate state which is created in hadron–hadron scattering can be thought of as a very hot gas from which hadrons successively evaporate (Fig. 16.9). Thus it is clear that the lightest hadrons will most frequently be evaporated. In general, one expects a sort of Boltzmann distribution for the mean multiplicity:

$$\langle n_a \rangle \propto e^{-m_a/kT}, \tag{16.90}$$

where k is the Boltzmann constant and T is a characteristic temperature. Experimentally, one finds that the characteristic "thermal" energy of the fire ball is of the same order of magnitude as the pion mass:

$$kT \cong m_\pi. \tag{16.91}$$

Such thermodynamical concepts have had a certain amount of success in describing the many-particle production in high energy hadron–hadron scattering (Hagedorn 1968, 1983 and references therein).

A further phenomenological aspect of hadron–hadron reactions has to do with the transverse and longitudinal momenta of the particles produced at high energies. As an example, we again take proton–proton scattering at a

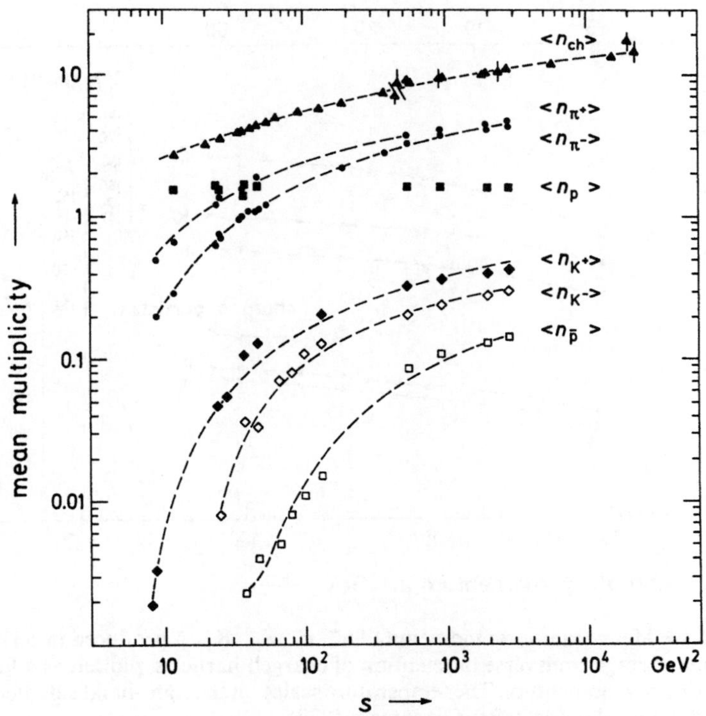

Figure 16.8 Mean multiplicity for π^+, π^-, K^+, K^-, p, $\bar{p}$ and for the total number of charged particles as a function of s in pp scattering (after Giacomelli 1979).

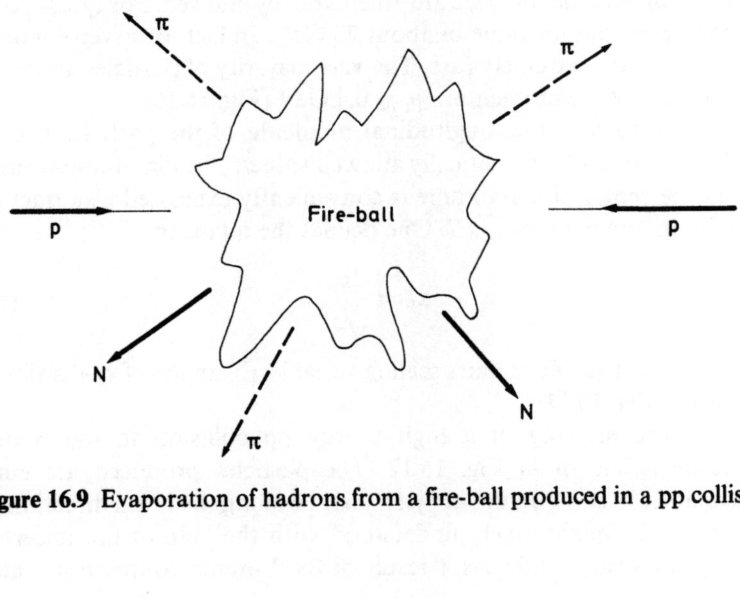

Figure 16.9 Evaporation of hadrons from a fire-ball produced in a pp collision.

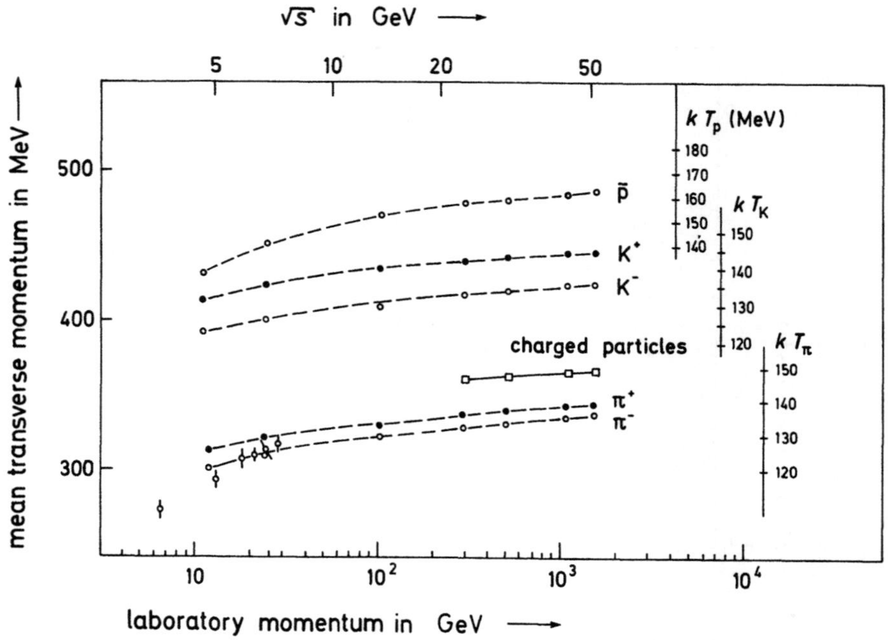

Figure 16.10 Mean transverse momenta of π^+, π^-, K^+, K^-, $\bar{p}$ produced in pp scattering and the average transverse momentum of charged particles, plotted as a function of the laboratory momentum. The temperature scales on the right-hand side stem from a thermodynamical model (after Giacomelli 1979).

center of mass energy $\sqrt{s} \gg m_p$. For the hadrons in the final state, transverse momenta p_t of the order of $\sqrt{s}/2$ are kinematically allowed. For $\sqrt{s} \cong 50$ GeV, p_t can take a maximum value of about 25 GeV. In fact, transverse momenta as large as this are extremely rare. The vast majority of particles appear with very small transverse momentum $p_t \lesssim 0.4$ GeV (Fig. 16.10).

In contrast to this, the longitudinal momenta of the particles produced cover the full range of kinematically allowed values. The longitudinal momentum p_l in the center of mass frame is conveniently expressed as a fraction of the maximum momentum $\sqrt{s}/2$. One defines the quantity

$$x_F = \frac{2p_l}{\sqrt{s}}, \tag{16.92}$$

which is referred to as Feynman's scaling variable. Examples of x_F-distributions are shown in Fig. 16.11.

A schematic drawing of a high energy pp collision in the center of mass frame is shown in Fig. 16.12. The particles produced are emitted predominantly in two bundles or "jets" that lie along the beam direction. This behavior can be qualitatively understood with the help of the uncertainty relation (Heisenberg 1952). As a result of the Lorentz contraction, the two

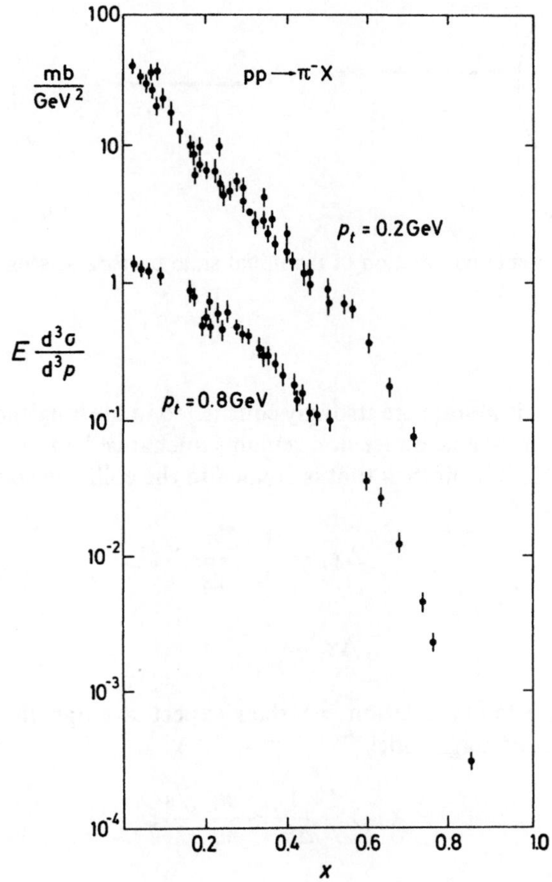

Figure 16.11 Invariant cross-sections for π^- production in pp collisions as a function of x_F for $p_t = 0.2$ GeV and 0.8 GeV. The data were obtained at various energies at the ISR (after Giacomelli 1979).

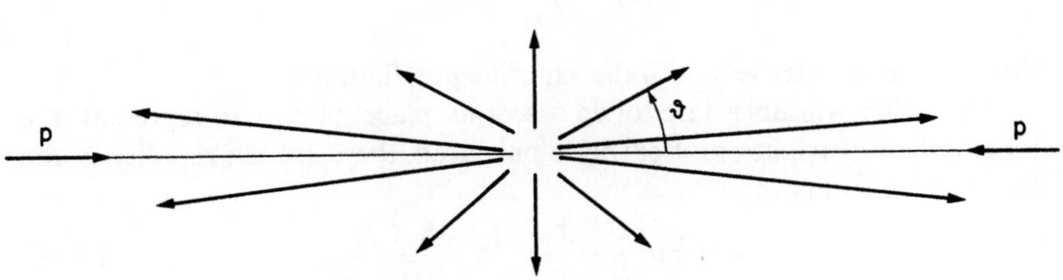

Figure 16.12 Schematic view of a proton–proton collision at high energy. The arrows represent the momenta of the particles produced.

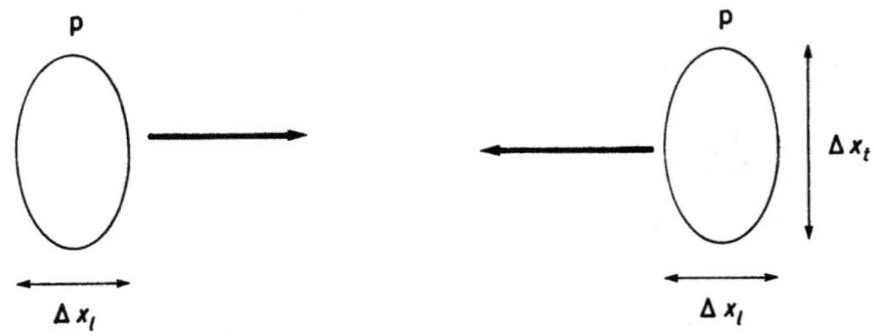

Figure 16.13 Lorentz contraction of the initial state protons as seen in the center of mass system.

protons in the initial state are strongly contracted in the longitudinal direction whereas their transverse dimension remains unchanged and is of the order of $1/m_\pi$ (Fig. 16.13). The fireball that is created in the collision should thus have the dimensions

$$\Delta x_1 \sim \frac{1}{m_\pi} \cdot \frac{2m_p}{\sqrt{s}}, \qquad \Delta x_t \sim \frac{1}{m_\pi}. \tag{16.93}$$

From the uncertainty relation we then expect average momenta of the following order of magnitude:

$$\langle p_l \rangle \sim \frac{1}{\Delta x_l} \sim \frac{m_\pi}{m_p} \frac{\sqrt{s}}{2},$$

and

$$\langle x_F \rangle = \left\langle \frac{2p_l}{\sqrt{s}} \right\rangle \sim \frac{m_\pi}{m_p} \sim 0.14, \qquad \langle p_t \rangle \sim \frac{1}{\Delta x_t} \sim m_\pi. \tag{16.94}$$

These agree qualitatively with the experimental findings.

Two other variables are often used in place of x_F to represent the longitudinal distribution of particles produced; these are the rapidity y, and the pseudorapidity η:

$$y = \tfrac{1}{2} \ln \frac{E + p_l}{E - p_l} = \ln \frac{E + p_l}{\sqrt{m^2 + p_t^2}}, \tag{16.95}$$

$$\eta = \tfrac{1}{2} \ln \frac{|\boldsymbol{p}| + p_l}{|\boldsymbol{p}| - p_l} = -\ln \tan \frac{\vartheta}{2}. \tag{16.96}$$

Here m is the mass of the particle produced and E, $\boldsymbol{p}$, ϑ and $p_l = |\boldsymbol{p}| \cos \vartheta$ are its energy, momentum, angle, and longitudinal momentum in the center of mass frame (Fig. 16.12). The distribution of charged particles produced in $\bar{p}p$

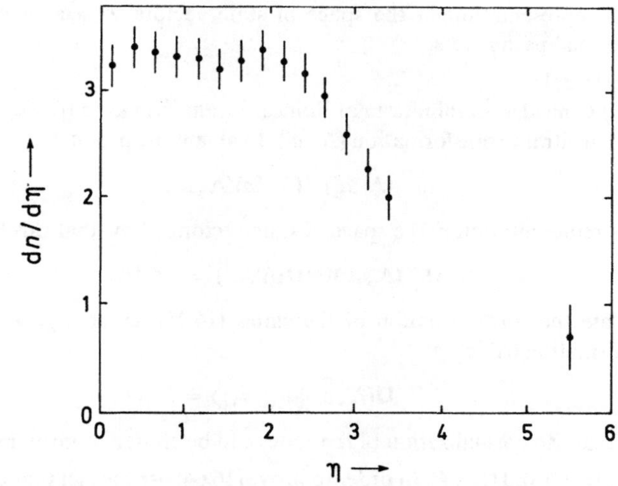

Figure 16.14 The distribution of charged particles produced in $\bar{p}p$ collisions at a center of mass energy of $\sqrt{s} = 540$ GeV, plotted against the pseudorapidity η. The figure shows only the region $\eta \geq 0$ since the distribution is symmetric about $\eta = 0$ due to charge-conjugation invariance. Single diffractive events have not been included (after Arnison 1983b).

collisions at center of mass energy $\sqrt{s} = 540$ GeV is shown in Fig. 16.14 as a function of η. We note the existence of a plateau in the central region with a particle density (Arnison 1983b)

$$\left.\frac{dn}{d\eta}\right|_{\eta=0} = 3.3 \pm 0.2. \tag{16.97}$$

The distributions observed in pp collisions for center of mass energies $\sqrt{s} \cong 50$–60 GeV show the same qualitative features but the height of the central plateau is somewhat reduced (Thomé 1977):

$$\left.\frac{dn}{d\eta}\right|_{\eta=0} \cong 1.9 \qquad \text{for} \quad \sqrt{s} = 63 \text{ GeV}. \tag{16.98}$$

To conclude this chapter we mention a special but rare phenomenon in hadron–hadron collisions, namely, the production of hadrons with very high transverse momenta. Such events are today interpreted as more or less direct manifestations of large angle scattering of the hadron constituents from one another. These reactions will be investigated in more detail in Chap. 20.

Exercises

16.1 Prove the commutation rules in (16.15).

Hint: Consider two infinitesimal Poincaré transformations $(\Lambda, \delta a)$ and $(\Lambda', \delta a')$ as in (16.12). Expand the product

$$(\Lambda', \delta a')^{-1}(\Lambda, \delta a)^{-1}(\Lambda', \delta a')(\Lambda, \delta a)$$

and its representation in the space of state vectors to second order in the infinitesimal parameters.

16.2 Prove (16.26).

Hint: Consider an infinitesimal Poincaré transformation $(\Lambda, \delta a)$ as in (16.12) and an arbitrary transformation (Λ', a'). Evaluate the product

$$(\Lambda', a')^{-1}(\Lambda, \delta a)(\Lambda', a')$$

and its representation in the space of state vectors. Show that this leads to

$$\mathbf{U}^{-1}(\Lambda', a')\mathbf{P}^{\mu}\mathbf{U}(\Lambda', a') = \Lambda'^{\mu}{}_{\nu}\mathbf{P}^{\nu}. \tag{16.99}$$

16.3 Calculate the transformation of the states (16.25) under a general Poincaré transformation (Λ', a')

$$\mathbf{U}(\Lambda', a')|a(\boldsymbol{p}, s_3)\rangle = ?.$$

Show that the normalization of the states can be chosen as given in (16.27).

16.4 Prove (16.63) and (16.64). In order to prove (16.64) use the fact that the spherical harmonics $Y_{l,l_3}(\hat{\boldsymbol{p}})$ form a complete orthonormal system of functions on the unit sphere.

16.5 Derive the expansion of $g(\vartheta)$ in terms of partial waves (16.77) using the formulas in Appendix C.

16.6 If we wish simply to derive the angular distribution for the decay of a resonance, it is possible to proceed as follows: We consider the expansion of the initial state (16.64) and choose the third coordinate axis to lie in the direction of $\boldsymbol{p}$. Making use of the properties of the spherical harmonics, show that the initial state then only contains terms with $l_3 = 0$ and $j_3 = r = \pm\frac{1}{2}$. A resonance is thus only produced in a spin state $j_3 = \pm\frac{1}{2}$. Construct, according to (16.60), the final states which appear in the decay of a resonance with $j^P = \frac{1}{2}^-, \frac{1}{2}^+, \frac{3}{2}^+$, or $\frac{5}{2}^+$. Derive the corresponding angular distribution for the pions. Compare this with the general results of (16.76)–(16.78).

16.7 According to (16.5), for s close to m_Δ^2, the amplitude $a_{1,3/2}(s)$ behaves as follows:

$$a_{1,3/2}(s) \propto \frac{1}{s - m_\Delta^2 + i m_\Delta \Gamma_\Delta}.$$

Determine, according to (16.84), the correct normalization of $a_{1,3/2}(s)$ and also the behavior of the phase shift $\delta_{1,3/2}(s)$ as a function of s.

16.8 Consider the collision of two particles a and b in frames of reference that are obtained from the center of mass system by Lorentz boosts along the axis defined by the momenta of the incoming particles. Define, in analogy to (16.95), the rapidity of a produced particle c in these frames of reference. Discuss the transformation behavior of the rapidity.

CHAPTER 17

Internal Symmetries of the Strong Interaction and the Quark Model

We now turn our attention to the internal symmetries of the hadrons. In Chap. 15 we have already met the isospin. In this chapter we want to have a closer look at the SU(3) flavor group of Gell-Mann and Ne'eman. To this end, we must familiarize ourselves with the mathematical properties of the SU(3) group.

17.1 Mathematics of the SU(3) Group

We consider a three-dimensional vector space. Let the states

$$|1\rangle, \quad |2\rangle, \quad |3\rangle,$$

be orthonormal basis states with

$$\langle i|j\rangle = \delta_{ij} \qquad (1 \le i, j \le 3).$$

We might identify these states with fictitious one-quark states:

$$|1\rangle \equiv |\mathrm{u}\rangle, \qquad |2\rangle \equiv |\mathrm{d}\rangle, \qquad |3\rangle \equiv |\mathrm{s}\rangle. \tag{17.1}$$

The SU(3) group is the group of unitary transformations $\mathbf{U}$ with determinant 1 in our three-dimensional vector space. In the chosen basis we have:

$$\begin{aligned} \mathbf{U}|i\rangle &= |j\rangle U_{ji}, \\ \mathbf{U}^\dagger \mathbf{U} &= 1, \\ \det \mathbf{U} &= 1. \end{aligned} \tag{17.2}$$

As is the case for rotations in ordinary space or in isospin space, it is useful to consider infinitesimal transformations which we write in the following form:

$$\mathbf{U} = 1 + \mathrm{i}\delta\varphi \cdot \mathbf{H}, \tag{17.3}$$

where the real parameter $\delta\varphi$ is treated as a first order quantity and where $\mathbf{H}$ is a matrix. Inserting this in (17.2) leads to

$$\begin{aligned} (1 + \mathrm{i}\delta\varphi\mathbf{H})(1 - \mathrm{i}\delta\varphi\mathbf{H}^\dagger) &= 1, \\ \det(1 + \mathrm{i}\delta\varphi\mathbf{H}) = 1 + \mathrm{i}\delta\varphi(\mathrm{Sp}\,\mathbf{H}) &= 1. \end{aligned} \tag{17.4}$$

It then follows that $\mathbf{H}$ must satisfy the conditions

$$\begin{aligned} \mathbf{H} &= \mathbf{H}^\dagger, \\ \mathrm{Tr}\,\mathbf{H} &= 0. \end{aligned} \tag{17.5}$$

We see that $\mathbf{H}$ is a Hermitian matrix with vanishing trace. Conversely, every such matrix $\mathbf{H}$ generates an infinitesimal SU(3) transformation.

The number of linearly independent Hermitian 3×3 matrices is 9 since we have in each case to specify three real numbers for the diagonal elements and three complex numbers for the elements above the diagonal. The condition of vanishing trace (17.5) reduces the number of linearly independent matrices to 8. A suitable basis for the 3×3 Hermitian matrices with trace zero is provided by the Gell-Mann λ-matrices (Gell-Mann 1964):

$$\lambda_1 = \begin{pmatrix} 0 & 1 & 0 \\ 1 & 0 & 0 \\ 0 & 0 & 0 \end{pmatrix}, \quad \lambda_2 = \begin{pmatrix} 0 & -\mathrm{i} & 0 \\ \mathrm{i} & 0 & 0 \\ 0 & 0 & 0 \end{pmatrix}, \quad \lambda_3 = \begin{pmatrix} 1 & 0 & 0 \\ 0 & -1 & 0 \\ 0 & 0 & 0 \end{pmatrix},$$

$$\lambda_4 = \begin{pmatrix} 0 & 0 & 1 \\ 0 & 0 & 0 \\ 1 & 0 & 0 \end{pmatrix}, \quad \lambda_5 = \begin{pmatrix} 0 & 0 & -\mathrm{i} \\ 0 & 0 & 0 \\ \mathrm{i} & 0 & 0 \end{pmatrix}, \quad \lambda_6 = \begin{pmatrix} 0 & 0 & 0 \\ 0 & 0 & 1 \\ 0 & 1 & 0 \end{pmatrix},$$

$$\lambda_7 = \begin{pmatrix} 0 & 0 & 0 \\ 0 & 0 & -\mathrm{i} \\ 0 & \mathrm{i} & 0 \end{pmatrix}, \quad \lambda_8 = \frac{1}{\sqrt{3}} \begin{pmatrix} 1 & 0 & 0 \\ 0 & 1 & 0 \\ 0 & 0 & -2 \end{pmatrix}. \tag{17.6}$$

These matrices are the generalization of the isospin matrices τ_1, τ_2, and τ_3 to the case of SU(3). The normalization is chosen such that

$$\mathrm{Tr}(\lambda_a \lambda_b) = 2\delta_{ab}. \tag{17.7}$$

Each infinitesimal transformation of the SU(3) group can then be written in the form

$$\mathbf{U} = 1 + \mathrm{i}\delta\varphi_a \frac{\lambda_a}{2}. \tag{17.8}$$

As usual, repeated indices must be summed over.

The matrices $\frac{1}{2}\lambda$ are known as the generators of the SU(3) group. They obey the following commutation rules:

$$\left[\frac{\lambda_a}{2}, \frac{\lambda_b}{2}\right] = \mathrm{i} f_{abc} \frac{\lambda_c}{2}. \tag{17.9}$$

The quantities f_{abc} are called the structure constants of the SU(3) group. They characterize the algebra of λ-matrices and therefore the finite transformations in a suitable neighborhood of the unit transformation. With the help of (17.7) we find for f_{abc}:

$$f_{abc} = \frac{1}{4\mathrm{i}} \operatorname{Tr}[\lambda_a, \lambda_b]\lambda_c. \tag{17.10}$$

From this it is easy to see that f_{abc} is real and totally antisymmetric. The explicit form of the structure constants, and other useful relationships for the λ-matrices are given in Appendix C.

A further interesting quantity is the anticommutator of two λ-matrices:

$$\{\lambda_a, \lambda_b\} \equiv \lambda_a\lambda_b + \lambda_b\lambda_a. \tag{17.11}$$

This anticommutator is itself a Hermitian matrix and thus it must be possible to represent it as a linear combination of the unit matrix and the 8 λ-matrices:

$$\{\lambda_a, \lambda_b\} = c_{ab} 1 + 2d_{abc}\lambda_c. \tag{17.12}$$

If we form the trace in (17.12) then we readily find

$$c_{ab} = \tfrac{4}{3}\delta_{ab}. \tag{17.13}$$

Multiplying (17.12) by λ_c and forming the trace gives

$$d_{abc} = \tfrac{1}{4} \operatorname{Tr}\{\lambda_a, \lambda_b\}\lambda_c. \tag{17.14}$$

From this it can easily be seen that d_{abc} is real and totally symmetric. Its explicit form is given in Appendix C.

We now come to the representations of the SU(3) group. A unitary representation of the SU(3) group is a mapping of the 3×3 matrices $\mathbf{U}$ into unitary matrices $\mathbf{D}(\mathbf{U})$ which, in general, act in a space of a different dimension,

$$\begin{aligned} &\mathbf{U}\mathbf{U} \to \mathbf{D}(\mathbf{U}),\\ &\mathbf{D}(\mathbf{U}) \colon n \times n \text{ Matrix},\\ &\mathbf{D}^\dagger(\mathbf{U})\mathbf{D}(\mathbf{U}) = 1. \end{aligned} \tag{17.15}$$

To be a representation the mapping has to preserve the group property, i.e., for arbitrary $\mathbf{U}, \mathbf{V} \in$ SU(3) we must have

$$\mathbf{D}(\mathbf{U})\mathbf{D}(\mathbf{V}) = \mathbf{D}(\mathbf{U}\mathbf{V}). \tag{17.16}$$

We can immediately list a few representations. The trivial representation assigns 1 to every matrix $\mathbf{U}$:

$$\mathbf{U} \to 1. \tag{17.17}$$

Since it is one-dimensional it is called the 1-representation. The matrices $\mathbf{U}$ themselves also form a representation:

$$\mathbf{U} \to \mathbf{U}. \tag{17.18}$$

This is known as the 3-representation since it acts in a three-dimensional space. A further three-dimensional representation is obtained by assigning to $\mathbf{U}$ its complex conjugate matrix:

$$\mathbf{U} \to \mathbf{U}^*. \tag{17.19}$$

This is the 3*-representation.

For arbitrary representations the infinitesimal SU(3) transformations (17.8) are represented by matrices $\mathbf{D}(\mathbf{U})$ where to first order in $\delta\varphi_a$:

$$\begin{aligned} \mathbf{U} &= 1 + \mathrm{i}\delta\varphi_a \frac{\lambda_a}{2} \to \mathbf{D}(\mathbf{U}), \\ \mathbf{D}(\mathbf{U}) &= 1 + \mathrm{i}\delta\varphi_a \mathbf{F}_a. \end{aligned} \tag{17.20}$$

The matrices $\mathbf{F}_a$ defined by the above expansion are known as the generators of the representation. They are Hermitian matrices and, as can easily be shown from the group property of a representation, they obey the same commutation rules as the matrices $\lambda_a/2$:

$$[\mathbf{F}_a, \mathbf{F}_b] = \mathrm{i} f_{abc} \mathbf{F}_c. \tag{17.21}$$

Conversely, a set of Hermitian matrices $\mathbf{F}_a$ that satisfies these commutation rules leads to a unitary representation of the SU(3) group.

We now wish to discuss a further representation. For this purpose we consider all complex 3×3 matrices with zero trace as a linear space. A basis in this space R is formed by the λ-matrices and the space therefore has dimension 8. A general element of this space can be written

$$\mathbf{C} = c_a \frac{\lambda_a}{2}, \tag{17.22}$$

where the c_a are arbitrary complex numbers. Considering the matrix $\mathbf{C}$ as a vector in the space R we denote it by $|\mathbf{C}\rangle$. The scalar product in the space R is

$$\langle \mathbf{C}' | \mathbf{C} \rangle = 2\,\mathrm{Tr}(\mathbf{C}'^\dagger \mathbf{C}). \tag{17.23}$$

We define the transformations $\mathbf{D}(\mathbf{U})$ of the adjoint or 8-representation by

$$\mathbf{D}(\mathbf{U})\colon \mathbf{C} \to \mathbf{U}\mathbf{C}\mathbf{U}^\dagger, \tag{17.24}$$

where $\mathbf{U} \in \mathrm{SU}(3)$. It is easy to convince oneself that $\mathbf{D}(\mathbf{U})$ is a unitary transformation in the space R, and that the group property (17.16) is satisfied. If we let $\mathbf{U}$ in (17.24) be an infinitesimal transformation (17.8) then we obtain the generators of the adjoint representation:

$$\begin{aligned} (1 + \mathrm{i}\delta\varphi_a \mathbf{F}_a)\colon \mathbf{C} &\to \left(1 + \mathrm{i}\delta\varphi_a \frac{\lambda_a}{2}\right) \mathbf{C} \left(1 - \mathrm{i}\delta\varphi_b \frac{\lambda_b}{2}\right), \\ \mathbf{F}_a\colon \mathbf{C} &\to \left[\frac{\lambda_a}{2}, \mathbf{C}\right]. \end{aligned} \tag{17.25}$$

In particular, for $\mathbf{C} = \lambda_c/2$ we find

$$\mathbf{F}_a\colon \frac{\lambda_c}{2} \to \left[\frac{\lambda_a}{2}, \frac{\lambda_c}{2}\right] = \mathrm{i} f_{acb} \frac{\lambda_b}{2}. \tag{17.26}$$

Expressed as a transformation in the space R (17.26) takes on the form

$$\mathbf{F}_a \left|\frac{\lambda_c}{2}\right\rangle = \left|\frac{\lambda_b}{2}\right\rangle (\mathbf{F}_a)_{bc} = \left|\frac{\lambda_b}{2}\right\rangle \mathrm{i} f_{acb}. \tag{17.27}$$

In the basis of the Gell-Mann matrices, the generators $\mathbf{F}_a$ of the adjoint representation are thus 8 × 8 matrices that are given by the struture constants of the group:

$$(\mathbf{F}_a)_{bc} = \mathrm{i} f_{acb} = \frac{1}{\mathrm{i}} f_{abc}. \tag{17.28}$$

The concepts of reducible and irreducible representations are explained in Appendix C. One can readily convince oneself that the representations 3, 3*, and 8 are irreducible.

We now close our mathematical treatment of the SU(3) group; for further discussion we refer the reader to the literature (see, e.g., Gell-Mann 1964; Lipkin 1966; Gourdin 1967).

17.2 The Quark Model and the Flavor-SU(3) Group

In this section we will discuss Gell-Mann and Ne'eman's theory of the "eightfold way". We shall immediately use the language of the quark model although, historically, the "eightfold way" preceded the quark model.

The fundamental assumption of the quark model for hadrons is that mesons are quark–antiquark bound states and that baryons are three-quark states. In a schematic notation where spin and spatial degrees of freedom are suppressed we thus write:

$$\begin{aligned} \text{Mesons} &\sim \mathrm{q}\bar{\mathrm{q}}, \\ \text{Baryons} &\sim \mathrm{qqq}. \end{aligned} \tag{17.29}$$

The observed hadrons are eigenstates of the Hamilton operator for the strong interaction $\mathbf{H}_{\mathrm{st}}$. The group-theoretical concepts that have been developed above will be very useful now since they allow us to make predictions about the hadron spectrum without actually needing to solve the Schrödinger equation. In fact, a satisfactory derivation of the ansatz (17.29) within the framework of quantum chromodynamics remains an unsolved problem.

We begin by considering an idealized world with only three quarks u, d, and s (Table 15.4) and we make the following assumptions:

(i) Flavor universality of the strong forces, i.e., the strong forces should act in the same way on quarks with different flavor. (This is to be understood in the same sense as the universality of the electromagnetic forces for electrons and muons; see Chap. 11.)

(ii) Equality of the masses of u-, d-, and s-quarks:

$$m_u = m_d = m_s. \tag{17.30}$$

The Hamilton operator of the strong interaction $\mathbf{H}_{st}$ is then invariant under arbitrary SU(3) transformations of the quarks u, d, and s. Explicitly, this means the following: The Hilbert space of the quark model consists of states with one quark, two quarks, one antiquark, a quark–antiquark pair, etc. In order to simplify our notation we shall write

$$q_1 \equiv u, \qquad q_2 \equiv d, \qquad q_3 \equiv s. \tag{17.31}$$

By means of a unitary transformation for the one-quark states

$$|q_i\rangle \to |q_j\rangle U_{ji}, \tag{17.32}$$

where $\mathbf{U} = (U_{ji})$ is from SU(3), we can define in the usual manner a representation $\mathbf{D}(\mathbf{U})$ of the SU(3) group in the full Hilbert space. The two-quark states, for example, are transformed as follows:

$$\mathbf{D}(\mathbf{U})\colon |q_i\rangle|q_j\rangle \to |q_{i'}\rangle|q_{j'}\rangle U_{i'i}U_{j'j}. \tag{17.33}$$

Antiquarks are transformed with the complex conjugate matrix

$$\mathbf{D}(\mathbf{U})\colon |\bar{q}_i\rangle \to |\bar{q}_j\rangle U_{ji}^{*}. \tag{17.34}$$

Under the assumptions (i) and (ii) these flavor-SU(3) transformations and their generating operators $\mathbf{F}_a$ commute with the Hamilton operator. We have

$$[\mathbf{H}_{st}, \mathbf{D}(\mathbf{U})] = 0 \tag{17.35}$$

for all $\mathbf{U} \in$ SU(3) or, for $a = 1, \ldots, 8$,

$$[\mathbf{H}_{st}, \mathbf{F}_a] = 0. \tag{17.36}$$

To the operators $\mathbf{F}_a$ there thus correspond eight additive conserved quantum numbers.

We have already met a few of these conserved quantities. In Table 15.4 the isospin and hypercharge quantum numbers of the quarks were given. The quarks u and d form an isodoublet and s an isosinglet. For the third component of the isospin we have

$$\begin{aligned} \mathbf{I}_3|u\rangle &= \tfrac{1}{2}|u\rangle, \\ \mathbf{I}_3|d\rangle &= -\tfrac{1}{2}|d\rangle, \\ \mathbf{I}_3|s\rangle &= 0. \end{aligned} \tag{17.37}$$

If we compare this with the generators $\mathbf{F}_a$ which, when applied to the one-quark states, give (17.32), (17.8), (17.20)

$$\mathbf{F}_a|q_i\rangle = |q_j\rangle\left(\frac{\lambda_a}{2}\right)_{ji}, \tag{17.38}$$

then we readily see from the explicit form of the λ-matrices (17.6) that we can identify $\mathbf{F}_3$ with $\mathbf{I}_3$. Analogously, we find that $\mathbf{F}_1$ and $\mathbf{F}_2$ can be identified with the isospin components $\mathbf{I}_1$ and $\mathbf{I}_2$ and the hypercharge $\mathbf{Y}$ with $2\mathbf{F}_8/\sqrt{3}$:

$$\mathbf{I}_1 = \mathbf{F}_1, \qquad \mathbf{I}_2 = \mathbf{F}_2, \qquad \mathbf{I}_3 = \mathbf{F}_3, \qquad \mathbf{Y} = \frac{2}{\sqrt{3}}\mathbf{F}_8. \tag{17.39}$$

This identification of course then also holds for all states that are constructed from quarks and antiquarks. From the commutation rules of the SU(3) algebra (17.21) we can easily check that

$$\begin{aligned} [\mathbf{I}_a, \mathbf{I}_b] &= \mathrm{i}\varepsilon_{abc}\mathbf{I}_c \qquad && (1 \le a, b, c \le 3), \\ [\mathbf{I}_a, \mathbf{Y}] &= 0 && (a = 1, 2, 3). \end{aligned} \tag{17.40}$$

The isospin generators and the hypercharge thus generate a subgroup SU(2) × U(1) of the flavor-SU(3) group:

$$\mathrm{SU}(3) \supset \mathrm{SU}(2) \times \mathrm{U}(1).$$

Therefore our model at least incorporates the observed conservation of isospin and hypercharge. From (17.40) we read off that in particular

$$[\mathbf{I}_3, \mathbf{Y}] = 0,$$

or

$$[\mathbf{F}_3, \mathbf{F}_8] = 0.$$

It is also possible to show that there are no other linearly independent generators of the SU(3) group that commute with $\mathbf{F}_3$ and $\mathbf{F}_8$. The SU(3) group possesses exactly two linearly independent simultaneously diagonalizable generators. The group is said to have rank 2. Thus, in every representation, we can choose basis vectors that are simultaneous eigenstates of $\mathbf{F}_3 = \mathbf{I}_3$ and $\mathbf{F}_8$ (or $\mathbf{Y}$). It is useful to plot the eigenvalues of $\mathbf{I}_3$ and $\mathbf{Y}$ that occur in a representation as points in the $\mathbf{I}_3$–$\mathbf{Y}$ plane. For convenience, we choose the unit length along the $\mathbf{Y}$-axis to be a factor $\sqrt{3}/2$ shorter than that along the $\mathbf{I}_3$-axis.

We start by plotting the eigenvalues of $\mathbf{I}_3$ and $\mathbf{Y}$ that occur for the quarks. As we see, this so-called weight diagram forms an equilateral triangle (Fig. 17.1).

Next we consider the antiquarks. Under the flavor-SU(3) group they transform with the complex conjugate matrix $\mathbf{U}^*$ as given in (17.34). For an infinitesimal transformation we have

$$\begin{aligned} \mathbf{U}^* &= \left(1 + \mathrm{i}\delta\varphi_a \frac{\lambda_a}{2}\right)^* = 1 + \mathrm{i}\delta\varphi_a\left(-\frac{\lambda_a^*}{2}\right) \\ &= 1 + \mathrm{i}\delta\varphi_a\left(-\frac{\lambda_a^{\mathrm{T}}}{2}\right). \end{aligned} \tag{17.41}$$

The generators of the antiquark representation 3* are thus

$$\mathbf{F}_a = -\frac{\lambda_a^{\mathrm{T}}}{2}. \tag{17.42}$$

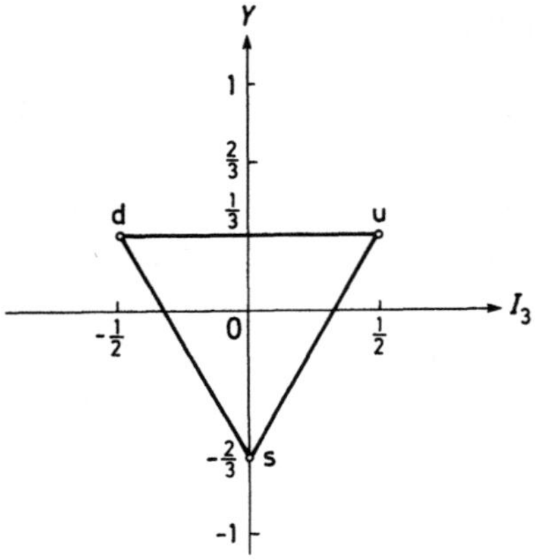

Figure 17.1 Weight diagram for the quark representation (3).

The eigenvalues of $\mathbf{I}_3$ and $\mathbf{Y}$ that appear in the antiquark representation are therefore obtained from those of the quark representation by a reflection at the origin (Fig. 17.2).

Now we turn to states with one quark and one antiquark. We have three quarks and three antiquarks and thus, when we hold all other variables such as the momentum and spin of the quarks and antiquarks fixed, the corresponding space is nine dimensional. The transformation of the basis states is as follows:

$$\mathbf{D}(\mathbf{U})(|q_i\rangle|\bar{q}_j\rangle) = |q_{i'}\rangle|\bar{q}_{j'}\rangle U_{i'i}U^*_{j'j}. \tag{17.43}$$

We can immediately write down one invariant state:

$$|1\rangle = \frac{1}{\sqrt{3}}|q_i\rangle|\bar{q}_i\rangle. \tag{17.44}$$

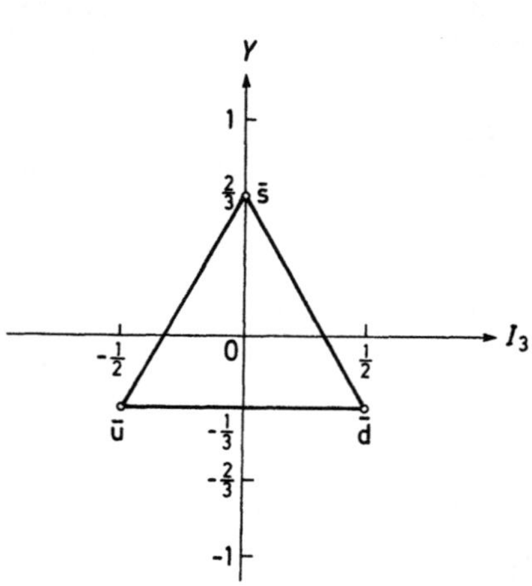

Figure 17.2 Weight diagram for the antiquark representation (3*).

In fact we have

$$\mathbf{D}(\mathbf{U})|1\rangle = \frac{1}{\sqrt{3}}|q_{i'}\rangle|\bar{q}_{j'}\rangle U_{i'i}U^*_{j'i}$$
$$= \frac{1}{\sqrt{3}}|q_{i'}\rangle|\bar{q}_{j'}\rangle(\mathbf{U}\mathbf{U}^\dagger)_{i'j'}$$
$$= \frac{1}{\sqrt{3}}|q_{i'}\rangle|\bar{q}_{i'}\rangle = |1\rangle. \tag{17.45}$$

We now investigate the states that are orthogonal to the state $|1\rangle$. A general state vector of this kind can be written as follows:

$$|\mathbf{C}\rangle = C_{ij}|q_i\rangle|\bar{q}_j\rangle, \tag{17.46}$$

where we set $\mathbf{C} = (C_{ij})$ and demand

$$\langle 1|\mathbf{C}\rangle = \frac{1}{\sqrt{3}}\,\mathrm{Tr}\,\mathbf{C} = 0, \tag{17.47}$$

i.e., $\mathbf{C}$ is a 3×3 matrix with zero trace. The transformation behavior is, according to (17.43),

$$\mathbf{D}(\mathbf{U})|\mathbf{C}\rangle = C_{ij}|q_{i'}\rangle|\bar{q}_{j'}\rangle U_{i'i}U^*_{j'j}$$
$$= (\mathbf{U}\mathbf{C}\mathbf{U}^\dagger)_{i'j'}|q_{i'}\rangle|\bar{q}_{j'}\rangle$$
$$= |\mathbf{C}'\rangle, \tag{17.48}$$

where

$$\mathbf{C}' = \mathbf{U}\mathbf{C}\mathbf{U}^\dagger. \tag{17.49}$$

This, however, is precisely the transformation law for the adjoint representation (17.24). With this we have reduced the product of the quark and antiquark representations to irreducible parts. Our result is

$$3 \times 3^* = 1 + 8. \tag{17.50}$$

What are the consequences of these results for the mesons, which, according to (17.29), are considered to be quark–antiquark bound states? With our assumptions (17.36) we have exact flavor-SU(3) invariance. It follows that all hadrons can be classified according to irreducible representations of the flavor-SU(3) group, and that all particles in an SU(3) multiplet must have the same mass. In order to prove this statement one uses the same methods with which one shows that rotational invariance in ordinary space guarantees that all spin orientations of a particle have the same mass (see (16.21) and the discussion of Schur's lemma in Appendix C).

As we have seen, from one quark and one antiquark, it is only possible to construct the trivial representation 1 and the adjoint representation 8. Thus, in our theory, all mesons must occur either as flavor-SU(3) singlets or as octets, i.e., in sets of eight all with the same mass, spin, and parity.

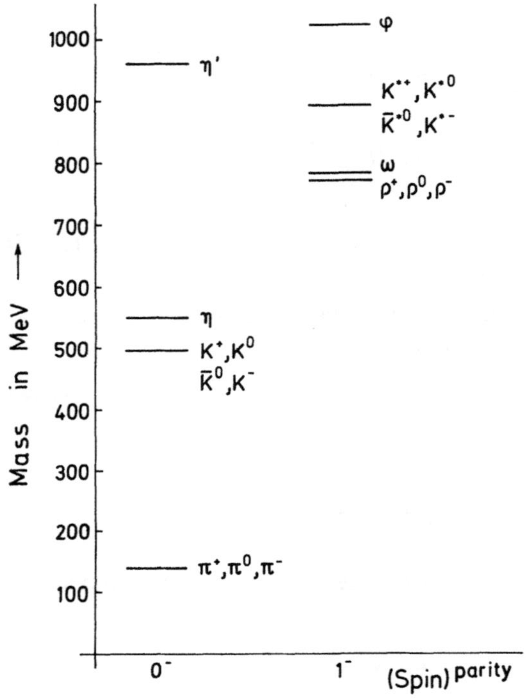

Figure 17.3 Level scheme for the lightest mesons.

Let us now see how well this agrees with experiment. From Table 15.1, we obtain for the lightest mesons the level scheme shown in Fig. 17.3. We indeed recognize a set of eight pseudoscalar mesons that are significantly lighter than the η'-meson, but neither for these 0^--mesons, nor for the 1^--mesons, can one claim a degeneracy of the masses. Let us now see whether at least the isospin and hypercharge quantum numbers correspond to those expected for flavor singlets and octets.

The values of hypercharge Y and isospin I occurring for mesons in the quark model are most easily determined with the help of the weight diagrams. The quantum numbers Y and I_3 are additive and thus their values for a $q\bar{q}$ state are obtained by simply adding the values for q and $\bar{q}$. The points in the weight diagram for the $3 \times 3^*$-representation, are thus obtained by taking every point of the quark diagram to be the origin of an antiquark diagram. In this way it is also easy to read off the multiplicities of the weights (Fig. 17.4). The hypercharge commutes with the isospin generators (17.40) and thus also with the square of the total isospin $\vec{T}^2$. The two states with $Y = +1$ must therefore form an isodoublet and likewise for the two states with $Y = -1$. The two states with $Y = 0$ and $I_3 = \pm 1$ must belong to an isotriplet. If we subtract the corresponding neutral member with $I_3 = 0$ from the states with $Y = I_3 = 0$, then there remain two linearly independent states. One of these is the SU(3) singlet state (17.44) and the other is an octet state with $Y = I_3 = 0$. Since this latter state has no isospin partner, it must be an isosinglet. We thus obtain

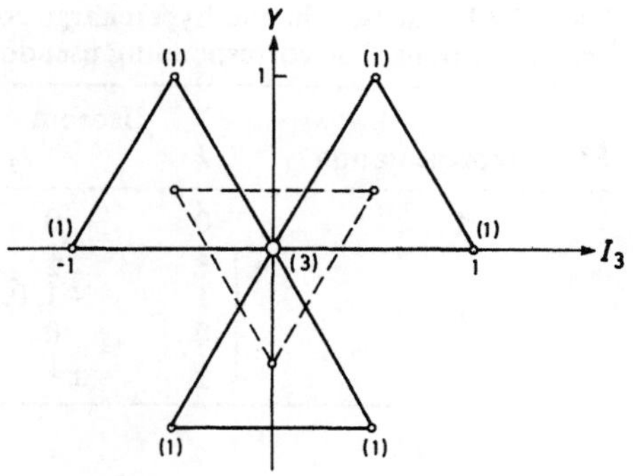

Figure 17.4 Weight diagram for the representation $3 \times 3^*$. The numbers in brackets are the multiplicities.

the isospin values given in Table 17.1 for the singlet and octet representations of SU(3).

This scheme agrees very well with the observed isospin and hypercharge values of the mesons as indicated in Table 17.1 for the pseudoscalar mesons. It therefore makes sense to arrange the mesons in weight diagrams for the singlet and octet representations (Fig. 17.5). Here we write η_1 and η_8 for the singlet and octet since, as we shall see shortly, the physical states η' and η are singlet–octet mixtures. From the weight diagrams in Fig. 17.5, it is also a straightforward matter to read off the quark content of the mesons:

$$
\begin{aligned}
\pi^+ &\sim \mathrm{u\bar{d}}, \\
\pi^0 &\sim \frac{1}{\sqrt{2}}(\mathrm{u\bar{u}} - \mathrm{d\bar{d}}), \\
\pi^- &\sim \mathrm{d\bar{u}}, \\
\mathrm{K}^+ &\sim \mathrm{u\bar{s}}, \\
\mathrm{K}^+ &\sim \mathrm{d\bar{s}}, \\
\bar{\mathrm{K}}^0 &\sim \mathrm{s\bar{d}}, \\
\mathrm{K}^- &\sim \mathrm{s\bar{u}}, \\
\eta_8 &\sim \frac{1}{\sqrt{6}}(\mathrm{u\bar{u}} + \mathrm{d\bar{d}} - 2\mathrm{s\bar{s}}), \\
\eta_1 &\sim \frac{1}{\sqrt{3}}(\mathrm{u\bar{u}} + \mathrm{d\bar{d}} + \mathrm{s\bar{s}}).
\end{aligned}
\tag{17.51}
$$

One of the recognized triumphs of the theory of the eightfold way is that all of the mesons so far observed that consist only of u-, d-, and s-quarks, fit into singlet and octet representations of the SU(3) group. Exceptions to this rule, so-called exotic mesons with, for example, isospin $= \frac{3}{2}$ and hypercharge $= 1$, or $I = 2$, and $Y = 0$, have never been observed.

Table 17.1 The isospin and hypercharge content of the SU(3) singlet (1) and octet (8) representations. The corresponding pseudoscalar mesons are also listed.

SU(3) representation	Isospin I	I_3	Hypercharge Y	0^- particles
1	0	0	0	η'
8	$\frac{1}{2}$	$\pm\frac{1}{2}$	1	K^+, K^0
	1	$+1, 0, -1$	0	π^+, π^0, π^-
	0	0	0	η
	$\frac{1}{2}$	$\pm\frac{1}{2}$	-1	$\bar{K}^0, K^-$

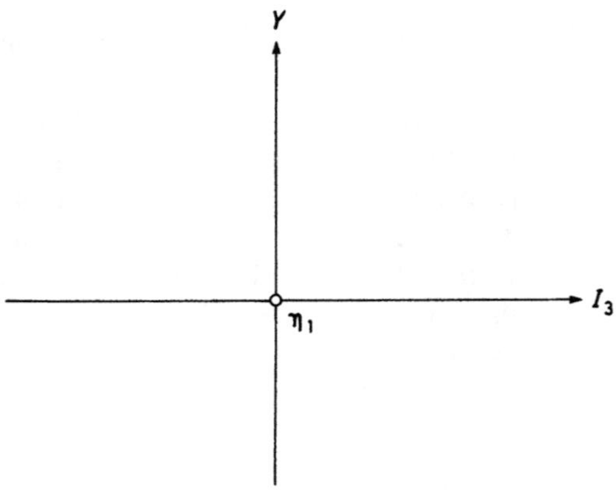

a)

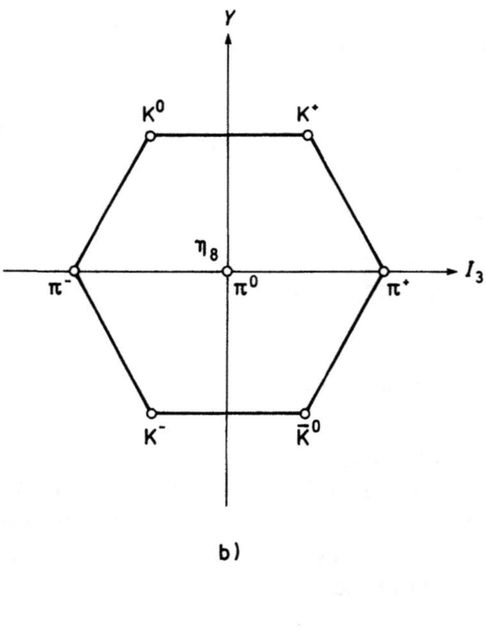

b)

Figure 17.5 Weight diagrams for the singlet (a) and octet (b) multiplets of the pseudo-scalar mesons.

Before we move on to discuss the origin of the mass splitting, we will turn our attention to the spectrum of the baryons in the limit of exact flavor-SU(3) invariance. In the quark model we consider baryons to be bound states of three quarks (17.29). We thus have to investigate which SU(3) states can be constructed from three quarks. We have 27 basis states of the form

$$|q_i\rangle|q_j\rangle|q_k\rangle,$$

where $1 \leq i, j, k \leq 3$. We first classify the two-quark states

$$|q_i\rangle|q_j\rangle$$

according to irreducible representations. As can easily be seen, we need only form the symmetric and antisymmetric combinations:

$$(|q_i\rangle|q_j\rangle + |q_j\rangle|q_i\rangle),$$

$$\varepsilon_{lij}|q_i\rangle|q_j\rangle \qquad (l = 1, 2, 3).$$

We have six symmetric states and the corresponding representation is the 6-representation. The three antisymmetric states can readily be seen to form a basis for the 3*-representation. Thus we have the reduction

$$3 \times 3 = 3^* + 6. \tag{17.52}$$

The weight diagram of the 6-representation can readily be found using the same methods that we have applied for the reduction of the 3 × 3*-representation.

We now add a third quark, i.e., we consider the representation

$$3 \times 3 \times 3 = (3^* + 6) \times 3. \tag{17.53}$$

The representation 3* × 3 has already been reduced out (17.50). If we combine the states of the 6-representation with a further quark, then we can first of all form the totally symmetric quark states:

$$|q_i\rangle|q_j\rangle|q_k\rangle + \text{permutations}.$$

It is a simple matter of counting to show that there are ten linearly independent states of this sort. These states form a basis for the irreducible 10-representation. In the product representation 6 × 3, there remain a further eight linearly independent states, which, as can easily be seen, belong to the octet representation. Indeed, the basis vectors can be chosen as follows:

$$|a\rangle = (|q_i\rangle|q_j\rangle + |q_j\rangle|q_i\rangle)|q_k\rangle\varepsilon_{jkl}\lambda^a_{il} \qquad (a = 1, \ldots, 8). \tag{17.54}$$

Therefore we have the reduction formulas

$$\begin{aligned} 6 \times 3 &= 8 + 10, \\ 3 \times 3 \times 3 &= 1 + 8 + 8 + 10. \end{aligned} \tag{17.55}$$

From this we now infer that the baryons in our model occur as SU(3) singlets, octets, and decuplets. The observed baryons again fit very well into this scheme. The lightest baryons with $(\text{spin})^{\text{parity}} = \frac{1}{2}^+$, which include the

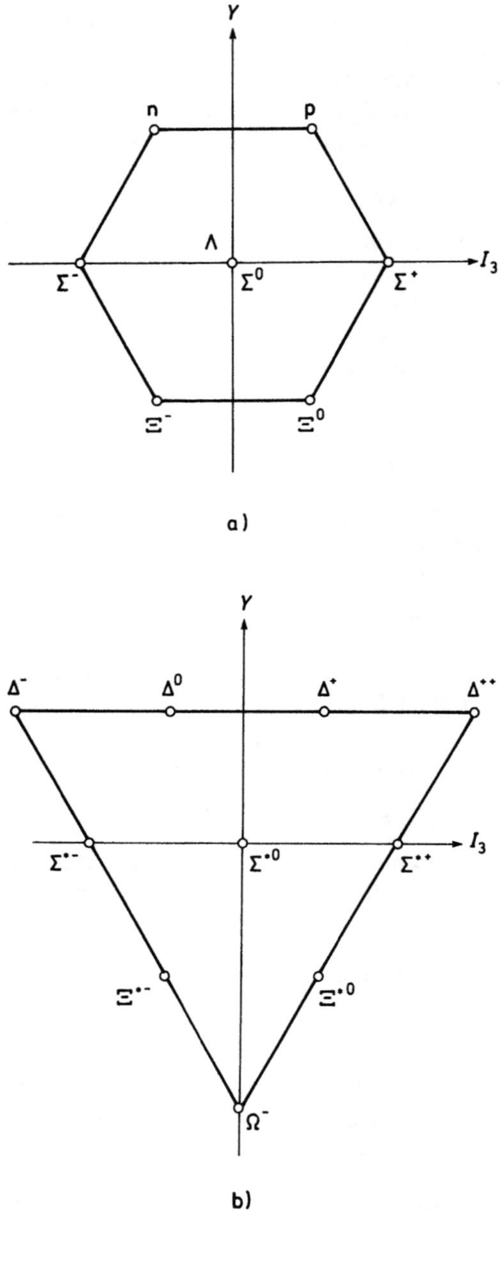

Figure 17.6 Weight diagrams for the lightest baryon octet (a) and for the lightest baryon decuplet (b).

nucleons, form an octet; the lightest $\frac{3}{2}^+$ baryons, to which the Δ-resonance and the Ω^- belong, form a decuplet. The weight diagrams are shown in Fig. 17.6. In this we have set (cf. Table 1.2) $\Delta \equiv \Delta(1232)$, $\Sigma^* \equiv \Sigma(1385)$ and $\Xi^* \equiv \Xi$ (1530).

As was the case for the mesons, one finds for the baryons too, that all states can be arranged into SU(3) multiplets, but that these states are certainly not degenerate in mass as our assumptions (i) and (ii) regarding the SU(3) symmetry would have led us to expect.

17.3 The Gell-Mann–Okubo Mass Formula

We must now discuss the origin of the mass splitting within an SU(3) multiplet. The assumptions (i) and (ii) at the beginning of Sect. 17.2, which have led to exact SU(3) invariance, are obviously too strong. Today it is believed that assumption (i) is correct, since the gluons, the carriers of the strong force, do not distinguish between quarks of different flavors. Assumption (ii) [eq. (17.30)] however, is too restrictive. As we shall see, one achieves agreement with experiment when one sets the masses of the u- and d-quarks to be equal, but makes the s-quark somewhat heavier. In place of (17.30) we adopt the hypothesis

$$m_{\mathrm{u}} = m_{\mathrm{d}} < m_{\mathrm{s}}. \tag{17.56}$$

Let us take the hypothetical u-, d-, and s-quark to be at rest. The Hamilton operator in the three-dimensional space defined by the quark flavor degrees of freedom would be

$$\begin{aligned}
(\langle \mathrm{q}_i | \mathbf{H}_{\mathrm{st}} | \mathrm{q}_j \rangle) &= \begin{pmatrix} m_{\mathrm{u}} & 0 & 0 \\ 0 & m_{\mathrm{d}} & 0 \\ 0 & 0 & m_{\mathrm{s}} \end{pmatrix} \\
&= \frac{2m_{\mathrm{u}} + m_{\mathrm{s}}}{3} \begin{pmatrix} 1 & 0 & 0 \\ 0 & 1 & 0 \\ 0 & 0 & 1 \end{pmatrix} + \frac{m_{\mathrm{u}} - m_{\mathrm{s}}}{3} \begin{pmatrix} 1 & 0 & 0 \\ 0 & 1 & 0 \\ 0 & 0 & -2 \end{pmatrix} \\
&= \frac{2m_{\mathrm{u}} + m_{\mathrm{s}}}{3} \mathbb{1} + \frac{m_{\mathrm{u}} - m_{\mathrm{s}}}{\sqrt{3}} \lambda_8 .
\end{aligned} \tag{17.57}$$

For this special case, the Hamilton operator thus consists of an SU(3)-invariant term and a term that transforms like the eighth component of an octet. With the help of a little field-theoretical formalism, one can see that this property of the Hamilton operator must hold in general as long as (17.56) is valid (see Appendix E). We thus write

$$\mathbf{H}_{\mathrm{st}} = \mathbf{H}_0 + \mathbf{H}_8 , \tag{17.58}$$

where $\mathbf{H}_0$ should be invariant and $\mathbf{H}_8$ should transform under the SU(3) group like the eighth component of an octet.

We now make the assumption that $\mathbf{H}_8$ is small in comparison to $\mathbf{H}_0$ such that we can apply perturbation theory with respect to $\mathbf{H}_8$. This assumption must, of course, be checked experimentally. To zeroth order we have degenerate SU(3) multiplets. We consider, for example, the baryon octet to which the nucleon belongs. For the unperturbed states of baryons at rest, $|\mathrm{B}^{(0)}\rangle$, we then have

$$\mathbf{H}_0 |\mathrm{B}^{(0)}\rangle = m^{(0)} |\mathrm{B}^{(0)}\rangle , \tag{17.59}$$

where B = p, n, Λ, Σ^+, Σ^0, Σ^-, Ξ^0, or Ξ^- and $m^{(0)}$ is the common eigenvalue. In first-order perturbation theory we find for the mass eigenvalues, m_{B}, of $\mathbf{H}_{\mathrm{st}}$:

$$m_{\mathrm{B}} = m^{(0)} + \langle \mathrm{B}^{(0)} | \mathbf{H}_8 | \mathrm{B}^{(0)} \rangle. \tag{17.60}$$

The matrix element of the octet operator $\mathbf{H}_8$ between octet states $|\mathrm{B}^{(0)}\rangle$ can be treated by means of the so-called Wigner–Eckart theorem. In essence, this says that we must search for all matrices in the space of the eight baryon states that transform like the eighth component of an octet. The matrix

$$(\langle \mathrm{B}'^{(0)} | \mathbf{H}_8 | \mathrm{B}^{(0)} \rangle)$$

is then a linear combination of these matrices.

We already know of one set of matrices that transform like an octet, namely the generators of the representation of the SU(3) group in the space of the baryon states:

$$\mathbf{F}_a \quad (a = 1, \dots, 8).$$

We can form a further set of octet matrices with the aid of the totally symmetric symbol d_{abc} (17.4):

$$\mathbf{D}_a = d_{abc} \mathbf{F}_b \mathbf{F}_c. \tag{17.61}$$

We will state without proof that, for an arbitrary irreducible representation, in particular for the octet representation, there are no other linearly independent octet matrices. However, for certain special representations, it is quite possible that even the matrices $\mathbf{F}_a$ and $\mathbf{D}_a$ are linearly dependent. This is the case, for example, for the decuplet representation. For our octet baryons we can thus write:

$$\langle \mathrm{B}'^{(0)} | \mathbf{H}_8 | \mathrm{B}^{(0)} \rangle = \langle \mathrm{B}'^{(0)} | \delta m'_1 \mathbf{F}_8 + \delta m'_2 d_{8ab} \mathbf{F}_a \mathbf{F}_b | \mathrm{B}^{(0)} \rangle, \tag{17.62}$$

where $\delta m'_1$ and $\delta m'_2$ are two constants.

We now make use of the explicit form of d_{abc} from Appendix C and find:

$$\begin{aligned} d_{8ab} \mathbf{F}_a \mathbf{F}_b &= \frac{1}{\sqrt{3}} (\mathbf{F}_1^2 + \mathbf{F}_2^2 + \mathbf{F}_3^2) - \frac{1}{2\sqrt{3}} (\mathbf{F}_4^2 + \mathbf{F}_5^2 + \mathbf{F}_6^2 + \mathbf{F}_7^2) - \frac{1}{\sqrt{3}} \mathbf{F}_8^2 \\ &= -\frac{1}{2\sqrt{3}} (\mathbf{F}_a \mathbf{F}_a) + \frac{3}{2\sqrt{3}} (\mathbf{F}_1^2 + \mathbf{F}_2^2 + \mathbf{F}_3^2) - \frac{1}{2\sqrt{3}} \mathbf{F}_8^2. \end{aligned} \tag{17.63}$$

If we use the isospin and hypercharge identification from (17.39), then we obtain in terms of redefined constants δm_1 and δm_2:

$$\langle \mathrm{B}'^{(0)} | \mathbf{H}_8 | \mathrm{B}^{(0)} \rangle = \langle \mathrm{B}'^{(0)} | \delta m_1 \mathbf{Y} + \delta m_2 [\vec{\mathbf{T}}^2 - \tfrac{1}{4} \mathbf{Y}^2 - \tfrac{1}{3} \mathbf{F}_a \mathbf{F}_a] | \mathrm{B}^{(0)} \rangle. \tag{17.64}$$

The quantity $\mathbf{F}_a \mathbf{F}_a$ is, as can easily be seen, an invariant under SU(3) rotations. Thus, in an irreducible representation it has a fixed value. This term from $\mathbf{H}_8$ can be absorbed into $m^{(0)}$. The perturbatively corrected mass eigenvalue (17.60) of a baryon B with hypercharge Y and isospin I is therefore given by:

$$m_{\mathrm{B}} = \bar{m} + \delta m_1 Y + \delta m_2 (I(I+1) - \tfrac{1}{4} Y^2). \tag{17.65}$$

Table 17.2 The hypercharge Y, isospin I, and average masses for the isomultiplets of the lightest baryon octet.

Baryons	Y	I	Average mass (MeV)
p, n	1	$\frac{1}{2}$	939
Λ	0	0	1116
$\Sigma^+, \Sigma^0, \Sigma^-$	0	1	1193
Ξ^0, Ξ^-	-1	$\frac{1}{2}$	1318

This is the Gell-Mann–Okubo formula. The three parameters $\bar{m}$, δm_1, and δm_2 remain undetermined by our purely group-theoretical considerations.

We now wish to know whether this formula is consistent with experiment. The observed masses of the $\frac{1}{2}^+$ baryons are given in Table 17.2. Since our assumptions, imply exact isospin symmetry, states within an isomultiplet remain degenerate. Experimentally, this is correct to within an accuracy of a few MeV (Table 15.1). According to our theory, we now have to describe the four mass values in Table 17.2 with the three-parameter Gell-Mann–Okubo formula, i.e., our theory yields a linear relationship between the four masses, namely

$$\underset{1129\text{ MeV (exp.)}}{\frac{m_N + m_\Xi}{2}} = \underset{1135\text{ MeV (exp.)}}{\frac{3m_\Lambda + m_\Sigma}{4}} \tag{17.66}$$

where $m_N = (m_p + m_n)/2$. The corresponding experimental values are written below the two terms. Using first-order perturbation theory we could hardly have hoped for a better agreement.

For the baryon decuplet, the predictive power of the Gell-Mann–Okubo formula is even greater. As can easily be seen by inserting the explicit values of I and Y from this representation (Fig. 17.6(b)), the third term on the right-hand side of (17.65) is, in this case, also proportional to Y. The Gell-Mann–Okubo formula then takes on the simpler form

$$m_B = \bar{m} + \delta m \cdot Y \qquad \text{(decuplet formula)} \tag{17.67}$$

This imples equal spacings between the masses of isomultiplets with different hypercharge (Table 17.3).

At the time when the theory of the eightfold way was developed, the Ω^--particle had not yet been discovered. Its existence was postulated since one place in the decuplet remained to be filled. Its mass could be predicted to within a few MeV using (17.67). The fact that the Ω^- was subsequently discovered by experiments has already been mentioned in Chap. 15.

From Tables 17.2 and 17.3 we see that the typical order of magnitude of the flavor-SU(3) symmetry breaking is given by

$$\frac{m_\Lambda - m_N}{m_N} \cong 20\%, \tag{17.68}$$

Table 17.3 The hypercharge Y, isospin I, and average masses for the isomultiplets of the lightest baryon decuplet. The last column gives the mass differences between successive isomultiplets.

Baryon isomultiplet	Y	I	Average mass (MeV)	Mass difference δm (MeV)
Δ	1	$\frac{3}{2}$	1232	
Σ^*	0	1	1385	153
Ξ^*	-1	$\frac{1}{2}$	1530	145
Ω^-	-2	0	1672	142

so that the use of first-order perturbation theory is reasonably justified. The order of the levels can also be readily understood. States with smaller hypercharge contain more s-quarks and from (17.56) are thus heavier.

We now move on to a brief discussion of the mass splitting for the pseudoscalar mesons. Since our Hamilton operator (17.58) contains an octet part, there will be nonvanishing matrix elements between octet and singlet states. If we consider the Hamilton operator in the space of the nine states (17.51), and neglect all transitions to other states, then the 9×9 mass matrix $\mathscr{M}$ that is to be diagonalized has the following form:

$$\mathscr{M} = (\langle\alpha|\mathbf{H}_{\mathrm{st}}|\beta\rangle), \tag{17.69}$$

in which α and β stand for the nine mesons $\pi^+, \pi^0, \ldots, \eta_8, \eta_1$, all of which are considered to be at rest. For the individual matrix elements we can make use of group theory in a manner analogous to that used for the baryons to derive the following relations and parametrizations:

$$\langle\eta_1|\mathbf{H}_{\mathrm{st}}|\eta_1\rangle = \langle\eta_1|\mathbf{H}_0 + \mathbf{H}_8|\eta_1\rangle = \langle\eta_1|\mathbf{H}_0|\eta_1\rangle \equiv m_1, \tag{17.70}$$

$$\langle\alpha|\mathbf{H}_{\mathrm{st}}|\eta_1\rangle = \langle\alpha|\mathbf{H}_8|\eta_1\rangle = 0 \qquad \text{for} \quad \alpha \neq \eta_8, \eta_1; \tag{17.71}$$

$$\langle\eta_8|\mathbf{H}_{\mathrm{st}}|\eta_1\rangle = \langle\eta_8|\mathbf{H}_8|\eta_1\rangle; \tag{17.72}$$

$$\langle\alpha|\mathbf{H}_{\mathrm{st}}|\beta\rangle = \langle\alpha|\mathbf{H}_0 + \mathbf{H}_8|\beta\rangle = \langle\alpha|\bar{m}_8 + \delta m(\vec{\mathbf{I}}^2 - \tfrac{1}{4}\mathbf{Y}^2)|\beta\rangle$$
$$\text{for} \quad \alpha \neq \eta_1, \beta \neq \eta_1. \tag{17.73}$$

In the case of the mesons, particles and their antiparticles are in the same SU(3) multiplet, for example, K^+ and K^- or K^0 and $\bar{K}^0$ in our meson octet. According to the CPT-theorem, particle and antiparticle have the same mass but opposite hypercharge Y. Thus, in contrast to the case of the baryon octet, there is now no term proportional to Y in the mass formula (17.73). Group theory therefore allows us to parametrize the mass matrix $\mathscr{M}$ for our nine mesons in terms of the quantities m_1, $\bar{m}_8$, δm, and $\langle\eta_8|\mathbf{H}_8|\eta_1\rangle$.

As is easily seen, the matrix $\mathscr{M}$ is almost diagonal in the above basis. The exception is made by the subspace spanned by η_8 and η_1. There we have to diagonalize a 2×2 matrix $\mathscr{M}'$ in order to obtain the mass eigenvalues. From (17.70)–(17.73) we find for the mass eigenvalues of the π- and K-mesons and for $\mathscr{M}'$:

$$
\begin{aligned}
m_\pi &= \bar{m}_8 + 2\delta m, \\
m_\mathrm{K} &= \bar{m}_8 + \tfrac{1}{2}\delta m, \\
\mathscr{M}' &= \begin{pmatrix} \bar{m}_8 & \langle \eta_8 | \mathbf{H}_8 | \eta_1 \rangle \\ \langle \eta_1 | \mathbf{H}_8 | \eta_8 \rangle & m_1 \end{pmatrix}.
\end{aligned}
\tag{17.74}
$$

By means of a suitable choice of the relative phase of the states $|\eta_8\rangle$ and $|\eta_1\rangle$ we can always ensure that

$$
\langle \eta_8 | \mathbf{H}_8 | \eta_1 \rangle \geq 0. \tag{17.75}
$$

The eigenstates and eigenvalues of $\mathscr{M}'$ yield the physical particles η and η' with their masses m_η and $m_{\eta'}$. With the convention (17.75) we can write

$$
\begin{aligned}
|\eta\rangle &= \cos\vartheta |\eta_8\rangle + \sin\vartheta |\eta_1\rangle, \\
|\eta'\rangle &= -\sin\vartheta |\eta_8\rangle + \cos\vartheta |\eta_1\rangle.
\end{aligned}
\tag{17.76}
$$

Here ϑ is the mixing angle which can easily be calculated from the observed masses of the nine pseudoscalar mesons (Exercise 17.7). We find a relatively small octet–singlet mixing

$$
\vartheta \cong -24^\circ. \tag{17.77}
$$

If the above considerations are applied to the square of the masses instead of the mass itself, then an even smaller mixing angle of -11° is obtained (Dumbrajs 1983).

For the vector mesons ρ, K*, ω, and ϕ, it turns out that in some sense the mixing is extreme (ideal). The quark content of the observed isoscalar vector mesons is, to a good approximation, as follows:

$$
\begin{aligned}
\omega &\sim \frac{1}{\sqrt{2}}(\mathrm{u\bar{u}} + \mathrm{d\bar{d}}), \\
\phi &= \mathrm{s\bar{s}}.
\end{aligned}
\tag{17.78}
$$

For the baryon octet we have not considered the mixing with a singlet. The reason for this will become clear in the next section where we shall explain the SU(6) theory.

17.4 The SU(6) Symmetry

If we look at the masses of the baryons in Tables 17.2 and 17.3, then we see that the splitting between the Δ and the nucleon is not much larger than that between the Λ and the nucleon. As we will now show, this fact is an indication that the forces between the quarks are, to a first approximation, independent of spin (Sakita 1964; Gürsey 1964; for a review see Dyson 1966).

As before we consider the case of three quark flavors. If we take the spin into account, then a quark has six degrees of freedom in spin × flavor space.

As a basis we choose

$$\overset{\uparrow}{u}, \overset{\downarrow}{u}, \overset{\uparrow}{d}, \overset{\downarrow}{d}, \overset{\uparrow}{s}, \overset{\downarrow}{s},$$

where the arrows indicate spin in the positive (↑) and negative (↓) z-direction. If the forces between quarks are spin independent and the u-, d-, and s-quarks all have the same mass, then the Hamiltonian must display a U(6) symmetry. The special case of a common phase transformation of the quark degrees of freedom has to do with baryon number. If we separate this out then the remaining symmetry is SU(6).

In the quark model for the hadron spectrum one assumes that the motion of the quarks is nonrelativistic. One assumes a particular form for the potential, for example, that corresponding to harmonic forces, and solves a Schröding equation. We will now discuss the baryon spectrum in these terms. For reasonable potentials the ground state has a symmetric spatial wavefunction with zero orbital angular momenta. Due to the fermionic nature of the quarks, their overall wavefunction has to be totally antisymmetric. As already mentioned in Chap. 15, the quarks also carry a color degree of freedom. For baryons, the quark wavefunction is totally antisymmetric with respect to color:

$$\mathrm{B} \sim \mathrm{q}_\alpha \mathrm{q}_\beta \mathrm{q}_\gamma \varepsilon_{\alpha\beta\gamma} \qquad (\alpha, \beta, \gamma\text{: Color indices}). \tag{17.79}$$

The Pauli principle therefore demands for baryons that the flavor × spin part of the quark wavefunction is totally symmetric, i.e., our states must be totally symmetric combinations of the six basis states $\overset{\uparrow}{u}, \ldots, \overset{\downarrow}{s}$. A simple enumeration shows that this gives 56 states for a three-quark system (Exercise 17.8).

We immediately see that there can be no flavor-SU(3) singlet among these states. The singlet SU(3) wavefunction in flavor space would be

$$(\mathrm{uds})_{\text{totally antisymmetrized}}.$$

Correspondingly, the spin wavefunction would also have to be totally antisymmetric. This however, is impossible for three particles with spin $\frac{1}{2}$. In the ground state there is thus no SU(3) singlet and for this reason we did not consider the mixing between octets and singlets in our discussion of the Gell-Mann–Okubo formula for baryons.

On taking a closer look at the 56 states, one finds that they contain an octet with spin $\frac{1}{2}$ and a decuplet with spin $\frac{3}{2}$. With the notation (spin multiplicity, SU(3)-representation) we have

$$56 = (2, 8) + (4, 10). \tag{17.80}$$

Experimentally one finds for the lowest baryon levels exactly these numbers. The SU(6) symmetry thus enables us to understand this fact. The degeneracy of the 56 states is lifted by the different quark masses (17.56), by the spin–spin coupling terms and, for higher states also by the spin–orbit coupling terms. Our understanding of these terms has progressed significantly with the development of quantum chromodynamics (De Rujula 1975). Today it can be claimed that we possess a very good phenomenological model for the baryon spectrum. Admittedly, this is not a model derived from first principles, but it is at least inspired by QCD (Isgur 1978, 1979; Gromes 1980).

We now present a few examples of wavefunctions in our simple SU(6) model. Only the flavor and spin parts are given explicitly [Notation: Hadron (spin, z-component of spin)]:

$$\Delta^{++}(\tfrac{3}{2},\tfrac{3}{2}) \sim \overset{\uparrow}{u}\overset{\uparrow}{u}\overset{\uparrow}{u},$$

$$\Omega^{-}(\tfrac{3}{2},\tfrac{3}{2}) \sim \overset{\uparrow}{s}\overset{\uparrow}{s}\overset{\uparrow}{s},$$

$$\begin{aligned} p(\tfrac{1}{2},\tfrac{1}{2}) &\sim \frac{1}{\sqrt{18}}\{\overset{\uparrow}{u}(\overset{\uparrow}{u}\overset{\downarrow}{d} - \overset{\downarrow}{u}\overset{\uparrow}{d}) + \text{permutations}\},\\ &= \frac{1}{\sqrt{18}}\{2\overset{\uparrow}{u}\overset{\uparrow}{u}\overset{\downarrow}{d} + 2\overset{\uparrow}{u}\overset{\downarrow}{d}\overset{\uparrow}{u}\\ &\quad + 2\overset{\downarrow}{d}\overset{\uparrow}{u}\overset{\uparrow}{u} - \overset{\uparrow}{u}\overset{\downarrow}{u}\overset{\uparrow}{d} - \overset{\downarrow}{u}\overset{\uparrow}{u}\overset{\uparrow}{d} - \overset{\downarrow}{u}\overset{\uparrow}{d}\overset{\uparrow}{u} - \overset{\uparrow}{u}\overset{\uparrow}{d}\overset{\downarrow}{u} - \overset{\uparrow}{d}\overset{\uparrow}{u}\overset{\downarrow}{u} - \overset{\uparrow}{d}\overset{\downarrow}{u}\overset{\uparrow}{u}\},\\ n(\tfrac{1}{2},\tfrac{1}{2}) &\sim \frac{1}{\sqrt{18}}\{2\overset{\uparrow}{d}\overset{\uparrow}{d}\overset{\downarrow}{u} + 2\overset{\uparrow}{d}\overset{\downarrow}{u}\overset{\uparrow}{d}\\ &\quad + 2\overset{\downarrow}{u}\overset{\uparrow}{d}\overset{\uparrow}{d} - \overset{\uparrow}{d}\overset{\downarrow}{d}\overset{\uparrow}{u} - \overset{\downarrow}{d}\overset{\uparrow}{d}\overset{\uparrow}{u} - \overset{\downarrow}{d}\overset{\uparrow}{u}\overset{\uparrow}{d} - \overset{\uparrow}{d}\overset{\uparrow}{u}\overset{\downarrow}{d} - \overset{\uparrow}{u}\overset{\uparrow}{d}\overset{\downarrow}{d} - \overset{\uparrow}{u}\overset{\downarrow}{d}\overset{\uparrow}{d}\}. \end{aligned} \tag{17.81}$$

An extensive table of such explicit wavefunctions is given by Thirring (1965). These wavefunctions can be used, for example, to calculate the static properties of baryons. We will illustrate this for the magnetic moment of the proton and the neutron.

For a baryon in the ground state, the magnetic moment must be the sum of the magnetic moments of the quarks since there is no orbital angular momentum. The operator of the magnetic moment is then

$$\boldsymbol{\mu} = \tfrac{2}{3}\mu_u\boldsymbol{\sigma}_u - \tfrac{1}{3}\mu_d\boldsymbol{\sigma}_d - \tfrac{1}{3}\mu_s\boldsymbol{\sigma}_s. \tag{17.82}$$

Here $\mu_u\mu_d$ and μ_s are the quark magnetons which, if the quarks only have a Dirac magnetic moment, have the form

$$\mu_q = \frac{e}{2m_q} \qquad (q = u, d, s). \tag{17.83}$$

For equal u- and d-quark masses we then have

$$\mu_u = \mu_d. \tag{17.84}$$

If we form the expectation values of the operator μ (17.82) using the quark wavefunctions (17.81) for the proton and neutron, then we readily obtain for the magnetic moments μ_p, μ_n of the proton and neutron

$$\begin{aligned} \mu_p &= \mu_u, \\ \mu_n &= -\tfrac{2}{3}\mu_u. \end{aligned} \tag{17.85}$$

The ratio is thus predicted as

$$\left.\frac{\mu_n}{\mu_p}\right|_{\text{theory}} = -\tfrac{2}{3}. \tag{17.86}$$

In fact, this agrees very well with experiment:

$$\begin{aligned} \mu_p &= 2.793\frac{e}{2m_p}, \\ \mu_n &= -1.913\frac{e}{2m_p}, \\ \left.\frac{\mu_n}{\mu_p}\right|_{\text{experiment}} &= \frac{-1.913}{2.793} = -0.685. \end{aligned} \tag{17.87}$$

This is one of the major achievements of the SU(6) theory.

If we consider this as a confirmation of the underlying ideas, we can also go on to determine the quark masses. It follows from (17.83)–(17.87) that

$$\begin{aligned} 2.793\frac{e}{2m_p} &= \frac{e}{2m_u}, \\ m_u &= \frac{m_p}{2.793} = 336\ \text{MeV}. \end{aligned} \tag{17.88}$$

This is just the order of magnitude for the quark masses that one requires for potential models in order to describe the spectrum. There one finds, very roughly

$$m_u \sim \tfrac{1}{3}m_p. \tag{17.89}$$

The SU(6) theory has, of course, also been applied to the mesons. We do not wish to go into details here, but simply mention that the 0^-- and 1^--mesons of Fig. 17.3 are ordered into a singlet and a 35-multiplet of the SU(6). For further discussion we refer the reader to the literature (see, for example, Rosner 1974).

Exercises

17.1 With the help of (17.10) and (17.14) demonstrate the total antisymmetry of f_{abc} and the total symmetry of d_{abc}.

17.2 Consider a representation of the SU(3) group as in (17.15), (17.16) with the generators $\mathbf{F}_a$ (17.20). Show that the commutation rules (17.21) follow from the group property (17.16) holding for representations and vice versa.

17.3 With the help of the Jacobi identity, calculate the commutator $[\mathbf{F}_a, \mathbf{F}_b]$ for the matrices $\mathbf{F}$ defined in (17.25).

17.4 This exercise is intended to clarify the physical meaning of the generators of the flavor-SU(3) group $\mathbf{F}_a$ ($a = 1, \ldots, 8$). Consider the operators

$$\begin{aligned} \mathbf{I}_\pm &= \mathbf{F}_1 \pm \mathrm{i}\,\mathbf{F}_2, \\ \mathbf{U}_\pm &= \mathbf{F}_6 \pm \mathrm{i}\,\mathbf{F}_7, \\ \mathbf{V}_\pm &= \mathbf{F}_4 \pm \mathrm{i}\,\mathbf{F}_5. \end{aligned} \tag{17.90}$$

What effect do these operators have when applied to the quark states $|u\rangle$, $|d\rangle$, and $|s\rangle$ as in (17.38)? What effect do they have in a general representation of the flavor-SU(3) group?

17.5 Construct the weight diagram for the 6- and the 10-representations of the SU(3) group.

17.6 With the help of the operators $\mathbf{I}_\pm$, $\mathbf{U}_\pm$, and $\mathbf{V}_\pm$ of (17.90), demonstrate the irreducibility of the SU(3) representations 3, 3*, 6, 8, and 10.

17.7 Determine the mixing angle of the 0^--mesons in (17.76). As experimental input use the masses of the physical particles η and η' and the mass $\bar{m}_8$ derived from the masses of the π- and the K-mesons. Carry out the same calculation for the 1^--mesons and verify (17.78).

17.8 Show that there are exactly 56 linearly independent three-quark states that are totally symmetric combinations of the six states $\overset{\uparrow}{\mathrm{u}} \ldots, \overset{\downarrow}{\mathrm{s}}$. Verify (17.80).

17.9 Calculate the magnetic moment of the proton and the neutron in the SU(6) model and verify (17.85).

CHAPTER 18

The Naïve Parton Model

In the last chapter we discussed the nonrelativistic quark model which has proved to be quite successful in describing the hadron spectrum. In this chapter, we will discuss the parton model to describe so-called hard processes. In this model, too, the hadrons are considered to be composed of constituents, the partons, and as it turns out some of these partons carry exactly the quark quantum numbers. In contrast to our previous approach, we will not however consider hadrons at rest; instead we shall be dealing with rapidly moving hadrons. As might be expected, the connection between the constituent wavefunctions of a particle at rest and those of a rapidly moving particle is far from simple. Thus it should come as no surprise that the picture of hadronic constituents which will be developed in this chapter, appears somewhat different to the one derived in the nonrelativistic approach of the previous chapter.

The parton model was introduced at the end of the 1960s by a number of authors (Feynman 1969, 1972; Bjorken 1969a; Drell 1971; Landshoff 1971, 1971a): The success of the model was phenomenal. Also, the arrival of quantum chromodynamics (QCD), has not really rendered it superfluous. In fact, almost the opposite is true; in many of the phenomenological applications of QCD, it is still necessary to assume the validty of the parton description of hadrons in order to make contact between theory and experiment.

We now give a few examples of reactions to which the parton model has been applied. Here X stands for hadrons in the final state.

— Electron–positron annihilation into hadrons at high center of mass energies:

$$e^+ + e^- \to X. \tag{18.1}$$

— Deep inelastic electron–nucleon scattering, i.e., scattering with large transfers of energy and momentum:

$$e + N \to e + X. \tag{18.2}$$

— Muon pair production in high energy nucleon–nucleon collisions, the so-called Drell–Yan process (Drell 1970):

$$N + N \to \mu^+ + \mu^- + X. \tag{18.3}$$

— Production of hadrons with large transverse momentum in high energy hadron–hadron collisions, for example:

$$p + \bar{p} \to \text{jets with large } p_T + X. \tag{18.4}$$

We now want to introduce the basic assumptions of the parton model and then see how these can be used to describe the above reactions.

Assumption I. A rapidly moving hadron appears as a jet of partons, all of which travel in more or less the same direction as the parent hadron. The three-momentum of the hadron is shared out among the partons (Fig. 18.1).

Assumption II. For hard reactions such as those listed above, we can make use of the *impulse approximation.* The rule for calculating reaction rates for hadrons is as follows: one calculates the reaction rate for the basic process with free partons and then sums incoherently over the contributions of all partons in the hadrons.

Unfortunately, until now, it has not been possible to deduce these rules from first principles. However, a number of plausibility arguments have been used to justify them (Feynman 1969; see also Nachtmann 1977). We will not go into further details here but will concentrate instead on the applications of these assumptions.

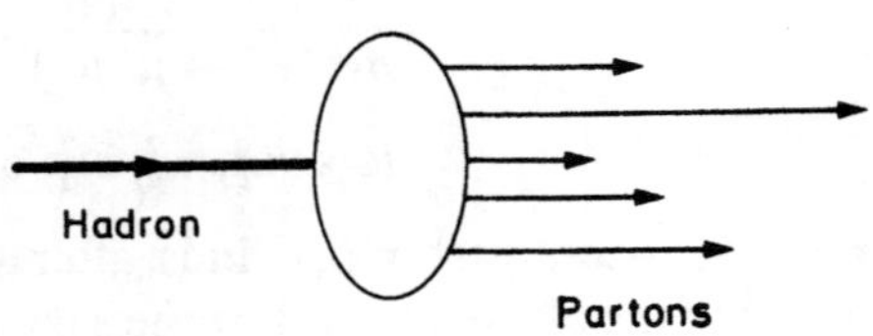

Figure 18.1 Partons as constituents of a rapidly moving hadron.

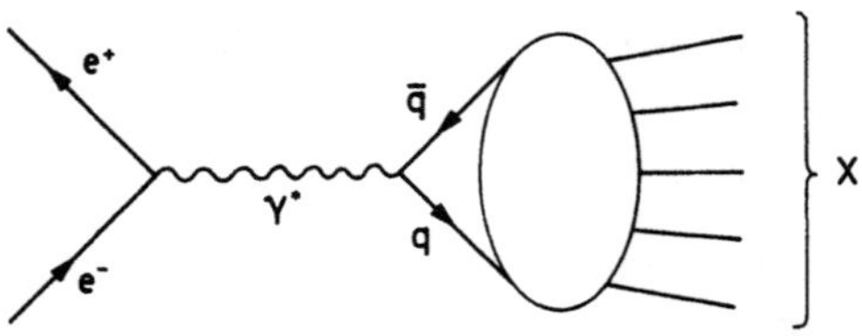

Figure 18.2 The reaction $e^+e^- \to$ hadrons in the parton model.

18.1 Electron–Positron Annihilation into Hadrons

In the parton model we picture this reaction in the following way (Fig. 18.2): An electron and a positron annihilate into a virtual photon that materializes primarily into a parton–antiparton (q, $\bar{q}$) pair. The partons are then assumed to transform into hadrons with a probability of one. Thus, in order to calculate the total scattering cross-section, we simply need to consider the reaction

$$e^+ + e^- \to q + \bar{q} \tag{18.5}$$

and sum over all contributing partons q (Cabibbo 1970).

The production of a parton pair (18.5) is very similar to the well-known muon pair production

$$e^+ + e^- \to \mu^+ + \mu^-, \tag{18.6}$$

which is a pure quantum electrodynamical process (Fig. 11.3). The scattering cross-section for the reaction (18.6) has already been calculated in Sect. 11.2. There we found for large center of mass energies $\sqrt{s}$:

$$\sigma(e^+e^- \to \mu^+\mu^-) = \frac{4\pi\alpha^2}{3}\frac{1}{s} \qquad (s \gg m_e^2, m_\mu^2). \tag{18.7}$$

This result can be readily understood. The power of the fine-structure constant $\alpha = e^2/4\pi$ follows from counting the vertices in Fig. 11.3. For high energies, where the masses can be neglected, only the length scale $1/\sqrt{s}$ remains and thus the cross-section must be proportional to $1/s$. Simple scale arguments of this type will be used frequently in our discussion of the parton model.

Switching to the parton process (18.5), we note that the cross-section for spin-$\frac{1}{2}$ partons, in the approximation where their masses are neglected, must be the same as in (18.7) but multiplied by the square of the parton charge Q_q since this can be different from the muon charge. For the ratio of cross-sections we thus find

$$\frac{\sigma(e^+e^- \to q\bar{q})}{\sigma(e^+e^- \to \mu^+\mu^-)} = Q_q^2, \qquad (s \gg m_e^2, m_\mu^2, m_q^2;\ \text{q: spin-}\tfrac{1}{2}\text{parton}). \tag{18.8}$$

For a spin-zero parton one finds, after a simple calculation (cf. Exercise 11.2), only $\frac{1}{4}$ of the above result. For the sake of simplicity, we will exclude charged partons with spin $> \frac{1}{2}$. We now apply Assumption II and obtain for the ratio

$R(s)$ of the hadronic to the muonic cross-section

$$R(s) = \frac{\sigma(e^+e^- \to \text{hadrons})}{\sigma(e^+e^- \to \mu^+\mu^-)} = \sum_q Q_q^2 + \frac{1}{4}\sum_{q'} Q_{q'}^2, \tag{18.9}$$

(q: spin-$\frac{1}{2}$ partons, q′: spin-0 partons).

As we shall see in the next section, the data for deep inelastic scattering strongly suggest that all charged partons have spin $\frac{1}{2}$. If we assume that the partons are identical to the familiar quarks u, d, and s, then for quarks *without color*, we find

$$R(s) = (\tfrac{2}{3})^2 + (-\tfrac{1}{3})^2 + (-\tfrac{1}{3})^2 = \tfrac{2}{3}. \tag{18.10}$$

For quarks with three colors, we have three times as many final states, corresponding to

$$e^+e^- \to q_\alpha \bar{q}_\alpha, \tag{18.11}$$

where $\alpha = 1, 2, 3$ is the color index and q = u, d, s. Thus for *three* colors we have

$$R(s) = 3 \cdot \tfrac{2}{3} = 2. \tag{18.12}$$

These theoretical predictions must be compared with experimental data for center of mass energies $\sqrt{s}$ that are large compared to the masses of u-, d-, and s-quarks, but which remain below the range where the charm quark, c, comes into play. Thus we are interested in energies

$$1\ \text{GeV} \lesssim \sqrt{s} \lesssim 3\ \text{GeV}. \tag{18.13}$$

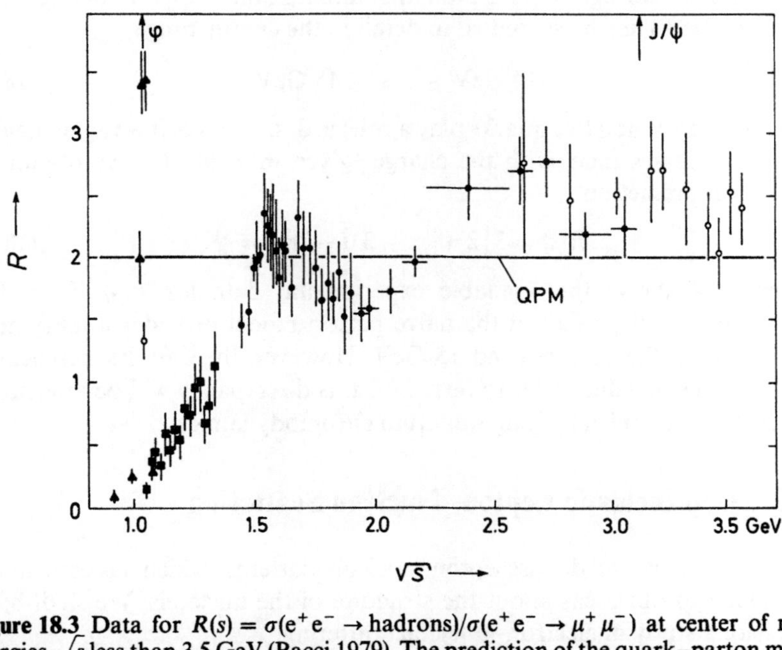

Figure 18.3 Data for $R(s) = \sigma(e^+e^- \to \text{hadrons})/\sigma(e^+e^- \to \mu^+\mu^-)$ at center of mass energies $\sqrt{s}$ less than 3.5 GeV (Bacci 1979). The prediction of the quark–parton model (QPM) for this energy region is $R(s) = 2$.

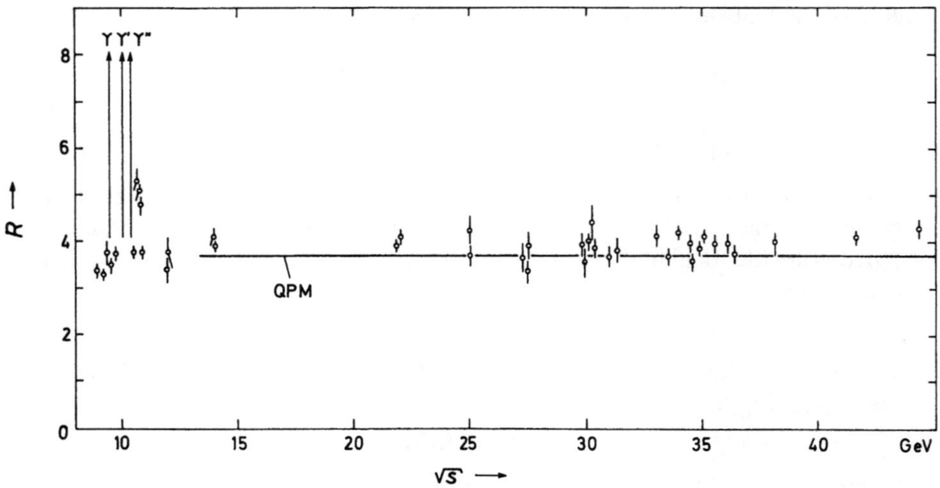

Figure 18.4 Experimental data for the ratio $R(s) = \sigma(e^+e^- \to \text{hadrons})/\sigma_{\mu\mu}$ where $\sigma_{\mu\mu} = 4\pi\alpha^2/(3s)$ for center of mass energies $8 \leq \sqrt{s} \leq 45$ GeV (Wu 1984). The prediction of the quark–parton model (QPM) for $\sqrt{s} \gtrsim 12$ GeV is $R(s) = 11/3$.

The data (Fig. 18.3) in fact show that for $\sqrt{s} \gtrsim 1.5$ GeV, $R(s)$ is approximately constant with a value of about 2. A value $R(s) \cong \frac{2}{3}$ can certainly be ruled out. These data are held to be one of the most decisive pieces of experimental evidence to support the hypothesis of three quark colors.

At the e^+e^- storage ring PETRA in Hamburg and at PEP in Stanford, the ratio $R(s)$ (18.9) has been studied in detail in the energy range

$$14 \text{ GeV} \lesssim \sqrt{s} \lesssim 45 \text{ GeV}. \tag{18.14}$$

In this energy region five quarks play a role: u, d, s, c, and b. If we again neglect all quark masses then, with the charges given in Table 1.3, we obtain the theoretical prediction

$$R(s) = 3\{2\cdot(\tfrac{2}{3})^2 + 3\cdot(-\tfrac{1}{3})^2\} = \tfrac{11}{3}. \tag{18.15}$$

Figure 18.4 shows the available experimental data for $R(s)$. To a first approximation the result of the naïve parton model provides a fairly good description of the data beyond 15 GeV. However, the data lie significantly above the naïve value of $\frac{11}{3}$. In Sect. 20.2 this discrepancy will be interpreted as a higher order effect within quantum chromodynamics.

18.2 Deep Inelastic Lepton–Nucleon Scattering

In this section we will discuss lepton–nucleon scattering which has contributed very much to our ideas about the structure of the nucleons. We shall begin here with a study of electron–nucleon scattering:

$$e + N \to e + X. \tag{18.16}$$

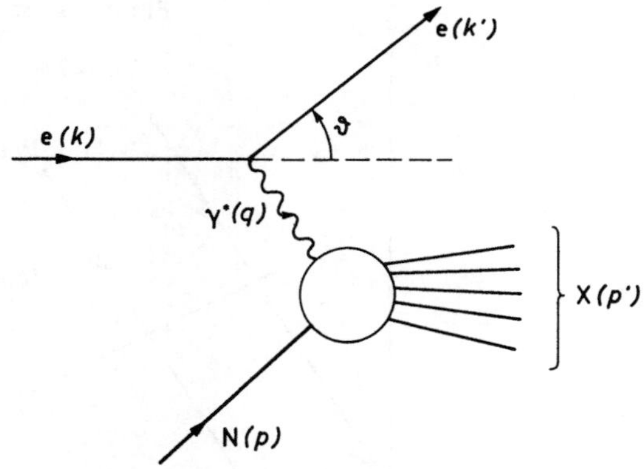

Figure 18.5 The diagram for electron–nucleon scattering.

The corresponding Feynman diagram is shown in Fig. 18.5 in which the four-momenta of the participating particles are also given. We consider only the leading order in the expansion in powers of the fine-structure constant α and ignore the electron mass. The electron emits a virtual photon that is absorbed by the nucleon. The nucleon thereupon breaks up, in general, into many hadrons. We are mainly interested in large energy and momentum transfers since, here, the constituent structure of the nucleons should become apparent. Indeed, it was in an analogous manner that the scattering of α-particles from atoms led to Rutherford's discovery of the atomic nucleus as a constituent of the atom.

The most important kinematic quantities for our reaction (18.16) are defined in Table 18.1.

Due to the conservation of baryon number, the invariant mass of the hadrons in the final state must be at least as large as the mass of the nucleon.

Table 18.1 Kinematic quantities of lepton–nucleon scattering.

Quantity	Description
M	Nucleon mass
$E = pk/M$	Energy of incoming lepton in laboratory system
$E' = pk'/M$	Energy of outgoing lepton in laboratory system
ϑ	Scattering angle of lepton in laboratory system
$q = k - k'$	Four-momentum transfer
$Q^2 = -q^2$ $= 4EE' \sin^2 \vartheta/2$	Square of four-momentum transfer
$\nu = pq/M$	Energy transfer in laboratory system
$W^2 = (p + q)^2$	Invariant squared mass of the hadrons in the final state
$x \equiv \omega^{-1} = Q^2/(2M\nu)$	Bjorken's scaling variable ($0 \le x \le 1$)
$y = \nu/E$	(Energy transfer)/(maximum energy transfer) ($0 \le y \le 1$)
$P = M\nu/Q = Q/(2x)$	Momentum of the nucleon in the Breit system

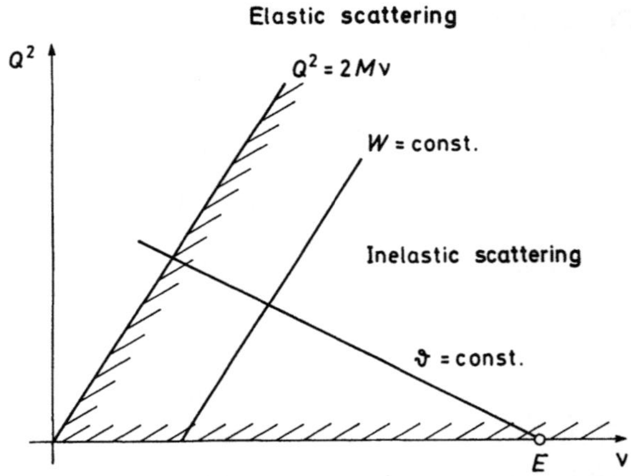

Figure 18.6 Kinematically allowed region in the ν–Q^2 plane. Line of constant invariant mass of the hadrons in the final state: W = const. Line of constant scattering angle at a particular energy E of the incoming electron: ϑ = const.

We thus have the kinematic bound

$$\begin{aligned} W^2 = (p+q)^2 = M^2 + 2M\nu - Q^2 &\geq M^2, \\ 2M\nu &\geq Q^2. \end{aligned} \tag{18.17}$$

The condition $2M\nu = Q^2$ corresponds to elastic scattering. The kinematically allowed region in the ν–Q^2 plane is shown in Fig. 18.6. The kinematic bounds for the variables x and y (Table 18.1) accordingly are:

$$\begin{aligned} 0 \leq x \leq 1, \\ 0 \leq y \leq 1. \end{aligned} \tag{18.18}$$

Deep inelastic electron–nucleon scattering was in the beginning investigated mainly at the linear accelerator at Stanford (SLAC). These studies involved electron energies and Q^2 values of

$$\begin{aligned} 1\ \mathrm{GeV} \lesssim E \lesssim 30\ \mathrm{GeV}, \\ 0 \lesssim Q^2 \lesssim 20\ \mathrm{GeV}^2. \end{aligned} \tag{18.19}$$

Further studies at FNAL and CERN made use of muons in place of electrons and employed higher energies. More recently energies of $E \sim 300$ GeV and Q^2 values of 200 GeV^2 have been reached (Drees 1983).

The S-matrix element for the reaction (18.16) has the following form:

$$\begin{aligned} \langle \mathrm{e}(k')\mathrm{X}(p')|\,\mathbf{S}\,|\mathrm{e}(k)\mathrm{N}(p)\rangle = -\mathrm{i}(2\pi)^4\delta(p'+k'-p-q) \\ \cdot 4\pi\alpha\, \bar{u}(k')\gamma_\mu u(k)\frac{1}{q^2}\langle \mathrm{X}(p')|\,\mathscr{J}^\mu(0)\,|N(p)\rangle. \end{aligned} \tag{18.20}$$

Here $e\mathscr{J}^{\mu}$ is the hadronic part of the electromagnetic current. The factor $1/q^2$ comes from the photon propagator in Fig. 18.5.

We shall be interested in experiments in which only the electron in the final state is observed and is analyzed according to its energy and scattering angle. The corresponding cross-section for unpolarized electrons can, as is easily deduced from (18.20), be represented as the product of a lepton and a hadron tensor:

$$\frac{\partial^2\sigma}{\partial E'\,\partial\Omega'} = \frac{2\alpha^2}{Q^4}\frac{E'}{E} l^{\lambda\rho} W^{(\mathrm{eN})}_{\rho\lambda}, \tag{18.21}$$

where

$$\begin{aligned} l^{\lambda\rho} &= k'^{\lambda}k^{\rho} + k^{\lambda}k'^{\rho} - g^{\lambda\rho}(kk'), \\ W^{(\mathrm{eN})}_{\rho\lambda} &= \sum_{\mathrm{X}} \frac{1}{2M}(2\pi)^3\delta(p'-p-q) \\ &\quad \cdot \sum_{\mathrm{Spins}}{}' \langle \mathrm{N}(p)|\mathscr{J}_{\rho}(0)|\mathrm{X}(p')\rangle\langle \mathrm{X}(p')|\mathscr{J}_{\lambda}(0)|\mathrm{N}(p)\rangle. \end{aligned} \tag{18.22}$$

Since the $\mathscr{J}_{\lambda}$ are Hermitian operators, we have

$$W^{(\mathrm{eN})}_{\rho\lambda} = (W^{(\mathrm{eN})}_{\lambda\rho})^{*}. \tag{18.23}$$

A few general principles help to further simplify the tensor $W^{(\mathrm{eN})}_{\rho\lambda}$. Due to the summation over all hadronic final states, and the use of unpolarized initial nucleous, $W^{(\mathrm{eN})}_{\rho\lambda}$ can only depend on the four-vectors p and q. Furthermore, $W^{(\mathrm{eN})}_{\rho\lambda}$ must be a Lorentz tensor that, as a result of the conservation of the electromagnetic current

$$\partial^{\rho}\mathscr{J}_{\rho}(x) = 0, \tag{18.24}$$

satisfies the conditions

$$\begin{aligned} q^{\rho}W^{(\mathrm{eN})}_{\rho\lambda}(p,q) &= 0, \\ q^{\lambda}W^{(\mathrm{eN})}_{\rho\lambda}(p,q) &= 0. \end{aligned} \tag{18.25}$$

We can easily write down three tensors that satisfy these conditions:

$$\begin{aligned} &\left(-g_{\rho\lambda} + \frac{q_{\rho}q_{\lambda}}{q^2}\right), \\ &\left(p_{\rho} - \frac{(pq)q_{\rho}}{q^2}\right)\left(p_{\lambda} - \frac{(pq)q_{\lambda}}{q^2}\right), \\ &\mathrm{i}\varepsilon_{\rho\lambda\sigma\tau}p^{\sigma}q^{\tau}. \end{aligned}$$

It can be shown that there is no other independent tensor that can be formed from p, q, $g_{\mu\nu}$ and $\varepsilon_{\mu\nu\rho\sigma}$ which satisfies (18.25). The third of the above tensors can be dropped in the present context because the electromagnetic current is a pure vector current, which means the no term with odd parity can appear in $W^{(\mathrm{eN})}_{\rho\lambda}$. Thus $W^{(\mathrm{eN})}_{\rho\lambda}$ can be represented as a sum of the two remaining tensors multiplied by unknown scalar functions. The invariant scalar functions can only depend on the invariants that we can form from p and q, namely, $Q^2 = -q^2$ and $\nu = pq/M$. Thus we can make the ansatz

$$
\begin{aligned}
W_{\rho\lambda}^{(\mathrm{eN})} = &\left(-g_{\rho\lambda} + \frac{q_\rho q_\lambda}{q^2}\right) W_1(\nu, Q^2) \\
&+ \left(p_\rho - \frac{(pq)q_\rho}{q^2}\right)\left(p_\lambda - \frac{(pq)q_\lambda}{q^2}\right)\frac{1}{M^2} W_2(\nu, Q^2).
\end{aligned}
\tag{18.26}
$$

The functions $W_1(\nu, Q^2)$ and $W_2(\nu, Q^2)$, which, on account of (18.23), are real, are known as the structure functions of the nucleon. They contain the information about the dynamics of the strong interaction part of the process.

Inserting (18.26) into (18.21) then yields, after a simple calculation:

$$
\frac{\partial^2 \sigma^{(\mathrm{eN})}}{\partial E' \, \partial \Omega'} = \frac{4\alpha^2 E'^2}{Q^4}\left\{2W_1^{(\mathrm{eN})}(\nu, Q^2)\sin^2\frac{\vartheta}{2} + W_2^{(\mathrm{eN})}(\nu, Q^2)\cos^2\frac{\vartheta}{2}\right\}. \tag{18.27}
$$

From this we see that for small scattering angles, one measures mainly W_2 and for large scattering angles mainly W_1. However, large scattering angles are strongly supressed by the factor $Q^4 \propto \sin^4 \vartheta/2$ in the denominator (cf. Table 18.1).

We will now attempt to apply the parton model to our reaction. For this we have to use a frame of reference in which the nucleon is moving very fast. A suitable system is the so-called *Breit frame* in which the electron has the same energy before and after the scattering. The energy transfer in this system is therefore zero and the virtual photon only transfers three-momentum. It can easily be seen that the Breit frame can be chosen to have the further property that the three-momenta of the nucleon and of the virtual photon are antiparallel and point in the positive and negative x^3-directions, respectively. The four-momenta p of the nucleon and q of the virtual photon are then of the form

$$
p = \begin{pmatrix} \sqrt{P^2 + M^2} \\ 0 \\ 0 \\ P \end{pmatrix}, \qquad q = \begin{pmatrix} 0 \\ 0 \\ 0 \\ -Q \end{pmatrix}. \tag{18.28}
$$

The quantities P and Q are as given in Table 18.1 since from (18.28) we indeed obtain:

$$
q^2 = -Q^2, \qquad pq = PQ = M\nu. \tag{18.29}
$$

Following Bjorken (1969) we consider the limit

$$
\begin{aligned}
&Q^2 \to \infty, \\
&\nu \to \infty, \\
&x = \frac{Q^2}{2M\nu} \quad \text{fixed.}
\end{aligned}
\tag{18.30}
$$

In this Bjorken limit the third component of the momentum of the nucleon in the Breit frame tends to infinity

$$P = \frac{Q}{2x} \to \infty .$$

We then have a rapidly moving nucleon and can treat this, according to *Assumption I* of the parton model, as a beam of partons all of which travel in approximately the same direction as the nucleon. If we neglect the transverse momenta of the partons entirely, then the three-momentum $\boldsymbol{p}_{\mathrm{i}}$ of the parton i is a fraction ξ of the three-momentum of the nucleon:

$$\boldsymbol{p}_{\mathrm{i}} = \xi \begin{pmatrix} 0 \\ 0 \\ P \end{pmatrix}, \tag{18.31}$$

where $0 \leq \xi \leq 1$. We shall describe the parton content of the nucleon by distribution functions. For the partons of type j let

$$N_j(\xi)\,\mathrm{d}\xi$$

be the number of partons with momentum fraction between ξ and $\xi + \mathrm{d}\xi$. Partons and antipartons are treated as separate types in this procedure.

We now apply *Assumption II* of the parton model. According to this we have to calculate the scattering of the electron from free partons and must then sum incoherently over all the parton contributions. The parton scattering diagram is shown in Fig. 18.7, and a diagrammatic representation of the total reaction is given in Fig. 18.8. Since we are observing the reaction in the Breit frame where all particles are in motion before the collision, it is convenient to work with reaction rates rather than cross-sections. According to (5.54) and (18.21), the reaction rate for electron nucleon scattering in the approximation $m_{\mathrm{e}} = 0$ is given by

$$\begin{aligned} \mathrm{d}\Gamma(\mathrm{e}(\boldsymbol{k}) + \mathrm{N}(\boldsymbol{p}) \to \mathrm{e}(\boldsymbol{k}') + \mathrm{X}) &= \frac{1}{V}\frac{w(s, M^2, 0)}{2k^0 p^0}\,\mathrm{d}\sigma(\mathrm{e}(\boldsymbol{k}) + \mathrm{N}(\boldsymbol{p}) \to \mathrm{e}(\boldsymbol{k}') + \mathrm{X}) \\ &= \frac{1}{V}\frac{w(s, M^2, 0)}{2k^0 p^0}\frac{\partial^2\sigma}{E'\,\partial E'\,\partial\Omega'}\frac{\mathrm{d}^3k'}{k'^0} \\ &= \frac{1}{V}\frac{1}{k^0}\frac{2\alpha^2}{Q^4}\,l^{\lambda\rho}\Gamma_{\rho\lambda}\frac{\mathrm{d}^3k'}{k'^0}, \end{aligned} \tag{18.32}$$

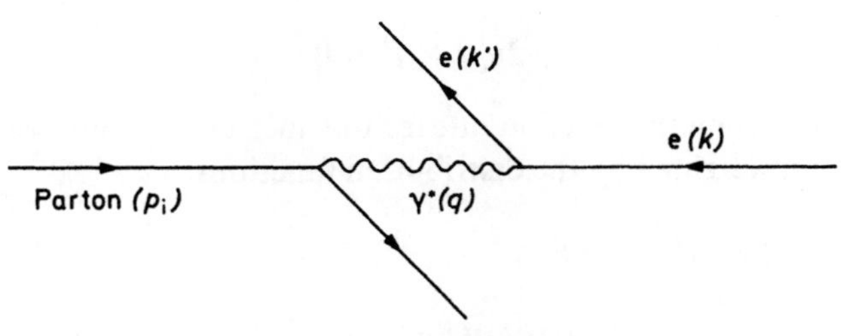

Figure 18.7 The diagram for electron–parton scattering.

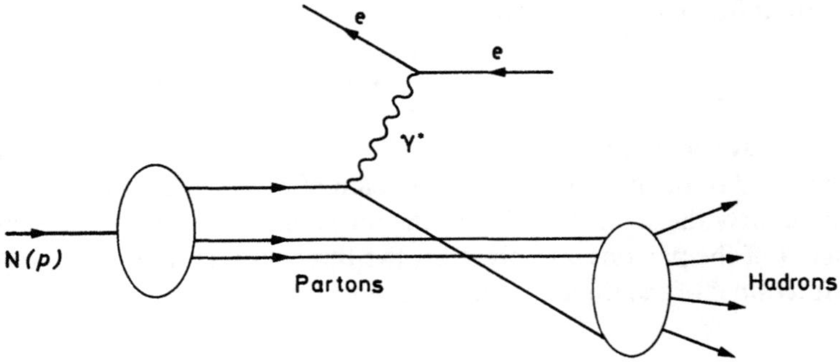

Figure 18.8 Electron–nucleon scattering in the parton model.

where we set

$$\Gamma_{\rho\lambda} = \frac{M}{p^0} W^{(\mathrm{eN})}_{\rho\lambda}.$$

The reaction rate for a parton a_i with four-momentum p_i can be defined in an analogous way. According to the diagram in Fig. 18.7, we have to calculate the elastic electron–parton scattering. For a spin-$\frac{1}{2}$ parton with charge Q_i and mass m_i, we find

$$\mathrm{d}\Gamma(\mathrm{e}(\boldsymbol{k}) + \mathrm{a}_i(\boldsymbol{p}_i) \to \mathrm{e}(\boldsymbol{k}') + \mathrm{a}_i(\boldsymbol{p}_i')) = \frac{1}{V}\frac{1}{k^0}\frac{2\alpha^2}{Q^4} l^{\lambda\rho}\Gamma^{(i)}_{\rho\lambda}\frac{\mathrm{d}^3 k'}{k'^0}, \qquad (18.33)$$

$$\Gamma^{(i)}_{\rho\lambda}(p_i, q) = \frac{Q_i^2}{p_i^0}\delta(2p_i q + q^2)$$
$$\cdot\left\{\left(-g_{\rho\lambda} + \frac{q_\rho q_\lambda}{q^2}\right)\left(-\frac{q^2}{2}\right) + \left(p_{i_\rho} - \frac{(p_i q)q_\rho}{q^2}\right)\left(p_{i_\lambda} - \frac{(p_i q)q_\lambda}{q^2}\right)2\right\}. \qquad (18.34)$$

The δ-function stems from the fact that, according to *Assumption II*, we treat the partons as free before and after the scattering; this means that their momenta must satisfy the mass-shell condition

$$m_i^2 = p_i^2 = p_i'^2 = (p_i + q)^2, \qquad (18.35)$$

implying

$$2p_i q + q^2 = 0.$$

In order to obtain the reaction rate for the nucleon, we sum over all parton contributions, weighted by the distribution functions:

$$\mathrm{d}\Gamma(\mathrm{e}(\boldsymbol{k}) + \mathrm{N}(\boldsymbol{p}) \to \mathrm{e}(\boldsymbol{k}') + \mathrm{X})$$
$$= \sum_i \int_0^1 \mathrm{d}\xi\; N_i(\xi)\,\mathrm{d}\Gamma(\mathrm{e}(\boldsymbol{k}) + \mathrm{a}_i(\boldsymbol{p}_i) \to \mathrm{e}(\boldsymbol{k}') + \mathrm{a}_i(\boldsymbol{p}_i'))|_{\boldsymbol{p}_i = \xi\boldsymbol{p}}. \qquad (18.36)$$

With (18.32)–(18.34) we then find

$$\Gamma_{\rho\lambda} = \frac{M}{p^0} W_{\rho\lambda}(p, q)$$
$$= \sum_i \int_0^1 \mathrm{d}\xi \, N_i(\xi) \Gamma^{(i)}_{\rho\lambda}(p_i, q)|_{p_i = \xi p}. \tag{18.37}$$

In the spirit of a high energy approximation, we now neglect the masses of the nucleons and partons and use the proportionality in ξ also for the energies in the Breit frame

$$p_i^0 \cong \xi p^0. \tag{18.38}$$

With the help of the δ-function in (18.34), the integral in (18.37) can now easily be carried out. For the case where all charged partons have spin $\frac{1}{2}$, we get in this way for the structure functions:

$$2MW_1(\nu, Q^2) = \sum_i Q_i^2 N_i(x),$$
$$\nu W_2(\nu, Q^2) = x \sum_i Q_i^2 N_i(x). \tag{18.39}$$

If charged, spin-zero partons are taken into account then this generalizes to:

$$2MW_1(\nu, Q^2) = \sum_i Q_i^2 N_i(x),$$
$$\nu W_2(\nu, Q^2) = x \sum_i Q_i^2 N_i(x) + x \sum_{i'} Q_{i'}^2 N_{i'}(x), \tag{18.40}$$

(i: spin-$\frac{1}{2}$ partons, i': spin-0 partons).

From (18.40) we see that for given ν and Q^2, exactly the partons with momentum fraction x contribute to the structure functions. This can readily be understood as a consequence of energy-momentum conservation in electron–parton scattering in the Breit frame (Exercise 18.1).

The naïve parton model yields the so-called Bjorken scaling behavior for the structure functions. Even before the construction of the parton model, Bjorken had shown from abstract considerations that the structure functions $2MW_1$ and νW_2 should, in the Bjorken limit (18.30), tend to nontrivial functions of the one variable x (Bjorken 1969). More precisely, he predicted:

$$2MW_1(\nu, Q^2) = F_1(x) + \mathrm{O}\left(\frac{1}{Q^2}\right),$$
$$\nu W_2(\nu, Q^2) = F_2(x) + \mathrm{O}\left(\frac{1}{Q^2}\right). \tag{18.41}$$

The parton model result (18.40) provides a simple interpretation for the scaling functions $F_{1,2}(x)$. When the masses and transverse momenta of the partons are taken into account, terms of the order of $1/Q^2$ also appear in the parton model; these, however, have been neglected in (18.39), (18.40).

The older data (Miller 1972; Bodek 1973) were fully compatible with (18.41) and it seemed that for $Q^2 \gtrsim 2\ \mathrm{GeV}^2$ one could neglect the terms of order $1/Q^2$.

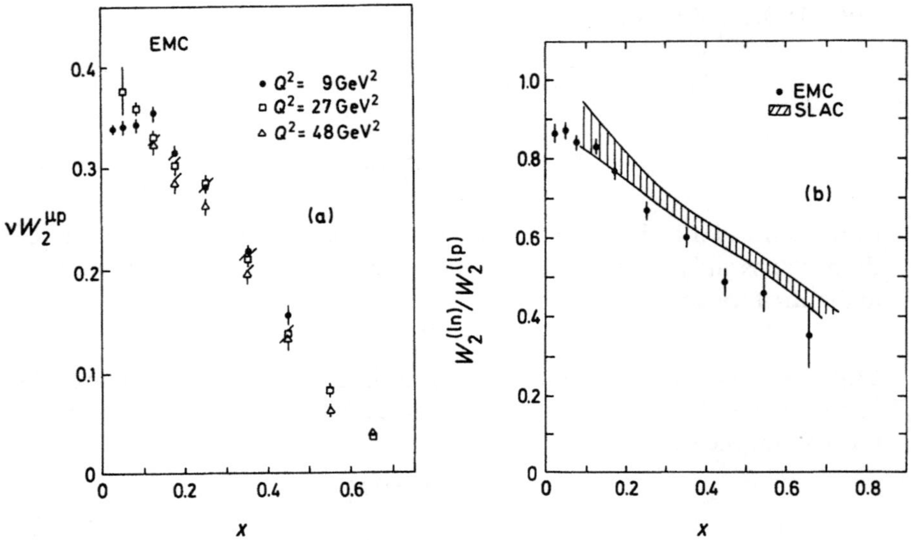

Figure 18.9 Data from the European Muon Collaboration (EMC) for the structure function $\nu W_2^{(\mu p)}(\nu, Q^2)$ of the proton as a function of $x = Q^2/(2M\nu)$ for various Q^2 values. Exact Bjorken scale invariance would demand that the data points for the same x but different Q^2 should lie on top of one another (a). Part (b) shows the ratio of the neutron and proton structure functions $W_2^{(\mathrm{ln})}(\nu, Q^2)$ and $W_2^{(\mathrm{lp})}(\nu, Q^2)$ ($\mathrm{l} = e, \mu$) as a function of x. The shaded band represents the SLAC data obtained from electron scattering in the interval $2 \le Q^2 \le 20$ GeV^2. The points correspond to preliminary EMC data from muon scattering in the interval $10 \le Q^2 \le 80$ GeV^2 (after Drees 1983 and Dydak 1983).

Even the recent data for deep inelastic muon–proton scattering show that Bjorken scale invariance represents a very good approximation to reality (Fig. 18.9(a)). For the structure functions of the neutron, which are extracted from the scattering data for a deuteron target with the help of a model for the deuteron, one also finds approximate validity of Bjorken scale invariance. Data for the ratio of the structure functions $W_2^{(\mu n)}$ and $W_2^{(\mu p)}$ are shown in Fig. 18.9(b).

Encouraged by this successful comparison of theory with experiment, we can now proceed to use (18.40) to determine the spin of the charged partons. Following Callan and Gross (Callan 1969), we consider the ratio

$$R(x) = \frac{F_2(x) - xF_1(x)}{F_2(x)}. \tag{18.42}$$

From (18.40) we see that $R(x)$ measures the contribution of the spin-zero partons, i.e.,

$$R(x) = \begin{cases} 0 \text{ if all charged partons have spin } \frac{1}{2}, \\ 1 \text{ if all charged partons have spin } 0. \end{cases}$$

Recent μ–p scattering experiments (Rith 1983) yield for the average of $R(x)$ over the whole x-range and a mean value $\langle Q^2 \rangle = 22.5\ \mathrm{GeV}^2$ the result:

$$\langle R \rangle = -0.01 \pm 0.11. \tag{18.43}$$

This is compatible with $R = 0$ and is strong evidence that the charged partons all have spin $\frac{1}{2}$. Henceforth we shall assume that to be the case.

18.3 The Flavor Quantum Numbers of the Partons

In this section we will discuss the charge and isospin quantum numbers of the partons. It is found that deep inelastic data, when interpreted in terms of the parton model, unambiguously lead to nonintegral charges for the partons.

One of the most important numbers in this connection is the ratio of the cross-sections for electron–nucleon and neutrino–nucleon scattering:

$$\begin{aligned} \nu_\mu + \mathrm{N} &\to \mu^- + \mathrm{X}, \\ \bar{\nu}_\mu + \mathrm{N} &\to \mu^+ + \mathrm{X}. \end{aligned} \tag{18.44}$$

The kinematics of these charged current processes is very similar to that of electron–nucleon scattering. The Feynman diagram for the neutrino scattering (Fig. 18.10) is interpreted as follows: A neutrino emits a virtual W^+-boson and thereby changes into a muon μ^-. The W-boson is absorbed by the nucleon. In contrast to the well-known massless and neutral photon, the W-boson is heavy and charged; its discovery in 1983 will be discussed in detail in Part IV. The mass of the W-boson was found, as predicted, to be $m_\mathrm{W} \cong 80\ \mathrm{GeV}$. At center of mass energies $\sqrt{s} \ll m_\mathrm{W}$, the exchange of a W-boson leads to a practically point-like coupling between lepton and hadron currents which can be described by the effective Lagrange density (see Part IV)

$$\mathscr{L}_{\mathrm{eff}}(x) = -\frac{G}{\sqrt{2}}(\ell^\lambda(x)\mathscr{J}_\lambda^+(x) + \ell^{\lambda\dagger}(x)\mathscr{J}_\lambda^-(x)). \tag{18.45}$$

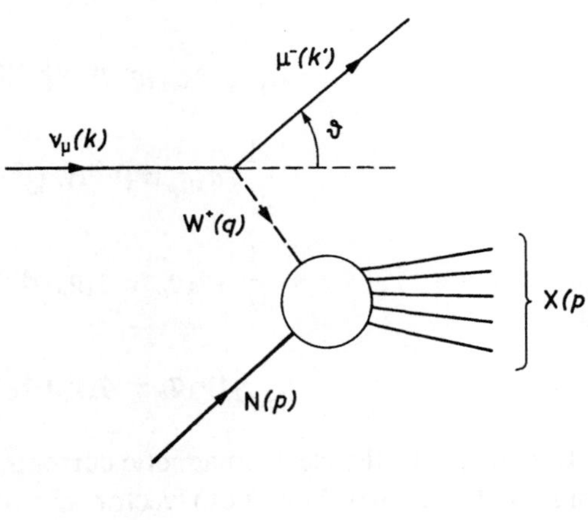

Figure 18.10 The Feynman diagram for neutrino–nucleon scattering via charged currents.

Here $G \cong 1.16 \times 10^{-5}\ \mathrm{GeV}^{-2}$ is Fermi's constant and the lepton current ℓ^λ is constructed from the fields μ and ν_μ of the muon and the neutrino as follows:

$$\ell^\lambda = \bar{\mu}\gamma^\lambda(1-\gamma_5)\nu_\mu .$$

The $\mathcal{J}_\lambda^\pm$ are the charged weak hadronic currents that were first investigated in the β-decay of the neutron and of atomic nuclei. They must satisfy the following relation:

$$(\mathcal{J}_\lambda^+)^\dagger = \mathcal{J}_\lambda^- ,$$

in order to guarantee that the effective Lagrange density is Hermitian.

The calculation of the transition matrix element and the cross-section for the reactions (18.44) now proceeds exactly according to the methods presented in Sect. 18.2. Neglecting the muon mass, we find for the neutrino and anti-neutrino scattering

$$\frac{\partial^2\sigma^{(\nu,\bar\nu)}}{\partial E'\,\partial\Omega'} = \frac{G^2}{2\pi^2}E'^2\left\{2W_1^{(\nu,\bar\nu)}(\nu,Q^2)\sin^2\frac{\vartheta}{2} + W_2^{(\nu,\bar\nu)}(\nu,Q^2)\cos^2\frac{\vartheta}{2} \right.$$
$$\left. \mp W_3^{(\nu,\bar\nu)}(\nu,Q^2)\frac{E+E'}{M}\sin^2\frac{\vartheta}{2}\right\}. \qquad (18.46)$$

Here $W_{1,2,3}^{(\nu,\bar\nu)}$ are the weak structure functions of the nucleon. They are defined completely analogously to the electromagnetic structure functions (18.26), via the expansion of the hadron tensors $W_{\lambda\rho}^{(\nu,\bar\nu)}$ in terms of a suitable tensor basis:

$$W_{\lambda\rho}^{(\nu,\bar\nu)}(p,q) = \sum_X \frac{1}{2M}(2\pi)^3\delta(p'-p-q)$$
$$\cdot \sum_{\mathrm{Spins}}{}' \langle \mathrm{N}(p)|\mathcal{J}_\lambda^\mp|\mathrm{X}(p')\rangle\langle \mathrm{X}(p')|\mathcal{J}_\rho^\pm|\mathrm{N}(p)\rangle$$
$$= \left(-g_{\lambda\rho} + \frac{q_\lambda q_\rho}{q^2}\right)W_1^{(\nu,\bar\nu)}(\nu,Q^2)$$
$$+ \left(p_\lambda - \frac{(pq)q_\lambda}{q^2}\right)\left(p_\rho - \frac{(pq)q_\rho}{q^2}\right)\frac{1}{M^2}W_2^{(\nu,\bar\nu)}(\nu,Q^2)$$
$$- \frac{\mathrm{i}}{2M^2}\varepsilon_{\lambda\rho\alpha\beta}p^\alpha q^\beta W_3^{(\nu,\bar\nu)}(\nu,Q^2)$$
$$+ \frac{1}{M^2}q_\lambda q_\rho W_4^{(\nu,\bar\nu)}(\nu,Q^2)$$
$$+ \frac{1}{M^2}(p_\lambda q_\rho + q_\lambda p_\rho)W_5^{(\nu,\bar\nu)}(\nu,Q^2)$$
$$+ \frac{\mathrm{i}}{M^2}(p_\lambda q_\rho - q_\lambda p_\rho)W_6^{(\nu,\bar\nu)}(\nu,Q^2). \qquad (18.47)$$

In contrast to the electromagnetic currents, the weak currents $\mathcal{J}_\lambda^\pm$ are not conserved and they have both vector and axial vector parts. Thus, in the expansion of the tensors $W_{\lambda\rho}^{(\nu,\bar{\nu})}$ more invariant functions appear than in (18.26); in particular, a term of odd parity proportional to the ε-tensor. The structure function W_6 vanishes if time-reversal invariance holds. In any case, W_6 and also W_5 and W_4 are almost impossible to measure since their contribution to the scattering cross-section is proportional to the muon mass divided by the energy of the incident neutrino. For neutrino energies in the GeV range, such terms can be neglected as was done in (18.46). By comparing the scattering cross-sections (18.27) and (18.46), we further see that we need to make essentially the following substitution:

$$\frac{\alpha}{Q^2} \to \frac{G}{2\pi\sqrt{2}}.$$

The fine-structure constant multiplied by the photon propagator is replaced by Fermi's constant when one goes from electron to neutrino scattering.

We will now apply the parton model to neutrino–nucleon scattering. For this we must make an assumption concerning the composition of the weak current in terms of parton fields. The most obvious approach is to try the quark model. If we consider only u-, d-, and s-quarks, and neglect the Cabibbo angle, then the phenomenology of weak meson and baryon decays suggests the ansatz (see Part IV):

$$\begin{aligned} \mathcal{J}_\lambda^+ &= \bar{\mathrm{u}}\gamma_\lambda(1-\gamma_5)\mathrm{d}, \\ \mathcal{J}_\lambda^- &= \bar{\mathrm{d}}\gamma_\lambda(1-\gamma_5)\mathrm{u}. \end{aligned} \tag{18.48}$$

With this, the fundamental parton processes underlying the reactions (18.44) are the following (Fig. 18.11):

$$\begin{aligned} \nu_\mu + \mathrm{d} &\to \mu^- + \mathrm{u}, \\ \nu_\mu + \bar{\mathrm{u}} &\to \mu^- + \bar{\mathrm{d}}, \end{aligned} \tag{18.49}$$

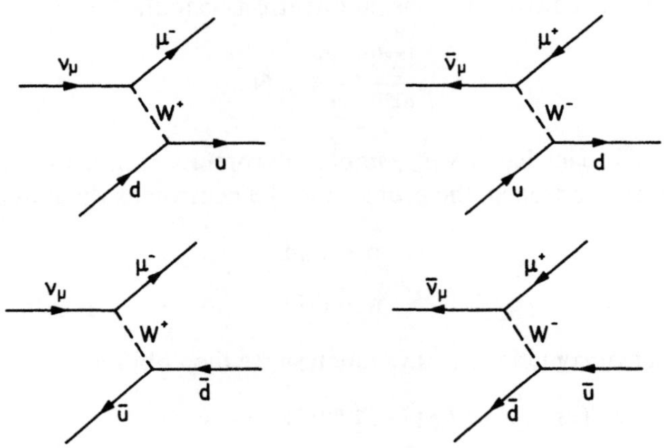

Figure 18.11 The diagrams for the fundamental parton processes (18.49), (18.50) underlying neutrino- and antineutrino–nucleon scattering.

$$\begin{aligned} &\bar{\nu}_\mu + \mathrm{u} \to \mu^+ + \mathrm{d}, \\ &\bar{\nu}_\mu + \bar{\mathrm{d}} \to \mu^+ + \bar{\mathrm{u}}. \end{aligned} \tag{18.50}$$

It is now a straightforward matter to apply the rules of the parton model in order to derive the predictions for charged current processes. In summary the results for the structure functions of neutrino–, antineutrino–, and electron–nucleon scattering in the Bjorken limit are:

$$\begin{aligned} 2MW_1^{(\nu)}(\nu, Q^2) &\to F_1^{(\nu)}(x) = 2N_\mathrm{d}(x) + 2N_{\bar{\mathrm{u}}}(x), \\ \nu W_2^{(\nu)}(\nu, Q^2) &\to F_2^{(\nu)}(x) = xF_1^{(\nu)}(x), \\ \nu W_3^{(\nu)}(\nu, Q^2) &\to F_3^{(\nu)}(x) = -2N_\mathrm{d}(x) + 2N_{\bar{\mathrm{u}}}(x); \end{aligned} \tag{18.51}$$

$$\begin{aligned} 2MW_1^{(\bar{\nu})}(\nu, Q^2) &\to F_1^{(\bar{\nu})}(x) = 2N_\mathrm{u}(x) + 2N_{\bar{\mathrm{d}}}(x), \\ \nu W_2^{(\bar{\nu})}(\nu, Q^2) &\to F_2^{(\bar{\nu})}(x) = xF_1^{(\bar{\nu})}(x), \\ \nu W_3^{(\bar{\nu})}(\nu, Q^2) &\to F_3^{(\bar{\nu})}(x) = -2N_\mathrm{u}(x) + 2N_{\bar{\mathrm{d}}}(x); \end{aligned} \tag{18.52}$$

$$\begin{aligned} 2MW_1^{(\mathrm{e})}(\nu, Q^2) &\to F_1^{(\mathrm{e})}(x) = \tfrac{4}{9}N_\mathrm{u}(x) + \tfrac{1}{9}N_\mathrm{d}(x) + \tfrac{1}{9}N_\mathrm{s}(x) \\ &\qquad + \tfrac{4}{9}N_{\bar{\mathrm{u}}}(x) + \tfrac{1}{9}N_{\bar{\mathrm{d}}}(x) + \tfrac{1}{9}N_{\bar{\mathrm{s}}}(x), \\ \nu W_2^{(\mathrm{e})}(\nu, Q^2) &\to F_2^{(\mathrm{e})}(x) = xF_1^{(\mathrm{e})}(x). \end{aligned} \tag{18.53}$$

These are fundamental equations of the quark–parton model. It is to be noted that the distribution functions for the quarks and the antiquarks depend on the particular nucleon target under consideration.

The deep inelastic neutrino–nucleon scattering was first studied in detail with the help of the bubble chamber "Gargamelle" at CERN (Deden 1975). In these experiments, liquid freon, CF_3Br, which contains approximately equal numbers of protons and neutrons, was used as a target. As can be seen in Fig. 18.12, the ratio of the structure function F_2 for neutrino and electron scattering averaged over protons and neutrons is found to be

$$\frac{F_2^{(\nu\mathrm{N})}(x)}{F_2^{(\mathrm{eN})}(x)} \cong 3.60.$$

This ratio in fact has a very simple interpretation. Let us describe, in a simple-minded approach, the proton and the neutron as three-quark states:

$$\begin{aligned} \mathrm{p} &\sim \mathrm{uud}, \\ \mathrm{n} &\sim \mathrm{ddu}. \end{aligned} \tag{18.54}$$

For the electromagnetic structure function we then obtain

$$\begin{aligned} F_2^{(\mathrm{eN})}(x) &= \tfrac{1}{2}(F_2^{(\mathrm{ep})}(x) + F_2^{(\mathrm{en})}(x)) \\ &= \tfrac{1}{2}x\{\tfrac{4}{9}N_\mathrm{u}^\mathrm{p}(x) + \tfrac{1}{9}N_\mathrm{d}^\mathrm{p}(x) + \tfrac{4}{9}N_\mathrm{u}^\mathrm{n}(x) + \tfrac{1}{9}N_\mathrm{d}^\mathrm{n}(x)\} \\ &= \tfrac{5}{18}x(N_\mathrm{u}^\mathrm{p}(x) + N_\mathrm{d}^\mathrm{p}(x)). \end{aligned} \tag{18.55}$$

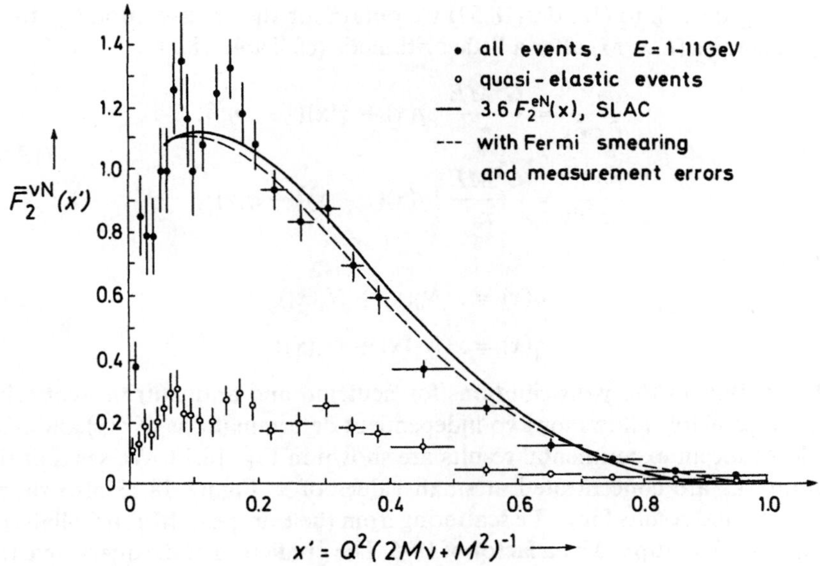

Figure 18.12 Data for the structure function $F_2^{(\nu N)}$ of neutrino–nucleon scattering, averaged over proton and neutron, as a function of $x' = Q^2/(2M\nu + M^2)$. For comparison the likewise averaged structure function $F_2^{(eN)}$ multiplied by 3.6 is also shown (solid and dashed lines) (after Deden 1975).

In deriving this we have made use of the generally valid isospin symmetry relations (Exercise 18.5):

$$\begin{aligned} N_u^p(x) &= N_d^n(x), \\ N_d^p(x) &= N_u^n(x). \end{aligned} \tag{18.56}$$

For the corresponding weak structure function we find from (18.51)

$$F_2^{(\nu N)}(x) = \tfrac{1}{2}x\{2N_d^p(x) + 2N_d^n(x)\} = x(N_d^p(x) + N_u^p(x)). \tag{18.57}$$

and thus for the ratio we obtain

$$\frac{F_2^{(\nu N)}(x)}{F_2^{(eN)}(x)} = \frac{18}{5} = 3.60, \tag{18.58}$$

in perfect agreement with experiment (Fig. 18.12).

In a more exact analysis, of course, the antiquarks in the nucleon must be taken into account. The antiquark distributions are obtained from measurements of the structure functions F_2 and F_3 according to (18.51), (18.52). Very detailed studies have been made of the scattering of neutrinos and antineutrinos from iron. We shall ignore the small excess of neutrons over protons in iron and regard this nucleus as an isoscalar target. For the quark distribution per nucleon we then have:

$$\begin{aligned} N_u(x) &= N_d(x), \\ N_{\bar{u}}(x) &= N_{\bar{d}}(x). \end{aligned} \tag{18.59}$$

Inserting these into (18.51), (18.52) we obtain for the cross-section (18.46) at high energies ($E \gg M$) after a little arithmetic (cf. Table 18.1):

$$\frac{\partial\sigma^{(\nu N)}}{\partial x\,\partial y} = \frac{G^2 ME}{\pi}\{q(x) + \bar{q}(x)(1-y)^2\},$$
$$\frac{\partial\sigma^{(\bar{\nu} N)}}{\partial x\,\partial y} = \frac{G^2 ME}{\pi}\{q(x)(1-y)^2 + \bar{q}(x)\}, \tag{18.60}$$

where

$$q(x) = x(N_u(x) + N_d(x)),$$
$$\bar{q}(x) = x(N_{\bar{u}}(x) + N_{\bar{d}}(x)).$$

The analysis of the y-distributions for neutrino and antineutrino scattering thus in principle allows for two independent determinations of $q(x)$ and $\bar{q}(x)$.

Some recent experimental results are shown in Fig. 18.13. We see that the antiquarks are concentrated at small values of x. Figure 18.13 also shows experimental results for μ–Fe scattering from the European Muon Collaboration (EMC) multiplied by a factor of 18/5. The prediction of the quark–parton model (18.58), which should be valid for x-values where antiquarks are negligible, is evidently well confirmed.

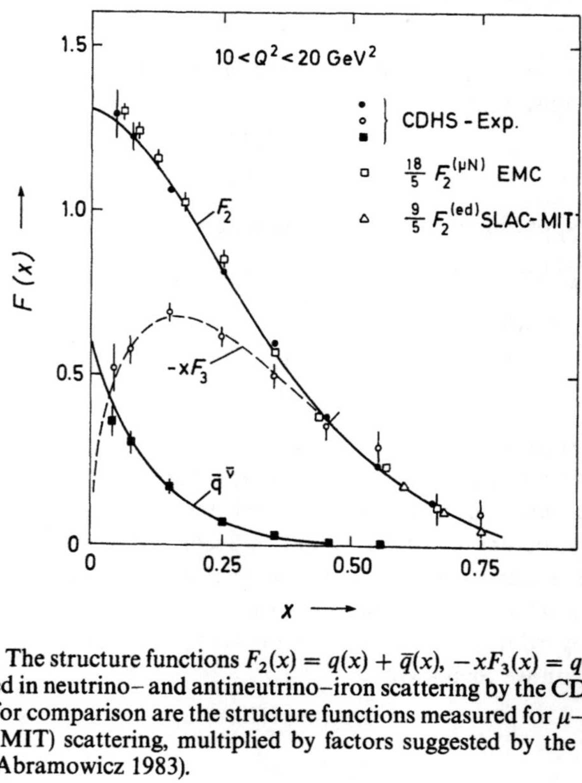

Figure 18.13 The structure functions $F_2(x) = q(x) + \bar{q}(x)$, $-xF_3(x) = q(x) - \bar{q}(x)$, and $\bar{q}(x)$, measured in neutrino- and antineutrino-iron scattering by the CDHS collaboration. Shown for comparison are the structure functions measured for μ–Fe (EMC) and e–d (SLAC–MIT) scattering, multiplied by factors suggested by the quark–parton model (after Abramowicz 1983).

The total cross-sections for neutrino– and antineutrino–nucleon scattering are obtained from (18.60) by integrating over x and y:

$$\sigma(\nu_\mu \mathrm{N} \to \mu^- \mathrm{X}) = \frac{G^2 ME}{\pi}\{\langle q\rangle + \tfrac{1}{3}\langle \bar{q}\rangle\},$$

$$\sigma(\bar{\nu}_\mu \mathrm{N} \to \mu^+ \mathrm{X}) = \frac{G^2 ME}{\pi}\{\tfrac{1}{3}\langle q\rangle + \langle \bar{q}\rangle\}, \tag{18.61}$$

$(E \gg M)$.

Here we have put

$$\langle q\rangle = \int_0^1 \mathrm{d}x\, x(N_\mathrm{u}(x) + N_\mathrm{d}(x)),$$

$$\langle \bar{q}\rangle = \int_0^1 \mathrm{d}x\, x(N_{\bar{\mathrm{u}}}(x) + N_{\bar{\mathrm{d}}}(x)). \tag{18.62}$$

The parton model thus predicts a linear increase of the total cross-section with neutrino energy E. This, too, is in good agreement with experiment (Fig. 18.14).

We also note that the factor 18/5 in (18.58) depends in an essential way on fractional quark charges. For comparison, let us suppose the partons to be bare nucleons, i.e., point-like particles p′ and n′ with the quantum numbers of the proton and the neutron. A simple model for the nucleon would then be

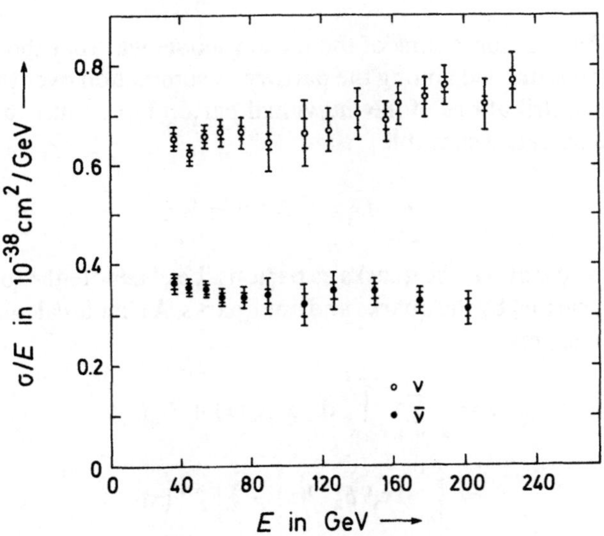

Figure 18.14 The total cross-sections for neutrino– and antineutrino–iron scattering. Shown are the cross-sections per nucleon divided by the neutrino energy $\sigma(\nu_\mu \mathrm{N} \to \mu^- \mathrm{X})/E$ and $\sigma(\bar{\nu}_\mu \mathrm{N} \to \mu^+ \mathrm{X})/E$ as a function of E. The average values are $\langle \sigma(\nu_\mu \mathrm{N} \to \mu^- \mathrm{X})/E\rangle = (0.669 \pm 0.003 \pm 0.024)\cdot 10^{-38}\ \mathrm{cm}^2/\mathrm{GeV}$ and $\langle \sigma(\bar{\nu}_\mu \mathrm{N} \to \mu^+ \mathrm{X})/E\rangle = (0.340 \pm 0.003 \pm 0.020)\cdot 10^{-38}\ \mathrm{cm}^2/\mathrm{GeV}$. The first error represents the statistical, the second the systematic uncertainty (after Blair 1983).

the following:

$$\mathrm{p} \sim \mathrm{p}' + \text{neutral partons},$$

$$\mathrm{n} \sim \mathrm{n}' + \text{neutral partons}.$$

It can readily be seen that this leads to the following result:

$$\frac{F_2^{(\nu N)}(x)}{F_2^{(eN)}(x)} = 2,$$

which is in contradiction with experiment.

These simple arguments can, in fact, be put into much more general form. It is possible to show that the experimental data on electron– and neutrino–nucleon scattering, together with that for electron–positron annihilation and π^0-decay, unambiguously exclude all models of partons with integer charge (Kühnelt 1975). On the other hand, partons with third integer charges, as in the quark model, are compatible with all experimental results. With this, the nature of the partons that take part in the weak and the electromagnetic interaction is made clear: They are one and the same as the quarks which we already know. In the next section we will discuss the evidence for gluons obtained from deep inelastic scattering experiments.

18.4 Sum Rules and Evidence for Flavor-Neutral Partons, Gluons

One of the basic assumptions of the parton model was that the momentum of the nucleon is divided among the partons. A summation over the fractional momentum contributions of the individual parton types must thus give one. We obtain a so-called sum rule

$$\int_0^1 \mathrm{d}x\, x \sum_j N_j(x) = 1. \tag{18.63}$$

We already know about the quarks as partons. Let I denote the total momentum fraction carried by the quarks and antiquarks. A simple calculation shows that I can be expressed as:

$$\begin{aligned} I &= \sum_{q=u,d,s} \int_0^1 \mathrm{d}x\, x(N_q(x) + N_{\bar{q}}(x)) \\ &= \int_0^1 \mathrm{d}x(9F_2^{(eN)}(x) - \tfrac{3}{2}F_2^{(\nu N)}(x)). \end{aligned} \tag{18.64}$$

This integral should be equal to one if the quarks and antiquarks carry the total momentum of the nucleon. From the experimental data however (Fig. 18.13), one finds

$$I \cong 0.50.$$

From this we must conclude that there are further partons in the nucleon besides quarks and antiquarks, and that these carry about half of the total momentum. With respect to the weak and electromagnetic interactions, these partons are neutral. The next obvious step was to assume that these partons, whose existence was concluded from indirect evidence, are responsible for the strong interaction between quarks. They thus became known as gluons. They are supposed to be responsible for the fact that the quarks cannot be extracted from the nucleon, but are permanently bound. In the next chapter we shall see that the natural theory of quarks, quantum chromodynamics, in fact, demands the existence of additional quanta which are vector particles. These can be identified with the gluons introduced here. Later we will discuss more recent and more direct evidence for gluons.

A further sum rule is obtained from the fact that the sum of the charge of the individual partons must give the total charge, Q_N, of the nucleon in question ($Q_N = 1$ for the proton and $Q_N = 0$ for the neutron)

$$\int_0^1 dx \sum_j Q_j N_j(x) = Q_N. \tag{18.65}$$

Finally, there is a sum rule stemming from the conservation of baryon number, which for quarks is $\frac{1}{3}$ and for antiquarks $-\frac{1}{3}$. The baryon number of the nucleon is 1 and thus we get

$$\int_0^1 dx \sum_q \tfrac{1}{3}[N_q(x) - N_{\bar{q}}(x)] = 1. \tag{18.66}$$

Equation (18.65) is Adler's sum rule written in the language of the parton model. Originaly this sum rule was derived by Adler before the development of the parton model using the techniques of current algebra. Historically, this provided important impetus for the development of the parton idea (Bjorken 1966). Equation (18.66) was first derived by Gross and Llewellyn Smith (Gross 1969). The experimental verification of these sum rules is not easy, since high quality data at small values of x are necessary. The result of detailed investigations is that all existing data are compatible with the sum rules (Eisele 1982; Bergsma et al. 1983; MacFarlane 1984).

18.5 The Drell–Yan Process

In this section we will discuss the production of $\mu^-\mu^+$ pairs in hadron–hadron collisions. We consider, for example, the proton–antiproton reaction

$$p(p_1) + \bar{p}(p_2) \to \mu^-(k_1) + \mu^+(k_2) + X, \tag{18.67}$$

where X stands for the other particles in the final state. Following Drell and Yan (Drell 1970), we imagine this reaction to proceed by the annihilation of a quark–antiquark pair into a virtual photon (Fig. 18.15):

$$q(p_1') + \bar{q}(p_2') \to \mu^-(k_1) + \mu^+(k_2). \tag{18.68}$$

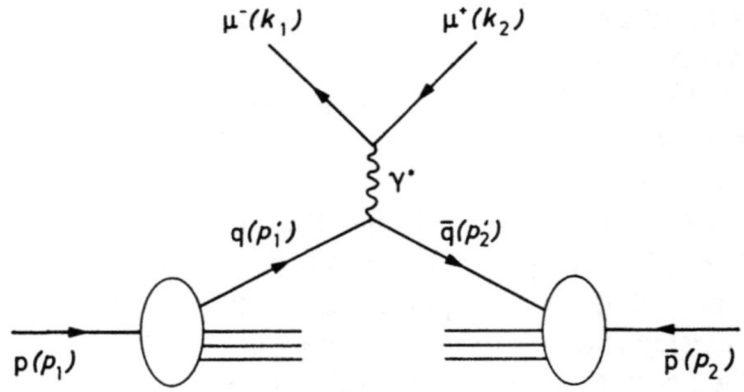

Figure 18.15 The diagram for the production of a $\mu^- \mu^+$ pair in a proton–antiproton collision by quark–antiquark annihilation.

In order to calculate the scattering cross-section for the reaction (18.67), we work in the center of mass frame of the $\mathrm{p\bar{p}}$ scattering and consider the limiting case of high energies:

$$s = (p_1 + p_2)^2 \to \infty. \tag{18.69}$$

Both hadrons in Fig. 18.15 then have high energy, and we can apply the rules of the parton model. In the following we will only be interested in the distribution for the invariant mass m of the muon pair.

The transition rate for the elementary parton process (18.68) is readily found to be

$$\frac{d\Gamma}{dm^2}(\mathrm{q}(p_1')+\bar{\mathrm{q}}(p_2') \to \mu^-(k_1)+\mu^+(k_2)) = \frac{1}{Vp_1'^0 p_2'^0}\frac{2\pi\alpha^2}{9}Q_\mathrm{q}^2\delta((p_1'+p_2')^2-m^2), \tag{18.70}$$

where

$$m^2 = (k_1 + k_2)^2.$$

Here Q_q is the charge of the quark q. It should be noted that the transition rate (18.70) is valid for quarks that are *unpolarized* with respect to color. The quark and antiquark in Fig. 18.15 can each come in three possible colors or, in other words, there are in each case three basis states in color space:

Quark color basis states: R, G, B,
Antiquark color basis states: $\bar{\mathrm{R}}$, $\bar{\mathrm{G}}$, $\bar{\mathrm{B}}$.

The three colors red, green, and blue (R, G, B), are, of course, only to be interpreted symbolically. When the quark and antiquark join together, nine possible color combinations result. Of these, only $\mathrm{R\bar{R}}$, $\mathrm{G\bar{G}}$, and $\mathrm{B\bar{B}}$ can lead to a color-neutral state like the virtual photon. On averaging over the "color spins" of the initial state, we obtain the factor $\frac{3}{9} = \frac{1}{3}$ which is taken into account in (18.70).

In the next step, we sum incoherently over the contributions of all parton reactions. The transition rate for the proton–antiproton reaction then becomes

$$\frac{d\Gamma}{dm^2}(\mathrm{p}(p_1) + \bar{\mathrm{p}}(p_2) \to \mu^- + \mu^+ + \mathrm{X})$$

$$= \sum_{\mathrm{q}} \left\{ \int_0^1 dx_1\, N_{\mathrm{q}}^{\mathrm{p}}(x_1) \int_0^1 dx_2\, N_{\bar{\mathrm{q}}}^{\bar{\mathrm{p}}}(x_2) \right.$$

$$\left. \cdot \frac{d\Gamma}{dm^2}(\mathrm{q}(x_1 p_1) + \bar{\mathrm{q}}(x_2 p_2) \to \mu^- + \mu^+) + (q \leftrightarrow \bar{\mathrm{q}}) \right\}. \quad (18.71)$$

Here $N_{\mathrm{q}}^{\mathrm{p}}$ and $N_{\bar{\mathrm{q}}}^{\bar{\mathrm{p}}}$ are the parton distribution functions for proton and antiproton. The term $(\mathrm{q} \leftrightarrow \bar{\mathrm{q}})$ takes into account the fact that of course also a quark from $\bar{\mathrm{p}}$ and an antiquark from p can annihilate giving a $\mu^-\mu^+$ pair. The cross-section is now obtained using (5.54) for $s \gg M^2$ as

$$\frac{d\sigma}{dm^2}(\mathrm{p} + \bar{\mathrm{p}} \to \mu^- + \mu^+ + \mathrm{X})$$

$$= \frac{4\pi\alpha^2}{9} \frac{1}{m^4} \cdot \sum_q \left\{ Q_{\mathrm{q}}^2 \int_0^1 dx_1 \int_0^1 dx_2\, x_1 N_{\mathrm{q}}^{\mathrm{p}}(x_1) x_2 N_{\bar{\mathrm{q}}}^{\bar{\mathrm{p}}}(x_2) \right.$$

$$\left. \cdot \delta\left(x_1 x_2 - \frac{m^2}{s}\right) + (q \leftrightarrow \bar{\mathrm{q}}) \right\}. \quad (18.72)$$

The generalization of this formula to Drell–Yan reactions for other hadrons is obvious.

According to this result, the dimensionless cross-section $m^4\, d\sigma/dm^2$ should be independent of s when the ratio m^2/s is held constant, i.e., it should display a scaling behavior. This is confirmed quite well by the experimental data (Cox 1982; Yan 1981). In order to perform a quantitative test of (18.72) one inserts the parton distributions as determined from the deep inelastic scattering experiments. As a result of charge-conjugation invariance, it naturally holds that

$$N_{\bar{\mathrm{q}}}^{\bar{\mathrm{p}}}(x) = N_{\mathrm{q}}^{\mathrm{p}}(x), \qquad N_{\mathrm{q}}^{\bar{\mathrm{p}}}(x) = N_{\bar{\mathrm{q}}}^{\mathrm{p}}(x). \quad (18.73)$$

One then finds that the shape of the distribution for the scaled invariant mass of the muon pair is well reproduced (Fig. 18.16) however not the absolute normalization. One thus defines a so-called "K-factor"

$$K = \frac{\left.\dfrac{d\sigma}{dm^2}\right|_{\mathrm{exp.}}}{\left.\dfrac{d\sigma}{dm^2}\right|_{\mathrm{theor.}}}. \quad (18.74)$$

For the antiproton–nucleon collisions a K-factor of about 2.3 was needed as described in Fig. 18.16. Similar results have been found for muon pair production in all hadron collisions studied so far (Cox 1982). It would appear that this K-factor must be understood as an effect of quantum chromodynamics

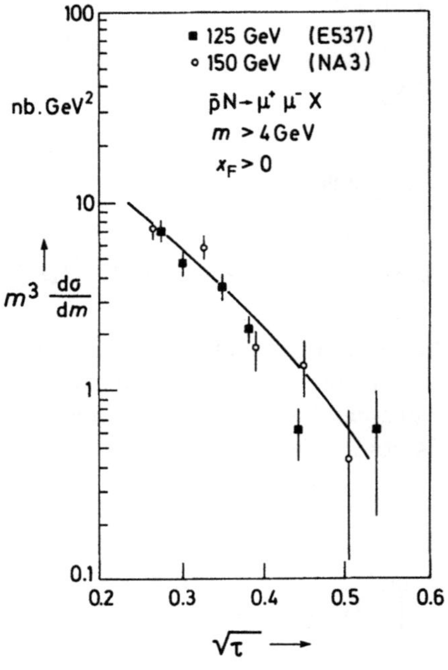

Figure 18.16 The dimensionless cross-section $m^3\, d\sigma/dm$ for the production of muon pairs of invariant mass m in anti-proton–nucleon scattering as a function of $\sqrt{\tau} = m/\sqrt{s}$. The curve is the prediction of the Drell–Yan model (18.72), multiplied by the factor $K = 2.3$ (after Cox 1982).

(Altarelli 1979; Kubar-André 1979; for an overview see Altarelli 1982, and for a less conventional interpretation of the K-factor see Nachtmann 1984).

The Drell–Yan process has attained particular importance because it is currently believed that the production of W- and Z-bosons occurs by means of a similar process (Fig. 18.17):

$$\begin{aligned} \mathrm{p} + \bar{\mathrm{p}} &\rightarrow \mathrm{W}^{\pm} + \mathrm{X}, \\ \mathrm{p} + \bar{\mathrm{p}} &\rightarrow \mathrm{Z} + X. \end{aligned} \tag{18.75}$$

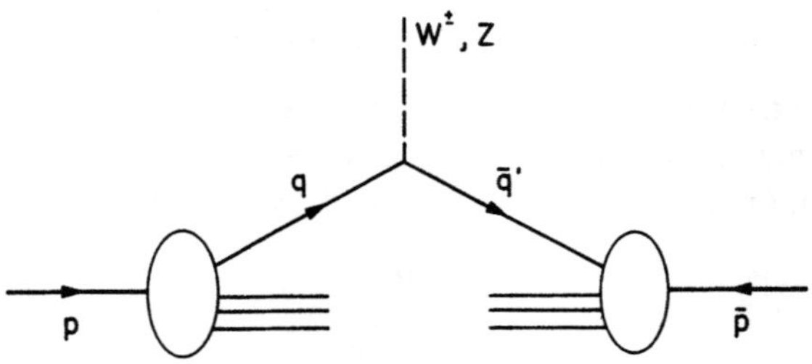

Figure 18.17 The production of W- and Z-bosons in $\mathrm{p\bar{p}}$ collisions. The main contributions are obtained from the annihilation of the following quark–antiquark pairs:

$$\mathrm{W}^+: \quad \mathrm{q} = \mathrm{u}, \bar{\mathrm{q}}' = \bar{\mathrm{d}};$$

$$\mathrm{W}^-: \quad \mathrm{q} = \mathrm{d}, \bar{\mathrm{q}}' = \bar{\mathrm{u}};$$

$$\mathrm{Z}: \quad \mathrm{q} = \mathrm{u}; \bar{\mathrm{q}}' = \bar{\mathrm{u}} \text{ and } \mathrm{q} = \mathrm{d}, \bar{\mathrm{q}}' = \bar{\mathrm{d}}.$$

These reactions, which led to the discovery of the intermediary bosons, will be discussed in detail in Part IV.

Exercises

18.1 Consider electron–nucleon scattering in the parton model (Fig. 18.18). Show that in the Breit frame (18.28), the electron scatters on partons whose fraction of the nucleon momentum is given by the Bjorken variable x, and that the longitudinal component of the parton momentum is reversed in the process. Use the conservation of energy and momentum for an elastic electron–parton collision.

18.2 Derive the formulas for the cross-section of ν–N and $\bar{\nu}$–N scattering (18.46) from the Lagrange density (18.45), for which the Hamilton operator for the interaction, $\mathbf{H}'$ is given by

$$\mathbf{H}'(t) = -\int \mathrm{d}^3x \, \mathscr{L}_{\mathrm{eff}}(\boldsymbol{x}, t). \tag{18.76}$$

18.3 Use the parton model in order to derive the relations (18.51), (18.52) from the expressions for the currents (18.48).

18.4 Assume a parton picture of the nucleon with bare nucleons p′ and n′ as partons. Write the weak current as

$$\mathscr{J}_\lambda^+ = \bar{\mathrm{p}}'\gamma_\lambda(1 - \gamma_5)\mathrm{n}'. \tag{18.77}$$

Let the charges of the isodoublets p′ and n′ be +1 and 0, respectively. What is the consequence of this for the structure functions $F_2^{(\nu\mathrm{N})}(x)$ and $F_2^{(\mathrm{eN})}(x)$ averaged over proton and neutron, and for their ratio?

18.5 The charge-symmetry operation is defined as a special isospin transformation. It consists of a rotation through an angle π about the second axis in isospin space. The corresponding operator in the space of state vectors is

$$\mathbf{P}_I = \exp(\mathrm{i}\pi \mathbf{I}_2). \tag{18.78}$$

How are proton, neutron, u-, and d-quarks transformed by $\mathbf{P}_I$? Show that charge-symmetry alone is sufficient to yield the relations (18.56).

18.6 Let $\langle q \rangle$ and $\langle \bar{q} \rangle$ (18.62) represent the momentum fractions of the nucleon carried by the quarks u and d or the antiquarks $\bar{\mathrm{u}}$ and $\bar{\mathrm{d}}$. Determine the numerical values of these quantities from the data of Fig. 18.14. According to Fig. 18.13, the s-quark and $\bar{\mathrm{s}}$-antiquark content of the nucleon is very small. Neglect s and $\bar{\mathrm{s}}$ in the nucleon altogether to give a further estimate for the integral I (18.64).

18.7 With the help of the results of Exercise 18.5, show that the structure functions $F_2^{(\mathrm{ep})}$ and $F_2^{(\mathrm{en})}$ of protons and neutrons in the framework of the quark–parton model can be represented by the distribution functions of the *proton* as follows:

$$\begin{aligned} F_2^{(\mathrm{ep})}(x) &= x\{\tfrac{4}{9}N_\mathrm{u}^\mathrm{p}(x) + \tfrac{1}{9}N_\mathrm{d}^\mathrm{p}(x) + \tfrac{1}{9}N_\mathrm{s}^\mathrm{p}(x) + \tfrac{4}{9}N_{\bar{\mathrm{u}}}^\mathrm{p}(x) \\ &\quad + \tfrac{1}{9}N_{\bar{\mathrm{d}}}^\mathrm{p}(x) + \tfrac{1}{9}N_{\bar{\mathrm{s}}}^\mathrm{p}(x)\}, \\ F_2^{(\mathrm{en})}(x) &= x\{\tfrac{1}{9}N_\mathrm{u}^\mathrm{p}(x) + \tfrac{4}{9}N_\mathrm{d}^\mathrm{p}(x) + \tfrac{1}{9}N_\mathrm{s}^\mathrm{p}(x) + \tfrac{1}{9}N_{\bar{\mathrm{u}}}^\mathrm{p}(x) \\ &\quad + \tfrac{4}{9}N_{\bar{\mathrm{d}}}^\mathrm{p}(x) + \tfrac{1}{9}N_{\bar{\mathrm{s}}}^\mathrm{p}(x)\}. \end{aligned} \tag{18.79}$$

Show that this result of the quark–parton model implies that the ratio $F_2^{(en)}(x)/F_2^{(ep)}(x)$ is bounded from below by the value $\frac{1}{4}$ (Nachtmann 1972; Majumdar 1971). From Fig. 18.9(b) it would appear that the measured ratio $W_2^{(en)}/W_2^{(ep)} = F_2^{(en)}/F_2^{(ep)}$ for $x \to 1$ does indeed approach $\frac{1}{4}$. What can one conclude from this about the behavior of the parton distributions of the proton for $x \to 1$?

18.8 Consider (18.65) for the proton and for the neutron and form the difference. Express the result in terms of the structure functions $W_2^{(\nu p)}$ and $W_2^{(\bar{\nu} p)}$ (see Adler 1966).

18.9 Consider (18.66) for the proton and for the neutron and form the sum. Express the result in terms of the structure functions $F_3^{(\nu)}$ and $F_3^{(\bar{\nu})}$ (see Gross 1969).

CHAPTER 19

The Basic Principles of Quantum Chromodynamics

19.1 The Lagrange Density of Quantum Chromodynamics (QCD)

We have learnt in the previous chapters that hadrons can be understood as bound states of constituents called partons. We found that the partons that take part in the electromagnetic and weak interactions have spin $\frac{1}{2}$ and carry the internal quantum numbers of the quarks; in particular, they carry fractional charges. We have further discovered that the quarks possess an additional degree of freedom, the color, and that experiments demand the existence of flavor-neutral partons, gluons, in the nucleon. The question now arises as to what forces are responsible for holding the partons together in the hadrons.

The answer to this question is known once we know the Lagrange density that describes the interaction between quarks and gluons. This Lagrange density should certainly contain a kinetic term for the quarks:

$$\mathcal{L}_{\mathrm{q}}^{(0)}(x) = \sum_{j=1}^{f} \bar{\mathbf{q}}^j(x)(\mathrm{i}\gamma^\lambda \partial_\lambda - m_j)\mathbf{q}^j(x). \tag{19.1}$$

Here we have numbered the different quark flavors with $j = 1, \ldots, f$ and written their masses as m_j. The quark fields $\mathbf{q}^j$ have three color components

$$\mathbf{q}^j = \begin{pmatrix} \mathbf{q}_1^j \\ \mathbf{q}_2^j \\ \mathbf{q}_3^j \end{pmatrix}, \qquad \mathbf{q}^1 \equiv \mathbf{u}, \quad \mathbf{q}^2 \equiv \mathbf{d}, \quad \mathbf{q}^3 \equiv \mathbf{s}, \quad \mathbf{q}^4 \equiv \mathbf{c}, \ldots . \tag{19.2}$$

The Lagrange density $\mathscr{L}_q^{(0)}$ (19.1) cannot be really satisfactory since $\mathscr{L}_q^{(0)}$ only describes the motion of free quarks. But we now show that there is another and more subtle physical reason why $\mathscr{L}_q^{(0)}$ is unacceptable as Lagrange density.

We learned previously that hadrons are color singlet states. A π^+-meson, for example, can be imagined as a bound state of a u- and a $\bar{\mathrm{d}}$-quark:

$$\pi^+ \sim \mathrm{u}_1\bar{\mathrm{d}}_1 + \mathrm{u}_2\bar{\mathrm{d}}_2 + \mathrm{u}_3\bar{\mathrm{d}}_3. \tag{19.3}$$

The physical π^+-meson is invariant under SU(3) rotations in color space. The same has been found for baryons [see (15.40)–(15.42) and Chap. 17]. When the bound states display an invariance, this should be a consequence of an invariance of the underlying Lagrange density. In fact, we find that the Lagrange density $\mathscr{L}_q^{(0)}$ is invariant under the following SU(3) color transformation

$$\mathbf{q}^j(x) \to \mathbf{U}\cdot\mathbf{q}^j(x) \qquad (j = 1, \dots, f), \tag{19.4}$$

where

$$\mathbf{U}\mathbf{U}^\dagger = 1,$$
$$\det \mathbf{U} = 1,$$
$$\mathbf{U} = \text{const., i.e., independent of } x.$$

An essential point here, is that $\mathbf{U}$ is a *constant* matrix. This means that with $\mathscr{L}_q^{(0)}$ as Lagrange density, we can choose the unobserved directions of the axes in color space in any manner we like. We must however choose them to be the same in Europe as in America or, for that matter, anywhere in the universe. This is a very unsatisfactory situation since it would give the relative orientation in color space at different space–time points an absolute significance although there are no corresponding observable physical effects.

A similar situation occurred in the theory of special relativity in which the relative state of motion of observers at different space–time points has an absolute significance. Let us take an inertial system, defined by a set of observers which are at rest relative to each other and another system where the observers are accelerated relative to each other. In these two systems, the relativistic equations of motion are different (Fig. 19.1). Einstein found this state of affairs to be highly unsatisfactory and made the postulate that for the

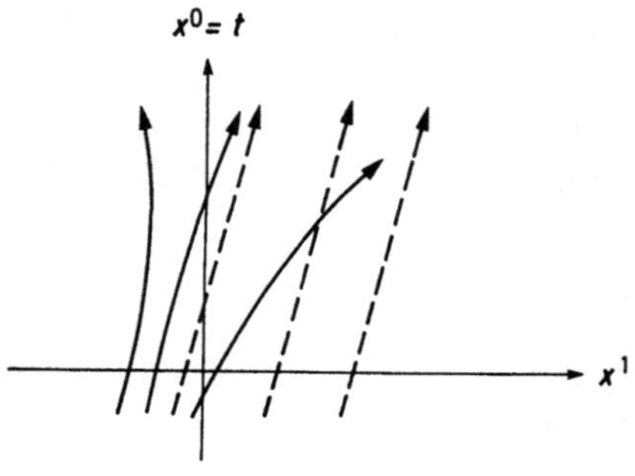

Figure 19.1 Observer systems in Minkowski space. Dashed lines: inertial system; Solid lines: observers accelerating relative to each other.

formulation of the basic equations of the theory, *all* observer systems should be equivalent. This principle led Einstein to the general theory of relativity (Einstein 1916). This theoy could only be formulated with the inclusion of gravity. We might regard the principle of general relativity as a "reason" for the existence of the gravitational field.

In the case of electrodynamics, H. Weyl (1929) discovered a principle that one can regard as a "reason" for the existence of the photon. The starting point is the Lagrange density for a free Dirac electron:

$$\mathscr{L}_e^{(0)}(x) = \bar{\psi}(x)(\mathrm{i}\gamma^\lambda \partial_\lambda - m_e)\psi(x). \tag{19.5}$$

Here ψ is the Dirac field, which itself is not observable—indeed, it changes sign upon rotation through 2π. Typical observable quantities are bilinear expressions such as the current density

$$\mathscr{J}^\lambda(x) = \bar{\psi}(x)\gamma^\lambda\psi(x).$$

These observables are unchanged when $\psi(x)$ is multiplied by a phase factor. The phase factor can have an arbitrary x-dependence, as we see by making the substitution

$$\psi(x) \to \mathrm{e}^{\mathrm{i}\alpha(x)}\psi(x) \tag{19.6}$$

which corresponds to

$$\bar{\psi}(x)\gamma^\lambda\psi(x) \to \bar{\psi}(x)\gamma^\lambda\psi(x).$$

It would seem reasonable to demand that the theory is invariant with respect to the unobservable phase transformation (19.6). We can see immediately that the Lagrange density (19.5) for the free Dirac field is not invariant. However, as discovered by H. Weyl, the coupled system of electrons and photons, which is described by the Maxwell–Dirac Lagrangian, is invariant. For this, the electromagnetic vector potential $A_\mu(x)$ must be appropriately transformed.

We have already met the Maxwell–Dirac Lagrange density in Chap. 9 (eq. (9.5)),

$$\mathscr{L}_{\mathrm{M.D.}}(x) = -\tfrac{1}{4}F_{\mu\nu}(x)F^{\mu\nu}(x) + \bar{\psi}(x)(\mathrm{i}\gamma^\mu D_\mu - m_e)\psi(x). \tag{19.7}$$

Here $F_{\mu\nu}$ is the field-strength tensor, D_μ is the covariant derivative:

$$\begin{aligned} F_{\mu\nu}(x) &= \partial_\mu A_\nu(x) - \partial_\nu A_\mu(x), \\ D_\mu\psi(x) &= (\partial_\mu - \mathrm{i}eA_\mu(x))\psi(x), \end{aligned} \tag{19.8}$$

and the electron charge is denoted by $(-e)$. It is easy to check that $\mathscr{L}_{\mathrm{M.D.}}$ is invariant under the following transformations, which H. Weyl named gauge transformations:

$$\begin{aligned} \psi(x) &\to \mathrm{e}^{\mathrm{i}\alpha(x)}\psi(x), \\ A_\mu(x) &\to A_\mu(x) + \frac{1}{e}\partial_\mu\alpha(x). \end{aligned} \tag{19.9}$$

Here $\alpha(x)$ is an arbitrary real function of x. In contemporary terminology, these transformations are said to form a U(1) gauge group.

H. Weyl interpreted his result to mean that the physical postulate of the invariance of the Dirac field under phase transformations (19.6) can be regarded as a "reason" for the existence of the photon.

We now return to quantum chromodynamics. We shall see that, the Lagrange density can be constructed along similar lines as in the Maxwell–Dirac case. To begin with, we impose the following physical postulate: It should be possible to choose the directions of the axes in color space completely freely at all space–time points since they are unobservable. Indeed, all hadrons are color singlets. In mathematical language, this means that we demand invariance under so-called local gauge transformations in color space:

$$\mathbf{q}^j(x) \to \mathbf{U}(x)\mathbf{q}^j(x) \qquad (j = 1, \dots f), \tag{19.10}$$

where $\mathbf{U}(x) \in SU(3)$ i.e., for all x it must hold that

$$\begin{aligned} \mathbf{U}(x)\mathbf{U}^\dagger(x) &= 1, \\ \det \mathbf{U}(x) &= 1. \end{aligned} \tag{19.11}$$

Apart from these conditions, $\mathbf{U}(x)$ can have an arbitrary x-dependence. It is easy to see that the Lagrange density of the free quark fields $\mathscr{L}_\mathrm{q}^{(0)}$ (19.1) is not invariant under the substitution (19.10):

$$\begin{aligned} \mathscr{L}_\mathrm{q}^{(0)}(x) &= \sum_{j=1}^{f} \bar{\mathbf{q}}^j(x)(\mathrm{i}\gamma^\lambda \partial_\lambda - m_j)\mathbf{q}^j(x) \\ &\to \sum_{j=1}^{f} \bar{\mathbf{q}}^j(x)(\mathrm{i}\gamma^\lambda \partial_\lambda - m_j + \mathrm{i}\gamma^\lambda \mathbf{U}^\dagger(x)\partial_\lambda \mathbf{U}(x))\mathbf{q}^j(x). \end{aligned} \tag{19.12}$$

The way to save the invariance in this case was discovered in 1954 in a somewhat different context (Yang 1954; Shaw 1954). The non-Abelian gauge theories, which were introduced at this time, now form the basis of the modern theory of elementary particles. As a historical note, we should mention that O. Klein in fact formulated a non-Abelian gauge theory as early as 1939 (for a historical review, see Veltman 1974).

In electrodynamics, the gauge principle is saved by the introduction of the photons. It turns out that the gauge principle of chromodynamics also requires the introduction of vector fields, which we shall call gluons. We need eight gluons corresponding to the eight linearly independent generators of the SU(3) color group. The eight real gluon vector potentials $G_\lambda^a(x)$ (with $a = 1, \dots, 8$) are combined into a 3×3 Hermitian traceless matrix

$$\begin{aligned} \mathbf{G}_\lambda(x) &= G_\lambda^a(x)\frac{\lambda_a}{2} = \mathbf{G}_\lambda^\dagger(x), \\ \mathrm{Tr}\, \mathbf{G}_\lambda(x) &= 0. \end{aligned} \tag{19.13}$$

The λ^a (with $a = 1, \dots, 8$) are the Gell-Mann λ-matrices (see Appendix C) now operating in color space.

For the coupling of the gluons to the quarks we attempt a "minimal substitution" as in electrodynamics, i.e., we replace all ordinary derivatives ∂_λ

as follows:

$$\partial_\lambda \to \mathbf{D}_\lambda = \partial_\lambda + \mathrm{i}g_s \mathbf{G}_\lambda(x). \tag{19.14}$$

The quantity $\mathbf{D}_\lambda$ is called the covariant derivative and g_s is a dimensionless coupling constant analogous to the electric charge e. With this substitution, we obtain from $\mathscr{L}_q^{(0)}$ the following expresion:

$$\mathscr{L}_q(x) = \sum_{j=1}^{f} \bar{\mathbf{q}}^j(x)(\mathrm{i}\gamma^\lambda \mathbf{D}_\lambda - m_j)\mathbf{q}^j(x). \tag{19.15}$$

It is easy to check that $\mathscr{L}_q$ is invariant under local gauge transformations (19.10), provided the gluon field is transformed as follows:

$$\mathbf{G}_\lambda(x) \to \mathbf{G}'_\lambda(x) = \mathbf{U}(x)\mathbf{G}_\lambda(x)\mathbf{U}^\dagger(x) - \frac{\mathrm{i}}{g_s}\mathbf{U}(x)\partial_\lambda \mathbf{U}^\dagger(x). \tag{19.16}$$

The matrix $\mathbf{G}'_\lambda(x)$ is also Hermitian and traceless for any $\mathbf{U}(x)$ that satisfies the conditions (19.11). This can easily be verified. From

$$\mathbf{U}(x)\mathbf{U}^\dagger(x) = 1$$

it follows that

$$\begin{aligned} &\mathbf{U}(x)(\partial_\lambda \mathbf{U}^\dagger(x)) + (\partial_\lambda \mathbf{U}(x))\mathbf{U}^\dagger(x) = 0, \\ &\mathbf{G}'^\dagger_\lambda(x) = \mathbf{G}'_\lambda(x). \end{aligned} \tag{19.17}$$

From

$$\det \mathbf{U}(x) = 1$$

we get in addition

$$\begin{aligned} \mathrm{Tr}(\mathbf{U}(x)\partial_\lambda \mathbf{U}^\dagger(x)) &= 0, \\ \mathrm{Tr}\, \mathbf{G}'_\lambda(x) &= 0. \end{aligned} \tag{19.18}$$

We have not, of course, gained anything so long as the vector potential $\mathbf{G}_\lambda$ of the gluon is treated as an external field. The gluon field itself must become a dynamical variable. In order to construct the part of the Lagrange density that contains the gluon dynamics, we will again allow electrodynamics to serve as a model. We introduce a gluon field-strength tensor $\mathbf{G}_{\lambda\rho}(x)$:

$$\mathbf{G}_{\lambda\rho}(x) = \partial_\lambda \mathbf{G}_\rho(x) - \partial_\rho \mathbf{G}_\lambda(x) + \mathrm{i}g_s[\mathbf{G}_\lambda(x), \mathbf{G}_\rho(x)], \tag{19.19}$$

which, for fixed x is cleary a Hermitian matrix with trace zero. Thus we define its components $G^a_{\lambda\rho}(x)$ (with $a = 1, \ldots, 8$) by

$$\mathbf{G}_{\lambda\rho}(x) = G^a_{\lambda\rho}(x)\frac{\lambda_a}{2},$$

and obtain from (19.19)

$$G^a_{\lambda\rho}(x) = \partial_\lambda G^a_\rho(x) - \partial_\rho G^a_\lambda(x) - g_s f_{abc} G^b_\lambda(x) G^c_\rho(x). \tag{19.20}$$

The f_{abc} are the structure constants of the SU(3) group. The term quadratic in the gluon field has no analogue in electrodynamics and is typical of

Table 19.1 Comparison between QED and QCD.

	QED	QCD
Quantum number	Electrical charge	Color
Fermions	Electrons	Quarks (color triplets)
Vector bosons	Photon (uncharged)	Gluons (color octet)
Gauge group	U(1) (Abelian)	SU(3) (non-Abelian)
Coupling constant	e	g_s
	$\alpha = \frac{e^2}{4\pi}$	$\alpha_s = \frac{g_s^2}{4\pi}$
Bound states	Positronium	Hadrons
Free particles (i.e., asymptotic states)	Electrons, positrons and photons	Only hadrons. Quarks and gluons are permanently bound

the non-Abelian character of the color group SU(3). This term is necessary in order to achieve a simple transformation behavior for the field-strength tensors under the gauge transformations. In fact, for a transformation of the gluon field according to (19.16), we find

$$\mathbf{G}_{\lambda\rho}(x) \to \mathbf{U}(x)\mathbf{G}_{\lambda\rho}(x)\mathbf{U}^{\dagger}(x). \tag{19.21}$$

It is now easy to write down a gauge-invariant Lagrange density for the quarks and gluons:

$$\begin{aligned} \mathscr{L}_{\mathrm{QCD}}(x) &= -\tfrac{1}{2}\,\mathrm{Tr}(\mathbf{G}_{\lambda\rho}(x)\mathbf{G}^{\lambda\rho}(x)) + \sum_{j=1}^{f} \bar{\mathbf{q}}^j(x)(\mathrm{i}\gamma^\lambda \mathbf{D}_\lambda - m_j)\mathbf{q}^j(x) \\ &= -\tfrac{1}{4}G^a_{\lambda\rho}(x)G^{\lambda\rho a}(x) \\ &\quad + \sum_{j=1}^{f} \bar{\mathbf{q}}^j(x)\left[\mathrm{i}\gamma^\lambda\left(\partial_\lambda + \mathrm{i}g_s G^a_\lambda(x)\frac{\lambda_a}{2}\right) - m_j\right]\mathbf{q}^j(x). \end{aligned} \tag{19.22}$$

This is the Lagrange density underlying quantum chromodynamics. Its structure is very similar to that of the Lagrange density (19.7) that provides the foundation of quantum electrodynamics.

In Table 19.1 a number of analogies between quantum electrodynamics and quantum chromodynamics are listed. Of course, there are also differences. The most obvious one is that electrons and photons are known to be free particles, whereas the quarks and gluons appear to be permanently confined inside the hadrons.

In order to describe the true hadron world, we must regard the Lagrange density (19.22) as the starting point of a quantum field theory of quarks and gluons. In an ideal case we would then solve this theory and look to see if it actually contains hadrons as color singlet bound states, and whether it reproduces all the properties that have been observed in the real world. It goes without saying that today we remain far removed from this ideal goal. Instead it has been necessary to develop approximation schemes. First of all, we can try to contruct a renormalized perturbation expansion for QCD in the knowledge that the analogous technique in QED is so successful. To carry out a consistent quantization and perturbation expansion for a non-Abelian gauge theory such as QCD is no easy matter. Nonetheless, a quite general solution has been found. Some particularly important contributions to this subject were made by t'Hooft (t'Hooft 1971; see also the review articles of t'Hooft 1973; Abers 1973; Politzer 1974; Marciano 1978). A perturbation expansion can, of course, only make sense when the expansion parameter is sufficiently small. In QCD we shall see that for certain reactions with large momentum transfer, this is actually the case. We will not present a detailed derivation of the perturbation series, but will simply give the Feynman rules of QCD in Appendix D.

The perturbation theory of QCD has, however, one severe shortcoming: To every finite order of the expansion, the asymptotic, i.e., free, particles of the theory are the quarks and the gluons. In the real world, however, the free particles are the hadrons whereas the quarks and the gluons are absent in the physical spectrum. Other approximation schemes have been suggested in order to treat this situation theoretically, for example, lattice gauge theories (Wilson 1974a, 1975).

In the numerical treatment of problems in lattice gauge theory, a major breakthrough has come in the last few years through the introduction of Monte-Carlo methods. Since this book is dedicated to phenomenological aspects of the subject, we will not go into further details about these methods but refer instead to Rebbi (1983). Here we shall place our emphasis on discussing the predictions of QCD for high energy reactions. We shall work with an extended version of the naïve parton model, which includes the effects of QCD and which sheds light on the physical content of the theory (Kogut 1974).

19.2 Violation of the Bjorken Scale Invariance in Deep Inelastic Scattering

In Chap. 18 we discussed the Bjorken scale invariance and the naive parton model. In this section we will present theoretical arguments which show that this scale invariance should be violated in a relativistic quantum field theory and, in particular, in QCD.

One of the most successful concepts in physics is the idea that large objects are built up from smaller ones. The large objects that are of interest here are the hadrons which we are trying to understand as bound states of smaller objects, the quarks and the gluons. In the naïve parton model we considered a rapidly moving hadron and described it in terms of a wavefunction for point-like constituents. In deep inelastic scattering, we only needed to know the distribution functions of the constituents to evaluate the reaction. For example, for the structure function νW_2 in deep inelastic electron–nucleon scattering we found (18.39)

$$\nu W_2(\nu, Q^2) = x \sum_i Q_i^2 N_i(x). \tag{19.23}$$

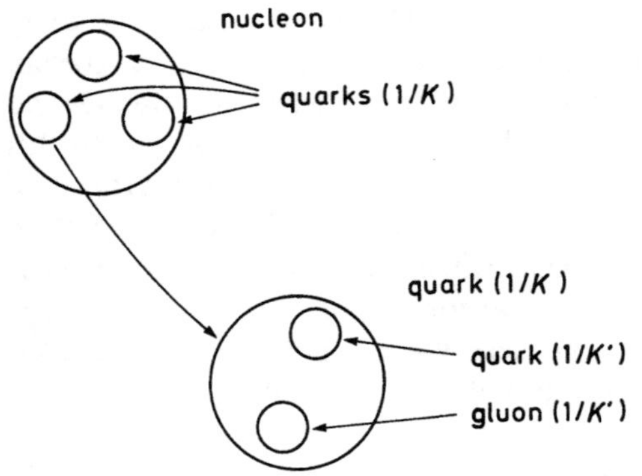

Figure 19.2 A nucleon composed of a hierarchy of constituents ($1/K > 1/K' \ldots$).

The distribution function $N_i(x)$ counts how many constituents of type i with longitudinal momentum fraction x there are inside the nucleon. It can be shown that the parton description of the nucleon follows essentially from quantum field theory if the assumption that the transverse momenta of the constituents are bounded is made (Feynman 1969).

In an honest relativistic quantum field theory, however, there are no cutoff parameters that can limit the transverse momenta, and the naïve parton description cannot be correct (Parisi 1973; Callan 1973). Kogut and Susskind (1974) have shown how the effect of a renormalizable quantum field theory can be understood within the framework of the parton picture. These authors describe a nucleon, not as a bound state of point-like constituents, but instead as a hierachy of smaller and smaller effective constituents. The constituents at the various levels of the description are each characterized by a length $1/K$ which can be roughly interpreted as their "size". If we observe a nucleon with a "microscope" whose resolving power is $1/K$, then we shall see the corresponding effective constituents. If we increase the resolving power we shall see the constituents of these consituents, etc. (Fig. 19.2).

In deep inelastic scattering we observe a nucleon with a "microscope" whose resolving power is proportional to $1/Q$, where $1/Q$ is the wavelength of the virtual photon in the Breit frame (18.28). If we describe the reaction in terms of effective constituents of size $1/K \sim 1/Q$, then we can apply the impulse approximation and take over all the formulas of the parton model. We simply have to remember that the distribution functions now refer to effective constituents and that they are thus dependent on the size of these constituents and therefore on Q. For νW_2, for example, we obtain

$$\nu W_2(\nu, Q^2) = x \sum_i Q_i^2 N_i(x, Q^2). \tag{19.24}$$

The distribution functions $N_i(x, Q^2)$ for effective constituents of various sizes will, in general, be different from one another. Thus we expect a Q^2-dependence in the structure functions for fixed x, i.e., a violation of the Bjorken scale invariance.

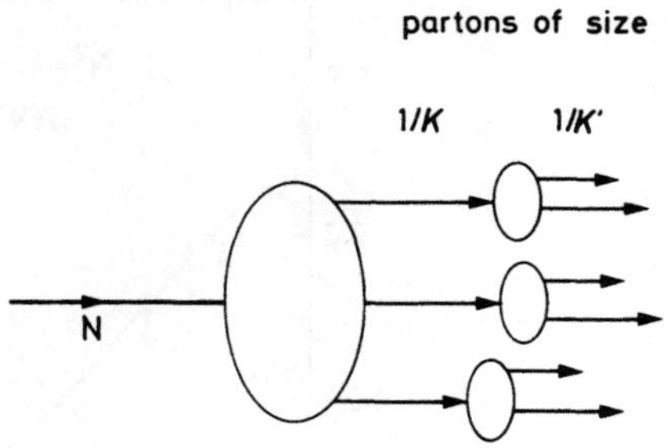

Figure 19.3 Distribution of the longitudinal momentum of a nucleon among the constituents of size $(1/K)$, and its redistribution among yet smaller constituents $(1/K > 1/K')$.

We can even predict the general features of this violation of scale invariance that is demanded by quantum field theory. If we resolve a large constituent into smaller ones, then the smaller constituents must share the longitudinal momentum of the larger one among themselves. This means that the smaller constituents must, on average, have a smaller longitudinal momentum. Their number becomes larger, however, and thus the total longitudinal momentum remains the same (Fig. 19.3). The pattern of scaling violations in the structure functions that follows from these arguments consists of an increase at small x and a decrease at large x with increasing Q^2 (Fig. 19.4).

This general pattern of scale breaking has, in fact, been seen in experiments. In Fig. 19.5 we show, as an example, data for muon–nucleon scattering. The question now arises as to whether we can quantitatively describe these data within the framework of a quantum field theory. At present there is only a single quantum field theoretical model for the strong interaction from which one can derive reliable results; this is QCD. The magic words that makes such calculations possible are *asymptotic freedom* (t'Hooft 1972; Politzer 1973, 1974; Gross 1973, 1973a, 1974).

In the picture that we have developed here we may formulate asymptotic freedom as follows: The coupling of quarks and gluons of "size" $1/K \sim 1/Q$ is characterized by an effective coupling strength $g_s(Q^2)$. The dependence of this coupling strength on Q^2 can be calculated perturbatively with the help of Feynman rules (see Appendix D). A very interesting result was obtained by the authors mentioned above, namely, that in QCD the effective coupling

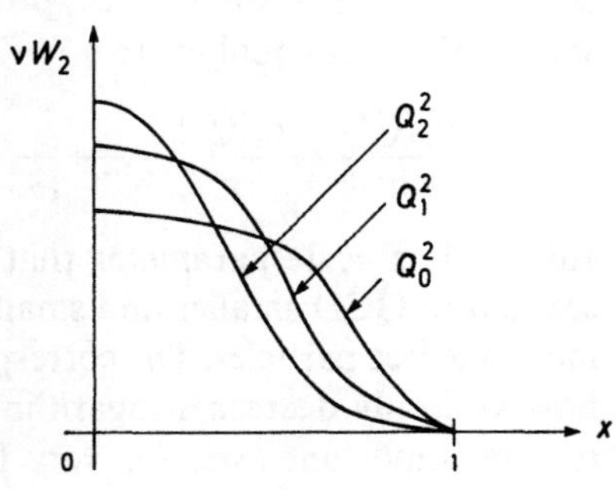

Figure 19.4 The general pattern expected for the violation of Bjorken scale invariance in quantum field theory. The structure function νW_2 is plotted for fixed Q^2 as a function of x (with $Q_0^2 < Q_1^2 < Q_2^2$).

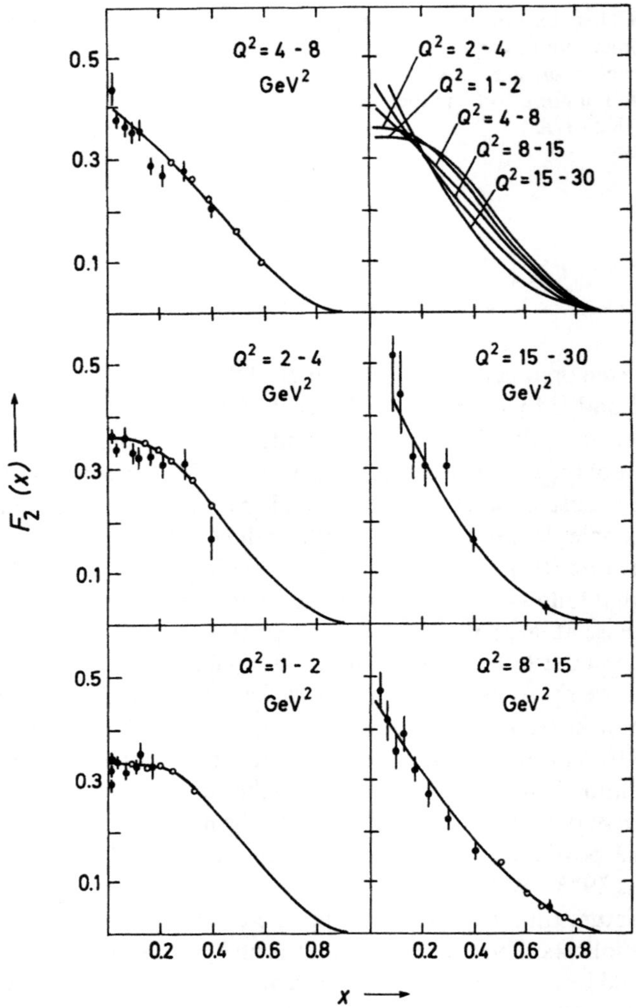

Figure 19.5 The structure function $F_2 = \nu W_2$ of the proton as a function of x for various Q^2 intervals (after Anderson 1977).

constant vanishes for $Q^2 \to \infty$, provided that the number of quark flavors is smaller than, or equal to, 16:

$$\frac{g_s^2(Q^2)}{4\pi^2} \equiv \frac{\alpha_s(Q^2)}{\pi} \xrightarrow[Q^2 \to \infty]{} \frac{12}{(33 - 2f)\ln(Q^2/\Lambda^2)} \qquad \text{for} \quad f \leq 16. \quad (19.25)$$

Here Λ is a scale parameter that must be determined experimentally. We see that in QCD smaller and smaller effective constituents behave more and more like free particles. The corresponding running coupling constant (19.25), however, only decreases logarithmically. Thus the effects of the interaction remain important even for very large Q^2. It is possible to show that the property of being asymptotically free is typical for non-Abelian gauge theories (Coleman 1973).

What predictions does asymptotic freedom yield for the distribution functions $N_i(x, Q^2)$ of the effective quarks and gluons? Until now, it has not been possible to derive the distribution functions themselves from first principles. However, since the effective coupling becomes small for $Q^2 \to \infty$ (19.25), we can apply perturbation theory in order to determine the *change* in the distribution functions for large Q^2. The evolution of the distribution functions is described by an integro-differential equation, which was first formulated by Altarelli and Parisi (Altarelli 1977). The general form of this equation can be derived from a few simple arguments:

(i) In first-order perturbation theory, the change in the distribution functions must be proportional to $g_s^2(Q^2)$.
(ii) Since the longitudinal momentum of a daughter parton must always be smaller than or equal to the longitudinal momentum of the parent parton (Fig. 19.3), the change in the distribution functions for longitudinal momentum fraction x can only be dependent on the distribution functions for momentum fractions y, where y is greater than or equal to x.
(iii) If we neglect the mass of the quarks, then the change in the distribution functions when measured on the dimensionless scale $\ln(Q^2)$ should, apart from the factor $g_s^2(Q^2)$, only depend on the ratios of longitudinal momenta.

According to these considerations, we can make the following ansatz for the evolution equation with initially undetermined splitting functions P_{ij}:

$$\frac{\partial N_i(x, Q^2)}{\partial \ln Q^2} = \frac{g_s^2(Q^2)}{8\pi^2} \sum_j \int_x^1 \frac{dy}{y} P_{ij}\left(\frac{x}{y}\right) N_j(y, Q^2). \qquad (19.26)$$

Here the indices i and j are labels for the different types of partons in the theory, i.e., quarks, antiquarks, and gluons:

$$i, j = \mathrm{u, d, s}, \ldots, \bar{\mathrm{u}}, \bar{\mathrm{d}}, \bar{\mathrm{s}}, \ldots, \mathrm{G}. \qquad (19.27)$$

Condition (ii) above has been taken into account in (19.26) by restricting the integration to $x \le y \le 1$, and condition (iii) is accounted for by writing the unknown functions P_{ij} as functions of the ratio of the longitudinal momenta x/y. The exact form of the functions P_{ij} cannot be derived from general principles. They are specific to the quantum field theoretical model in question. The calculation of these functions for the case of QCD will be discussed in the next section. Prior to this we want to introduce the moments of the distribution functions:

$$M_i(n, Q^2) = \int_0^1 dx\, x^{n-1} N_i(x, Q^2), \qquad (19.28)$$

where $n = 2, 3, 4, \ldots$. These moments are useful because they enable us to reduce the integro-differential equation (19.26) to a system of ordinary differential equations. After a simple calculation we find

$$\frac{\partial M_i(n, Q^2)}{\partial \ln Q^2} = -\frac{1}{\ln(Q^2/\Lambda^2)} \sum_j d_{ij}^n M_j(n, Q^2) \qquad (n = 2, 3, \ldots). \quad (19.29)$$

Here we have made use of (19.25) for $g_s^2(Q^2)$ and we define the matrix (d_{ij}^n), as the matrix of the *anomalous dimensions*, as follows:

$$d_{ij}^n = -\frac{6}{33 - 2f} \int_0^1 dz\, z^{n-1} P_{ij}(z). \quad (19.30)$$

Equation (19.29) also establishes a connection to the more abstract methods that use the operator product expansion (Wilson 1969, 1972) and the renormalization group analysis of the coefficient functions (Christ 1972). These abstract methods lead directly to predictions for the moments of the structure functions (cf. Peterman 1979; Altarelli 1982), rather than to predictions for the structure functions themselves.

19.3 The Calculation of Anomalous Dimensions in QCD

The calculation of the functions P_{ij} in the Altarelli–Parisi equation (19.26) and of the anomalous dimensions d_{ij}^n (19.30) can be carried out using a wide variety of methods. Here we will discuss a simple method which relies on the physical picture of Kogut and Susskind described in Sect. 19.2.

For the sake of simplicity, we consider a model with massless quarks of f different flavors. We then have SU(f) flavor symmetry. Together with charge-conjugation invariance, this reduces the number of independent functions P_{ij} (19.26) to four:

$$\begin{aligned}
&P_{\mathrm{qq}} \equiv P_{\mathrm{uu}} = P_{\mathrm{dd}} = \cdots = P_{\bar{\mathrm{u}}\bar{\mathrm{u}}} = P_{\bar{\mathrm{d}}\bar{\mathrm{d}}} = \cdots, \\
&P_{\mathrm{qG}} \equiv P_{\mathrm{uG}} = P_{\mathrm{dG}} = \cdots = P_{\bar{\mathrm{u}}\mathrm{G}} = P_{\bar{\mathrm{d}}\mathrm{G}} = \cdots, \\
&P_{\mathrm{Gq}} \equiv P_{\mathrm{Gu}} = P_{\mathrm{Gd}} = \cdots = P_{\mathrm{G}\bar{\mathrm{u}}} = P_{\mathrm{G}\bar{\mathrm{d}}} = \cdots, \\
&P_{\mathrm{GG}}.
\end{aligned} \quad (19.31)$$

In order to calculate these functions, we shall, for the most part, only apply concepts that are well known from ordinary quantum mechanics.

We consider a rapidly moving nucleon with momentum $\boldsymbol{p}$ ($|\boldsymbol{p}| \gg M$) described by the state vector $|N(\boldsymbol{p})\rangle$. It would be nice if we could determine the composition of this state from its property of being an eigenstate of the true Hamilton operator $\mathbf{H}$ of QCD. However, the true Hamiltonian of QCD is far too complicated. It couples all momentum components of the quark and gluon fields with one another. Since it is not possible to solve the whole problem in one go, we attempt it step by step. We construct a series of Hamilton operators $\mathbf{H}_K$ with a cutoff parameter K for the transverse momenta. These Hamilton operators are then divided into a potential term $\mathbf{V}$, supposed to describe the permanent binding of the quarks, and a "kinetic" term $\mathbf{H}_K^{\mathrm{kin}}$, supposed to describe the coupling between momenta whose transverse components extend to the cutoff parameter K, which we will always assume to be sufficiently large:

$$\mathsf{H}_K = \mathsf{H}_K^{\text{kin}} + \mathsf{V}. \tag{19.32}$$

The effective quarks and gluons of Sect. 19.2 are *defined* as eigenstates of the kinetic term $\mathsf{H}_K^{\text{kin}}$. For massless quarks and gluons, we thus have

$$\begin{aligned} \mathsf{H}_K^{\text{kin}} |q_K(\boldsymbol{p})\rangle &= |\boldsymbol{p}|\,|q_K(\boldsymbol{p})\rangle, \\ \mathsf{H}_K^{\text{kin}} |G_K(\boldsymbol{k})\rangle &= |\boldsymbol{k}|\,|G_K(\boldsymbol{k})\rangle. \end{aligned} \tag{19.33}$$

The idea is to represent the nucleon state as a superposition of such quark and gluon states.

We now discuss the appropriate choice for the cutoff parameter K in deep inelastic scattering. We work in the Breit frame (18.28). As can easily be seen, the maximum transverse momentum of a hadron in the final state in deep inelastic electron–nucleon scattering is of the order of Q:

$$|\boldsymbol{k}_\perp|_{\max} \cong \frac{Q}{2}\left(\frac{1-x}{x}\right)^{1/2}. \tag{19.34}$$

However, the transverse momentum distributions fall off very steeply. Thus to describe the reaction, it should be sufficient to choose a transverse cutoff parameter $K \ll Q$. We therefore write

$$K = \varepsilon Q, \tag{19.35}$$

where $0 < \varepsilon \ll 1$. The quantity ε will be held fixed throughout the calculation. This can be justified by an estimate of the higher orders.[1] The distribution functions $N_i(x, Q^2)$ introduced previously (19.24) will be identified with the distribution functions for the eigenstates of $\mathsf{H}_K^{\text{kin}}$ (19.33).

In order to calculate, for example, P_{Gq} from (19.31), we consider quarks of transverse "size" $1/K$ (19.33), i.e., a state

$$|q_K(\boldsymbol{p})\rangle.$$

This state has the following distribution functions:

$$\begin{aligned} N_{\text{q}}(y, Q^2) &= \delta(1-y), \\ N_{\text{G}}(y, Q^2) &= 0. \end{aligned} \tag{19.36}$$

We now increase the momentum transfer from Q to $Q' > Q$ and the cutoff parameter correspondingly from K to $K' > K$. The state $|q_K(\boldsymbol{p})\rangle$, which is, by definition, an eigenstate of $\mathsf{H}_K^{\text{kin}}$, is now expanded in terms of eigenstates of $\mathsf{H}_{K'}^{\text{kin}}$. We write

$$\begin{aligned} K' &= K + \Delta K, \\ \mathsf{H}_{K'}^{\text{kin}} &= \mathsf{H}_K^{\text{kin}} + \mathsf{H}'_{\Delta K}. \end{aligned} \tag{19.37}$$

[1] In second order, for instance, we encounter terms of the form $(\alpha_s(Q^2))^2 \ln\varepsilon \propto \ln\varepsilon/(\ln(Q^2/\Lambda^2))^2$. For constant ε this leads to a decent expansion in powers of $1/\ln(Q^2/\Lambda^2)$ whereas this is in general not the case if we let ε vary with Q^2. Take as an example $\varepsilon \propto Q^{-\rho}$ with $\rho > 0$.

The term $\mathbf{H}'_{\Delta K}$ contains the coupling of the momenta whose transverse components lie between K and K'.

Appendix E contains a derivation of the Hamilton operator from the Lagrange density (19.22) using the gauge conditions

$$G_0^a(x) = 0 \qquad (a = 1, \ldots, 8). \tag{19.38}$$

For time $t = 0$ we infer

$$\mathbf{H}'_{\Delta K} = -g_s(Q^2) \int_{K \le |\boldsymbol{k}_\perp| \le K'} \mathrm{d}^3x \left\{ \sum_{j=1}^{f} \bar{\mathbf{q}}^j(\boldsymbol{x}) \vec{\mathbf{G}}^a(\boldsymbol{x}) \gamma \frac{\lambda_a}{2} \mathbf{q}^j(\boldsymbol{x}) \right.$$
$$\left. - f_{abc} \mathbf{G}^{\alpha a}(\boldsymbol{x}) \mathbf{G}^{\beta b}(\boldsymbol{x}) \nabla_\alpha \mathbf{G}^{\beta c}(\boldsymbol{x}) \right\} + \mathrm{O}(g_s^2(Q^2)), \tag{19.39}$$

where α and β $(1 \le \alpha, \beta \le 3)$ denote spatial indices. The first-order term in g_s contains the quark–gluon coupling and the three-gluon coupling as expected. We have identified the coupling constant with the corresponding running coupling strength (19.25). This can be justified *a posteriori* since, in this way, we shall find the same results as are obtained from more abstract methods (see Peterman 1979; Altarelli 1982).

We now apply first-order perturbation theory in order to decompose the quark state $|\mathrm{q}_K(\boldsymbol{p})\rangle$ into eigenstates of $\mathbf{H}^{\mathrm{kin}}_{K'}$. The "large" quark will contain, with a certain amplitude, a "small" quark of the same momentum, and also an admixture of quark–gluon states (Fig. 19.6). We accordingly write

$$|\mathrm{q}_K(\boldsymbol{p})\rangle = C_\mathrm{q} |\mathrm{q}_{K'}(\boldsymbol{p})\rangle + \sum_{\boldsymbol{k}, \boldsymbol{p}'} C(\boldsymbol{k}, \boldsymbol{p}') |\mathrm{G}_{K'}(\boldsymbol{k}), \mathrm{q}_{K'}(\boldsymbol{p}')\rangle, \tag{19.40}$$

where

$$C(\boldsymbol{k}, \boldsymbol{p}') = \frac{\langle \mathrm{G}_{K'}(\boldsymbol{k}), \mathrm{q}_{K'}(\boldsymbol{p}') | \mathbf{H}'_{\Delta K} | \mathrm{q}_K(\boldsymbol{p}) \rangle}{|\boldsymbol{k}| + |\boldsymbol{p}'| - |\boldsymbol{p}|} = \mathrm{O}(g_s(Q^2)),$$
$$C_\mathrm{q} = 1 + \mathrm{O}(g_s^2(Q^2)).$$

From this we obtain for the distribution function of a gluon $\mathrm{G}_{K'}$ in the quark q_K:

$$N_\mathrm{G}(x, Q'^2) = \sum_{\substack{\boldsymbol{k}, \boldsymbol{p}' \\ K \le |\boldsymbol{k}_\perp| \le K'}} \delta\left(x - \frac{(\boldsymbol{k}\boldsymbol{p})}{|\boldsymbol{p}|^2} \right) \cdot |C(\boldsymbol{k}, \boldsymbol{p}')|^2. \tag{19.41}$$

If, on the other hand, we insert the initial conditions (19.36) into the Altarelli–Parisi equation, we obtain

$$N_\mathrm{G}(x, Q'^2) = N_\mathrm{G}(x, Q^2) + \Delta N_\mathrm{G}(x, Q^2)$$
$$= \frac{g_s^2(Q^2)}{8\pi^2} P_{\mathrm{Gq}}(x) \Delta \ln Q^2. \tag{19.42}$$

In order to determine P_{Gq} we now merely need to carry out the phase-space integration in (19.41) and to compare this with (19.42). This simple, but rather

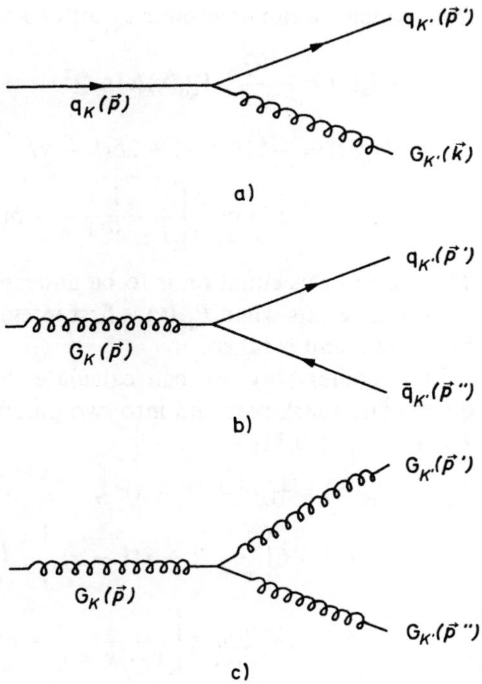

Figure 19.6 The splitting of partons of transverse "size" $1/K$ into two partons of "size" $1/K'$ $(K' > K)$. Quark into quark and gluon (a); gluon into quark and antiquark (b); gluon into two gluons (c).

lengthy calculation is presented in Appendix E. The result is

$$P_{\mathrm{Gq}}(x) = \frac{4}{3}\frac{1 + (1-x)^2}{x}. \tag{19.43}$$

The factor $1/x$ is typical for a bremsstrahlung spectrum of vector particles as we know from QED (Sect. 12.2).

From (19.43) we can determine the change in the quark distribution function $\Delta N_{\mathrm{q}}(x)$ without a long calculation. Since momentum is conserved, we have the condition

$$\boldsymbol{p}' + \boldsymbol{k} = \boldsymbol{p}$$

in the state (19.40). From this it follows that a quark $\mathrm{q}_{K'}$ with longitudinal momentum fraction $x < 1$ is always accompanied by a gluon with momentum fraction $1 - x$. For $x = 1$ the term $|\mathrm{q}_{K'}(\boldsymbol{p})\rangle$ in (19.40) contributes a δ-function. Thus in total we have

$$\Delta N_{\mathrm{q}}(x) = \Delta N_{\mathrm{G}}(1-x) - c_{\mathrm{q}}\delta(1-x), \tag{19.44}$$

where $c_{\mathrm{q}} = 1 - |C_{\mathrm{q}}|^2$ measures the decrease of the probability for the single quark state in (19.40). From the conservation of longitudinal momentum, we also have the following sum rule:

$$\int_0^1 \mathrm{d}x\, x[\Delta N_{\mathrm{q}}(x) + \Delta N_{\mathrm{G}}(x)] = 0, \tag{19.45}$$

from which we can determine c_q. After a simple calculation we find

$$\Delta N_q(x) = \frac{g_s^2(Q^2)}{8\pi^2} P_{qq}(x) \Delta \ln Q^2,$$

$$P_{qq}(x) = -\tfrac{4}{3}(1+x) + 2\delta(1-x) + \lim_{\eta \to +0} \frac{8}{3}\left[\frac{1}{1-x+\eta} - \delta(1-x)\int_0^1 dz \frac{1}{1-z+\eta}\right]. \qquad (19.46)$$

The limit in this equation is to be understood in the sense of a distribution, i.e., it only exists when $P_{qq}(x)$ is first integrated with a well-behaved function before the limit is taken.

In a similar way, we can calculate the splitting of a gluon G_K into a quark–antiquark pair and into two gluons. From this we find the functions P_{qG} and P_{GG} (19.31):

$$P_{qG}(x) = \tfrac{1}{2}[x^2 + (1-x)^2], \qquad (19.47)$$

$$P_{GG}(x) = 6\left[\frac{1}{x} - 2 + x(1-x)\right] + \left(\frac{11}{2} - \frac{f}{3}\right)\delta(1-x) + \lim_{\eta \to +0} 6\left[\frac{1}{1-x+\eta} - \delta(1-x)\int_0^1 dz \frac{1}{1-z+\eta}\right]. \qquad (19.48)$$

By simple integration we finally obtain for the anomalous dimensions d_{ij}^n (19.30):

$$d_{qq}^n = \frac{4}{33-2f}\left\{1 - \frac{2}{n(n+1)} + 4\sum_{j=2}^{n}\frac{1}{j}\right\},$$

$$d_{Gq}^n = -\frac{8}{33-2f}\frac{n^2+n+2}{n(n^2-1)},$$

$$d_{qG}^n = -\frac{3}{33-2f}\frac{n^2+n+2}{n(n+1)(n+2)},$$

$$d_{GG}^n = \frac{9}{33-2f}\left\{\frac{1}{3} + \frac{2f}{9} - \frac{4}{n(n-1)} - \frac{4}{(n+1)(n+2)} + 4\sum_{j=2}^{n}\frac{1}{j}\right\}. \qquad (19.49)$$

With this we now have all the results that we need to make a first-order comparison of the deep inelastic scattering experiments with QCD.

19.4 Comparison of Deep Inelastic Scattering Data with QCD

A whole range of methods have been, and are, used in order to compare the Altarelli–Parisi equation (19.26) or the equation for the moments (19.29) with experimental data. To begin with we will discuss a few results obtained from analyzing moments.

We consider the structure function $F_3^{(\nu N)}$ for neutrino–nucleon scattering (18.51) averaged over proton and neutron and form its moments $\tilde{M}(n, Q^2)$:

$$F_3^{(\nu N)} = \tfrac{1}{2}(F_3^{(\nu p)} + F_3^{(\nu n)}), \tag{19.50}$$

$$\tilde{M}(n, Q^2) = \int_0^1 \mathrm{d}x\, x^{n-2}[-xF_3^{(\nu N)}(x, Q^2)] \qquad (n = 2, 3, \ldots). \tag{19.51}$$

With expression (18.51) for $F_3^{(\nu N)}(x, Q^2)$ in terms of quark distribution functions which now depend on Q^2, and, using the relations (18.56), we obtain

$$-F_3^{\nu N}(x, Q^2) = N_{\mathrm{u}}^{\mathrm{p}}(x, Q^2) + N_{\mathrm{d}}^{\mathrm{p}}(x, Q^2) - N_{\bar{\mathrm{u}}}^{\mathrm{p}}(x, Q^2) - N_{\bar{\mathrm{d}}}^{\mathrm{p}}(x, Q^2). \tag{19.52}$$

If we form the moments (19.51) and use the differential equation (19.29), then, on account of the relations (19.31), which of course carry over to the anomalous dimensions d_{ij}^n, we find

$$\frac{\partial \tilde{M}(n, Q^2)}{\partial \ln Q^2} = -\frac{d_{\mathrm{qq}}^n}{\ln(Q^2/\Lambda^2)} \tilde{M}(n, Q^2). \tag{19.53}$$

The gluon distributions play no role here since F_3 is proportional to the difference between quark and antiquark distributions.

For the moments $\tilde{M}$ we thus obtain a simple differential equation whose solution can readily be given

$$\tilde{M}(n, Q^2) = C_n (\ln Q^2/\Lambda^2)^{-d_{\mathrm{qq}}^n}, \tag{19.54}$$

where the C_n are integration constants. Equation (19.54) can also be written as follows:

$$(\tilde{M}(n, Q^2))^{-1/d_{\mathrm{qq}}^n} = C_n' (\ln Q^2 - \ln \Lambda^2) \tag{19.55}$$

with different constants C_n'. Thus first-order perturbation theory in QCD predicts a linear dependence of the quantities on the left-hand side of (19.55) on $\ln Q^2$. The experiments confirm this very well as can be seen in Fig. 19.7, which shows results of neutrino scattering from iron. Note that these data refer to averages over protons and neutrons bound in the Fe nucleus. It has been found that even at Q^2 values of 200 GeV^2 the binding effects in nuclei are not negligible (Aubert 1983). However, this should have no influence on the behavior predicted by (19.55). The moments shown in Fig. 19.7 are, in fact, evaluated from more complicated formulas than those given in (19.51); they contain corrections for the finite nucleon mass (Nachtmann 1973; De Rujula 1977; Wandzura 1977).

The intersection of the straight lines in Fig. 19.7 with the Q^2-axis should, according to (19.55), determine the parameter Λ. We find

$$\begin{aligned} \Lambda^2 &\cong 0.1\ \mathrm{GeV}^2, \\ \Lambda &\cong 300\ \mathrm{MeV}. \end{aligned} \tag{19.56}$$

For the coupling strength at say $Q^2 = 5$ GeV^2, with the number of flavors

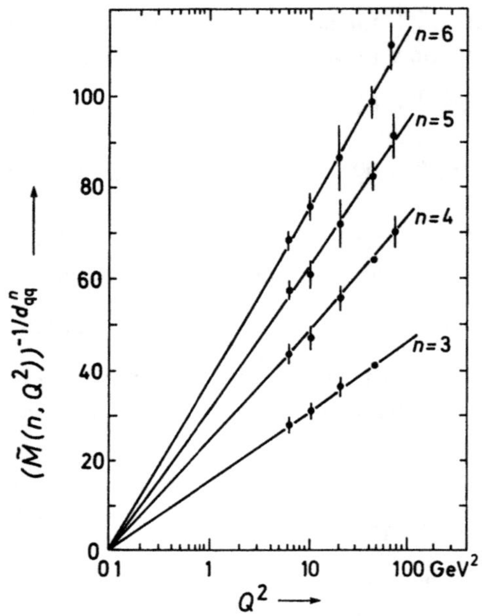

Figure 19.7 The moments of the structure function $F_3^{(\nu N)}$ measured in neutrino–iron scattering (after de Groot 1979).

being equal to four ($f = 4$), we then get from (19.25)

$$\frac{\alpha_s}{\pi}(5\ \mathrm{GeV}^2) \cong 0.12. \tag{19.57}$$

This is small enough to justify the use of first-order perturbation theory in QCD.

The parameter Λ is, in fact, extremely tricky (Baće 1978) and, strictly speaking, may not be used in first-order perturbation theory. This rather subtle point will not be discussed further here; we simply note that even in first-order perturbation theory it is both meaningful and unproblematic to talk about an effective coupling strength as in (19.57). In order to unambiguously define a parameter Λ, one has to choose a renormalization scheme and carry out the theoretical calculations for the process in question up to second order to α_s. Today, one often uses the so-called $\overline{\mathrm{MS}}$ scheme (Bardeen 1978) in which the running coupling strength has the form

$$\frac{\alpha_s(Q^2)}{\pi} = \frac{12}{(33-2f)t}\left\{1 - \frac{918-114f}{(33-2f)^2}\frac{\ln t}{t} + \mathrm{O}\left(\frac{1}{t^2}\right)\right\}, \tag{19.58}$$

where

$$t = \ln Q^2/\Lambda^2_{\overline{\mathrm{MS}}}.$$

In the limiting case $t \to \infty$ we of course retrieve (19.25).

Another method to compare (19.55) with the data is to plot the moments for various values of n against one another. From (19.55) we get

$$\ln \tilde{M}(n, Q^2) = \ln C_n - d_{qq}^n \ln(\ln Q^2/\Lambda^2),$$

$$\ln \tilde{M}(m, Q^2) = \ln C_m - d_{qq}^m \ln(\ln Q^2/\Lambda^2),$$

$$\frac{1}{d_{qq}^n} \ln \tilde{M}(n, Q^2) - \frac{1}{d_{qq}^m} \ln \tilde{M}(m, Q^2) = \text{const.} \tag{19.59}$$

First-order perturbation theory in QCD thus predicts that the logarithms of the moments give a straight line when plotted against one another with a slope that can be calculated. From (19.49) we find, for example:

$$\begin{aligned} d_{qq}^5/d_{qq}^3 &= 1.456, \\ d_{qq}^6/d_{qq}^4 &= 1.290, \\ d_{qq}^7/d_{qq}^3 &= 1.761. \end{aligned} \tag{19.60}$$

The comparison with the data is shown in Fig. 19.8. The agreement between theory and experiment is impressive.

The analysis of the structure function F_2 and its moments is somewhat more complicated. An analysis of neutrino scattering data using the Altarelli–Parisi equation (19.26) is shown in Fig. 19.9. As parameters that must be fitted to the observed scaling violations, one has not only Λ or $\alpha_s(Q^2)$, but also the entire distribution function of the gluons in the nucleon. In order to arrive at

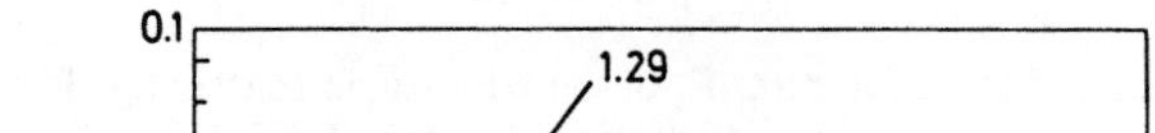

Figure 19.8 Logarithms of moments of the structure function F_3 plotted against one another. The predictions of QCD are straight lines whose slopes can be calculated to have the values given in the figure (after Bosetti 1978).

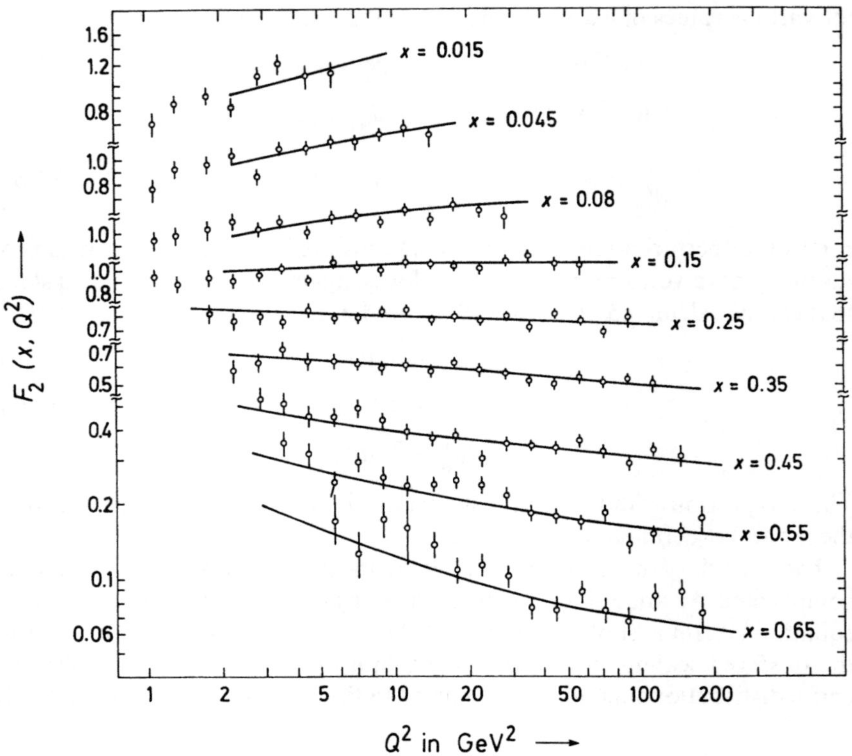

Figure 19.9 The structure function F_2 of neutrino–iron scattering. The curves are fits made on the basis of first-order perturbation theory in QCD (after Abramowicz 1983).

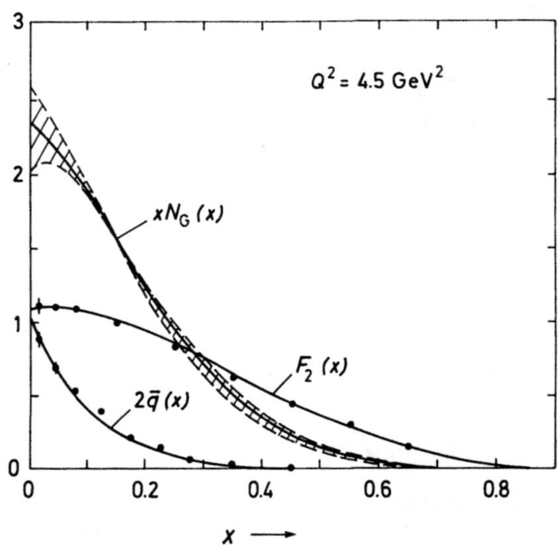

Figure 19.10 The structure functions $F_2(x)$ and $2\bar{q}(x)$ for iron, calculated per nucleon. All data have been converted to $Q^2 = 4.5\ \mathrm{GeV}^2$. The curves are QCD fits. The gluon distribution $xN_G(x)$ calculated from the scaling violations is also shown. The shaded area indicates the uncertainty corresponding to one standard deviation (after Abramowicz 1983).

meaningful results, one must simultaneously fit the antiquark distributions, which can be extracted by combining the structure functions F_2 and F_3. The gluon distribution obtained in this way is shown in Fig. 19.10.

From the QCD fits to the data for deep inelastic lepton–nucleon scattering, one finds the following value for the QCD parameter $\Lambda_{\overline{MS}}$ Dydak 1983):

$$\Lambda_{\overline{MS}} = 250 \pm 150 \text{ MeV}. \tag{19.61}$$

The large error bar for $\Lambda_{\overline{MS}}$ arises mainly from the uncertainty in the theoretical input to the analysis of the data. Besides the logarithmic deviations from Bjorken scaling behavior that are determined by $\Lambda_{\overline{MS}}$, one also expects additive terms in the structure functions which behave as powers of $1/Q^2$ for $Q^2 \to \infty$. Neither theoretically nor experimentally does one have much control over such "higher twist" terms and they represent the main source of uncertainty in many QCD analyses.

What have we learned about the structure of the nucleon from deep inelastic scattering? As an essential result we found that nucleons are built up from quarks, antiquarks, and gluons whose momentum distributions are described

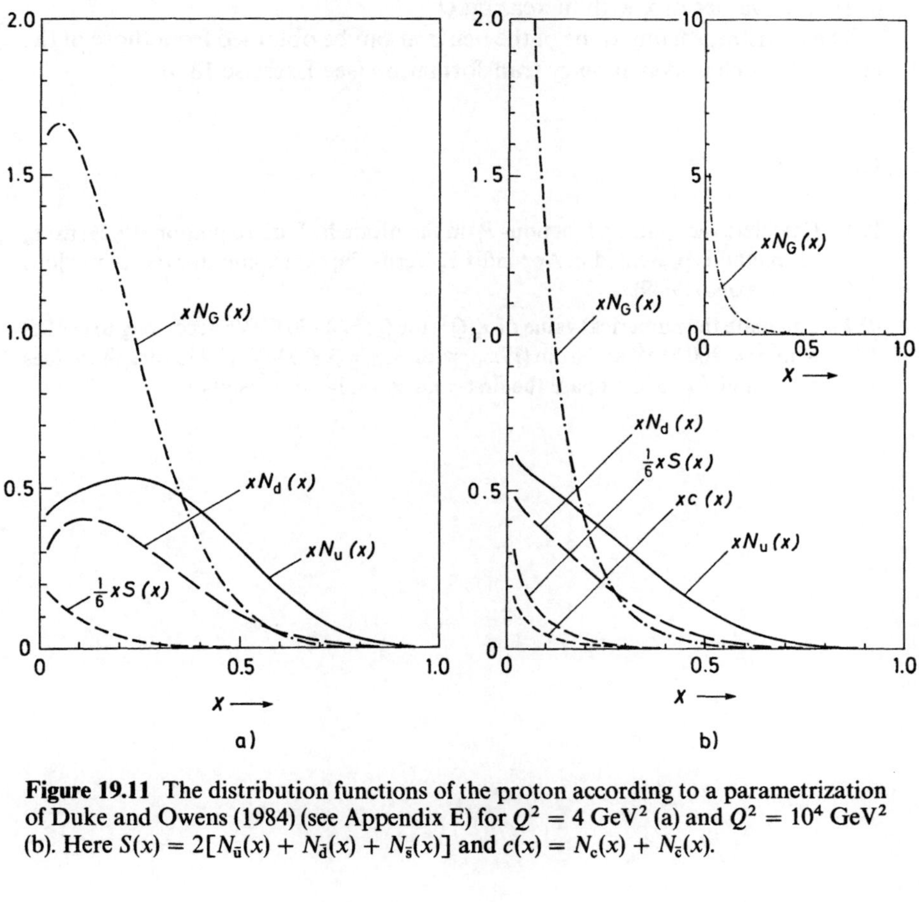

Figure 19.11 The distribution functions of the proton according to a parametrization of Duke and Owens (1984) (see Appendix E) for $Q^2 = 4$ GeV^2 (a) and $Q^2 = 10^4$ GeV^2 (b). Here $S(x) = 2[N_{\bar{u}}(x) + N_{\bar{d}}(x) + N_{\bar{s}}(x)]$ and $c(x) = N_c(x) + N_{\bar{c}}(x)$.

by Q^2-dependent distribution functions. It would seem an obvious next step to determine these distribution functions from a general fit to the available data in the region $4 \leq Q^2 \leq 200\ \mathrm{GeV}^2$ and to calculate by means of (19.26) the QCD prediction for the distribution functions at still higher Q^2. This has been done, for example, by Duke and Owens (Duke 1984). These authors also presented suitable parametrizations of the distribution functions of the proton, which are useful for many purposes. One of their parametrizations is discussed in Appendix E. The resulting quark, antiquark, and gluon distributions of the proton for $Q^2 = 4\ \mathrm{GeV}^2$ and $Q^2 = 10^4\ \mathrm{GeV}^2$ are shown in Fig. 19.11. We note that the d-quark distribution is softer than that of the u-quark, i.e., it decreases more strongly for $x \to 1$. This result is based mainly on the data shown in Fig. 18.9(b) (see Exercise 18.7). The antiquark distribution is softer than the gluon distribution which, in turn, is softer than the quark distribution. This pattern can be understood by assuming that for some relatively low value of Q^2($Q^2 \lesssim 1\ \mathrm{GeV}^2$) the proton is composed of only three quarks. With increasing Q^2 gluons then emerge due to the evolution of the quark distributions (Fig. 19.6(a)) and antiquarks emerge from the evolution of the thus generated gluon distribution (Fig. 19.6(b)). The comparison of Fig. 19.11(a) with Fig. 19.11(b) finally shows that as a general consequence of (19.26) the distributions shift to smaller values of x with increasing Q^2.

The distribution functions of the neutron can be obtained from those of the proton by a charge-symmetry transformation (see Exercise 18.5).

Exercises

19.1 Calculate the splitting functions P_{ij} in the Altarelli–Parisi equation (19.26) using the methods presented in Appendix E. Verify the expressions for the anomalous dimensions (19.49).

19.2 Calculate the numerical value of $\alpha_s(Q^2)$ for $Q^2 = 1170\ \mathrm{GeV}^2$ according to (19.25) with $\Lambda = 300$ MeV and from (19.58) with $\Lambda_{\overline{\mathrm{MS}}} = 300$ MeV. Use in turn the values $f = 4$ and $f = 5$. Compare the first- and second-order results.

CHAPTER 20

Jet and Quarkonium Physics

20.1 The Naïve Jet Model

In this chapter we wish to discuss the predictions that can be derived about the hadronic final states of high energy reactions from the concepts of quarks and gluons. Let us take, for example, the electron–positron annihilation into hadrons discussed in Sect. 18.1. In the quark picture (Fig. 18.2), the virtual photon produces a quark–antiquark pair. In the laboratory, however, we only observe hadrons. How are the hadrons created from the quarks?

Theoretical ideas about this problem were developed relatively early (Drell 1969, 1970a; Polyakov 1971; Cabibbo 1970; Feynman 1972). In the naïve parton model we have regarded a rapidly moving nucleon as a *jet* of partons, all traveling in more or less the same direction as the nucleon. The transverse momentum of the partons in the nucleon was assumed to be small ($|\boldsymbol{p}_T| \lesssim$ 300 GeV). We now reverse this picture and assume that a rapidly moving parton (quark or gluon) fragments into a jet of hadrons (Fig. 20.1). This is known as the *fragmentation hypothesis.* The hadrons in the jet are again supposed to share the total momentum of the parton, and have limited transverse momentum relative to the direction of motion of the original parton.

Today it is possible to underpin the validity of this jet picture with impressive experimental evidence. In electron–positron annihilation, the hadrons produced at center of mass energies $\sqrt{s} \gtrsim 10$ GeV do indeed appear in the form of jets. In Fig. 20.2(b), (c) we show examples of two- and three-jet events that have been observed with the detector JADE at the large storage ring PETRA. The way of observing these events is illustrated in Fig. 20.2(a), in

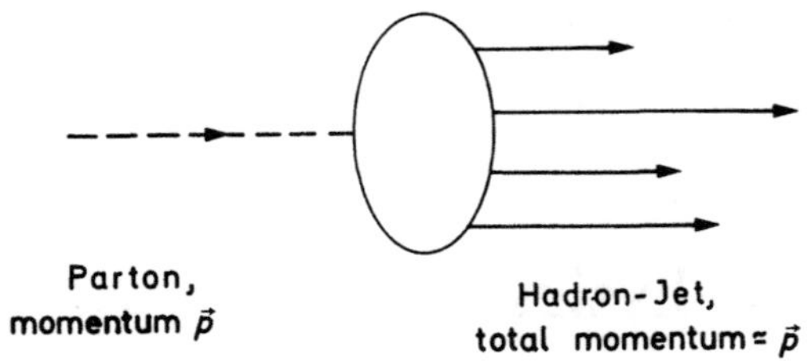

Figure 20.1 Fragmentation of a parton (quark or gluon) into a hadron jet.

which two essential elements of the JADE detector are shown: The jet chamber and the ring of lead-glass counters. The former is a wire chamber that serves to detect the charged particles. The axial magnetic field present in this chamber causes the curvature of the tracks of the charged particles in Fig. 20.2(b), (c). By means of the lead–glass counters, the neutral particles can also be detected (see Bartel (1979) for a description of the detector). Two-jet events are the most frequent type of event occurring in e^+e^- annihilation into hadrons.

We now wish to give a quantitative description of these jets. In the naïve jet model one makes the following hypotheses:

(i) The form of a jet is independent of the reaction in which the parent parton (quark or gluon) is produced.
(ii) The distribution functions for the individual types of hadrons in a jet obey scaling laws which are analogous to the Bjorken scaling law (18.41). The transverse momentum distribution should be independent of the momentum of the parent parton and again has a typical width of a few hundred MeV ($|\boldsymbol{p}_T| \lesssim 300$ MeV). The distributions in longitudinal momentum should scale with the momentum of the parent parton, i.e., they depend only on the longitudinal momentum *fraction* of the hadron considered.
(iii) If the transverse width is neglected, the hadron distributions should be described by *fragmentation* functions, that represent the analogue of the quark distribution functions in deep inelastic scattering. If $\boldsymbol{p}$ is the momentum of the original parton, for example, that of a quark q, then the longitudinal momentum fraction of a hadron h with momentum $\boldsymbol{p}_h$ is

$$z = \frac{\boldsymbol{p}_h \cdot \boldsymbol{p}}{|\boldsymbol{p}|^2}. \tag{20.1}$$

The probability of finding the hadron h with momentum fraction between z and $z + dz$ should then be given by

$$D_q^h(z)\, dz$$

where $D_q^h(z)$ is the fragmentation function, which should be dependent only on the momentum fraction z and not on the absolute magnitude of the momentum.

We first discuss the production of hadrons in e^+e^- annihilation. A simple observable is the inclusive cross-section for the reaction

$$e^+ + e^- \to h + X, \tag{20.2}$$

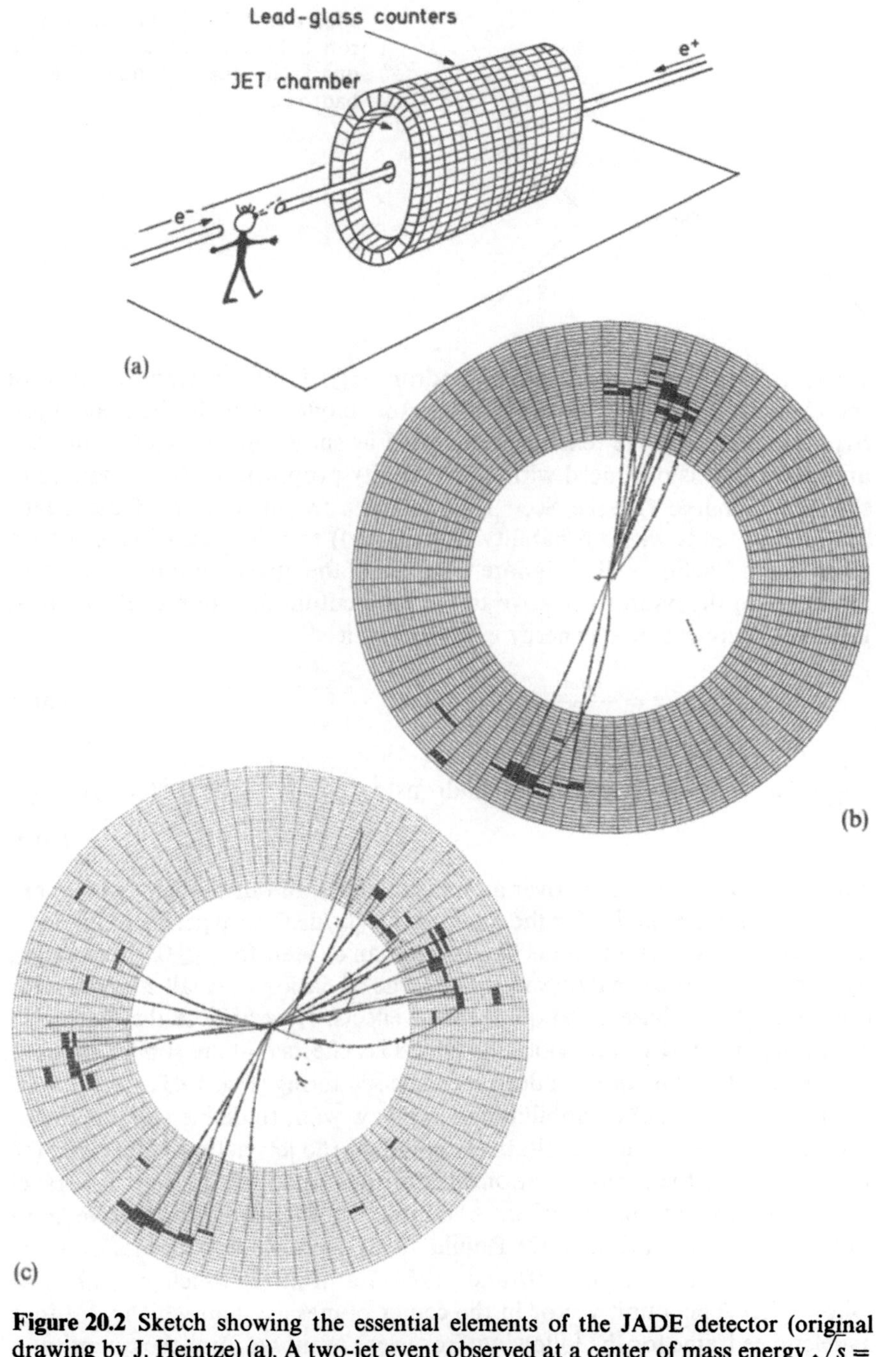

Figure 20.2 Sketch showing the essential elements of the JADE detector (original drawing by J. Heintze) (a). A two-jet event observed at a center of mass energy $\sqrt{s} =$ 30 GeV (b). A three-jet event at $\sqrt{s} = 31$ GeV (c). Visible are the tracks of charged particles in the central JET chamber and the signals from the lead–glass counters. The latter can also detect neutral particles that produce no track in the JET chamber. (The event pictures were made available by J. von Krogh.)

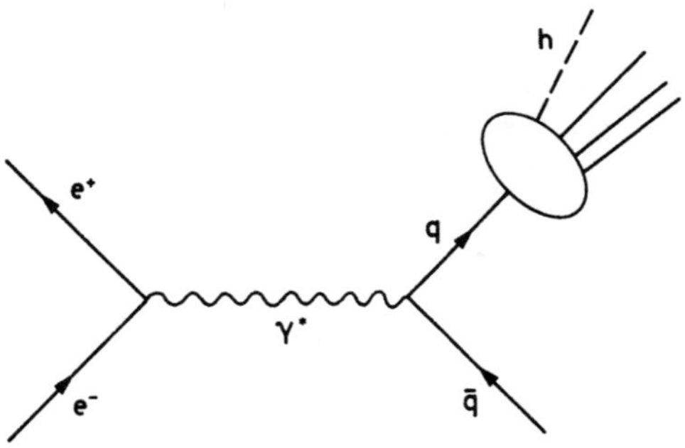

Figure 20.3 Production of a hadron h by the fragmentation of a quark in the reaction $e^+e^- \to$ hadrons.

where h stands for any kind of hadron, e.g., for π^+-mesons or any of the charged hadrons. In the quark–parton model with the fragmentation hypotheses (i)–(iii), the reaction is pictured as shown in Fig. 20.3. A quark–antiquark pair is produced with a probability proportional to the square of the quark charge Q_q (see Sect. 18.1) and the fragmentation of the quark (antiquark) leads with probability $D_q^h(z)$ ($D_{\bar{q}}^h(z)$) to a hadron of type h with momentum fraction z. If we ignore the mass of the quarks and the transverse momenta of the hadrons relative to the momentum direction of the quarks, then, with center of mass energy $\sqrt{s}$, we obtain

$$z = \frac{2|\boldsymbol{p}_h|}{\sqrt{s}}, \tag{20.3}$$

$$\frac{d\sigma}{dz}(e^+e^- \to h + X) = \sigma_{tot}(e^+e^- \to \text{hadrons}) \cdot \left(\sum_q Q_q^2\right)^{-1} \cdot \sum_q Q_q^2 (D_q^h(z) + D_{\bar{q}}^h(z)), \tag{20.4}$$

where the summation runs over all quark flavors. In our model we therefore expect a scaling behavior for the quantity $\sigma_{tot}^{-1}\, d\sigma/dz$. The experimental results for this quantity are plotted in Fig. 20.4. As can be seen, for $z \gtrsim 0.2$, the scaling hypothesis is approximately valid. The large deviation at small z values is, at least in part, attributable to quark mass effects. Nevertheless there are also clear signs of QCD scaling violation effects i.e. effects that are similar to those for the structure functions in deep inelastic scattering (Fig. 19.5).

Proceeding with e^+e^- annihilation, we now want to define suitable quantities that enable us to quantitatively compare the jets actually observed (cf. Fig. 20.2) with the approximation of ideal jets in which all hadrons travel strictly parallel to one another. A number of such quantities have been suggested (see the review of De Rujula 1978). Here we will discuss only the quantity known as "thrust" (Brandt 1964; Farhi 1977) which is defined as follows: Let $\hat{\boldsymbol{n}}$ be a unit vector in the center of mass system which we adopt as reference frame for the following. For every event we form the quantity

$$t(\hat{\boldsymbol{n}}) = \frac{\sum_i |\boldsymbol{p}_i \cdot \hat{\boldsymbol{n}}|}{\sum_i |\boldsymbol{p}_i|}. \tag{20.5}$$

Figure 20.4 The inclusive cross-section for the production of charged particles in e^+e^- annihilation into hadrons at center of mass energies $\sqrt{s} = 14$ GeV and $\sqrt{s} = 34$ GeV (after Althoff 1984a).

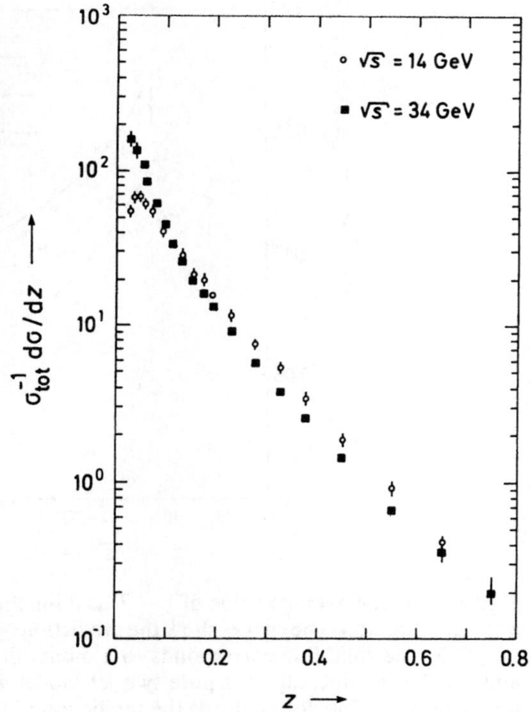

where $\boldsymbol{p}_i$ are the hadron momenta. The quantity "thrust" is the maximum of $t(\hat{\boldsymbol{n}})$ with respect to $\hat{\boldsymbol{n}}$, where $\hat{\boldsymbol{n}}$ can take all values on the unit sphere:

$$T = \max_{\hat{\boldsymbol{n}}} t(\hat{\boldsymbol{n}}). \tag{20.6}$$

The direction of $\hat{\boldsymbol{n}}$ for which t is maximal is called the thrust axis. It is the axis along which the longitudinal momenta are maximized.

Let us now examine the values of the thrust that we obtain in some special cases. For two ideal jets (Fig. 20.5(a)) we find

$$T(2 \text{ ideal jets}) = 1. \tag{20.7}$$

Here the thrust axis is the same as the jet axis. For a completely isotropic

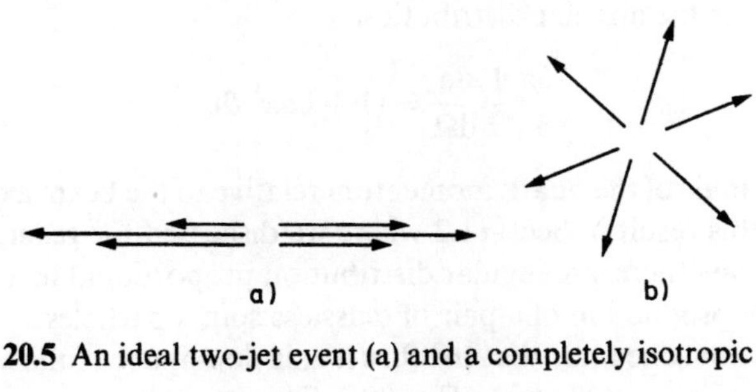

Figure 20.5 An ideal two-jet event (a) and a completely isotropic event (b).

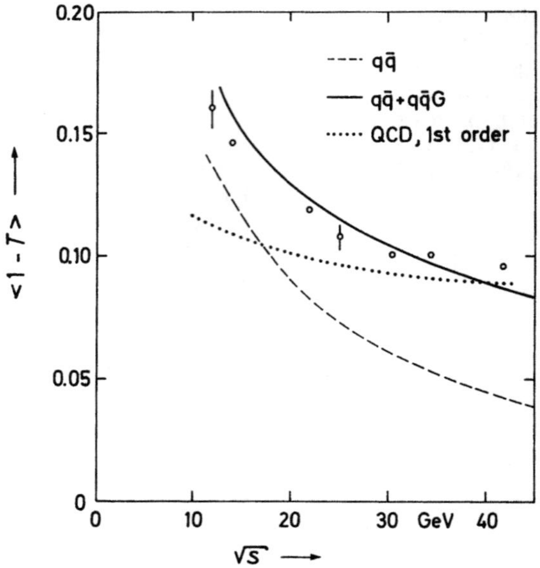

Figure 20.6 The average value of $1 - T$ as a function of the center of mass energy in e^+e^- annihilation. The curves show the predictions of jet models within the framework of QCD. The solid line corresponds to a model that includes gluon bremsstrahlung and the dashed line, $q\bar{q}$, to a pure two-jet model, in each case with the inclusion of fragmentation. The dotted line is the prediction of QCD to first order (20.15) without fragmentation effects (after Althoff 1984a).

event, on the other hand, we obtain

$$T(\text{isotropic event}) = \frac{\int d\Omega_p |\boldsymbol{p} \cdot \hat{\boldsymbol{n}}|}{\int d\Omega_p |\boldsymbol{p}|} = \frac{1}{2}. \tag{20.8}$$

If the events in e^+e^- annihilation actually have two-jet structure, then the average value of $(1 - T)$ should be close to zero. As can be seen in Fig. 20.6, for increasing center of mass energy, this quantity does indeed tend to zero.

The distribution of the thrust axes is also of considerable interest. If the thrust axis is identical to the axis defined by the momenta of the primary quark–antiquark pair, then both should display the same angular dependence. Neglecting in Fig. 20.3 the effect of the hadronization we obtain for massless quarks the angular distribution

$$\frac{16\pi}{3} \frac{1}{\sigma} \frac{d\sigma}{d\Omega} = (1 + \cos^2 \vartheta). \tag{20.9}$$

Here ϑ is the angle of the quark momentum relative to the beam axis. We have already met this result in Sect. 11.2 where we discussed the reaction $e^+e^- \to \mu^+\mu^-$. As we saw there, an angular distribution proportional to $1 + \cos^2 \vartheta$ is typical for the production of a pair of massless spin-$\frac{1}{2}$ particles.

In experiments, one finds that the thrust axis does indeed follow a distribution corresponding to (20.9); see Fig. 20.7. This provides strong support for

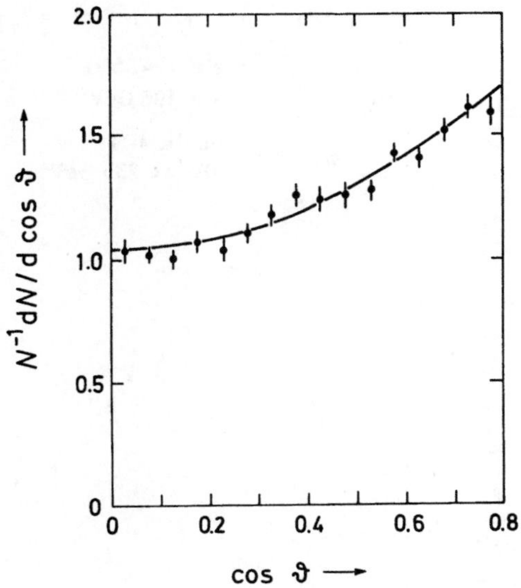

Figure 20.7 Angular distribution of the thrust axis in e^+e^- annihilation at a center of mass energy $\sqrt{s} = 34$ GeV. The curve is proportional to $(1 + \cos^2\theta)$ and corresponds to the theoretical expectation for spin-$\frac{1}{2}$ partons (20.9). The data and the curve are normalized to one over the interval $0 \leq \cos\vartheta \leq 0.8$ (after Althoff 1984a).

our assumption that e^+e^- annihilation leads primarily to the production of a pair of spin-$\frac{1}{2}$ particles.

It is now possible to investigate these events further; for example, one can study the behavior of the transverse momentum relative to the thrust axis. As expected, one finds an average transverse momentum of the order of 300 MeV (see the review article by Wu (1984)).

The production of hadrons in deep inelastic lepton–nucleon scattering can be treated in a similar way to the e^+e^- annihilation. Let us consider the case of muon–proton scattering in which a hadron h is detected in the final state, i.e., the inclusive reaction

$$\mu + \mathrm{p} \rightarrow \mu + \mathrm{h} + \mathrm{X}. \tag{20.10}$$

We see from Figs. 18.7 and 18.8 that this reaction consists primarily of the collision with a quark which should then, according to the naïve jet model, fragment into a hadron jet. Let ν be the energy transferred by the muon (Table 18.1) and E_h the energy of the hadron h in the laboratory system.

We now consider the whole reaction in the Breit frame (18.28) where the quark picture applies, to calculate the momentum fraction z carried by the hadron. Ignoring masses and transverse momenta of quarks and hadrons in this system we readily find

$$z = \frac{E_\mathrm{h}}{\nu}. \tag{20.11}$$

The z-distribution of the hadrons h is obtained in analogy to (20.4) from (18.27), (18.39) as

$$\frac{1}{\sigma}\frac{\mathrm{d}\sigma}{\mathrm{d}z} = \frac{\sum_\mathrm{q} Q_\mathrm{q}^2(\langle N_\mathrm{q}\rangle D_\mathrm{q}^\mathrm{h}(z) + \langle N_{\bar{\mathrm{q}}}\rangle D_{\bar{\mathrm{q}}}^\mathrm{h}(z))}{\sum_\mathrm{q} Q_\mathrm{q}^2(\langle N_\mathrm{q}\rangle + \langle N_{\bar{\mathrm{q}}}\rangle)}. \tag{20.12}$$

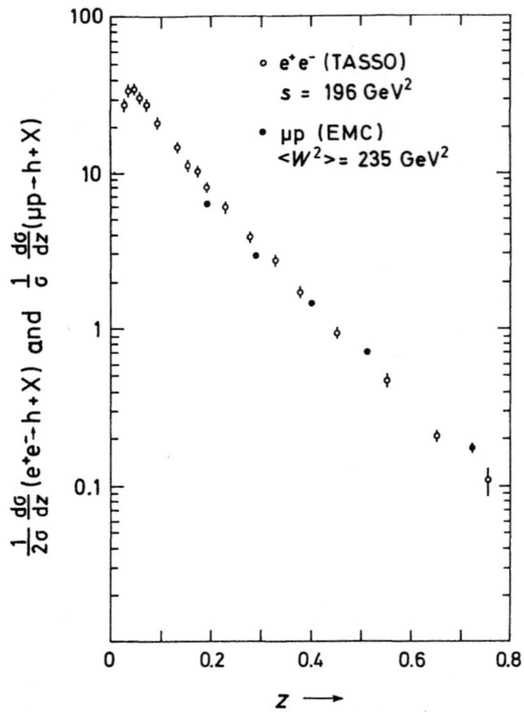

Figure 20.8 The z-distribution of charged hadrons h in e^+e^- annihilation at $s = 196\ \text{GeV}^2$ and in μp scattering at $\langle W^2 \rangle = 235\ \text{GeV}^2$. The quantities plotted correspond to weighted averages of the fragmentation functions $D_q^h(z)$ of the produced quarks [see equations (20.4) (20.12)]. The e^+e^- data are those from the TASSO group (Althoff 1984a) and the μp data are from the EMC group (J. Dress and A. Schneider, private communication). The μp data are preliminary. Their statistical error is visible on the plot only for the point with the highest z value, and the systematic error, which is not shown, is less than 15% for all points.

The sum runs over all contributing quark flavors and $\langle N_q \rangle$, $\langle N_{\bar{q}} \rangle$ denote the average values of the number of quarks and antiquarks on which a scattering occurred. These averages depend, of course, on the kinematic region in x, Q^2, and ϑ covered by the experiment in question.

An experimental z-distribution for charged hadrons is shown in Fig. 20.8. For comparison with hypothesis (i) of the naïve jet model, this figure also shows the corresponding data for e^+e^- annihilation from Fig. 20.4. One can claim quite good confirmation of the hypothesis that the fragmentation is independent of the reaction in which the quarks are produced.[1] With this we conclude our discussion of the naïve fragmentation model.

20.2 Jets and QCD Effects in Electron–Positron Annihilation into Hadrons

In this section we will present a few applications of QCD to the physics of jets. The problem is again that to all finite orders of perturbation theory QCD yields final states with quarks and gluons whereas, in the real world, only

[1] We are excluding here the unlikely possibility that the difference in the weights with which the various quark flavors contribute in the two reactions and a dependence of the fragmentation on the reaction conspire to give identical distributions for the hadrons.

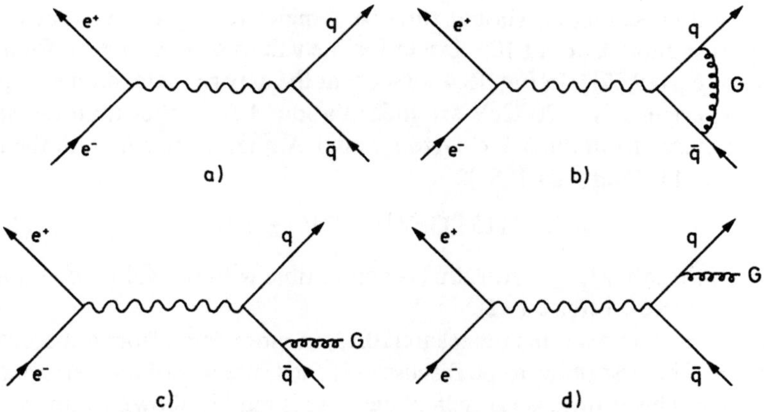

Figure 20.9 Diagrams that contribute to the total e^+e^- annihilation rate up to order g_s^2: zero order diagram (a) and the corresponding gluon correction (b); the diagrams for the emission of real gluons (c, d).

hadrons are observed. An important hypothesis relating to this was formulated by Sterman and Weinberg (Sterman, 1977). According to this, it should be possible to directly use certain cross-sections and distributions that are calculated for quarks and gluons for the corresponding hadronic quantities.

We have already met the simplest example of such a quantity. This is the total cross-section for e^+e^- annihilation into hadrons. As was argued in Sect. 18.1, it should suffice to calculate the parton production rate since the conversion of partons into hadrons does not change the total rate. In QCD, however, alongside the production of a quark–antiquark pair, which in the naïve model was the only contribution, we must now also consider the production of more complicated final states of quarks and gluons. We expect, for example, final states that consist of a quark, an antiquark, and a gluon. The corresponding lowest order diagrams are shown in Fig. 20.9 and can be evaluated using the Feynman rules of QCD (Appelquist 1973; Zee 1973). Today we know part of the corrections to the total cross-section up to order α_s^3 (Dine 1979; Chetyrkin 1979; Celmaster 1979; Gorishny 1987). In the $\overline{\text{MS}}$ scheme, with α_s as in (19.58), one obtains up to order α_s^2

$$\frac{\sigma_{\text{tot}}(e^+e^- \to \text{hadrons})}{\sigma(e^+e^- \to \mu^+\mu^-)} = 3\sum_{q=1}^{f} Q_q^2\left(1 + \frac{\alpha_s(s)}{\pi} + C_2\left(\frac{\alpha_s(s)}{\pi}\right)^2 + O(\alpha_s^3)\right), \tag{20.13}$$

where

$$C_2 = 1.98 - 0.115 \cdot f$$

and s is the square of the center of mass energy. This result holds for massless quarks and the sum runs over all quark flavors that contribute in the energy region considered.

The total cross-section should thus be somewhat larger than that of the naïve parton model; about 10% larger for the values of α_s/π that we found in Sect. 19.4 [eq. (19.57)]. In Fig. 18.4 we see that the experimental data for center of mass energies $\sqrt{s} \geq 20$ GeV are indeed about 10% higher than the value of 11/3 deduced from the naïve parton model. A quantitative fit with the help of (20.13) yields (Naroska 1987):

$$\alpha_s(s = 1156\ \text{GeV}^2) = 0.19 \pm 0.07.^2 \tag{20.14}$$

To within the fairly large error, this is compatible with the values of α_s found in Sect. 19.4 (see Exercise 19.2).

We now wish to examine the result (20.13) in more detail from a theoretical viewpoint. The first point to be discussed is the finiteness of the perturbative corrections. This feature is not self-evident. As is readily shown by an explicit calculation, the diagram in Fig. 20.9(b), for example, yields an infinite result, a so-called infrared singularity, since the gluons are massless. The infrared catastrophe has already been discussed in relation to quantum electrodynamics. Infrared singularities occur whenever we ask an unphysical question, for example, when we ask about the production of quark–antiquark pairs without allowing for any accompanying gluons. However, all measuring equipment has a finite energy resolution, ΔE. An accompanying gluon with an energy smaller than ΔE certainly cannot be excluded and it only makes sense to ask about a quark–antiquark pair together with gluons of energy smaller than ΔE. It turns out that physically meaningful questions of this type always yield an answer *without* infrared singularities. An inquiry into the total annihilation rate is certainly physically meaningful and, correspondingly, in (20.13) no infrared singularities occur.

The second point concerns the fact that the cross-section (20.13) is finite although massless quarks are used. This observation is also far from trivial and needs some discussion. If we inquire about the production rate for a quark and an antiquark with a gluon moving parallel to it (Fig. 20.10), then the result is certainly not finite for massless quarks. If we look at the contribution of the diagram in Fig. 20.9(c) to the amplitude of this process, then, from the virtual

[2] Gorishny (1987) claims that the coefficient of the α_s^2 term in (20.13) is sizable and that this contribution should not be neglected in the analysis.

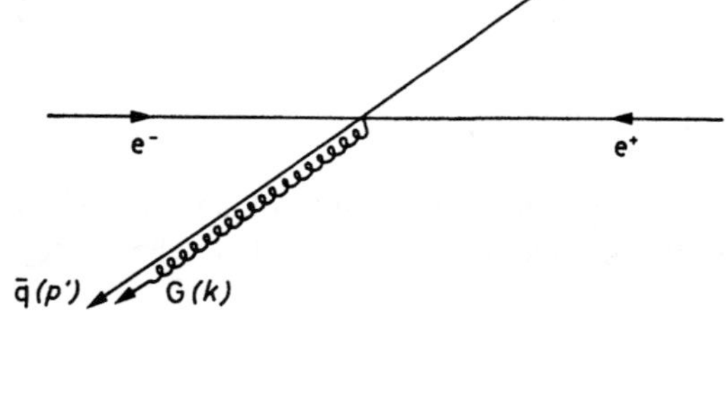

Figure 20.10 Production in e^+e^- annihilation of a quark q and an antiquark $\bar{q}$ with a gluon emitted in a collinear direction.

quark line, we obtain a factor

$$\frac{1}{(p'+k)^2 - m_q^2} = \frac{1}{2p' \cdot k}.$$

The relevant four-momenta are given in Fig. 20.10. However, for massless quarks and gluons, p' and k are light-like four-vectors which are parallel and we obtain a so-called mass singularity in the amplitude:

$$\frac{1}{2p' \cdot k} \to \infty \qquad \text{for} \quad m_q \to 0.$$

In the total rate (20.13) no singularity appears. This can be traced back to the fact that in (20.13) a sum is taken over all collinear states.

It can be shown that mass singularities always appear when we pose a question which would distinguish between one parton and several collinear partons of the same total momentum. Our measuring devices can only detect hadrons, and it is difficult to imagine that one could determine from looking at the hadrons, whether they have been produced from one parton or from two collinear partons. Thus it seems plausible to demand that quantities that one wishes to carry directly over from quarks and gluons to hadrons should not distinguish between one parton and several collinear partons of the same total momentum. In mathematical language this means that they must be free of mass singularities.

These arguments were combined by Sterman and Weinberg in the form of a hypothesis (Sterman 1977). According to this hypothesis, all quantities that are free of infrared and mass singularities are directly transferable from quarks and gluons to hadrons. More precisely, the statement is that such quantities only receive correction terms from the hadronization that decrease with an inverse power of the center of mass energy.

The quantity thrust, already introduced in (20.6), is a "good" quantity in the sense proposed by Sterman and Weinberg. Since the numerator and denominator in (20.5) are linear in the momenta, it is clearly immaterial whether or not one momentum is divided into two collinear parts. We thus expect no infrared or mass singularities and this is confirmed by the explicit calculation. It should thus be possible, for example, to calculate the average value of $1 - T$ for quarks and gluons and then to make a direct comparison with the hadronic data (Fig. 20.6). To lowest order, in which only the diagrams in Fig. 20.9 contribute, the calculation yields (De Rujula 1978)

$$\langle 1 - T \rangle = 1.57 \cdot \frac{2\alpha_s(s)}{\pi}. \tag{20.15}$$

If we take (19.25) for α_s with $\Lambda = 300$ MeV and $f = 5$ (since five quarks contribute at the energies used at PETRA, then we find the curve labelled "QCD" in Fig. 20.6. As can be seen, for center of mass energies $\sqrt{s} \lesssim 30$ GeV, the terms of order $1/\sqrt{s}$, which are neglected in (20.15), must contribute strongly. Higher orders of α_s, which are also neglected in (20.15), have been

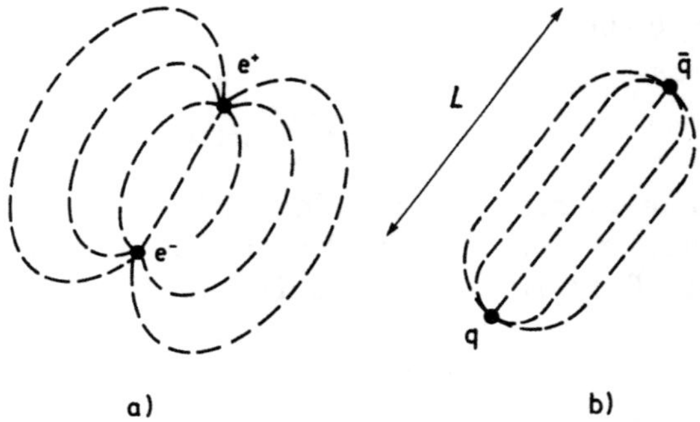

Figure 20.11 Electric dipole field around an e^+e^- pair (a) and the gluon field around a $q\bar{q}$ pair (b).

calculated (Ellis 1981; Fabrizius 1981; Vermaseren 1981) but do not significantly change the form of the QCD curve in Fig. 20.6.

If we want to compare the currently available data with QCD, we must therefore find a way to take the hadronization into account; we need a model. A simple description of how hadrons are created from the quark–antiquark pair produced in e^+e^- annihilation, is the following: When the quark separates from the antiquark, a gluon field is produced, in exactly the same way as an electromagnetic field arises when an electron–positron pair separates. An e^+e^- pair produces an electric dipole field which extends in all directions in space (Fig. 20.11(a)). It is well known that such a dipole field does not prevent us from separating the electron and the positron at the cost of a finite amount of energy. At small separations, the situation should be similar for quarks on account of asymptotic freedom. However, at large distances a radically different behavior is necessary; one which prevents the occurrence of free quarks. One imagines that a thin flux tube forms between the quark and the antiquark. If the quark and antiquark are sufficiently far apart ($\gtrsim 10^{-13}$ cm) then it is supposed that the energy of the flux tube is proportional to its length L (Fig. 20.11(b)). Such a linearly increasing potential was also suggested by the lattice gauge theories mentioned previously. It clearly yields permanently bound quarks.

According to this picture, the kinetic energy of the primary $q\bar{q}$ pair produced in e^+e^- annihilation is converted into potential energy of the flux tube. The flux tube is then broken at some point, at which another quark–antiquark pair is created from the vacuum. The two flux tubes, which are now separated from one another, can be further split if their energy content is high enough, or they may directly yield mesons (Fig. 20.12). This scheme also enables us to understand why the hadrons that are formed from a fast quark all travel in approximately the same direction and form a jet. Detailed models to describe the fragmentation of quarks and gluons into hadrons that are founded on the idea of such recursive processes have been proposed by a number of authors (Field 1978; Hoyer 1979; Andersson 1980, 1983; Jones 1981; Webber 1984). These models naturally contain a few free parameters that must be fitted to

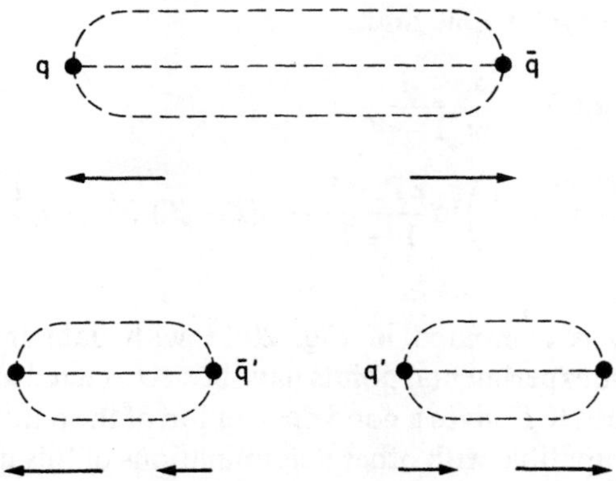

Figure 20.12 Breaking of a gluon flux tube by the creation of a quark–antiquark pair $q'\bar{q}'$. The arrows represent the momenta of the quarks and antiquarks.

experiment; however, they give a complete description of the fragmentation process. Here we shall not explore them further. We only point out that these models allow us to generate hadron distributions in full detail from the distributions for quarks and gluons calculated by QCD perturbation theory. The jet model curves in Fig. 20.6 were obtained in this way.

Perhaps the most important application to date of the fragmentation model is in the discussion of three-jet events in e^+e^- annihilation. If we look at the diagrams in Fig. 20.9, we see that there is a certain probability that a bremsstrahlung gluon is produced at a large angle with respect to the quark–antiquark pair. If our picture of the fragmentation is correct, then such an event would yield three jets. In experiments, this has actually been observed; in Fig. 20.2(c) we have shown an example. However, one should note that, due to statistical fluctuations, it is possible for a two-jet event to look like a three-jet event. To clearly demonstrate the existence of three-jet events one is forced, for better or for worse, to use a fragmentation model in order to determine the number of apparent three-jet events which are expected from statistical fluctuations occurring in two-jet events. Following many detailed studies of the experimental data, it has been concluded that for center of mass energies $\sqrt{s} \gtrsim 25$ GeV true three-jet events do occur (see Wu (1984) for a review).

If we assume that the existence of three-jet events is proved, then according to Fig. 20.9 the three jets must derive in the framework of QCD, from a quark, an antiquark, and a gluon. Since we are already familiar with the properties of quarks, it becomes possible to empirically determine the properties of the gluons from the three-jet events.

We will discuss the determination of the spin of the gluon. For this purpose we again consider the quantity thrust (20.6). The thrust distribution of the three partons in the final state of the reaction

$$e^+ + e^- \rightarrow q + \bar{q} + G \tag{20.16}$$

can easily be calculated from the diagrams of Fig. 20.9(c), (d) by means of the

Feynman rules for QCD. One finds

$$\frac{1}{\sigma_{tot}}\frac{d\sigma}{dT}(e^+e^- \to q\bar{q}G) = \frac{2\alpha_s}{3\pi}\frac{1}{1-T}$$

$$\cdot\left\{\left(\frac{4}{T} - 6 + 6T\right)\ln\frac{2T-1}{1-T} - 3(2-T)(3T-2)\right\} \qquad (\tfrac{2}{3} \leq T \leq 1). \tag{20.17}$$

This distribution is compared in Fig. 20.13 with data from the TASSO collaboration. The experimental points have been corrected for fragmentation effects. We see that QCD gives a good description of the data. The fitted value of $\alpha_s \cong 0.17$ is compatible with other determinations of this quantity [(20.14), see also Exercise 19.2]. A calculation based on the radiation of "scalar gluons", however, does not yield agreement with experiment.

This and similar investigations can be considered as direct evidence for the existence of gluons with a spin value of 1 as required by theory. What we are still lacking is direct evidence that the gluons also carry color, i.e., that they have the self-coupling represented by the three- and four-gluon vertices of

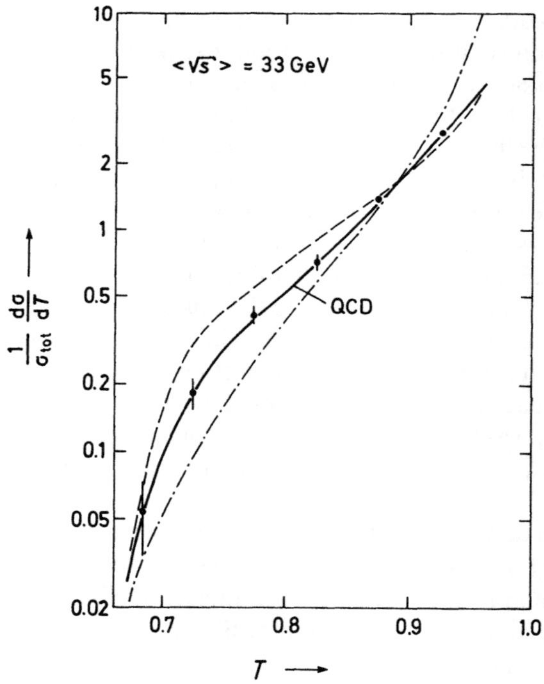

Figure 20.13 The thrust distribution of three-jet events observed in e^+e^- annihilation corrected for the effects of fragmentation. The data are from the TASSO group. The solid line is the first order prediction of QCD for $\alpha_s = 0.17$ (20.17). The dashed line is the result obtained for the emission of scalar gluons, and the dash–dotted line is the result for the so-called "constituent interchange" model (after Wolf 1982).

QCD. Possible means of demonstrating this have been discussed by, among others, Körner (1981) and Nachtmann (1982).

Recent determinations of α_s from e^+e^- annihilation into hadrons typically yield values in the range

$$0.10 \leq \alpha_s(s) \leq 0.17$$

for $\sqrt{s} = 35$ GeV (see, e.g., Naroska (1987)), where the smaller values tend to be favored (J. von Krogh, private communication). The considerable uncertainty in the value of α_s stems from the theoretical uncertainties in the description of the fragmentation process.

20.3 Quarkonium

In this section we will discuss the bound states of a heavy quark–antiquark pair. To date we are familiar with $c\bar{c}$ and $b\bar{b}$ states; in the future it should also be possible to study $t\bar{t}$ states. The symbols c, b, and t stand for the quarks charm, bottom and top. The masses of c and b are approximately

$$\begin{aligned} m_c &\cong 1.3 \text{ GeV}, \\ m_b &\cong 4.5 \text{ GeV}. \end{aligned} \tag{20.18}$$

The top quark has not been found yet. Experiments at the TRISTAN ring in Japan give a lower bound for the t-mass of $m_t > 26$ GeV (Takasaki 1988).

The first quarkonium state to be discovered was the J/ψ-particle, which was found in the reactions (Aubert 1974; Augustin 1974)

$$p + p \to J/\psi + X, \quad J/\psi \hookrightarrow e^+e^- \tag{20.19}$$

$$e^+ + e^- \to J/\psi. \tag{20.20}$$

Today we know from experiments that the J/ψ is a particle with spin 1 and with negative parity and charge conjugation. Its mass, total decay rate, and leptonic branching ratios are

$$\begin{aligned} m_{J/\psi} &= 3096.9 \pm 0.1 \text{ MeV}, \\ \Gamma(J/\psi \to \text{anything}) &= 0.063 \pm 0.009 \text{ MeV}, \\ \frac{\Gamma(J/\psi \to e^+e^-)}{\Gamma(J/\psi \to \text{anything})} &= \frac{\Gamma(J/\psi \to \mu^+\mu^-)}{\Gamma(J/\psi \to \text{anything})} = 7.4 \pm 1.2\%. \end{aligned} \tag{20.21}$$

This particle presented physicists with a big puzzle; namely, to understand how a particle with a mass of 3 GeV could have a width of only 63 keV. This is much smaller than the typical hadronic width of about 100 MeV (16.8). The solution to this puzzle was provided by QCD as we shall see.

The J/ψ-particle is today believed to be a $c\bar{c}$ bound state, exactly analogous to the e^-e^+ bound state orthopositronium, which indeed also has spin 1 and negative parity (Chap. 13). The electron and positron are held together by the

Coulomb force. The force that binds the quark and antiquark together should stem from the gluons. From our discussion in Sect. 20.2 we expect that the corresponding potential $V(r)$ is Coulomb-like for small quark–antiquark separations, but that it grows linearly for large separations (Fig. 20.11). For small separations, we can calculate the potential from the one-gluon exchange (Exercise 20.4). We find

$$V(r) \to -\frac{4}{3}\frac{\alpha_s(r^{-2})}{r} = \frac{4}{3}\frac{1}{r}\frac{12\pi}{(33-2f)\ln\Lambda^2 r^2} \qquad \text{for} \quad r \to 0. \tag{20.22}$$

The factor 4/3 that appears here comes from the color-SU(3) group. For large separations we assume a linear form

$$V(r) \to k \cdot r \qquad \text{for} \quad r \to \infty, \tag{20.23}$$

where k is the so-called string tension. A good theory for the potential at intermediate separations does not exist. However, there have been some interesting attempts to interpolate between the limiting cases (20.22) and (20.23), for example, the Richardson potential which was further improved by Buchmüller and Tye (Richardson 1979; Buchmüller 1981). Other methods to treat the charmonium system that bypass the use of a quark-antiquark potential have also been suggested (Novikov 1978; Reinders 1981; for reviews, see Shifman 1981; Gottfried 1983; Reinders 1985).

In the following we shall assume that a potential exists between quark and antiquark of the general form shown in Fig. 20.14. Its parameters must

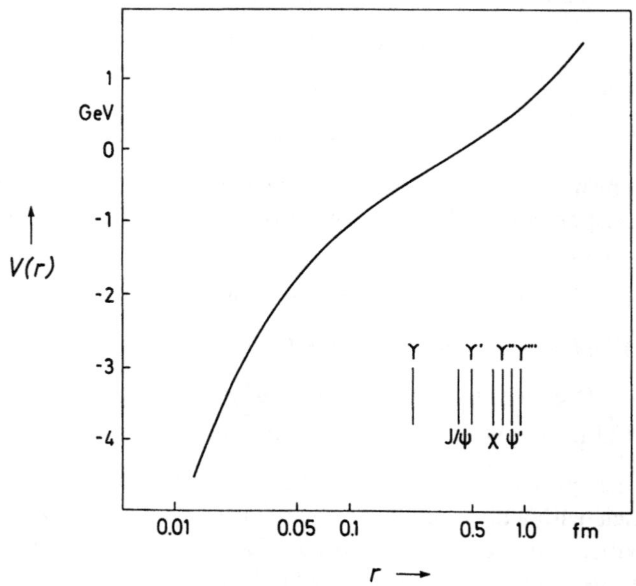

Figure 20.14 Example of a quark-antiquark potential satisfying (20.22), (20.23) for $r \to 0$ and $r \to \infty$. The mean radii $(\langle r^2 \rangle)^{1/2}$ of individual quarkonium states are indicated by the vertical lines marked Υ, J/Ψ, etc. (after Buchmüller 1981).

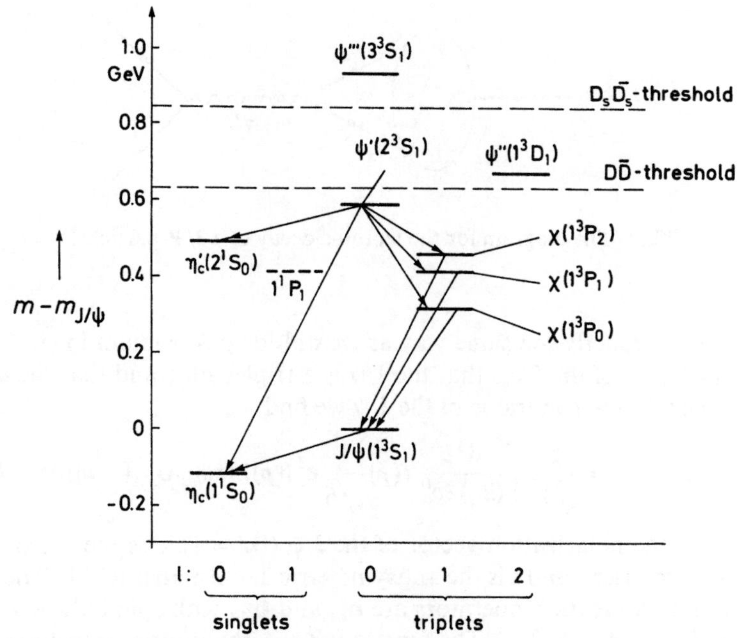

Figure 20.15 The level scheme observed for the charmonium system and the observed photon transitions. The as yet undiscovered $1\,{}^1P_1$ level is shown by a dashed line at the position predicted by Reinders (1981).

be determined individually by comparison with experimental data. In this potential, heavy quarks will be nonrelativistic and so we can make use of a Schrödinger equation to describe the states.

For a potential model of this kind we expect a whole spectrum of quarkonium states with triplets and singlets, S-, P-, and D-waves, radial excitations, etc., just like the case of positronium (Figs. 13.2 and 13.3). There should also be electromagnetic transitions between the various levels. All of these expectations have been brilliantly confirmed by experiment. In Fig. 20.15 we show the observed level scheme of the $c\bar{c}$ system. The theory predicts the existence of exactly one further state, the $1\,{}^1P_1$ level, that lies below the $D\bar{D}$ threshold (the latter will be explained later). The $1\,{}^1P_1$ level, however, is difficult to produce and so it is not surprising that it has not yet been observed. The observed photon transitions are also indicated in Fig. 20.15 (see Scharre (1981) for a review).

We now turn to the decay of the J/ψ-particle. It is possible for a quark and an antiquark to annihilate into a virtual photon which can then rematerialize into an e^+e^- or a $\mu^+\mu^-$ pair (Fig. 20.16). In this way we obtain the leptonic decay processes

$$J/\psi \to e^+e^-,\, \mu^+\mu^- .$$

In order to calculate these decay processes we write down the wavefunction

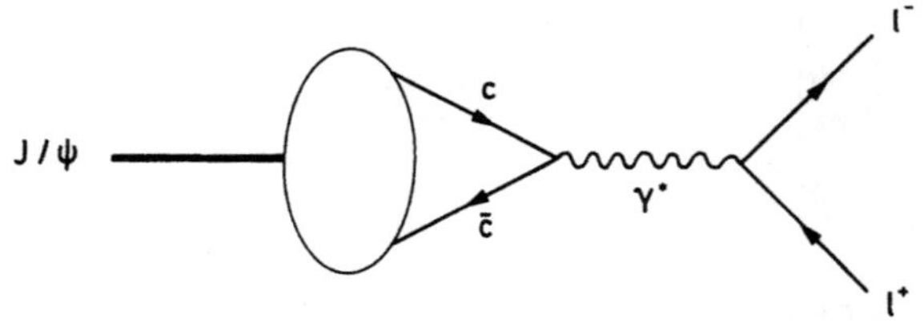

Figure 20.16 The diagram for the leptonic decay of a J/Ψ-particle ($\ell = e, \mu$).

of the J/ψ in exactly the same way as we did for positronium in (13.17). We must take note of the facts that the J/ψ is a triplet state and that the quarks carry color. In the rest frame of the J/ψ we find

$$|J/\psi(\boldsymbol{\lambda})\rangle = \frac{1}{\sqrt{V}}\int \frac{d^3p}{(2\pi)2p^0} f(\boldsymbol{p}) \frac{1}{\sqrt{6}} \mathbf{a}^\dagger_{sA}(\boldsymbol{p})\boldsymbol{\lambda}(\boldsymbol{\sigma}\varepsilon)_{ss'}\mathbf{b}^\dagger_{s'A}(-\boldsymbol{p})|0\rangle, \quad (20.24)$$

where $\boldsymbol{\lambda}$ is the polarization vector of the J/ψ ($|\boldsymbol{\lambda}| = 1$); $\boldsymbol{\sigma}$ is the vector of the Pauli spin matrices, and ε is the antisymmetric 2×2 matrix (C.11). The quark and antiquark creation operators are $\mathbf{a}^\dagger_{sA}$ and $\mathbf{b}^\dagger_{s'A}$ with spin indices s, s' and color index A ($A = 1, 2, 3$). The function $f(\boldsymbol{p})$ is the Fourier transform of the ordinary Schrödinger wavefunction $\psi(\boldsymbol{x})$,

$$\psi(\boldsymbol{x}) = \int \frac{d^3p}{(2\pi)^3} e^{i\boldsymbol{p}\boldsymbol{x}} f(\boldsymbol{p}). \quad (20.25)$$

With this, the state (20.24) is normalized to one:

$$\langle J/\psi(\boldsymbol{\lambda})|J/\psi(\boldsymbol{\lambda})\rangle = |\boldsymbol{\lambda}|^2 = 1.$$

The procedure now exactly follows the calculation of positronium decay. We first calculate the amplitude for the transition

$$c + \bar{c} \to \ell^- + \ell^+,$$

where $\ell = e, \mu$ and convolute this with the wavefunction of the $c\bar{c}$ system in the charmonium state J/ψ. Neglecting the masses of the leptons and using the nonrelativistic approximation for the motion of c and $\bar{c}$, we then obtain:

$$\Gamma(J/\psi \to \ell^-\ell^+) = \frac{16\pi\alpha^2 Q_c^2 |\psi(0)|^2}{(m_{J/\psi})^2}, \quad (20.26)$$

where $Q_c = 2/3$ is the charge of the c-quark. We will consider the wavefunction at the origin $\psi(0)$ as a free parameter since we have not specified the exact form of the potential.

What do the hadronic decay processes of the J/ψ look like? Since the J/ψ represents the lowest state of the $c\bar{c}$ system with parallel spins, for kinematical reasons it should not be possible for it to decay into a pair of "free" c, $\bar{c}$ quarks. Free quarks do not exist anyhow but there are mesons that consist of a c-quark and a $\bar{u}$-, $\bar{d}$-, or $\bar{s}$-antiquark (Table 15.1). These are

$$\begin{aligned} D^0 &\sim c\bar{u}, \\ D^+ &\sim c\bar{d}, \\ D_s^+ &\sim c\bar{s}. \end{aligned} \tag{20.27}$$

Thus we expect that the thresholds for $D\bar{D}$ and $D_s\bar{D}_s$ production lie higher than the J/ψ mass and this is indeed the case (Fig. 20.15). If the J/ψ-particle is to decay hadronically, the c- and $\bar{c}$-quarks must annihilate.

A possible annihilation process that gives hadrons in the final state is one that proceeds via a virtual photon. We simply need to replace the lepton pair $\ell^-\ell^+$ in Fig. 20.16 by a quark pair $u\bar{u}$, $d\bar{d}$, or $s\bar{s}$. Due to the color degree of freedom of the quarks we obtain (cf. Sect. 18.1)

$$\frac{\Gamma(J/\psi \to \gamma^* \to \text{hadrons})}{\Gamma(J/\psi \to e^-e^+)} \cong 3 \sum_{q=u,d,s} Q_q^2 = 2. \tag{20.28}$$

A second possibility is the annihilation of the $c\bar{c}$ pair into gluons. Since heavy quarkonia are relatively small objects (Fig. 20.14), this annihilation should occur at small separations. Thus we will call upon the asymptotic freedom property of QCD and calculate the decay rate as if the gluons were free massless particles.

We first demonstrate that as a consequence of the Landau–Yang theorem (Chap. 13) the spin-1 quarkonium state J/ψ cannot decay into two gluons:

$$J/\psi \not\to 2 \text{ gluons}. \tag{20.29}$$

To do so we consider the color-state of the two gluons. Their wavefunction must form a color singlet since the J/ψ is color neutral. The only color-singlet combination of two gluons with color indices a and b $(a, b = 1, \ldots, 8)$ is proportional to δ_{ab}, and therefore symmetric with respect to the exchange of the two gluons. The conditions for the rest of the wavefunction of the two gluons are thus the same as those for two photons (13.11)–(13.13), i.e., we have the situation referred to in the Landau–Yang theorem.

The J/ψ-particle must therefore decay into at least three gluons (Fig. 20.17) since, due to color conservation, an annihilation via a single virtual gluon is also impossible:

$$J/\psi \to 3 \text{ gluons}. \tag{20.30}$$

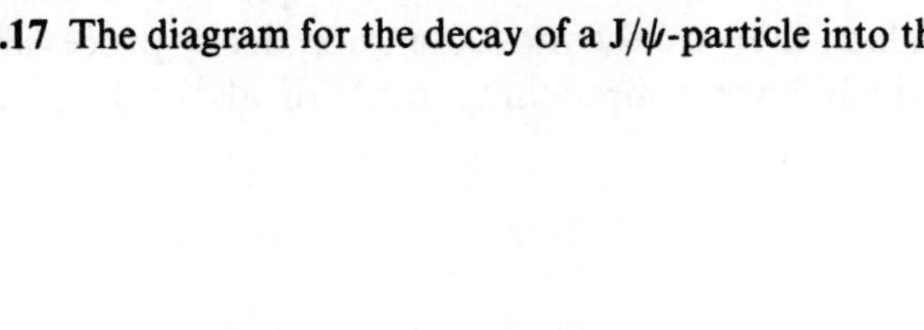

Figure 20.17 The diagram for the decay of a J/ψ-particle into three gluons.

This process can be calculated analogously to the decay of positronium into three photons and one finds

$$\Gamma(\mathrm{J}/\psi \to 3\mathrm{G}) = \frac{160}{81}(\pi^2 - 9)(\alpha_s(m_{\mathrm{J}/\psi}^2))^3 \frac{|\psi(0)|^2}{(m_{\mathrm{J}/\psi})^2}. \tag{20.31}$$

The decisive point which explains the puzzle of the small decay width of the J/ψ is the observation that the process takes place at small separations. The corresponding coupling parameter α_s for the gluons should thus be small on account of asymptotic freedom. An estimate, similar to that for positronium decay (13.32), tells us that the c- and $\bar{\mathrm{c}}$-quark can in fact only annihilate when their separation is smaller than about $1/m_c$. Thus in (20.31) we have written $m_{\mathrm{J}/\psi}^2 \sim (2m_c)^2$ as the argument of α_s and, according to (19.25), (19.56), this yields a small value for α_s. A process of order α_s^3 is then strongly suppressed and the lifetime of the J/ψ-particle is relatively long. These ideas were originally put forward by Appelquist and Politzer before the discovery of the J/ψ-particle (Appelquist 1975). Subsequently, they have been fully confirmed by experiment.

Equations (20.26) and (20.31) clearly furnish a means of determining α_s since in the ratio of the rates, the wavefunction at the origin cancels. Taking (20.28) into account as well, we find

$$\Gamma(\mathrm{J}/\psi \to \text{anything}) = \Gamma(\mathrm{J}/\psi \to 3\mathrm{G}) + 4\Gamma(\mathrm{J}/\psi \to \mathrm{e}^+\mathrm{e}^-), \tag{20.32}$$

$$\frac{\Gamma(\mathrm{J}/\psi \to 3\mathrm{G})}{\Gamma(\mathrm{J}/\psi \to \mathrm{e}^+\mathrm{e}^-)} = \frac{1 - 4B_{\mathrm{ee}}}{B_{\mathrm{ee}}} = \frac{10(\pi^2 - 9)\alpha_s^3}{81\pi\alpha^2 Q_c^2},$$

where B_{ee} is the electronic branching ratio,

$$B_{\mathrm{ee}} = \frac{\Gamma(\mathrm{J}/\psi \to \mathrm{e}^+\mathrm{e}^-)}{\Gamma(\mathrm{J}/\psi \to \text{anything})}.$$

If we insert the value of B_{ee} from (20.21) we obtain

$$\alpha_s(m_{\mathrm{J}/\psi}^2) \cong 0.19. \tag{20.33}$$

This is significantly smaller than the values of α_s extracted from deep inelastic scattering (19.57). Even a more exact analysis which takes into account corrections to the next order in α_s produces the same result (Mackenzie 1981). It is probably the relativistic corrections, which have been neglected in this description of the J/ψ as a $\mathrm{c}\bar{\mathrm{c}}$ bound state, that are responsible for this discrepancy.

The ϕ-meson is regarded in the quark model as an $\mathrm{s}\bar{\mathrm{s}}$ state (17.78), and with a mass $m_\phi = 1020$ MeV, it represents the limiting case of a "heavy" quarkonium state. Many of the ideas in this section remain qualitatively valid for the ϕ-meson. The decay processes $\phi \to \mathrm{K}^+\mathrm{K}^-$ and $\phi \to \mathrm{K}^0\bar{\mathrm{K}}^0$, in which each of the K-mesons contains a "heavy" s- or $\bar{\mathrm{s}}$-quark, are energetically just possible and, in fact, they are the most common decay processes. The decays $\phi \to \pi^+\pi^-\pi^0$, which have a branching ratio of about 15%, can—at least qualitatively—be understood as $s\bar{s}$ annihilation into gluons followed by fragmentation into pions.

In 1977 the first levels of the $b\bar{b}$ quarkonium system were discovered: the Υ-particle and its radial excitations (Herb 1977). Since then, these states have been studied in detail, particularly in electron–positron annihilation (see Berkelman 1983). The theoretical description of the bottonium system is very similar to that of the charmonium system. From the gluonic and leptonic branching ratios of the Υ, it is again possible to determine a value for α_s and for $\Lambda_{\overline{MS}}$. One finds (Mackenzie 1981)

$$\alpha_s(20.6\ \mathrm{GeV}^2) = 0.158 \begin{cases} +0.012 \\ -0.010, \end{cases}$$
$$\Lambda_{\overline{MS}} = 100 \begin{cases} +34 \\ -25 \end{cases} \mathrm{MeV}, \tag{20.34}$$

where the number of active quark flavors is four.[3] This value of $\Lambda_{\overline{MS}}$ lies at the lower limit of the interval allowed by the data on deep inelastic scattering (19.61).

According to our ideas about the fragmentation of partons into hadrons, the decay of a heavy $^3S_1$ quarkonium state into three gluons should lead to three hadron jets. The J/ψ-particle with a mass of 3.1 GeV, is too light for a three-jet structure to be noticeable in the hadronic final states of the decay. For the decay of the Υ-particle, on the other hand, with a mass of 9.5 GeV, some evidence of a three-jet structure has been observed (Berger 1978; see also Berkelman 1983). However, the mass of the Υ is still relatively small and thus it is hardly possible to distinguish three isolated jets. For the $^3S_1$-state of the toponium system, with a mass of more than 50 GeV this should be possible.

20.4 Jets in Hadron–Hadron Collisions

Hadron–hadron collisions at high energies are best analyzed in terms of collisions between the hadronic constituents, the quarks and the gluons. Most of the cross-section is due to a scattering at small angles between the constituents. Occasionally, however, a large-angle scattering of two energetic constituents takes place. As an example, Fig. 20.18 shows a proton–antiproton collision in which a quark and an antiquark scatter from one another by means of the exchange of a gluon. This process is exactly analogous to the Bhabha scattering in QED (Sect. 10.2). In the final state of the collision shown in Fig. 20.18 we expect, as a fragmentation product of the scattered quarks, two jets of hadrons with large transverse momentum and from the fragmentation of the spectator partons, two jets along the beam axis. Events

[3] The parameter $\Lambda_{\overline{MS}}$ is dependent on the number of active quark flavors (f). Therefore we should really write $\Lambda^{(4)}_{\overline{MS}}$ for the value in (20.34) and $\Lambda^{(5)}_{\overline{MS}}$ for the relevant Λ-parameter at PETRA energies where $f = 5$. The relationship between the different Λ-parameters can be calculated theoretically (Weinberg 1980; Bernreuther 1982).

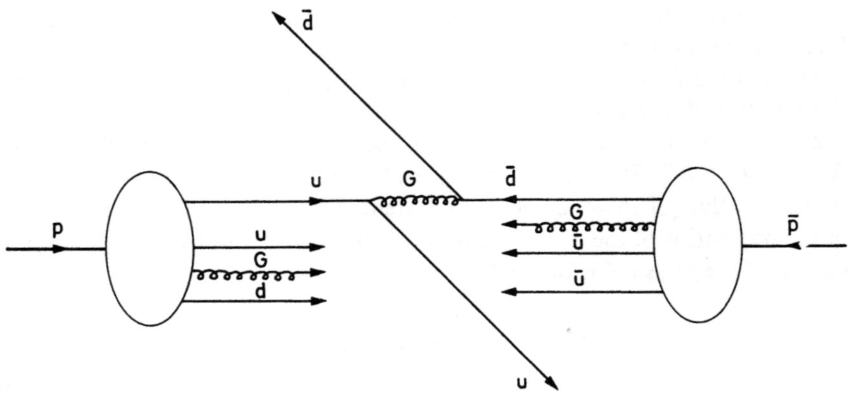

Figure 20.18 Diagram for the scattering of a quark u from an antiquark d̄ with the exchange of a gluon in a pp̄ collision.

of this type have in fact been observed in pp collisions in the storage ring ISR at center of mass energies $\sqrt{s} \gtrsim 60$ GeV (Åkesson 1984; Breakstone 1984a), and in pp̄ collisions at $\sqrt{s} = 540$ GeV (Banner 1982; Arnison 1983f, g, 1984b; Bagnaia 1984a–d).

In Fig. 20.19 we show an example of such an event observed in the UA2 detector and represented by a so-called "Lego plot". The Lego plot is a two-dimensional histogram with bins in the azimuth φ and polar angle ϑ with respect to the beam axis in the center of mass system. The height of an individual tower is proportional to the transverse energy deposited in the respective bin. This transverse energy is related to the energy E by $E_T = E \sin \vartheta$. The particles of the spectator jets have small transverse energies ($E_T \lesssim 2$ GeV) and so, on the scale of this picture, only the two jets with large p_T are clearly visible. Noteworthy is the clearly isolated and highly collimated nature of these jets.

In order to describe these reactions theoretically, we first consider the "hard" parton–parton scattering processes. To leading order in α_s the possible reactions with two partons in the final state are those listed in Table 20.1. This

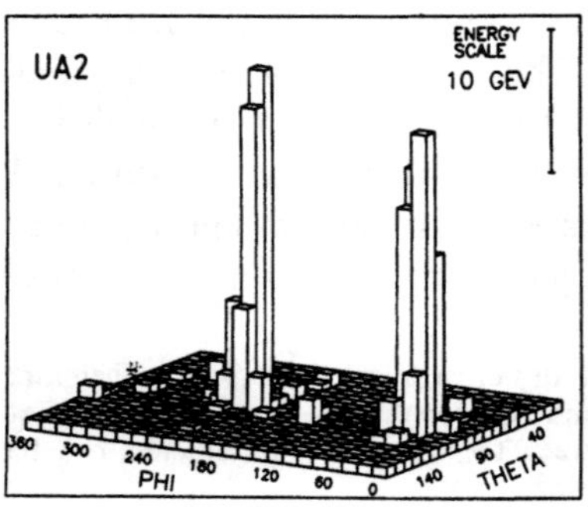

Figure 20.19 Lego plot of an event with two jets with large transverse momentum in a pp̄ collision at center of mass energy $\sqrt{s} = 540$ GeV, observed with the UA2 detector at the storage ring Sp$\bar{\text{p}}$S. Plotted is the observed transverse energy E_T per bin of polar angle ϑ (THETA) and azimuth φ (PHI). The energy scale is marked in GeV. (This plot was made available by E. Putzer of the UA2 collaboration.)

Table 20.1 The cross-sections for the possible parton-parton two-particle reactions to leading order in α_s. The various quark flavors are labeled by i, j $(1 \le i, j \le f)$ [see (19.2)].

Reaction	Cross-section $\frac{\hat{s}^2}{\pi\alpha_s^2}\frac{d\sigma}{d\hat{t}}$
$q^i + q^j \to q^i + q^j\ (i \neq j)$ $q^i + \bar{q}^j \to q^i + \bar{q}^j\ (i \neq j)$	$\frac{4}{9}\frac{\hat{s}^2 + \hat{u}^2}{\hat{t}^2}$
$q^i + q^i \to q^i + q^i$	$\frac{4}{9}\left(\frac{\hat{s}^2 + \hat{u}^2}{\hat{t}^2} + \frac{\hat{s}^2 + \hat{t}^2}{\hat{u}^2}\right) - \frac{8}{27}\frac{\hat{s}^2}{\hat{u}\hat{t}}$
$q^i + \bar{q}^i \to q^j + \bar{q}^j\ (i \neq j)$	$\frac{4}{9}\frac{\hat{t}^2 + \hat{u}^2}{\hat{s}^2}$
$q^i + \bar{q}^i \to q^i + \bar{q}^i$	$\frac{4}{9}\left(\frac{\hat{s}^2 + \hat{u}^2}{\hat{t}^2} + \frac{\hat{t}^2 + \hat{u}^2}{\hat{s}^2}\right) - \frac{8}{27}\frac{\hat{u}^2}{\hat{s}\hat{t}}$
$q^i + \bar{q}^i \to G + G$	$\frac{32}{27}\frac{\hat{u}^2 + \hat{t}^2}{\hat{u}\hat{t}} - \frac{8}{3}\frac{\hat{u}^2 + \hat{t}^2}{\hat{s}^2}$
$G + G \to q^i + \bar{q}^i$	$\frac{1}{6}\frac{\hat{u}^2 + \hat{t}^2}{\hat{u}\hat{t}} - \frac{3}{8}\frac{\hat{u}^2 + \hat{t}^2}{\hat{s}^2}$
$q^i + G \to q^i + G$ $\bar{q}^i + G \to \bar{q}^i + G$	$-\frac{4}{9}\frac{\hat{u}^2 + \hat{s}^2}{\hat{u}\hat{s}} + \frac{\hat{u}^2 + \hat{s}^2}{\hat{t}^2}$
$G + G \to G + G$	$\frac{9}{2}\left(3 - \frac{\hat{u}\hat{t}}{\hat{s}^2} - \frac{\hat{u}\hat{s}}{\hat{t}^2} - \frac{\hat{s}\hat{t}}{\hat{u}^2}\right)$

table also contains the corresponding cross-sections valid for massless partons that are unpolarized with respect to spin and color (Combridge 1977). We use as variables the Lorentz invariants $\hat{s}$, $\hat{t}$, and $\hat{u}$ which, for a general parton reaction

$$a(k_1) + b(k_2) \to a'(k) + b'(k'), \tag{20.35}$$

are defined as follows

$$\begin{aligned} \hat{s} &= (k_1 + k_2)^2, \\ \hat{t} &= (k_1 - k)^2, \\ \hat{u} &= (k_2 - k)^2. \end{aligned} \tag{20.36}$$

The diagram for the reaction $u + \bar{d} \to u + \bar{d}$ is shown in Fig. 20.18. The diagrams for the other reactions listed in Table 20.1 can be readily constructed from the Feynman rules of QCD (Appendix D; see Exercise 20.7).

In order to calculate the cross-section for the production of jets in a hadron–hadron collision, we have to convolute the parton scattering cross-sections with the corresponding parton distribution functions (Chap. 19). As an example we calculate the inclusive cross-section for the production of jets

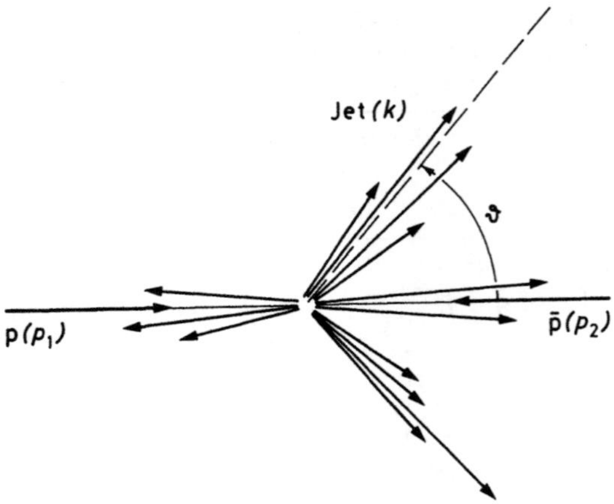

Figure 20.20 Schematic view of a proton–antiproton collision in which two jets with large E_T and two jets along the beam axis are produced. The four-momentum of a jet is denoted by k, and the production angle in the center of mass system by ϑ.

with transverse energy E_T and pseudorapidity $\eta = -\ln \tan(\vartheta/2)$ in $p\bar{p}$ collisions (Fig. 20.20)

$$p(p_1) + \bar{p}(p_2) \to \text{jet}(E_T, \eta) + X. \tag{20.37}$$

In the following we shall work in the center of mass system of the $p\bar{p}$ collisions. The center of mass energy is $\sqrt{s}$. Let a be a parton of the proton with momentum fraction x_1, and b a parton of the antiproton with momentum fraction x_2. We consider a reaction $a + b \to a' + b'$ corresponding to (20.35). The four-momenta in the initial state are then

$$\begin{aligned} k_1 &= x_1 p_1, \\ k_2 &= x_2 p_2. \end{aligned} \tag{20.38}$$

From the fragmentation of the parton a′ we obtain a jet with four-momentum k. We neglect the invariant masses of the jets and of the proton and antiproton. With $E \equiv k^0$ we then have the following relations

$$\begin{aligned} E_T &= E \sin \vartheta, \\ \frac{d^3k}{2k^0} &= \frac{E_T}{2 \sin \vartheta} dE_T \, d\vartheta \, d\varphi = \frac{E_T}{2} dE_T \, d\eta \, d\varphi, \end{aligned} \tag{20.39}$$

where ϑ and φ are the polar and azimuthal angles of $\boldsymbol{k}$. The invariants of the parton–parton reaction are given by

$$\begin{aligned} \hat{s} &= (x_1 p_1 + x_2 p_2)^2 = x_1 x_2 s, \\ \hat{t} &= (x_1 p_1 - k)^2 = -x_1 \sqrt{s} E_T \frac{1 - \cos \vartheta}{\sin \vartheta}, \\ \hat{u} &= (x_2 p_2 - k)^2 = -x_2 \sqrt{s} E_T \frac{1 + \cos \vartheta}{\sin \vartheta}. \end{aligned} \tag{20.40}$$

We now express the transition rate $d\Gamma \equiv dw/T$ (5.54) for the parton–parton reaction in terms of the differential cross-section $d\sigma/d\hat{t}$. From (5.49)–(5.54) we obtain

$$
\begin{aligned}
&\mathrm{d}\Gamma(\mathrm{a}(k_1) + \mathrm{b}(k_2) \rightarrow \mathrm{a}'(k) + \mathrm{b}'(k')) \\
&\quad = \frac{\hat{s}^2}{V\pi k_1^0 k_2^0} \frac{\mathrm{d}^3k\, \mathrm{d}^3k'}{2k^0\, 2k'^0} \delta(k_1 + k_2 - k - k') \frac{\mathrm{d}\sigma}{\mathrm{d}\hat{t}}(\mathrm{ab} \rightarrow \mathrm{a'b'}). \qquad (20.41)
\end{aligned}
$$

The reaction rate for the $\mathrm{p\bar{p}}$ collision (20.37) is obtained by summing up the contributions of all partons a, b with arbitrary momentum fractions x_1, x_2 in the initial state, and all partons a′ in the final state. A summation over b′ is not necessary since, given a, b, and a′, the type of parton b is already determined (see Table 20.1). We thus have

$$
\begin{aligned}
&\mathrm{d}\Gamma(\mathrm{p}(p_1) + \bar{\mathrm{p}}(p_2) \rightarrow \mathrm{jet}(k) + \mathrm{X}) = \sum_{\mathrm{a,b,a'}} \int_0^1 \mathrm{d}x_1\, N_\mathrm{a}^\mathrm{p}(x_1, E_\mathrm{T}^2) \\
&\quad \cdot \int_0^1 \mathrm{d}x_2\, N_\mathrm{b}^{\bar{\mathrm{p}}}(x_2, E_\mathrm{T}^2) \mathrm{d}\Gamma(\mathrm{a}(x_1p_1) + \mathrm{b}(x_2p_2) \rightarrow \mathrm{a}'(k) + \mathrm{b}'(k')), \qquad (20.42)
\end{aligned}
$$

where $N_\mathrm{a}^\mathrm{p}(x_1, E_\mathrm{T}^2)$ and $N_\mathrm{b}^{\bar{\mathrm{p}}}(x_2, E_\mathrm{T}^2)$ are the distribution functions for partons of type a in the proton and for partons of type b in the antiproton, respectively. The resolution scale is set to E_T^2 in the spirit of the discussion in Sect. 19.2. As a result of charge-conjugation invariance it holds that

$$
N_\mathrm{b}^{\bar{\mathrm{p}}}(x, E_\mathrm{T}^2) = N_{\bar{\mathrm{b}}}^\mathrm{p}(x, E_\mathrm{T}^2), \qquad (20.43)
$$

so that all the distribution functions that appear here are known from deep inelastic lepton–proton scattering. A suitable parametrization of these distribution functions is given by Duke 1984 (see Appendix E).

We now insert the transition rate for the parton reaction (20.41), and, since we are interested in the one-jet inclusive cross-section, we integrate over the momentum directions of the parton b′. If we also integrate over the azimuthal angle φ of the observed jet, then from (20.42) and (5.54) we obtain for the scattering cross-section

$$
\begin{aligned}
&\frac{\partial^2 \sigma}{\partial E_\mathrm{T}\, \partial \eta}(\mathrm{p}(p_1) + \bar{\mathrm{p}}(p_2) \rightarrow \mathrm{jet}(E_\mathrm{T}, \eta) + \mathrm{X}) \\
&\quad = \sum_{\mathrm{a,b,a'}} \int_0^1 \mathrm{d}x_1\, x_1 N_\mathrm{a}^\mathrm{p}(x_1, E_\mathrm{T}^2) \int_0^1 \mathrm{d}x_2\, x_2 N_\mathrm{b}^{\bar{\mathrm{p}}}(x_2, E_\mathrm{T}^2) \\
&\qquad \cdot \theta\left(x_1 + x_2 - \frac{2E_\mathrm{T}}{\sqrt{s}\sin\vartheta} \right) \\
&\qquad \cdot \delta\left(x_1 x_2 - \frac{E_\mathrm{T}}{\sqrt{s}} \frac{x_1(1 - \cos\vartheta) + x_2(1 + \cos\vartheta)}{\sin\vartheta} \right) \\
&\qquad \cdot 2E_\mathrm{T} \frac{\mathrm{d}\sigma}{\mathrm{d}\hat{t}}[\mathrm{a}(x_1p_1) + \mathrm{b}(x_2p_2) \rightarrow \mathrm{a}'(k) + \mathrm{b}'(k')]. \qquad (20.44)
\end{aligned}
$$

In Fig. 20.21 we show data for the one-jet inclusive cross-section at center of mass energy $\sqrt{s} = 540$ GeV, together with the result of a numerical evalua-

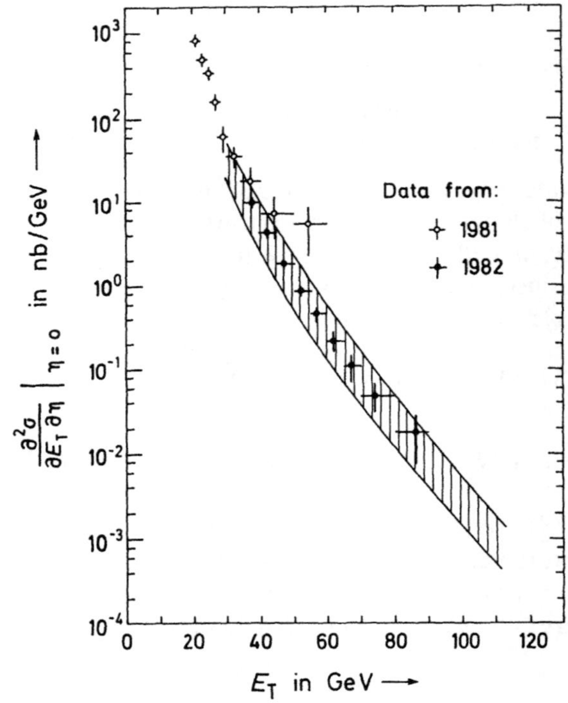

Figure 20.21 The one-jet inclusive cross-section $\partial^2\sigma/\partial E_T\,\partial\eta$ for $\eta = 0$ as a function of E_T for $p\bar{p}$ collisions at a center of mass energy $\sqrt{s} = 540$ GeV. The shaded band corresponds to the QCD prediction (after Arnison 1983f; for the calculation of the theoretical curves see Horgan 1981 and Kunszt 1983).

tion of (20.44). The error band of the theoretical result stems from uncertainties in the experimental determination of the parton distribution functions and of the coupling constant α_s and from the neglect of corrections of higher orders in α_s. One can definitely conclude from this comparison, that the QCD calculation agrees with experiment to within the error limits. The theory also predicts that jets with large E_T should always occur in pairs (Fig. 20.18). In 80–90% of the observed events this is indeed the case (Banner 1982; Arnison 1983f). In the remaining 10–20% of events more than two jets with large E_T appear. This can be understood as a manifestation of processes of higher order in α_s.

This concludes our discussion of the production of jets in hadron–hadron collisions. As a testing ground for QCD, these reactions should be very important in the future.

Exercises

20.1 Calculate the minimal value of the quantity thrust (20.6) for planar events with arbitrarily many hadrons and for events with only three hadrons. Neglect the hadron masses for this consideration.

20.2 Calculate the cross-section for the inclusive production of a hadron in deep inelastic muon–proton scattering (20.10). What do the expressions for $\langle N_q \rangle$ and $\langle N_{\bar{q}} \rangle$ (20.12) look like if the average value over a certain region in x, Q^2, and ϑ is taken.

20.3 Calculate the cross-section for the reaction $e^+ + e^- \to q + \bar{q} + G$ to lowest order in α_s. Verify the results in (20.17) and (20.15). What are the corresponding equations for scalar gluons $\tilde{G}$ where the interaction Lagrange density is taken as $\mathscr{L}'(x) = \tilde{g}\bar{q}(x)q(x)\tilde{G}(x)$?

20.4 Consider the scattering of a heavy quark from a heavy antiquark. According to the rules in Appendix D, to lowest order in α_s only the one-gluon exchange contributes. In the limit of very slow quarks and antiquarks, the scattering matrix yields the expression for the nonrelativistic potential. This can be seen by considering, for example, a wave packet for the antiquark that is concentrated at the origin. Then the antiquark can be regarded as a fixed scattering center and comparing with (5.14) we can read off the expression for the potential. Show that for a $q\bar{q}$ color singlet state this potential is given by

$$V(r) = -\frac{4}{3}\frac{\alpha_s}{r}. \tag{20.45}$$

The strong coupling strength α_s depends, however, on the momentum scale of the process considered. What would you suggest should be used as momentum scale for α_s in (20.45)?

20.5 The quark–antiquark potential suggested by Richardson reads

$$V(r) = \int \frac{d^3q}{(2\pi)^3} e^{i\boldsymbol{q}\boldsymbol{x}} \tilde{V}(\boldsymbol{q}), \tag{20.46}$$

where

$$\tilde{V}(\boldsymbol{q}) = -\frac{4}{3}\frac{12\pi}{33-2f}\frac{4\pi}{\boldsymbol{q}^2 \ln(1+\boldsymbol{q}^2/\Lambda^2)}. \tag{20.47}$$

Show that this potential possesses the limiting properties given in (20.22), (20.23).

20.6 Calculate to lowest order of α_s the decay of a $^3S_1$ quarkonium state into three gluons, using the results of Exercise 13.1.

20.7 Calculate the cross-sections for the reactions given in Table 20.1 to leading order in α_s according to the Feynman rules of QCD.

PART IV

The Electroweak Interaction

CHAPTER 21

From β-Decay to the W-Boson. A Historical Survey

21.1 The Early Days, the Neutrino Hypothesis, Four-Fermion Coupling

The history of weak interactions could be said to have begun on the 1st of March 1896. This was the day on which Henri Becquerel discovered radioactivity (Becquerel 1896). We today know that the majority of the rays that he discovered stem from the β-decay of heavy nuclei. The distinction between α- and β-rays was first made by Rutherford in 1899. For a long time the nature and the origin of the radioactive rays remained a puzzle. This is not surprising, since it was not until 1911 that Rutherford discovered the atomic nucleus (Rutherford 1911, 1911a). It was Niels Bohr who first attributed the origin of β-rays to the nucleus (Bohr 1913). James Chadwick, who later discovered the neutron, showed in 1914 that the β-rays possess a continuous energy spectrum (Chadwick 1914). This fact actually concealed a major problem although this was only demonstrated 13 years later by Ellis and Wooster. They studied a β-transition leading from an initial nucleus to a final nucleus of well-defined energy:

$$^{210}_{83}\mathrm{Bi} \xrightarrow{\beta^-} {}^{210}_{84}\mathrm{Po}.$$

(In older literature the notation RaE is used for the isotope $^{210}_{83}$Bi.). They found that the energy release in the decay, which was measured with a calorimeter, corresponded not to the *maximum*, but to the *average* energy of the β-rays. The maximum β-energy was known to be $\Delta E \cong 1050$ keV, and the average energy $\langle E_\beta \rangle \cong 390$ keV. The average energy per decay, measured in the calorimeter, was $\langle E_{\mathrm{cal}} \rangle \cong 350$ keV (Ellis 1927).

The consequences seemed to be clear. Either energy was not conserved or the energy was being carried away by another particle that was not absorbed in the calorimeter. Niels Bohr supported the first of these alternatives. It was another three years before W. Pauli took the daring step of postulating the existence of a new particle, today known as the neutrino. He did this in 1930 in a famous letter to a group of physicists who were gathered for a conference in Tübingen. In the same letter he sent his apologies for not being able to attend in person—the conference unfortunately conflicted with a ball taking place in Zürich! (Pauli 1964). The neutrino that he proposed was supposed to have spin $\frac{1}{2}$. (Pais (1977) gives an account of many of the historical events of this time.)

The next milestone in the history of the weak interaction is the monumental work by Fermi, in which he developed the first theory of β-decay (Fermi 1933, 1934). His theory is constructed along the same lines as quantum electrodynamics. At that time, and for a long time afterwards, the proton and the newly discovered neutron were regarded as elementary particles. The interaction of the proton with the electromagnetic field is then given, according to QED (9.9), by

$$\mathbf{H}'_{\mathrm{em}} = e \int \mathrm{d}^3x\, \bar{\mathbf{p}}(x)\gamma^\mu \mathbf{p}(x)\mathbf{A}_\mu(x). \tag{21.1}$$

Here $\mathbf{A}_\mu(x)$ is the electromagnetic vector potential and $\mathbf{p}(x)$ is the Dirac field operator of the proton. In order to describe the β-decay of nuclei, Fermi made an ansatz for the elementary β-decay of the neutron

$$\mathrm{n} \to \mathrm{p} + \mathrm{e}^- + \bar{\nu}_\mathrm{e} \tag{21.2}$$

which was motivated by the structure of (21.1). The vector potential $\mathbf{A}_\mu(x)$ was replaced by a vector field constructed from the electron and neutrino fields. The transition of a neutron into a proton could easily be built in by regarding the proton and the neutron, as Heisenberg did, as two isospin states of one and the same particle. The interaction postulated by Fermi reads[1]

$$\mathbf{H}'_\beta = G \int \mathrm{d}^3x (\bar{\mathbf{p}}(x)\gamma^\mu \mathbf{n}(x))(\bar{\mathbf{e}}(x)\gamma_\mu \nu(x)) + \mathrm{h.c.} \tag{21.3}$$

In this G was a new fundamental constant, which is known today as the Fermi constant. Its dimensions are $(\mathrm{mass})^{-2}$. By comparing his theory with the experimental data for β-decay, Fermi concluded that the neutrino must have zero, or very small, mass. He also obtained an estimate for the coupling constant G:

$$G \cong 4 \cdot 10^{-50}\,\mathrm{cm}^3\,\mathrm{erg} \mathrel{\hat{=}} 0.3 \cdot 10^{-5}\,\mathrm{GeV}^{-2},$$

which agrees in order of magnitude with the present-day value of

$$G \cong 1.1 \cdot 10^{-5}\,\mathrm{GeV}^{-2}.$$

[1] Here, and in the following, we denote the field operators of the particles p, n, e, ... (with Latin symbols) by $\mathbf{p}(x)$, $\mathbf{n}(x)$, $\mathbf{e}(x)$, ...

Direct experimental proof of the existence of the antineutrino as a free particle was only obtained much later by Reines and co-workers (Reines 1953, 1956, 1959). They made use of a large reactor as a source of antineutrinos. The products formed when nuclei are split are rich in neutrons and often undergo β-decay. The antineutrinos that are produced in this decay leave the reactor and can be detected outside. A large tank of water mixed with various salts was used as detector to observe the inverse reaction to β-decay

$$\bar{\nu}_e + \mathrm{p} \rightarrow \mathrm{n} + \mathrm{e}^+. \tag{21.4}$$

Fermi's ansatz for the weak interaction as a local coupling of four spin-$\frac{1}{2}$ fields was extremely successful. With some additions and modifications, which we shall discuss explicitly below, the four-fermion coupling describes nearly all experiments made so far in the field of weak interactions!

First, however, we continue to trace the historical development. It was very soon recognized that Fermi's ansatz needed to be generalized in order to describe all observed β-decay processes (Gamov 1936). If we remain with the four-fermion coupling, and for simplicity exclude couplings involving derivatives then we face the problem of determining the number of ways in which we can form a Lagrange density $\mathscr{L}'$ of the weak interaction from the four fermion fields. According to the rules of Sect. 3.4, we obtain $\mathbf{H}'_\beta$ as

$$\mathbf{H}'_\beta = -\int d^3x\, \mathscr{L}'(x).$$

We must require that the Lagrange density behaves like a scalar under Lorentz transformations. Our problem can thus easily be solved with the help of the Dirac covariants (Sect. 4.3). The electron and neutrino fields together can give a scalar, vector, tensor, axial vector, or pseudoscalar current (4.80) of the general form

$$\bar{\mathbf{e}}(x)\mathbf{M}\nu(x),$$

where

$$\mathbf{M} = \mathbb{1}, \gamma^\mu, \sigma^{\mu\nu}, \gamma^\mu\gamma_5, \gamma_5. \tag{21.5}$$

The same situation holds for the proton-neutron pair. One of the candidates for the Lagrange density $\mathscr{L}'$ is Fermi's ansatz, which corresponds to contracting the p–n vector with the e–ν vector:

$$\mathscr{L}'(x) \propto (\bar{\mathbf{p}}(x)\gamma^\mu\mathbf{n}(x))(\bar{\mathbf{e}}(x)\gamma_\mu\nu(x)). \tag{21.6}$$

The most general Lagrange density that behaves as a scalar under proper orthochronous Lorentz transformations is obtained by contracting the pn scalar with the eν scalar or pseudoscalar, the pn vector with the eν vector or axial vector, etc.:

$$\begin{aligned}\mathscr{L}'(x) = \sum_{j=1}^{5} \{ & g_j\bar{\mathbf{p}}(x)\mathbf{M}_j\mathbf{n}(x)\bar{\mathbf{e}}(x)\mathbf{M}'_j\nu(x) \\ & + g'_j\bar{\mathbf{p}}(x)\mathbf{M}_j\mathbf{n}(x)\bar{\mathbf{e}}(x)\mathbf{M}'_j\gamma_5\nu(x)\} + \text{h.c.}\end{aligned} \tag{21.7}$$

Here $j = 1, \ldots, 5$, enumerate the possibilities

$$\mathbf{M}_j \otimes \mathbf{M}'_j = \mathbb{1} \otimes \mathbb{1}, \gamma^\mu \otimes \gamma_\mu, \sigma^{\mu\nu} \otimes \sigma_{\mu\nu}, \gamma^\mu\gamma_5 \otimes \gamma_\mu\gamma_5, \gamma_5 \otimes \gamma_5, \tag{21.8}$$

and g_j, g'_j are arbitrary complex coupling constants. This corresponds to 20 real parameters. One might ask whether a different combination of the fields, for example,

$$(\bar{\mathbf{p}}(x)\nu(x))(\bar{\mathbf{e}}(x)\mathbf{n}(x))$$

yields anything new. It turns out that this coupling can also be brought into the form (21.7) by means of a so-called Fierz transformation (Fierz 1937). The explicit form of the Fierz transformation is given in Appendix F.

We now discuss the behavior of the Lagrange density (21.7) under parity and time-reversal transformations. Using the formulas of Sect. 4.5, one readily finds that time-reversal invariance demands real coupling constants g_j and g'_j. If we further require parity invariance, then couplings of the form (scalar ⊗ pseudoscalar), (vector ⊗ axial vector), etc., are excluded. In other words, all constants g'_j must vanish.

For a long time it was believed that parity is a conserved quantum number also for the weak interaction, and thus g'_j was set to zero. The combination of vector (V), axial vector (A), scalar (S), tensor (T), and pseudoscalar (P) couplings that constitute the weak interaction Hamiltonian can be experimentally determined by studying β-decays. As a consequence of some incorrect experimental results, it initially seemed that tensor currents (T) played an important role.

21.2 Parity Violation and the (V − A) Theory

It was the discovery of parity violation and the subsequent theoretical and experimental developments which clarified the structure of the interaction responsible for β-decay. The impetus for this came, however, not from β-decay itself, but from the physics of strange particles. Two particles had been discovered, which were then known as τ and ϑ, that decayed into three and two pions, respectively:

$$\begin{aligned} \tau^+ &\to \pi^+\pi^+\pi^-, \\ \vartheta^+ &= \pi^+\pi^0. \end{aligned} \tag{21.9}$$

The experimental analysis showed that the final state of the τ-decay had negative parity, whereas that of the ϑ-decay had positive parity. Curiously, however, the experiments indicated that τ and ϑ both have the same mass and lifetime. This puzzle was brilliantly solved by Lee and Yang, who raised the possibility that parity is not conserved in weak processes (Lee 1956). The τ- and ϑ-particles then simply correspond to two different decay channels of one and the same particle. Today this is called the K^+-meson:

$$K^+ \to \pi^+\pi^+\pi^-, \pi^+\pi^0. \tag{21.10}$$

Lee and Yang were also able to show that none of the previous experiments on β-decay could exclude a possible parity violation. They suggested some new experiments to clarify the question of parity violation. It indeed turned out that in β-decay, and also in other weak processes, parity is not conserved. Nature distinguishes between right-handed and left-handed systems (Wu 1957; Garwin 1957).

These developments rapidly helped to clarify the structure of the β-decay interaction. The Lagrange density for β-decay is given by a product of a V − A current formed by the electron and the neutrino fields with V and A currents formed by the proton and neutron fields:

$$\mathscr{L}'_\beta(x) = -\frac{G_\beta}{\sqrt{2}}\bar{\mathsf{p}}(x)\gamma^\lambda\left(1 - \frac{g_A}{g_V}\gamma_5\right)\mathsf{n}(x)\bar{\mathsf{e}}(x)\gamma_\lambda(1-\gamma_5)\nu(x) + \text{h.c.} \tag{21.11}$$

The corresponding Hamiltonian of the interaction is

$$\hat{H}'_\beta = -\int d^3x\, \mathscr{L}'_\beta(x).$$

The parameters G_β and g_A/g_V are coupling constants that have to be determined experimentally. Their currently accepted values are

$$\begin{aligned} G_\beta &= (1.14730 \pm 0.00064)\cdot 10^{-5}\,\text{GeV}^{-2}, \\ g_A/g_V &= 1.255 \pm 0.006. \end{aligned} \tag{21.12}$$

We see that Fermi's original ansatz (21.3) came very close to the true description, although it contained no axial-vector currents and no parity violating terms. The Fermi constant that appears in (21.11) has been given the index β for reasons that will be discussed in detail below. The form of the lepton current was postulated on the basis of theoretical considerations by Salam, Landau and Lee and Yang (Salam 1957; Landau 1956a, b; Lee 1957). The combination of vector and axial vector for the neutron–proton current was suggested by Feynman and Gell-Mann, Sudarshan and Marshak, and Theis (Feynman 1958; Sudarshan 1957; Theis 1958). These authors postulated a pure (V − A) structure, not only for the lepton currents, but even for the hadron currents. As we shall see, at the quark level this is actually the case.

Readers who would like to know more about the history of the weak interaction up to 1957 are recommended to read the article written at that time by W. Pauli (Pauli 1984).

We will now take a closer look at the lepton current. This current is invariant when the neutrino field (or electron field) is multiplied by $-\gamma_5$. Making the replacement

$$\nu(x) \to (-\gamma_5)\nu(x), \tag{21.13}$$

which today is known as a chiral transformation, we have

$$\bar{\mathsf{e}}(x)\gamma_\lambda(1-\gamma_5)\nu(x) \to \bar{\mathsf{e}}(x)\gamma_\lambda(1-\gamma_5)\nu(x).$$

Invariances under γ_5 transformations were first considered by Stech and Jensen, although only when applied to the electron and neutrino fields simultaneously (Stech 1955). It is easy to see that fermion mass terms are not invariant under γ_5 transformations. It was an important step to consider an invariance which is only present in the theory if mass terms are neglected. In the subsequent development of weak interaction theory this point of view has played an essential role as we shall see.

The strict (V − A) form of the lepton current has drastic consequences for the neutrinos, which we wish to regard as massless. This we will now show with the help of the above γ_5 transformation. We consider the Dirac equation without mass term, first for a classical Dirac spinor $\psi(x)$:

$$i\gamma^\mu \partial_\mu \psi(x) = 0. \tag{21.14}$$

We readily see that with $\psi(x)$ also $\gamma_5\psi(x)$ is a solution of this Dirac equation:

$$i\gamma^\mu \partial_\mu \gamma_5 \psi(x) = -\gamma_5 i\gamma^\mu \partial_\mu \psi(x) = 0. \tag{21.15}$$

For zero mass we can thus classify the solutions according to their behavior upon multiplication with γ_5. Since

$$(\gamma_5)^2 = 1$$

the only eigenvalues are ± 1. The corresponding solutions are denoted by $\psi_\pm(x)$:

$$\gamma_5 \psi_\pm(x) = \pm\psi_\pm(x). \tag{21.16}$$

As we will now show, these solutions represent particles or antiparticles whose spins are parallel or antiparallel to the momentum direction. The eigenvalue of the spin in the momentum direction is known as the *helicity*. States with positive and negative helicity can be represented pictorically as right- and left-handed screws (cf. Fig. 2.7).

Let us look at the plane wave solutions of the Dirac equation for zero mass. The general form of a negative frequency solution in terms of an arbitrary two-component spinor ξ is [cf. (4.38)]:

$$\psi(x) = e^{-ipx}\, u(p), \tag{21.17}$$

where

$$u(p) = \sqrt{|\boldsymbol{p}|}\begin{pmatrix} \xi \\ \boldsymbol{\sigma}\cdot\hat{\boldsymbol{p}}\xi \end{pmatrix}, \qquad \hat{\boldsymbol{p}} = \frac{\boldsymbol{p}}{|\boldsymbol{p}|}. \tag{21.18}$$

The spin eigenstates with respect to the direction of the momentum are obtained by choosing for ξ the eigenvectors of $\boldsymbol{\sigma}\hat{\boldsymbol{p}}$. They are denoted by $\xi_{R,L}$ and obey

$$(\boldsymbol{\sigma}\hat{\boldsymbol{p}})\xi_{R,L} = \pm\xi_{R,L}. \tag{21.19}$$

The Dirac spinors with $\xi = \xi_{R,L}$ in (21.18) are denoted by $u_\pm(p)$. It is easily verified that

$$u_\pm(p) = \sqrt{|\boldsymbol{p}|}\begin{pmatrix} \xi_{R,L} \\ \pm\xi_{R,L} \end{pmatrix}, \qquad \gamma_5 u_\pm(p) = \pm u_\pm(p). \tag{21.20}$$

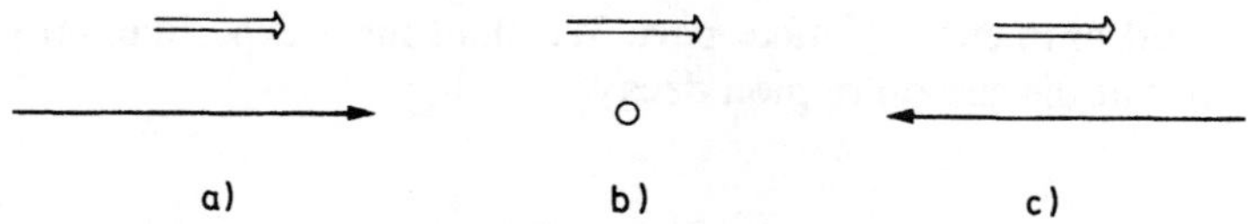

Figure 21.1 Applying a Lorentz boost to transform a massive particle with helicity $+\frac{1}{2}$ (a) via its rest frame (b) to a frame where it moves in the opposite direction (c) and thus has reversed helicity. The single arrows represent the momenta and the double arrows the spin direction.

The spinors $u_\pm(p)$ are thus eigenstates of γ_5. For the positive frequency solutions we find from (4.41) that the eigenstates $v_\pm(p)$ of γ_5 have the form

$$v_\pm(p) = -\sqrt{|\boldsymbol{p}|}\begin{pmatrix} \pm\varepsilon\xi^*_{\mathrm{L,R}} \\ \varepsilon\xi^*_{\mathrm{L,R}} \end{pmatrix}, \tag{21.21}$$

where $\pm$ again refer to the convention

$$\gamma_5 v_\pm(p) = \pm v_\pm(p).$$

According to our discussion in Sect. 4.4, the solution involving u_- corresponds to a particle whose spin points in the direction opposite to the direction of motion; it is thus a *left-handed* particle. The spinor v_- contains, according to (21.21), the quantity ξ^*_{R}. The associated solution therefore corresponds to an antiparticle with spin in the direction of motion, i.e., to a right-handed antiparticle. For the solutions involving u_+ and v_+ with γ_5-eigenvalue $+1$, exactly the reverse holds.

We also want to show briefly that for massless particles, the helicity is a Lorentz-invariant property, but not for particles with finite mass. A massive particle can be "caught up" with, that is, transformed to its rest frame, or even "overtaken" so that it moves in the opposite direction. In this sequence the helicity changes sign if the transformation is a pure Lorentz boost (Fig. 21.1). A *massless* particle, on the other hand, can never be overtaken since it travels at the speed of light and therefore its helicity is conserved.

We now return to the Lagrange density for β-decay (21.11) and consider the neutrino field operator expanded in terms of creation and annihilation operators. In place of the basis u_s, v_s of Chap. 4, we make use of the helicity basis $u_\pm$, $v_\pm$. Clearly the factor $(1-\gamma_5)$ then projects out u_- and v_- and we obtain:

$$\begin{aligned}\tfrac{1}{2}(1-\gamma_5)\nu(x) &\equiv \nu_{\mathrm{L}}(x) \\ &= \int \frac{d^3p}{(2\pi)^3}\frac{1}{2p_0}\{\mathrm{e}^{ipx}\,v_-(p)\mathbf{b}^\dagger_{\mathrm{R}}(p) + \mathrm{e}^{-ipx}\,u_-(p)\mathbf{a}_{\mathrm{L}}(p)\}. \end{aligned}\tag{21.22}$$

Here we have denoted the creation and annihilation operators for right-handed antineutrinos and left-handed neutrinos by $\mathbf{b}^\dagger_{\mathrm{R}}$ and $\mathbf{a}_{\mathrm{L}}$. The corresponding field operator is conventionally denoted by ν_{L}. Thus in β-decay, according to the Lagrange density (21.11), only left-handed neutrinos and right-handed antineutrinos take part. We shall meet experimental proof of this fact in our discussion of pion decay.

21.3 The Universality of the Weak Interaction and the Cabibbo Theory

We now wish to discuss another property of the weak interaction, namely, its *universality*. The muon was discovered in 1937 and was at first believed to be the Yukawa particle responsible for nuclear forces. In about 1947, however, it became clear that the muon did not take part in the strong interaction and that it decayed via a weak process. The main decay mode of the μ^- is

$$\mu^- \to e^- + \bar{\nu}_e + \nu_\mu .$$

Here we have introduced two types of neutrino, ν_e and ν_μ. In 1962, the existence of two types of neutrino was demonstrated experimentally in Brookhaven, USA (Danby 1962).

From the lifetime of the muon, Puppi (1948) inferred an approximate equality of the coupling constants for μ-decay and for β-decay, and thus an approximate universality of the weak interaction. Today it is believed that μ-decay is described by the following Lagrange density

$$\mathscr{L}'_\mu(x) = -\frac{G_\mu}{\sqrt{2}}\bar{\nu}_\mu(x)\gamma^\lambda(1-\gamma_5)\mu(x)\bar{e}(x)\gamma_\lambda(1-\gamma_5)\nu_e(x) + \text{h.c.} \quad (21.23)$$

The experimental value of the coupling constant G_μ has in fact turned out to be almost identical to the value of G_β in (21.12). A comprehensive analysis yields (Bég 1982)

$$\begin{aligned} G_\mu &= (1.16632 \pm 0.00002)\cdot 10^{-5}\,\text{GeV}^{-2}, \\ \frac{G_\beta}{G_\mu} &\cong 0.98. \end{aligned} \quad (21.24)$$

The fact that G_μ and G_β are so close to each other may seem surprising at first since the muon is a point particle, whereas the neutron is a hadron with an inner structure and an extent of about 10^{-13} cm. However, we are already familiar with an analogous situation concerning the electric charge. The muon, electron and proton all have the same charge, except in sign, although the charge distributions are completely different from one another. This analogy led to the idea that the hadronic vector current in β-decay has similar properties as the electromagnetic current, i.e., to the hypothesis of conserved vector current or CVC hypothesis (Gershtein 1956; Feynman 1958). These developments will not be discussed in greater detail here but we shall instead address the problem of universality directly from the modern standpoint.

In the 1950s and 1960s a host of new hadrons were discovered and among these were many that decay via the weak interaction. We will give a few examples.

(i) Weak decays of hadrons in which leptons occur (semileptonic processes)

$$\begin{aligned}
\pi^+ &\to \mu^+ + \nu_\mu, \quad e^+ + \nu_e, \\
\pi^+ &\to \pi^0 + e^+ + \nu_e, \\
K^+ &\to \mu^+ + \nu_\mu, \quad e^+ + \nu_e, \\
\Lambda &\to p + e^- + \bar{\nu}_e, \\
\Sigma^+ &\to \Lambda + e^+ + \nu_e.
\end{aligned} \tag{21.25}$$

(ii) Weak decays of hadrons into hadrons (nonleptonic processes)

$$\begin{aligned}
K^+ &\to \pi^+ + \pi^+ + \pi^-, \quad \pi^+ + \pi^0, \\
K^0 &\to \pi^+ + \pi^-, \\
\Lambda &\to p + \pi^-, \quad n + \pi^0, \\
\Sigma^+ &\to p + \pi^0, \quad n + \pi^+.
\end{aligned} \tag{21.26}$$

The question now arises as to whether, for every new hadron, we ought to introduce a new term into the β-decay interaction containing the corresponding hadron fields. This would obviously yield a vast number of coupling constants. In fact, it was realized quite early on, that in the weak interaction, only a few hadron *currents* with particular quantum numbers, play a role (Feynman 1958). At that time, however, the existence and properties of these currents had to be postulated since no fundamental theory of hadrons had been developed. The real progress in our understanding of weak processes involving hadrons came only with the idea of regarding hadrons as bound states of constituents, i.e., with the quark hypothesis. It was then possible to attribute the β-decay of hadrons to elementary quark reactions in a very similar way as the β-decay of a nucleus can be described in terms of the decay of the bound nucleons. According to this picture, the elementary quark reaction that is responsible, for example, for the β-decay of the neutron, is the transition of a d-quark into a u-quark with the emission of an electron–antineutrino pair

$$d \to u + e^- + \bar{\nu}_e. \tag{21.27}$$

In the decay of the neutron it is a bound d-quark that undergoes this reaction

$$n \sim (ddu) \to (p \sim (duu)) + e^- + \bar{\nu}_e. \tag{21.28}$$

The quark process (21.27), after crossing the u-quark to the initial state, also describes the decay $\pi \to e^- + \bar{\nu}_e$,

$$\pi^- \sim (\bar{u}d) \to e^- + \bar{\nu}_e. \tag{21.29}$$

Today, the Lagrange density for these β-processes is well known from detailed studies of decays and, in particular, from neutrino reactions. It has the form

$$\begin{aligned}
\mathscr{L}'_\beta(x) = &-\frac{G_\beta}{\sqrt{2}}\bar{\mathbf{u}}(x)\gamma^\lambda(1-\gamma_5)\mathbf{d}(x) \\
&\cdot[\bar{\mathbf{e}}(x)\gamma_\lambda(1-\gamma_5)\nu_e(x) + \bar{\mu}(x)\gamma_\lambda(1-\gamma_5)\nu_\mu(x)] + \text{h.c.}
\end{aligned} \tag{21.30}$$

Here we have also included the coupling to the muon and its neutrino. It turns out that the muonic and electronic currents couple with the same strength (μ–e universality).

The quark picture thus has enabled us to write down a simple interaction which describes a variety of weak hadronic decays and which can directly be compared with the interaction for muon decay (21.23). Apart from a small discrepancy in the coupling constants (21.24) the analogy is, in fact, complete.

Our interaction (21.30) cannot represent the whole story since it does not allow for changes of the strangeness. For example, the decay of the K^+-meson

$$K^+ \to \mu^+ + \nu_\mu \tag{21.31}$$

cannot be explained with the elementary process (21.27). Following a suggestion of Cabibbo, the decay of strange particles is taken care of—in present-day terminology—by replacing the d-quark field in (21.30) by a linear combination of d- and s-quark fields:

$$\mathbf{d} \to \mathbf{d}' = \cos \vartheta_C \mathbf{d} + \sin \vartheta_C \mathbf{s}. \tag{21.32}$$

Here ϑ_C is the Cabibbo angle, which must be determined experimentally (Cabibbo 1963). (Cabibbo, of course, formulated the above hypothesis in terms of the currents; indeed, the quark model was only introduced in 1964.) The Cabibbo-rotated quark d′ should then, in its decay, couple with the same strength as the muon. Thus we can write

$$\mathscr{L}'(x) = -\frac{G_\mu}{\sqrt{2}} \bar{\mathbf{u}}(x)\gamma^\lambda(1-\gamma_5)[\cos \vartheta_C \mathbf{d}(x) + \sin \vartheta_C \mathbf{s}(x)]$$
$$\cdot [\bar{\mathbf{e}}(x)\gamma_\lambda(1-\gamma_5)\nu_e(x) + \bar{\mu}(x)\gamma_\lambda(1-\gamma_5)\nu_\mu(x)] + \text{h.c.} \tag{21.33}$$

As a new elementary process, we now have the transition of an s-quark into a u-quark with the emission of a lepton pair and with an amplitude proportional to $\sin \vartheta_C$

$$\mathrm{s} \to \mathrm{u} + \ell^- + \bar{\nu}_\ell \qquad (\ell = \mathrm{e}, \mu). \tag{21.34}$$

This coupling allows, for example, the following K-decay:

$$K^- \sim (\mathrm{s\bar{u}}) \to \mu^- + \bar{\nu}_\mu. \tag{21.35}$$

By means of the Cabibbo ansatz (21.33) a connection was established between the decay of strange particles and the discrepancy in the values of the constants G_β and G_μ. This connection is well supported by experiment. From (21.24) we obtain:

$$\frac{G_\beta}{G_\mu} = \cos \vartheta_C \cong 0.98. \tag{21.36}$$

The analysis of K-meson and strange baryon decays yields, as will be shown in Sect. 23.4,

$$\begin{aligned} &\sin \vartheta_C \cong 0.21, \\ &\cos \vartheta_C = \sqrt{1 - \sin^2 \vartheta_C} \cong 0.98, \end{aligned} \tag{21.37}$$

which is in very good agreement with (21.36).

21.4 Neutral Currents, the W- and Z-Bosons, and the Glashow–Weinberg–Salam Theory

The Cabibbo form of the interaction (21.33) was very successful. It described all semileptonic reactions and decay processes known up to 1973. Whether it was equally successful in describing nonleptonic decay processes became a topic of lively debate. Since reliable calculations of these processes are not easy to perform, even nowadays, this question is still an open one. In 1973, however, the so-called *neutral currents* were discovered in neutrino reactions (Hasert 1973a, b, 1974). The reactions observed were a purely leptonic process

$$\bar{\nu}_\mu + e^- \rightarrow \bar{\nu}_\mu + e^-, \tag{21.38}$$

and also semileptonic processes which, in the quark picture, corresponded to the elastic scattering of a neutrino or an antineutrino form a quark,

$$\begin{aligned} \nu_\mu + u &\rightarrow \nu_\mu + u, \\ \nu_\mu + d &\rightarrow \nu_\mu + d, \\ \bar{\nu}_\mu + u &\rightarrow \bar{\nu}_\mu + u, \\ \bar{\nu}_\mu + d &\rightarrow \bar{\nu}_\mu + d \end{aligned} \tag{21.39}$$

and the analogous reactions for antiquarks. In these reactions, the quark charge and, more generally, the flavor of the participating quarks and leptons, remains unchanged. Thus one speaks of reactions via neutral currents.

In contrast to this, in the reactions considered previously (21.27), (21.34) the quark charge is changed. These are correspondingly referred to as reactions via charged currents. A short time after the discovery of neutral currents, the J/ψ-particle (Aubert 1974; Augustin 1974) and charmed particles (Cazzoli 1975; Goldhaber 1976) were discovered. The quark picture had thus gained its fourth quark, the c-quark.

Both neutral currents and charm were, in fact, predicted by the theory and their experimental discovery led to major progress in our understanding of the forces that act between elementary particles. In particular, it was possible to arrive at a unified picture of the weak and electromagnetic interactions.

The starting point of the theoretical considerations dates back over many decades. In the weak interaction we have a current–current coupling of the form that was originally considered by Fermi (21.3). In QED, on the other hand, the current couples to a Bose field (21.1). It was Yukawa who first discussed the possibility that the weak interaction could likewise be described by the coupling of a current to a boson. This boson must, in contrast to QED, be massive (Yukawa 1935). His argument ran approximately as follows: Consider electron–electron scattering. According to QED, this is described by the exchange of a photon (Fig. 21.2(a)). For nonrelativistic electrons, the one-photon exchange yields the Coulomb potential whose range, as is well known, is infinite

$$V_{\mathrm{Coul.}}(r) = \frac{e^2}{4\pi r}. \tag{21.40}$$

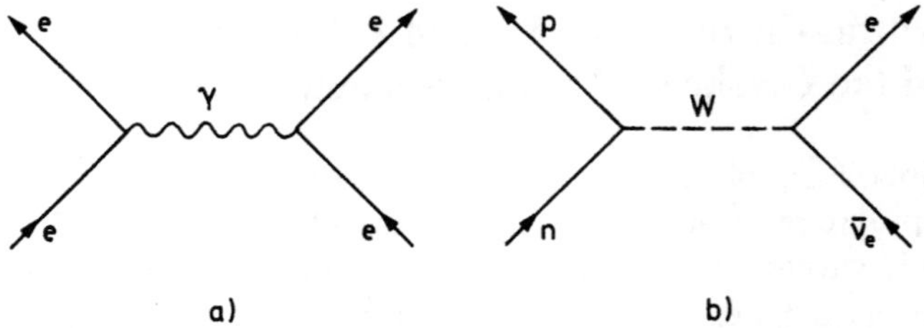

Figure 21.2 Diagrams for electron–electron scattering via the exchange of a photon (a), and for the β-decay of the neutron via the exchange of a W-boson (b).

For the weak interaction we replace the photon by a massive boson, which we will call the W-boson (Fig. 21.2(b)), and then, in place of the Coulomb potential, we obtain the Yukawa potential

$$V_{\mathrm{Yuk.}}(r) = \frac{g^2}{4\pi r} \mathrm{e}^{-m_{\mathrm{W}} r}. \tag{21.41}$$

Here g is again a dimensionless coupling constant analogous to the electromagnetic coupling constant e. If the W-boson is made very heavy, then we can approximate the Yukawa potential by a δ-function:

$$\frac{g^2}{4\pi r} \mathrm{e}^{-m_{\mathrm{W}} r} \xrightarrow[m_{\mathrm{W}} \to \infty]{} \frac{g^2}{m_{\mathrm{W}}^2} \delta^3(\boldsymbol{x}). \tag{21.42}$$

The diagram in Fig. 21.2(b) then becomes a point coupling of two currents, which was exactly the case considered by Fermi. The Fermi constant is given by

$$G \propto \frac{g^2}{m_{\mathrm{W}}^2}. \tag{21.43}$$

If we further assume that the fundamental coupling for the W-boson and for the photon is of the same magnitude,

$$g^2 \cong e^2 = 4\pi\alpha, \tag{21.44}$$

then we also obtain an estimate for the mass of the heavy boson:

$$m_{\mathrm{W}} \cong \sqrt{4\pi\alpha/G} \cong 90\ \mathrm{GeV}. \tag{21.45}$$

It is in fact known today that the W-boson has a mass of about 80 GeV. The importance of Yukawa's idea is not in any way diminished by the fact that he wanted to describe both nuclear forces and the weak force in terms of the exchange of the same boson. The extent of his insight can be judged by the fact that the search for the bosons of the weak interaction was only successful in 1983 with experiments performed at the proton–antiproton storage ring Sp$\bar{\mathrm{p}}$S at CERN, as will be described in detail.

For a long time, the development of the idea that bosons are responsible for the weak interaction remained a rather esoteric occupation for theoreticians. It is noteworthy that as early as 1938, O. Klein wrote down a theory of β-processes that made use of the same self-coupled vector fields that we today know as Yang–Mills fields (Klein 1939). The modern version of the theory was formulated in the work by Glashow (1961), Weinberg (1967), and Salam (1968). Many other physicists also contributed important ideas to this work. The inclusion of quarks in the theory was achieved by Glashow, Iliopuolos, and Maiani (Glashow 1970; Weinberg 1972a). It is due to these developments that we today have a renormalizable theory that simultaneously describes both the weak and the electromagnetic interactions; this is the theory of quantum flavor dynamics (QFD). The theory of the strong interaction, QCD, has been introduced in Part III. QFD and QCD together form a theory of all known interactions with the exception of gravity. We shall call this theory the *standard model* of elementary particle physics. One of the aims of this part of the book is to present the standard model and to illustrate its ability to describe the experimental results.

Exercise

21.1 Derive the conditions on the coupling constants g_j, g_j' in (21.7) that result from assuming invariance of the weak interaction under the parity, charge-conjugation, and time-reversal transformations, respectively. Show that CPT invariance is always valid, and thus imposes no conditions on g_j and g_j'.

CHAPTER 22

The Lagrange Densities of Quantum Flavor Dynamics and of the Standard Model

22.1 The Gauge Group of the Electroweak Interaction

Today we regard leptons and quarks as fundamental fermions and we arrange them into three families (Table 22.1). The first experimental evidence for the top quark (t), with a mass of about 40 GeV, was reported in 1984 (Arnison 1984e) but this evidence has not been corroborated in the meantime. The leptons and quarks listed in Table 22.1 are called the different flavor species of the fermions. In addition to flavor the quarks possess the color degree of freedom. The masses of these fundamental particles are also given in the table. For possible ways of defining a mass for quarks, which is not an obvious procedure since the quarks are not observable as free particles, the reader is referred to Gasser (1982). The neutrino mass will be assumed to be zero although the upper experimental limit is, apart from the case of the electron neutrino, not particularly low.

We now remind ourselves of the gauge principles that were made use of in Part III to construct the Lagrange densities of QED and QCD. Here we shall proceed in a very similar manner. To begin with, we just consider the electron and its neutrino. For each of these particles there is a corresponding Dirac field operator. For the neutrino we already know that in β-decay only the left-handed part ν_{e_L} couples (21.22). The Dirac field operator for the electron can likewise be split into a "left-handed" and a "right-handed" part by setting

$$\mathbf{e}(x) = \mathbf{e}_L(x) + \mathbf{e}_R(x), \tag{22.1}$$

Table 22.1 The fundamental particles of quantum flavor dynamics. The leptons and quarks, which are the fundamental fermions, are arranged in three families. The fermions in each of the three columns form one family. The masses of the particles (or their upper or lower limits) are also given. Values for the leptons and gauge bosons are taken from Particle Data Group 1986. The values for the quarks u, d, and s correspond to the "running" quark masses at a scale of 1 GeV; For the c- and b-quarks the respective quark mass itself is taken as renormalization point (Gasser 1982). The lower limits for the top quark and the Higgs boson masses are quoted according to Takasaki (1988) and Lee-Franzini (1988).

	Particle, mass in MeV						Charge
Leptons	ν_e	<0.000046	ν_μ	<0.50	ν_τ	<164	0
	e	0.5110034 ± 0.0000014	μ	105.65932 ± 0.00029	τ	1784.2 ± 3.2	-1
Quarks	u	4.5 ± 1.4	c	1270 ± 50	t	$>26\,000$	$+\frac{2}{3}$
	d	7.9 ± 2.4	s	155 ± 50	b	4250 ± 100	$-\frac{1}{3}$
Gauge bosons	γ			$<3 \cdot 10^{-33}$			0
	$W^\pm$			81 800 $\pm 1\,500$			± 1
	Z			92 600 $\pm 1\,700$			0
Higgs boson	?			$>3\,200$			0

where

$$\begin{aligned} \mathbf{e}_L(x) &= \tfrac{1}{2}(1-\gamma_5)\mathbf{e}(x), \\ \mathbf{e}_R(x) &= \tfrac{1}{2}(1-\gamma_5)\mathbf{e}(x). \end{aligned} \tag{22.2}$$

Due to the electron mass term $\mathbf{e}_L$ and $\mathbf{e}_R$ are clearly *not* solutions to the Dirac equation and a strict connection between helicity and the eigenvalue of γ_5 does not hold.

However, the electron mass is very small in comparison with the energies of a few GeV which are common in high energy physics. It would seem a good idea to consider an approximation to the real world in which the electron is assumed to be massless. We will furthermore imagine that the electromagnetic and weak coupling are "switched off" at first. There then remain the three fields

$$\nu_{eL}, \quad \mathbf{e}_L, \quad \mathbf{e}_R.$$

The Lagrange density for these Dirac fields is

$$\mathscr{L}_0(x) = (\bar{\nu}_{eL}(x), \bar{\mathbf{e}}_L(x))(\mathrm{i}\gamma^\lambda \partial_\lambda)\begin{pmatrix} \nu_{eL}(x) \\ \mathbf{e}_L(x) \end{pmatrix} + \bar{\mathbf{e}}_R(x)\mathrm{i}\gamma^\lambda \partial_\lambda \mathbf{e}_R(x). \tag{22.3}$$

This expression exhibits a complete symmetry between the left-handed neutrino and electron fields, whereas the right-handed electron field leads an independent existence. In the considered approximation to the real world, it is not possible to physically distinguish between a left-handed neutrino and a left-handed electron. Expressed mathematically, this means that the Lagrange density (22.3) is invariant under arbitrary SU(2) rotations in the space of ν_{eL} and $\mathbf{e}_{\mathrm{L}}$. If we make the substitution

$$\begin{pmatrix} \nu_{\mathrm{eL}}(x) \\ \mathbf{e}_{\mathrm{L}}(x) \end{pmatrix} \to \mathbf{U} \begin{pmatrix} \nu_{\mathrm{eL}}(x) \\ \mathbf{e}_{\mathrm{L}}(x) \end{pmatrix}, \tag{22.4}$$

where $\mathbf{U}$ is any x-independent matrix from the SU(2) group, then we have

$$\mathscr{L}_0(x) \to \mathscr{L}_0(x).$$

We now again let the idea of locality enter into the picture. The Lagrange density $\mathscr{L}_0$ is only invariant if we redefine the neutrino and the electron in the same way at all places and at all times. This would imply that the basic equations for the electron and the neutrino on Earth would be dependent on which particle an observer in the Andromeda nebula chooses to call an electron and which a neutrino. The way to avoid this situation, which is unappealing from a physical point of view, has already been discussed within QCD. We postulate the invariance of the theory under *local* SU(2) transformations:

$$\begin{pmatrix} \nu_{\mathrm{eL}}(x) \\ \mathbf{e}_{\mathrm{L}}(x) \end{pmatrix} \to \mathbf{U}(x) \begin{pmatrix} \nu_{\mathrm{eL}}(x) \\ \mathbf{e}_{\mathrm{L}}(x) \end{pmatrix}, \tag{22.5}$$

where $\mathbf{U}(x) \in \mathrm{SU}(2)$ for all x. The Lagrange density $\mathscr{L}_0$ (22.3) is no longer invariant, but the invariance can be rescued by introducing vector fields. We need the same number of real vector fields as there are generators in the invariance group; thus in the case of the SU(2) group, we must introduce three vector fields. The resulting Lagrange density can be taken over from Sect. 19.1, replacing the SU(3) by the SU(2) group.

For the fundamental representation of the SU(2) group we have as generators the Pauli spin matrices τ_1, τ_2, and τ_3. The corresponding vector fields will be denoted as $\mathbf{W}^1_\lambda$, $\mathbf{W}^2_\lambda$, and $\mathbf{W}^3_\lambda$ and we combine them into a Hermitian 2×2 matrix with zero trace:

$$\mathbf{W}_\lambda(x) = \mathbf{W}^a_\lambda(x)\frac{\tau_a}{2}. \tag{22.6}$$

We define the matrix of field strengths as

$$\begin{aligned} \mathbf{W}_{\lambda\rho}(x) &= \partial_\lambda \mathbf{W}_\rho(x) - \partial_\rho \mathbf{W}_\lambda(x) + \mathrm{i}g[\mathbf{W}_\lambda(x), \mathbf{W}_\rho(x)] \\ &= \mathbf{W}^a_{\lambda\rho}(x)\frac{\tau_a}{2}, \end{aligned} \tag{22.7a}$$

where

$$\mathbf{W}^a_{\lambda\rho}(x) = \partial_\lambda \mathbf{W}^a_\rho(x) - \partial_\rho \mathbf{W}^a_\lambda(x) - g\varepsilon_{abc}\mathbf{W}^b_\lambda(x)\mathbf{W}^c_\rho(x). \tag{22.7b}$$

Here ε_{abc} are the structure constants of the SU(2) group and g is a gauge coupling constant. The Lagrange density for the electron, the neutrino and the W-fields would then be [see (19.22)]

$$\mathscr{L}(x) = \tfrac{1}{2}\mathrm{Tr}(\mathbf{W}_{\lambda\rho}(x)\mathbf{W}^{\lambda\rho}(x)) + (\bar{\nu}_{eL}(x), \bar{\mathbf{e}}_L(x))\mathrm{i}\gamma^\lambda(\partial_\lambda + \mathrm{i}g\mathbf{W}_\lambda)\begin{pmatrix}\nu_{eL}(x)\\ \mathbf{e}_L(x)\end{pmatrix} + \bar{\mathbf{e}}_R(x)\mathrm{i}\gamma^\lambda\partial_\lambda\mathbf{e}_R(x). \tag{22.8}$$

Exactly as in Part III, one can convince oneself that this Lagrange density is invariant under SU(2) gauge transformations. In other words, for all substitutions of the form

$$\begin{aligned}&\mathbf{W}_\lambda(x) \to \mathbf{U}(x)\mathbf{W}_\lambda(x)\mathbf{U}^\dagger(x) - \frac{\mathrm{i}}{g}\mathbf{U}(x)\partial_\lambda\mathbf{U}^\dagger(x),\\ &\begin{pmatrix}\nu_{eL}(x)\\ \mathbf{e}_L(x)\end{pmatrix} = \mathbf{U}(x)\begin{pmatrix}\nu_{eL}(x)\\ \mathbf{e}_L(x)\end{pmatrix},\\ &\mathbf{e}_R(x) \to \mathbf{e}_R(x),\end{aligned} \tag{22.9}$$

where $\mathbf{U}(x) \in \mathrm{SU}(2)$ for all x, we have

$$\mathscr{L}(x) \to \mathscr{L}(x).$$

The gauge group that we have just introduced will be referred to as the weak isospin group. The fields ν_{eL} and $\mathbf{e}_L$ form a weak doublet, whereas $\mathbf{e}_R$ represents a singlet. We stress that the physics of this group has *nothing* to do with the isospin symmetry of the strong interaction.

Let us now take a closer look at the ν–e–W coupling which originated from the requirement of SU(2) invariance. With the definitions

$$\mathbf{W}_\lambda^\pm = \frac{1}{\sqrt{2}}(\mathbf{W}_\lambda^1 \mp \mathrm{i}\mathbf{W}_\lambda^2) \tag{22.10}$$

we obtain from (22.8) as coupling term

$$\begin{aligned}\mathscr{L}_{\nu e\mathrm{W}} &= -g(\bar{\nu}_{eL}, \bar{\mathbf{e}}_L)\gamma^\lambda\mathbf{W}_\lambda^a\frac{\tau_a}{2}\begin{pmatrix}\nu_{eL}\\ \mathbf{e}_L\end{pmatrix}\\ &= -g(\bar{\nu}_{eL}, \bar{\mathbf{e}}_L)\gamma^\lambda\frac{1}{2}\begin{pmatrix}\mathbf{W}_\lambda^3 & \sqrt{2}\mathbf{W}_\lambda^+\\ \sqrt{2}\mathbf{W}_\lambda^- & -\mathbf{W}_\lambda^3\end{pmatrix}\begin{pmatrix}\nu_{eL}\\ \mathbf{e}_L\end{pmatrix}\\ &= -\frac{g}{2}\{\mathbf{W}_\lambda^3(\bar{\nu}_{eL}\gamma^\lambda\nu_{eL} - \bar{\mathbf{e}}_L\gamma^\lambda\mathbf{e}_L) + \sqrt{2}\mathbf{W}_\lambda^+\bar{\nu}_{eL}\gamma^\lambda\mathbf{e}_L\\ &\quad + \sqrt{2}\mathbf{W}_\lambda^-\bar{\mathbf{e}}_L\gamma^\lambda\nu_{eL}\}.\end{aligned} \tag{22.11}$$

We shall first interpret the terms with $\mathbf{W}_\lambda^\pm$. The fields $\mathbf{W}_\lambda^a$ ($a = 1, 2, 3$) are real and thus correspond to Hermitian operators in the quantized theory. Therefore

$$\mathbf{W}_\lambda^+ = (\mathbf{W}_\lambda^-)^\dagger. \tag{22.12}$$

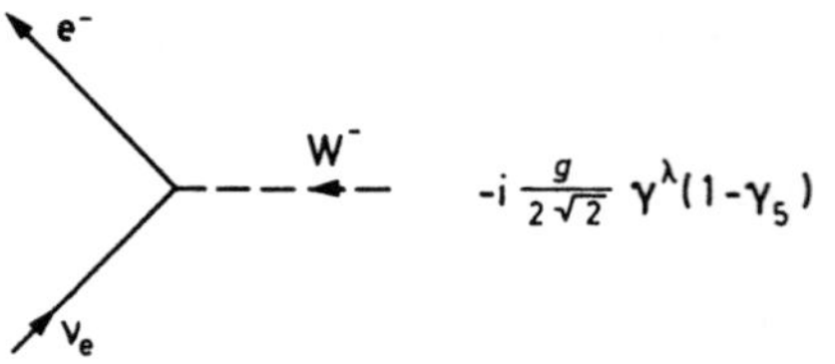

Figure 22.1 The elementary coupling between electron, electron neutrino, and W-boson as given in (22.11); Feynman diagram and analytical expression.

Expressed in terms of annihilation and creation operators, the field $\mathbf{W}_\lambda^-$ ($\mathbf{W}_\lambda^+$) has the effect of annihilating $W^-(W^+)$ particles and creating $W^+(W^-)$ particles. In the language of Feynman diagrams, the coupling (22.11) describes the transformation of a neutrino into an electron with the absorption of a W^- particle and also all those elementary processes which are obtained therefrom by crossing (Fig. 22.1). We note that the ν_e–e current that occurs here already has the $\gamma^\lambda(1-\gamma_5)$ structure which we found in the discussion of β- and μ-decay (21.11), (21.33). If we further assume electron–muon universality, then the muon and its neutrino couple to the W-boson in exactly the same way as the electron and its neutrino. To lowest order in g we then obtain for μ-decay the diagram shown in Fig. 22.2.

At first sight this already looks quite promising. The SU(2) gauge principle seems to lead, in a natural way, to weak processes in which the participating currents have the correct (V − A) structure. We have only to make the W-boson very heavy and then, along the lines of Yukawa [(21.42)], we effectively obtain, at low energies, a point-like four-fermion coupling.

Unfortunately, in the Lagrange density (22.8) there is nothing that could generate a mass for the W-boson. Electromagnetism is also missing. According to (22.11), the W_λ^3-boson couples to the left-handed neutrino ν_{e_L} and to the left-handed electron e_L, but not to the right-handed electron e_R. Thus it is *not* possible to identify W_λ^3 with the photon field. Indeed, the photon does *not* couple to the neutrino but to the left- and right-handed electrons. According to (9.5) the Lagrange density for this interaction is

$$\begin{aligned}\mathscr{L}'_{\text{em}} &= e\cdot\bar{\mathbf{e}}(x)\gamma^\lambda\mathbf{e}(x)\mathbf{A}_\lambda(x)\\ &= e\cdot\{\bar{\mathbf{e}}_R(x)\gamma^\lambda\mathbf{e}_R(x)+\bar{\mathbf{e}}_L(x)\gamma^\lambda\mathbf{e}_L(x)\}\mathbf{A}_\lambda(x),\end{aligned}\tag{22.13}$$

where e is the positron charge ($e > 0$).

We first address the question of how the theory developed so far can be extended to include electromagnetism. For this purpose we return to the

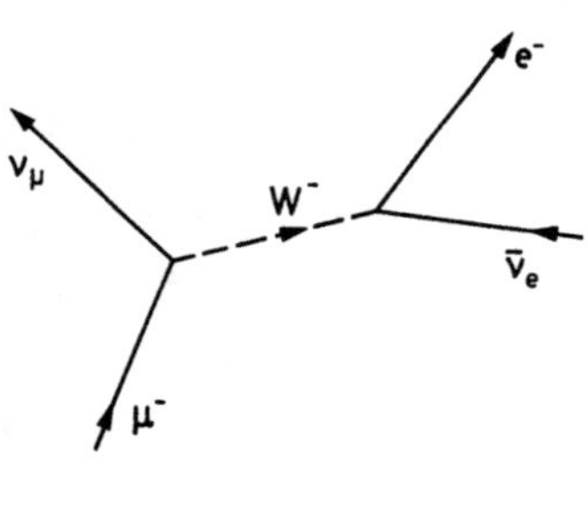

Figure 22.2 The diagram for the decay $\mu^- \to e^- + \bar{\nu}_e + \nu_\mu$ via the intermediate boson W.

Lagrange density of the free fields (22.3). Until now we have only considered invariance under SU(2) transformations. However, $\mathscr{L}_0(x)$ is also invariant under two phase transformations corresponding to two U(1) groups. To see this we make the substitutions

$$\begin{aligned} \begin{pmatrix} \nu_{\mathrm{eL}}(x) \\ \mathbf{e}_{\mathrm{L}}(x) \end{pmatrix} &\to \mathrm{e}^{+\mathrm{i}\varphi'} \cdot \begin{pmatrix} \nu_{\mathrm{eL}}(x) \\ \mathbf{e}_{\mathrm{L}}(x) \end{pmatrix}, \\ \mathbf{e}_{\mathrm{R}}(x) &\to \mathrm{e}^{+\mathrm{i}\varphi}\, \mathbf{e}_{\mathrm{R}}(x), \end{aligned} \tag{22.14}$$

where φ and φ' are constant phases, and find that $\mathscr{L}_0(x) \to \mathscr{L}_0(x)$. It would now be most satisfactory to gauge these two U(1) groups in analogy to Weyl's approach in quantum electrodynamics. However, this would yield *two* further massless gauge bosons. We would then have two photon-like bosons in the theory—in contradiction to experiment. Since only one photon is found in nature, it will be sufficient to gauge a special combination of the phase transformations (22.14). We therefore consider U(1) transformations of the form

$$\begin{aligned} \begin{pmatrix} \nu_{\mathrm{eL}}(x) \\ \mathbf{e}_{\mathrm{L}}(x) \end{pmatrix} &\to \mathrm{e}^{+\mathrm{i}y_{\mathrm{L}}\chi} \begin{pmatrix} \nu_{\mathrm{eL}}(x) \\ \mathbf{e}_{\mathrm{L}}(x) \end{pmatrix}, \\ \mathbf{e}_{\mathrm{R}}(x) &\to \mathrm{e}^{+\mathrm{i}y_{\mathrm{R}}\chi}\, \mathbf{e}_{\mathrm{R}}(x), \end{aligned} \tag{22.15}$$

where y_{L} and y_{R} are fixed numbers whose values will become known later. The operator generating this group will be referred to as the *weak hypercharge* $\mathbf{Y}$. Again, this has nothing to do with the hypercharge of the strong interaction. The arbitrary way in which the above phase transformation is selected is a very unsatisfactory aspect of flavor dynamics. Only within the framework of Grand Unified Theories does one attempt to implement the idea of local gauge invariance in a radical way, i.e., for as big a transformation group as possible.

We thus give the fields $\nu_{\mathrm{e_L}}$ and $\mathbf{e}_{\mathrm{L}}$ the hypercharge y_{L}, and the field $\mathbf{e}_{\mathrm{R}}$, the hypercharge y_{R}. If these fermions are combined into a spinor $\psi(x)$, then the transformation of the U(1) hypercharge group can be written as

$$\psi(x) = \begin{pmatrix} \nu_{\mathrm{eL}}(x) \\ \mathbf{e}_{\mathrm{L}}(x) \\ \mathbf{e}_{\mathrm{R}}(x) \end{pmatrix} \to \mathrm{e}^{\mathrm{i}\chi\mathbf{Y}} \begin{pmatrix} \nu_{\mathrm{eL}}(x) \\ \mathbf{e}_{\mathrm{L}}(x) \\ \mathbf{e}_{\mathrm{R}}(x) \end{pmatrix} = \mathrm{e}^{\mathrm{i}\chi\mathbf{Y}} \cdot \psi(x), \tag{22.16}$$

where

$$\mathbf{Y} = \begin{pmatrix} y_{\mathrm{L}} & 0 & 0 \\ 0 & y_{\mathrm{L}} & 0 \\ 0 & 0 & y_{\mathrm{R}} \end{pmatrix}.$$

In order to make the U(1) group into a gauge group one proceeds in exactly the same way as in quantum electrodynamics. In place of the charge $\mathbf{Q}$ we have the hypercharge $\mathbf{Y}$. We have to introduce a real vector field $\mathbf{B}_\lambda$ and an accompanying gauge coupling constant g'. The field strength tensor for this Abelian gauge field is exactly as in electrodynamics

$$\mathbf{B}_{\lambda\rho} = \partial_\lambda \mathbf{B}_\rho - \partial_\rho \mathbf{B}_\lambda . \tag{22.17}$$

The Lagrange density with gauged SU(2) group and gauged U(1) hypercharge group is given by

$$\mathscr{L}(x) = -\tfrac{1}{2}\,\mathrm{Tr}(\mathbf{W}_{\lambda\rho}(x)\mathbf{W}^{\lambda\rho}(x)) - \tfrac{1}{4}\mathbf{B}_{\lambda\rho}(x)\mathbf{B}^{\lambda\rho}(x) + \bar{\psi}(x)\mathrm{i}\gamma^{\lambda}\mathbf{D}_{\lambda}\psi(x), \tag{22.18}$$

where

$$\mathbf{D}_{\lambda}\psi(x) = (\partial_{\lambda} + \mathrm{i}g\mathbf{W}^{a}_{\lambda}(x)\mathbf{T}_a + \mathrm{i}g'\mathbf{B}_{\lambda}(x)\mathbf{Y})\psi(x)$$

is the covariant derivative as in Part III. The matrices $\mathbf{T}_a$ $(a = 1, 2, 3)$ have the form

$$\mathbf{T}_a = \left(\begin{array}{c|c} \frac{1}{2}\tau_a & 0 \\ \hline 0 & 0 \end{array}\right) \tag{22.19}$$

and, together with the hypercharge matrix $\mathbf{Y}$, they form a representation of the generators of the group SU(2) × U(1) since they obey the following commutation rules:

$$\begin{aligned} [\mathbf{T}_a, \mathbf{T}_b] &= \mathrm{i}\varepsilon_{abc}\mathbf{T}_c, \\ [\mathbf{T}_a, \mathbf{Y}] &= 0. \end{aligned} \tag{22.20}$$

Written in full detail the coupling terms in (22.18) are:

$$\begin{aligned} \mathscr{L}' &= -\bar{\psi}\gamma^{\lambda}(g\mathbf{W}^{a}_{\lambda}\mathbf{T}_a + g'\mathbf{B}_{\lambda}\mathbf{Y})\psi \\ &= -\frac{g}{\sqrt{2}}(\mathbf{W}^{+}_{\lambda}\bar{\nu}_{e\mathrm{L}}\gamma^{\lambda}\mathbf{e}_{\mathrm{L}} + \mathbf{W}^{-}_{\lambda}\bar{\mathbf{e}}_{\mathrm{L}}\gamma^{\lambda}\nu_{e\mathrm{L}}) \\ &\quad - \tfrac{1}{2}(g\mathbf{W}^{3}_{\lambda} + 2y_{\mathrm{L}}g'\mathbf{B}_{\lambda})\bar{\nu}_{e\mathrm{L}}\gamma^{\lambda}\nu_{e\mathrm{L}} \\ &\quad + \tfrac{1}{2}(g\mathbf{W}^{3}_{\lambda} - 2y_{\mathrm{L}}g'\mathbf{B}_{\lambda})\bar{\mathbf{e}}_{\mathrm{L}}\gamma^{\lambda}\mathbf{e}_{\mathrm{L}} - y_{\mathrm{R}}g'\mathbf{B}_{\lambda}\bar{\mathbf{e}}_{\mathrm{R}}\gamma^{\lambda}\mathbf{e}_{\mathrm{R}}. \end{aligned} \tag{22.21}$$

We have already discussed a possible interpretation of the terms containing $\mathbf{W}^{\pm}_{\lambda}$. We now want to arrange the terms containing $\mathbf{W}^{3}_{\lambda}$ and $\mathbf{B}_{\lambda}$ in such a way that the electromagnetic coupling (22.13) can be identified in the resulting expression. We further note that one of the constants y_{L} or y_{R} can be freely chosen since we also have g' as a free parameter. It is conventional to choose

$$y_{\mathrm{L}} = -\tfrac{1}{2}. \tag{22.22}$$

At the level discussed so far, the two vector fields $\mathbf{W}^{3}_{\lambda}$ and $\mathbf{B}_{\lambda}$ are on an equal footing as they are both neutral and both massless. Any two orthogonal linear combinations of these fields form an equivalent basis. It is suggestive to use as one element of such a new basis the linear combination that couples to the neutrino current. The corresponding normalized field is denoted by $\mathbf{Z}_{\lambda}$:

$$\mathbf{Z}_{\lambda} = \frac{1}{\sqrt{g^2 + g'^2}}(g\mathbf{W}^{3}_{\lambda} - g'\mathbf{B}_{\lambda}). \tag{22.23}$$

Since the neutrino has no electric charge, $\mathbf{Z}_{\lambda}$ cannot contain any component of the photon field. A candidate for the latter is thus given by the linear combination that is orthogonal to $\mathbf{Z}_\lambda$:

$$\mathbf{A}_\lambda = \frac{1}{\sqrt{g^2 + g'^2}} (g' \mathbf{W}_\lambda^3 + g \mathbf{B}_\lambda). \tag{22.24}$$

If we write

$$\begin{aligned} \sin \vartheta_W &= \frac{g'}{\sqrt{g^2 + g'^2}}, \\ \cos \vartheta_W &= \frac{g}{\sqrt{g^2 + g'^2}}, \end{aligned} \tag{22.25}$$

then we can re-express (22.23), (22.24) as

$$\begin{aligned} \mathbf{Z}_\lambda &= \cos \vartheta_W \mathbf{W}_\lambda^3 - \sin \vartheta_W \mathbf{B}_\lambda, \\ \mathbf{A}_\lambda &= \sin \vartheta_W \mathbf{W}_\lambda^3 + \cos \vartheta_W \mathbf{B}_\lambda. \end{aligned} \tag{22.26}$$

The angle ϑ_W is known as the weak angle (Glashow 1961). If we replace $\mathbf{W}_\lambda^3$ and $\mathbf{B}_\lambda$ in $\mathscr{L}'$ (22.21) by $\mathbf{Z}_\lambda$ and $\mathbf{A}_\lambda$ then we obtain:

$$\begin{aligned} \mathscr{L}' = &-\frac{g}{\sqrt{2}} (\mathbf{W}_\lambda^+ \bar{\nu}_{eL} \gamma^\lambda \mathbf{e}_L + \mathbf{W}_\lambda^- \bar{\mathbf{e}}_L \gamma^\lambda \nu_{eL}) \\ &- \sqrt{g^2 + g'^2}\, \mathbf{Z}_\lambda \{ \tfrac{1}{2} \bar{\nu}_{eL} \gamma^\lambda \nu_{eL} - \tfrac{1}{2} \bar{\mathbf{e}}_L \gamma^\lambda \mathbf{e}_L \\ &- \sin^2 \vartheta_W \cdot (-\bar{\mathbf{e}}_L \gamma^\lambda \mathbf{e}_L + y_R \bar{\mathbf{e}}_R \gamma^\lambda \mathbf{e}_R) \} \\ &- \frac{g g'}{\sqrt{g^2 + g'^2}} \mathbf{A}_\lambda (-\bar{\mathbf{e}}_L \gamma^\lambda \mathbf{e}_L + y_R \bar{\mathbf{e}}_R \gamma^\lambda \mathbf{e}_R). \end{aligned} \tag{22.27}$$

We can thus attain the correct form of the electromagnetic coupling (22.13) if we set

$$y_R = -1, \qquad \frac{g g'}{\sqrt{g^2 + g'^2}} = e. \tag{22.28}$$

With this, and with (22.25), we obtain the relations

$$e = g \sin \vartheta_W = g' \cos \vartheta_W = \sqrt{g^2 + g'^2} \sin \vartheta_W \cos \vartheta_W. \tag{22.29}$$

The Lagrange density $\mathscr{L}'$ can be written in more compact form as follows:

$$\begin{aligned} \mathscr{L}' = -e \Bigg\{ &\mathbf{A}_\lambda \mathscr{J}_{em}^\lambda + \frac{1}{\sqrt{2} \sin \vartheta_W} (\mathbf{W}_\lambda^+ \bar{\nu}_{eL} \gamma^\lambda \mathbf{e}_L + \mathbf{W}_\lambda^- \bar{\mathbf{e}}_L \gamma^\lambda \nu_{eL}) \\ &+ \frac{1}{\sin \vartheta_W \cos \vartheta_W} \mathbf{Z}_\lambda \mathscr{J}_{NC}^\lambda \Bigg\}, \end{aligned} \tag{22.30}$$

where

$$\mathscr{J}_{em}^\lambda = -\bar{\mathbf{e}}_L \gamma^\lambda \mathbf{e}_L - \bar{\mathbf{e}}_R \gamma^\lambda \mathbf{e}_R = -\bar{\mathbf{e}} \gamma^\lambda \mathbf{e}, \tag{22.31}$$

$$\mathscr{J}_{NC}^\lambda = \tfrac{1}{2} \bar{\nu}_{eL} \gamma^\lambda \nu_{eL} - \tfrac{1}{2} \bar{\mathbf{e}}_L \gamma^\lambda \mathbf{e}_L - \sin^2 \vartheta_W \mathscr{J}_{em}^\lambda. \tag{22.32}$$

The electromagnetic current is denoted here by $\mathscr{J}^{\lambda}_{\text{em}}$ and $\mathscr{J}^{\lambda}_{\text{NC}}$ is the neutral current, which couples to the Z-boson.

Let us discuss this result. By gauging the group SU(2) × U(1) we have attained a coupling structure that can describe the electromagnetic interaction and the charged current processes of the weak interaction. However, we have also obtained an additional neutral boson Z. This is not unwelcome, because, since 1973, we know of the existence of neutral weak currents. The necessity for the Z-boson stemmed essentially from the requirement that the charged weak current with its (V − A) structure and the electromagnetic current with its pure vector structure, should be brought into a unified system. This "unification" of the weak and the electromagnetic interactions was, however, somewhat contrived. Indeed, it was necessary to choose a suitable value of y_{R}. The neutral Bose field $\mathbf{Z}_\lambda$ and the photon field $\mathbf{A}_\lambda$ are linear combinations of $\mathbf{W}^3_\lambda$ and $\mathbf{B}_\lambda$. The fields $\mathbf{W}^3_\lambda$ and $\mathbf{B}_\lambda$ in turn are the third, isospin partners of the charged fields $\mathbf{W}^\pm_\lambda$, and the gauge field to the hypercharge transformation. The weak mixing angle ϑ_{W} remains a free parameter in this theory, whose value can only be determined by experiment.

This is essentially the state of the art that was reached by Glashow in 1961. However, the theory is not yet realistic in this form. We still have to give masses to the electron and to the bosons $\mathrm{W}^\pm$ and Z. The masses of the bosons $\mathrm{W}^\pm$ and Z must, in fact, be very large so that we obtain an almost point-like four-fermion coupling at low energies. One could simply try to add explicit mass terms for the electron and the W- and Z-bosons to the Lagrange density. However, this violates gauge invariance and, as it turns out, leads to a very nasty theory that is not renormalizable. Such a theory contains an infinity of parameters that must be fitted to experiment and hardly deserves to be called a theory.

A possible way out of these difficulties was presented in the work of Weinberg (1967) and Salam (1968). They made use of the phenomenon of spontaneous symmetry breaking; the topic of the next section.

22.2 The Higgs Field and Spontaneous Symmetry Breaking

One speaks of spontaneous symmetry breaking when the *fundamental equations* of a system possess a symmetry that is not displayed by the *ground state.* This phenomenon plays an important role in many branches of physics.

A simple example from mechanics can be used to demonstrate the essential features. We consider a point particle moving along an axis ρ in a one-dimensional potential $V(\rho)$ given by

$$V(\rho) = -\tfrac{1}{2}\mu^2\rho^2 + \tfrac{1}{4}\lambda\rho^4, \tag{22.33}$$

where $\mu^2 > 0$ and $\lambda > 0$ (Fig. 22.3). This potential is obviously symmetric with respect to exchange of ρ and $-\rho$, i.e.,

$$V(-\rho) = V(\rho). \tag{23.34}$$

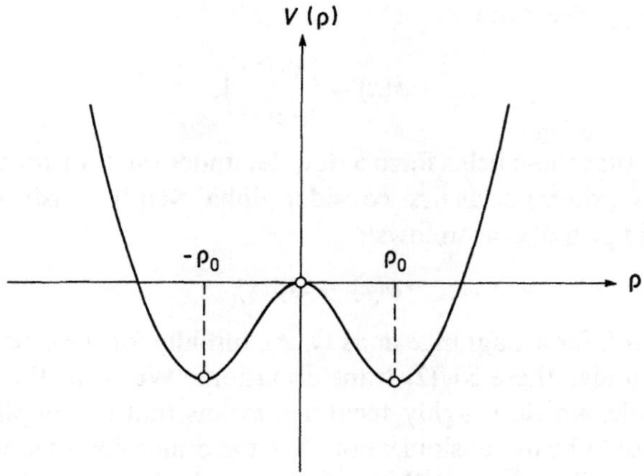

Figure 22.3 A potential $V(\rho)$ with two stable equilibrium positions ρ_0 and $-\rho_0$, as a model showing spontaneous symmetry breaking.

The equilibrium positions of the point particle are obtained from the condition

$$\frac{\partial V(\rho)}{\partial \rho} = -\mu^2\rho + \lambda\rho^3 = 0, \tag{22.35}$$

which is satisfied by $\rho = 0$ and $\rho = \pm\rho_0$ where

$$\rho_0 = \sqrt{\frac{\mu^2}{\lambda}}. \tag{22.36}$$

As is immediately evident from Fig. 22.3, the points of *stable* equilibrium lie at $\rho = \pm\rho_0$. In the ground state the particle is situated *either* at $\rho = \rho_0$ *or* at $\rho = -\rho_0$. Neither of these equilibrium positions, when taken alone, shows the symmetry of the potential under the exchange of ρ and $-\rho$. The symmetry is spontaneously broken. However, from the symmetry of the potential, we are quite entitled to conclude that if $\rho = \rho_0$ is an equilibrium position, then $-\rho_0$ must be too. It is easy to see that spontaneous symmetry breaking is, quite generally, associated with a degeneracy of the ground state.

Other well-known phenomena which are understood in terms of spontaneous symmetry breaking are the spontaneous magnetization of a solid and the condensation of water vapor.

A possible way to generate masses for the bosons $W^{\pm}$ and Z by means of a spontaneous symmetry breaking, was discussed by Weinberg (1967) and Salam (1968). Their ideas drew upon several earlier publications (Higgs 1964, 1964a, 1966; Englert 1964; Guralnik 1964; Kibble 1967). The mechanism suggested by Weinberg and Salam requires, alongside the fermion and vector fields considered in Sect. 22.1, additional scalar fields, the so-called Higgs fields. In the simplest version, it is sufficient to introduce two complex scalar fields ϕ_1 and ϕ_2. We write

$$\phi(x) = \begin{pmatrix} \phi_1(x) \\ \phi_2(x) \end{pmatrix} \tag{22.37}$$

and demand that these fields form a doublet under the transformations of the weak isospin group. Let us first consider global SU(2) transformations $\mathbf{U}$ as in (22.4), acting on $\phi(x)$ as follows:

$$\phi(x) \to \mathbf{U}\phi(x). \tag{22.38}$$

We now look for a Lagrange density $\mathscr{L}_\phi$, initially for the ϕ-field alone, that is invariant under these SU(2) transformations. We want the theory to be renormalizable, which, roughly speaking, means that all coupling constants occurring should be dimensionless or have the dimension of a positive power of mass. As is easily seen, with this requirement the Lagrange density $\mathscr{L}_\phi$ must have the following form:

$$\mathscr{L}_\phi = (\partial_\mu \phi^\dagger)(\partial^\mu \phi) - V(\phi), \tag{22.39}$$

where

$$V(\phi) = \kappa\phi^\dagger\phi + \lambda(\phi^\dagger\phi)^2.$$

For the theory to be stable we must require $\lambda > 0$; the sign of κ however, is not determined *ab initio*. We shall choose $\kappa < 0$ since this, as we shall see, leads to spontaneous symmetry breaking and to massive W- and Z-bosons. Thus our assumptions are:

$$\begin{aligned} \kappa = -\mu^2 < 0, \\ \lambda > 0. \end{aligned} \tag{22.40}$$

We now regard the field $\phi(x)$ as a classical field and inquire about the state of lowest energy. Applying the methods of Sect. 3.4 (3.104)–(3.107) to the Lagrange density (22.39), we obtain for the total energy H of the field the expression:

$$H = \int \mathrm{d}^3x\, \{\dot\phi^\dagger(x)\dot\phi(x) + \nabla\phi^\dagger(x)\nabla\phi(x) + V(\phi(x))\}. \tag{22.41}$$

Since the derivative terms are positive definite, we obtain the lowest energy state for a constant field $\phi(x) \equiv \phi = \text{const.}$ that minimizes the potential function $V(\phi)$. If we write $\rho/\sqrt{2}$ for the "length" of the field ϕ,

$$\frac{\rho}{\sqrt{2}} = +\sqrt{\phi^\dagger\phi}, \tag{22.42}$$

then we have to seek the minimum of the function

$$V(\phi) = -\tfrac{1}{2}\mu^2\rho^2 + \tfrac{1}{4}\lambda\rho^4. \tag{22.43}$$

The function $V(\phi)$ is identical to the restriction to ($\rho \geq 0$) of the potential shown in Fig. 22.3. The minimum of the potential, and therefore of the total energy (22.41), is thus reached for

$$\rho = \rho_0 = \sqrt{\frac{\mu^2}{\lambda}}. \tag{22.44}$$

This is only a condition on the "length" of the field ϕ. The orientation of the ground-state field in the two-dimensional isospin space is *not* determined. Reminding ourselves of the ordinary rotations and their representation in the two-dimensional spin space, we see that the fields ϕ with length $\rho_0/\sqrt{2}$ can be parametrized as follows:

$$\phi = e^{i(\boldsymbol{\tau}/2)\boldsymbol{\varphi}} \begin{pmatrix} 0 \\ \frac{1}{\sqrt{2}}\rho_0 \end{pmatrix}. \tag{22.45}$$

Here $\boldsymbol{\varphi}$ is a vector in isospin space with $|\boldsymbol{\varphi}| < 2\pi$. The ground state is therefore infinitely degenerate. Each of the individual equivalent ground states, for example, the field configuration

$$\phi = \begin{pmatrix} 0 \\ \frac{1}{\sqrt{2}}\rho_0 \end{pmatrix} \tag{22.46}$$

is *not* invariant under SU(2) transformations. The SU(2) group, which is a symmetry of the Lagrange density, is spontaneously broken by the ground state.

If we go over to the quantum theory of the field $\phi(x)$, then it turns out that the classical ground state corresponds, in a first approximation, to the vacuum expectation value of the quantum field. A possible vacuum expectation value would be, according to (22.46)

$$\langle 0|\phi(x)|0\rangle = \begin{pmatrix} 0 \\ \frac{1}{\sqrt{2}}\rho_0 \end{pmatrix}. \tag{22.47}$$

The analogy with the spontaneous magnetization of a solid body should help us to understand this result. In the theory of magnets the magnetization $\boldsymbol{M}$ corresponds to our field ϕ. In the case of spontaneous magnetization, $\boldsymbol{M}$ assumes a nonzero value. However, only the magnitude $|\boldsymbol{M}|$ is determined, not the direction of the magnetization. Thus although the theory possesses rotational invariance, the system settles in a state which is not invariant as the magnetization points in a particular direction.

How can perturbation theory be developed in the case of spontaneous symmetry breaking? The standard trick here is to define a new field $\phi'(x)$ by means of

$$\phi'(x) = \phi(x) - \langle 0|\phi(x)|0\rangle, \tag{22.48}$$

whose vacuum expectation value is zero:

$$\langle 0|\phi'(x)|0\rangle = 0.$$

It can be shown that by using the field $\phi'(x)$, perturbation theory may be formulated in the usual manner.

We will now allow the Higgs field to interact with the gauge bosons and the fermions. The structure of the coupling should, of course, respect the gauge invariance under the SU(2) × U(1) group of weak isospin and weak hypercharge. We would also like to have a renormalizable theory. The field $\phi(x)$ was a weak isodoublet. We can thus form an isospin invariant coupling between ϕ, the left-handed fermion doublet, and the right-handed electron singlet. We thus write the corresponding term in the Lagrange density as follows:

$$\begin{aligned}\mathscr{L}_{\text{Yuk}} &= -c_e \bar{\mathbf{e}}_R \phi^\dagger \begin{pmatrix} \nu_{eL} \\ \mathbf{e}_L \end{pmatrix} + \text{h.c.} \\ &= -c_e(\phi_1^\dagger \bar{\mathbf{e}}_R \nu_{e_L} + \phi_2^\dagger \bar{\mathbf{e}}_R \mathbf{e}_L) + \text{h.c.} \end{aligned} \tag{22.49}$$

Here c_e is a coupling constant, which initially may be complex. We note that this coupling of the Higgs bosons to the fermions is renormalizable and of the same type as that considered by Yukawa for the coupling of the pions to the nucleon.

It is obvious that the coupling (22.49) is also invariant under local isospin transformations. We now further demand that the coupling is invariant under hypercharge transformations. For this we need to assign a suitable hypercharge y_H to the Higgs field ϕ. The elementary processes described by (22.49), for example the reaction

$$e_L \to e_R + \phi_2, \tag{22.50}$$

should conserve hypercharge. This yields the condition

$$y_H = y_L - y_R = \tfrac{1}{2}. \tag{22.51}$$

With this, everything is now determined. If we add to our previous expression (22.18) the Yukawa term (22.49) and the Lagrange density for the Higgs field (22.39), then we obtain a gauge-invariant Lagrange density if, in addition, we also replace the derivatives of the Higgs field by covariant derivatives:

$$\partial_\lambda \phi \to \mathbf{D}_\lambda \phi = \left(\partial_\lambda + \mathrm{i}g \mathbf{W}_\lambda^a \frac{\tau_a}{2} + \mathrm{i}g' \mathbf{B}_\lambda y_H \right) \phi. \tag{22.52}$$

This rule always yields gauge-invariant expressions. For the total Lagrange density we obtain in this way:

$$\begin{aligned}\mathscr{L} = &-\tfrac{1}{2}\,\mathrm{Tr}(\mathbf{W}_{\lambda\rho}\mathbf{W}^{\lambda\rho}) - \tfrac{1}{4}\mathbf{B}_{\lambda\rho}\mathbf{B}^{\lambda\rho} + (\bar{\nu}_{eL}, \bar{\mathbf{e}}_L)\mathrm{i}\gamma^\lambda \mathbf{D}_\lambda \begin{pmatrix} \nu_{eL} \\ \mathbf{e}_L \end{pmatrix} \\ &+ \bar{\mathbf{e}}_R \mathrm{i}\gamma^\lambda \mathbf{D}_\lambda \mathbf{e}_R - c_e \bar{\mathbf{e}}_R \phi^\dagger \begin{pmatrix} \nu_{eL} \\ \mathbf{e}_L \end{pmatrix} - c_e^*(\bar{\nu}_{eL}, \bar{\mathbf{e}}_L)\phi \mathbf{e}_R \\ &+ (\mathbf{D}_\lambda \phi)^\dagger (\mathbf{D}^\lambda \phi) - V(\phi). \end{aligned} \tag{22.53}$$

Note that right- and left-handed Fermi fields are only coupled together in the Yukawa term. We now make use of the freedom to redefine the field $\mathbf{e}_R$ by multiplying it by a constant phase factor as follows:

$$\mathbf{e}_R \to e^{i\varphi}\, \mathbf{e}_R . \tag{22.54}$$

This obviously is equivalent to replacing the parameter c_e in (22.53) by $c_e \exp(-i\varphi)$. The phase of c_e is thus arbitrary and has no physical significance. Without loss of generality we can demand that $c_e \geq 0$. With this the number of free parameters present in our Lagrange density amounts to a total of five, namely:

$$g,\ g',\ c_e,\ \mu^2,\ \lambda .$$

The Lagrange density (22.53) is invariant under gauge transformations of the SU(2) × U(1) group. We again list the explicit form of these transformations since we shall shortly be needing them.

SU(2) gauge transformations:

$$\begin{aligned}
&\mathbf{W}_\lambda(x) \to \mathbf{U}(x)\mathbf{W}_\lambda(x)\mathbf{U}^\dagger(x) - \frac{i}{g}\mathbf{U}(x)\partial_\lambda \mathbf{U}^\dagger(x),\\
&\mathbf{B}_\lambda(x) \to \mathbf{B}_\lambda(x),\\
&\begin{pmatrix} \nu_{eL}(x) \\ \mathbf{e}_L(x) \end{pmatrix} \to \mathbf{U}(x)\begin{pmatrix} \nu_{eL}(x) \\ \mathbf{e}_L(x) \end{pmatrix},\\
&\mathbf{e}_R(x) \to \mathbf{e}_R(x),\\
&\begin{pmatrix} \phi_1(x) \\ \phi_2(x) \end{pmatrix} \to \mathbf{U}(x)\begin{pmatrix} \phi_1(x) \\ \phi_2(x) \end{pmatrix}.
\end{aligned} \tag{22.55}$$

Here

$$\mathbf{U}(x) = e^{i(\tau_a/2)\varphi^a(x)}, \tag{22.56}$$

with $\varphi^a(x)$ $(a = 1, 2, 3)$ being arbitrary real functions of x.

U(1) gauge transformations:

$$\begin{aligned}
&\mathbf{W}_\lambda(x) \to \mathbf{W}_\lambda(x),\\
&\mathbf{B}_\lambda(x) \to \mathbf{B}_\lambda(x) - \frac{1}{g'}\partial_\lambda \chi(x),\\
&\begin{pmatrix} \nu_{eL}(x) \\ \mathbf{e}_L(x) \end{pmatrix} \to e^{iy_L\chi(x)}\begin{pmatrix} \nu_{eL}(x) \\ \mathbf{e}_L(x) \end{pmatrix},\\
&\mathbf{e}_R(x) \to e^{iy_R\chi(x)}\, \mathbf{e}_R(x),\\
&\phi(x) \to e^{iy_H\chi(x)}\, \phi(x).
\end{aligned} \tag{22.57}$$

Here $\chi(x)$ is again an arbitrary real function of x.

By means of a gauge transformation we can obviously rotate the Higgs field ϕ into any direction in isospin space, and furthermore, this can be done independently at all points in space and time. Therefore, according to (22.45) there is always an SU(2) gauge transformation $\mathbf{U}(x)$ which achieves the following:

$$\mathbf{U}(x)\phi(x) = \begin{pmatrix} 0 \\ \frac{1}{\sqrt{2}}\rho(x) \end{pmatrix}, \tag{22.58}$$

where $\rho(x) \geq 0$. In other words, we can demand as gauge condition that the Higgs field has the form of the right-hand side of (22.58) (unitary gauge).

We would now like to know the vacuum expectation value of the radial field $\rho(x)$. It can easily be seen that all the points that we made concerning the Higgs field alone are also valid here. Thus, to a first approximation, the vacuum expectation value of the Higgs field is given by the minimum of the potential V and from (22.47) we have

$$\begin{aligned} \langle 0|\rho(x)|0\rangle &= \rho_0 \\ \langle 0|\phi(x)|0\rangle &= \begin{pmatrix} 0 \\ \frac{1}{\sqrt{2}}\rho_0 \end{pmatrix}. \end{aligned} \tag{22.59}$$

This vacuum expectation value is obviously not invariant under the full SU(2) × U(1) gauge group. Only the U(1) subgroup, which is generated by $\mathbf{T}_3 + \mathbf{Y}$, leaves the vacuum expectation value of ϕ invariant:

$$e^{i\chi(x)((\tau_3/2)+y_H)} \begin{pmatrix} 0 \\ \frac{1}{\sqrt{2}}\rho_0 \end{pmatrix} = \begin{pmatrix} 0 \\ \frac{1}{\sqrt{2}}\rho_0 \end{pmatrix}. \tag{22.60}$$

This U(1) gauge group is not identical to the hypercharge group. As we shall see shortly, it exactly corresponds to the gauge group of electromagnetism.

The way to proceed further is now clear. We shift the field $\rho(x)$ by an amount equal to its vacuum expectation value ρ_0; i.e., we introduce a new field ρ' with a vacuum expectation value of zero:

$$\begin{aligned} \rho'(x) &= \rho(x) - \rho_0, \\ \langle 0|\rho'(x)|0\rangle &= 0. \end{aligned} \tag{22.61}$$

With the help of the field ρ' we can apply field theory in the usual manner, in particular perturbation theory, and we can derive Feynman rules. When we rewrite the Lagrange density in terms of the field ρ', we obtain from the covariant derivative of the Higgs field (22.52), (22.53) the following quadratic form in the gauge fields:

$$\langle 0|\phi^\dagger|0\rangle\left(-ig\mathbf{W}^a_\lambda\frac{\tau_a}{2}-ig'\mathbf{B}_\lambda y_{\mathrm{H}}\right)\left(ig\mathbf{W}^{\lambda a}\frac{\tau_a}{2}+ig'\mathbf{B}^\lambda y_{\mathrm{H}}\right)\langle 0|\phi|0\rangle$$

$$=\left(0,\frac{1}{\sqrt{2}}\rho_0\right)\cdot\begin{pmatrix}\dfrac{2gg'\mathbf{A}_\lambda+(g^2-g'^2)\mathbf{Z}_\lambda}{2\sqrt{g^2+g'^2}} & \dfrac{g}{\sqrt{2}}\mathbf{W}^+_\lambda\\ \dfrac{g}{\sqrt{2}}\mathbf{W}^-_\lambda & -\dfrac{\sqrt{g^2+g'^2}}{2}\mathbf{Z}_\lambda\end{pmatrix}$$

$$\cdot\begin{pmatrix}\dfrac{2gg'\mathbf{A}^\lambda+(g^2-g'^2)\mathbf{Z}^\lambda}{2\sqrt{g^2+g'^2}} & \dfrac{g}{\sqrt{2}}\mathbf{W}^{\lambda+}\\ \dfrac{g}{\sqrt{2}}\mathbf{W}^{\lambda-} & -\dfrac{\sqrt{g^2+g'^2}}{2}\mathbf{Z}^\lambda\end{pmatrix}\begin{pmatrix}0\\ \dfrac{1}{\sqrt{2}}\rho_0\end{pmatrix}$$

$$=\frac{g^2\rho_0^2}{4}\mathbf{W}^-_\lambda\mathbf{W}^{\lambda+}+\frac{(g^2+g'^2)\rho_0^2}{8}\mathbf{Z}_\lambda\mathbf{Z}^\lambda. \tag{22.62}$$

The terms clearly signify that the bosons $\mathrm{W}^\pm$ and Z have become massive whereas the photon remains massless, exactly as it should be. We can read off the boson masses as:

$$\begin{aligned}m_{\mathrm{W}}^2&=\frac{g^2\rho_0^2}{4}=\frac{e^2\rho_0^2}{4\sin^2\vartheta_{\mathrm{W}}},\\ m_{\mathrm{Z}}^2&=\frac{(g^2+g'^2)\rho_0^2}{4}=\frac{e^2\rho_0^2}{4\sin^2\vartheta_{\mathrm{W}}\cos^2\vartheta_{\mathrm{W}}}.\end{aligned} \tag{22.63}$$

Thus if the vacuum expectation value ρ_0 of the Higgs field is large enough, the W- and Z-bosons become arbitrarily heavy.

What about the fermion masses? From the Yukawa coupling of the Higgs bosons to the fermions in (22.53) we obtain a term of the form

$$\begin{aligned}&-c_{\mathrm{e}}\left[\bar{\mathbf{e}}_{\mathrm{R}}\langle 0|\phi^\dagger|0\rangle\begin{pmatrix}\nu_{\mathrm{eL}}\\ \mathbf{e}_{\mathrm{L}}\end{pmatrix}+(\bar{\nu}_{\mathrm{eL}},\bar{\mathbf{e}}_{\mathrm{L}})\langle 0|\phi|0\rangle\mathbf{e}_{\mathrm{R}}\right]\\ &\qquad=-c_{\mathrm{e}}\left[\bar{\mathbf{e}}_{\mathrm{R}}\frac{1}{\sqrt{2}}\rho_0\mathbf{e}_{\mathrm{L}}+\bar{\mathbf{e}}_{\mathrm{L}}\frac{1}{\sqrt{2}}\rho_0\mathbf{e}_{\mathrm{R}}\right]\\ &\qquad=-c_{\mathrm{e}}\frac{\rho_0}{\sqrt{2}}(\bar{\mathbf{e}}_{\mathrm{R}}\mathbf{e}_{\mathrm{L}}+\bar{\mathbf{e}}_{\mathrm{L}}\mathbf{e}_{\mathrm{R}})\\ &\qquad=-c_{\mathrm{e}}\frac{\rho_0}{\sqrt{2}}\bar{\mathbf{e}}\mathbf{e}.\end{aligned} \tag{22.64}$$

This has exactly the form of a mass term for the electron with m_e given by

$$m_e = c_e \frac{\rho_0}{\sqrt{2}}. \tag{22.65}$$

With this we have succeeded in giving a mass to the electron and at the same time the neutrino remains massless although the left-handed components ν_{eL} and $\mathbf{e}_L$ enter into the basic equations in a fully symmetric way. This is achieved via the spontaneous symmetry breaking. The ground state, which, in a field theory, is the vacuum state, points in a particular direction in the isospin space spanned by ν_{eL} and $\mathbf{e}_L$.

The total Lagrange density, after making the transition to the field $\rho'(x)$ (22.61) has the following form:

$$\begin{aligned}\mathcal{L} = &-\tfrac{1}{2}\,\mathrm{Tr}(\mathbf{W}_{\lambda\rho}\mathbf{W}^{\lambda\rho}) - \tfrac{1}{4}\mathbf{B}_{\lambda\rho}\mathbf{B}^{\lambda\rho} + \bar{\nu}_{eL}\mathrm{i}\gamma^\lambda\partial_\lambda\nu_{eL} + \bar{\mathbf{e}}\mathrm{i}\gamma^\lambda\partial_\lambda\mathbf{e} \\ &+ \mathbf{W}_\lambda^+\mathbf{W}^{-\lambda}m_W^2\left(1+\frac{\rho'}{\rho_0}\right)^2 + \tfrac{1}{2}\mathbf{Z}_\lambda\mathbf{Z}^\lambda m_Z^2\left(1+\frac{\rho'}{\rho_0}\right)^2 \\ &- m_e\bar{\mathbf{e}}\mathbf{e}\left(1+\frac{\rho'}{\rho_0}\right) + \tfrac{1}{2}\partial_\lambda\rho'\,\partial^\lambda\rho' \\ &- \tfrac{1}{2}m_{\rho'}^2(\rho')^2\left[1+\frac{\rho'}{\rho_0}+\frac{1}{4}\left(\frac{\rho'}{\rho_0}\right)^2\right] + \mathcal{L}'.\end{aligned} \tag{22.66}$$

Here $\mathcal{L}'$ is the coupling term between the fermions and the vector bosons as in (22.30)–(22.32). In the first two terms of $\mathcal{L}$ we imagine that $\mathbf{W}_\lambda^3$ and $\mathbf{B}_\lambda$ are expressed in terms of $\mathbf{Z}_\lambda$ and $\mathbf{A}_\lambda$ according to (22.26). The parameter $m_{\rho'}$ is defined as

$$m_{\rho'}^2 = 2\lambda\rho_0^2. \tag{22.67}$$

By examining the Lagrange density (22.66) we can read off the particle content of our theory. We obtain: a massless vector boson, the photon with field $\mathbf{A}_\lambda$; three massive vector bosons, $W^\pm$ and Z; a massless left-handed fermion, the neutrino; a massive fermion, the electron; and, finally, a massive neutral boson with spin zero and mass $m_{\rho'}$, the physical Higgs particle. The field belonging to the Higgs particle is ρ'. According to (22.66), the Higgs particle possesses a self-coupling. It also couples to the W- and Z-bosons and to the electron, with coupling strengths proportional to the masses of the respective particles.

Until 1982, of the particles in this theory, only the photon, the neutrino and the electron were known. The discovery of the vector bosons $W^\pm$ and Z was made in 1983 in experiments at the proton–antiproton storage ring at CERN, as will be discussed later on. What is still lacking, however, is experimental proof of the existence of the Higgs particle.

The form of the Lagrange density in (22.66) suggests that, instead of the original parameters of the model (22.53), we take, for example, the following five quantities as new parameters

$$e,\ \sin\vartheta_W,\ m_e,\ m_W^2,\ m_{\rho'}^2.$$

For the Z mass and the vacuum expectation value ρ_0 we then have, according to (22.63):

$$m_Z^2 = \frac{m_W^2}{\cos^2 \vartheta_W}, \tag{22.68}$$

$$\rho_0 = 2m_W \sqrt{\frac{\sin^2 \vartheta_W}{e^2}}. \tag{22.69}$$

In the next section we shall see how the Fermi constant G is expressed in terms of these parameters and what this implies for the numerical values of the W and Z masses and for ρ_0.

22.3 The Extension of Quantum Flavor Dynamics to Other Leptons and to Quarks, and the Effective Lagrange Density at Low Energies

In Nature, not only the electron and the electron neutrino are known, but also a whole series of other fundamental fermions (Table 22.1). All of these leptons and quarks take part in the weak interaction. In QFD, all left-handed fermions are arranged into weak isodoublets and all right-handed fermions form singlets. The hypercharge quantum numbers are chosen by hand as in the case of the electron in such a way that the experimentally observed electrical charges of the fermions are reproduced. The flavor quantum numbers of the leptons and the quarks are listed in Table 22.2. As we shall see, the quarks d′, s′, and b′, which, together with u, c, and t, form left-handed doublets, are not identical to the quarks d, s, and b of well-defined mass, but are linear combinations of the latter set. We also see that the standard scheme demands the existence of a top quark, t.

Table 22.2 The flavor quantum numbers of the leptons and the quarks. T is the weak isospin, T_3 its third component, Y the weak hypercharge, and Q the electrical charge. The weak isodoublets are written as bi-spinors. The quarks d′, s′, and b′ are obtained from the quarks d, s, and b of defined mass by means of a generalized Cabibbo rotation.

			T	T_3	Y	Q
$\begin{pmatrix} \nu_{eL} \\ e_L \end{pmatrix}$	$\begin{pmatrix} \nu_{\mu L} \\ \mu_L \end{pmatrix}$	$\begin{pmatrix} \nu_{\tau L} \\ \tau_L \end{pmatrix}$	$\begin{matrix} \frac{1}{2} \\ \frac{1}{2} \end{matrix}$	$\begin{matrix} \frac{1}{2} \\ -\frac{1}{2} \end{matrix}$	$\begin{matrix} -\frac{1}{2} \\ -\frac{1}{2} \end{matrix}$	$\begin{matrix} 0 \\ -1 \end{matrix}$
e_R	μ_R	τ_R	0	0	-1	-1
$\begin{pmatrix} u_L \\ d'_L \end{pmatrix}$	$\begin{pmatrix} c_L \\ s'_L \end{pmatrix}$	$\begin{pmatrix} t_L \\ b'_L \end{pmatrix}$	$\begin{matrix} \frac{1}{2} \\ \frac{1}{2} \end{matrix}$	$\begin{matrix} \frac{1}{2} \\ -\frac{1}{2} \end{matrix}$	$\begin{matrix} \frac{1}{6} \\ \frac{1}{6} \end{matrix}$	$\begin{matrix} \frac{2}{3} \\ -\frac{1}{3} \end{matrix}$
u_R	c_R	t_R	0	0	$\frac{2}{3}$	$\frac{2}{3}$
d_R	s_R	b_R	0	0	$-\frac{1}{3}$	$-\frac{1}{3}$

We now wish to include all these fermions in the Lagrange density of QFD. To do this combine all the Fermi fields into a total spinor ψ:

$$\psi = \begin{pmatrix} \nu_{eL} \\ \mathbf{e}_L \\ \mathbf{e}_R \\ \vdots \\ \mathbf{b}_R \end{pmatrix}. \tag{22.70}$$

The corresponding representation matrices of the SU(2) × U(1) algebra are again denoted by $\mathbf{T}_a$ (with $a = 1, 2, 3$) and $\mathbf{Y}$. The covariant derivative of the Fermi fields is then given, according to the rule of minimal substitution, by

$$\mathbf{D}_\lambda \psi = (\partial_\lambda + \mathrm{i}g\mathbf{W}^a_\lambda \mathbf{T}_a + \mathrm{i}g'\mathbf{B}_\lambda \mathbf{Y})\psi. \tag{22.71}$$

In the standard version of the model, one contents oneself with a Higgs doublet ϕ. The coupling of the Higgs field to itself, and to the gauge bosons, remains as in (22.53). However, the Yukawa coupling of the fermions to the Higgs field becomes more complicated. The most general expression for this coupling that is compatible with the conservation of fermion number reads

$$\mathscr{L}_{\mathrm{Yuk}}(x) = \bar{\psi}(x)\phi_i(x)\mathbf{C}_i\psi(x) + \mathrm{h.c.}, \tag{22.72}$$

where $\mathbf{C}_i$ (with $i = 1, 2$) are any complex matrices acting on the flavor index of ψ. We have, of course, to demand that the Yukawa coupling is invariant under SU(2) × U(1) transformations. The ensuing restrictions for the matrices $\mathbf{C}_i$ will be discussed below.

After these preparations, we are ready to write down the Lagrange density of QFD as a generalization of (22.53):

$$\begin{aligned}\mathscr{L} = &-\tfrac{1}{2}\operatorname{Tr}(\mathbf{W}_{\lambda\rho}\mathbf{W}^{\lambda\rho}) - \tfrac{1}{4}\mathbf{B}_{\lambda\rho}\mathbf{B}^{\lambda\rho} + \bar{\psi}\mathrm{i}\gamma^\lambda \mathbf{D}_\lambda\psi + \mathscr{L}_{\mathrm{Yuk}} \\ &+ (\mathbf{D}_\lambda\phi)^\dagger(\mathbf{D}^\lambda\phi) - V(\phi).\end{aligned} \tag{22.73}$$

With the gauge condition (22.55), and after introducing the field ρ' (22.61), this becomes:

$$\begin{aligned}\mathscr{L}_{\mathrm{QFD}} = &-\tfrac{1}{2}\operatorname{Tr}(\mathbf{W}_{\lambda\rho}\mathbf{W}^{\lambda\rho}) - \tfrac{1}{4}\mathbf{B}_{\lambda\rho}\mathbf{B}^{\lambda\rho} \\ &+ \mathbf{W}^+_\lambda \mathbf{W}^{-\lambda} m_W^2\left(1 + \frac{\rho'}{\rho_0}\right)^2 + \tfrac{1}{2}\mathbf{Z}_\lambda \mathbf{Z}^\lambda m_Z^2\left(1 + \frac{\rho'}{\rho_0}\right)^2 \\ &+ \bar{\psi}\mathrm{i}\gamma^\lambda \mathbf{D}_\lambda\psi - \bar{\psi}\mathbf{M}\psi\left(1 + \frac{\rho'}{\rho_0}\right) \\ &+ \tfrac{1}{2}\partial_\lambda\rho'\partial^\lambda\rho' - \tfrac{1}{2}m_{\rho'}^2\rho'^2\left[1 + \frac{\rho'}{\rho_0} + \frac{1}{4}\left(\frac{\rho'}{\rho_0}\right)^2\right],\end{aligned} \tag{22.74}$$

where

$$\mathbf{M} = -(\mathbf{C}_2 + \mathbf{C}_2^\dagger)\frac{\rho_0}{\sqrt{2}} \tag{22.75}$$

is a mass matrix for the fermions in generalization of (22.65). We shall see, that this mass matrix gives rise to Cabibbo-like angles in a natural manner.

From the Lagrange density (22.74) we can easily read off the particle content of QFD, which has already been given in Table 22.1. The fermions, and the specific representations of the group SU(2) × U(1) in which they occur were chosen by hand in the beginning. Thus we have to regard this as "input" to the theory. The physical postulate of local gauge invariance then predicted the existence of the gauge bosons as an "output" of the theory. The spontaneous symmetry breaking, whose necessity in fact only arose from experiment, and for which more fundamental justification could not be given, predicted the existence of the Higgs boson. All of the particles demanded by the theory, with the exception of the top quark and the Higgs boson, have already been discovered.

We will now take a closer look at some of the terms in the Lagrange density (22.74). The coupling of the fermions to the gauge bosons is given by the covariant derivative of the fermions (22.71). Thus, apart from the coupling constants, it is fixed by the gauge group SU(2) × U(1) and by the flavor quantum numbers of the fermions. The corresponding term in the Lagrange density will be denoted $\mathscr{L}_{\text{int}}$, i.e.,

$$\begin{aligned} \bar{\psi} \mathrm{i}\gamma^\lambda \mathbf{D}_\lambda \psi &= \bar{\psi} \mathrm{i}\gamma^\lambda \partial_\lambda \psi + \mathscr{L}_{\text{int}} \\ \mathscr{L}_{\text{int}} &= -\bar{\psi}\gamma^\lambda (g \mathbf{W}_\lambda^a \mathbf{T}_a + g' \mathbf{B}_\lambda \mathbf{Y})\psi . \end{aligned} \tag{22.76}$$

Expressed in terms of the physical boson fields $\mathbf{W}_\lambda^\pm$, $\mathbf{Z}_\lambda$, and $\mathbf{A}_\lambda$ this gives

$$\begin{aligned} \mathscr{L}_{\text{int}} = -e \Bigg\{ \mathbf{A}_\lambda \mathscr{J}_{\text{em}}^\lambda &+ \frac{1}{\sin \vartheta_{\text{W}} \cos \vartheta_{\text{W}}} \mathbf{Z}_\lambda \mathscr{J}_{\text{NC}}^\lambda \\ &+ \frac{1}{\sqrt{2} \sin \vartheta_{\text{W}}} (\mathbf{W}_\lambda^+ \mathscr{J}_{\text{CC}}^\lambda + \mathbf{W}_\lambda^- \mathscr{J}_{\text{CC}}^{\lambda\dagger}) \Bigg\}, \end{aligned} \tag{22.77}$$

where

$$\begin{aligned} \mathscr{J}_{\text{em}}^\lambda &= \bar{\psi}\gamma^\lambda (\mathbf{T}_3 + \mathbf{Y})\psi , \\ \mathscr{J}_{\text{NC}}^\lambda &= \bar{\psi}\gamma^\lambda (\mathbf{T}_3 - \sin^2 \vartheta_{\text{W}} (\mathbf{T}_3 + \mathbf{Y}))\psi = \bar{\psi}\gamma^\lambda \mathbf{T}_3 \psi - \sin^2 \vartheta_{\text{W}} \mathscr{J}_{\text{em}}^\lambda , \\ \mathscr{J}_{\text{CC}}^\lambda &= \bar{\psi}\gamma^\lambda (\mathbf{T}_1 + \mathrm{i}\mathbf{T}_2)\psi . \end{aligned} \tag{22.78}$$

The currents which appear here—the electromagnetic current $\mathscr{J}_{\text{em}}$, the neutral current $\mathscr{J}_{\text{NC}}$, and the charged current $\mathscr{J}_{\text{CC}}$, are the fundamental objects in terms of which the electroweak interaction is expressed at low energies. The matrix of the electric charge $\mathbf{Q}$ of the fermions is given by

$$\mathbf{Q} = \mathbf{T}_3 + \mathbf{Y}. \tag{22.79}$$

From this we see that the gauge group of electromagnetism is exactly the U(1) gauge group that, according to (22.60), leaves the vacuum expectation value of ϕ invariant, and is therefore not broken. The hypercharge values of the fermions in Table 22.2 are *chosen* such that the electromagnetic current is a

Table 22.3 The fermion–gauge-boson vertices of quantum flavor dynamics.

Vertex	Analytical expression
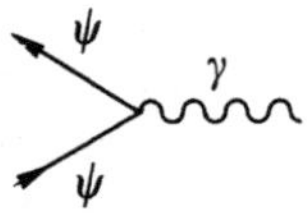	$-\mathrm{i}e\mathbf{Q}\,\gamma^\lambda$
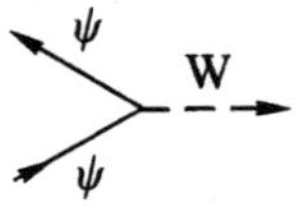	$-\mathrm{i}\dfrac{e}{\sqrt{2}\sin\vartheta_{\mathrm{W}}}(\mathbf{T}_1+\mathrm{i}\mathbf{T}_2)\,\gamma^\lambda$
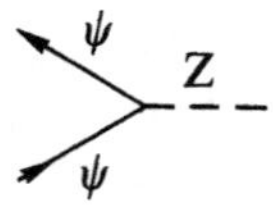	$-\mathrm{i}\dfrac{e}{\sin\vartheta_{\mathrm{W}}\cos\vartheta_{\mathrm{W}}}(\mathbf{T}_3-\sin^2\vartheta_{\mathrm{W}}\mathbf{Q})\,\gamma^\lambda$

pure vector current and yields the experimentally observed charges. This arranging of quantum numbers by hand instead of having recourse to fundamental physical postulates is one of the few unsatisfactory aspects of QFD.

It is now possible—as for QED—to develop Feynman rules for QFD. The coupling terms in (22.77) correspond to fermion–boson vertices (Table 22.3). Gauge bosons in the initial and final state, and the exchanges of virtual gauge bosons are represented by external and internal gauge boson lines, respectively. In Table 22.4 we give the analytical expressions for the internal lines; these are readily derived by generalizing the photon propagator of QED. The complete Feynman rules of QFD, in the so-called unitary gauge, are presented in Appendix G.

We now consider fermion–fermion scattering and the corresponding crossed reactions, i.e., processes of the type

$$\begin{aligned} \mathrm{e}^- + \mathrm{e}^- &\to \mathrm{e}^- + \mathrm{e}^-,\\ \mathrm{e}^- + \mathrm{e}^+ &\to \mu^- + \mu^+,\\ \nu_{\mathrm{e}} + \mathrm{e} &\to \nu_{\mathrm{e}} + \mathrm{e},\\ \nu_\mu + \mathrm{e} &\to \nu_\mu + \mathrm{e},\\ \nu_\mu + \mathrm{d} &\to \mu^- + \mathrm{u},\\ \mu^- &\to \mathrm{e}^- + \bar{\nu}_{\mathrm{e}} + \nu_\mu \quad \text{(etc.)} \end{aligned} \tag{22.80}$$

Of these, the first two are already familiar to us from QED. Neutrino reactions have been discussed in Part III. In our present theory, reactions of this type

Table 22.4 The internal lines representing the propagation of the gauge bosons in quantum flavor dynamics and the corresponding analytical expressions. The four-momentum flowing along the lines is denoted by q.

Propagator	Analytical expression
$\gamma(q)$ 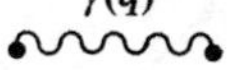	$\dfrac{\mathrm{i}g_{\rho\lambda}}{q^2}$
$\mathrm{W}(q)$ 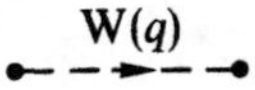	$\dfrac{-\mathrm{i}\left(g_{\rho\lambda}-\dfrac{q_\rho q_\lambda}{m_\mathrm{W}^2}\right)}{q^2-m_\mathrm{W}^2} \xrightarrow[q^2 \ll m_\mathrm{W}^2]{} \dfrac{\mathrm{i}g_{\rho\lambda}}{m_\mathrm{W}^2}$
$\mathrm{Z}(q)$ 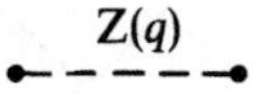	$\dfrac{-\mathrm{i}\left(g_{\rho\lambda}-\dfrac{q_\rho q_\lambda}{m_\mathrm{Z}^2}\right)}{q^2-m_\mathrm{Z}^2} \xrightarrow[q^2 \ll m_\mathrm{Z}^2]{} \dfrac{\mathrm{i}g_{\rho\lambda}}{m_\mathrm{Z}^2}$

are described to lowest order by the exchange of the bosons γ, $\mathrm{W}^\pm$, and Z (Fig. 22.4).

If the momentum transfers q are very small in comparison to the W and Z masses, then we can replace the W and Z propagators of Table 22.4 by constants. A constant propagator in momentum space corresponds to a δ-function in real space. The corresponding exchange diagrams in Fig. 22.4 thus reduce to diagrams with a point-like coupling of the four fermions. As can easily be seen, these diagrams can be generated from an *effective* Lagrange density of current × current form:

$$\begin{aligned}\mathscr{L}_\mathrm{eff} = &-\frac{e^2}{2m_\mathrm{Z}^2 \sin^2\vartheta_\mathrm{W}\cos^2\vartheta_\mathrm{W}}\mathscr{J}_\mathrm{NC}^\lambda\, g_{\lambda\rho}\mathscr{J}_\mathrm{NC}^\rho \\ &-\frac{e^2}{2m_\mathrm{W}^2\sin^2\vartheta_\mathrm{W}}\mathscr{J}_\mathrm{CC}^\lambda\, g_{\lambda\rho}\mathscr{J}_\mathrm{CC}^{\rho\dagger}.\end{aligned} \tag{22.81}$$

This should be understood as follows: In the limit $q^2 \ll m_\mathrm{W}^2, m_\mathrm{Z}^2$, the *second order* diagrams in Fig. 22.4 with W and Z exchange give analytical expressions for the S-matrix elements that are identical to the matrix elements of the S operator

$$\mathbf{S} = 1 + \mathrm{i}\int \mathrm{d}x\, \mathscr{L}_\mathrm{eff}(x), \tag{22.82}$$

where according to the rules of perturbation theory $\mathscr{L}_\mathrm{eff}$ is used in first order.

With this form of the weak coupling, we have made the connection to the phenomenological four-fermion coupling that was discussed in Chap. 21. The Lagrange density $\mathscr{L}_\mu'$ for muon decay is contained in the coupling of the charged currents in (22.81). We merely have to single out the term that couples

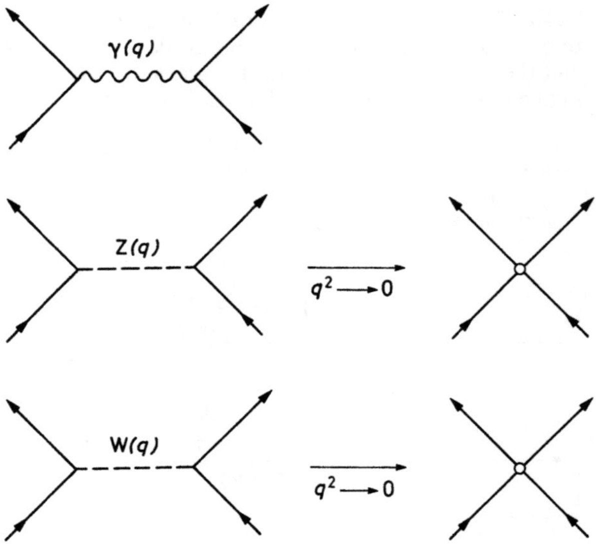

Figure 22.4 Diagrams for fermion–fermion scattering in the standard model. The interpretation of the various fermion lines depends on the reaction considered; for example, in the case of electron–electron scattering they represent incoming and outgoing electron lines. In the low energy region ($q^2 \ll m_W^2, m_Z^2$), the exchange of W and Z effectively act as point-like four-fermion couplings.

the muon, electron, and neutrino fields. This gives

$$\begin{aligned}\mathscr{L}'_\mu &= -\frac{e^2}{2m_W^2 \sin^2 \vartheta_W}[(\bar{\nu}_{\mu L}\gamma^\lambda \mu_L)(\bar{e}_L\gamma_\lambda \nu_{eL}) + \text{h.c.}] \\ &= -\frac{e^2}{8m_W^2 \sin^2 \vartheta_W}[\bar{\nu}_\mu \gamma^\lambda (1-\gamma_5)\mu\, \bar{e}\gamma_\lambda(1-\gamma_5)\nu_e + \text{h.c.}]. \qquad (22.83)\end{aligned}$$

If we compare this with (21.23), we find for the Fermi constant $G \equiv G_\mu$:

$$\frac{1}{\sqrt{2}}G = \frac{e^2}{8m_W^2 \sin^2 \vartheta_W}, \qquad (22.84a)$$

or by solving for m_W^2:

$$m_W^2 = \frac{e^2}{4\sqrt{2}G \cdot \sin^2 \vartheta_W}. \qquad (22.84b)$$

As we shall see in the next chapter, G and $\sin^2 \vartheta_W$ can be determined from low energy experiments. One finds:

$$G \cong 1.17 \cdot 10^{-5}\ \text{GeV}^{-2},$$
$$\sin^2 \vartheta_W \cong 0.23.$$

Using (22.68), (22.69) one then obtains the following predictions for the mass values of the W- and Z-bosons and for the vacuum expectation value, ρ_0, of the Higgs field:

$$\begin{aligned} m_W &\cong 77.7 \text{ GeV}, \\ m_Z &\cong 88.6 \text{ GeV}, \\ \rho_0 &\cong 2^{-1/4} G^{-1/2} = 246 \text{ GeV}. \end{aligned} \tag{22.85}$$

In terms of the Fermi constant the effective Lagrange density (22.81) can be written in the compact form:

$$\mathscr{L}_{\text{eff}} = -\frac{4G}{\sqrt{2}} \{ \mathscr{J}^{\lambda}_{\text{CC}} \mathscr{J}^{\dagger}_{\text{CC}\lambda} + \mathscr{J}^{\lambda}_{\text{NC}} \mathscr{J}_{\text{NC}\lambda} \}. \tag{22.86}$$

In subsequent chapters we shall discuss a few tests of this form of the weak interaction coupling. Here we will simply mention that the *universality* of the weak interaction (Chap. 21) is automatically fulfilled in QFD: In the diagram with W exchange (Fig. 22.4) *every* fermion doublet must couple to the W-boson with the same strength. This is a consequence of the gauged isospin invariance. Furthermore, since there is only *one* $W^{\pm}$-boson in QFD, the effective four-fermion coupling via W exchange must be the same for any pair of doublets.

In the next section, before discussing the phenomenological applications of QFD, we will take a closer look at the Yukawa coupling.

22.4 The Mass Matrix and the Cabibbo Angles

In this section we will study the Yukawa coupling (22.72). In QFD, the right-handed fermions are all singlets, and the left-handed all doublets. In order to construct an SU(2) × U(1)-invariant Yukawa coupling, we begin by forming singlets from the Higgs doublet and the left-handed fermion doublets. Let

$$\begin{pmatrix} \psi_{1\text{L}} \\ \psi_{2\text{L}} \end{pmatrix}$$

be an arbitrary fermion doublet. An isospin-invariant combination can be formed by using either ϕ or its adjoint $\phi^{\dagger}$, since both represent isodoublets:

$$\begin{aligned} \phi^{\dagger} \begin{pmatrix} \psi_{1\text{L}} \\ \psi_{2\text{L}} \end{pmatrix} &= \phi_1^{\dagger} \psi_{1\text{L}} + \phi_2^{\dagger} \psi_{2\text{L}}, \\ \phi^{\text{T}} \varepsilon \begin{pmatrix} \psi_{1\text{L}} \\ \psi_{2\text{L}} \end{pmatrix} &= \phi_1 \psi_{2\text{L}} - \phi_2 \psi_{1\text{L}}. \end{aligned} \tag{22.87}$$

Here ε is the antisymmetric 2×2 matrix (C.11). These invariants must be combined in all possible ways with the right-handed singlets. We must be careful, however, to ensure that the hypercharge is conserved. The resulting coupling is

$$\mathscr{L}_{\text{Yuk}} = -(\bar{\mathbf{e}}_R, \bar{\mu}_R, \bar{\tau}_R)\mathbf{C}_\ell \begin{pmatrix} \phi^\dagger \begin{pmatrix} \nu_{eL} \\ \mathbf{e}_L \end{pmatrix} \\ \phi^\dagger \begin{pmatrix} \nu_{\mu L} \\ \mu_L \end{pmatrix} \\ \phi^\dagger \begin{pmatrix} \nu_{\tau L} \\ \tau_L \end{pmatrix} \end{pmatrix} + (\bar{\mathbf{u}}_R, \bar{\mathbf{c}}_R, \bar{\mathbf{t}}_R)\mathbf{C}'_q \begin{pmatrix} \phi^T \varepsilon \begin{pmatrix} \mathbf{u}_L \\ \mathbf{d}'_L \end{pmatrix} \\ \phi^T \varepsilon \begin{pmatrix} \mathbf{c}_L \\ \mathbf{s}'_L \end{pmatrix} \\ \phi^T \varepsilon \begin{pmatrix} \mathbf{t}_L \\ \mathbf{b}'_L \end{pmatrix} \end{pmatrix}$$

$$- (\bar{\mathbf{d}}'_R, \bar{\mathbf{s}}'_R, \bar{\mathbf{b}}'_R)\mathbf{C}_q \begin{pmatrix} \phi^\dagger \begin{pmatrix} \mathbf{u}_L \\ \mathbf{d}'_L \end{pmatrix} \\ \phi^\dagger \begin{pmatrix} \mathbf{c}_L \\ \mathbf{s}'_L \end{pmatrix} \\ \phi^\dagger \begin{pmatrix} \mathbf{t}_L \\ \mathbf{b}'_L \end{pmatrix} \end{pmatrix} + \text{h.c.} \tag{22.88}$$

Here $\mathbf{C}_\ell$, $\mathbf{C}'_q$, and $\mathbf{C}_q$ are initially arbitrary complex 3×3 matrices. We shall soon see, however, that without loss of generality, we can write them in a certain standard form. It should also be noted that for the quarks, both types of coupling to the Higgs field occur, but for the leptons only one type. This stems from the fact that we have assumed the neutrinos to be massless and have not included right-handed neutrino fields.

Not all choices for the various matrices $\mathbf{C}_\ell$, $\mathbf{C}'_q$, and $\mathbf{C}_q$ lead to different physical theories. For example, we can carry out a change of basis for the right-handed lepton fields, i.e., make the substitution

$$\begin{pmatrix} \mathbf{e}_R \\ \mu_R \\ \tau_R \end{pmatrix} \to \mathbf{U}_1 \begin{pmatrix} \mathbf{e}_R \\ \mu_R \\ \tau_R \end{pmatrix}, \tag{22.89}$$

where $\mathbf{U}_1$ is any constant unitary 3×3 matrix $[\mathbf{U}_1 \in \mathrm{U}(3)]$. Since the fields $\mathbf{e}_R$, μ_R, and τ_R all carry the same quantum numbers, all terms in the Lagrange density (22.73) apart from the Yukawa term $\mathscr{L}_{\text{Yuk}}$ remain unchanged. For the latter the matrix $\mathbf{C}_\ell$ undergoes the transformation

$$\mathbf{C}_\ell \to \mathbf{U}_1^\dagger \mathbf{C}_\ell. \tag{22.90}$$

In an analogous way, we can carry out a basis change for the lepton doublets, the quark doublets, and the quark singlets since in each category the quantum numbers are the same. The matrices $\mathbf{C}_\ell$, $\mathbf{C}'_q$, and $\mathbf{C}_q$ are thereby transformed as follows:

$$\begin{aligned} \mathbf{C}_\ell &\to \mathbf{U}_1^\dagger \mathbf{C}_\ell \mathbf{V}_1, \\ \mathbf{C}'_q &\to \mathbf{U}_2^\dagger \mathbf{C}'_q \mathbf{V}_2, \\ \mathbf{C}_q &\to \mathbf{U}_3^\dagger \mathbf{C}_q \mathbf{V}_2, \end{aligned} \tag{22.91}$$

where $\mathbf{U}_1, \mathbf{U}_2, \mathbf{U}_3, \mathbf{V}_1, \mathbf{V}_2 \in \mathrm{U}(3)$ must hold. We therefore have the freedom to multiply the matrices $\mathbf{C}_\ell$, $\mathbf{C}'_q$ and $\mathbf{C}_q$ by constant unitary 3×3 matrices as given above. This does not change the physics since we are simply changing the basis within sets of equivalent fields. It should be noted that the matrices $\mathbf{C}'_q$ and $\mathbf{C}_q$ for the quarks must be multiplied from the right by the *same* matrix $\mathbf{V}_2$ since in (22.88) the same quark doublets appear to the right of $\mathbf{C}'_q$ and $\mathbf{C}_q$.

We next consider the lepton matrix $\mathbf{C}_\ell$. The matrix $\mathbf{C}_\ell\mathbf{C}_\ell^\dagger$ is Hermitian and nonnegative. Under the transformations (22.91) we have

$$\mathbf{C}_\ell\mathbf{C}_\ell^\dagger \to \mathbf{U}_1^\dagger\mathbf{C}_\ell\mathbf{C}_\ell^\dagger\mathbf{U}_1 . \tag{22.92}$$

By making a suitable choice of $\mathbf{U}_1$ we can arrange that $\mathbf{C}_\ell\mathbf{C}_\ell^\dagger$ becomes diagonal, i.e., transforms into

$$\mathbf{C}_\ell\mathbf{C}_\ell^\dagger = \begin{pmatrix} c_e^2 & 0 & 0 \\ 0 & c_\mu^2 & 0 \\ 0 & 0 & c_\tau^2 \end{pmatrix}, \tag{22.93}$$

where c_e, c_μ, and $c_\tau \geq 0$. Thus it clearly holds that

$$\mathbf{C}_\ell = \begin{pmatrix} c_e & 0 & 0 \\ 0 & c_\mu & 0 \\ 0 & 0 & c_\tau \end{pmatrix}\mathbf{W}, \tag{22.94}$$

where $\mathbf{W}$ is a unitary matrix [$\mathbf{W} \in \mathrm{U}(3)$]. If in (22.91) we now set

$$\mathbf{V}_1 = \mathbf{W}^\dagger, \tag{22.95}$$

then we can bring $\mathbf{C}_\ell$ into diagonal form with nonnegative diagonal elements:

$$\mathbf{C}_\ell = \begin{pmatrix} c_e & 0 & 0 \\ 0 & c_\mu & 0 \\ 0 & 0 & c_\tau \end{pmatrix}. \tag{22.96}$$

For the quarks we can begin in complete analogy by bringing the matrix $\mathbf{C}'_q$ into diagonal form:

$$\mathbf{C}'_q = \begin{pmatrix} c_u & 0 & 0 \\ 0 & c_c & 0 \\ 0 & 0 & c_t \end{pmatrix}, \tag{22.97}$$

where c_u, c_c, $c_t \geq 0$. According to (22.91), however, the matrix $\mathbf{C}_q$ can then only be multiplied from the left by an arbitrary matrix $\mathbf{U}_3^\dagger$. In analogy with (22.94) we can arrive at

$$\mathbf{C}_q = \begin{pmatrix} c_d & 0 & 0 \\ 0 & c_s & 0 \\ 0 & 0 & c_b \end{pmatrix}\mathbf{V}^\dagger, \tag{22.98}$$

where c_d, c_s, $c_b \geq 0$ and $\mathbf{V}^\dagger \in \mathrm{U}(3)$. If, corresponding to (22.91), we then transform again with

$$\mathbf{U}_3 = \mathbf{V}^\dagger$$

from the left, we can ensure that $\mathbf{C}_q$ becomes Hermitian. We can therefore demand that $\mathbf{C}_q$ takes on the canonical form

$$\mathbf{C}_q = \mathbf{V}\begin{pmatrix} c_d & 0 & 0 \\ 0 & c_s & 0 \\ 0 & 0 & c_b \end{pmatrix}\mathbf{V}^\dagger. \tag{22.99}$$

The matrix $\mathbf{V}$ is the so-called *Kobayashi–Maskawa* matrix (Kobayashi 1973).

We have now taken advantage of virtually all the freedom which the transformations (22.91) allow us. The forms of $\mathbf{C}'_q$ and $\mathbf{C}_q$ (22.97), (22.99) only remain unchanged under a transformation with a unitary diagonal matrix. We can see this if, in (22.91), we put

$$\mathbf{V}_2 = \mathbf{U}_2 = \mathbf{U}_3 = \mathbf{U}_\varphi \equiv \begin{pmatrix} e^{i\varphi_1} & 0 & 0 \\ 0 & e^{i\varphi_2} & 0 \\ 0 & 0 & e^{i\varphi_3} \end{pmatrix},$$

where φ_1, φ_2, and φ_3 are arbitrary phases. We then have

$$\begin{aligned} \mathbf{C}'_q &= \begin{pmatrix} c_u & 0 & 0 \\ 0 & c_c & 0 \\ 0 & 0 & c_t \end{pmatrix} \rightarrow \mathbf{U}^\dagger_\varphi \mathbf{C}'_q \mathbf{U}_\varphi = \mathbf{C}'_q, \\ \mathbf{C}_q &= \mathbf{V}\begin{pmatrix} c_d & 0 & 0 \\ 0 & c_s & 0 \\ 0 & 0 & c_b \end{pmatrix}\mathbf{V}^\dagger \rightarrow \mathbf{V}'\begin{pmatrix} c_d & 0 & 0 \\ 0 & c_s & 0 \\ 0 & 0 & c_b \end{pmatrix}\mathbf{V}'^\dagger, \end{aligned} \tag{22.100}$$

where $\mathbf{V}' = \mathbf{U}^\dagger_\varphi \mathbf{V}$. The form of $\mathbf{C}_q$ (22.99) clearly also remains unchanged if we multiply the matrix $\mathbf{V}$ from the right with a unitary diagonal matrix $\mathbf{U}_\chi$

$$\mathbf{U}_\chi = \begin{pmatrix} e^{i\chi_1} & 0 & 0 \\ 0 & e^{i\chi_2} & 0 \\ 0 & 0 & e^{i\chi_3} \end{pmatrix}, \tag{22.101}$$

where χ_1, χ_2, and χ_3 are again arbitrary phases. Thus the canonical forms that we have demanded for $\mathbf{C}'_q$ and $\mathbf{C}_q$ (22.97), (22.99) still allow us the freedom to replace the unitary matrix $\mathbf{V}$ as follows:

$$\mathbf{V} \rightarrow \mathbf{U}^\dagger_\varphi \mathbf{V} \mathbf{U}_\chi, \tag{22.102}$$

where $\mathbf{U}_\varphi$ and $\mathbf{U}_\chi$ must be diagonal matrices of U(3). We can use this to bring $\mathbf{V}$ itself into a canonical form.

We first consider the simpler case of two families instead of three (cf. Table 22.1). The argument runs exactly as before, except that $\mathbf{V}$ is now merely a 2×2 matrix for which, according to (22.102), we have the following freedom of transformations:

$$\begin{aligned} \mathbf{V} &= \begin{pmatrix} V_{11} & V_{12} \\ V_{21} & V_{22} \end{pmatrix} \rightarrow \begin{pmatrix} e^{-i\varphi_1} & 0 \\ 0 & e^{-i\varphi_2} \end{pmatrix}\begin{pmatrix} V_{11} & V_{12} \\ V_{21} & V_{22} \end{pmatrix}\begin{pmatrix} e^{i\chi_1} & 0 \\ 0 & e^{i\chi_2} \end{pmatrix} \\ &= \begin{pmatrix} e^{-i(\varphi_1-\chi_1)}\, V_{11} & e^{-i(\varphi_1-\chi_2)}\, V_{12} \\ e^{-i(\varphi_2-\chi_1)}\, V_{21} & e^{-i(\varphi_2-\chi_2)}\, V_{22} \end{pmatrix}. \end{aligned} \tag{22.103}$$

Three of the four phase differences, for example,

$$\varphi_1 - \chi_1, \quad \varphi_1 - \chi_2, \quad \varphi_2 - \chi_1$$

can be chosen at will, whereas the fourth is then fixed since

$$\varphi_2 - \chi_2 = (\varphi_2 - \chi_1) + (\varphi_1 - \chi_2) - (\varphi_1 - \chi_1).$$

We can make use of this to demand as canonical form for **V**

$$\begin{aligned} V_{11} &\geq 0, \quad \text{(a)} \\ V_{12} &\geq 0, \quad \text{(b)} \\ V_{21} &\leq 0. \quad \text{(c)} \end{aligned} \tag{22.104}$$

The unitarity of **V** further requires

$$\begin{aligned} |V_{11}|^2 + |V_{12}|^2 &= 1, \quad \text{(a)} \\ V_{11}^* V_{21} + V_{12}^* V_{22} &= 0, \quad \text{(b)} \\ |V_{21}|^2 + |V_{22}|^2 &= 1. \quad \text{(c)} \end{aligned} \tag{22.105}$$

The equations (22.104a), (22.104b) and (22.105a) can be identically fulfilled if we set

$$\begin{aligned} V_{11} &= \cos \vartheta_C, \\ V_{12} &= \sin \vartheta_C, \end{aligned} \tag{22.106}$$

where $0 \leq \vartheta_C \leq \pi/2$. The remaining equalities and inequalities then readily yield that the canonical form for **V** is given for two families by

$$\mathbf{V} = \begin{pmatrix} \cos \vartheta_C & \sin \vartheta_C \\ -\sin \vartheta_C & \cos \vartheta_C \end{pmatrix}. \tag{22.107}$$

The angle ϑ_C is, as we shall see shortly, the usual Cabibbo angle. It should be noted that in the case of two families, it has been possible to rotate away *all* the complex phases in **V**.

For the case of three families which, as far as we know today, is that realized in Nature, an analogous procedure shows that the matrix **V** can be brought into a canonical form that can be parametrized in terms of three angles ϑ_i $(i = 1, 2, 3)$ and a phase δ:

$$\mathbf{V} = \begin{pmatrix} c_1 & s_1 c_3 & s_1 s_3 \\ -s_1 c_2 & c_1 c_2 c_3 - s_2 s_3 \, e^{i\delta} & c_1 c_2 s_3 + s_2 c_3 \, e^{i\delta} \\ -s_1 s_2 & c_1 s_2 c_3 + c_2 s_3 \, e^{i\delta} & c_1 s_2 s_3 - c_2 c_3 \, e^{i\delta} \end{pmatrix}. \tag{22.108}$$

Here the abbreviations

$$\begin{aligned} c_i &= \cos \vartheta_i, \\ s_i &= \sin \vartheta_i \qquad (i = 1, 2, 3), \end{aligned}$$

have been used and the ranges for the parameters are

$$\begin{aligned} 0 \leq \vartheta_i \leq \pi/2, \\ 0 \leq \delta \leq 2\pi. \end{aligned}$$

The derivation of this form for the Kobayashi–Maskawa matrix is discussed in Appendix H. It turns out that for the case of three families, it is *not* generally possible to rotate away all the complex phases in the matrix $\mathbf{V}$. This will be an important factor in the discussion of CP violation. A special case occurs if two quark masses are equal and/or one of the angles $\vartheta_i = 0$ or $\pi/2$. In this case one can always obtain a real matrix $\mathbf{V}$ with suitable transformations (22.91).

We now apply the preceding results to reduce the Yukawa coupling (22.88) to a canonical form. In the unitary gauge (22.58), with a physical Higgs field ρ' as defined in (22.61), we end up with:

$$\begin{aligned}
\mathcal{L}_{\mathrm{Yuk}} = \Bigg\{ &-(\bar{\mathbf{e}}_{\mathrm{R}}, \bar{\mu}_{\mathrm{R}}, \bar{\tau}_{\mathrm{R}}) \begin{pmatrix} c_{\mathrm{e}} & 0 & 0 \\ 0 & c_\mu & 0 \\ 0 & 0 & c_\tau \end{pmatrix} \begin{pmatrix} \mathbf{e}_{\mathrm{L}} \\ \mu_{\mathrm{L}} \\ \tau_{\mathrm{L}} \end{pmatrix} \\
&- (\bar{\mathbf{u}}_{\mathrm{R}}, \bar{\mathbf{c}}_{\mathrm{R}}, \bar{\mathbf{t}}_{\mathrm{R}}) \begin{pmatrix} c_{\mathrm{u}} & 0 & 0 \\ 0 & c_{\mathrm{c}} & 0 \\ 0 & 0 & c_{\mathrm{t}} \end{pmatrix} \begin{pmatrix} \mathbf{u}_{\mathrm{L}} \\ \mathbf{c}_{\mathrm{L}} \\ \mathbf{t}_{\mathrm{L}} \end{pmatrix} \\
&-(\bar{\mathbf{d}}'_{\mathrm{R}}, \bar{\mathbf{s}}'_{\mathrm{R}}, \bar{\mathbf{b}}'_{\mathrm{R}})\mathbf{V} \begin{pmatrix} c_{\mathrm{d}} & 0 & 0 \\ 0 & c_{\mathrm{s}} & 0 \\ 0 & 0 & c_{\mathrm{b}} \end{pmatrix} \mathbf{V}^\dagger \begin{pmatrix} \mathbf{d}'_{\mathrm{L}} \\ \mathbf{s}'_{\mathrm{L}} \\ \mathbf{b}'_{\mathrm{L}} \end{pmatrix} + \text{h.c.} \Bigg\} \frac{\rho_0}{\sqrt{2}} \left(1 + \frac{\rho'}{\rho_0}\right) \\
= \Bigg\{ &-(\bar{\mathbf{e}}, \bar{\mu}, \bar{\tau}) \begin{pmatrix} m_{\mathrm{e}} & 0 & 0 \\ 0 & m_\mu & 0 \\ 0 & 0 & m_\tau \end{pmatrix} \begin{pmatrix} \mathbf{e} \\ \mu \\ \tau \end{pmatrix} \\
&- (\bar{\mathbf{u}}, \bar{\mathbf{c}}, \bar{\mathbf{t}}) \begin{pmatrix} m_{\mathrm{u}} & 0 & 0 \\ 0 & m_{\mathrm{c}} & 0 \\ 0 & 0 & m_{\mathrm{t}} \end{pmatrix} \begin{pmatrix} \mathbf{u} \\ \mathbf{c} \\ \mathbf{t} \end{pmatrix} \\
&- (\bar{\mathbf{d}}', \bar{\mathbf{s}}', \bar{\mathbf{b}}')\mathbf{V} \begin{pmatrix} m_{\mathrm{d}} & 0 & 0 \\ 0 & m_{\mathrm{s}} & 0 \\ 0 & 0 & m_{\mathrm{b}} \end{pmatrix} \mathbf{V}^\dagger \begin{pmatrix} \mathbf{d}' \\ \mathbf{s}' \\ \mathbf{b}' \end{pmatrix} \Bigg\} \left(1 + \frac{\rho'}{\rho_0}\right) .
\end{aligned} \tag{22.109}$$

By definition:

$$m_{\mathrm{e}} = c_{\mathrm{e}} \frac{\rho_0}{\sqrt{2}}, \ldots, m_{\mathrm{b}} = c_{\mathrm{b}} \frac{\rho_0}{\sqrt{2}} . \tag{22.110}$$

The quantities $m_{\mathrm{e}}, m_\mu, m_\tau, m_{\mathrm{u}}, m_{\mathrm{c}}$, and m_{t} can immediately be identified as the masses of the corresponding leptons and quarks. However, for a Kobayashi–Maskawa matrix $\mathbf{V}$ not equal to the unit matrix, we see that the quark fields $\mathbf{d}'$, $\mathbf{s}'$, and $\mathbf{b}'$ do *not* correspond to particles of well-defined mass. The fields $\mathbf{d}'$, $\mathbf{s}'$, and $\mathbf{b}'$ were defined as isospin partners of the fields $\mathbf{u}$, $\mathbf{c}$, and $\mathbf{t}$. The fields $\mathbf{d}$, $\mathbf{s}$, and $\mathbf{b}$, which do correspond to quarks of well-defined mass are given according to (22.109) by

$$\begin{pmatrix} \mathbf{d} \\ \mathbf{s} \\ \mathbf{b} \end{pmatrix} = \mathbf{V}^\dagger \begin{pmatrix} \mathbf{d}' \\ \mathbf{s}' \\ \mathbf{b}' \end{pmatrix} . \tag{22.111}$$

What are the physical consequences of this generalized Cabibbo rotation? To answer this question we consider the charged and neutral currents (22.78). Since the charged current contains the isospin raising operator, the fields **u**, **c**, and **t** are connected here with their isospin partners **d**′, **s**′, and **b**′, and we find

$$\begin{aligned}\mathscr{J}^{\lambda}_{\mathrm{CC}} &= (\bar{\nu}_{\mathrm{eL}}, \bar{\nu}_{\mu\mathrm{L}}, \bar{\nu}_{\tau\mathrm{L}})\gamma^{\lambda}\begin{pmatrix}\mathbf{e}_{\mathrm{L}}\\ \mu_{\mathrm{L}}\\ \tau_{\mathrm{L}}\end{pmatrix} + (\bar{\mathbf{u}}_{\mathrm{L}}, \bar{\mathbf{c}}_{\mathrm{L}}, \bar{\mathbf{t}}_{\mathrm{L}})\gamma^{\lambda}\begin{pmatrix}\mathbf{d}'_{\mathrm{L}}\\ \mathbf{s}'_{\mathrm{L}}\\ \mathbf{b}'_{\mathrm{L}}\end{pmatrix}\\ &= (\bar{\nu}_{\mathrm{eL}}, \bar{\nu}_{\mu\mathrm{L}}, \bar{\nu}_{\tau\mathrm{L}})\gamma^{\lambda}\begin{pmatrix}\mathbf{e}_{\mathrm{L}}\\ \mu_{\mathrm{L}}\\ \tau_{\mathrm{L}}\end{pmatrix} + (\bar{\mathbf{u}}_{\mathrm{L}}, \bar{\mathbf{c}}_{\mathrm{L}}, \bar{\mathbf{t}}_{\mathrm{L}})\gamma^{\lambda}\mathbf{V}\begin{pmatrix}\mathbf{d}_{\mathrm{L}}\\ \mathbf{s}_{\mathrm{L}}\\ \mathbf{b}_{\mathrm{L}}\end{pmatrix}.\end{aligned} \tag{22.112}$$

Thus the matrix **V** occurs in the charged current and produces, for example, the following series of transitions into the u-quark, with amplitudes that are proportional to the matrix element of **V** indicated in the brackets:

$$\begin{aligned} \mathrm{d} &\to \mathrm{u} + \mathrm{W}^- \quad (V_{11}),\\ \mathrm{s} &\to \mathrm{u} + \mathrm{W}^- \quad (V_{12}),\\ \mathrm{b} &\to \mathrm{u} + \mathrm{W}^- \quad (V_{13}).\end{aligned} \tag{22.113}$$

The transition d → u, for example, is responsible for the decay of the neutron, and the transition s → u for the decay of the Λ-hyperon (Fig. 22.5). We also note that, according to (22.112), in the lepton part of the charged current no mixing matrix occurs. This stems, as we have seen, from the fact that we have assumed all neutrinos to be massless. For massive neutrinos, one would, in general, also obtain a mixing matrix for the leptons. This would lead, among other things, to neutrino oscillations (Langacker 1989).

We will also look at the neutral current. On account of the *unitarity* of the

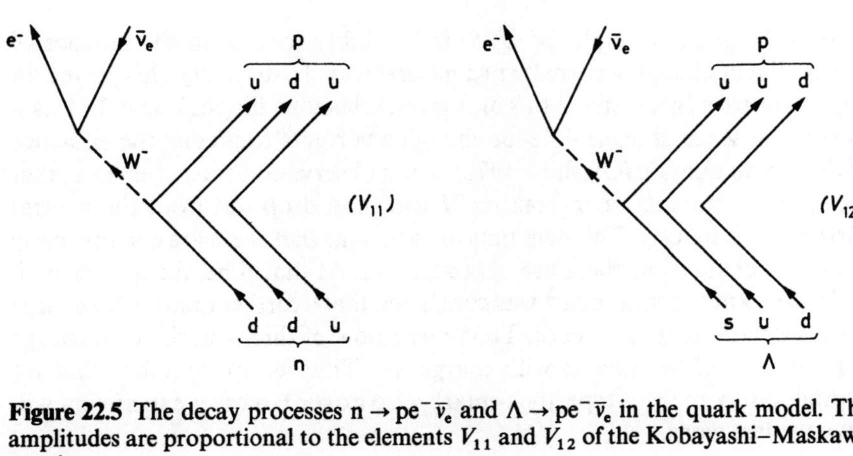

Figure 22.5 The decay processes $n \to pe^-\bar{\nu}_e$ and $\Lambda \to pe^-\bar{\nu}_e$ in the quark model. The amplitudes are proportional to the elements V_{11} and V_{12} of the Kobayashi–Maskawa matrix.

Kobayashi–Maskawa matrix **V**, we find from (22.78)

$$\begin{aligned}\mathscr{J}^{\lambda}_{\mathrm{NC}} = (\bar{\nu}_e, \bar{\nu}_\mu, \bar{\nu}_\tau)\gamma^\lambda \frac{1}{2}\frac{1-\gamma_5}{2}\begin{pmatrix}\nu_e\\ \nu_\mu\\ \nu_\tau\end{pmatrix} \\ + (\bar{\mathbf{e}}, \bar{\mu}, \bar{\tau})\gamma^\lambda\left(-\frac{1}{2}\frac{1-\gamma_5}{2} + \sin^2\vartheta_{\mathrm{W}}\right)\begin{pmatrix}\mathbf{e}\\ \mu\\ \tau\end{pmatrix} \\ + (\bar{\mathbf{u}}, \bar{\mathbf{c}}, \bar{\mathbf{t}})\gamma^\lambda\left(\frac{1}{2}\frac{1-\gamma_5}{2} - \tfrac{2}{3}\sin^2\vartheta_{\mathrm{W}}\right)\begin{pmatrix}\mathbf{u}\\ \mathbf{c}\\ \mathbf{t}\end{pmatrix} \\ + (\bar{\mathbf{d}}, \bar{\mathbf{s}}, \bar{\mathbf{b}})\gamma^\lambda\left(-\frac{1}{2}\frac{1-\gamma_5}{2} + \tfrac{1}{3}\sin^2\vartheta_{\mathrm{W}}\right)\begin{pmatrix}\mathbf{d}\\ \mathbf{s}\\ \mathbf{b}\end{pmatrix}.\end{aligned} \tag{22.114}$$

It is noteworthy that no flavor changing terms appear here. For example, there is no transition

$$\mathrm{s} \to \mathrm{d} + \mathrm{Z}. \tag{22.115}$$

Such a transition could give rise to reactions such as the following decay of the long-lived K-meson:

$$\mathrm{K_L} \sim (\mathrm{s\bar{d}} + \mathrm{\bar{s}d}) \to \mu^+ + \mu^-. \tag{22.116}$$

Experimentally, this decay is very strongly suppressed. If we compare this with the kinematically similar decay of the positive K-meson

$$\mathrm{K}^+ \to \mu^+ + \nu_\mu, \tag{22.117}$$

then we find

$$\frac{\Gamma(\mathrm{K_L} \to \mu^+\mu^-)}{\Gamma(\mathrm{K}^+ \to \mu^+\nu_\mu)} = (3.42 \pm 0.71)\cdot 10^{-9}. \tag{22.118}$$

One of the great triumphs of QFD is its ability to explain the absence of neutral flavor changing currents in a natural way. Historically, this played an important role: By reversing this argument, Glashow, Iliopoulos, and Maiani used the absence of neutral flavor changing currents to predict the existence of the charm quark c (Glashow 1970). Their observation was, in essence, that the charged current mixing matrix **V** can only drop out from the neutral current if it is unitary. This demands in particular that there are equally many quarks of charge $\frac{2}{3}$ as there are of charge $-\frac{1}{3}$. At that time, the quarks u, d, and s were known, and thus it was concluded that a further quark with charge $\frac{2}{3}$, the c-quark, must also exist. Today we know of three quarks with charge $-\frac{1}{3}$, but only of two quarks with charge $+\frac{2}{3}$. Thus we are confident that yet another quark of the u-type, the t-quark, also exists. However experiment will have the last word.

22.5 The Lagrange Density of the Standard Model

So far in this chapter, we have only been considering the electroweak interaction and have ignored the strong force. We now want to combine the corresponding theories, QFD and QCD, into the *standard model.*

We begin this discussion by again arranging the fundamental fermions of Table 22.2 into a total spinor ψ as indicated in (22.70). Along with the SU(2) × U(1) group of the electroweak interaction, we now also consider the color-SU(3) group of the strong interaction (Chap. 19). With respect to this SU(3) group, all leptons are singlets since they do not participate in the strong interaction, and all quarks are triplets. The space of the spinors ψ thus also carries a nontrivial representation of the color-SU(3) group. The generating operators of this—highly reducible—representation will be denoted by $\mathbf{F}_a$ ($a = 1, \ldots, 8$). When $\mathbf{F}_a$ is applied to the spinor ψ (22.70) it multiplies all lepton fields by zero and all quark fields by $\lambda_a/2$:

$$\mathbf{F}_a \begin{pmatrix} \nu_{eL} \\ \vdots \\ \tau_R \\ \mathbf{u}_L \\ \vdots \\ \mathbf{b}_R \end{pmatrix} = \begin{pmatrix} 0 \\ \vdots \\ 0 \\ \frac{\lambda_a}{2}\mathbf{u}_L \\ \vdots \\ \frac{\lambda_a}{2}\mathbf{b}_R \end{pmatrix}. \tag{22.119}$$

The generators of the color-SU(3) group obviously commute with the generators $\mathbf{T}_a$ ($a = 1, 2, 3$) and $\mathbf{Y}$ of the SU(2) × U(1) group of the electroweak interaction. Together $\mathbf{F}_a$, $\mathbf{T}_a$, and $\mathbf{Y}$ generate the group

$$\mathscr{F} = \mathrm{SU}(3) \times \mathrm{SU}(2) \times \mathrm{U}(1), \tag{22.120}$$

which is the direct product of the color group SU(3), the group SU(2) of the weak isospin, and the group U(1) of the weak hypercharge.

This group $\mathscr{F}$ is the fundamental group that is gauged in the standard model, whereby the strong and electroweak forces are generated. Since we are dealing here with a direct product, we do not encounter any complications; however, neither do we get any interrelations between the strong and the electroweak forces. By applying the rules of Sect. 19.1 we can immediately extend the Lagrange density of QFD (22.73), (22.74) to become a gauge theory also with respect to the color-SU(3) group. The resulting Lagrange density, written in manifestly gauge invariant form, is as follows:

$$\begin{aligned}\mathscr{L} = &-\tfrac{1}{2}\,\mathrm{Tr}(\mathbf{G}_{\lambda\rho}\mathbf{G}^{\lambda\rho}) - \tfrac{1}{2}\,\mathrm{Tr}(\mathbf{W}_{\lambda\rho}\mathbf{W}^{\lambda\rho}) - \tfrac{1}{4}\mathbf{B}_{\lambda\rho}\mathbf{B}^{\lambda\rho} \\ &+ \bar{\psi}\mathrm{i}\gamma^\lambda \mathbf{D}_\lambda\psi + \mathscr{L}_{\mathrm{Yuk}} + (\mathbf{D}_\lambda\phi)^\dagger(\mathbf{D}^\lambda\phi) - V(\phi).\end{aligned} \tag{22.121}$$

Here $\mathbf{G}_{\lambda\rho}$ is the gluon field-strength tensor (19.19), and, according to (19.14) and (22.71), the covariant derivative of the spinor reads:

$$\mathbf{D}_\lambda \psi = (\partial_\lambda + \mathrm{i} g_s \mathbf{G}_\lambda^a \mathbf{F}_a + \mathrm{i} g \mathbf{W}_\lambda^a \mathbf{T}_a + \mathrm{i} g' \mathbf{B}_\lambda \mathbf{Y})\psi. \tag{22.122}$$

We regard the Higgs field ϕ as a color singlet. Thus the Higgs part of $\mathscr{L}$ and the Yukawa term $\mathscr{L}_{\mathrm{Yuk}}$ are identical to those of (22.73).

The mechanism of the spontaneous symmetry breaking of the SU(2) × U(1) part of the gauge group $\mathscr{F}$ is exactly the same as for QFD alone. The gluons are not involved and the SU(3) part of $\mathscr{F}$ remains unbroken. In terms of the physical fields the Lagrange density takes on the explicit form:

$$\begin{aligned}\mathscr{L} = &-\tfrac{1}{2}\operatorname{Tr}(\mathbf{G}_{\lambda\rho}\mathbf{G}^{\lambda\rho}) - \tfrac{1}{2}\operatorname{Tr}(\mathbf{W}_{\lambda\rho}\mathbf{W}^{\lambda\rho}) - \tfrac{1}{4}\mathbf{B}_{\lambda\rho}\mathbf{B}^{\lambda\rho} \\ &+ \mathbf{W}_\lambda^+ \mathbf{W}^{-\lambda} m_{\mathrm{W}}^2 \left(1 + \frac{\rho'}{\rho_0}\right)^2 + \tfrac{1}{2}\mathbf{Z}_\lambda \mathbf{Z}^\lambda m_{\mathrm{Z}}^2 \left(1 + \frac{\rho'}{\rho_0}\right)^2 \\ &+ \sum_\ell \left\{ \overline{\nu}_{\ell\mathrm{L}} \mathrm{i}\gamma^\lambda \partial_\lambda \nu_{\ell\mathrm{L}} + \overline{\ell}\left[\mathrm{i}\gamma^\lambda\partial_\lambda - m_\ell\left(1 + \frac{\rho'}{\rho_0}\right)\right]\ell \right\} \\ &+ \sum_{\mathbf{q}} \overline{\mathbf{q}}\left[\mathrm{i}\gamma^\lambda\left(\partial_\lambda + \mathrm{i} g_s \mathbf{G}_\lambda^a \frac{\lambda_a}{2}\right) - m_{\mathbf{q}}\left(1 + \frac{\rho'}{\rho_0}\right)\right]\mathbf{q} \\ &+ \mathscr{L}_{\mathrm{int}} + \tfrac{1}{2}\partial_\lambda\rho'\partial^\lambda\rho' - \tfrac{1}{2}m_{\rho'}^2\rho'^2\left[1 + \frac{\rho'}{\rho_0} + \frac{1}{4}\left(\frac{\rho'}{\rho_0}\right)^2\right].\end{aligned} \tag{22.123}$$

The indices $\ell = \mathrm{e}, \mu, \tau$ and $\mathrm{q} = \mathrm{u, c, t, d, s, b}$ are to be summed over. The term $\mathscr{L}_{\mathrm{int}}$ is given in (22.77). The currents $\mathscr{J}_{\mathrm{CC}}^\lambda$ and $\mathscr{J}_{\mathrm{NC}}^\lambda$ that occur there are to be taken from (22.112), (22.114) and the electromagnetic current is

$$\mathscr{J}_{\mathrm{em}}^\lambda = \sum_\ell - \overline{\ell}\gamma^\lambda\ell + \sum_{\mathbf{q}} Q_{\mathbf{q}}\overline{\mathbf{q}}\gamma^\lambda\mathbf{q}, \tag{22.124}$$

where Q_{q} are the quark charges. For convenience we continued in (22.123) to make reference to the field-strength tensors $\mathbf{W}_{\lambda\rho}$ and $\mathbf{B}_{\lambda\rho}$. It is understood that they too are expressed in terms of the physical boson fields $\mathbf{W}_\lambda^\pm$, $\mathbf{Z}_\lambda$, and $\mathbf{A}_\lambda$.

The gauge symmetry that is manifest in (22.123) corresponds to the unbroken part SU(3) × $\mathrm{U_{em}}(1)$ of the full gauge group $\mathscr{F}$. The typical terms that we have obtained from the spontaneous symmetry breaking are the mass terms for the W- and Z-bosons. However, for very large energies $E \gg m_{\mathrm{W}} \cong m_{\mathrm{Z}} \cong 100$ GeV, we can neglect these boson masses. Apart from small corrections of the order of m_{W}/E, the scattering cross-sections will then satisfy the symmetry relations demanded by the full gauge group $\mathscr{F}$. Thus at high energies the underlying gauge group is directly manifest. In contrast to this, for energies $E \ll m_{\mathrm{W}}, m_{\mathrm{Z}}$, we only see the unbroken part of $\mathscr{F}$ as manifest symmetry group in the scattering cross-sections and decay rates. This state of affairs can be represented schematically as follows:

$$\mathscr{F} = \mathrm{SU}(3) \times \mathrm{SU}(2) \times \mathrm{U}(1) \xrightarrow[m_{\mathrm{Z}} \cong 100\,\mathrm{GeV}]{} \mathrm{SU}(3) \times \mathrm{U_{em}}(1), \tag{22.125}$$

where the arrow indicates spontaneous symmetry breaking and the typical mass scale connected with this is written underneath.

We have now encountered all the properties of the standard model that we need to pursue the phenomenology. One of our tasks will be to show how those parameters of the standard model which have not already been treated in Part III, are determined experimentally. For reference we list here a complete set of independent parameters. There are

3 coupling constants

$$g_s, e, \sin\vartheta_W$$

2 boson masses

$$m_W, m'_\rho$$

3 lepton masses

$$m_e, m_\mu, m_\tau$$

6 quark masses

$$m_u, m_d, m_c, m_s, m_t, m_b$$

and 4 parameters of the Kobayashi–Maskawa matrix

$$\vartheta_1, \vartheta_2, \vartheta_3, \delta.$$

The mass of the Z-boson is given by (22.68). Thus there are a total of 18 parameters, a figure which does not include the neutrino masses which are assumed to be zero. Such a number of independent parameters would seem to be rather high for a truly fundamental theory. If it should be experimentally shown that the neutrinos also have mass, then this number will substantially increase because in this case one needs to consider a Kobayashi–Maskawa matrix in the lepton sector, and also the possibility of so-called "Majorana mass terms" for the neutrinos (for a review see Frampton (1982)).

Exercises

22.1 Consider the Lagrange density $\mathscr{L}_\phi$ (22.39) for the classical field ϕ. With the help of (3.104)–(3.107) calculate the corresponding Hamiltonian.

22.2 Consider the gauge transformations that correspond to the generator $\mathbf{T}_3 + \mathbf{Y}$ of the SU(2) × U(1) group [see (22.60)]. Show that these transformations yield the U(1) gauge group of electromagnetism (9.7).

22.3 Consider the Lagrange density (22.77) and the Feynman rules of Tables 22.3 and 22.4 that follow from it. Show that the diagrams for fermion–fermion scattering in Fig. 22.4 which involve the exchange of a W- or Z-boson lead, at low energies, to the following S-matrix:

$$\mathbf{S} = 1 + \mathrm{i}\int \mathrm{d}x\, \mathscr{L}_{\mathrm{eff}}(x) \tag{22.126}$$

where $\mathscr{L}_{\mathrm{eff}}(x)$ is given in (22.81).

CHAPTER 23

Decay Processes in the Standard Model and the Determination of the Quark Mixing Angles in the Charged Current

23.1 The Decay of the Muon

The muon decays with a branching ratio of almost 100 % into an electron, an electron antineutrino, and a muon neutrino:

$$\mu^- \to e^- + \bar{\nu}_e + \nu_\mu .$$

To leading order of perturbation theory in the standard model the muon can only decay by the process, which is described by the diagram in Fig. 23.1. If this diagram is evaluated according to the Feynman rules of QFD (Appendix G), we obtain for the S-matrix element

$$\begin{aligned} S_{fi} &= (2\pi)^4 \delta(p_1 - p_2 - p_3 - p_4)\left(-\mathrm{i}\frac{e}{\sqrt{2}\sin\vartheta_W}\right)^2 \\ &\quad \cdot \bar{u}(p_2)\frac{1+\gamma_5}{2}\gamma^\lambda\frac{1-\gamma_5}{2}u(p_1)\frac{\mathrm{i}g_{\lambda\rho}}{m_W^2} \\ &\quad \cdot \bar{u}(p_3)\frac{1+\gamma_5}{2}\gamma^\rho\frac{1-\gamma_5}{2}v(p_4) \\ &= (2\pi)^4 \delta(p_1 - p_2 - p_3 - p_4)\left(-\mathrm{i}\frac{G}{\sqrt{2}}\right)\bar{u}(p_2)\gamma^\lambda(1-\gamma_5)u(p_1) \\ &\quad \cdot \bar{u}(p_3)\gamma_\lambda(1-\gamma_5)v(p_4). \end{aligned} \tag{23.1}$$

Here we have used the low energy approximation for the W-boson propagator

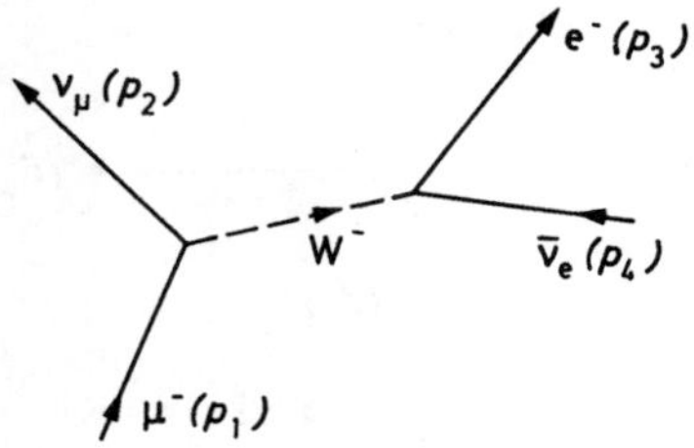

Figure 23.1 The lowest order diagram for muon decay. The four-momenta of the particles are p_j (with $j = 1, \ldots, 4$).

(Table 22.4). Since all momentum transfers that occur are at most of the order of the muon mass, this amounts to neglecting terms of order m_μ^2/m_W^2.

Using the formulas of Sect. 5.4, we obtain from (23.1) the following result for the inverse lifetime τ_μ^{-1} of the muon:

$$\tau_\mu^{-1} = \Gamma(\mu^- \to e^- + \bar{\nu}_e + \nu_\mu) = \frac{G^2 m_\mu^5}{192\pi^3}. \tag{23.2a}$$

Here we have neglected the electron mass in comparison to the muon mass. The more exact formula, in which electromagnetic radiative corrections to order α are also taken into account reads (see Bég 1982)

$$\tau_\mu^{-1} = \frac{G^2 m_\mu^5}{192\pi^3}\left(1 - 8\frac{m_e^2}{m_\mu^2}\right)\left(1 + \frac{\alpha}{2\pi}\left(\frac{25}{4} - \pi^2\right)\right). \tag{23.2b}$$

What about even higher order radiative corrections? Here the question of the precise definition of G arises. We may *define* G by means of (23.2b), i.e., by postulating this equation to get *no* higher order radiative corrections at all. Of course, other relations involving G, for instance (22.84), will then acquire (computable) corrections. Proceeding in this way one obtains from the measured lifetime of the muon

$$\tau_\mu = 2.19709(5) \cdot 10^{-6}\ \mathrm{s}$$

the following high precision value for the Fermi constant:

$$G = 1.16632(2) \cdot 10^{-5}\ \mathrm{GeV}^{-2}. \tag{23.3}$$

We will now study the energy distribution of the emitted electron in the rest frame of the decaying muon (Fig. 23.2). Simple kinematics yields the following relationship between the invariant mass W of the neutrino pair and the electron energy E_e

$$W^2 = (p_2 + p_4)^2,$$
$$E_e = \frac{m_\mu^2 - W^2}{2m_\mu}. \tag{23.4}$$

The maximum electron energy thus occurs for $W = 0$, corresponding to the configuration of Fig. 23.2(b), and is given by

$$E_{e\,\max} = \frac{m_\mu}{2}. \tag{23.5}$$

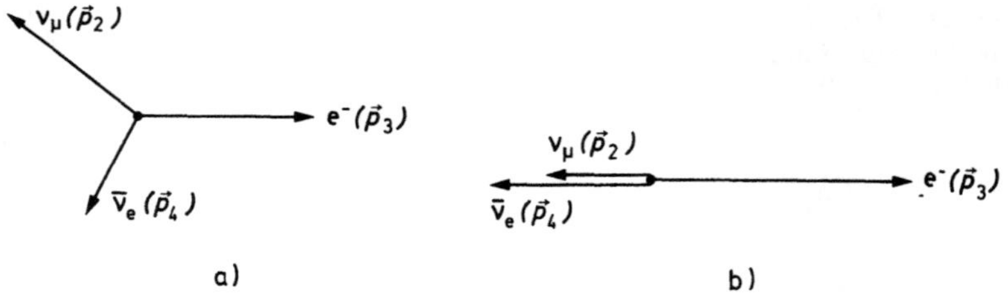

Figure 23.2 A general momentum configuration (a) and the configuration with maximum electron energy (b) in the decay of a muon at rest.

We now define

$$x = \frac{E_e}{E_{e\,\max}} = \frac{2E_e}{m_\mu}, \tag{23.6}$$

such that $0 \leq x \leq 1$. The distribution in x can also be readily calculated from the S-matrix element (23.1). One finds

$$\begin{aligned} \frac{1}{\Gamma}\frac{d\Gamma}{dx} &= 12x^2[1 - x + \tfrac{2}{3}\rho(\tfrac{4}{3}x - 1)], \\ \rho &= \tfrac{3}{4}. \end{aligned} \tag{23.7}$$

Here ρ is the Michel parameter. As Michel (1950) showed for parity-conserving couplings and, together with Bouchiat (Bouchiat 1957) for the general case, every four-fermion coupling with V, A, S, P, and T terms, as in (21.7), leads to an energy spectrum of the form (23.7), but in general with different values for the ρ parameter.

Let us assume, for example, that the muon couples to its neutrino via a right-handed (V + A) current, and the electron to its neutrino with a (V − A) current as before. A simple calculation then yields $\rho = 0$ for the Michel parameter. In this case the energy spectrum (23.7) has a zero for $x = 1$, i.e., the configuration with maximum electron energy (Fig. 23.2(b)) would be forbidden. This can be readily understood as a consequence of angular momentum conservation. For the above coupling, ν_μ and $\bar{\nu}_e$ would have helicity $+\frac{1}{2}$ and the electron helicity $-\frac{1}{2}$. In the configuration of Fig. 23.2(b) this would lead to an angular momentum of $-\frac{3}{2}$ with respect to the direction of motion of the electron. This, however, is forbidden, since the decaying muon only has spin $\frac{1}{2}$.

The measurement of the Michel parameter thus provides a good test of the standard theory, in which only left-handed currents couple to the W-boson. The experimental result is

$$\rho = 0.752 \pm 0.003 \tag{23.8}$$

in very good agreement with the value of $\frac{3}{4}$ demanded by the standard model.

With this we conclude our brief discussion of muon decay. Further discussion and other references can be found in Scheck (1978). Higher order corrections have been calculated by Sirlin (1980).

23.2 The Decay of the τ-Lepton

The heavy lepton τ was discovered in 1975 at the storage ring SPEAR in Stanford and shortly afterwards was also observed at the DORIS ring in Hamburg (Perl 1975; Feldman 1976; Burmester 1977, 1977a). The mass of the τ is

$$m_\tau = 1784.2 \pm 3.2 \text{ MeV}. \tag{23.9}$$

Today we know of a large number of decay modes of the τ, purely leptonic ones and semihadronic modes:

$$\tau^- \to \nu_\tau + e^- + \bar{\nu}_e, \qquad \nu_\tau + \mu^- + \bar{\nu}_\mu, \tag{23.10a}$$

$$\tau^- \to \nu_\tau + \text{hadrons}. \tag{23.10b}$$

In the standard model these decay processes are described by the diagrams shown in Fig. 23.3. The hadrons are imagined in the sense of the parton model to result from the fragmentation of the quarks.

In the standard model the τ-lepton couples to the W-boson in exactly the same way as the muon and the electron. Thus, for purely leptonic decay processes, we can immediately take over the decay rate from Sect. 23.1. Neglecting the masses of the muon and the electron in comparison to the τ mass, we have

$$\Gamma(\tau^- \to \nu_\tau + e^- + \bar{\nu}_e) = \Gamma(\tau^- \to \nu_\tau + \mu^- + \bar{\nu}_\mu) = \frac{G^2 m_\tau^5}{192\pi^3}. \tag{23.11}$$

As was the case for muon decay, a measurement of the electron or muon spectrum provides us with information about whether the τ couples to its neutrino via a (V − A) current. For the Michel parameter of the τ, the

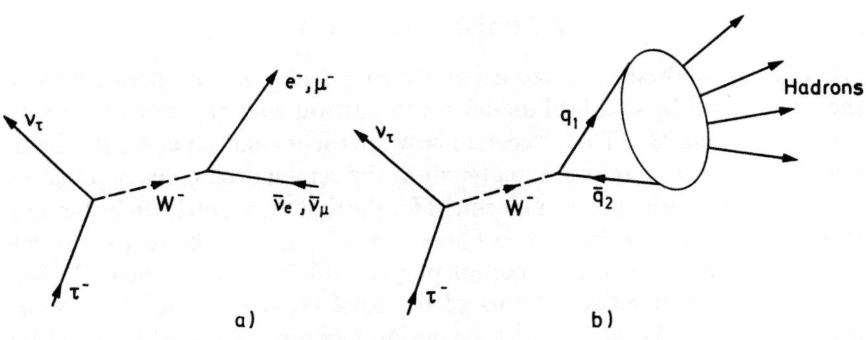

Figure 23.3 The diagrams for the decay of the τ-lepton into leptons (a) and into ν_τ + hadrons (b) in the standard model to leading order in the perturbation expansion of the electroweak interaction.

experimental value combined from the electronic and muonic decays was found to be

$$\rho = 0.71 \pm 0.08. \tag{23.12}$$

This is nicely compatible with the standard model and excludes a pure right-handed $\tau - \nu_\tau - \mathrm{W}$ coupling leading to $\rho = 0$.

We now estimate the rate for the decay of the τ into a neutrino and hadrons (Fig. 23.3(b)). For this we make use of the parton model. We argue that the transformation of the quark pair into hadrons occurs with a probability of one. The rate for the decay $\tau^- \to \nu_\tau +$ hadrons is then given by the rate for the decay $\tau^- \to \nu_\tau +$ quark–antiquark pairs,

$$\tau^- \to \nu_\tau + \mathrm{q}_1 + \bar{\mathrm{q}}_2. \tag{23.13}$$

Since the c- and t-quarks are too heavy, the only energetically allowed processes are

$$\begin{aligned} &\tau^- \to \nu_\tau + \mathrm{d} + \bar{\mathrm{u}} \qquad (V_{11} \cong \cos \vartheta_C); \\ &\tau^- \to \nu_\tau + \mathrm{s} + \bar{\mathrm{u}} \qquad (V_{12} \cong \sin \vartheta_C). \end{aligned} \tag{23.14}$$

The amplitudes for these two reactions are proportional to the matrix elements V_{11} and V_{12} of the Kobayashi–Maskawa matrix. These are well approximated by $\cos \vartheta_C$ and $\sin \vartheta_C$. The quark final state $\mathrm{d}\bar{\mathrm{u}}$ ($\mathrm{s}\bar{\mathrm{u}}$) leads to a hadron state with strangeness $S = 0$ ($S = 1$). If we now neglect the masses of the u-, d-, and s-quarks compared to the mass of the τ, then the matrix element for the decay of the τ into $\nu_\tau +$ quarks can be calculated exactly as for the purely leptonic decay. We must take into account, however, that the quarks are colored, i.e., that we have to sum over three different final states. For the decay rates this yields

$$\begin{aligned} &\Gamma(\tau^- \to \nu_\tau + \mathrm{d} + \bar{\mathrm{u}} \to \nu_\tau + \text{hadrons } (S = 0)) \\ &\qquad \cong 3 \cos^2 \vartheta_C \cdot \Gamma(\tau^- \to \nu_\tau + \mathrm{e}^- + \bar{\nu}_\mathrm{e}), \\ &\Gamma(\tau^- \to \nu_\tau + \mathrm{s} + \bar{\mathrm{u}} \to \nu_\tau + \text{hadrons } (S = -1)) \\ &\qquad \cong 3 \sin^2 \vartheta_C \cdot \Gamma(\tau^- \to \nu_\tau + \mathrm{e}^- + \bar{\nu}_\mathrm{e}). \end{aligned} \tag{23.15}$$

Gathering all these results together, we obtain a set of naïve predictions for the τ-lepton in the standard model; a comparison with experimental results is given in Table 23.1. The agreement between theory and experiment is really quite good. From our crude treatment of the semihadronic decay processes we can hardly expect perfect agreement for the first three entries in Table 23.1. However, the leptonic decay *rates* $\Gamma(\tau \to \nu_\tau \ell \bar{\nu}_\ell)$ ($\ell = \mathrm{e}, \mu$) can be very accurately predicted, since the strong interaction plays no role here. Experimentally, they are obtained from measurements of the total lifetime τ_τ and the leptonic branching ratios. As can be seen from the last two rows of Table 23.1, to within the rather large experimental error, the measurements are compatible with the theory.

At this point we close our discussion of the τ-lepton.

Table 23.1 Theoretical predictions and experimental values for selected decay parameters of the τ-lepton in the standard model. Both the electroweak corrections and the gluonic corrections when hadrons are involved have been neglected in the theoretical values.

	Theory	Experiment
Lifetime τ_τ	$\left[5\cdot\frac{G^2 m_\tau^5}{192\pi^3}\right]^{-1} = 3.19\cdot 10^{-13}\,\mathrm{s}$	$(3.4 \pm 0.5)\cdot 10^{-13}\,\mathrm{s}$
$B_e \equiv \frac{\Gamma(\tau\to\nu_\tau e\bar{\nu}_e)}{\Gamma(\tau\to\text{anything})}$	20 %	$(16.2 \pm 0.9)\%$
$B_\mu \equiv \frac{\Gamma(\tau\to\nu_\tau \mu\bar{\nu}_\mu)}{\Gamma(\tau\to\text{anything})}$	20 %	$(18.5 \pm 1.1)\%$
$\Gamma^{-1}(\tau\to\nu_\tau e\bar{\nu}_e) = B_e^{-1}\tau_\tau$	$\left[\frac{G^2 m_\tau^5}{192\pi^3}\left(1-8\frac{m_e^2}{m_\tau^2}\right)\right]^{-1} = (1.593 \pm 0.014)\cdot 10^{-12}\,\mathrm{s}$	$(2.06 \pm 0.32)\cdot 10^{-12}\,\mathrm{s}$
$\Gamma^{-1}(\tau\to\nu_\tau \mu\bar{\nu}_\mu) = B_\mu^{-1}\tau_\tau$	$\left[\frac{G^2 m_\tau^5}{192\pi^3}\left(1-8\frac{m_\mu^2}{m_\tau^2}\right)\right]^{-1} = (1.638 \pm 0.014)\cdot 10^{-12}\,\mathrm{s}$	$(1.84 \pm 0.29)\cdot 10^{-12}\,\mathrm{s}$

23.3 The β-Decay of the Neutron and the Determination of the Kobayashi–Maskawa Matrix Element V_{11}

As an example for the decays of conventional hadrons, we discuss in this section the β-decay of the neutron:

$$\mathrm{n}(p_1) \to \mathrm{p}(p_2) + \mathrm{e}^-(p_3) + \bar{\nu}_e(p_4). \tag{23.16}$$

The momenta of the particles are indicated in brackets. This process offers us a possible means of determining the element V_{11} of the Kobayashi–Maskawa matrix.

The diagram for neutron decay in the standard model has already been given in Fig. 22.5. In order to find the S-matrix element, we start from the effective Lagrange density (22.86) with the charged current $\mathcal{J}_{\mathrm{CC}}^{\lambda}$ as in (22.112). We then obtain:

$$S_{fi} = (2\pi)^4\delta(p_1 - p_2 - p_3 - p_4)\left(-\mathrm{i}\frac{G}{\sqrt{2}}V_{11}\right)$$
$$\cdot\bar{u}_e(p_3)\gamma^\lambda(1-\gamma_5)v_{\nu_e}(p_4)\langle \mathrm{p}(p_2)|\bar{\mathbf{u}}(0)\gamma_\lambda(1-\gamma_5)\mathbf{d}(0)|\mathrm{n}(p_1)\rangle. \tag{23.17}$$

The Kobayashi–Maskawa matrix element V_{11} appears in this expression because a d-quark changes into a u-quark with the emission of a W^--boson (Fig. 22.5). The initially unknown matrix element of the quark current between nucleon states also appears in (23.17). Fortunately, the vector part of this hadronic matrix element can be obtained by symmetry arguments.

The energy release in the β-decay of the neutron is very small in comparison to the rest mass of the nucleons:

$$m_n - m_p = 1.29 \text{ MeV}. \tag{23.18}$$

Thus the nucleon only receives a small recoil in the process. For the calculation of the matrix element we can neglect the recoil and set

$$p_1 \cong p_2 \cong p. \tag{23.19}$$

We further assume exact isospin invariance of the strong interaction. We divide the hadronic matrix element into vector and axial-vector parts. Quite generally, we can write:

$$\langle \mathrm{p}(p)|\bar{\mathbf{u}}(0)\gamma^\lambda \mathbf{d}(0)|\mathrm{n}(p)\rangle = \bar{u}_p(p)\Gamma^\lambda_V(p)u_n(p), \tag{23.20}$$

$$\langle \mathrm{p}(p)|\bar{\mathbf{u}}(0)\gamma^\lambda\gamma_5 \mathbf{d}(0)|\mathrm{n}(p)\rangle = \bar{u}_p(p)\Gamma^\lambda_A(p)u_n(p). \tag{23.21}$$

Here $u_{p,n}$ are the Dirac spinors of the proton and the neutron and $\Gamma^\lambda_V(p)$ and $\Gamma^\lambda_A(p)$ are matrices in Dirac space. As a consequence of the parity invariance of the *strong* interaction, Γ^λ_V must be a polar four-vector and Γ^λ_A an axial four-vector. Simple possibilities for these would be

$$\begin{aligned} \Gamma^\lambda_V(p) &= g_V\gamma^\lambda, \\ \Gamma^\lambda_A(p) &= g_A\gamma^\lambda\gamma_5, \end{aligned} \tag{23.22}$$

where g_V and g_A are constants. It turns out that these already exhaust all the possibilities available under the approximation (23.19). For example, if we were to write

$$\Gamma^\lambda_V(p) = p^\lambda, \tag{23.23}$$

this would bring nothing new since for the Dirac spinors we have

$$\bar{u}(p)p^\lambda u(p) = m\bar{u}(p)\gamma^\lambda u(p). \tag{23.24}$$

We now show that the vector coupling constant g_V is determined by the isospin symmetry of the strong interaction. We work in the rest frame of the nucleons and recall that the zeroth component of the vector current, when integrated over space, gives the isospin charge:

$$\mathbf{I}_1 + i\mathbf{I}_2 = \int d^3x\, \bar{\mathbf{u}}(x)\gamma_0 \mathbf{d}(x). \tag{23.25}$$

Since the neutron and the proton form an isospin doublet, it follows, in the rest system of the nucleons, that

$$\langle \mathrm{p}(p,s_2)|\mathbf{I}_1 + i\mathbf{I}_2|\mathrm{n}(p,s_1)\rangle = \langle \mathrm{p}(p,s_2)|\mathrm{p}(p,s_1)\rangle,$$

$$\int d^3x\, \langle \mathrm{p}(p,s_2)|\bar{\mathbf{u}}(0)\gamma_0\mathbf{d}(0)|\mathrm{n}(p,s_1)\rangle = \delta_{s_2s_1}2p^0V,$$

$$V g_V\bar{u}(p,s_2)\gamma_0u(p,s_1) = \delta_{s_2s_1}2p^0V,$$

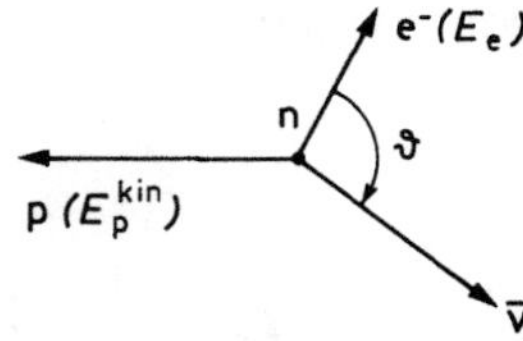

Figure 23.4 Decay of a neutron in its rest frame: $n \to p + e^- + \bar{\nu}_e$.

where s_1 and s_2 are the spin indices. From this we obtain the important relation

$$g_V = 1. \tag{23.26}$$

As in previous instances we have used a normalization volume V in order to be able to work with states of well-defined momentum. The normalization of the states is as in (5.24).

It is not possible to derive a value for the coupling constant g_A from simple symmetry arguments. By means of certain postulates valid in QCD, concerning the algebra of currents[1] (Gell-Mann 1964a, b) and the nature of the pions, this constant can however be related to other experimentally measurable quantities by the so-called Adler–Weisberger sum rule (Adler 1965; Weisberger 1966). We will not go into further details on this point, and instead we leave g_A as a free parameter.

If we assume that the Fermi constant is known from muon decay (23.3), then the matrix element for the neutron β-decay has one other unknown parameter, V_{11}. If we substitute the expressions (23.20)–(23.22) into (23.17), we obtain for the S-matrix element

$$S_{fi} = (2\pi)^4 \delta(p_1 - p_2 - p_3 - p_4)\left(-i\frac{G}{\sqrt{2}}V_{11}\right)$$
$$\cdot \bar{u}_e(p_3)\gamma^\lambda(1-\gamma_5)v_{\nu_e}(p_4)\bar{u}_p(p_2)(g_V\gamma_\lambda - g_A\gamma_\lambda\gamma_5)u_n(p_1), \tag{23.27}$$

where according to (23.26), we can set $g_V = 1$.

In neutron decay there are enough independently measurable quantities to enable both the parameters g_A and V_{11} to be determined. Let us consider the decay in the rest frame of the neutron. One can measure, for example, the energy of the electron (E_e) and the kinetic energy (E_p^{kin}) of the recoil proton (Fig. 23.4). The conventional treatment is to calculate from E_p^{kin} the angle ϑ between the electron and the unobserved neutrino. One then takes E_e and ϑ as independent variables. Neglecting the recoil energy of the proton in the energy balance, we obtain from energy and momentum

[1] With the help of the postulates of current algebra, many interesting relationships were derived in the 1960s. We refer to the excellent introduction to this field and the collection of original papers presented in Adler (1968). Many of the results of current algebra are also exactly valid in QCD; for others there are corrections which can be calculated perturbatively.

conservation

$$E_e + E_\nu = E_0 \equiv m_n - m_p, \tag{23.28}$$

$$\boldsymbol{p}_e + \boldsymbol{p}_\nu = \boldsymbol{p}_p, \tag{23.29}$$

$$E_p^{kin} = \frac{1}{2m_p}\boldsymbol{p}_p^2 = \frac{1}{2m_p}\{E_e^2 - m_e^2 + (E_0 - E_e)^2 + 2\sqrt{E_e^2 - m_e^2}(E_0 - E_e)\cos\vartheta\}. \tag{23.30}$$

The S-matrix element (23.27) leads to a differential decay rate given by

$$\begin{aligned} &d\Gamma(n \to p + e^- + \bar{\nu}_e) \\ &= (G \cdot V_{11})^2 \frac{1}{4\pi^3}(E_0 - E_e)^2 E_e \sqrt{E_e^2 - m_e^2}\left(1 + 3\left(\frac{g_A}{g_V}\right)^2\right) \\ &\cdot \left\{1 + \frac{\sqrt{E_e^2 - m_e^2}}{E_e} a \cos\vartheta\right\} dE_e \, d\cos\vartheta, \end{aligned} \tag{23.31}$$

where

$$a = \frac{\left(1 - \dfrac{g_A}{g_V}\right)^2}{1 + 3\left(\dfrac{g_A}{g_V}\right)^2}.$$

Since the Fermi constant G is known from muon decay (23.3), we need to measure two quantities in order to determine V_{11} and g_A/g_V, for example, the angular correlation parameter a and the total lifetime τ_n of the neutron, which is the inverse of the total decay rate. In actually carrying out such an analysis one must, however, take various corrections into account. In (23.31), for instance, the fact that the emitted electron moves in the Coulomb field of the proton has been ignored. One also has to give up the approximation in (23.19) and take account of further terms in the expansion of Γ_V^λ and Γ_A^λ (23.20)–(23.22), the so-called weak magnetism and the induced pseudoscalar. A detailed discussion of these corrections is given by Wilkinson (1982) where further references can also be found.

The experimental value of the lifetime, τ_n, of the neutron is

$$\tau_n = 898 \pm 16 \text{ s} \tag{23.32a}$$

and the ratio of the constants g_A and g_V is found to be

$$\frac{g_A}{g_V} = 1.254 \pm 0.006. \tag{23.32b}$$

Since the uncertainty in these quantities is of the order of one percent, the decay of the free neutron only allows a determination of V_{11} to the same precision. A more exact determination can be made from so-called superallowed nuclear β-decay processes, where bound neutrons decay (Bég 1982; Wilkinson 1978, 1980; Paschos 1982a):

$$V_{11} = 0.973 \pm 0.002. \tag{23.33}$$

The fact that V_{11} turns out to be less than one is not trivial, and is a necessary condition for the concept of the generalized Cabibbo rotation to make sense phenomenologically.

23.4 Hyperon Decay Processes and the Determination of V_{12}

A whole series of hyperons, i.e., baryons that possess strangeness, are known today. In the quark picture, these particles contain at least one s-quark (Chap. 17). Many of the hyperons undergo semileptonic decay processes (Table 23.2), which can be made use of to determine the Kobayashi–Maskawa matrix element V_{12}.

As an example we consider the decay $\Lambda \to \mathrm{pe}^- \bar{\nu}_e$. The kinematics of this process is very similar to that of neutron β-decay. The respective quark diagram has already been given in Fig. 22.5. The s-quark transforms into a u-quark under the emission of a W^--boson; thus the amplitude is proportional to V_{12}. Starting from the effective Lagrange density (22.86) we find

$$\begin{aligned}\langle \mathrm{p}(p_2)\mathrm{e}^-(p_3)\bar{\nu}_e(p_4)|\mathbf{S}|\Lambda(p_1)\rangle \\ &= (2\pi)^4\delta(p_1 - p_2 - p_3 - p_4)\left(-\mathrm{i}\frac{G}{\sqrt{2}}V_{12}\right) \\ &\quad \cdot \bar{u}(p_3)\gamma^\lambda(1-\gamma_5)v_{\nu_e}(p_4) \\ &\quad \cdot \langle \mathrm{p}(p_2)|\bar{\mathbf{u}}(0)\gamma_\lambda(1-\gamma_5)\mathbf{s}(0)|\Lambda(p_1)\rangle .\end{aligned} \tag{23.34}$$

Table 23.2 The semileptonic decay processes of hyperons allowed in the standard model. All of these decay processes apart from $\Xi^- \to \Sigma^0 + \mu^- + \bar{\nu}_\mu$ have actually been observed. For the latter decay, the experimental upper limit for the decay rate is nonetheless compatible with the prediction of the standard model. To date, no semileptonic hyperon decay has been observed that is forbidden by the standard model, e.g., the decay $\Sigma^+ \to n + e^+ + \nu_e$.

Hyperon	Quark Content	Observed semileptonic decay modes
Λ	sud	$\mathrm{p} + \mathrm{e}^- + \bar{\nu}_e$
		$\mathrm{p} + \mu^- + \bar{\nu}_\mu$
Σ^+	suu	$\Lambda + \mathrm{e}^+ + \nu_e$
Σ^-	sdd	$\mathrm{n} + \mathrm{e}^- + \bar{\nu}_e$
		$\mathrm{n} + \mu^- + \bar{\nu}_\mu$
		$\Lambda + \mathrm{e}^- + \bar{\nu}_e$
Ξ^-	ssd	$\Lambda + \mathrm{e}^- + \bar{\nu}_e$
		$\Lambda + \mu^- + \bar{\nu}_\mu$
		$\Sigma^0 + \mathrm{e}^- + \bar{\nu}_e$
		$\Sigma^0 + \mu^- + \bar{\nu}_\mu$
Ω^-	sss	$\Xi^0 + \mathrm{e}^- + \bar{\nu}_e$

In order to determine the quantity V_{12} we again need some information about the hadronic matrix element. The energy released in the reaction is still relatively small in comparison to the mass of the nucleons:

$$m_\Lambda - m_\mathrm{p} = 177.3\ \mathrm{MeV}. \tag{23.35}$$

In a first approximation the recoil of the proton can therefore be neglected. We thus have again

$$\begin{gathered} p_2 \cong p_1 \equiv p, \\ \langle \mathrm{p}(p)|\bar{\mathbf{u}}(0)\gamma_\lambda(1-\gamma_5)\mathbf{s}(0)|\Lambda(p)\rangle = \bar{u}(p)(\tilde{g}_\mathrm{V}\gamma_\lambda - \tilde{g}_\mathrm{A}\gamma_\lambda\gamma_5)u(p), \end{gathered} \tag{23.36}$$

where $\tilde{g}_\mathrm{V}$ and $\tilde{g}_\mathrm{A}$ are constants.

In place of the isospin symmetry we can now make use of the flavor-SU(3) symmetry of Gell-Mann and Ne'eman in order to determine the matrix element $\tilde{g}_\mathrm{V}$ of the vector current. As was discussed in Chap. 17, p and Λ are members of the baryon octet, and the charge

$$\mathbf{F}_4 + \mathrm{i}\mathbf{F}_5 = \int \mathrm{d}^3x\, \bar{\mathbf{u}}(x)\gamma_0\mathbf{s}(x)$$

is a generating operator of the SU(3) group. In the limiting case of exact SU(3) symmetry one finds, in analogy to (23.26), the result

$$\tilde{g}_\mathrm{V} = \sqrt{\tfrac{3}{2}}. \tag{23.37}$$

With this we can determine the constant V_{12} in a very similar way as V_{11} was determined from neutron decay. In reality, of course, one must take into account that the SU(3) symmetry is relatively strongly broken. Indeed, the mass difference $m_\Lambda - m_\mathrm{p}$ is not especially small. The analysis of the data from hyperon decays yields (Bourquin 1983):

$$V_{12} = 0.231 \pm 0.003. \tag{23.38}$$

An alternative approach to the determination of V_{12} is offered by the decays of the K-mesons. In this case, the analysis of the data, using the best available theoretical methods for the calculation of the hadronic matrix elements, yields $V_{12} = 0.220 \pm 0.002$ (Leutwyler 1984). This is slightly lower than (23.38) but in Leutwyler (1984) it is shown that the error in (23.38) should be increased to 0.005, making the discrepancy statistically insignificant. The combined value from hyperon and K decays given there is

$$V_{12} = 0.221 \pm 0.002. \tag{23.39}$$

We can now use the unitarity relation to see what the theory predicts for the size of the coupling between the b-quark and the u-quark. From the relation

$$|V_{11}|^2 + |V_{12}|^2 + |V_{13}|^2 = 1 \tag{23.40}$$

we find, using the values of V_{11} and V_{12} (23.33), (23.39)

$$|V_{13}|^2 \leq 0.010. \tag{23.41}$$

If we had used (23.38) for V_{12}, the upper limit for V_{13} would have been even smaller. Thus, according to the theory, the transition $b \to u$ should be strongly suppressed and may even be completely absent. Investigations that directly deal with particles containing b-quarks are needed to clarify this (cf. Sect. 23.6).

23.5 The Decays of Charged Pions

The decay processes of the $\pi^\pm$-mesons have played an important historical role because they provided evidence for the (V − A) structure of the leptonic contribution to the charged current. We will thus discuss them in some detail.

The charged pions decay with an average lifetime of

$$\tau = (2.6030 \pm 0.0023) \cdot 10^{-8}\ \text{s}.$$

The known decay channels and their branching ratios are given in Table 23.3.

We look now at the leptonic decay processes, for example, that of the π^-:

$$\pi^-(p) \to \ell^-(p_1) + \bar{\nu}_\ell(p_2), \tag{23.42}$$

where $\ell = \text{e}, \mu$. The four-momenta of the particles are p, p_1, and p_2. The striking feature of Table 23.3 is that the electronic channel is very strongly suppressed relative to the muonic one although the phase space for the former is larger.

The diagram describing this decay in the standard model is shown in Fig. 23.5. According to the effective Lagrange density (22.86), the transition amplitude for our reaction is given by

$$\begin{aligned}
&\langle \ell^-(p_1)\bar{\nu}_\ell(p_2)|\mathbf{S}|\pi^-(p)\rangle \\
&= (2\pi)^4\delta(p - p_1 - p_2)\left(-\mathrm{i}\frac{G}{\sqrt{2}}V_{11}\right) \\
&\quad \cdot \bar{u}_\ell(p_1)\gamma^\lambda(1-\gamma_5)v_{\nu_\ell}(p_2)\langle 0|\bar{\mathbf{u}}(0)\gamma_\lambda(1-\gamma_5)\mathbf{d}(0)|\pi^-(p)\rangle.
\end{aligned} \tag{23.43}$$

It is again impossible to calculate the hadronic matrix element from first principles. We must be satisfied with a general analysis which relies on the invariance of the strong interaction under Lorentz and parity transformations.

Table 23.3 The observed decay modes of the $\pi^\pm$ mesons and their measured branching ratios.

Decay channel	Branching ratio
$\pi^\pm \to \mu^\pm + \nu_\mu(\bar{\nu}_\mu)$	100 %
$\text{e}^\pm + \nu_e(\bar{\nu}_e)$	$(1.232 \pm 0.024) \cdot 10^{-4}$
$\mu^\pm + \nu_\mu(\bar{\nu}_\mu) + \gamma$	$(1.24 \pm 0.25) \cdot 10^{-4}$
$\text{e}^\pm + \nu_e(\bar{\nu}_e) + \gamma$	$(5.6 \pm 0.7) \cdot 10^{-8}$
$\pi^0 + \text{e}^\pm + \nu_e(\bar{\nu}_e)$	$(1.033 \pm 0.034) \cdot 10^{-8}$

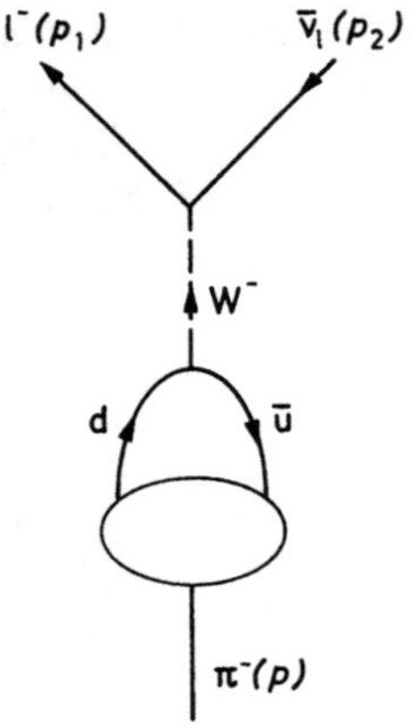

Figure 23.5 Diagram for the decay processes $\pi^- \to \ell^- + \bar{\nu}_\ell$ in the standard model ($\ell = \mu$, e).

The matrix element of the quark current in (23.43) transforms like a four-vector. The only four-vector at our disposal is the momentum of the pion. As a consequence of this and the normalization (3.69) of our states the matrix elements of the vector and axial vector currents, must have the form

$$\langle 0|\bar{\mathbf{u}}(0)\gamma_\lambda \mathbf{d}(0)|\pi^-(p)\rangle = \mathrm{i}p_\lambda \tilde{f}_\pi, \tag{23.44}$$

$$\langle 0|\bar{\mathbf{u}}(0)\gamma_\lambda\gamma_5 \mathbf{d}(0)|\pi^-(p)\rangle = \mathrm{i}p_\lambda f_\pi, \tag{23.45}$$

where $\tilde{f}_\pi$ and f_π are constants with the dimension of mass. Since the pion has negative intrinsic parity, the matrix element of the vector current must transform as a pseudoscalar and that of the axial-vector current as a polar vector. The momentum p_λ is a polar vector and therefore the matrix element of the vector current vanishes, i.e., we have

$$\tilde{f}_\pi \equiv 0. \tag{23.46}$$

The decay constant f_π in (23.45) can be nonzero and must be determined experimentally.

We now insert the hadronic matrix element into (23.43). Using the following relationship for the lepton spinors:

$$\begin{aligned}\bar{u}_\ell(p_1)\not{p}(1-\gamma_5)v_{\nu_\ell}(p_2) &= \bar{u}_\ell(p_1)(\not{p}_1+\not{p}_2)(1-\gamma_5)v_{\nu_\ell}(p_2)\\ &= m_\ell \bar{u}_\ell(p_1)(1-\gamma_5)v_{\nu_\ell}(p_2)\end{aligned} \tag{23.47}$$

we obtain for the S-matrix element

$$\begin{aligned}&\langle \ell^-(p_1)\bar{\nu}_\ell(p_2)|\mathbf{S}|\pi^-(p)\rangle\\ &\quad = (2\pi)^4\delta(p-p_1-p_2)\left(-\mathrm{i}\frac{G}{\sqrt{2}}V_{11}\right)\cdot f_\pi\cdot \mathrm{i}m_\ell \bar{u}_\ell(p_1)(1-\gamma_5)v_{\nu_\ell}(p_2).\end{aligned} \tag{23.48}$$

From this we can readily calculate that the decay rate is given by

$$\Gamma(\pi^- \to \ell^- + \bar{\nu}_\ell) = (GV_{11}f_\pi)^2\frac{m_\pi m_\ell^2}{8\pi}\left(1-\frac{m_\ell^2}{m_\pi^2}\right)^2. \tag{23.49}$$

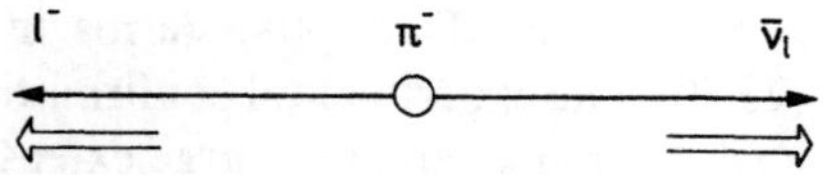

Figure 23.6 The decay $\pi^- \to \ell^- + \overline{\nu}_\ell$ in the center of mass system where $\ell = \mathrm{e}$ or μ. For massless neutrinos that couple via a (V − A) current, the helicity of the decay products must be as shown by the double arrows.

What is noteworthy about these results is that the S-matrix element is proportional to the lepton mass m_ℓ, or that correspondingly the decay rate is proportional to m_ℓ^2. This can actually be understood quite easily by considering the decay in the pion rest frame. Here the ℓ^- and the $\overline{\nu}_\ell$ travel away from one another in diametrically opposite directions. The massless antineutrino must always have positive helicity for (V − A) coupling as was discussed in Chap. 21. The pion has spin zero and thus, due to angular momentum conservation, the lepton ℓ^- must also be emitted with *positive* helicity (Fig. 23.6). Only in this way can we obtain a final state with a zero angular momentum component in the direction of motion of the leptons. If the lepton ℓ^- were also massless, then within (V − A) theory it would always have to have *negative* helicity. The total angular momentum component in the direction of motion of the neutrino would then be +1. However, it is not possible for the spin-zero pion to decay into such a state. Consequently, the amplitude for $m_\ell = 0$ must vanish; this is confirmed by the result of our explicit calculation (23.48). Thus the theory makes a noteworthy prediction for the ratio of the electronic to the muonic decay rates:

$$\left.\frac{\Gamma(\pi^- \to \mathrm{e}^- \overline{\nu}_\mathrm{e})}{\Gamma(\pi^- \to \mu^- \overline{\nu}_\mu)}\right|_{\text{theor.}} = \frac{m_\mathrm{e}^2}{m_\mu^2} \frac{\left(1 - \dfrac{m_\mathrm{e}^2}{m_\pi^2}\right)^2}{\left(1 - \dfrac{m_\mu^2}{m_\pi^2}\right)^2} = 1.284 \cdot 10^{-4}. \qquad (23.50)$$

This theoretical value is slightly modified by the inclusion of radiative corrections (Marciano 1976; Goldman 1977):

$$\left.\frac{\Gamma(\pi^- \to \mathrm{e}^- \overline{\nu}_\mathrm{e})}{\Gamma(\pi^- \to \mu^- \overline{\nu}_\mu)}\right|_{\text{theor.}} = 1.233 \cdot 10^{-4}. \qquad (23.51)$$

The agreement with the experimental value (Table 23.3) is now excellent. From the muonic decay rate and from the knowledge of G and V_{11} one also obtains an experimental value for the pion decay constant f_π (Dumbrajs 1983):

$$f_\pi = 0.94 m_\pi. \qquad (23.52)$$

A simple analysis shows that in order to derive (23.50) we merely need the following assumptions:

(i) massless neutrinos;
(ii) lepton currents of vector or axial vector type only, but with arbitrary relative strengths of V and A;
(iii) electron–muon universality.

S-, P-, and T-coupling terms in the lepton current invalidate equation (23.50). The experimental confirmation of this equation is therefore a sign that such couplings are, to a large extent, absent.

A specific test for the (V − A) structure of the lepton current emerges from Fig. 23.6. The lepton $\ell^-(\ell^+)$ always has positive (negative) helicity in $\pi^-(\pi^+)$ decay. To within the error bounds, this has been experimentally confirmed (Roesch 1982, Abela 1983). Combining the direct helicity measurements reported there with an analysis by Fetscher (1984) yields, for example, for the helicity of the μ^-

$$0.9959 \leq 2h(\mu^-) \leq 1.0 \qquad (23.53)$$

in excellent agreement with the expected value of +1.

23.6 The Decay of Particles Containing a Heavy Quark c or b

In Chap. 17 we studied hadrons, which are composed of the "light" quarks u, d, and s. Since the gluons, which are responsible for the binding of the quarks, feel nothing of the quark flavor, there should also be bound states of the light with the "heavy" quarks c, b, and t, and between two heavy quarks. Bound states of the type $c\bar{c}$ and $b\bar{b}$ have already been discussed in Sect. 20.3. Here we shall be concerned with particles that contain a heavy quark, c or b, and a light quark. We will limit the discussion to pseudoscalar meson states (Table 23.4).

The strong and the electromagnetic interaction do not alter the quark flavor. Thus the ground states listed in Table 23.4 of mesons containing a c- or a b-quark can only decay via a weak process. If we had free quarks instead

Table 23.4 The pseudoscalar mesons containing a heavy quark c or b and a light antiquark $\bar{u}$, $\bar{d}$, or $\bar{s}$. The corresponding antiparticles D^-, $\bar{D}^0$, etc., contain a heavy antiquark, $\bar{c}$ or $\bar{b}$, and a light quark u, d, or s. All of these particles, apart from $\bar{B}_s^0$ and B_s^0, have actually been observed. The mass values quoted here are those of Particle Data Group (1984). The lifetimes are the average values calculated by Schubert (1984) from the experimental results. The lifetimes of B^- and $\bar{B}^0$ are not known separately and the value given is to be interpreted as an average of the two lifetimes with weights which presumably are about equal.

Particle	Quark Content	Mass in MeV	Lifetime τ in s
D^+	$c\bar{d}$	1869.4 ± 0.6	$(8.3 \pm 1.0)\cdot 10^{-13}$
D^0	$c\bar{u}$	1864.7 ± 0.6	$(3.60 \pm 0.35)\cdot 10^{-13}$
D_s^+	$c\bar{s}$	1971 ± 6	$2.80\left\{\begin{matrix}+1.1\\-0.7\end{matrix}\right.\cdot 10^{-13}$
B^-	$b\bar{u}$	5270.8 ± 3.0	$(11 \pm 4)\cdot 10^{-13}$ (common value for B^- and $\bar{B}^0$)
$\bar{B}^0$	$b\bar{d}$	5274.2 ± 2.8	
$\bar{B}_s^0$	$b\bar{s}$	?	

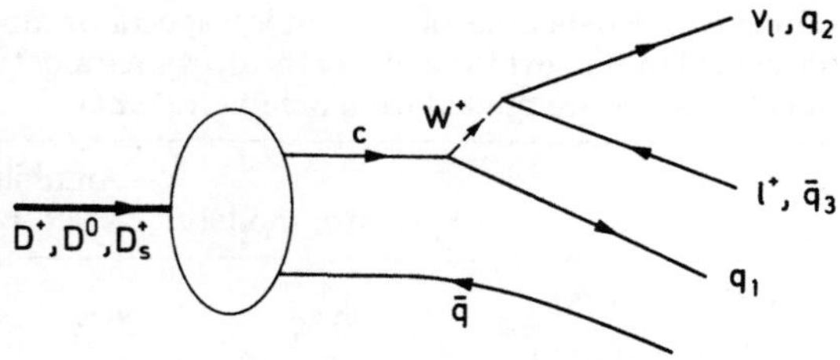

Figure 23.7 The decay processes of charmed mesons in the spectator model. Here $\bar{q}$ stands for the quark $\bar{d}(\bar{u}, \bar{s})$ when we consider the meson $D^+(D^0, D_s^+)$. According to (23.54) we must set $\ell = e, \mu$; $q_1 = s, d$; $q_2 = u$, and $q_3 = \bar{d}, \bar{s}$.

of mesons, then for the c-quark, for example, the following decay reactions would be possible:

$$\begin{aligned} c \to\ & s + \ell^+ + \nu_\ell && (V_{22}^*), \\ & d + \ell^+ + \nu_\ell && (V_{21}^*), \\ & s + \bar{d} + u && (V_{22}^* V_{11}), \\ & d + \bar{d} + u && (V_{21}^* V_{11}), \\ & s + \bar{s} + u && (V_{22}^* V_{12}), \\ & d + \bar{s} + u && (V_{21}^* V_{12}). \end{aligned} \tag{23.54}$$

Here $\ell = e, \mu$, and we have given in brackets the Kobayashi–Maskawa matrix elements that appear in the amplitudes. These decay processes of free quarks would be exactly analogous to the μ- and τ-decay processes of Sects. 23.1 and 23.2. They would give us an ideal means of determining many of the matrix elements V_{ij}. In reality, of course, things are not so simple since the quarks are always bound in hadrons.

We turn first to a discussion of the mesons with charm quarks D^+, D^0, and D_s^+. In the simplest model proposed it was assumed that the decay of the c-quark was essentially unchanged by the presence of the $\bar{u}$, $\bar{d}$, or $\bar{s}$ in the meson. It was supposed that the antiquarks in the meson were merely spectators, as shown in Fig. 23.7 (Gaillard 1975; Ellis 1975). The predictions of this spectator model can be readily deduced from the results of Sects. 23.1 and 23.2. We neglect the masses of the light quarks u, d, and s and take account of the unitarity relation (23.40) which is already approximately fulfilled by V_{11} and V_{12}. We then obtain, for example, the following relations:

$$\begin{aligned} \Gamma(D^+ \to \ell^+ \nu_\ell X) &\cong \Gamma(D^0 \to \ell^+ \nu_\ell X) \\ &\cong (|V_{22}|^2 + |V_{21}|^2) \frac{G^2 m_c^5}{192\pi^3}, \\ \Gamma(D^+ \to \text{hadrons}) &\cong \Gamma(D^0 \to \text{hadrons}) \\ &\cong 3(|V_{22}|^2 + |V_{21}|^2) \frac{G^2 m_c^5}{192\pi^3}. \end{aligned} \tag{23.55}$$

A few other typical predictions of the spectator model are given in Table 23.5.

Table 23.5 Predictions of the simplest spectator model and of the annihilation model discussed in the text for a few of the decay parameters of mesons with charm. Experimental values are quoted from Schubert (1984).

	Spectator model	Annihilation model	$r = 2.2$	Experiment
$\frac{\Gamma(D^+ \to \ell^+ \nu_\ell X)}{\Gamma(D^+ \to \text{anything})}$	20 %	20 %	20 %	(14 ± 3) %
$\frac{\Gamma(D^0 \to \ell^+ \nu_\ell X)}{\Gamma(D^0 \to \text{anything})}$	20 %	$\frac{1}{5+3r}$	8.6 %	(6 ± 1) %
τ_{D^+}/τ_{D^0}	1	$\frac{5+3r}{5}$	2.3	2.3 ± 0.3
$\frac{\Gamma(D_s^+ \to \ell^+ \nu_\ell X)}{\Gamma(D_s^+ \to \text{anything})}$	20 %	20 %	20 %	?
$\tau_{D^+}/\tau_{D_s^+}$	$\cong 1$	$\cong 1 + r$	3.2	$3.0\begin{cases}+0.8\\-1.2\end{cases}$

It came as a surprise that the experimental results deviated considerably from the predictions of the spectator model (Table 23.5). The cause of the discrepancy between the naïve theory and the experimental results is today attributed to the effects of quark binding in the hadrons, i.e., to the strong interaction. Evidently the antiquarks in the mesons take part more actively in the decay process, and the picture of a meson as a pure quark–antiquark state is also too naïve. Indeed, from Sect. 18.4, we know that the gluons play an important role in the hadrons. However, the possibility that the weak interaction is not identical to that demanded by the standard model can also not be completely excluded.

We will now briefly discuss the annihilation model, which provides a satisfactory description of the inclusive data (Bander 1980; Bernreuther 1980; Fritzsch 1980; Rosen 1980). For this discussion we will set the Kobayashi–Maskawa matrix **V** equal to the unit matrix. As we shall see, this is in fact a good approximation. For the decay of the c-quark one then has only the possibilities

$$\begin{aligned} c &\to s + \ell^+ + \nu_\ell, \\ c &\to s + \bar{d} + u, \end{aligned} \tag{23.56}$$

where $\ell = e$ or μ. By crossing over a quark from the final to the initial state, we obtain two-particle reactions that can contribute to the decay of the D^0 and D_s^+, but not to the D^+-mesons (Fig. 23.8).

$$\begin{aligned} D^0 &\sim c\bar{u} \to \bar{d} + s, \\ D_s^+ &\sim c\bar{s} \to \ell^+ + \nu_\ell, \\ D_s^+ &\sim c\bar{s} \to \bar{d} + u. \end{aligned} \tag{23.57}$$

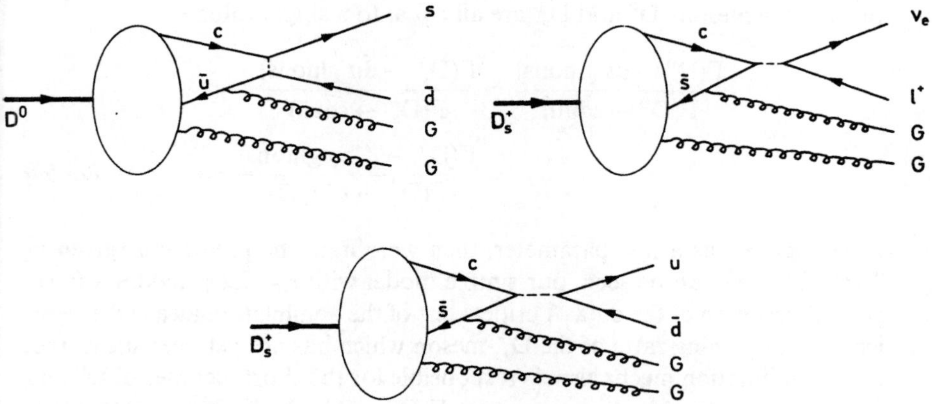

Figure 23.8 Typical diagrams with the emission of gluons G that arise as corrections to the decay of D^0 and D_s^+ mesons by the annihilation mechanism, (23.58). Similar diagrams exist for the emission of any number (0, 1, 2, 3, ...) of gluons. The diagrams without gluon emission correspond to the reactions (23.57). The exchanged W-boson is illustrated in each diagram by a short dashed line.

These possibilities were discussed at the time when the spectator model was developed, but their contributions to the decay rate were held to be insignificant. We are already familiar with the argument from our discussion of π^--decay (Sect. 23.5). We consider the example of the decay of the spin-zero particle D^0 into an antiquark–quark pair $\bar{d}$ and s. If we neglect the quark masses and all binding and gluon effects, then, as a result of the (V − A) structure of the charged current, this decay is forbidden, exactly like the decay $\pi^- \to e^- \bar{\nu}_e$ for massless electrons. The authors of the annihilation model correctly pointed out that the suppression of this decay, which is based on angular momentum conservation, can be lifted by the gluons in the mesons, since these can also carry away angular momentum. Thus one expects contributions to the decay of D^0 and D_s^+ from the following reactions:

$$\begin{aligned} D^0 &\sim c\bar{u} \to \bar{d} + s + \text{gluons}, \\ D_s^+ &\sim c\bar{s} \to \ell^+ + \nu_\ell + \text{gluons}, \\ D_s^+ &\sim c\bar{s} \to \bar{d} + u + \text{gluons}. \end{aligned} \tag{23.58}$$

Some typical diagrams for these processes are shown in Fig. 23.8. The effects of hard gluons are, however, only of the order of α_s and are therefore small. In order to obtain large effects one must assume that soft gluons are emitted. For these we still have no theoretical treatment based on first principles.

Here we will merely discuss a simple ansatz in which we assume that the ratios of annihilation to spectator contributions in the corresponding decay rates of the mesons D^0 and D_s^+ are all equal to a single value r:

$$\frac{\Gamma(D^0 \to \bar{d}s \text{ gluons})}{\Gamma(D^0 \to s\bar{d}u\bar{u})} = \frac{\Gamma(D_s^+ \to \bar{d}u \text{ gluons})}{\Gamma(D_s^+ \to s\bar{d}u\bar{s})} = \frac{\Gamma(D_s^+ \to \ell^+ \nu_\ell \text{ gluons})}{\Gamma(D_s^+ \to \ell^+ \nu_\ell s\bar{s})} = r. \tag{23.59}$$

If we regard r as a free parameter, then we obtain the predictions given in Table 23.5. As can be seen, our simple model with $r = 2.2$ provides a fairly good description of the data. A critical test of the annihilation idea is the semi-leptonic branching ratio of the D_s^+-meson which has not yet been measured. If the annihilation mechanism is responsible for the short lifetimes of D^0 and D_s^+ compared to D^+, then we expect $\Gamma(D_s^+ \to \ell^+\nu_\ell X)/\Gamma(D_s^+ \to \text{anything}) \cong 20\,\%$ as recorded in Table 23.5. If, however, the short lifetimes of D^0 and D_s^+ are due to other effects that only affect nonleptonic channels, we instead expect

$$\frac{\Gamma(D_s^+ \to \ell^+\nu_\ell X)}{\Gamma(D_s^+ \to \text{anything})} \cong \frac{\Gamma(D^0 \to \ell^+\nu_\ell X)}{\Gamma(D^0 \to \text{anything})} \cong 6\,\%.$$

In recent years a great deal of effort has gone into theoretical calculations of charmed particle decays. Despite this, it still has not been possible to get results from first principles. However, extensive calculations using various plausible assumptions have led to a qualitative understanding of the charmed particle decays. For inclusive decays, the magnitudes of various effects due to gluons have been estimated, and annihilation, as discussed above, does not seem to be as important as was previously thought. On the other hand, many exclusive decay channels have been calculated in a sophisticated version of the spectator model using model wavefunctions for the hadrons (Bauer (1987) and references therein). We refer the reader to Shifman (1988) for a review of these developments.

We also want to briefly discuss the decay of mesons containing a b-quark. Should we here too expect similarly large deviations from the simple spectator model? To answer this question we consider the extent to which the rates for annihilation and spectator processes (Figs. 23.7, and 23.8) will depend on the mass of the quarks. A classical annihilation process has been treated in Sect. 23.5 with the decay $\pi^- \to \ell^-\bar{\nu}_\ell$. We found that the hadronic matrix element is determined by the decay constant f_π. In essence, f_π represents the probability amplitude for finding the d- and the ū-quark in the π-meson at the same point in space (Fig. 23.5). Let f_D and f_B be the analogous decay constants for D- and B-mesons. Then, from dimensional arguments

$$\Gamma(\text{ann.}; D, D_s) \propto G^2 m_c^3 f_D^2,$$

$$\Gamma(\text{spect.}; D, D_s) \propto G^2 m_c^5,$$

$$\Gamma(\text{ann.}; B) \propto G^2 m_b^3 f_B^2,$$

$$\Gamma(\text{spect.}; B) \propto G^2 m_b^5.$$

Plausible arguments lead to the expectation $f_B/f_D \cong (m_c/m_b)^{1/2}$ (cf. Exercise 23.8). From this we obtain the following estimate:

$$\frac{\Gamma(\text{ann.; B})}{\Gamma(\text{spect.; B})} \cong \frac{\Gamma(\text{ann.; D, D}_s)}{\Gamma(\text{spect.; D, D}_s)}\left(\frac{m_c}{m_b}\right)^3 \cong r \cdot 0.03 \cong 0.06. \qquad (23.60)$$

For B-mesons the spectator model should thus be a good approximation. The possible decay processes for the b-quark are

$$\begin{aligned} &b \to c\ell^-\bar{\nu}_\ell, \qquad c q \bar{q}', \\ &b \to u\ell^-\bar{\nu}_\ell, \qquad u q \bar{q}', \end{aligned} \qquad (23.61)$$

where $\ell = e, \mu, \tau$; $q = d, s$, and $q' = u, c$. The total decay rate of the B-mesons is therefore mainly determined by the strength of the transitions $b \to c$ and $b \to u$, i.e., by the Kobayashi–Maskawa matrix elements V_{23} and V_{13}. If the difference in the c- and u-quark masses is taken into account, one finds the following theoretical prediction for the lifetime of the B-meson (Gaillard 1980)

$$\tau_B = \frac{1}{\Gamma(B \to \text{anything})} = (3.68\,|V_{23}|^2 + 7.8\,|V_{13}|^2)^{-1} \cdot 10^{-14}\ \text{s}. \qquad (23.62)$$

Here for the phase space factors the numerical values quoted by Kleinknecht (1984) have been used.

The measurement of the B-meson lifetime was reported for the first time in Hanson (1983) and Chadwick (1983). The presently available results lead to the following average value:

$$\tau_B = (14.2 \pm 2.7) \cdot 10^{-13}\ \text{s}. \qquad (23.63)$$

From (23.62) we then find

$$(|V_{23}|^2 + 2.1\,|V_{13}|^2)^{1/2} = 0.04 \pm 0.01. \qquad (23.64)$$

Independently of (23.41), we conclude from this that neither of the transitions $b \to u$ and $b \to c$ is particularly strong. In the Kobayashi–Maskawa scheme with six quarks, the transition $b \to t$, which of course does not contribute to the decay of the B-mesons, must have a strength of approximately 1:

$$|V_{33}| \cong 1. \qquad (23.65)$$

Independent information about V_{13} and V_{23} is obtained, for example, from the semileptonic decay processes of the b-quark:

$$\begin{aligned} &b \to c\ell^-\bar{\nu}_\ell \qquad (V_{23}), \\ &b \to u\ell^-\bar{\nu}_\ell \qquad (V_{13}). \end{aligned} \qquad (23.66)$$

The amplitudes are again proportional to the matrix elements in brackets. At the quark level, these two decay processes can be distinguished via the kinematics of the reaction since the mass of the c-quark is significantly larger than that of the u-quark (Table 22.1). This has the consequence that the lepton energy spectrum is softer for the transition $b \to c$ than for the transition $b \to u$

Table 23.6 The absolute value $|V_{ij}|$ of the elements of the Kobayashi–Maskawa matrix as determined by Kleinknecht (1983) in a general fit to experimental results. The range of values given corresponds to one standard deviation.

i (q) \ j (q′)	1 (d)	2 (s)	3 (b)
1 (u)	0.9723–0.9737	0.228–0.234	0.000–0.008
2 (c)	0.228–0.234	0.9704–0.9726	0.042–0.067
3 (t)	0.003–0.016	0.041–0.066	0.9977–0.9991

(Exercise 23.6). A study of the leptonic energy spectrum in the semileptonic decay of the B-mesons

$$\mathrm{B} \sim \mathrm{b}\bar{\mathrm{q}} \to \ell^- + \bar{\nu}_\ell + \mathrm{X} \tag{23.67}$$

thus allows, in principle, a determination of the ratio V_{13}/V_{23}. One must bear in mind, of course, that binding and gluon effects might lead to a modification of our naïve estimate which is valid for free quarks. This question has been studied in detail by, among others, Altarelli and co-workers (Altarelli 1982a), who showed that the lepton energy spectrum is, indeed, very sensitive to the ratio V_{13}/V_{23}. The existing data have been analyzed in this way and show that the transition $\mathrm{b} \to \mathrm{u}$ is strongly suppressed in comparison to $\mathrm{b} \to \mathrm{c}$.

Given the experimental information about the Kobayashi–Maskawa matrix the obvious next step is to attempt a general fit of the independent parameters. This has been done, for example, by Kleinknecht and Renk (1983).[2] Besides data discussed in this chapter, they have also used information from neutrino reactions and from the K^0–$\bar{\mathrm{K}}^0$ system. Their result for the absolute magnitudes $|V_{ij}|$ is given in Table 23.6. For the phase δ they conclude that a large range of values remains possible, namely

$$2^0 < \delta < 178^0. \tag{23.68}$$

A noticeable feature of the experimental results is that the quarks of different families do not mix strongly with one another. The standard model itself cannot provide an explanation for this. Theoretical attempts to calculate the Kobayashi–Maskawa matrix necessarily require information from a deeper level.

Exercises

23.1 Calculate the decay rate of the muon and the energy spectrum of the decay electron for a four-fermion interaction of the general form

[2] The parametrization of the Kobayashi–Maskawa matrix used by Kleinknecht and Renk corresponds to our equation (22.108). Their Table 1 contains misprints (K. Kleinknecht, private communication).

$$\mathscr{L}'_\mu = -\frac{G}{\sqrt{2}}\bar{\nu}_\mu \gamma^\lambda (a - b\gamma_5)\mu\, \bar{\mathrm{e}}\gamma_\lambda(1-\gamma_5)\nu_{\mathrm{e}} + \text{h.c.}, \tag{23.69}$$

where

$$|a|^2 + |b|^2 = 2.$$

How does the Michel parameter ρ depend on a and b?

23.2 We consider the Lagrange density of QCD (19.22) for two flavors u and d whose masses we set equal:

$$m_{\mathrm{u}} = m_{\mathrm{d}}. \tag{23.70}$$

This Lagrange density is invariant under the strong isospin group

$$\begin{pmatrix} \mathbf{u}(x) \\ \mathbf{d}(x) \end{pmatrix} \to \mathbf{U} \begin{pmatrix} \mathbf{u}(x) \\ \mathbf{d}(x) \end{pmatrix}, \tag{23.71}$$

where $\mathbf{U} \in \mathrm{SU}(2)$. Construct the corresponding currents and charges with the help of the Noether theorems of Sect. 3.3.

23.3 Consider the isospin charges of Exercise 23.2

$$\mathbf{I}_a = \int \mathrm{d}^3x\, (\bar{\mathbf{u}}(x), \bar{\mathbf{d}}(x))\, \tfrac{1}{2}\tau_a \gamma^0 \begin{pmatrix} \mathbf{u}(x) \\ \mathbf{d}(x) \end{pmatrix} \qquad (a = 1, 2, 3). \tag{23.72}$$

Derive the commutation rules of the isospin charges with the help of the canonical anticommutation rules of the quark fields [see (4.92)].

23.4 Calculate the constant $\tilde{g}_{\mathrm{V}}$ in (23.36) in the limiting case of exact flavor-SU(3) invariance.

23.5 Calculate the rate for the leptonic decays of the charged pion (23.42) assuming the following Lagrange densities for the interaction ($\ell = \mathrm{e}, \mu$):

$$\text{(i)} \quad \mathscr{L}' = -\frac{G}{\sqrt{2}}\bar{\ell}\gamma^\lambda(a - b\gamma_5)\nu_\ell \mathscr{J}_\lambda + \text{h.c.}, \tag{23.73}$$

$$\text{(ii)} \quad \mathscr{L}' = -\frac{G}{\sqrt{2}}\bar{\ell}(a - b\gamma_5)\nu_\ell \mathscr{J} + \text{h.c.}. \tag{23.74}$$

Here a and b are arbitrary with

$$|a|^2 + |b|^2 = 2 \tag{23.75}$$

and $\mathscr{J}_\lambda(\mathscr{J})$ is a hadronic axial vector (pseudoscalar) current.

23.6 Calculate the rate for the decay processes in (23.66) for $\ell = \mathrm{e}, \mu$ in the aproximation $m_\ell = m_{\mathrm{u}} = 0$ but with $m_{\mathrm{c}} \neq 0$. Discuss the resulting energy spectrum of the leptons ℓ^-.

23.7 Calculate the total decay width of the B-meson in the spectator model, taking into account the decay processes (23.61). In doing so set

$$m_{\mathrm{e}} = m_\mu = m_{\mathrm{u}} = m_{\mathrm{d}} = m_{\mathrm{s}} = 0$$

but $m_{\mathrm{c}} \neq 0$ and $m_\tau \neq 0$.

23.8 Consider a pseudoscalar meson M composed of a heavy quark Q of mass m_{Q} and a light antiquark $\bar{\mathrm{q}}$ of mass m_{q}. Assume that M can be described as a nonrelativistic $\mathrm{Q}\bar{\mathrm{q}}$ bound state with the binding potential being independent of the quark masses as expected in QCD (Sect. 20.3). Define the decay constant f_{M}

in analogy to (23.45) and calculate the rate for the decay $M \to \mu\nu$ as for $\pi \to \mu\nu$, assuming a corresponding coupling of $Q\bar{q}$ to $\mu\nu$. Calculate, on the other hand, $M \to \mu\nu$ as was discussed for $J/\psi \to \ell^+\ell^-$ in Sect. 20.3. Show that for $m_Q \to \infty$ and m_q fixed, the wavefunction at the origin $\psi(0)$ becomes independent of m_Q and that the decay constant f_M becomes proportional to $m_Q^{-1/2}$.

CHAPTER 24

The Neutral Current and the Determination of $\sin^2 \vartheta_W$

In this chapter we consider a number of possible methods of obtaining the weak angle ϑ_W that determines the composition of the neutral current in the standard model (22.78). For this purpose, one could, in principle, use any reaction to which the neutral current contributes. Here we shall confine ourselves to a discussion of neutrino–electron and neutrino–nucleon scattering and electron–positron annihilation. A review that also considers other reactions is given, for example, by Kim (1981).

24.1 Neutrino–Electron Scattering

To date, the following neutrino–electron scattering processes have been observed in experiments with accelerators:

$$\begin{aligned} \nu_\mu + e &\to \nu_\mu + e, \\ \bar{\nu}_\mu + e &\to \bar{\nu}_\mu + e, \\ \nu_e + e &\to \nu_e + e. \end{aligned} \tag{24.1}$$

At reactors providing an intense source of electron antineutrinos, the reaction

$$\bar{\nu}_e + e \to \bar{\nu}_e + e \tag{24.2}$$

has also been observed.

To perform a theoretical analysis of these processes we use (22.86) as our

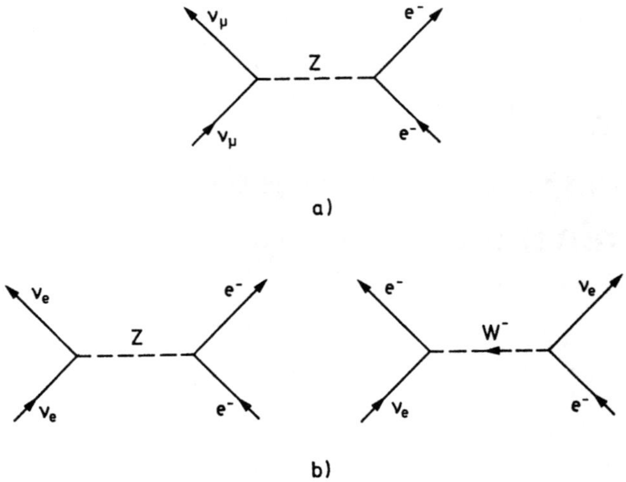

Figure 24.1 Diagrams for ν_μe scattering (a) and ν_ee scattering (b). The diagrams for $\bar{\nu}_\mu$e and $\bar{\nu}_e$e scattering are obtained by reinterpreting the neutrino lines in (a) and (b) according to a crossing transformation (Chap. 2).

starting point. The relevant part of the effective Lagrange density is

$$\mathscr{L}_{\text{eff}} = -\frac{4G}{\sqrt{2}}\{\bar{\nu}_{eL}\gamma^\lambda \mathbf{e}_L \bar{\mathbf{e}}_L \gamma_\lambda \nu_{eL} + \mathscr{J}^\lambda_{NC}\mathscr{J}_{NC\,\lambda}\}. \tag{24.3}$$

The neutral current has the form

$$\mathscr{J}^\lambda_{NC} = \tfrac{1}{2}\bar{\nu}_{eL}\gamma^\lambda \nu_{eL} + \tfrac{1}{2}\bar{\nu}_{\mu L}\gamma^\lambda \nu_{\mu L} + \tfrac{1}{2}\bar{\mathbf{e}}(g^e_V \gamma^\lambda - g^e_A \gamma^\lambda \gamma_5)\mathbf{e} + \cdots, \tag{24.4}$$

where the dots (...) stand for contributions that play no role in neutrino–electron scattering. In the standard model we have

$$\begin{aligned} g^e_V &= -\tfrac{1}{2} + 2\sin^2\vartheta_W, \\ g^e_A &= -\tfrac{1}{2}. \end{aligned} \tag{24.5}$$

In this section we leave g^e_V *and* g^e_A as free parameters that will be fitted to experiment. It should be noted that in ν_μ–e and $\bar{\nu}_\mu$–e scattering, only the neutral current contributes, whereas for ν_e–e and $\bar{\nu}_e$–e scattering both the neutral and the charged current play a role (Fig. 24.1).

For calculational purposes it is useful to rewrite the product of the charged currents in (24.3) by means of a Fierz transformation, such that it, too, appears as a product of neutral currents. Using the formulas of Appendix F we obtain

$$\begin{aligned} \mathscr{L}_{\text{eff}} = -\frac{G}{\sqrt{2}}\{&\bar{\nu}_e\gamma^\lambda(1-\gamma_5)\nu_e \bar{\mathbf{e}}(G_V\gamma_\lambda - G_A\gamma_\lambda\gamma_5)\mathbf{e} \\ &+ \bar{\nu}_\mu\gamma^\lambda(1-\gamma_5)\nu_\mu \bar{\mathbf{e}}(g^e_V\gamma_\lambda - g^e_A\gamma_\lambda\gamma_5)\mathbf{e}\}, \end{aligned} \tag{24.6}$$

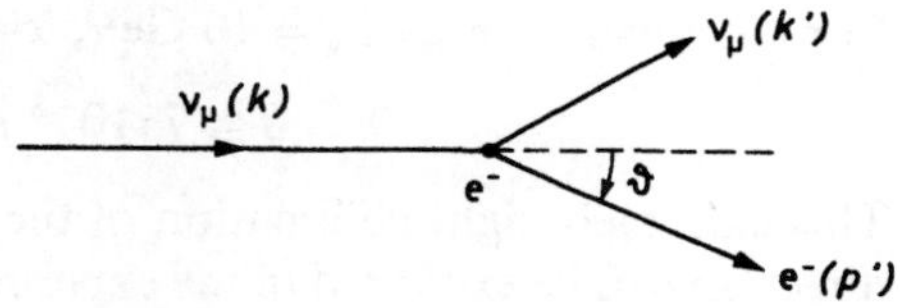

Figure 24.2 ν_μe scattering in the laboratory frame where ϑ is the scattering angle of the electron.

where

$$G_V = g_V^e + 1, \qquad G_A = g_A^e + 1.$$

As before terms in $\mathscr{L}_{\text{eff}}$ that are unimportant for our purposes have been omitted.

We now study ν_μ–e scattering (Fig. 24.2)

$$\nu_\mu(k) + e^-(p) \to \nu_\mu(k') + e^-(p') \tag{24.7}$$

in more detail. We define the following kinematical variables:

$$\begin{aligned} s &= (k+p)^2 = (k'+p')^2, \\ E_\nu &= \frac{pk}{m_e}, \\ E'_e &= \frac{pp'}{m_e}, \\ y &= \frac{E'_e - m_e}{E_\nu}. \end{aligned} \tag{24.8}$$

Here $\sqrt{s}$ is the center of mass energy, E_ν is the initial energy of the neutrino, and E'_e is the energy of the scattered electron in the laboratory frame. The dimensionless variable y gives the fraction of the neutrino energy that is converted into kinetic energy of the scattered electron. The kinematic limits for y turn out to be

$$0 \le y \le \left(1 + \frac{m_e}{2E_\nu}\right)^{-1}. \tag{24.9}$$

For the scattering angle ϑ of the electron in the laboratory frame, we find the relationship

$$\cos\vartheta = \left(1 + \frac{m_e}{E_\nu}\right)\left(\frac{E'_e - m_e}{E'_e + m_e}\right)^{1/2}. \tag{24.10}$$

In the limiting case of high energies, i.e., for $E_\nu, E'_e \gg m_e$, this becomes

$$\vartheta^2 \cong 2m_e \frac{E_\nu - E'_e}{E_\nu E'_e} = \frac{2m_e}{E_\nu}\frac{1-y}{y}. \tag{24.11}$$

Typical scattering angles at high energies are thus of the order of

$$\vartheta \cong \sqrt{\frac{m_e}{E_\nu}}. \tag{24.12}$$

For a neutrino energy $E_\nu = 10$ GeV, for example, this implies

$$\vartheta \cong 7 \cdot 10^{-3} \text{ rad} \cong 0.4^\circ. \tag{24.13}$$

This extremely tight collimation of the scattered electrons into the forward direction can be exploited in the experiment to distinguish neutrino–electron scattering from background reactions.

The S-matrix element and the scattering cross-section for the reaction (24.7) are obtained from the effective Lagrange density (24.6) as

$$\begin{aligned} \langle \nu_\mu(k')\mathrm{e}(p')|\mathbf{S}|\nu_\mu(k)\mathrm{e}(p)\rangle &= -\mathrm{i}\frac{G}{\sqrt{2}}(2\pi)^4\delta(p'+k'-p-k) \\ &\quad \cdot \bar{u}_\nu(k')\gamma^\lambda(1-\gamma_5)u_\nu(k)\,\bar{u}_\mathrm{e}(p')\gamma_\lambda(g_\mathrm{V}^\mathrm{e} - g_\mathrm{A}^\mathrm{e}\gamma_5)u_\mathrm{e}(p), \end{aligned} \tag{24.14}$$

$$\begin{aligned} \frac{\mathrm{d}\sigma^{(\nu_\mu \mathrm{e})}}{\mathrm{d}y} &= \frac{G^2}{\pi}2m_\mathrm{e}E_\nu \\ &\quad \cdot \left\{\left(\frac{g_\mathrm{V}^\mathrm{e}+g_\mathrm{A}^\mathrm{e}}{2}\right)^2 + \left(\frac{g_\mathrm{V}^\mathrm{e}-g_\mathrm{A}^\mathrm{e}}{2}\right)^2(1-y)^2 - \tfrac{1}{4}[(g_\mathrm{V}^\mathrm{e})^2-(g_\mathrm{A}^\mathrm{e})^2]\frac{m_\mathrm{e}}{E_\nu}y\right\}. \end{aligned} \tag{24.15}$$

For the $\bar{\nu}_\mu$–e scattering, we find analogously

$$\begin{aligned} \frac{\mathrm{d}\sigma^{(\bar{\nu}_\mu \mathrm{e})}}{\mathrm{d}y} &= \frac{G^2}{\pi}2m_\mathrm{e}E_{\bar{\nu}} \\ &\quad \cdot \left\{\left(\frac{g_\mathrm{V}^\mathrm{e}-g_\mathrm{A}^\mathrm{e}}{2}\right)^2 + \left(\frac{g_\mathrm{V}^\mathrm{e}+g_\mathrm{A}^\mathrm{e}}{2}\right)^2(1-y)^2 - \tfrac{1}{4}[(g_\mathrm{V}^\mathrm{e})^2-(g_\mathrm{A}^\mathrm{e})^2]\frac{m_\mathrm{e}}{E_{\bar{\nu}}}y\right\}. \end{aligned} \tag{24.16}$$

For the ν_e–e and $\bar{\nu}_\mathrm{e}$–e scattering we simply need to replace g_V^e and g_A^e by G_V and G_A. Equations (24.15), (24.16) were originally derived by t'Hooft (1971a).

Integrating over y yields the total elastic cross-sections. In the high energy limit ($E_\nu \gg m_\mathrm{e}$) one finds

$$\begin{aligned} \sigma(\nu_\mu \mathrm{e}) &= \frac{G^2}{\pi}2m_\mathrm{e}E_\nu\left\{\left(\frac{g_\mathrm{V}^\mathrm{e}+g_\mathrm{A}^\mathrm{e}}{2}\right)^2 + \frac{1}{3}\left(\frac{g_\mathrm{V}^\mathrm{e}-g_\mathrm{A}^\mathrm{e}}{2}\right)^2\right\}, \\ \sigma(\bar{\nu}_\mu \mathrm{e}) &= \frac{G^2}{\pi}2m_\mathrm{e}E_{\bar{\nu}}\left\{\left(\frac{g_\mathrm{V}^\mathrm{e}-g_\mathrm{A}^\mathrm{e}}{2}\right)^2 + \frac{1}{3}\left(\frac{g_\mathrm{V}^\mathrm{e}+g_\mathrm{A}^\mathrm{e}}{2}\right)^2\right\}. \end{aligned} \tag{24.17}$$

The order of magnitude of the cross-sections is given by

$$\frac{G^2}{\pi}m_\mathrm{e}E_\nu = 8.62 \cdot 10^{-42}\left(\frac{E_\nu}{1\text{ GeV}}\right)\text{cm}^2.$$

Despite their small size, these cross-sections have been measured in a number of experiments. The results obtained, for example, by Bergsma (1984), are

$$\begin{aligned} \frac{\sigma(\nu_\mu \mathrm{e})}{E_\nu} &= [1.9 \pm 0.4\text{ (stat.)} \pm 0.4\text{ (syst.)}] \cdot 10^{-42}\frac{\text{cm}^2}{\text{GeV}}, \\ \frac{\sigma(\bar{\nu}_\mu \mathrm{e})}{E_{\bar{\nu}}} &= [1.5 \pm 0.3\text{ (stat.)} \pm 0.4\text{ (syst.)}] \cdot 10^{-42}\frac{\text{cm}^2}{\text{GeV}}, \end{aligned} \tag{24.18}$$

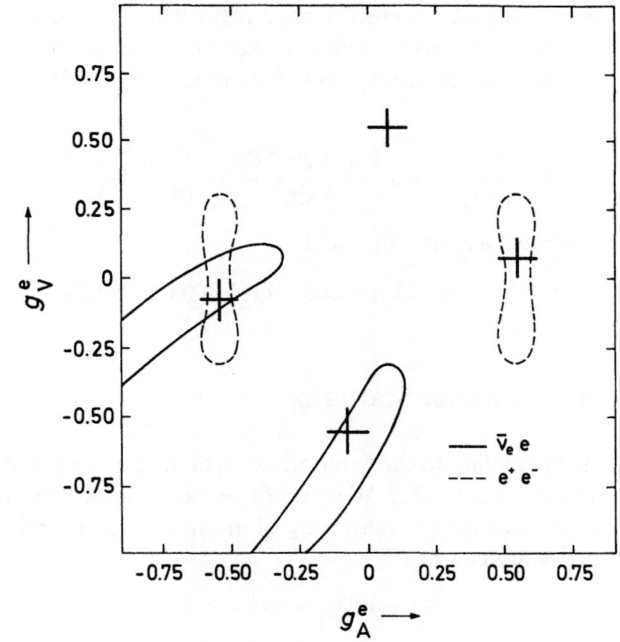

Figure 24.3 The experimental determination of the weak coupling constants g_V^e and g_A^e of the electron. The values for $\sigma(\nu_\mu e)$ and $\sigma(\bar{\nu}_\mu e)$ given in (24.18) yield the four solutions marked with crosses. The allowed regions deduced from the results of $\bar{\nu}_e e$ scattering (Reines 1976) and e^+e^- annihilation (Panman 1984) are bounded by solid and dashed lines, respectively. Together, the three types of experiment lead to a unique solution for g_V^e and g_A^e (after Bergsma 1984).

Here both statistical and systematic errors are shown. These results still admit four solutions for g_V^e and g_A^e (Fig. 24.3). This is understandable since (24.17) is invariant under the substitutions

$$g_V^e \to -g_V^e, \qquad g_A^e \to -g_A^e$$

and

$$g_V^e \to g_A^e, \qquad g_A^e \to g_V^e.$$

Experimental studies of $\bar{\nu}_e$–e and ν_e–e scattering can, in principle, remove this ambiguity. However, the current precision of these experiments is such that in order to obtain a unique solution one also has to include into the analysis results for e^+e^- annihilation into leptons (Sect. 24.3). In this way one obtains (Bergsma 1984)

$$\begin{aligned} g_V^e &= -0.08 \pm 0.07\,(\text{stat.}) \pm 0.03\,(\text{syst.}), \\ g_A^e &= -0.54 \pm 0.05\,(\text{stat.}) \pm 0.06\,(\text{syst.}). \end{aligned} \tag{24.19}$$

The value of g_A^e in (24.19) agrees very well with the standard model which demands $g_A^e = -0.5$. If we presuppose the validity of the standard model, then only one parameter needs to be determined from the data, namely $\sin^2 \vartheta_W$.

For this it is advantageous to consider the ratio $R = \sigma(\nu_\mu e)/\sigma(\overline{\nu}_\mu e)$ since R is less affected by systematic errors and can thus be measured more accurately than the individual cross-sections. According to (24.5), (24.17) the theoretical expectation is

$$R = \frac{\sigma(\nu_\mu e)}{\sigma(\overline{\nu}_\mu e)} = 3\frac{1 - 4\sin^2\vartheta_W + \frac{16}{3}\sin^4\vartheta_W}{1 - 4\sin^2\vartheta_W + 16\sin^4\vartheta_W}. \tag{24.20}$$

The above-mentioned experiment yields

$$\sin^2\vartheta_W = 0.215 \pm 0.032\,(\text{stat.}) \pm 0.012\,(\text{syst.}). \tag{24.21}$$

24.2 Neutrino–Nucleon Scattering

One of the best approaches to the determination of the weak angle ϑ_W is offered by neutrino–nucleon scattering. We confine our discussion here to the total inclusive cross-sections for the scattering of muon neutrinos and muon antineutrinos from nucleons, i.e.:

$$\begin{aligned} \nu_\mu(k) + \mathrm{N}(p) &\to \nu_\mu(k') + \mathrm{X}, \\ \overline{\nu}_\mu(k) + \mathrm{N}(p) &\to \overline{\nu}_\mu(k') + \mathrm{X}. \end{aligned} \tag{24.22}$$

The notation to be used for the momenta is as indicated here. The kinematics of this type of reaction is very similar to that of electron–nucleon scattering via the charged current, which was discussed in Chap. 18. We will again make use of the variables defined in Table 18.1.

The reactions (24.22) were first observed by the Gargamelle collaboration (Hasert 1973b). Today we possess good experimental data for these reactions, particularly in the deep inelastic region (see Geweniger (1984) for a review).

The theoretical treatment of the deep inelastic neutrino–nucleon scattering makes use of the parton model. This model has certainly proved its worth for the analogous charged current reactions treated in Chap. 18. In Fig. 24.4 we present the diagrams for ν_μ–N scattering via charged and neutral currents in the quark–parton model.

We next remind ourselves of the rules of the parton model (Chap. 18). In order to determine the transition rate for neutrino scattering from nucleons, we must first calculate the transition rates for the individual point-like partons and then sum incoherently over the contributions of all the partons in the nucleon. For the sake of simplicity, we will assume that as far as the quark and antiquark content of the nucleon is concerned only u- and d-quarks are present. We then have to take account of the following elementary parton reactions:

For scattering via neutral currents:

$$\begin{aligned} \nu_\mu(\overline{\nu}_\mu) + \mathrm{u} &\to \nu_\mu(\overline{\nu}_\mu) + \mathrm{u}, \\ \nu_\mu(\overline{\nu}_\mu) + \mathrm{d} &\to \nu_\mu(\overline{\nu}_\mu) + \mathrm{d}. \end{aligned} \tag{24.23}$$

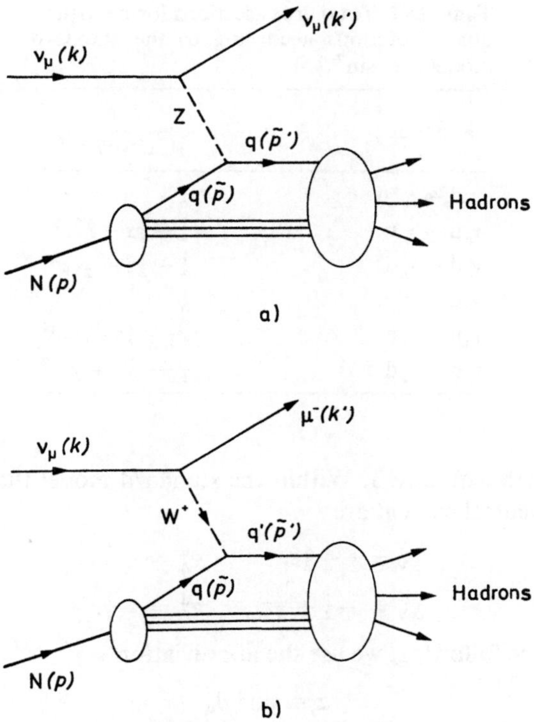

Figure 24.4 Neutrino–nucleon scattering in the quark–parton model via neutral currents (a) and charged currents (b). In place of a quark q, it is of course possible for an antiquark $\bar{q}$ to act as elementary scattering partner of the leptons.

For scattering via charged currents:

$$\begin{aligned} \nu_\mu + \mathrm{d} &\to \mu^- + \mathrm{u}, \\ \bar{\nu}_\mu + \mathrm{u} &\to \mu^+ + \mathrm{d}. \end{aligned} \tag{24.24}$$

The cross-sections for these parton reactions can be taken directly from neutrino–electron scattering since the quark–partons are point particles like the electron and since the muon mass is neglected in the calculation. We must, of course, replace g_V^e and g_A^e in the charged and neutral current by the coupling constants appropriate for quarks. The part of the effective Lagrange density (22.86) relevant for our reactions is

$$\begin{aligned} \mathscr{L}_{\mathrm{eff}} = -\frac{G}{\sqrt{2}} \{ &\bar{\nu}_\mu \gamma^\lambda (1-\gamma_5)\nu_\mu [\bar{\mathbf{u}}\gamma_\lambda (g_V^u - g_A^u \gamma_5)\mathbf{u} + \bar{\mathbf{d}}\gamma_\lambda (g_V^d - g_A^d\gamma_5)\mathbf{d}] \\ &+ \bar{\mu}\gamma^\lambda (1-\gamma_5)\nu_\mu\, \bar{\mathbf{u}}\gamma_\lambda(1-\gamma_5)\mathbf{d} \\ &+ \bar{\nu}_\mu \gamma^\lambda (1-\gamma_5)\mu\, \bar{\mathbf{d}}\gamma_\lambda (1-\gamma_5)\mathbf{u}\}. \end{aligned} \tag{24.25}$$

In the charged current we have set the element V_{11} of the Kobayashi–

Table 24.1 Total cross-sections for neutrino–quark reactions according to the standard model ($z \equiv \sin^2 \vartheta_W$).

Reaction	$\dfrac{\sigma}{2G^2(k\tilde{p})\pi^{-1}}$
$\nu_\mu d \to \mu^- u$	1
$\nu_\mu u \to \nu_\mu u$	$\frac{1}{4} - \frac{2}{3}z + \frac{16}{27}z^2$
$\nu_\mu d \to \nu_\mu d$	$\frac{1}{4} - \frac{1}{3}z + \frac{4}{27}z^2$
$\bar{\nu}_\mu u \to \mu^+ d$	$\frac{1}{3}$
$\bar{\nu}_\mu u \to \bar{\nu}_\mu u$	$\frac{1}{12} - \frac{2}{9}z + \frac{16}{27}z^2$
$\bar{\nu}_\mu d \to \bar{\nu}_\mu d$	$\frac{1}{12} - \frac{1}{9}z + \frac{4}{27}z^2$

Maskawa matrix equal to 1. Within the standard model the coupling constants in the neutral current are

$$g_V^u = \tfrac{1}{2} - \tfrac{4}{3}z, \qquad g_A^u = +\tfrac{1}{2},$$
$$g_V^d = -\tfrac{1}{2} + \tfrac{2}{3}z, \qquad g_A^d = -\tfrac{1}{2}. \tag{24.26}$$

Here, and in the following, we use the abbreviation

$$z \equiv \sin^2 \vartheta_W. \tag{24.27}$$

For high energies we can neglect the quark masses and with the appropriate substitutions, we obtain from (24.17) the neutrino–quark cross-sections given in Table 24.1. Here $\tilde{p}$ is the four-momentum of the quark before the scattering (cf. Fig. 24.4).

To calculate the cross-section for neutrino–nucleon scattering we must sum incoherently over the contributions of the quarks in the nucleon. As in Chap. 18, we denote the momentum fraction of a quark by x, and the distribution functions of the u- and d-quarks in the nucleon by $N_{u,d}(x)$. According to Chap. 18 we get, for example, for the reaction

$$\nu_\mu + N \to \mu^- + X \tag{24.28}$$

the following result for the total cross-section

$$\sigma(\nu_\mu N \to \mu^- X) = \int_0^1 dx\, N_d(x) \sigma(\nu_\mu d \to \mu^- u)|_{\tilde{p}=xp}$$
$$= \frac{G^2}{\pi} ME \cdot 2 \int_0^1 dx\, x N_d(x). \tag{24.29}$$

Here E is the energy of the incident neutrino in the laboratory system. The other neutrino–nucleon cross-sections are calculated in an analogous way.

Experimental studies have concentrated on the scattering from heavy nuclei. As target material one uses, for example, iron or marble. The measured cross-sections are therefore essentially the average value for proton and neutron. We write $N_u(x)$ and $N_d(x)$ for these averages of the distribution functions for u- and d-quarks respectively, and $N(x)$ for their sum. Then, by definition, and as a consequence of isospin invariance, we have

$$N(x) = N_u(x) + N_d(x) = 2N_u(x) = 2N_d(x). \tag{24.30}$$

With this we obtain from Table 24.1 the following cross-sections averaged over proton and neutron:

$$\begin{aligned}
\sigma(\nu_\mu \mathrm{N} \to \mu^- \mathrm{X}) &= \frac{G^2}{\pi} ME \cdot \int_0^1 \mathrm{d}x\, xN(x), \\
\sigma(\nu_\mu \mathrm{N} \to \nu_\mu \mathrm{X}) &= \frac{G^2}{\pi} ME \cdot \int_0^1 \mathrm{d}x\, xN(x) \cdot \{\tfrac{1}{2} - z + \tfrac{20}{27} z^2\}, \\
\sigma(\bar{\nu}_\mu \mathrm{N} \to \mu^+ \mathrm{X}) &= \frac{G^2}{\pi} ME \cdot \int_0^1 \mathrm{d}x\, xN(x) \tfrac{1}{3}, \\
\sigma(\bar{\nu}_\mu \mathrm{N} \to \bar{\nu}_\mu \mathrm{X}) &= \frac{G^2}{\pi} ME \cdot \int_0^1 \mathrm{d}x\, xN(x) \cdot \tfrac{1}{3} \{\tfrac{1}{2} - z + \tfrac{20}{9} z^2\}.
\end{aligned} \tag{24.31}$$

For the ratios R_ν and $R_{\bar{\nu}}$ of the total cross-sections for neutrino–nucleon and antineutrino–nucleon scattering via neutral and charged currents, we then obtain

$$\begin{aligned}
R_\nu &= \frac{\sigma(\nu_\mu \mathrm{N} \to \nu_\mu \mathrm{X})}{\sigma(\nu_\mu \mathrm{N} \to \mu^- \mathrm{X})} = \tfrac{1}{2} - z + \tfrac{20}{27} z^2, \\
R_{\bar{\nu}} &= \frac{\sigma(\bar{\nu}_\mu \mathrm{N} \to \bar{\nu}_\mu \mathrm{X})}{\sigma(\bar{\nu}_\mu \mathrm{N} \to \mu^+ \mathrm{X})} = \tfrac{1}{2} - z + \tfrac{20}{9} z^2.
\end{aligned} \tag{24.32}$$

Here $z \equiv \sin^2 \vartheta_W$ can vary between zero and one. If we look at these reactions in the R_ν–$R_{\bar{\nu}}$ plane, then according to the standard model, the experimental points must lie on a particular curve, the so-called "Weinberg nose" (Fig. 24.5). This assumes, however, the validity of the simplest quark–parton model for the nucleons. The position of the experimental points on the "nose" should determine the parameter $\sin^2 \vartheta_W$.

The experimental results were summarized by Geweniger (1984). As an example we quote the values presented by the CDHS and the CHARM collaborations:

$$\begin{aligned}
&\left.\begin{aligned} R_\nu &= 0.301 \pm 0.007 \\ R_{\bar{\nu}} &= 0.363 \pm 0.015 \end{aligned}\right\} \text{CDHS, Fe target (Geweniger 1984),} \\
&\left.\begin{aligned} R_\nu &= 0.320 \pm 0.010 \\ R_{\bar{\nu}} &= 0.377 \pm 0.020 \end{aligned}\right\} \text{CHARM, marble target (Jonker 1981).}
\end{aligned} \tag{24.33}$$

As can be seen from Fig. 24.5, with a value of about 0.23 for $\sin^2 \vartheta_W$, the standard model with the most naïve parton model, gives a surprisingly good description of the data. For a precise determination of $\sin^2 \vartheta_W$ one must, of course, take a number of corrections into account, for example, the presence

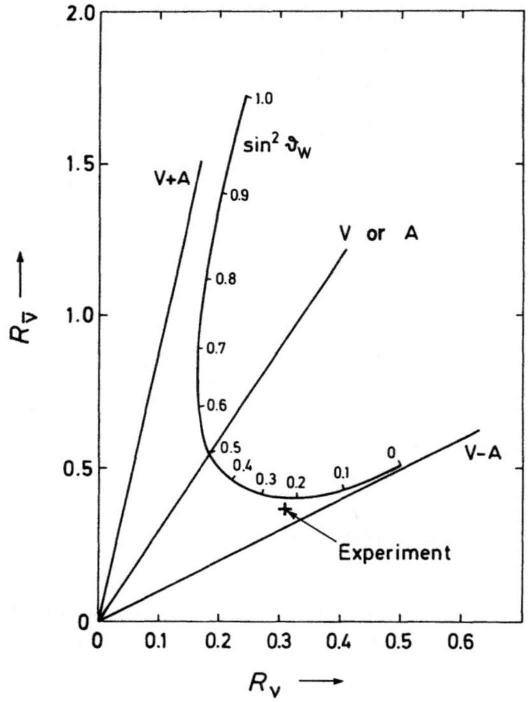

Figure 24.5 The predictions of the standard model for $R_\nu = \sigma(\nu_\mu N \to \nu_\mu X)/\sigma(\nu_\mu N \to \mu^- X)$ and $R_{\bar{\nu}} = \sigma(\bar{\nu}_\mu N \to \bar{\nu}_\mu X)/\sigma(\bar{\nu}_\mu N \to \mu^+ X)$ as functions of $\sin^2 \vartheta_W$. The straight lines refer to a pure (V − A), V, A, and (V + A) form of the neutral quark current. As experimental values, the average values for R_ν and $R_{\bar{\nu}}$ from (24.33) are given.

of the antiquarks in the nucleon (Chap. 19) and electroweak radiative corrections (Sirlin 1981). It must also be exactly specified how $\sin^2 \vartheta_W$ is defined in higher orders of perturbation theory. One possibility, which we will make use of in what follows, is to define $\sin^2 \vartheta_W$ such that (22.68) holds to all orders in terms of the physical masses of the W- and Z-bosons (Sirlin 1980)

$$\sin^2 \vartheta_W = 1 - \frac{m_W^2}{m_Z^2}. \tag{24.34}$$

With this convention, the neutrino–nucleon scattering experiments give, according to Geweniger (1984), a value of

$$\sin^2 \vartheta_W = 0.225 \pm 0.007\,(\text{exp.}) \pm 0.006\,(\text{theor.}), \tag{24.35}$$

where both experimental and theoretical uncertainties are indicated.

24.3 Effects of the Weak Interaction in Electron–Positron Annihilation

The reaction

$$e^+ + e^- \to \mu^+ + \mu^-$$

via the exchange of a virtual photon (Fig. 24.6(a)) has already been discussed in Chap. 11. In the standard model of the electroweak interaction there is a further lowest order diagram that contributes to this reaction, namely the

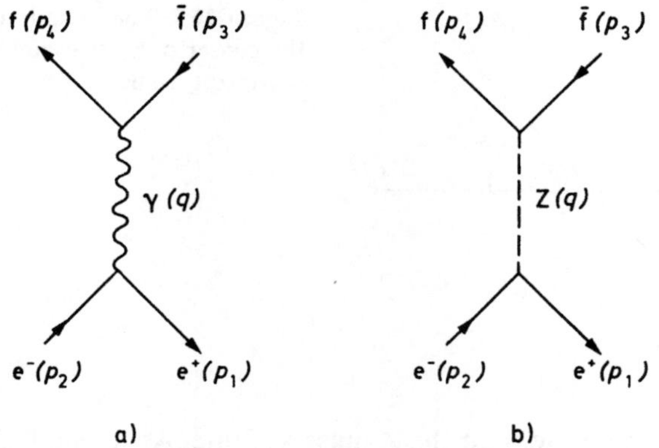

Figure 24.6 The annihilation of an e^+e^- pair with the production of a fermion–antifermion pair $f\,\bar{f}$ ($f \neq e$). In lowest order, the reaction occurs via a virtual photon γ(a) or a virtual Z-boson (b).

diagram where the role of the virtual photon is taken over by a virtual Z-boson (Fig. 24.6(b)). In the following we will discuss the more general case of e^+e^- annihilation into a fermion–antifermion pair:

$$e^+(p_1) + e^-(p_2) \to \bar{f}(p_3) + f(p_4) \qquad (f = \mu, \tau, u, d, s, c, b, \ldots; f \neq e). \tag{24.36}$$

In this f can be any fundamental fermion, be it lepton or quark, but not the electron. The corresponding diagrams are shown in Fig. 24.6. The reaction

$$e^+ + e^- \to e^+ + e^- \tag{24.37}$$

must be treated separately since here bosons can also be exchanged in the t-channel (see Sect. 10.2).

If a quark–antiquark pair is emitted, then we observe two hadron jets in the final state as described in Chap. 20. The momentum direction of the quarks is to be identified with the jet axes, determined, for example, with the help of the quantity thrust (20.6).

The kinematic variables for our reaction are the square of the total energy and the scattering angle in the center of mass frame (Fig. 24.7):

$$s = (p_1 + p_2)^2, \qquad \cos\vartheta = \frac{\boldsymbol{p}_1 \cdot \boldsymbol{p}_3}{|\boldsymbol{p}_1||\boldsymbol{p}_3|}. \tag{24.38}$$

The theoretical consideration starts from the interaction Lagrange density of the gauge bosons with the fermions (22.77)

$$\mathcal{L}_{\text{int}} = -e\Bigg\{\mathbf{A}_\lambda \mathcal{J}^\lambda_{\text{em}} + \frac{1}{\sin\vartheta_W \cos\vartheta_W}\mathbf{Z}_\lambda \mathcal{J}^\lambda_{\text{NC}} + \frac{1}{\sqrt{2}\sin\vartheta_W}(\mathbf{W}^+_\lambda \mathcal{J}^\lambda_{\text{CC}} + \mathbf{W}^-_\lambda \mathcal{J}^{\lambda\dagger}_{\text{CC}})\Bigg\}. \tag{24.39}$$

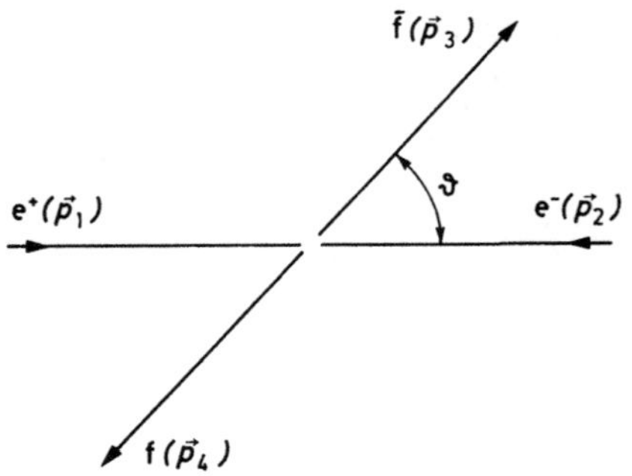

Figure 24.7 The reaction $e^+e^- \to f\,\bar{f}$ in the center of mass system where ϑ is the scattering angle.

Unlike before we now treat the Z mass as finite. As we shall shortly see, the Z also has a nonzero width Γ_Z—it is an unstable particle. We will again write the neutral current in the general form:

$$\mathscr{J}^{\lambda}_{NC} = \sum_f \tfrac{1}{2}\bar{\mathbf{f}}(g^f_V\gamma^\lambda - g^f_A\gamma^\lambda\gamma_5)\mathbf{f}, \tag{24.40}$$

where g^f_V and g^f_A are regarded as free parameters that must be determined from experiment. The summation in (24.40) runs over all fundamental fermions including the electron. In the standard model it is possible to express all the coupling constants in terms of the single parameter $\sin^2 \vartheta_W$:

$$\begin{aligned} g^f_V &= T^f_{3L} + T^f_{3R} - 2\sin^2\vartheta_W Q_f, \\ g^f_A &= T^f_{3L} - T^f_{3R}. \end{aligned} \tag{24.41}$$

Here $T^f_{3L,R}$ are the third components of the weak isospin of the left- and right-handed parts of the fermion f, and Q_f is its electrical charge. The values for the coupling constants of the fermions that result from (24.41) are summarized in Table 24.2. In evaluating the diagrams in Fig. 24.6 we neglect all fermion masses. Taking account of the finite width Γ_Z of the Z-boson in its propagator (cf. Sect. 16.1), we obtain for the S-matrix element:

$$\begin{aligned} S_{fi} = {} & ie^2(2\pi)^4\delta(p_1+p_2-p_3-p_4) \\ & \cdot\Bigg\{\bar{u}_f(p_4)Q_f\gamma^\lambda v_f(p_3)\frac{1}{s}\bar{v}_e(p_1)(-\gamma_\lambda)u_e(p_2) \\ & + \frac{1}{4}\left(\frac{1}{\sin\vartheta_W\cos\vartheta_W}\right)^2\bar{u}_f(p_4)(g^f_V\gamma^\lambda - g^f_A\gamma^\lambda\gamma_5)v_f(p_3) \\ & \cdot\frac{1}{s-m_Z^2+im_Z\Gamma_Z}\bar{v}_e(p_1)(g^e_V\gamma_\lambda - g^e_A\gamma_\lambda\gamma_5)u_e(p_2)\Bigg\}. \end{aligned} \tag{24.42}$$

From this it is simple to calculate the differential cross-section, for the case where unpolarized electrons are assumed for the initial state and where the polarization in the final state is left unobserved:

$$\frac{d\sigma}{d\Omega_3} = \frac{\alpha^2}{4s}\{R_f(s)(1+\cos^2\vartheta) + I_f(s)\cos\vartheta\}. \tag{24.43}$$

Table 24.2 Values of the coupling constants g_V^f and g_A^f of the neutral current for the fundamental fermions f in the standard model ($z \equiv \sin^2 \vartheta_W$).

f	g_V^f	g_A^f
ν_e, ν_μ, ν_τ	$\frac{1}{2}$	$\frac{1}{2}$
e, μ, τ	$-\frac{1}{2} + 2z$	$-\frac{1}{2}$
u, c, t	$\frac{1}{2} - \frac{4}{3}z$	$\frac{1}{2}$
d, s, b	$-\frac{1}{2} + \frac{2}{3}z$	$-\frac{1}{2}$

Here Ω_3 is the solid angle corresponding to $\boldsymbol{p}_3$ and

$$\begin{aligned} R_f(s) = N_c^f\{Q_f^2 &+ 2Q_f g_V^e g_V^f \operatorname{Re} K(s) \\ &+ [(g_V^e)^2 + (g_A^e)^2][(g_V^f)^2 + (g_A^f)^2]|K(s)|^2\}, \end{aligned} \tag{24.44}$$

$$I_f(s) = N_c^f\{4Q_f g_A^e g_A^f \operatorname{Re} K(s) + 8 g_V^e g_A^e g_V^f g_A^f |K(s)|^2\},$$

$$K(s) = \frac{-s}{4(\sin\vartheta_W \cos\vartheta_W)^2(s - m_Z^2 + i m_Z \Gamma_Z)}, \tag{24.45}$$

$$N_c^f = \begin{cases} 1 & \text{for leptons,} \\ 3 & \text{for quarks.} \end{cases} \tag{24.46}$$

The factor N_c^f takes account of the fact that the quarks have three possible colors. With the help of (22.68), (22.84a) we can also write $K(s)$ in the following form

$$K(s) = \frac{Gs}{2\sqrt{2}\pi\alpha} \frac{-m_Z^2}{s - m_Z^2 + i m_Z \Gamma_Z}. \tag{24.47}$$

For $s \ll m_Z^2$ one can thus express $K(s)$ in terms of the Fermi constant G and the fine-structure constant α alone:

$$K(s) \xrightarrow[s \ll m_Z^2]{} \frac{Gs}{2\sqrt{2}\pi\alpha} = 0.220\left(\frac{\sqrt{s}}{35\ \text{GeV}}\right)^2. \tag{24.48}$$

The numerical value of G is taken from (23.3).

We will now discuss our result. For $s \to 0$ only the one-photon exchange contribution remains, which corresponds to the quantum electrodynamical result (Chap. 11):

$$\begin{aligned} R_f(s) &\xrightarrow[s \to 0]{} N_c^f Q_f^2, \\ I_f(s) &\xrightarrow[s \to 0]{} 0. \end{aligned} \tag{24.49}$$

The angular distribution is then, according to (24.43), symmetric with respect to the forward and backward hemisphere, so that exactly as many fermions are produced with $\vartheta < 90°$ as with $\vartheta > 90°$. For the integrated cross-section we obtain the standard result

$$\sigma(e^+e^- \to \bar{f}f) = \frac{4\pi\alpha^2}{3s} N_c^f Q_f^2. \tag{24.50}$$

For nonzero values of s, where the Z-boson also contributes, the angular distribution becomes asymmetric with respect to $\vartheta = 90°$. We define the (forward–backward) asymmetry by

$$A_f(s) = \frac{\sigma(\vartheta < 90°) - \sigma(\vartheta > 90°)}{\sigma(\vartheta < 90°) + \sigma(\vartheta > 90°)}. \tag{24.51}$$

A simple calculation yields

$$A_f(s) = \frac{3}{8}\frac{I_f(s)}{R_f(s)}. \tag{24.52}$$

The integrated cross-section is also modified:

$$\begin{aligned} \sigma(e^+e^- \to \bar{f}f) &= \frac{4\pi\alpha^2}{3s} R_f(s), \\ R_f(s) &= N_c^f Q_f^2 + \delta R_f(s). \end{aligned} \tag{24.53}$$

For center of mass energies much smaller than the Z mass, and neglecting Γ_Z and terms of the order of $K^2(s)$, we obtain from (24.44):

$$\begin{aligned} \delta R_f(s) &= 2N_c^f Q_f g_V^e g_V^f \cdot \frac{Gs}{2\sqrt{2}\pi\alpha} \cdot \frac{m_Z^2}{m_Z^2 - s}, \\ A_f(s) &= \frac{3}{2} \cdot \frac{g_A^e g_A^f}{Q_f} \cdot \frac{Gs}{2\sqrt{2}\pi\alpha} \cdot \frac{m_Z^2}{m_Z^2 - s}. \end{aligned} \tag{24.54}$$

At low energies, the asymmetry and the deviation of the integrated cross-section from the standard value increase essentially linearly with the square of the center of mass energy. At $\sqrt{s} = 35$ GeV we find numerically, using (24.48), a predicted asymmetry of the order of 10–20%.

We now discuss the reaction $e^+e^- \to \mu^+\mu^-$, which is the reaction that has been most thoroughly studied experimentally. The results from the PETRA experiments for the integrated cross-section, or more precisely for $R_\mu(s)$, have already been presented in Fig. 11.4. Right up to the highest center of mass energies which have been available at PETRA, there is no discernible deviation in $R_\mu(s)$ from the prediction of QED. We conclude from Fig. 11.4 that

$$|\delta R_\mu(s)| \lesssim 0.1 \qquad \text{for} \quad s < 2000\ \text{GeV}^2. \tag{24.55}$$

According to (24.48) and (24.54) the product of the vector coupling constants of e and μ is therefore restricted to

$$|g_V^e g_V^\mu| \lesssim 0.11. \tag{24.56}$$

We next look at the angular distribution of the muons in the reaction $e^+e^- \to \mu^+\mu^-$. Data from the JADE group have already been shown in Fig. 11.5. At

Table 24.3 Values for the asymmetry parameter A_μ in the reaction $e^+e^- \to \mu^+\mu^-$ (24.51). The experimental values are averages obtained by combining the results of several experiments [for $\sqrt{s} = 29$ GeV and 34.5 GeV, from Naroska (1983); for $\sqrt{s} = 42.5$ GeV, from Bowdery (1984)]. The theoretical values of the standard model are calculated from (24.54).

	A_μ in %	
$\sqrt{s}$ in GeV	Experimental value	Theoretical value according to the standard model (Equation (24.54), $g_A^e = g_A^\mu = -\frac{1}{2}$)
29.0	-6.3 ± 0.9	-6.3
34.5	-10.8 ± 1.1	-9.3
42.5	-17.6 ± 2.7	-15.4

center of mass energies $\sqrt{s} \gtrsim 30$ GeV, one recognizes a significant deviation from the symmetric QED distribution. The average values of the observed asymmetries are presented in Table 24.3. By comparing these with the theoretical expression in (24.54) we obtain a value for the product of the axial-vector coupling constants:

$$g_A^e g_A^\mu = \begin{cases} 0.25 \pm 0.04 & (\sqrt{s} = 29 \text{ GeV}), \\ 0.29 \pm 0.03 & (\sqrt{s} = 34.5 \text{ GeV}), \\ 0.28 \pm 0.03 & (\text{average value}). \end{cases} \tag{24.57}$$

Here we have used $m_Z = 92.9$ GeV (Table 22.1).

We now apply these results in order to obtain conditions on the coupling constants g_V^e and g_A^e of the electron. To do this it is necessary, however, to assume e–μ universality, i.e., to *presuppose* equal coupling constants for e and μ:

$$g_V^e = g_V^\mu, \qquad g_A^e = g_A^\mu. \tag{24.58}$$

Equations (24.56), (24.57) then demand

$$|g_V^e| \lesssim 0.33, \qquad |g_A^e| = 0.52 \pm 0.02. \tag{24.59}$$

If, in addition, the results of the reaction $e^+e^- \to e^+e^-$ are used, one obtains the allowed regions marked with e^+e^- in Fig. 24.3. We see that there is compatibility with only one (24.19) of the four possibilities for g_V^e and g_A^e allowed by the results of ν_μ–e, $\bar{\nu}_\mu$–e, and $\bar{\nu}_e$–e scattering.

Measurements of the asymmetry parameter have also been made for $\tau^+\tau^-$ production and for the production of the heavy quark–antiquark pairs $c\bar{c}$ and $b\bar{b}$. In the latter cases, of course, only the two jets can be observed in the final state (Chap. 20) and the problem arises as to how to identify the jet as fragmentation product of a particular quark flavor. This can be achieved, however, by observing, for example, jets that are accompanied by a muon (Fig. 24.8). Muons with large transverse momentum relative to the jet axis stem

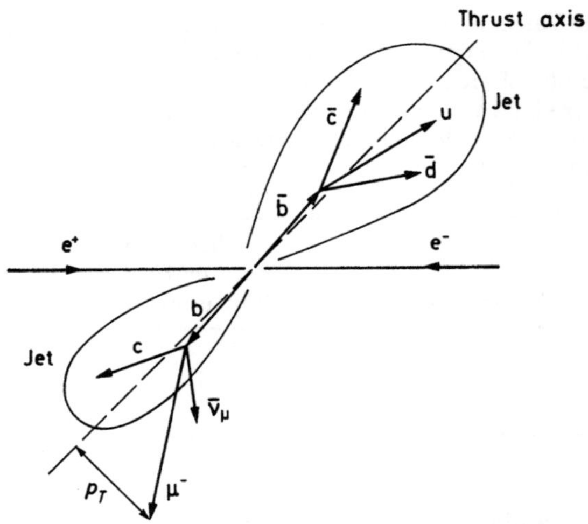

Figure 24.8 Schematic view of a reaction $e^+e^- \to \bar{b}b$ in which the b-quark decays semileptonically and the $\bar{b}$-antiquark nonleptonically. A large transverse momentum p_T of the μ^- relative to the thrust axis can serve as an indication for the production of the $b\bar{b}$ pair.

principally from the semileptonic decay of a heavy quark, which is of course, bound in a corresponding hadron (Sect. 23.6). Possible reactions are:

$$\begin{aligned} b &\to \mu^- + \bar{\nu}_\mu + c, \\ \bar{b} &\to \mu^+ + \nu_\mu + \bar{c}, \\ c &\to \mu^+ + \nu_\mu + s, \\ \bar{c} &\to \mu^- + \bar{\nu}_\mu + \bar{s}. \end{aligned} \tag{24.60}$$

The transverse momentum of the muon is, for simple kinematical reasons, of the order of half the mass of the decaying quark, and thus, from Table 22.1, about 2 GeV for b-decay and 0.6 GeV for c-decay. From the transverse momentum one can therefore infer—at least on a statistical basis—the flavor of the original quark.

The experiments at PETRA with $\sqrt{s} = 34.5$ GeV gave the following average values for the asymmetries (Bowdery 1984; Wicklund 1984)

$$\begin{aligned} A_\tau &= -7.7 \pm 1.9\%, \\ A_c &= -14.2 \pm 5.2\%, \\ A_b &= -23.5 \pm 5.5\%. \end{aligned} \tag{24.61}$$

With the help of (24.54) we can use these values to determine g_A^τ, g_A^c, and g_A^b, since we already know g_A^e. In Table 24.4 we list the corresponding results. This table summarizes all available information about the couplings of the fundamental fermions to the Z-boson. To within the error margin, the values agree

Table 24.4 A summary of the results for the coupling constants of the neutral current. We have made use of the following information: for g_V^e, g_A^e equations (24.19), (24.59); for g_A^μ equation (24.57) and e–μ universality; for g_A^τ, g_A^c, g_A^b equations (24.61), (24.54) and g_A^e as determined above; for g_V^u, g_A^u, g_V^d, g_A^d the values quoted by Panman (1984) for the so-called chiral coupling constants q_L, q_R ($q = u, d$) satisfying $g_V^q = q_L + q_R$, $g_A^q = q_L - q_R$, which are obtained from experiments on νN and eD scattering and on parity violation in atoms.

Fermion	Experiment		Standard model $\sin^2\vartheta_W = 0.23$	
f	g_V^f	g_A^f	g_V^f	g_A^f
e	-0.08 ± 0.08	-0.52 ± 0.02		
μ		-0.52 ± 0.02	-0.04	$-\frac{1}{2}$
τ		-0.40 ± 0.10		
u	0.22 ± 0.02	0.56 ± 0.02		
c		0.49 ± 0.18	0.19	$\frac{1}{2}$
t				
d	-0.33 ± 0.03	-0.41 ± 0.03		
s			-0.35	$-\frac{1}{2}$
b		-0.40 ± 0.09		

with those of the standard model. The table makes evident the lack of information about the s-quark and that the information about the vector coupling constants is sparse.

For the remainder of this section we shall assume the validity of the standard model. With $K(s)$ given by (24.45) the asymmetry obtained from (24.52) then only depends on m_Z, Γ_Z, and $\sin^2\vartheta_W$.

Since m_Z is known from direct measurements and Γ_Z can be calculated in the standard model (Chap. 25), the only remaining free parameter is $\sin^2\vartheta_W$. By measuring the asymmetry—for example $A_\mu(s)$—we can thus determine $\sin^2\vartheta_W$ (Böhm, A. 1984). The value obtained, together with other precision determinations of $\sin^2\vartheta_W$, is given in Table 24.5. Typical energy scales for the experiments are also listed there. The agreement between the values obtained at quite different energies is impressive, and represents a major achievement of the standard model. The next step is to check the electroweak interaction at the level of diagrams with one loop. For the calculation of such electroweak radiative corrections we refer the reader to the review of Bég (1982). The typical order of magnitude of the electroweak corrections to $\sin^2\vartheta_W$ is

$$\frac{\delta\sin^2\vartheta_W}{\sin^2\vartheta_W} \cong \alpha\ln\left(\frac{m_Z}{\mu}\right)^2 . \tag{24.62}$$

Here μ is the energy scale of the measurement except in experiments dealing with parity violation in atoms where it is a typical hadronic scale, $\mu \cong 0.3$ GeV. The radiative shift of the values of $\sin^2\vartheta_W$ in Table 24.5 is thus about 5%. To verify these shifts one needs to perform at least *two* experiments both

Table 24.5 A summary of the determinations of the parameter $\sin^2 \vartheta_W$ by experiments at a variety of energies. We use here the definition $\sin^2 \vartheta_W = 1 - m_W^2/m_Z^2$ (24.34). The energy scales refer to typical center of mass energies or to energy and momentum transfers. For experiments on parity violation in atoms the binding energy of the electron in the ground state of the hydrogen atom is shown as the energy scale. Of the two errors quoted, the first is experimental (quadratic sum of statistical and systematic errors), and the second is an estimate of the theoretical uncertainty. For the determinations which rely on the reactions $\nu e \to \nu e$ and $e^+e^- \to \mu^+\mu^-$ and (by definition) for that which comes from m_W and m_Z, there is negligible theoretical uncertainty or none at all. Electroweak radiative corrections have been included (see, Marciano 1980, 1981, 1983; Beg 1982; Llewellyn Smith 1982; Paschos 1982; Wetzel 1983, 1983a; Böhm, M. 1984, and references therein).

Measured quantity or reaction	Typical energy scale in GeV	$\sin^2 \vartheta_W$	Reference
Parity violation in atoms	10^{-8}	$0.207 \pm 0.043 \pm 0.045$	Piketty (1984)
$\nu e \to \nu e$	10^{-1}	0.215 ± 0.034	Eq. (24.21)
$eD \to eX$	1	$0.223 \pm 0.015 \pm 0.020$	Prescott (1978, 1979)
$\nu N \to \nu X, \mu X$	5	$0.225 \pm 0.007 \pm 0.006$	Eq. (24.35)
$\mu N \to \mu X$	10	$0.23 \pm 0.08 \pm 0.02$	Argento (1983)
$e^+e^- \to \mu^+\mu^-$	35	0.18 ± 0.02	Böhm, A. (1984)
$1 - \dfrac{m_W^2}{m_Z^2}$	85	0.24 ± 0.06	m_W and m_Z from Table 22.1

of which allow $\sin^2 \vartheta_W$ to be determined to an accuracy of 1–2%. This aim should be reached in the near future.

Exercises

24.1 Determine the kinematic limits for the variable y in (24.8).

24.2 Calculate the differential cross-section $d\sigma/d\Omega_3$ and the asymmetry A_f for the reaction (24.36) assuming the following interaction involving two Z-bosons (see, e.g., de Groot 1980):

$$\mathscr{L}_{int} = -e\left\{\mathbf{A}_\lambda \mathscr{J}^\lambda_{em} + \frac{1}{\sin \vartheta_W \cos \vartheta_W}\left(\mathbf{Z}_\lambda \mathscr{J}^\lambda_{NC} + \frac{\sqrt{C} m_{Z'}}{m_Z}\mathbf{Z}'_\lambda \mathscr{J}^\lambda_{em}\right)\right\}, \quad (24.63)$$

where $C \geq 0$ is a constant. What is the form of the effective Lagrange density at low energies ($s \ll m_Z^2, m_{Z'}^2$)? Estimate an upper limit for C from the experimental data of Fig. 11.4.

24.3 Calculate the differential cross-section for the reaction $e^+e^- \to e^+e^-$ taking into account the diagrams of lowest order for the exchange of a photon or a Z-boson in the s- or t-channel.

CHAPTER 25

The Physics of the Z-, W-, and Higgs Bosons

With the proton–antiproton storage ring Sp$\bar{\text{p}}$S at CERN, the Tevatron at FNAL, SLC at SLAC, and the new accelerators under construction LEP and HERA), we are venturing into energy ranges where the production of Z- and W-bosons is possible. These bosons were actually discovered in 1983 in experiments at the Sp$\bar{\text{p}}$S ring at CERN. In this chapter we will first delineate the basic properties of the Z- and W- bosons and then move on to discuss their production in p$\bar{\text{p}}$ collisions. For the Higgs boson, which has not yet been discovered, we will present a few of the predictions of the standard model.

25.1 The Z-Boson

The standard model predicts the existence of a Z-boson whose mass according to (22.84) and (22.68) is given by:

$$m_Z = \left(\frac{\pi\alpha}{\sqrt{2}G}\right)^{1/2} \frac{1}{\sin\vartheta_W \cdot \cos\vartheta_W}. \tag{25.1a}$$

With G from (23.3) and $\sin^2\vartheta_W = 0.225$ from (24.35) this yields

$$m_Z = 89.3 \text{ GeV}. \tag{25.1b}$$

However, the Z-boson is not a stable particle. As we saw in Chap. 22, it couples, for example, to all fundamental fermions f (24.39). Via this coupling the Z-boson can decay into a fermion–antifermion pair:

$$Z \to f + \bar{f} \tag{25.2}$$

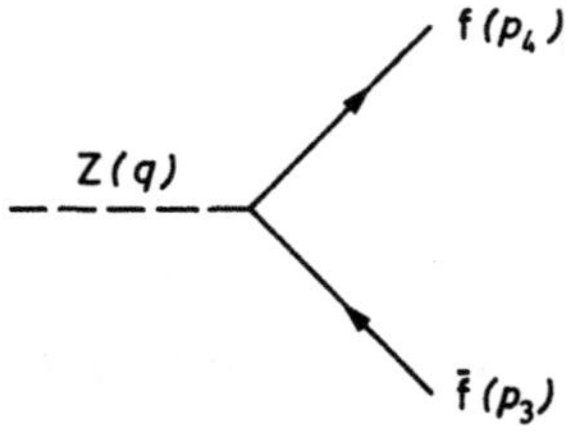

Figure 25.1 Lowest order diagram for the decay of the Z-boson into a fermion–antifermion pair. Here f is one of the fundamental fermions f = ν_e, ν_μ, ν_τ, e, μ, τ, u, c, t, d, s, b.

and therefore has a finite width Γ_Z. In Fig. 25.1 we show the corresponding lowest order diagram. It is easy to see that in the standard model all other decay processes of the Z-boson are of higher order in the gauge coupling of the electroweak or strong interaction.

Our first aim is to calculate, to leading order, the total width and the branching ratios of the Z. For an order of magnitude estimate of the decay width we can argue as follows. Since the coupling in (24.39) is proportional to e, the width must be proportional to α. If we neglect the fermion masses, the only remaining quantity possessing dimensions is the Z mass. Consequently, we expect

$$\Gamma_Z \cong \alpha m_Z \cong \frac{90\ \text{GeV}}{137} \cong 1\ \text{GeV}. \tag{25.3}$$

The actual calculation starts from determining the S-matrix element for the decay Z → f + f̄ from the Lagrange density (24.39). Writing the neutral current $\mathscr{J}_{NC}$ again in the general form (24.40) we get

$$\langle \bar{\mathrm{f}}(p_3)\mathrm{f}(p_4)|\mathbf{S}|\mathrm{Z}(q)\rangle = -\mathrm{i}e\frac{1}{2\sin\vartheta_W\cos\vartheta_W}(2\pi)^4\delta(q-p_3-p_4)$$
$$\cdot\,\bar{u}(p_4)\gamma^\lambda(g_V^f - g_A^f\gamma_5)v(p_3)\varepsilon_\lambda. \tag{25.4}$$

Here ε is the polarization vector of the Z-boson. From this the width for the decay Z → f + f̄ is found to be

$$\Gamma(\mathrm{Z}\to\mathrm{f}\bar{\mathrm{f}}) = \frac{\alpha m_Z(1-4m_f^2/m_Z^2)^{1/2}}{12\sin^2\vartheta_W\cos^2\vartheta_W}$$
$$\cdot N_c^f\left\{(g_V^f)^2\left(1+\frac{2m_f^2}{m_Z^2}\right)+(g_A^f)^2\left(1-\frac{4m_f^2}{m_Z^2}\right)\right\} \tag{25.5}$$

where, as previously, $N_c^f = 1$ for leptons and $N_c^f = 3$ for quarks. Quarks as decay products will, of course, fragment into hadrons.

We now substitute the values for g_V^f and g_A^f from the standard model (Table 24.2). For the decay into a $\nu_e\bar{\nu}_e$ pair, for example, we find

$$\Gamma(\mathrm{Z}\to\nu_e+\bar{\nu}_e) = \frac{\alpha m_Z}{24\sin^2\vartheta_W\cos^2\vartheta_W}. \tag{25.6}$$

The numerical value of this rate is given in Table 25.1 together with the decay

Table 25.1 The decay widths of the Z-boson calculated to lowest order from (25.5). Here $z \equiv \sin^2 \vartheta_W$ and all lepton and quark masses apart from m_t have been neglected. The numerical values are for $m_Z = 89.3$ GeV [eq. (25.1b)], $m_t = 40$ GeV and $\sin^2 \vartheta_W = 0.225$ [eq. (24.35)].

f	$\Gamma(Z \to f\bar{f})/\Gamma(Z \to \nu_e\bar{\nu}_e)$		$\Gamma(Z \to f\bar{f})$ (MeV)	$\Gamma(Z \to f\bar{f})/\Gamma_Z$
			$z = 0.225$; $m_t = 40$ GeV	
ν_e, ν_μ, ν_τ	1	1	156	6.7 %
e, μ, τ	$\frac{1}{2}[1 + (1 - 4z)^2]$	0.51	79	3.4 %
u, c	$\frac{3}{2}[1 + (1 - \frac{8}{3}z)^2]$	1.74	271	11.6 %
t	$\frac{3}{2}\left(1 - \frac{4m_t^2}{m_Z^2}\right)^{1/2}\left[1 - \frac{4m_t^2}{m_Z^2} + (1 - \frac{8}{3}z)^2\left(1 + \frac{2m_t^2}{m_Z^2}\right)\right]$	0.28	44	1.9 %
d, s, b	$\frac{3}{2}[1 + (1 - \frac{4}{3}z)^2]$	2.24	348	14.9 %

widths for the other channels. For the total decay width we obtain

$$\Gamma_Z = 2.3 \text{ GeV}. \tag{25.7}$$

To make precise comparisons of theory and experiment, which should, in the near future, become possible, it is important to take electroweak and gluonic radiative corrections into account (Marciano 1979; Albert 1980; Wetzel 1983; Böhm, M. 1984; Bég 1982 and references therein). These corrections have the effect that the Z-mass is no longer given precisely by (25.1a) but instead by a modified relationship which is rather complicated and also involves the still unknown masses of the Higgs particle and the top quark. Numerically the results shown in Fig. 25.2 are obtained. With the value of $\sin^2 \vartheta_W$ from (24.35),

$$\sin^2 \vartheta_W = 0.225 \pm 0.009, \tag{25.8}$$

where the error is now the quadratic sum of the experimental and theoretical errors, we find the following prediction of the standard model for the Z mass

$$m_Z = 92.6 \begin{cases} +1.9 \\ -1.6 \end{cases} \text{GeV}. \tag{25.9}$$

We assume here that the unknown Higgs mass lies between 10 and 1,000 GeV and $m_t = 40$ GeV. Comparing this result with (25.1b) we see that the radiative corrections lead to a shift of about 3 GeV in the Z mass. This is a readily measurable amount.

The partial widths of the Z-boson are also influenced by radiative corrections. The magnitude of these corrections is dependent upon which parameters we use to express the partial widths in lowest order (25.5). In place of the set α, m_Z, and ϑ_W we can, by exploiting (22.84), (24.34), for example, use the set

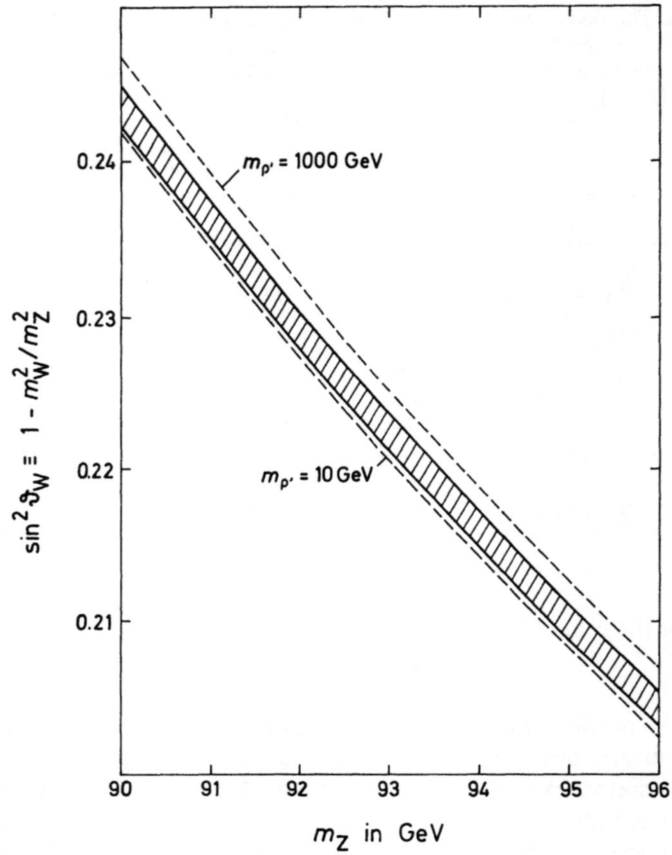

Figure 25.2 The relationship between $\sin^2 \vartheta_W \equiv 1 - m_W^2/m_Z^2$ and the Z mass m_Z in the standard model with inclusion of electroweak and hadronic corrections to (25.1a). The shaded band corresponds to a Higgs mass $m_{\rho'} = 40$ GeV and a t-quark mass $m_t = 40 \pm 10$ GeV. An additional uncertainty in m_Z of ± 0.1 GeV from the hadronic corrections has been taken into account. Curves for $m_t = 40$ GeV and $m_{\rho'} = 10$ GeV and 1000 GeV are also shown (W. Wetzel, private communication).

consisting of the Fermi constant G together with m_Z and ϑ_W. More precisely, G and ϑ_W should be *defined* by (23.2b) and (24.34), respectively; i.e., by definition, no radiative corrections appear in these relationships. With this new choice of parameters, even the electroweak radiative corrections to the partial widths of the Z turn out to be very small (Wetzel 1983). For leptons, one finds to a precision of about 1 ‰:

$$\Gamma(\mathrm{Z} \to \mathrm{f}\bar{\mathrm{f}}) = \frac{G m_Z^3}{12\sqrt{2}\pi}\left(1 - \frac{4m_f^2}{m_Z^2}\right)^{1/2} \cdot \frac{1}{2}\left[1 - \frac{4m_f^2}{m_Z^2} + (1 - 4\sin^2\vartheta_W |Q_f|)^2 \left(1 + \frac{2m_f^2}{m_Z^2}\right)\right]$$

$$(\mathrm{f} = \nu_e, \nu_\mu, \nu_\tau, e, \mu, \tau). \quad (25.10)$$

The Z-decay into quarks (i.e., hadrons) is a process analogous to the hadronic "decay" of a virtual photon produced in an e^+e^- collision (Chaps. 18 and 20). The decay

$$Z \to q + \bar{q}, \tag{25.11}$$

where q stands for one of the quarks u, d, s, c, b, for which $m_q \ll m_Z$, will yield two hadron jets in the final state. In addition, in QCD one also expects processes yielding 3, 4, ... jets:

$$\begin{aligned} Z \to\; & q + \bar{q} + G, \\ & q + \bar{q} + G + G, \\ & q + \bar{q} + q' + q', \quad \text{etc.} \end{aligned} \tag{25.12}$$

The evaluation of these processes again proceeds in analogy to that for e^+e^- annihilation via a virtual photon. The gluon corrections to order α_s, in fact modify the partial width for the decay into a particular quark channel in exactly the same way as in the case of e^+e^- annihilation into hadrons, cf. (20.13). The result is thus

$$\Gamma(Z \to q\bar{q} \text{ and } q\bar{q}G) = \frac{Gm_Z^3}{12\sqrt{2}\pi}\left(1 - \frac{4m_q^2}{m_Z^2}\right)^{1/2}$$

$$\cdot \frac{3}{2}\left[1 - \frac{4m_q^2}{m_Z^2} + (1 - 4\sin^2\vartheta_W|Q_q|)^2\left(1 + 2\frac{m_q^2}{m_Z^2}\right)\right]\left(1 + \frac{\alpha_s(m_Z^2)}{\pi}\right). \tag{25.13}$$

The use of the parameters G, m_Z, and ϑ_W guarantees that also for the quark channels the electroweak radiative corrections to (25.13) as a rule are very small ($\lesssim 1$ ‰) (Wetzel 1983; Akhundov 1986).

The Z decay into hadrons via a $t\bar{t}$ pair could be significantly more complicated if the threshold for the production of two particles each containing a t-quark is close to the Z mass. Mixing between the Z and the bound states of the $t\bar{t}$ system can become important in this case (see, e.g., Kühn 1985). There are also speculations that the Born estimate (25.5) is then drastically modified by strong interaction corrections (Horgan 1982). Nonetheless, we will use (25.13) as an estimate for $\Gamma(Z \to t\bar{t}$ and $t\bar{t}G)$. The partial widths and branching ratios calculated from (25.10), (25.13) are summarized in Table 25.2. For the total width Γ_Z we thus obtain the following prediction:

$$\Gamma_Z = 2.7 \pm 0.2 \text{ GeV}. \tag{25.14}$$

Rare decay processes of the Z-boson, which appear at higher orders of the electroweak interaction, have also been calculated (see Marciano 1979; Guberina 1980; Pietschmann 1982).

Having gained an overview of the decay processes of the Z-boson, we will now discuss the production of the Z in e^+e^- annihilation. As in Sect. 24.3 we begin by considering the reaction

$$e^+ + e^- \to \mu^+ + \mu^-. \tag{25.15}$$

The corresponding diagrams have already been shown in Fig. 24.6. If the

Table 25.2 The predictions of the standard model (25.10), (25.13) for the partial widths of the Z-boson. We have taken the values $m_Z = 92.6 \pm 1.9$ GeV, $m_t = 40 \pm 10$ GeV, $\sin^2\vartheta_W = 0.225 \pm 0.009$ and $\alpha_s(m_Z^2)/\pi = 0.04 \pm 0.01$. The error in the partial widths stems, other than for $Z \to t\bar{t} + t\bar{t}G$, mainly from the uncertainty in m_Z. Electroweak and gluonic radiative corrections are taken into account as described in the text.

Final state X	$\Gamma(Z \to X)$ (MeV)	$\Gamma(Z \to X)/\Gamma_Z$ (%)
$\nu_e\bar{\nu}_e$, $\nu_\mu\bar{\nu}_\mu$, $\nu_\tau\bar{\nu}_\tau$	174 ± 11	6.5 ± 0.3
e^-e^+, $\mu^-\mu^+$, $\tau^-\tau^+$	88 ± 6	3.3 ± 0.2
$u\bar{u} + u\bar{u}G$, $c\bar{c} + c\bar{c}G$	314 ± 21	11.7 ± 0.6
$t\bar{t} + t\bar{t}G$	$63\left\{\begin{matrix}+100\\-63\end{matrix}\right.$	$2.3\left\{\begin{matrix}+4.2\\-2.3\end{matrix}\right.$
$d\bar{d} + d\bar{d}G$, $s\bar{s} + s\bar{s}G$	404 ± 29	15.1 ± 0.7
$b\bar{b} + b\bar{b}G$	400 ± 29	14.9 ± 0.7
Total	2685 ± 200	

center of mass energy $\sqrt{s}$ approaches the Z mass m_Z, then the contribution of the virtual photon becomes very small compared to the Z contribution and the cross-section shows typical resonance behavior (cf. Sect. 16.1). From (24.43)–(24.51) and (25.5) the contribution of the Z-boson is found to be

$$\sigma(e^+e^- \to Z \to \mu^+\mu^-) = 12\pi \frac{\Gamma(Z \to e^+e^-)\Gamma(Z \to \mu^+\mu^-)}{(s - m_Z^2)^2 + m_Z^2\Gamma_Z^2} . \tag{25.16}$$

Since the electronic (muonic) branching ratio is sizable, the Z-boson will show up in electron–positron annihilation as a prominent resonance at about 90 GeV (Fig. 25.3).

We now consider the production of the Z-boson in e^+e^- annihilation and its subsequent decay into a general final state X. As can easily be deduced from Fig. 25.4, provided no one spatial direction is favored by the final state X, the cross-section obeys

$$\sigma(e^+e^- \to Z \to X) = 12\pi \frac{\Gamma(Z \to e^+e^-)\Gamma(Z \to X)}{(s - m_Z^2)^2 + m_Z^2\Gamma_Z^2} . \tag{25.17}$$

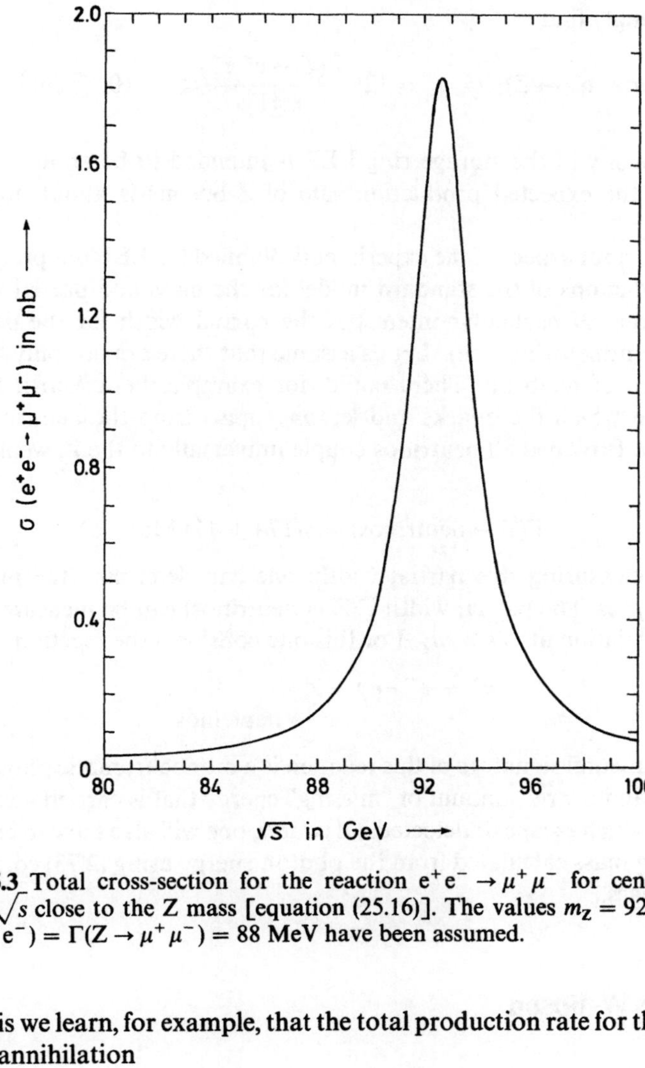

Figure 25.3 Total cross-section for the reaction $e^+e^- \to \mu^+\mu^-$ for center of mass energies $\sqrt{s}$ close to the Z mass [equation (25.16)]. The values $m_Z = 92.6$ GeV and $\Gamma(Z \to e^+e^-) = \Gamma(Z \to \mu^+\mu^-) = 88$ MeV have been assumed.

From this we learn, for example, that the total production rate for the Z-boson in e^+e^- annihilation

$$\sigma(e^+e^- \to Z) = 12\pi \frac{\Gamma(Z \to e^+e^-)\Gamma_Z}{(s - m_Z^2)^2 + m_Z^2\Gamma_Z^2}. \tag{25.18}$$

For the maximum of the production cross-section, which occurs at $\sqrt{s} = m_Z$,

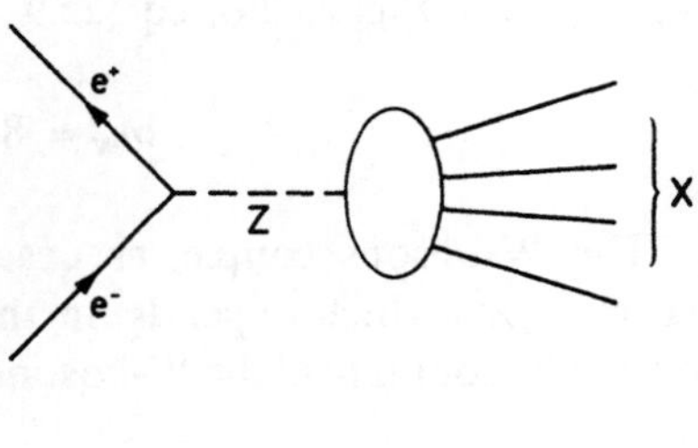

Figure 25.4 Diagram for the production of a Z-boson in e^+e^- annihilation with its subsequent decay into a final state X.

we accordingly find

$$\sigma(e^+e^- \to Z)|_{\sqrt{s}=m_Z} = 12\pi \frac{\Gamma(Z \to e^+e^-)}{m_Z^2 \Gamma_Z} \cong 5 \cdot 10^{-32}\ \mathrm{cm}^2. \tag{25.19}$$

The luminosity of the storage ring LEP is intended to be about 10^{31} cm^{-2} s^{-1}. Thus the expected production rate of Z-bosons is about one in two seconds.

An important aspect of the experiments planned for LEP is a precise check of the predictions of the standard model for the mass and partial widths of the Z-boson. Of particular interest is the partial width for the decay into neutrino–antineutrino pairs. Let us assume that there are not only three, but $n \geq 3$ types of neutrino. There could, for example, be a fourth family of fermions in which the quarks and leptons, apart from the neutrino, are all very heavy. Provided all neutrinos couple universally to the Z, we have from Table 25.2

$$\Gamma(Z \to \text{neutrinos}) = n(174 \pm 11)\ \text{MeV}. \tag{25.20}$$

Thus, by measuring this partial width, one can determine the number of neutrino types. The partial width $\Gamma(Z \to \text{neutrinos})$ can be measured, e.g., in e^+e^- annihilation at $\sqrt{s} > m_Z$. For this one considers the reaction

$$e^+ + e^- \to \gamma + \underset{\hookrightarrow \text{neutrinos}}{Z} \quad . \tag{25.21}$$

The experimental signature of this reaction is a monochromatic photon in the final state and a large amount of "missing" energy that is carried away by the neutrinos which escape undetected. Of course, one will also have to check that the missing mass calculated from the photon energy using (2.73) corresponds to the Z-mass.

25.2 The W-Boson

In the standard model we expect from (22.84) to find $W^\pm$-bosons at a mass of

$$m_W = \left(\frac{\pi\alpha}{\sqrt{2}G}\right)^{1/2} \frac{1}{\sin \vartheta_W} = 78.6\ \text{GeV} \tag{25.22}$$

for $\sin^2 \vartheta = 0.225$. Radiative corrections again lead to a slight change in this value. From Fig. 25.2 or eq. (25.9) the corrected prediction for m_W is:

$$m_W = 81.5 \begin{Bmatrix} +2.2 \\ -1.9 \end{Bmatrix} \text{GeV}. \tag{25.23}$$

The W-bosons couple, as described in (22.77), to the charged fermion current $\mathscr{J}^\lambda_{CC}$ which depends on the weak isospin raising operator $\mathbf{T}_1 + \mathrm{i}\mathbf{T}_2$ only. The coupling of the W-bosons to the fermions is thus universal, i.e., of the

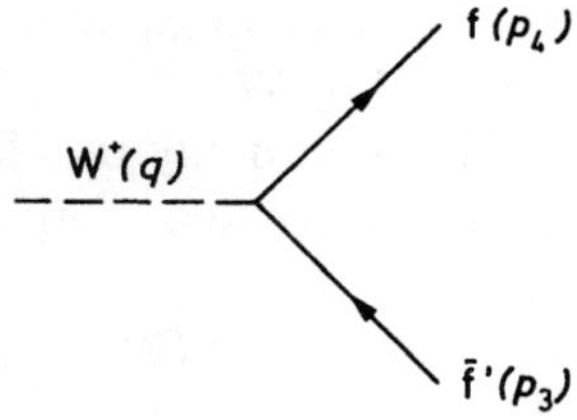

Figure 25.5 Lowest order diagram for the decay of a W^+-boson. Here (f, f′) stands either for (ν_ℓ, ℓ) where $\ell = e, \mu, \tau$ or for (q_i, q_j) where $q_i = u, c, t$ and $q_j = d, s, b$.

same strength for all weak isodoublets (Table 22.2):

$$\begin{pmatrix} \nu_{eL} \\ e_L \end{pmatrix}, \quad \begin{pmatrix} \nu_{\mu L} \\ \mu_L \end{pmatrix}, \quad \begin{pmatrix} \nu_{\tau L} \\ \tau_L \end{pmatrix}, \quad \begin{pmatrix} u_L \\ d'_L \end{pmatrix}, \quad \begin{pmatrix} c_L \\ s'_L \end{pmatrix}, \quad \begin{pmatrix} t_L \\ b'_L \end{pmatrix}. \tag{25.24}$$

As we have seen in Sect. 22.4, the quarks d′, s′, and b′—the eigenstates of the weak isospin—and the quarks d, s, and b of defined mass are related to each other via the Kobayashi–Maskawa matrix. Taking account of this, we obtain, to leading order in the gauge coupling, the following decay processes for the W^+ (Fig. 25.5):

$$W^+ \to \nu_\ell + \ell^+, \tag{25.25}$$

$$W^+ \to q_i + \bar{q}_j, \tag{25.26}$$

where $\ell = e, \mu, \tau$; $q_i = u, c, t$, and $q_j = d, s, b$. For (25.26) the amplitude is proportional to the corresponding element V_{ij} of the Kobayashi–Maskawa matrix. The decay processes of the W^- are obtained analogously.

To exemplify the procedure of calculating these processes we consider the leptonic W-decay:

$$W^\pm(q, \varepsilon) \to \ell^\pm(p_3) + \nu_\ell(p_4), \tag{25.27}$$

where $\nu = \nu_\ell(\bar{\nu}_\ell)$ for $\ell^+(\ell^-)$. Here ε is the polarization vector of the $W^\pm$ and q, p_3 and p_4 are four-momenta. The S-matrix element is given according to the Feynman rules of QFD (Appendix G) by

$$\langle \ell^\pm(p_3), \nu(p_4)|\mathbf{S}|W^\pm(q, \varepsilon)\rangle$$

$$= -\mathrm{i}\frac{e}{\sqrt{2}\sin\vartheta_W}(2\pi)^4\delta(q - p_3 - p_4)\bar{u}(p_{4,3})\not{\varepsilon}\frac{1-\gamma_5}{2}v(p_{3,4}). \tag{25.28}$$

For later use, we will calculate the decay rate of a polarized W-boson, where the lepton ℓ with momentum $\boldsymbol{p}_3$ is emitted in a solid angle $d\Omega_3$, all quantities being referred to the rest frame of the W. The lepton polarization is not observed. Ignoring the lepton masses, we find after a simple calculation

$$d\Gamma(W^\pm(q, \varepsilon) \to \ell^\pm(p_3) + \nu(p_4))$$

$$= d\Omega_3 \frac{\alpha m_W}{32\pi \sin^2\vartheta_W}\{1 - (\hat{\boldsymbol{p}}_3\boldsymbol{\varepsilon})(\hat{\boldsymbol{p}}_3\boldsymbol{\varepsilon}^*) \pm \mathrm{i}(\boldsymbol{\varepsilon} \times \boldsymbol{\varepsilon}^*)\cdot\hat{\boldsymbol{p}}_3\} \qquad (\hat{\boldsymbol{p}}_3 = \boldsymbol{p}_3/|\boldsymbol{p}_3|). \tag{25.29}$$

Note the appearance of the parity violating term $(\boldsymbol{\varepsilon} \times \boldsymbol{\varepsilon}^*)\cdot\hat{\boldsymbol{p}}_3$ which stems from the (V–A) coupling of the W-boson, and which changes its sign when W^+ is replaced by W^-.

The total decay rate for $W \to \ell\nu_\ell$ is obtained from (25.29) by integration over all directions of $\hat{\boldsymbol{p}}_3$ as

$$\Gamma(W^+ \to \nu_\ell \ell^+) = \Gamma(W^- \to \bar{\nu}_\ell \ell^-) = \frac{\alpha m_W}{12 \sin^2 \vartheta_W}. \tag{25.30}$$

For the W decay into quarks we find in a similar way

$$\Gamma(W^+ \to q_i \bar{q}_j) = \Gamma(W^- \to \bar{q}_i q_j) = \frac{\alpha m_W}{4 \sin^2 \vartheta_W} |V_{ij}|^2, \tag{25.31}$$

for $q_i = u, c$; $q_j = d, s, b$ and

$$\Gamma(W^+ \to t\bar{q}_j) = \Gamma(W^- \to \bar{t} q_j)$$
$$= \frac{\alpha m_W}{4 \sin^2 \vartheta_W} |V_{3j}|^2 \left(1 - \frac{m_t^2}{m_W^2}\right)\left(1 - \frac{m_t^2}{2m_W^2} - \frac{m_t^4}{2m_W^4}\right) \tag{25.32}$$

for $q_j = d, s, b$. Here we have neglected all quark masses apart from m_t.

For precision comparisons of theory and experiment one would again need to take electroweak and gluonic radiative corrections into account. The gluonic corrections are found by employing the same methods as for the decay of the Z-boson. A calculation of the electroweak corrections has not yet been carried out for the W-boson. However, for similar reasons as in the case of the Z-decay, we expect that these corrections will be small when we make use of the parameters G, m_W and ϑ_W in the lowest order formula (W. Wetzel, private communication). We then obtain the predictions

$$\Gamma(W^+ \to \nu_\ell \ell^+) = \Gamma(W^- \to \bar{\nu}_\ell \ell^-) = \frac{G m_W^3}{6\sqrt{2}\pi}, \tag{25.33}$$

$$\Gamma(W^+ \to q_i\bar{q}_j \text{ and } q_i\bar{q}_j G) = \Gamma(W^- \to \bar{q}_i q_j \text{ and } \bar{q}_i q_j G)$$
$$= \frac{G m_W^3}{2\sqrt{2}\pi} |V_{ij}|^2 \left(1 - \frac{(m_{q_i} + m_{q_j})^2}{m_W^2}\right)^{1/2}$$
$$\cdot \left(1 - \frac{(m_{q_i} - m_{q_j})^2}{m_W^2}\right)^{1/2}$$
$$\cdot \left(1 - \frac{m_{q_i}^2 + m_{q_j}^2}{2m_W^2} - \frac{(m_{q_i}^2 - m_{q_j}^2)^2}{2m_W^4}\right)\left(1 + \frac{\alpha_s(m_W^2)}{\pi}\right). \tag{25.34}$$

The numerical values of these decay rates are given in Table 25.3. A similar problem as for the Z-decay could occur for decays into final states containing a t-quark if the threshold for the production of a $t\bar{d}$, $t\bar{s}$ or $t\bar{b}$ pair happens to be near the W mass.

An exact experimental determination of the partial widths for the W-bosons is expected to be much more difficult than for the Z-boson.

Table 25.3 The partial widths and the branching ratios of the W^+ according to the standard model (25.33), (25.34). We have used the values $m_W = 81.5 \pm 2.1$ GeV, $m_t = 40 \pm 10$ GeV, $\alpha_s(m_W^2)/\pi = 0.04 \pm 0.01$ and the Kobayashi–Maskawa matrix elements V_{ij} as given in Table 23.6.

Final state X	$\Gamma(W^+ \to X)$ (MeV)	$\Gamma(W^+ \to X)/\Gamma_W$
$e^+\nu_e$	237 ± 18	$(8.9 \pm 0.4)\,\%$
$\mu^+\nu_\mu$	237 ± 18	$(8.9 \pm 0.4)\,\%$
$\tau^+\nu_\tau$	237 ± 18	$(8.9 \pm 0.4)\,\%$
$u\bar{d} + u\bar{d}G$	700 ± 55	$(26.4 \pm 1.2)\,\%$
$c\bar{s} + c\bar{s}G$	698 ± 54	$(26.3 \pm 1.2)\,\%$
$t\bar{b} + t\bar{b}G$	475 ± 120	$(18 \pm 5)\,\%$
$u\bar{s} + u\bar{s}G$	40 ± 3	$(1.5 \pm 0.1)\,\%$
$c\bar{d} + c\bar{d}G$	40 ± 3	$(1.5 \pm 0.1)\,\%$
$c\bar{b} + c\bar{b}G$	2 ± 1	$(9 \pm 4) \cdot 10^{-4}$
$u\bar{b} + u\bar{b}G$	0.02 ± 0.02	$(1 \pm 1) \cdot 10^{-5}$
$t\bar{d} + t\bar{d}G$	0.1 ± 0.1	$(3 \pm 3) \cdot 10^{-5}$
$t\bar{s} + t\bar{s}G$	1.4 ± 1.2	$(5 \pm 4) \cdot 10^{-4}$
Total	2653 ± 240	

25.3 The Production of W- and Z-Bosons in p$\bar{p}$ Collisions

How can one produce W- and Z-bosons? The possibility of creating the Z-boson in e^+e^- annihilation via the reaction

$$e^+ + e^- \to Z \tag{25.35}$$

has already been discussed in Sect. 25.1. The appropriate storage rings—SLC and LEP—are, however, still under construction or at least not yet fully operational. The analogous reactions

$$\begin{aligned} \nu_e + e^+ &\to W^+ , \\ \bar{\nu}_e + e^- &\to W^- , \end{aligned} \tag{25.36}$$

could only be used for the production of W-bosons if one could build neutrino–electron storage rings, but this, of course, is impossible. We can, however, reverse the decay processes of Z- and W-bosons into quarks and use these as production reactions. If we consider as an example only the u- and d-quarks, then we have the processes

$$\begin{aligned} u + \bar{u} &\to Z, \qquad u + \bar{d} \to W^+ , \\ d + \bar{d} &\to Z, \qquad d + \bar{u} \to W^- , \end{aligned} \tag{25.37}$$

At this point, one might raise the objection that we have no quark storage rings, but as we will now discuss, for practical purposes, one can regard hadron storage rings as just that.

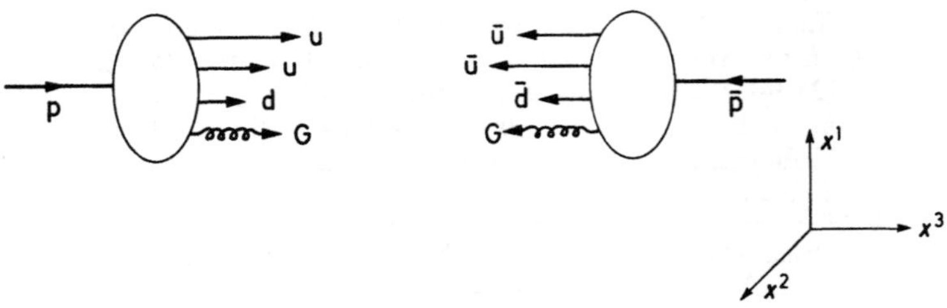

Figure 25.6 Parton picture of a proton and an antiproton colliding at high energy, e.g., in a $p\bar{p}$ storage ring. Most of the proton's momentum is carried by two u-quarks, a d-quark, and (an unknown number of) gluons G. For the antiproton it is carried by two ū-quarks, a d̄-quark, and gluons G. The choice of coordinates to be used later is also shown.

Using the naïve parton model, which will play an important role here, we regard a rapidly moving hadron as a "beam" of partons, i.e., quarks, antiquarks, and gluons (Chaps. 18 and 19). For example, in a high energy proton–antiproton storage ring, the proton corresponds essentially to a beam of u- and d-quarks and gluons, and the antiproton to a beam of ū- and d̄-quarks and gluons (Fig. 25.6). We know the quark and gluon distributions for the proton from deep inelastic lepton–nucleon scattering (Sect. 19.4). The antiquark distributions for the antiproton are, as a result of the invariance of the strong interaction under charge conjugation, identical to the respective quark distributions for the proton. For the same reason the gluon distributions are also identical. From Sect. 18.4, we know that the quarks in the proton and the antiquarks in the antiproton each carry about 50% of the momentum of the parent particle. If we assume three quarks in the proton, then each carries a typical momentum fraction $x_{\mathrm{q}} \cong \frac{1}{6}$; similarly, for an antiquark in the antiproton, $x_{\bar{\mathrm{q}}} \cong \frac{1}{6}$. We now consider a $p\bar{p}$ collision at a center of mass energy $\sqrt{s} \gg m_{\mathrm{p}}$. The typical center of mass energy $\sqrt{\hat{s}}$ of a quark–antiquark collision is then (see Sect. 18.5)

$$\sqrt{\hat{s}} = \sqrt{x_{\mathrm{q}} x_{\bar{\mathrm{q}}}\, s} \cong \tfrac{1}{6}\sqrt{s}. \tag{25.38}$$

Thus in order to produce W- and Z-bosons with a mass of about 100 GeV, one needs a center of mass energy of $\sqrt{s} \cong 600$ GeV in the $p\bar{p}$ collision.

In 1976, having arrived at this conclusion, C. Rubbia suggested that the existing proton accelerator SPS at CERN should be modified into a $p\bar{p}$ storage ring with a beam energy of 270 GeV, corresponding to a center of mass energy of 540 GeV (Rubbia 1977). A prerequisite for the realization of this project was the possibility of producing intense beams of antiprotons by means of the "stochastic cooling method" invented by S. van der Meer (van der Meer 1972; Möhl 1980). In 1983, with this storage ring ($Sp\bar{p}S$), the W- and Z-bosons were actually discovered (Arnison 1983a, c; Banner 1983; Bagnaia 1983).

We now wish to discuss the theory of Z and W production in $p\bar{p}$ collisions. These reactions are analogous to the Drell–Yan process (Sect. 18.5). The diagrams for Z and W production were also given previously in Fig. 18.17.

Let us start with the production of the Z-boson in a p$\bar{\text{p}}$ collision:

$$\mathrm{p}(p_1) + \bar{\mathrm{p}}(p_2) \to \mathrm{Z}(q, \varepsilon) + \mathrm{X}. \tag{25.39}$$

Here X represents the other particles in the final state and ε is the polarization vector of the Z. The elementary parton reactions are the quark–antiquark annihilations

$$\mathrm{q_A}(p_1') + \bar{\mathrm{q}}_\mathrm{B}(p_2') \to \mathrm{Z}(q, \varepsilon), \tag{25.40}$$

where q = u, d, ... and A, B ($1 \leq \mathrm{A}, \mathrm{B} \leq 3$) are the color indices. Using the expressions for the coupling of the Z to the quarks (24.39), (24.40), we obtain the S-matrix element for this reaction:

$$\begin{aligned}\langle \mathrm{Z}(q, \varepsilon)|\mathbf{S}|\mathrm{q_A}(p_1'), \bar{\mathrm{q}}_\mathrm{B}(p_2')\rangle = &-\mathrm{i}\frac{e}{\sin \vartheta_\mathrm{W} \cos \vartheta_\mathrm{W}}(2\pi)^4\delta(q - p_1' - p_2') \\ &\cdot \delta_\mathrm{AB}\, \tfrac{1}{2}\, \bar{v}(p_2')\not{\varepsilon}^*(g_\mathrm{V}^\mathrm{q} - g_\mathrm{A}^\mathrm{q}\gamma_5)u(p_1').\end{aligned} \tag{25.41}$$

We regard the Z here as a "stable" particle, i.e., we neglect its width. We are justified in doing this since we shall later be integrating over very broad quark and antiquark momentum distributions (Exercise 25.5). As is usual for calculations in the parton model, it is advantageous to consider the transition *rates*. For color-unpolarized quarks and antiquarks in the initial state, whose masses we neglect, we find from (25.41) using (25.5):

$$\begin{aligned}\mathrm{d}\Gamma(\mathrm{q}(p_1') + \bar{\mathrm{q}}(p_2') \to \mathrm{Z}(q, \varepsilon)) = &\frac{1}{V p_1'^0 p_2'^0}\frac{\mathrm{d}^3 q}{q^0} \cdot m_\mathrm{Z}\Gamma(\mathrm{Z} \to \mathrm{q}\bar{\mathrm{q}})\frac{\pi^2}{3}\delta(q - p_1' - p_2') \\ &\cdot \frac{1}{m_\mathrm{Z}^2}\Bigg\{p_{2\mu}' p_{1\nu}' + p_{1\mu}' p_{1\nu}' - g_{\mu\nu}(p_1' p_2') \\ &- \frac{2 g_\mathrm{V}^\mathrm{q} g_\mathrm{A}^\mathrm{q}}{(g_\mathrm{V}^\mathrm{q})^2 + (g_\mathrm{A}^\mathrm{q})^2}\mathrm{i}\varepsilon_{\alpha\mu\beta\nu} p_2'^\alpha p_1'^\beta\Bigg\}\varepsilon^{*\mu}\varepsilon^\nu.\end{aligned} \tag{25.42}$$

The factor of $\frac{1}{3}$ arises, as in the Drell–Yan process, from counting up the color combinations of q and $\bar{\text{q}}$ that can lead to the production of the Z, which is a color singlet. Summing over the spin polarizations of the Z, we obtain for the total production rate

$$\begin{aligned}&\mathrm{d}\Gamma(\mathrm{q}(p_1') + \bar{\mathrm{q}}(p_2') \to \mathrm{Z}(q)) \\ &\quad = \frac{1}{V p_1'^0 p_2'^0}\frac{\mathrm{d}^3 q}{q^0} m_\mathrm{Z}\Gamma(\mathrm{Z} \to \mathrm{q}\bar{\mathrm{q}})\frac{\pi^2}{3}\delta(q - p_1' - p_2').\end{aligned} \tag{25.43}$$

The remainder of the calculation exactly follows that for the Drell–Yan process in Sect. 18.5. We convolute the result for the parton reaction with the distribution functions of the quarks and antiquarks in the proton and the antiproton. For simplicity, we only take into account u- and d-quarks in the proton and ū- and d̄-antiquarks in the antiproton. This was shown in Sect.

19.4 to be a good approximation. We now write

$$\begin{aligned} p_1' &= x_1 p_1, \\ p_2' &= x_2 p_2, \\ \hat{s} &= (p_1' + p_2')^2 = x_1 x_2 s. \end{aligned} \tag{25.44}$$

The distribution functions of the quarks q in the proton and of the antiquarks $\bar{\mathrm{q}}$ in the antiproton are denoted, as in Chap. 18, by $N_{\mathrm{q}}^{\mathrm{p}}(x)$ and $N_{\bar{\mathrm{q}}}^{\bar{\mathrm{p}}}(x)$. As was mentioned previously, we have

$$N_{\mathrm{q}}^{\mathrm{p}}(x) = N_{\bar{\mathrm{q}}}^{\bar{\mathrm{p}}}(x). \tag{25.45}$$

With these relations, we find for the Z production cross-section in $\mathrm{p\bar{p}}$ collisions

$$\begin{aligned} \mathrm{d}\sigma(\mathrm{p}(p_1) + \bar{\mathrm{p}}(p_2) \to \mathrm{Z}(q) + \mathrm{X}) &= \frac{2\pi^2}{3s} \frac{\mathrm{d}^3 q}{q^0} \sum_{\mathrm{q=u,d}} m_{\mathrm{Z}} \Gamma(\mathrm{Z} \to \mathrm{q\bar{q}}) \\ &\cdot \int_0^1 \mathrm{d}x_1\, N_{\mathrm{q}}^{\mathrm{p}}(x_1) \int_0^1 \mathrm{d}x_2\, N_{\bar{\mathrm{q}}}^{\bar{\mathrm{p}}}(x_2) \frac{1}{x_1 x_2} \delta(q - x_1 p_1 - x_2 p_2) \end{aligned} \tag{25.46}$$

where X stands for the fragmentation products of the spectator partons.

We now discuss this result. Like the experimenters at the $\mathrm{Sp\bar{p}S}$ storage ring, we consider the $\mathrm{p\bar{p}}$ collision in its center of mass frame. We take the x^3-axis to be in the direction of the proton beam (Fig. 25.6). From the δ-function for energy-momentum conservation in (25.46) we find for the four-momentum of the Z-boson produced in the reaction

$$q = \begin{pmatrix} q^0 \\ q^1 \\ q^2 \\ q^3 \end{pmatrix} = \begin{pmatrix} (x_1 + x_2)\frac{\sqrt{s}}{2} \\ 0 \\ 0 \\ (x_1 - x_2)\frac{\sqrt{s}}{2} \end{pmatrix}. \tag{25.47}$$

According to our calculation, the Z should thus be produced with zero transverse momentum. This, of course, stems from the fact that we have neglected the transverse momenta of the quarks in the naïve parton model. The longitudinal momentum q^3 and the energy q^0 of the Z enable us, according to (25.47), to reconstruct the momentum fractions x_1 and x_2 of the partons from which the Z was produced.

For the distribution in q^3 we find from (25.46):

$$\begin{aligned} &\frac{\mathrm{d}\sigma}{\mathrm{d}q^3}(\mathrm{p}(p_1) + \bar{\mathrm{p}}(p_2) \to \mathrm{Z}(q) + \mathrm{X}) \\ &\quad = \frac{4\pi^2}{3s} \sum_{\mathrm{q=u,d}} \frac{\Gamma(\mathrm{Z} \to \mathrm{q\bar{q}})}{m_{\mathrm{Z}}\sqrt{m_{\mathrm{Z}}^2 + (q^3)^2}} N_{\mathrm{q}}^{\mathrm{p}}(x_1) N_{\bar{\mathrm{q}}}^{\bar{\mathrm{p}}}(x_2), \end{aligned} \tag{25.48}$$

where $x_{1,2} = (q^0 \pm q^3)/\sqrt{s}$. The total production cross-section is

$$\sigma(\mathrm{p}(p_1) + \bar{\mathrm{p}}(p_2) \to \mathrm{Z} + \mathrm{X}) = \frac{4\pi^2}{3s} \sum_{\mathrm{q=u,d}} m_{\mathrm{Z}}^{-1} \Gamma(\mathrm{Z} \to \mathrm{q\bar{q}})$$

$$\cdot \int_0^1 \mathrm{d}x_1\, N_{\mathrm{q}}^{\mathrm{p}}(x_1) \int_0^1 \mathrm{d}x_2\, N_{\bar{\mathrm{q}}}^{\bar{\mathrm{p}}}(x_2) \delta\left(x_1 x_2 - \frac{m_{\mathrm{Z}}^2}{s}\right). \tag{25.49}$$

For the production of $\mathrm{W}^{\pm}$-bosons in $p\bar{p}$ collisions, the calculation proceeds in a very similar way. If we again consider only the dominant reactions (25.37), then we obtain, e.g., for the total production cross-sections:

$$\sigma(\mathrm{p}(p_1) + \bar{\mathrm{p}}(p_2) \to \mathrm{W}^+ + \mathrm{X}) = \frac{4\pi^2}{3s} m_{\mathrm{W}}^{-1} \Gamma(\mathrm{W}^+ \to \mathrm{u\bar{d}})$$

$$\cdot \int_0^1 \mathrm{d}x_1\, N_{\mathrm{u}}^{\mathrm{p}}(x_1) \int_0^1 \mathrm{d}x_2\, N_{\bar{\mathrm{d}}}^{\bar{\mathrm{p}}}(x_2) \delta\left(x_1 x_2 - \frac{m_{\mathrm{W}}^2}{s}\right), \tag{25.50}$$

$$\sigma(\mathrm{p}(p_1) + \bar{\mathrm{p}}(p_2) \to \mathrm{W}^- + \mathrm{X}) = \frac{4\pi^2}{3s} m_{\mathrm{W}}^{-1} \Gamma(\mathrm{W}^- \to \mathrm{d\bar{u}})$$

$$\cdot \int_0^1 \mathrm{d}x_1\, N_{\mathrm{d}}^{\mathrm{p}}(x_1) \int_0^1 \mathrm{d}x_2\, N_{\bar{\mathrm{u}}}^{\bar{\mathrm{p}}}(x_2) \delta\left(x_1 x_2 - \frac{m_{\mathrm{W}}^2}{s}\right), \tag{25.51}$$

For a numerical evaluation, we make use of the parametrization given in Appendix E for the distribution functions, where we set $Q^2 \cong m_{\mathrm{W}}^2 \cong m_{\mathrm{Z}}^2 \cong 10^4\ \mathrm{GeV}^2$. If we take a center of mass energy $\sqrt{s} = 540$ GeV, corresponding to $m_{\mathrm{Z}}^2/s = 0.03$ and $m_{\mathrm{W}}^2/s = 0.02$, we find

$$\int_0^1 \mathrm{d}x_1\, N_{\mathrm{q}}^{\mathrm{p}}(x_1) \int_0^1 \mathrm{d}x_2\, N_{\bar{\mathrm{q}}}^{\bar{\mathrm{p}}}(x_2) \delta\left(x_1 x_2 - \frac{m_{\mathrm{Z}}^2}{s}\right) = \begin{cases} 12.18 & \text{for q = u,} \\ 4.18 & \text{for q = d,} \end{cases} \tag{25.52}$$

$$\int_0^1 \mathrm{d}x_1\, N_{\mathrm{u}}^{\mathrm{p}}(x_1) \int_0^1 \mathrm{d}x_2\, N_{\bar{\mathrm{d}}}^{\bar{\mathrm{p}}}(x_2) \delta\left(x_1 x_2 - \frac{m_{\mathrm{W}}^2}{s}\right) = 11.67. \tag{25.53}$$

With these we obtain production cross-sections of

$$\sigma(\mathrm{p} + \bar{\mathrm{p}} \to \mathrm{Z} + \mathrm{X}) = 1.0\ \text{nbarn},$$

$$\sigma(\mathrm{p} + \bar{\mathrm{p}} \to \mathrm{W}^+ + \mathrm{X}) = \sigma(\mathrm{p} + \bar{\mathrm{p}} \to \mathrm{W}^- + \mathrm{X}) = 1.7\ \text{nbarn}. \tag{25.54}$$

These results correspond to the naïve Drell–Yan model. For a more exact calculation one must naturally take into account corrections which arise in QCD. For example, we again expect the appearance of a "K-factor" (Sect. 18.4). As a result of QCD effects, the W- and Z-bosons produced in $p\bar{p}$ collisions will generally have a nonzero transverse momentum. For a more detailed discussion of these effects we refer the reader to, e.g., Altarelli (1984). There the following total production cross-sections are given for

$\sqrt{s} = 540$ GeV:

$$\sigma(\mathrm{p} + \bar{\mathrm{p}} \to \mathrm{Z} + \mathrm{X}) = \left(1.3 \begin{cases} +0.4 \\ -0.2 \end{cases}\right) \text{nbarn},$$

$$\sigma(\mathrm{p} + \bar{\mathrm{p}} \to \mathrm{W}^+ + \mathrm{X}) = \sigma(\mathrm{p} + \bar{\mathrm{p}} \to \mathrm{W}^- + \mathrm{X}) = \left(2.1 \begin{cases} +0.7 \\ -0.3 \end{cases}\right) \text{nbarn}. \quad (25.55)$$

These are not so very different from our simple estimate (25.54).

As already mentioned, W- and Z-bosons were first produced in 1983 in $\mathrm{p\bar{p}}$ collisions and detected via the following leptonic decay processes (Arnison 1983a, c, e, 1984a, d; Banner 1983; Bagnaia 1983, 1984e):

$$\mathrm{p} + \bar{\mathrm{p}} \to \mathrm{W}^+ + \mathrm{X} \quad , \qquad \mathrm{W}^+ \longrightarrow \mathrm{e}^+ \nu_\mathrm{e}, \mu^+ \nu_\mu \quad (25.56)$$

$$\mathrm{p} + \bar{\mathrm{p}} \to \mathrm{W}^- + \mathrm{X} \quad , \qquad \mathrm{W}^- \longrightarrow \mathrm{e}^- \bar{\nu}_\mathrm{e}, \mu^- \bar{\nu}_\mu \quad (25.57)$$

$$\mathrm{p} + \bar{\mathrm{p}} \to \mathrm{Z} + \mathrm{X} \quad . \qquad \mathrm{Z} \longrightarrow \mathrm{e}^+ \mathrm{e}^-, \mu^+ \mu^- \quad (25.58)$$

Here X stands for the fragmentation products of the spectators, which merely comprise hadrons with small transverse momenta $|\boldsymbol{p}_\mathrm{T}| \lesssim 1$ GeV). In W production one should see an isolated electron or muon and "missing" energy that is carried away by the corresponding neutrino. Such events have actually been observed (Fig. 25.7(a)). If one had an ideal detector, it would also be possible to deduce the energy and momentum of the neutrino from the energy-momentum balance. With the actual detectors, however, only the

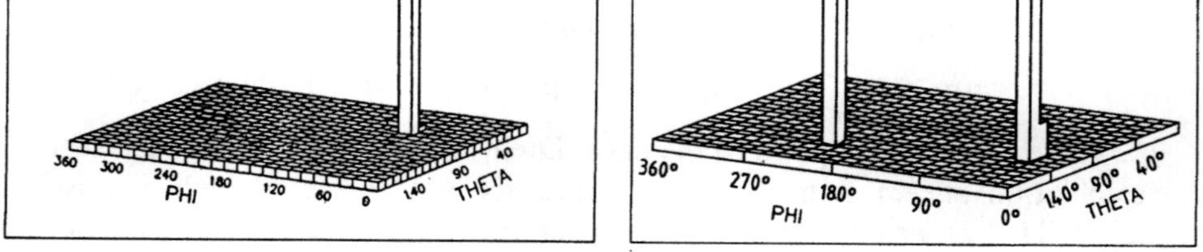

Figure 25.7 Lego plot of an event in which a W-boson was produced in a $\mathrm{p\bar{p}}$ collision at $\sqrt{s} = 540$ GeV and then decayed into an $\mathrm{e}\nu$ pair (a). An event that produced a Z-boson which then decayed into an $\mathrm{e}^+\mathrm{e}^-$ pair (b). The representation is the same as that of Fig. 20.19. (These plots were made available by A. Putzer of the UA2 collaboration.)

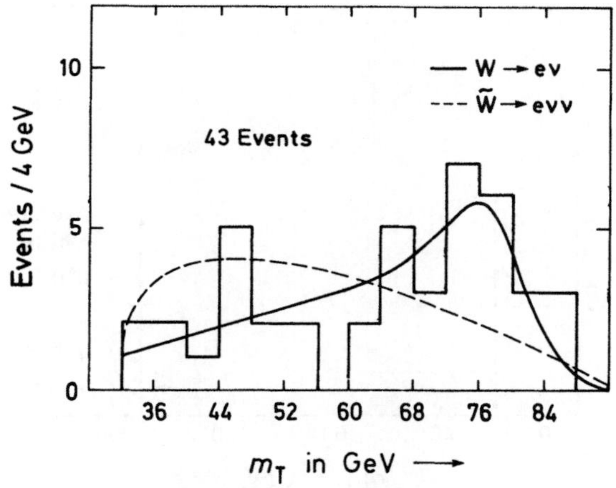

Figure 25.8 Distribution of the transverse mass $m_T = [2(|p_T^e||p_T^\nu| - p_T^e \cdot p_T^\nu)]^{1/2}$ of the electron–neutrino pair of W decay. The solid curve is the theoretical expectation with fitted W mass $m_W = 80.3$ GeV. For comparison, the dashed curve shows the theoretical expectation for the production of a hypothetical particle $\tilde{W}$ that undergoes a three-body decay (after Arnison 1983e).

transverse momentum p_T^ν of the neutrino can be accurately determined. The energy and longitudinal momentum of the neutrino are not so readily measurable since particles can escape undetected along the beam direction. (For a description of the UA1 and UA2 detectors, see Radermacher 1985, Banner 1982, and references cited therein.) To determine the mass of the W-boson, it is thus not possible in practice to make use of the invariant mass of the $\ell\nu$ pair ($\ell = e, \mu$). Instead of this, one considers, for example, the distribution of the "transverse" mass m_T:

$$m_T \equiv [2(|p_T^\ell||p_T^\nu| - p_T^\ell \cdot p_T^\nu)]^{1/2} \tag{25.59}$$

and fits the theoretical distribution to the data by treating m_W as a free parameter (Fig. 25.8). The average value of m_W obtained from all available data is (Particle Data Group 1986):

$$m_W = 81.8 \pm 1.5 \text{ GeV}, \tag{25.60}$$

in excellent agreement with the theoretical prediction (25.23).

In Z production one sees both charged leptons e^+e^- and $\mu^+\mu^-$ (Fig. 25.7(b)) and can thus reconstruct the invariant mass directly. As can be seen in Fig. 25.9, the Z can be detected in this way with almost no background. The measured mass is (Particle Data Group 1986):

$$m_Z = 92.6 \pm 1.7 \text{ GeV}, \tag{25.61}$$

again in excellent agreement with the theory (25.9). The production cross sections for the Z- and W-bosons, multiplied by the leptonic branching ratios,

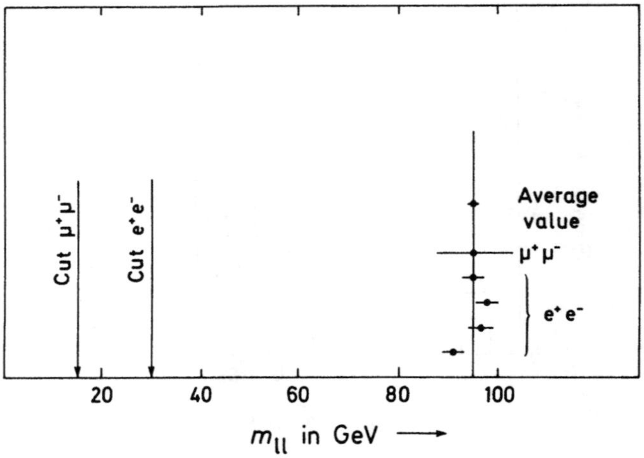

Figure 25.9 The invariant mass $m_{\ell\ell}$ of lepton pairs $\ell^+\ell^-$ (ℓ = e, μ) from Z decay processes observed at UA1 (after Arnison 1983c).

Table 25.4 The production cross-sections for W- and Z-bosons in $p\bar{p}$ collisions at $\sqrt{s} =$ 540 GeV multiplied by the leptonic branching ratios. The experimental values are from Eggert (1984) (UA1) and Bagnaia (1984e) (UA2). The first error quoted is statistical, and the second systematic. The theoretical values are calculated from (25.55) and Tables 25.2 and 25.3. All production cross-sections are given in picobarns (1 pbarn = 10^{-36} cm²).

	Experiment		
	UA1	UA2	Theory
$\sigma(p + \bar{p} \to W^\pm + X)\dfrac{\Gamma(W^\pm \to e\nu_e)}{\Gamma_{W^\pm}}$	$530 \pm 80 \pm 90$	$530 \pm 100 \pm 100$	$370\begin{cases}+100\\-60\end{cases}$
$\sigma(p + \bar{p} \to W^\pm + X)\dfrac{\Gamma(W^\pm \to \mu\nu_\mu)}{\Gamma_{W^\pm}}$	$670 \pm 170 \pm 150$		$370\begin{cases}+110\\-60\end{cases}$
$\sigma(p + \bar{p} \to Z + X)\dfrac{\Gamma(Z \to e^+e^-)}{\Gamma_Z}$	$41 \pm 21 \pm 7$	$110 \pm 40 \pm 20$	$43\begin{cases}+13\\-7\end{cases}$
$\sigma(p + \bar{p} \to Z + X)\dfrac{\Gamma(Z \to \mu^+\mu^-)}{\Gamma_Z}$	$100 \pm 50 \pm 15$		$43\begin{cases}+13\\-7\end{cases}$

have also been measured. These too, show good agreement with the theory (Table 25.4).

25.4 The Spin of the W-Boson

In this section we will show that the W- and Z-bosons produced in $p\bar{p}$ collisions are polarized, which explains why the angular distributions of the leptons from the decays $W \to \ell\nu_\ell$ and $Z \to \ell^+\ell^-$ ($\ell = e, \mu, \tau$) are asymmetric.

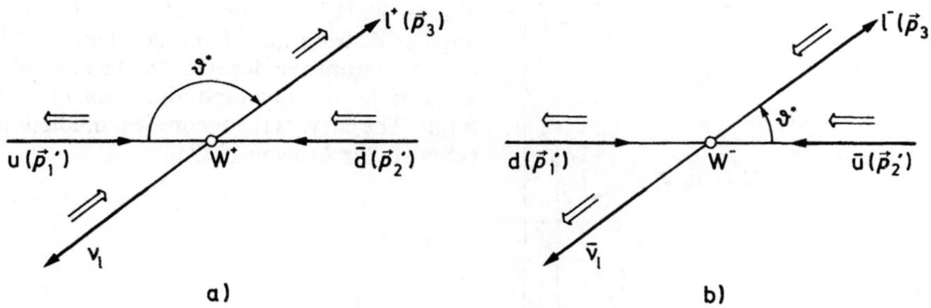

Figure 25.10 Production of a W^+-boson in a $u\bar{d}$ annihilation (a) and of a W^--boson in a $d\bar{u}$ annihilation (b) with subsequent decay into a lepton pair, observed in the rest frames of the W^+ and W^-, respectively. The helicities of the particles are indicated by the double arrows. The angle ϑ^* used in (25.64) is also shown.

The effect has been observed for the W-boson and gave evidence both for the spin of the W and for the parity violation in its production and decay.

We first consider the production of a W^+-boson through the annihilation of a u-quark from the proton with a $\bar{d}$-antiquark from the antiproton. We work in the rest frame of the W (Fig. 25.10). In the standard model, the charged current $\mathscr{J}^{\lambda}_{CC}$, to which the W couples, has a pure (V–A) structure; it is composed only of the left-handed quark and lepton fields (22.77), (22.112). As we already know from Sect. 21.3, when the fermion masses are neglected, the left-handed fields correspond to left-handed particles and right-handed antiparticles. In order for a W^+ to be produced, the helicities of the u-quark and the $\bar{d}$-antiquark must be those given in Fig. 25.10(a). From this it follows immediately that the W^+ is produced with a helicity of -1 with respect to the direction of flight of the u-quark, which, if the transverse momentum of the quark is neglected, is in the same direction as the proton momentum. For a coordinate system as depicted in Fig. 25.6 the polarization vector of the W^+ is thus

$$\varepsilon_{-1} = \frac{1}{\sqrt{2}} \begin{pmatrix} 1 \\ -i \\ 0 \end{pmatrix}. \tag{25.62}$$

In an analogous way, one sees that the W^- is likewise always produced with a helicity of -1 with respect to the proton direction (Fig. 25.10(b)). Provided the small contribution from annihilation of an antiquark in the proton with a quark in the antiproton is neglected, these results are valid for all W-bosons produced in $p\bar{p}$ collisions.

We now consider the leptonic decays of these polarized W-bosons, with reference to their rest frame:

$$W^{\pm}(q, \varepsilon_{-1}) \to \ell^{\pm}(p_3) + \nu \qquad (\ell = e, \mu, \tau) \tag{25.63}$$

where $\nu = \nu_\ell(\bar{\nu}_\ell)$ for $\ell^+(\ell^-)$. The angle between the momentum of the emitted lepton $\ell^-(\ell^+)$ and the proton (antiproton) beam is denoted by ϑ^* (Fig. 25.10).

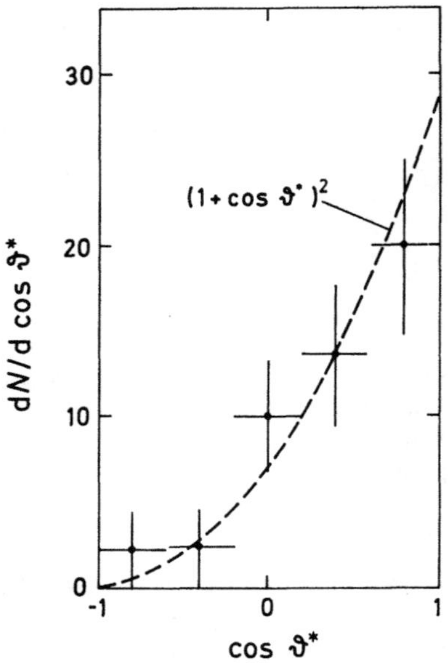

Figure 25.11 Experimental data for the angular distribution of charged leptons $\ell^\pm$ ($\ell = \mathrm{e}, \mu$) from the decay of the bosons $\mathrm{W}^\pm$ as seen in the corresponding boson rest frame. The curve is the theoretical prediction (25.64) (after Arnison 1983e).

From (25.29) we then obtain in both cases an angular distribution of the form

$$\frac{\mathrm{d}\Gamma}{\mathrm{d}\Omega_3}(\mathrm{W}^\pm(q, \varepsilon_{-1}) \to \ell^\pm(p_3) + \nu) \propto (1 + \cos\vartheta^*)^2. \tag{25.64}$$

In Fig. 25.11 we show experimental data for this angular distribution. As we see, the prediction of the standard model is confirmed.

In order to gain a better understanding of the implications of this result we make a general analysis of the spin state of the vector bosons created in $\mathrm{p\bar{p}}$ collisions. We begin by considering a Z-boson in its rest frame. The Z is a vector particle. In order to characterize its state we need to know its polarization. A pure state of the Z is characterized by a polarization vector $\boldsymbol{\varepsilon}$. Such a state is obtained, for example, when the reaction (25.39) proceeds from fully polarized protons and antiprotons, and when, along with the Z, a completely defined final state X is selected. In general, however, our knowledge, both of the initial and of the final state, is incomplete. One works with unpolarized protons and antiprotons and sums over final states X. The spin state of the Z is then described by a so-called spin density matrix ρ. We define ρ as the 3×3 matrix whose matrix element, for any chosen polarization vector $\boldsymbol{\varepsilon}$, gives the probability that this polarization vector is present in the state:

$$\rho = (\rho_{ij}),$$

$$\varepsilon^{*i}\rho_{ij}\varepsilon^j = \text{probability of finding } \boldsymbol{\varepsilon}. \tag{25.65}$$

It can easily be seen that this gives an unambiguous definition of ρ. The probability of finding any spin state at all must be equal to unity, i.e., the probabilities for every orthonormal basis in the space of polarization vectors must add up to one. If we choose as our basis the unit vectors in the coordinate directions, we obtain the condition;

$$\rho_{ii} = \mathrm{Tr}\,\rho = 1, \tag{25.66}$$

i.e., the trace of the density matrix is one. In the following, we shall also make use of the basis formed by the eigenvectors of the angular momentum component $\mathbf{J}^3$ (see Sect. 16.2 and Appendix C)

$$|1, \pm 1\rangle \equiv \boldsymbol{\varepsilon}_{\pm 1} = \mp \frac{1}{\sqrt{2}} \begin{pmatrix} 1 \\ \pm \mathrm{i} \\ 0 \end{pmatrix}, \qquad |1, 0\rangle \equiv \boldsymbol{\varepsilon}_0 = \begin{pmatrix} 0 \\ 0 \\ 1 \end{pmatrix}. \tag{25.67}$$

We have

$$\mathbf{J}^3 |1, m\rangle = m |1, m\rangle \qquad (m = 0, \pm 1). \tag{25.68}$$

For an unpolarized Z-boson, the probability of finding a polarization vector $\boldsymbol{\varepsilon}$ is, by definition, the same for all $\boldsymbol{\varepsilon}$ and the corresponding density matrix is thus proportional to the unit matrix. Using the condition (25.66) on the trace yields:

$$\rho(\text{unpol. Z}) = \tfrac{1}{3}\mathbb{1}. \tag{25.69}$$

We now consider the production of a Z-boson in $\mathrm{q}\bar{\mathrm{q}}$ annihilation (25.40). For unpolarized quarks the spin density matrix of the Z can be readily deduced from (25.42):

$$\rho_{ij} = \frac{1}{2}\left\{\delta_{ij} - e_3^i e_3^j + \frac{2 g_\mathrm{V}^\mathrm{q} g_\mathrm{A}^\mathrm{q}}{(g_\mathrm{V}^\mathrm{q})^2 + (g_\mathrm{A}^\mathrm{q})^2} \mathrm{i}\varepsilon_{ijk} e_3^k\right\}. \tag{25.70}$$

Here $\boldsymbol{e}_3$ is the unit vector in the x^3-direction, i.e., in the direction of the proton beam (Fig. 25.6). A simple calculation yields ρ in the basis of (25.67):

$$\langle 1, m|\rho|1, n\rangle = \begin{cases} \dfrac{1}{2}\dfrac{(g_\mathrm{V}^\mathrm{q} \mp g_\mathrm{A}^\mathrm{q})^2}{(g_\mathrm{V}^\mathrm{q})^2 + (g_\mathrm{A}^\mathrm{q})^2} & \text{for} \quad m = n = \pm 1, \\ 0 & \text{otherwise.} \end{cases} \tag{25.71}$$

Thus the Z produced by $\mathrm{q}\bar{\mathrm{q}}$ annihilation is polarized. We now assume that this Z decays leptonically:

$$\mathrm{Z} \to \ell^+(p_3) + \ell^-(p_4).$$

A straightforward calculation like that for the W-boson (25.29), (25.64) then yields for the angular distribution of the ℓ^+ in the rest frame of the Z:

$$\frac{1}{\Gamma}\frac{\mathrm{d}\Gamma}{\mathrm{d}\Omega_3}(\mathrm{Z} \to \ell^+(p_3) + \ell^-(p_4))$$

$$= \frac{3}{16\pi}\left\{1 + \cos^2\vartheta^* - \frac{8 g_\mathrm{V}^\mathrm{q} g_\mathrm{A}^\mathrm{q} g_\mathrm{V}^\ell g_\mathrm{A}^\ell}{[(g_\mathrm{V}^\mathrm{q})^2 + (g_\mathrm{A}^\mathrm{q})^2][(g_\mathrm{V}^\ell)^2 + (g_\mathrm{A}^\ell)^2]}\cos\vartheta^*\right\}. \tag{25.72}$$

Here ϑ^* denotes the polar angle of $\boldsymbol{p}_3$:

$$\cos\vartheta^* = \frac{\boldsymbol{p}_3 \cdot \boldsymbol{e}_3}{|\boldsymbol{p}_3|}. \tag{25.73}$$

For Z-bosons produced in $p\bar{p}$ collisions the formulas (25.71), (25.72) have to be suitably averaged over a $u\bar{u}$ and $d\bar{d}$ contribution.

Equation (25.72) shows that the ℓ^+ distribution in Z decay is only asymmetric with respect to $\vartheta^* = \pi/2$ when *all* coupling constants g_V^q, g_A^q, g_V^ℓ and g_A^ℓ are nonzero, i.e., when parity violation occurs both in the production and in the decay of the Z. The same conclusion is true for W-bosons. From Fig. 25.11 we thus infer that parity is violated by the coupling W–u–d, W–e–ν_e, and W–μ–ν_μ. However, from (25.72) we also see that the distribution in $\cos\vartheta^*$ remains unchanged when we reverse the signs of g_A^q and g_A^ℓ. This, too, holds for W-bosons as well. The observed distribution in Fig. 25.11 is thus compatible both with a (V–A) and with a (V + A) coupling of the quarks and leptons to the W-boson. To distinguish experimentally between these alternatives, one can, for example, use polarization measurements of τ-leptons produced in the decay $W \to \tau\nu_\tau$.

We finish by noting that the considerations of this section are closely related to those of Sect. 24.3.

25.5 The Higgs Boson

In the standard model there is a physical Higgs boson ρ', which represents a neutral spin-zero particle. This particle is *essential* in order to bring about the spontaneous symmetry breaking (Sect. 22.2). Experimental evidence of the existence of the Higgs boson is still lacking, From the above, it is clear that the Higgs boson plays a decisive role in confirming the validity of the standard model. In this section we will discuss a few predictions for the decay processes of the Higgs boson and possible ways for producing it.

According to the Lagrange density of the standard model (22.123), the Higgs boson couples to the W- and Z-bosons, to all fermions, and to itself. Thus at the level of the tree diagrams we expect the following decay processes for the ρ', provided they are energetically possible:

$$\begin{aligned} &\rho' \to W^+W^- && \left(\frac{2m_W^2}{\rho_0}\right); \\ &\rho' \to ZZ && \left(\frac{m_Z^2}{\rho_0}\right); \\ &\rho' \to f\bar{f} && \left(\frac{m_f}{\rho_0}\right). \end{aligned} \tag{25.74}$$

The respective coupling strengths are given in brackets and the quantity ρ_0 is given by

$$\rho_0 = 2^{-1/4} G^{-1/2} = 246 \text{ GeV}. \tag{25.75}$$

The coupling strengths are proportional to the masses or to the square of the masses. As we learned in Chap. 22, this is a necessary consequence of mass creation via spontaneous symmetry breaking. As a general rule, we thus expect

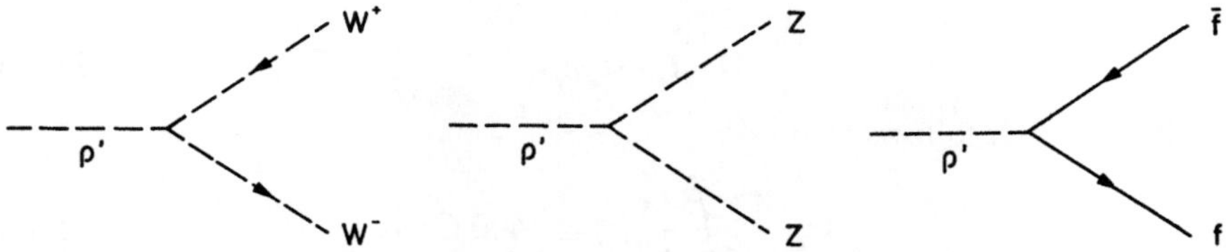

Figure 25.12 Diagrams for the decay processes of the Higgs boson to leading order in the couplings ($\mathrm{f} = \nu_e, \nu_\mu, \nu_\tau, e, \mu, \tau, u, c, t, d, s, b$).

that the Higgs particle will mainly decay into the heaviest energetically allowed particle. For its creation, too, it is best to begin with particles that are as heavy as possible.

The diagrams for the decay processes (25.74) are shown in Fig. 25.12. A straightforward calculation yields the following results for the decay rates:

$$\Gamma(\rho' \to \mathrm{W}^+\mathrm{W}^-) = \frac{m_{\rho'}}{16\pi}\left(\frac{m_{\rho'}}{\rho_0}\right)^2\left(1 - \frac{4m_\mathrm{W}^2}{m_{\rho'}^2}\right)^{1/2} \times \left(1 - 4\frac{m_\mathrm{W}^2}{m_{\rho'}^2} + 12\frac{m_\mathrm{W}^4}{m_{\rho'}^4}\right) \quad \text{for} \quad m_{\rho'} > 2m_\mathrm{W}; \tag{25.76}$$

$$\Gamma(\rho' \to \mathrm{ZZ}) = \frac{m_{\rho'}}{32\pi}\left(\frac{m_{\rho'}}{\rho_0}\right)^2\left(1 - \frac{4m_\mathrm{Z}^2}{m_{\rho'}^2}\right)^{1/2} \times \left(1 - 4\frac{m_\mathrm{Z}^2}{m_{\rho'}^2} + 12\frac{m_\mathrm{Z}^4}{m_{\rho'}^4}\right) \quad \text{for} \quad m_{\rho'} > 2m_\mathrm{Z}; \tag{25.77}$$

$$\Gamma(\rho' \to \mathrm{f}\bar{\mathrm{f}}) = \frac{m_{\rho'}}{8\pi}\left(\frac{m_\mathrm{f}}{\rho_0}\right)^2\left(1 - \frac{4m_\mathrm{f}^2}{m_{\rho'}^2}\right)^{3/2} \quad \text{for} \quad m_{\rho'} > 2m_\mathrm{f}. \tag{25.78}$$

Here f is one of the fundamental fermions. Thus, if the Higgs boson is lighter than two W-bosons ($m_{\rho'} < 2m_\mathrm{W}$), then it decays mainly into the heaviest fermion f for which $2m_\mathrm{f} < m_{\rho'}$. For $m_{\rho'} > 2m_\mathrm{W}$ or $m_{\rho'} > 2m_\mathrm{Z}$ the decay processes into Z- and W-bosons play the leading part, provided there are no further quarks or leptons that are heavier than the t-quark.

In the standard model, the Higgs mass $m_{\rho'}$ at first seems to be completely arbitrary. Upon a closer look, however, there are simple physical arguments in favor of an upper limit for $m_{\rho'}$. For this purpose we consider the case $m_{\rho'} \gg m_\mathrm{W}, m_\mathrm{Z}$. From (25.76), (25.77) we then find for the total width $\Gamma_{\rho'}$ of the Higgs boson:

$$\Gamma_{\rho'} \cong \Gamma(\rho' \to \mathrm{W}^+\mathrm{W}^-) + \Gamma(\rho' \to \mathrm{ZZ}) \cong \frac{3}{32\pi}\frac{m_{\rho'}^3}{\rho_0^2}. \tag{25.79}$$

It only makes sense to speak of a "particle" if its width is smaller than its mass, i.e., if

$$\Gamma_{\rho'} < m_{\rho'}. \tag{25.80}$$

With (25.79) it follows that

$$m_{\rho'} < \sqrt{\frac{32\pi}{3}}\rho_0 \cong 1400 \text{ GeV}. \tag{25.81}$$

A glance at the Lagrange density (22.123) shows that for a Higgs mass as large as this, the self-coupling will also be very large and so the application of perturbation theory cannot be justified. If we demand that the coupling constant of the $(\rho')^4$ term in (22.123) should not be larger than one, then we obtain a limit similar to (25.81) (Veltman 1977; Lee 1977):

$$m_{\rho'} < \sqrt{8}\rho_0 \cong 700 \text{ GeV}. \tag{25.82}$$

For larger Higgs masses we therefore have something like a *strong* interaction in the Higgs sector. Instead of a single Higgs particle, we might then expect a whole spectrum of states including perhaps resonances with very large width. In any case, we could no longer obtain the particle content of the standard model by simply looking at the Lagrangian (22.123).

It is also possible to derive a lower limit for the Higgs mass (Weinberg 1976; Linde 1976). If the mass $m_{\rho'}$ is too small, then quantum corrections to the Higgs potential $V(\phi)$ (22.43) become more important than the original terms. In this way, one finds that $m_{\rho'}$ should be greater than about 7 GeV:

$$m_{\rho'} \geq \frac{(6m_{\mathrm{W}}^4 + 3m_{\mathrm{Z}}^4)^{1/2}}{4\pi\rho_0} = 7.1 \text{ GeV}. \tag{25.83}$$

The experimental lower limit, at $m_{\rho'} > 3.2$ GeV (Table 22.1), is still smaller.

As the last topic of this section we discuss the possibility that the Higgs particle might be observed in the decay of a $^3\mathrm{S}_1$ quarkonium state which in the following is denoted generically by V:

$$\mathrm{V} \equiv {}^3\mathrm{S}_1(\mathrm{q\bar{q}}).$$

The quark q can be a c-, b-, or t-quark. Examples of such particles V are the J/ψ and the Υ, which were discussed in Sect. 20.3. It has been suggested that one should search for the following decay (Wilczek 1977):

$$\mathrm{V} \to \gamma + \rho'. \tag{25.84}$$

The monochromatic photons would serve as a signature for this reaction. The corresponding diagrams are shown in Fig. 25.13. The calculation proceeds in analogy to that for the decay of positronium into two photons (Sect. 13.2) and yields for the decay rate:

$$\Gamma(\mathrm{V} \to \gamma\rho') = 2\sqrt{2}G\alpha Q_{\mathrm{q}}^2\left(1 - \frac{m_{\rho'}^2}{m_{\mathrm{V}}^2}\right)|\psi(0)|^2. \tag{25.85}$$

Here m_{V} is the mass of the particle V and Q_{q} the charge of the quark q. Comparing this with the decay rate of the quarkonium state into a lepton pair

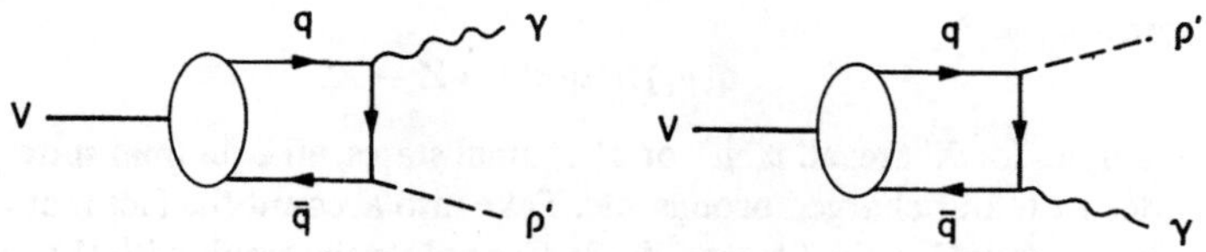

Figure 25.13 The diagrams for the decay of a $^3S_1$ quarkonium state V into a photon and a Higgs boson.

(20.26) we find

$$\frac{\Gamma(V \to \gamma\rho')}{\Gamma(V \to \ell^+\ell^-)} = \frac{Gm_V^2}{4\sqrt{2}\pi\alpha}\left(1 - \frac{m_{\rho'}^2}{m_V^2}\right). \tag{25.86}$$

For the Υ, whose measured mass and electronic branching ratio are

$$m_\Upsilon = 9.46 \text{ GeV},$$
$$\frac{\Gamma(\Upsilon \to e^+e^-)}{\Gamma(\Upsilon \to \text{anything})} = 2.5\%, \tag{25.87}$$

(25.86) yields the following numerical estimate for its decay into a Higgs particle and a photon:

$$\frac{\Gamma(\Upsilon \to \gamma\rho')}{\Gamma(\Upsilon \to \text{anything})} = 2.0 \cdot 10^{-4}\left(1 - \frac{m_{\rho'}^2}{m_V^2}\right). \tag{25.88}$$

In recent experiments, a sensitivity of this order of magnitude has been reached; however, no sign of a Higgs boson has shown up so far. Taking into account radiative corrections, which lower the expected rate (25.88) somewhat, one finds $m_{\rho'} \geq 3.2$ GeV as quoted in Table 22.1 (Lee-Franzini 1988).

With this we close our discussion of the Higgs boson. Further investigations of the anticipated properties of this particle are presented, e.g., by Ellis (1976).

Exercises

25.1 Evaluate numerically the contribution of one-photon exchange to the total cross-section $\sigma(e^+e^- \to \mu^+\mu^-)$ for $\sqrt{s} \cong m_Z$.

25.2 For the amplitude of the reaction $e^+e^- \to \mu^+\mu^-$ (24.42) carry out a partial wave analysis, as described in Sect. 16.3. In which partial waves does the Z-boson appear?

25.3 Draw the lowest order diagrams for the reaction (25.21) and calculate the cross-section and the angular distribution of the photon.

25.4 Calculate $\alpha_s(m_Z^2)$ and $\alpha_s(m_W^2)$ from (19.25) for $f = 5$. Assume $\alpha_s((35 \text{ GeV})^2) = 0.10$ to 0.17, as found from e^+e^- data (Sect. 20.2).

25.5 Calculate in a similar way as for the reaction $e^+e^- \to \mu^+\mu^-$ in Sect. 24.3 the rate for the production of a Z-boson in $q\bar{q}$ annihilation with the subsequent decay into any set of states X′ which remains unchanged under rotations in the rest frame of the Z

$$q(p_1') + \bar{q}(p_2') \to Z \to X'. \qquad (25.89)$$

Examples for X′ are: all $\mu^+\mu^-$ or e^+e^- final states, all 2-jet final states, all final states with four charged prongs, etc. Take into account the fact that the Z has a finite width Γ_Z as in (25.16), (25.17). Convolute the result with the quark and antiquark distributions of the proton and antiproton respectively to determine the cross-section for the reaction:

$$p(p_1) + \bar{p}(p_2) \to \underset{\llcorner\!\!\longrightarrow X'}{Z} + X \quad . \qquad (25.90)$$

Compare this expression with (25.46), which corresponds to $\Gamma_Z = 0$, and estimate the size of the correction due to the finite Z width.

25.6 Calculate the expectation value of the angular momentum operator $\vec{\mathbf{J}}$ in the state defined by the density matrix (25.70) for the Z in its rest frame.

25.7 Verify (25.72).

25.8 Consider the reaction

$$q(p_1') + \bar{q}(p_2') \to Z \to \ell^+(p_3) + \ell^-(p_4)$$

Calculate for fixed direction $\hat{\boldsymbol{p}}_3$ in the rest frame of the Z the expectation value of the angular momentum component $(\hat{\boldsymbol{p}}_3 \cdot \vec{\mathbf{J}})$ of the lepton pair. Calculate the average value of $\cos\vartheta^*$ for the distribution (25.72) and verify that

$$\langle \cos\vartheta^* \rangle = \frac{\langle \mathbf{J}^3 \rangle_Z \langle \hat{\boldsymbol{p}}_3 \vec{\mathbf{J}} \rangle_{\ell^+\ell^-}}{s(s+1)}. \qquad (25.91)$$

Here $\langle \mathbf{J}^3 \rangle_Z$ is the expectation value of $\mathbf{J}^3$ for the Z and $s = 1$ is the spin of the Z. This result is of general validity for all spin values (Jacob 1958).

25.9 Calculate the expected distribution in $\cos\vartheta^*$ for Z production in $p\bar{p}$ collisions. This requires suitable averaging over the contributions of u- and d-quarks in (25.72).

25.10 Show that a matrix ρ from a complex vector space $\mathbb{R}$ is unambiguously determined when the diagonal matrix elements $\langle a|\rho|a\rangle$ for all vectors $|a\rangle$ in the space $\mathbb{R}$ are given.

25.11 Verify (25.76)–(25.78) and (25.85), (25.86).

CHAPTER 26

The System of Neutral K-Mesons and CP Violation

Since 1957 we have known that the weak interaction possesses neither parity nor charge-conjugation invariance. Neither the parity operator **P** nor the charge-conjugation operator **C** are conserved.[1] Until 1964, however, it was believed that the product operator **CP** is a conserved quantity. As a result of the CPT theorem, the time-reversal operator **T** would then also be conserved. From the fundamental work of Christensen et al. (1964, 1965) we know that CP symmetry is broken in the decay of the neutral K-mesons. In the course of the further experimental and theoretical investigations of the K-mesons it was shown that the CP violation is connected with a violation of time-reversal invariance, whereas there was no evidence for a violation of CPT invariance (Schubert 1970; see also Casella 1969). In Nature, past and future are thus distinguishable even on a *microscopic* level.

Until now, the effects of CP and T violation have only been observed in the system of neutral K-mesons; thus we shall begin with a discussion of this system. We shall presuppose CPT invariance as demanded by the CPT theorem.

26.1 Phenomenology of the Neutral K-Mesons

The neutral K-mesons were discovered in strong reactions, in the associated production (Chap. 15). A typical production reaction was:

$$\pi^- + p \rightarrow K^0 + \Lambda. \tag{26.1}$$

[1] In this chapter and in Appendix I to shorten the notation we write **P**, **C**, and **T** for the parity, charge-conjugation, and time-reversal operators instead of **U**(P), **U**(C), and **V**(T) used in Chaps. 4 and 16.

Since a strangeness $S = +1$ was assigned to the K^0, it was not possible for the K^0 to be its own antiparticle:

$$\bar{K}^0 \neq K^0 \tag{26.2}$$

and thus there had to be two neutral K-meson states. This was predicted theoretically by Gell-Mann and Pais (1955) and experimentally confirmed by Landé et al. (1956). The states $|K^0\rangle$ and $|\bar{K}^0\rangle$ are unambiguously defined by the strong processes. For example, a K^0-meson produced in the reaction (26.1) can undergo a further strong interaction very close to its production point, leading, e.g., to the reaction:

$$K^0 + p \rightarrow n + K^+. \tag{26.3}$$

However, for the $\bar{K}^0$, this reaction is not possible due to the conservation of strangeness:

$$\bar{K}^0 + p \nrightarrow n + K^+. \tag{26.4}$$

On the other hand, by charge exchange with a nucleon, the $\bar{K}^0$ can produce a K^-:

$$\bar{K}^0 + n \rightarrow p + K^-. \tag{26.5}$$

From a study of the strong processes it was possible to deduce that the K^0 and $\bar{K}^0$ are pseudoscalar particles. For the parity operator **P** we thus have

$$\begin{aligned} \mathbf{P}|K^0\rangle &= -|K^0\rangle, \\ \mathbf{P}|\bar{K}^0\rangle &= -|\bar{K}^0\rangle. \end{aligned} \tag{26.6}$$

A charge-conjugation transformation **C** turns a K^0 into a $\bar{K}^0$. We can choose the relative phases of K^0 and $\bar{K}^0$ such that

$$\begin{aligned} \mathbf{C}|K^0\rangle &= |\bar{K}^0\rangle, \\ \mathbf{C}|\bar{K}^0\rangle &= |K^0\rangle. \end{aligned} \tag{26.7}$$

For the product **CP** it follows that

$$\begin{aligned} \mathbf{CP}|K^0\rangle &= -|\bar{K}^0\rangle, \\ \mathbf{CP}|\bar{K}^0\rangle &= -|K^0\rangle. \end{aligned} \tag{26.8}$$

The decay of the K-mesons is a weak process. We know that the weak interaction does not conserve strangeness S, parity P, or charge-conjugation C. The most common decay of the neutral K-mesons leads to two pions. In this *both* K^0 *and* $\bar{K}^0$ can yield the same final state:

$$\begin{aligned} K^0 &\rightarrow \pi^+\pi^-, \pi^0\pi^0, \\ \bar{K}^0 &\rightarrow \pi^+\pi^-, \pi^0\pi^0. \end{aligned} \tag{26.9}$$

It is therefore clear that at higher orders the weak interaction can also induce transitions between K^0 and $\bar{K}^0$ (Fig. 26.1). Thus the decay of K^0 and that of $\bar{K}^0$ cannot be considered separately—we must treat the two-state system K^0–$\bar{K}^0$ as a whole.

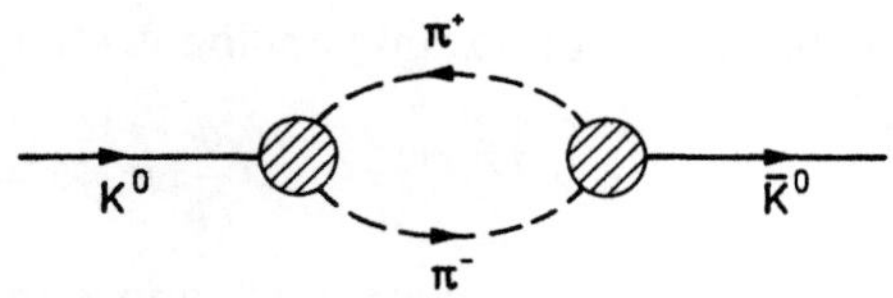

Figure 26.1 The transition $K^0 \to \bar{K}^0$, via an intermediary $\pi^+\pi^-$ pair. The two pions can be either real or virtual.

A suitable formalism for describing the decay of a multi-state system was developed by Weisskopf and Wigner (Weisskopf 1930). Here we will content ourselves with plausibility arguments. A more thorough discussion is presented in Appendix I.

We consider an unstable particle in its rest frame, for example, a π^+-meson. Let us assume that at time $t = 0$ exactly one pion is present. We write the state vector of the pion at $t = 0$ as

$$|\psi(0)\rangle = |\pi^+\rangle, \tag{26.10}$$

with the normalization condition:

$$\langle\psi(0)|\psi(0)\rangle = 1. \tag{26.11}$$

A plausible ansatz for the state of the *undecayed* pion at a later time $t > 0$ is then:

$$|\psi(t)\rangle = \mathrm{e}^{-\mathrm{i}(m_\pi - \mathrm{i}(1/2)\Gamma_\pi)t}\,|\psi(0)\rangle. \tag{26.12}$$

where m_π and Γ_π are the mass and total decay rate of the particle and t is the proper time. With this we obtain, on the one hand, an exponential decrease in the probability of finding an undecayed particle:

$$\langle\psi(t)|\psi(t)\rangle = \mathrm{e}^{-\Gamma_\pi t}. \tag{26.13}$$

On the other hand, we have that the phase of the state $|\psi(t)\rangle$ is given by the mass, i.e., the energy eigenvalue in the rest frame. The state vector of the undecayed particle thus satisfies an effective Schrödinger equation with the complex "mass" $(m_\pi - \mathrm{i}\Gamma_\pi/2)$:

$$\mathrm{i}\frac{\partial}{\partial t}|\psi(t)\rangle = \left(m_\pi - \frac{\mathrm{i}}{2}\Gamma_\pi\right)|\psi(t)\rangle. \tag{26.14}$$

Moving on now to the K-mesons, we have to allow for the fact that transitions can occur between K^0 and $\bar{K}^0$. The state vector of a neutral K-meson at a particular time t is given, in general, by the superposition of a K^0 and a $\bar{K}^0$ particle. The corresponding state vector must be expressed as follows:

$$|\psi(t)\rangle = \psi_1(t)|K^0\rangle + \psi_2(t)|\bar{K}^0\rangle. \tag{26.15}$$

Here $\psi_1(t)$ and $\psi_2(t)$ are the amplitudes, and $|\psi_1(t)|^2$ and $|\psi_2(t)|^2$ the probabilities, of encountering a K^0 or $\bar{K}^0$, respectively, at time t. The space of states for a neutral K-meson at rest is two-dimensional. As is shown in Appendix I, the state vector of (26.15) satisfies an effective Schrödinger equation that is exactly analogous to (26.14):

$$\mathrm{i}\frac{\partial}{\partial t}|\psi(t)\rangle = \mathcal{M}|\psi(t)\rangle. \tag{26.16}$$

Here $\mathcal{M}$ is an operator in the above-mentioned two-dimensional space; for a given basis, $\mathcal{M}$ is thus represented by a 2×2 matrix—which in general is non-Hermitian. However, as for every matrix, $\mathcal{M}$ can be split into a Hermitian and an anti-Hermitian part. We write these as follows:

$$\mathcal{M} = \mathbf{M} - \frac{\mathrm{i}}{2}\Gamma, \tag{26.17}$$

where $\mathbf{M} = \mathbf{M}^\dagger$ and $\Gamma = \Gamma^\dagger$. Obviously we then have

$$\mathbf{M} = \tfrac{1}{2}(\mathcal{M} + \mathcal{M}^\dagger), \tag{26.18a}$$

$$\Gamma = \mathrm{i}(\mathcal{M} - \mathcal{M}^\dagger). \tag{26.18b}$$

The matrix $\mathbf{M}$ is called the Hermitian mass matrix and Γ the decay matrix. These matrices are the generalization of the mass and decay width of an unstable particle that does not mix, cf. (26.14).

We now consider the eigenvalue problem for $\mathcal{M}$. We denote the eigenvectors by $|K_S\rangle$ and $|K_L\rangle$, and the eigenvalues by M_S and M_L:

$$\begin{aligned} \mathcal{M}|K_S\rangle &= M_S|K_S\rangle, \\ \mathcal{M}|K_L\rangle &= M_L|K_L\rangle. \end{aligned} \tag{26.19}$$

Since $\mathcal{M}$ is non-Hermitian, the eigenvalues are, in general, complex, and the eigenstates not orthogonal to one another. We thus write

$$\begin{aligned} M_S &= \left(m_S - \frac{\mathrm{i}}{2}\Gamma_S\right), \\ M_L &= \left(m_L - \frac{\mathrm{i}}{2}\Gamma_L\right), \end{aligned} \tag{26.20}$$

where m_S, m_L, Γ_S, and Γ_L are real. In Appendix I we show that $\Gamma_S \geq 0$ and $\Gamma_L \geq 0$. We will choose the normalization and relative phase of $|K_S\rangle$ and $|K_L\rangle$ such that

$$\langle K_S|K_S\rangle = \langle K_L|K_L\rangle = 1, \tag{26.21}$$

$$\langle K_S|K_L\rangle = \langle K_S|K_L\rangle^* \geq 0. \tag{26.22}$$

Since it has been experimentally demonstrated that $\langle K_S|K_L\rangle$ is not equal to zero, the relative phase of $|K_S\rangle$ and $|K_L\rangle$ is indeed fixed by requiring (26.22).

In an experiment, at time $t = 0$ a particular superposition of the particles K^0 and $\bar{K}^0$ will be present. The state vector is then

$$|\psi(0)\rangle = \psi_1|K^0\rangle + \psi_2|\bar{K}^0\rangle. \tag{26.23}$$

We obtain the time development of this state by solving the differential equation (26.16). It is convenient here to expand $|\psi(0)\rangle$ in terms of the states

$|K_S\rangle$ and $|K_L\rangle$:

$$|\psi(0)\rangle = c_S|K_S\rangle + c_L|K_L\rangle. \tag{26.24}$$

where c_S and c_L are certain constants. As can be shown with the help of (26.19), the required solution is then

$$|\psi(t)\rangle = e^{-i(m_S-(i/2)\Gamma_S)t}\, c_S|K_S\rangle + e^{-i(m_L-(i/2)\Gamma_L)t}\, c_L|K_L\rangle. \tag{26.25}$$

The decrease with time in the probability of finding an undecayed K-meson is given by

$$\langle\psi(t)|\psi(t)\rangle = |c_S|^2\, e^{-\Gamma_S t} + |c_L|^2\, e^{-\Gamma_L t} + 2\,\mathrm{Re}[c_S^* c_L \langle K_S|K_L\rangle\, e^{-i(m_L-m_S)t}]\, e^{-(1/2)(\Gamma_S+\Gamma_L)t}. \tag{26.26}$$

This relatively complicated expression for the t dependence contains information about the two decay rates Γ_S and Γ_L and about the mass difference $m_L - m_S$. A simple exponential decay law is obtained for the states $|K_S\rangle$ and $|K_L\rangle$, i.e., by putting $c_S = 1$, $c_L = 0$ or $c_L = 1$, $c_S = 0$. By convention, $|K_S\rangle$ represents the short-lived state and $|K_L\rangle$ the long-lived state, i.e., we choose the notation such that $\Gamma_S \geq \Gamma_L$.

What predictions are obtained if CP invariance is valid? From (26.8) we can readily write down the eigenstates of the CP operator. They are

$$|K_1^0\rangle = \frac{1}{\sqrt{2}}(|K^0\rangle - |\bar{K}^0\rangle),$$
$$|K_2^0\rangle = \frac{1}{\sqrt{2}}(|K^0\rangle + |\bar{K}^0\rangle), \tag{26.27}$$

where

$$\mathbf{CP}|K_1^0\rangle = |K_1^0\rangle,$$
$$\mathbf{CP}|K_2^0\rangle = -|K_2^0\rangle. \tag{26.28}$$

Thus if CP invariance holds, the mass matrix $\mathcal{M}$ cannot produce transitions between K_1^0, and K_2^0 and these states have a pure exponential decay law. Prior to 1964 it was already known that it is the short-lived neutral K-meson that decays into two pions

$$K_S \to \pi^+\pi^-, \pi^0\pi^0. \tag{26.29}$$

The two pions must be in a state with angular momentum $l = 0$ since the K-mesons have spin zero. It then follows immediately that the two-pion state is a CP eigenstate with eigenvalue $+1$:

$$\mathbf{CP}|\pi\pi(l=0)\rangle = |\pi\pi(l=0)\rangle. \tag{26.30}$$

Thus it was concluded that

$$K_S = K_1^0. \tag{26.31}$$

From this it follows that if CP invariance is valid then

$$K_L = K_2^0 \nrightarrow \pi\pi. \tag{26.32}$$

Indeed, the long-lived neutral K-meson mainly decays as follows

$$K_L \to \pi^+\pi^-\pi^0, \quad \pi^0\pi^0\pi^0, \quad \pi^\pm\mu^\mp\nu, \quad \pi^\pm e^\perp\nu. \tag{26.33}$$

The lifetimes are

$$\begin{aligned} \tau_S &= \Gamma_S^{-1} = (0.8923 \pm 0.0022)\cdot 10^{-10}\ \mathrm{s}, \\ \tau_L &= \Gamma_L^{-1} = (5.183 \pm 0.040)\cdot 10^{-8}\ \mathrm{s}. \end{aligned} \tag{26.34}$$

In the important experiment of Christensen et al. it was found, however, that the long-lived K-meson also decays into two pions, although the branching ratio for this is very small. The currently accepted values are

$$\begin{aligned} \frac{\Gamma(K_L \to \pi^+\pi^-)}{\Gamma_L} &= (2.03 \pm 0.05)\cdot 10^{-3}, \\ \frac{\Gamma(K_L \to \pi^0\pi^0)}{\Gamma_L} &= (0.94 \pm 0.18)\cdot 10^{-3}. \end{aligned} \tag{26.35}$$

This was the discovery of CP violation, a result of fundamental significance. CP violation plays an important role, for example, in the modern theories of cosmology which aim to explain the observed baryon–antibaryon asymmetry in the Universe (see Gibbons 1983; Linde 1984 and references therein).

A particularly clear piece of evidence for CP violation is found in the decay processes

$$K_L \to \pi^\pm \ell^\mp \nu, \tag{26.36}$$

where $\ell = e, \mu$. If the long-lived K-meson is an eigenstate of the CP operator, then the total rates for the ℓ^- and the ℓ^+ decay must be equal:

$$\Gamma(K_L \to \pi^+\ell^-\nu) = \Gamma(K_L \to \pi^-\ell^+\nu) \quad \text{(for CP invariance).} \tag{26.37}$$

Experimentally, one finds instead

$$\delta = \frac{\Gamma(K_L \to \pi^-\ell^+\nu) - \Gamma(K_L \to \pi^+\ell^-\nu)}{\Gamma(K_L \to \pi^-\ell^+\nu) + \Gamma(K_L \to \pi^+\ell^-\nu)} = (0.330 \pm 0.012)\cdot 10^{-2}. \tag{26.38}$$

Here we quote the average of the δ-values for $\ell = e$ and $\ell = \mu$ which are compatible with one another to within experimental error.

We now analyze the decay processes

$$K_{S,L} \to \pi^+\pi^-, \pi^0\pi^0 \tag{26.39}$$

in more detail. As already mentioned, the two pions must be in a state with orbital angular momentum $l = 0$ with respect to their center of mass system. Such a state can be constructed using the methods of Sect. 16.3. For a general center of mass energy $\sqrt{s}$, we define [see (16.58)]:

$$|\pi^+\pi^-; s; l = 0\rangle = \int d\Omega_k \frac{1}{\sqrt{4\pi}} |\pi^+(\boldsymbol{k})\pi^-(-\boldsymbol{k})\rangle, \tag{26.40}$$

$$|\pi^0\pi^0; s; l = 0\rangle = \int d\Omega_k \frac{1}{\sqrt{4\pi}} \frac{1}{\sqrt{2}} |\pi^0(\boldsymbol{k})\pi^0(-\boldsymbol{k})\rangle, \tag{26.41}$$

where $\sqrt{s} = 2k^0 = 2\sqrt{k^2 + m_\pi^2}$. The states (26.40), (26.41) both have the same normalization. The extra factor of $1/\sqrt{2}$ on the right-hand side of (26.41) accounts for the indistinguishability of the two π^0-mesons. We now analyze these states with respect to their (strong) isospin properties. A single pion has $I = 1$, and thus two pions can form states with $I = 0$, 1, and 2. For angular momentum $l = 0$, the two pions are in a symmetric momentum state. Because of the Bose symmetry, their state must then also be symmetric with respect to isospin and thus only the total isospin values $I = 0$ and $I = 2$ are possible. The isospin eigenstates for two pions with opposite momentum directions are readily constructed with the help of the Clebsch–Gordan coefficients. We are interested in states with $I = 0$ or $I = 2$ and a third component of $I_3 = 0$:

$$\begin{aligned} &|\pi(\boldsymbol{k})\pi(-\boldsymbol{k}); I = 0, I_3 = 0\rangle \\ &\qquad = \frac{1}{\sqrt{3}}\{|\pi^+(\boldsymbol{k})\pi^-(-\boldsymbol{k})\rangle - |\pi^0(\boldsymbol{k})\pi^0(-\boldsymbol{k})\rangle + |\pi^-(\boldsymbol{k})\pi^+(-\boldsymbol{k})\rangle\}, \\ &|\pi(\boldsymbol{k})\pi(-\boldsymbol{k}); I = 2, I_3 = 0\rangle \\ &\qquad = \frac{1}{\sqrt{6}}\{|\pi^+(\boldsymbol{k})\pi^-(-\boldsymbol{k})\rangle + 2|\pi^0(\boldsymbol{k})\pi^0(-\boldsymbol{k})\rangle + |\pi^-(\boldsymbol{k})\pi^+(-\boldsymbol{k})\rangle\}. \end{aligned} \tag{26.42}$$

The corresponding states with angular momentum $l = 0$ are defined as

$$|\pi\pi; s; l = 0; I, I_3 = 0\rangle = \int d\Omega_k \frac{1}{\sqrt{4\pi}} \frac{1}{\sqrt{2}} |\pi(\boldsymbol{k})\pi(-\boldsymbol{k}); I, I_3 = 0\rangle. \tag{26.43}$$

The normalization of these states is the same as that of the states (26.40), (26.41). In the following we use the abbreviated notation

$$\begin{aligned} |\pi^+\pi^-; s = m_K^2; l = 0\rangle &\equiv |\pi^+\pi^-\rangle, \\ |\pi^0\pi^0; s = m_K^2; l = 0\rangle &\equiv |\pi^0\pi^0\rangle, \\ |\pi\pi; s = m_K^2; l = 0; I, I_3 = 0\rangle &\equiv |\pi\pi, I\rangle. \end{aligned} \tag{26.44}$$

From (26.40)–(26.42) we then obtain

$$\begin{aligned} |\pi^+\pi^-\rangle &= \sqrt{\tfrac{2}{3}}|\pi\pi, 0\rangle + \sqrt{\tfrac{1}{3}}|\pi\pi, 2\rangle, \\ |\pi^0\pi^0\rangle &= -\sqrt{\tfrac{1}{3}}|\pi\pi, 0\rangle + \sqrt{\tfrac{2}{3}}|\pi\pi, 2\rangle. \end{aligned} \tag{26.45}$$

We now consider the amplitudes, i.e., the T-matrix elements, for the decays $K_{S,L} \to \pi\pi$ and define the following quantities:

$$\omega = \frac{\langle\pi\pi, 2|\mathbf{T}|K_S\rangle}{\langle\pi\pi, 0|\mathbf{T}|K_S\rangle}, \tag{26.46}$$

$$\eta_{+-} = \frac{\langle\pi^+\pi^-|\mathbf{T}|K_L\rangle}{\langle\pi^+\pi^-|\mathbf{T}|K_S\rangle}, \tag{26.47}$$

$$\eta_{00} = \frac{\langle\pi^0\pi^0|\mathbf{T}|K_L\rangle}{\langle\pi^0\pi^0|\mathbf{T}|K_S\rangle}. \tag{26.48}$$

We shall also set

$$\varphi_{+-} \equiv \arg \eta_{+-},$$
$$\varphi_{00} \equiv \arg \eta_{00}. \tag{26.49}$$

As a result of the convention (26.22), all phases are unambiguously defined and are observable quantities.

The amplitude ratio ω has nothing to do with CP violation but measures instead the fraction of $I = 2$ relative to $I = 0$ in the final state of the decay $K_S \to \pi\pi$. In Appendix I we show how ω can be determined from the ratio $\Gamma(K_S \to \pi^+\pi^-)/\Gamma(K_S \to \pi^0\pi^0)$. The experimental value of ω turns out to be small (Table 26.1). This fact, together with certain results from nonleptonic hyperon decay processes, forms the content of the so-called $\Delta I = \frac{1}{2}$ rule, whose theoretical derivation within the standard model is still an open problem.

The quantities η_{+-} and η_{00} are, according to (26.32), a direct measure of the CP violation. Their absolute magnitudes are obtained from the relations

$$\frac{\Gamma(K_L \to \pi^+\pi^-)}{\Gamma(K_S \to \pi^+\pi^-)} = |\eta_{+-}|^2,$$
$$\frac{\Gamma(K_L \to \pi^0\pi^0)}{\Gamma(K_S \to \pi^0\pi^0)} = |\eta_{00}|^2. \tag{26.50}$$

In order to determine the phases φ_{+-} and φ_{00}, one has to analyze the entire course of the $\pi^+\pi^-$ or $\pi^0\pi^0$ decay of a K state that is specified at a time $t = 0$. It is usual to introduce two further parameters, ε and ε', which, however, are dependent on the above quantities:

$$\varepsilon = \frac{\langle \pi\pi, 0|\mathbf{T}|K_L\rangle}{\langle \pi\pi, 0|\mathbf{T}|K_S\rangle}, \tag{26.51}$$

$$\varepsilon' = \frac{1}{\sqrt{2}}\left\{\frac{\langle \pi\pi, 2|\mathbf{T}|K_L\rangle}{\langle \pi\pi, 0|\mathbf{T}|K_L\rangle} - \frac{\langle \pi\pi, 2|\mathbf{T}|K_S\rangle}{\langle \pi\pi, 0|\mathbf{T}|K_S\rangle}\right\}\varepsilon. \tag{26.52}$$

A simple calculation yields

$$\varepsilon = \tfrac{1}{3}\{2\eta_{+-} + \eta_{00} + \sqrt{2}\omega(\eta_{+-} - \eta_{00})\}, \tag{26.53}$$

$$\varepsilon' = \tfrac{1}{3}(\eta_{+-} - \eta_{00})\left(1 - \frac{1}{\sqrt{2}}\omega - \omega^2\right). \tag{26.54}$$

The experimental values for these and some other parameters of interest are listed in Table 26.1. For the methods of determining these parameters the reader is referred to Appendix I and to the relevant review articles (e.g., Lee 1966; Kleinknecht 1976; Cronin 1981; Buras 1984).

Shortly after the discovery of CP violation, a simple phenomenological model known as the "superweak" model was proposed as an explanation of the observations (Wolfenstein 1964). This postulated the existence of a superweak force that mediates transitions between the CP eigenstates $|K_1^0\rangle$ and $|K_2^0\rangle$ (26.27), but which appears nowhere else. To analyze the conse-

Table 26.1 Experimental values for parameters of the K^0–$\bar{K}^0$ system. The value of ε'/ε is from Burkhardt (1988). The value of ω is calculated according to the methods of Appendix I (see Schubert 1970).

Quantity	Experimental value
$m_{K^0} = (m_L + m_S)/2$	497.67 ± 0.13 MeV
$m_L - m_S$	$(3.521 \pm 0.014)\cdot 10^{-12}$ MeV $\hat{=}$ $(0.5349 \pm 0.0022)\cdot 10^{10}\,\mathrm{s}^{-1}$
$\Gamma_S \equiv \Gamma(K_S \to \text{anything})$	$(1.121 \pm 0.003)\cdot 10^{10}\,\mathrm{s}^{-1}$
$\Gamma_L \equiv \Gamma(K_L \to \text{anything})$	$(1.929 \pm 0.015)\cdot 10^{7}\,\mathrm{s}^{-1}$
$\Gamma(K_S \to \pi^+\pi^-)/\Gamma_S$	$(68.61 \pm 0.24)\,\%$
$\Gamma(K_S \to \pi^0\pi^0)/\Gamma_S$	$(31.39 \pm 0.24)\,\%$
$\Gamma(K_L \to \pi^0\pi^0\pi^0)/\Gamma_L$	$(21.5 \pm 1.0)\,\%$
$\Gamma(K_L \to \pi^+\pi^-\pi^0)/\Gamma_L$	$(12.39 \pm 0.20)\,\%$
$\Gamma(K_L \to \pi^\pm\mu^\mp\nu)/\Gamma_L$	$(27.1 \pm 0.4)\,\%$
$\Gamma(K_L \to \pi^\pm e^\mp\nu)/\Gamma_L$	$(38.7 \pm 0.5)\,\%$
$\lvert\eta_{+-}\rvert$	$(2.274 \pm 0.022)\cdot 10^{-3}$
φ_{+-}	$(44.6 \pm 1.2)^0$
$\lvert\eta_{00}\rvert$	$(2.33 \pm 0.08)\cdot 10^{-3}$
φ_{00}	$(54 \pm 5)^0$
$(\varepsilon'/\varepsilon)e^{i0.05\pi}$	0.0033 ± 0.0011
ω	$(0.044 \pm 0.013)\exp i(-39^0 \pm 18^0)$

quences of this assumption we consider the mass matrix $\mathcal{M}$ for the neutral K-mesons in the basis $|K_1^0\rangle$, $|K_2^0\rangle$:

$$\mathcal{M} = \begin{pmatrix} \mathcal{M}_{11} & \mathcal{M}_{12} \\ \mathcal{M}_{21} & \mathcal{M}_{22} \end{pmatrix}, \tag{26.55}$$

where $\mathcal{M}_{ij} = \langle K_i^0|\mathcal{M}|K_j^0\rangle$. If $\mathbf{H}'_{sw}$ is the Hamiltonian of the superweak interaction, then on neglecting terms quadratic in $\mathbf{H}' \equiv \mathbf{H}'_{sw}$, we have from (I.37), (I.38) in Appendix I

$$\mathcal{M}_{12} = \langle K_1^0|\,\mathbf{H}'_{sw}|K_2^0\rangle. \tag{26.56}$$

The Hermiticity of $\mathbf{H}'_{sw}$ and the CPT invariance (I.66) yield:

$$\mathcal{M}_{12} = \mathcal{M}_{21}^*, \tag{26.57}$$

$$\mathcal{M}_{12} + \mathcal{M}_{21} = 0. \tag{26.58}$$

Thus, in the superweak model, $\mathcal{M}_{12}$ must be purely imaginary, $\mathcal{M}_{12} = \pm i|\mathcal{M}_{12}|$. It is now easy to determine the eigenvalues and eigenstates of $\mathcal{M}$. If we content ourselves with the leading terms of an expansion in the parameter $\mathcal{M}_{12}/(\mathcal{M}_{11} - \mathcal{M}_{22})$, we find

$$\begin{aligned} M_S &= \mathcal{M}_{11}, \\ M_L &= \mathcal{M}_{22}, \end{aligned} \tag{26.59}$$

$$\begin{aligned} |K_S\rangle &= |K_1^0\rangle + \varepsilon_S|K_2^0\rangle, \\ |K_L\rangle &= |K_2^0\rangle + \varepsilon_L|K_1^0\rangle. \end{aligned} \tag{26.60}$$

Here

$$\varepsilon_S = \varepsilon_L = \frac{\mathcal{M}_{12}}{M_L - M_S} = \frac{\pm i|\mathcal{M}_{12}|}{m_L - m_S - \frac{i}{2}(\Gamma_L - \Gamma_S)} \tag{26.61}$$

and the relative phase of the states $|K_L\rangle, |K_S\rangle$ is chosen such that (26.22) holds (see Appendix I).

In the superweak model, the only source of CP violation is, by definition, the CP "impurity" of the states $|K_S\rangle$ and $|K_L\rangle$ given in (26.60). The decay rates themselves are not supposed to induce any CP violation. In particular, this model predicts that the CP eigenstate $|K_2^0\rangle$, with eigenvalue -1, should not decay into two pions:

$$K_2^0 \not\to \pi^+\pi^-, \pi^0\pi^0.$$

Neglecting terms of order ε_L^2, it then follows that

$$\begin{aligned} \langle \pi^+\pi^-|\mathbf{T}|K_L\rangle &= \varepsilon_L\langle \pi^+\pi^-|\mathbf{T}|K_1^0\rangle = \varepsilon_L\langle \pi^+\pi^-|\mathbf{T}|K_S\rangle, \\ \langle \pi^0\pi^0|\mathbf{T}|K_L\rangle &= \varepsilon_L\langle \pi^0\pi^0|\mathbf{T}|K_1^0\rangle = \varepsilon_L\langle \pi^0\pi^0|\mathbf{T}|K_S\rangle. \end{aligned} \tag{26.62}$$

The predictions of the superweak model for the CP violation parameters in the decay $K \to \pi\pi$ are thus

$$\eta_{+-} = \eta_{00} = \varepsilon = \varepsilon_L = \frac{\mathcal{M}_{12}}{m_L - m_S - \frac{i}{2}(\Gamma_L - \Gamma_S)}, \tag{26.63}$$

$$\frac{\varepsilon'}{\varepsilon} = 0, \tag{26.64}$$

$$\varphi_{+-} = \varphi_{00} = \arctan\left(2\frac{m_L - m_S}{\Gamma_S - \Gamma_L}\right) + \arg \mathcal{M}_{12} - \frac{\pi}{2}, \tag{26.65}$$

$$\left(\arg \mathcal{M}_{12} = \pm\frac{\pi}{2}\right).$$

If we insert the experimental values for $m_L - m_S$ and $\Gamma_S - \Gamma_L$, we find

$$\varphi_{+-} = \varphi_{00} = 43.7° \pm 0.1° + \arg \mathcal{M}_{12} - 90°. \tag{26.66}$$

Comparing these predictions with the measured values of $|\eta_{+-}|$, $|\eta_{00}|$, φ_{+-}, φ_{00} and ε'/ε (Table 26.1) we see that provided $\mathcal{M}_{12}$ is positive imaginary the superweak model agrees with experiment except for ε'/ε. Setting for the moment ε'/ε to zero, and fitting the rest of the data in the framework of the superweak model, we would find for the only free parameter in the model, $\mathcal{M}_{12}$, the following:

$$\begin{aligned} \mathcal{M}_{12} &= i|\eta_{+-}|[(m_L - m_S)^2 + \tfrac{1}{4}(\Gamma_S - \Gamma_L)^2]^{1/2}, \\ |\mathcal{M}_{12}| &\cong 3\cdot 10^{-3}(m_L - m_S) \cong 10^{-14}\ \text{MeV}. \end{aligned} \tag{26.67}$$

The superweak interaction energy, $\mathcal{M}_{12}$, would thus be approximately a factor of 1000 smaller than the mass difference $m_L - m_S$, which stems from the weak interaction. It is indeed still possible that the observed CP violation is caused in part by such a superweak force.

26.2 CP Violation and CPT Invariance in the Standard Model

26.2.1 CP Violation in the Lagrange Density

We proceed from the Lagrange density $\mathcal{L}$ of the standard model in the form (22.123), and ask the question: When is this invariant under a CP transformation? A CP transformation must surely, transform Dirac fields of defined mass into themselves. We thus make the following ansatz in the basis of the physical fields (cf. Sect. 4.5):

$$\begin{aligned} \text{CP: } \nu_{\ell \mathrm{L}}(x) &\to \mathrm{e}^{\mathrm{i}\varphi_{\nu_\ell}} \gamma_0 \mathbf{S}(\mathrm{C}) \bar{\nu}^{\mathrm{T}}_{\ell \mathrm{L}}(x'), \\ \ell(x) &\to \mathrm{e}^{\mathrm{i}\varphi_\ell} \gamma_0 \mathbf{S}(\mathrm{C}) \bar{\ell}^{\mathrm{T}}(x'), \\ \mathbf{q}(x) &\to \mathrm{e}^{\mathrm{i}\varphi_{\mathrm{q}}} \gamma_0 \mathbf{S}(\mathrm{C}) \bar{\mathbf{q}}^{\mathrm{T}}(x'), \end{aligned} \tag{26.68}$$

where $\ell = \mathrm{e}, \mu, \tau$ and $\mathrm{q} = \mathrm{u, c, t, d, s, b}$. Moreover

$$x = \begin{pmatrix} t \\ \boldsymbol{x} \end{pmatrix}, \quad x' = \begin{pmatrix} t \\ -\boldsymbol{x} \end{pmatrix}$$

and $\mathbf{S}(\mathrm{C})$ is the charge conjugation matrix (4.124). We admit the occurrence of separate phases φ_{ν_ℓ}, φ_ℓ, and φ_{q} for the various fundamental fermions. As we already know from Sect. 4.5, the kinetic and mass terms for the Dirac fields in $\mathcal{L}$ remain invariant under the substitutions (26.68).

We now investigate whether these transformations of the Dirac fields can be supplemented by a suitable transformation of the Bose fields, such that the Lagrange density (22.123) is invariant.

Invariance of the Yukawa couplings, and of the Higgs part of $\mathcal{L}$, is achieved if we set

$$\text{CP: } \rho'(x) \to \rho'(x'). \tag{26.69}$$

We next consider the interaction term $\mathcal{L}_{\mathrm{int}}$ (22.77), in which the currents are given by (22.112), (22.114), (22.124). Applying the transformation (26.68) we obtain for the electromagnetic current:

$$\text{CP: } \mathcal{J}^{\lambda}_{\mathrm{em}}(x) \to -\mathcal{J}_{\lambda,\mathrm{em}}(x'). \tag{26.70}$$

In this we use the fact that the charge matrix $\mathbf{Q}$ for the Dirac fields is diagonal in the physical basis. Analogously, we find for the neutral current (22.114), which is also diagonal in the physical basis:

$$\text{CP: } \mathcal{J}^{\lambda}_{\mathrm{NC}}(x) \to -\mathcal{J}_{\lambda,\mathrm{NC}}(x'). \tag{26.71}$$

For the charged weak current (22.112), on the other hand, we have

$$
\begin{aligned}
\text{CP: } \mathscr{J}^{\lambda}_{\mathrm{CC}}(x) &= (\bar{\nu}_{\mathrm{eL}}(x), \bar{\nu}_{\mu\mathrm{L}}(x), \bar{\nu}_{\tau\mathrm{L}}(x))\gamma^{\lambda}\begin{pmatrix} \mathbf{e}_{\mathrm{L}}(x) \\ \mu_{\mathrm{L}}(x) \\ \tau_{\mathrm{L}}(x) \end{pmatrix} \\
&\quad + (\bar{\mathbf{u}}_{\mathrm{L}}(x), \bar{\mathbf{c}}_{\mathrm{L}}(x), \bar{\mathbf{t}}_{\mathrm{L}}(x))\gamma^{\lambda}\mathbf{V}\begin{pmatrix} \mathbf{d}_{\mathrm{L}}(x) \\ \mathbf{s}_{\mathrm{L}}(x) \\ \mathbf{b}_{\mathrm{L}}(x) \end{pmatrix} \\
&\longrightarrow -(\mathrm{e}^{\mathrm{i}\varphi_{\mathrm{e}}}\,\bar{\mathbf{e}}_{\mathrm{L}}(x'), \mathrm{e}^{\mathrm{i}\varphi_{\mu}}\,\bar{\mu}_{\mathrm{L}}(x'), \mathrm{e}^{\mathrm{i}\varphi_{\tau}}\,\bar{\tau}_{\mathrm{L}}(x'))\gamma_{\lambda}\begin{pmatrix} \mathrm{e}^{-\mathrm{i}\varphi_{\nu_{\mathrm{e}}}}\,\nu_{\mathrm{eL}}(x') \\ \mathrm{e}^{-\mathrm{i}\varphi_{\nu_{\mu}}}\,\nu_{\mu\mathrm{L}}(x') \\ \mathrm{e}^{-\mathrm{i}\varphi_{\nu_{\tau}}}\,\nu_{\tau\mathrm{L}}(x') \end{pmatrix} \\
&\quad -(\mathrm{e}^{\mathrm{i}\varphi_{\mathrm{d}}}\,\bar{\mathbf{d}}_{\mathrm{L}}(x'), \mathrm{e}^{\mathrm{i}\varphi_{\mathrm{s}}}\,\bar{\mathbf{s}}_{\mathrm{L}}(x'), \mathrm{e}^{\mathrm{i}\varphi_{\mathrm{b}}}\,\bar{\mathbf{b}}_{\mathrm{L}}(x'))\gamma_{\lambda}\mathbf{V}^{\mathrm{T}}\begin{pmatrix} \mathrm{e}^{-\mathrm{i}\varphi_{\mathrm{u}}}\,\mathbf{u}_{\mathrm{L}}(x') \\ \mathrm{e}^{-\mathrm{i}\varphi_{\mathrm{c}}}\,\mathbf{c}_{\mathrm{L}}(x') \\ \mathrm{e}^{-\mathrm{i}\varphi_{\mathrm{t}}}\,\mathbf{t}_{\mathrm{L}}(x') \end{pmatrix}.
\end{aligned}
\tag{26.72}
$$

If this CP transformation is to lead to an invariance of the Lagrange density $\mathscr{L}$ of the standard model, then the transformed charged current must either itself be present in $\mathscr{L}$, or be equivalent to another current appearing there. The only possible current in this respect is the Hermitian conjugated current $\mathscr{J}^{\dagger}_{\mathrm{CC}}$. Thus we must demand

$$
\text{CP: } \mathscr{J}^{\lambda}_{\mathrm{CC}}(x) \rightarrow -\mathrm{e}^{\mathrm{i}\chi}\,\mathscr{J}^{\dagger}_{\mathrm{CC}\lambda}(x'), \tag{26.73}
$$

where χ is an as yet arbitrary phase. Can we satisfy this requirement by making an appropriate choice of the phases φ_{ν_ℓ}, φ_ℓ, and φ_{q} in (26.68)? From (26.72), (26.73) we obtain the conditions:

$$
\begin{aligned}
\varphi_{\mathrm{e}} - \varphi_{\nu_{\mathrm{e}}} &= \chi, \\
\varphi_{\mu} - \varphi_{\nu_{\mu}} &= \chi, \\
\varphi_{\tau} - \varphi_{\nu_{\tau}} &= \chi,
\end{aligned}
\tag{26.74}
$$

$$
\begin{pmatrix} \mathrm{e}^{\mathrm{i}\varphi_{\mathrm{d}}} & 0 & 0 \\ 0 & \mathrm{e}^{\mathrm{i}\varphi_{\mathrm{s}}} & 0 \\ 0 & 0 & \mathrm{e}^{\mathrm{i}\varphi_{\mathrm{b}}} \end{pmatrix} \mathbf{V}^{\mathrm{T}} \begin{pmatrix} \mathrm{e}^{-\mathrm{i}\varphi_{\mathrm{u}}} & 0 & 0 \\ 0 & \mathrm{e}^{-\mathrm{i}\varphi_{\mathrm{c}}} & 0 \\ 0 & 0 & \mathrm{e}^{-\mathrm{i}\varphi_{\mathrm{t}}} \end{pmatrix} = \mathrm{e}^{\mathrm{i}\chi}\,\mathbf{V}^{\dagger}. \tag{26.75}
$$

Equation (26.75) can also be written

$$
\begin{pmatrix} \mathrm{e}^{-\mathrm{i}\varphi_{\mathrm{u}}} & 0 & 0 \\ 0 & \mathrm{e}^{-\mathrm{i}\varphi_{\mathrm{c}}} & 0 \\ 0 & 0 & \mathrm{e}^{-\mathrm{i}\varphi_{\mathrm{t}}} \end{pmatrix} \mathbf{V} \begin{pmatrix} \mathrm{e}^{\mathrm{i}(\varphi_{\mathrm{d}}-\chi)} & 0 & 0 \\ 0 & \mathrm{e}^{\mathrm{i}(\varphi_{\mathrm{s}}-\chi)} & 0 \\ 0 & 0 & \mathrm{e}^{\mathrm{i}(\varphi_{\mathrm{b}}-\chi)} \end{pmatrix} = \mathbf{V}^{*}. \tag{26.76}
$$

In Sect. 22.4 and Appendix H, exactly this type of transformation has been considered in order to bring $\mathbf{V}$ into a canonical form. Thus as a *necessary condition* for CP invariance, we find that the Kobayashi–Maskawa matrix and its complex conjugate $\mathbf{V}^{*}$ must have the same canonical form, i.e., in its canonical form, **V** must be real (Kobayashi 1973):

$$\mathbf{V} = \mathbf{V}^*. \tag{26.77}$$

For *two* families, it follows from (22.107) that this condition is *always* fulfilled: For three families, it is *only* fulfilled if the phase in the parametrization of **V** (22.108) has the value

$$\delta = 0 \quad \text{or} \quad \pi \tag{26.78}$$

Here we exclude from our discussion the experimentally unrealistic case of two quark masses being equal and/or of one of the angles ϑ_i parametrizing **V** being zero or $\pi/2$.

It remains to investigate whether the condition $\mathbf{V} = \mathbf{V}^*$ is also sufficient to ensure CP invariance in the standard model. This is, in fact, the case. If $\mathbf{V} = \mathbf{V}^*$, the Lagrange density $\mathscr{L}$ of (22.123) is invariant under the following transformation, apart from the addition of a total divergence:

$$\begin{aligned}
\text{CP:}\ \psi(x) &\to \gamma_0 \mathbf{S}(\mathrm{C})\bar{\psi}^{\mathrm{T}}(x') \qquad \text{(for all Dirac fields } \psi = \nu_{\ell \mathrm{L}}, \ell, \mathbf{q}\text{)},\\
\rho'(x) &\to \rho'(x'),\\
\mathbf{G}^{\mu}(x) &\to -\mathbf{G}_{\mu}^{\mathrm{T}}(x'),\\
\mathbf{W}^{\mu}(x) &\to -\mathbf{W}_{\mu}^{\mathrm{T}}(x'),\\
\mathbf{B}^{\mu}(x) &\to -\mathbf{B}_{\mu}(x').
\end{aligned} \tag{26.79}$$

From this result, it follows, for example, that it is impossible in the standard model to break the CP invariance if only *two* families and a single Higgs doublet are present. From the observed CP violation, Kobayashi and Maskawa (1973) concluded that there must either be more Higgs particles or more than two families. The subsequent discovery of the τ-lepton and the b-quark provided a striking confirmation of their prediction.

A phase $\delta \neq 0$ or π in the Kobayashi–Maskawa matrix is thus a source of CP violation in the standard model. No other possible sources of CP violation can be directly seen in the Lagrange density $\mathscr{L}$ (22.123). However, it has been shown by a number of authors that nonperturbative effects in QCD can lead to a CP violation (t'Hooft 1976a, b; Jackiw 1976; Callan 1976). What this amounts to is the following: The Lagrange density $\mathscr{L}$ (22.123) is replaced by an effective Lagrange density $\mathscr{L}_{\text{eff}}$ that contains an additional free parameter, the dimensionless angle ϑ:

$$\mathscr{L}_{\text{eff}}(x) = \mathscr{L}(x) + \mathscr{L}_{\vartheta}(x), \tag{26.80}$$

where

$$\mathscr{L}_{\vartheta}(x) = \frac{\vartheta\, g_s^2}{32\pi^2} \operatorname{Tr}\{\varepsilon^{\mu\nu\rho\lambda} \mathbf{G}_{\mu\nu}(x) \mathbf{G}_{\rho\lambda}(x)\}. \tag{26.81}$$

The additional term $\mathscr{L}_{\vartheta}$ obviously has odd parity and, as can easily be shown, is also odd with respect to the CP transformation (26.79). Thus $\mathscr{L}_{\vartheta}$ is even under the charge-conjugation transformation C, but on account of the CPT theorem (see below) it violates time reversal invariance T for $\vartheta \neq 0$. It can be shown that $\mathscr{L}_\vartheta$ induces an electric dipole moment d_n for the neutron—a quantity that can only be nonzero if there is P and T violation. The experimentally determined upper limit for the dipole moment of the neutron is, however, extremely small (Altarev 1981):

$$|d_n| < 6 \cdot 10^{-25}\, e \cdot \text{cm}. \tag{26.82}$$

From this one can conclude that $|\vartheta| < 10^{-9}$ (Baluni 1979; Crewther 1980). Thus the angle ϑ, which *a priori* can have any value, must in reality be very small. This fact is known as the "strong CP problem". The solution of this problem, i.e., the discovery of a natural reason for the small size of ϑ, must probably be sought beyond the standard model. A possible solution can be obtained by introducing additional Higgs particles and an additional symmetry (Peccei 1977). This led to the prediction of very light pseudoscalar particles, "axions" (Weinberg 1978; Wilczek 1978), which, however, have sofar not been observed experimentally in spite of considerable effort. It would seem that these particles, if they exist in Nature, can only couple to matter extremely weakly. They have to be what is called "invisible axions" (Kim 1979; Shifman 1980; Dine 1981; Wise 1981).

For the CP violation observed in the system of neutral K-mesons, $\mathscr{L}_\vartheta$ is almost certainly irrelevant. If we first imagine that the usual weak interaction is "switched off", then the CP eigenstates $|K_1^0\rangle$ and $|K_2^0\rangle$ (26.28) are also eigenstates of C with eigenvalues -1 and $+1$. Since $\mathscr{L}_\vartheta$ is even under charge conjugation, there can be no transitions between K_1^0 and K_2^0 and therefore no state mixing as in (26.60) can be induced. This could only occur by means of a combination of the weak interaction with $\mathscr{L}_\vartheta$. However, we would then expect

$$|\langle K_1^0|\mathscr{M}|K_2^0\rangle| \sim |\vartheta|(m_L - m_S) \lesssim 10^{-9}(m_L - m_S), \tag{26.83}$$

with which it is impossible to explain the observed strength of the CP violation (26.67).

26.2.2 CPT Invariance

The CPT theorem has already been mentioned on a number of occasions. In this section we will explicitly verify the CPT invariance of the Lagrange densities $\mathscr{L}$ (22.123) and $\mathscr{L}_{\text{eff}}$ (26.80). For convenience we write $\Theta \equiv \text{CPT}$. From the results of Sect. 4.5 we expect, at best, to be able to represent Θ by means of an *antiunitary* operator $\mathbf{V}(\Theta)$. We further recall from (4.142) that the antiunitary transform $\mathbf{A}'$ of an operator $\mathbf{A}$ is defined as

$$\mathbf{A}' = (\mathbf{V}(\Theta)\mathbf{A}\mathbf{V}^{-1}(\Theta))^{\dagger}. \tag{26.84}$$

For the Hermitian conjugate operator $\mathbf{A}^{\dagger}$ and for the product of two operators $\mathbf{A}$ and $\mathbf{B}$, it follows that

$$\begin{aligned}(\mathbf{A}^{\dagger})' &= (\mathbf{A}')^{\dagger},\\ (\mathbf{AB})' &= \mathbf{B}'\mathbf{A}'.\end{aligned} \tag{26.85}$$

We thus seek a transformation of the fields that appear in the Lagrange density $\mathscr{L}$, which, when we take note of the relations (26.85), leaves $\mathscr{L}$ invariant. For the Dirac fields we write this transformation as follows (cf. Exercise 4.5):

$$\Theta: \psi(x) \to \psi'(x) = -\gamma_5 \psi(-x) \qquad \text{for} \quad \psi = \nu_{\ell \mathrm{L}}, \ell, \mathbf{q}. \tag{26.86}$$

For $\bar{\psi}$ and for the currents we then get

$$\begin{aligned} \Theta: \bar{\psi}(x) &\to \bar{\psi}(-x)\gamma_5, \\ \mathscr{J}^{\lambda}_{\mathrm{em}}(x) &\to -\mathscr{J}^{\lambda}_{\mathrm{em}}(-x), \\ \mathscr{J}^{\lambda}_{\mathrm{NC}}(x) &\to -\mathscr{J}^{\lambda}_{\mathrm{NC}}(-x), \\ \mathscr{J}^{\lambda}_{\mathrm{CC}}(x) &\to -\mathscr{J}^{\lambda}_{\mathrm{CC}}(-x). \end{aligned} \tag{26.87}$$

If we now transform the Bose fields as follows:

$$\begin{aligned} \Theta: \rho'(x) &\to \rho'(-x), \\ \mathbf{G}^{\mu}(x) &\to -\mathbf{G}^{\mu}(-x), \\ \mathbf{W}^{\mu}(x) &\to -\mathbf{W}^{\mu}(-x), \\ \mathbf{B}^{\mu}(x) &\to -\mathbf{B}^{\mu}(-x), \end{aligned} \tag{26.88}$$

then we indeed find that the Lagrange density $\mathscr{L}_{\mathrm{eff}}$ (26.80) is invariant

$$\Theta: \mathscr{L}_{\mathrm{eff}}(x) \to \mathscr{L}_{\mathrm{eff}}(-x). \tag{26.89}$$

We still have to check whether, for the transformation defined in (26.86), (26.88), there exists an antiunitary operator $\mathbf{V}(\Theta)$ in the space of states such that

$$\begin{aligned} (\mathbf{V}(\Theta)\psi(x)\mathbf{V}^{-1}(\Theta))^{\dagger} &= -\gamma_5 \psi(-x), \\ (\mathbf{V}(\Theta)\rho'(x)\mathbf{V}^{-1}(\Theta))^{\dagger} &= \rho'(-x), \end{aligned} \tag{26.90}$$

etc.

With some very general mathematical assumptions, this turns out to be the case (see Streater 1964). From (26.86), (26.88) it is now a simple matter to obtain the transformation behavior of all operators that are constructed from the fundamental fields. Let us consider, for example, scalar or pseudoscalar currents $\mathbf{S}(x)$, vector or axial vector currents $\mathscr{J}_{\lambda}(x)$ and tensor currents $\mathbf{T}_{\lambda\rho}(x)$ which are constructed from the fields and their derivatives at the same point x. We find

$$\begin{aligned} \Theta: \mathbf{S}(x) &\to \mathbf{S}(-x), \\ \mathscr{J}_{\lambda}(x) &\to -\mathscr{J}_{\lambda}(-x), \\ \mathbf{T}_{\lambda\rho}(x) &\to \mathbf{T}_{\lambda\rho}(-x). \end{aligned} \tag{26.91}$$

This implies that every generalized charge operator $\mathbf{E}(0)$ of the form

$$\mathbf{E}(0) = \int_{t=0} \mathrm{d}^3 x \, \mathscr{J}_0(\boldsymbol{x}, t), \tag{26.92}$$

where $\mathscr{J}_\lambda(x)$ is a conserved or nonconserved current, is transformed as follows:

$$\Theta: \mathbf{E}(0) \to (\mathbf{V}(\Theta)\mathbf{E}(0)\mathbf{V}^{-1}(\Theta))^\dagger = -\mathbf{E}(0). \tag{26.93}$$

Analogously, for the four-momentum operator $\mathbf{P}_\mu$, which, according to (3.100), can be written as an integral over the energy-momentum tensor, we find the transformation behavior:

$$\Theta: \mathbf{P}_\mu \to (\mathbf{V}(\Theta)\mathbf{P}_\mu\mathbf{V}^{-1}(\Theta))^\dagger = \mathbf{P}_\mu. \tag{26.94}$$

From (26.93), (26.94) we can now state the following about particle physics: If any (conserved or nonconserved) Hermitian charge operator $\mathbf{E}(0) = \mathbf{E}^\dagger(0)$ has an eigenvalue $\kappa \neq 0$, then $-\kappa$ is also an eigenvalue, and moreover, it has the same multiplicity. To demonstrate this we let $|\kappa\rangle$ be the eigenvector with eigenvalue κ

$$\mathbf{E}(0)|\kappa\rangle = \kappa|\kappa\rangle. \tag{26.95}$$

From (26.93) it follows that

$$\mathbf{E}(0)\mathbf{V}(\Theta)|\kappa\rangle = -\kappa\mathbf{V}(\Theta)|\kappa\rangle. \tag{26.96}$$

Thus

$$|-\kappa\rangle \equiv \mathbf{V}(\Theta)|\kappa\rangle \tag{26.97}$$

is an eigenvector with eigenvalue $-\kappa$. For the matrix elements of the four-momentum operator between eigenstates $|\kappa_1\rangle$ and $|\kappa_2\rangle$ of $\mathbf{E}(0)$ it further follows from (26.94) that

$$\langle\kappa_1|\mathbf{P}_\mu|\kappa_2\rangle = \langle-\kappa_2|\mathbf{P}_\mu|-\kappa_1\rangle. \tag{26.98}$$

If we take, for example, the electric charge $\mathbf{Q}$ for the operator $\mathbf{E}$ and assume the existence of the proton with charge $+1$, then we conclude from the above that there must also be an antiproton with charge -1. If the proton is stable, i.e., if it is an eigenstate of $\mathbf{P}_\mu$, then from (26.98) the antiproton must also be stable and must have exacty the same rest mass as the proton. The extent to which this has been experimentaly confirmed has already been given in (16.50). The consequences of CPT invariance for unstable particles, in particular for the K-mesons, are discussed in Appendix I.

26.2.3 CP Violation in the Neutral K-Meson System in the Standard Model

It still remains to discuss whether the CP violation due to a phase $\delta \neq 0$ or π in the Kobayashi–Maskawa matrix is sufficient to explain the CP violation in the K^0–$\bar{K}^0$ system. Corresponding calculations within the standard model have been carried out by a number of authors (e.g., Ellis 1976a; Hagelin 1981; Ginsparg 1983; see Buras 1984 for a review). Unfortunately, the results still contain many uncertainties that stem from our lack of knowledge about the description of the K-mesons in terms of the quark and gluon constituents.

Figure 26.2 The diagrams of lowest order in the gauge couplings that bring about transitions $\bar{s}d \leftrightarrow \bar{d}s$, i.e., the transitions $K^0 \leftrightarrow \bar{K}^0$. Quarks of the u-type are represented here by $u_1 \equiv u$, $u_2 \equiv c$, $u_3 \equiv t$.

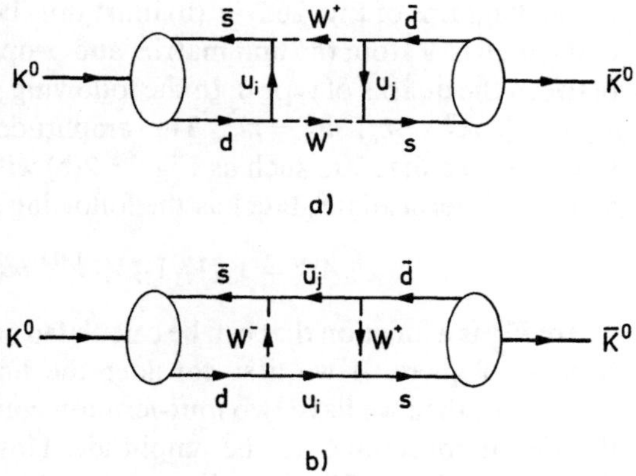

Until now it has not been possible to derive a detailed wavefunction for the K-mesons or other light hadrons from the basic principles of QCD. We therefore restrict the discussion to a few qualitative remarks about the calculations quoted above.

We consider the mass matrix $\mathcal{M}$ of the K-mesons as in (26.55). From (26.27) and using (I.66a) from Appendix I we obtain

$$\begin{aligned} \mathcal{M}_{22} - \mathcal{M}_{11} &= \langle K^0|\mathcal{M}|\bar{K}^0\rangle + \langle \bar{K}^0|\mathcal{M}|K^0\rangle, \\ \mathcal{M}_{12} &= \tfrac{1}{2}(\langle K^0|\mathcal{M}|\bar{K}^0\rangle - \langle \bar{K}^0|\mathcal{M}|K^0\rangle). \end{aligned} \tag{26.99}$$

In order to calculate the mass splitting $m_L - m_S$, which is given essentially by $\mathcal{M}_{22} - \mathcal{M}_{11}$, and the "false" CP admixtures to the states $|K_S\rangle$ and $|K_L\rangle$, which are determined by $\mathcal{M}_{12}$, we must thus look for diagrams that bring about transitions $K^0 \leftrightarrow \bar{K}^0$. In the naïve quark model, K^0 and $\bar{K}^0$ are quark–antiquark bound states of the following type (see Chap. 17):

$$K^0 \sim d\bar{s}, \qquad \bar{K}^0 \sim s\bar{d}. \tag{26.100}$$

To lowest order in the gauge couplings, the transitions $d\bar{s} \leftrightarrow s\bar{d}$ are brought about by the "box diagrams" of Fig. 26.2 (Gaillard 1974). If the quarks in the K-mesons were only weakly bound, we could content ourselves with these diagrams. In reality, however, the quarks are strongly bound and gluonic corrections to these diagrams (Fig. 26.2) are very important, although unfortunately difficult to calculate.

We now estimate the order of magnitude of the box diagrams. First, we note that these diagrams would yield no contribution if the Kobayashi–Maskawa matrix **V** were equal to the unit matrix. The d-quark would then couple only to the u-quark, and the s-quark only to the c-quark; hence the diagrams of Fig. 26.2 would give zero. The same would be true if the masses of the u-, c-, and t-quarks (i.e., of the u-type quarks) were equal. We would then be able, exactly as in the lepton sector, to set the matrix **V** equal to the unit matrix (Sect. 22.4) and thus have the above argument at hand. The sum of the diagrams of Fig. 26.2(a), (b) must thus be proportional to the deviation of the matrix **V** from the unit matrix, and proportional to the mass differences between the quarks of type u. In the following we write $u_1 \equiv u$, $u_2 \equiv c$, $u_3 \equiv t$, $m_1^2 = m_u^2$, $m_2^2 = m_c^2$, $m_3^2 = m_t^2$. The amplitude $A_{ij}^{(b)}$, which corresponds, for example, to a diagram such as Fig. 26.2(b) with a quark u_i and an antiquark $\bar{u}_j$, in the intermediate state has the following structure:

$$A_{ij}^{(b)} = V_{i2}^* V_{i1} V_{j2}^* V_{j1} F^{(b)}(m_W^2, m_i^2, m_j^2), \tag{26.101}$$

where $F^{(b)}$ is a function that can be calculated and whose behavior for $m_W \to \infty$ is easy to guess. If we first consider the limiting case of infinitely heavy W-bosons, then we have two four-fermion couplings, contributing as a factor the Fermi constant G to the amplitude. However, the loop integral of the diagram in Fig. 26.2(b) now diverges quadratically, as can easily be seen by counting the powers of the momenta in the propagators. Again by power counting we find that for a finite W mass, the integral is convergent. Thus the quadratic divergence is effectively cut off at m_W. Since the quark masses play no role in the divergent part, we therefore expect for $m_i, m_j \ll m_W$, an expansion of the form

$$\begin{aligned} &F^{(b)}(m_W^2, m_i^2, m_j^2) \\ &\quad = m_K^3 G^2 \{ f_0 m_W^2 + f_1 m_i^2 + f_2 m_j^2 + f_3 m_i m_j + O(m_W^{-2}) \}, \end{aligned} \tag{26.102}$$

where f_0, f_1, f_2, and f_3 are dimensionless constants.[2] The overall dimension has been made correct by multiplication with a suitable power of m_K. A second available quantity with the dimension of mass would be the inverse radius of the K-meson, but in fact, this is of the same order of magnitude as m_K. If we now sum over the contributions of the quarks u_i and $\bar{u}_j$, then, using the unitarity relations for **V**, we obtain

$$\begin{aligned} \sum_{i,j} A_{ij}^{(b)} &= \sum_{i,j} V_{i2}^* V_{i1} V_{j2}^* V_{j1} m_K^3 G^2 \\ &\quad \cdot \{ f_0 m_W^2 + f_1 m_i^2 + f_2 m_j^2 + f_3 m_i m_j + O(m_W^{-2}) \} \\ &= m_K^3 G^2 \sum_{i,j} V_{i2}^* V_{i1} V_{j2}^* V_{j1} \{ f_3 m_i m_j + O(m_W^{-2}) \}. \end{aligned} \tag{26.103}$$

An exactly analogous estimate is valid for the diagrams of Fig. 26.2(a). With the explicit parametrization of **V** from (22.108) we find that all terms in (26.103) are proportional to $\sin^2 \vartheta_1$.

We now consider the K_L–K_S mass difference. This quantity is nonzero even for the case of two families. Therefore we ignore the third generation, which, in any case, hardly mixes with the first two generations (Table 23.6) and, since $m_u \cong 0$, we obtain the estimate

$$\frac{m_L - m_S}{m_K} \sim \frac{\langle K^0 | \mathcal{M} | \bar{K}^0 \rangle}{m_K} \sim \sin^2 \vartheta_1 G^2 m_K^2 m_c^2 \cong 2 \cdot 10^{-12}. \tag{26.104}$$

[2] This is a highly simplified discussion which is only valid when all the m_i are approximately equal. In reality one must put $m_u \cong 0$ and $m_c \ll m_t$, which leads to the appearance of logarithms in the expansion of the amplitudes.

Considering the crudeness of our calculation this does not agree too badly with the experimental value (Table 26.1).

We turn now to the parameters related to CP violation. As has already been discussed, it is essential for the occurrence of a CP violation within the standard model, that *all three* families of quarks mix, and that $\delta \neq 0$ or π. For angles $\vartheta_2 = 0$ or $\vartheta_3 = 0$ one readily sees from (22.108) that the phase δ can be rotated away by a transformation of the type (22.102). Thus for the matrix element $\mathscr{M}_{12}$ that causes the CP mixing, we expect, in comparison to $m_L - m_S$, additional factors of $\sin\vartheta_2$, $\sin\vartheta_3$ and $\sin\delta$:

$$\frac{\mathscr{M}_{12}}{m_L - m_S} \sim \sin\vartheta_2 \cdot \sin\vartheta_3 \cdot \sin\delta. \tag{26.105}$$

The experimental values of the angles ϑ_2 and ϑ_3 are very small as can be seen from the values of $|V_{13}|$ and $|V_{31}|$ in Table 23.6. According to Kleinknecht (1983) one has

$$\begin{aligned} 0.015 < \sin\vartheta_2 < 0.09, \\ \sin\vartheta_3 < 0.04. \end{aligned} \tag{26.106}$$

With these we obtain the following estimate for η_{+-} and η_{00} (26.47), (26.48):

$$|\eta_{+-}| \sim |\eta_{00}| \sim |\varepsilon| \sim \left|\frac{\mathscr{M}_{12}}{m_L - m_S}\right| \sim \sin\vartheta_2 \cdot \sin\vartheta_3 \cdot \sin\delta \lesssim 3 \cdot 10^{-3} \cdot \sin\delta \tag{26.107}$$

The observed strength of the CP violation $|\eta_{+-}| \cong |\eta_{00}| \cong 2 \cdot 10^{-3}$ could thus be "explained" in the standard model by the weak mixing of the three quark families. The CP violating phase δ certainly does not have to be close to 0 or π. In fact, it would appear that δ must be close to $\pi/2$.

A further important point is that in the standard model there is no special reason why the parameter ε'/ε should be equal to zero. In this respect the standard model displays an essential difference from the superweak model of CP violation (26.62). Initially, theoretical calculations in the standard model led to a value of about 10^{-2} for ε'/ε, however with large uncertainties. The recent experimental result for ε'/ε (Table 26.1) is somewhat smaller than these original theoretical expectations but certainly no claim of incompatibility with the standard model can be made. The result $\varepsilon'/\varepsilon \neq 0$ does, however, rule out the model where a superweak interaction is the only source of CP violation [cf. (26.64)].

To conclude this chapter we mention that similar phenomena to those in the K^0–$\bar{K}^0$ system should also occur in the following systems of mesons with c-, b-, or t-quarks:

$$\begin{gathered} D^0(\sim c\bar{u}) - \bar{D}^0(\sim u\bar{c}), \\ B^0(\sim d\bar{b}) - \bar{B}^0(\sim b\bar{d}), \\ B^0_s(\sim s\bar{b}) - \bar{B}^0_s(\sim b\bar{s}), \\ T^0_u(\sim t\bar{u}) - \bar{T}^0_u(\sim u\bar{t}), \\ T^0_c(\sim t\bar{c}) - \bar{T}^0_c(\sim c\bar{t}). \end{gathered}$$

Simple considerations show that in comparison to K-mesons the effects should be smaller for these systems. As an example we consider the D^0–$\bar{D}^0$ system. What final states can occur in the decay both of D^0 and of $\bar{D}^0$? If we work with the simple spectator model (Sect. 23.6) at the quark level, then the decay processes of the c-quark given in (23.54) induce a whole series of possible decay processes for the D^0-meson. Of these, only the following lead to final states that also occur in $\bar{D}^0$ decay:

$$\begin{aligned} D^0 &\sim c\bar{u} \rightarrow d\bar{d}u\bar{u},\ s\bar{s}u\bar{u}, \\ \bar{D}^0 &\sim u\bar{c} \rightarrow d\bar{d}u\bar{u},\ s\bar{s}u\bar{u}. \end{aligned} \tag{26.108}$$

However, the Kobaysahi–Maskawa matrix elements appearing in (23.54) cause these channels to be strongly suppressed. The "communication" between D^0 and $\bar{D}^0$ should thus be very slight. Therefore we expect the decay time of a D^0 or $\bar{D}^0$ to be much shorter than the oscillation time for the transitions $D^0 \leftrightarrow \bar{D}^0$. In fact, no experiments have yet succeeded in observing D^0–$\bar{D}^0$ oscillations. This would, however, be a prerequisite for investigating whether the eigenstates of the mass matrix of the D^0–$\bar{D}^0$ system are also CP eigenstates. Even the direct comparison of the types of D^0 decay with the corresponding CP-transformed decay processes of the $\bar{D}^0$, has not yet produced any evidence of a CP violation.

For the B-mesons, one expects larger state mixing effects than for the D-mesons (see, e.g., Paschos 1983, 1984; Wolfenstein 1984). The B_s^0- and $\bar{B}_s^0$-mesons, for example, have, on the quark level, the following decay channels in common:

$$\begin{aligned} B_s^0 &\sim s\bar{b} \rightarrow s\bar{c}c\bar{s}, \\ \bar{B}_s^0 &\sim b\bar{s} \rightarrow c\bar{c}s\bar{s}. \end{aligned} \tag{26.109}$$

which are among the most important decay processes of these particles [see (23.61)]. The decay times and oscillation times of these mesons should thus be of approximately the same size. Indeed, recent experiments indicate sizable mixing effects for B_d and B_s mesons (Albajar 1987, Albrecht 1987). Despite this, it is believed that it will be very difficult to observe effects of CP violation in these systems.

Exercises

26.1 Calculate the normalization of the states defined in (26.40), (26.41), (26.43) [see also (16.63)].

26.2 The system of neutral K-mesons can be used to study the question of whether K^0 and $\bar{K}^0$ fall at the same rate in a gravitational field as is demanded by the equivalence principle of general relativity (Good 1961; for a detailed analysis and further references see Nachtmann 1969). Assume that a K^0 particle gains, as usual, an additional energy of $m_K U$ in a gravitational potential U, but that a $\bar{K}^0$ particle gains an energy of $(1-\eta)m_K U$ where η is a parameter. For $\eta = 2$, the potential energy of a $\bar{K}^0$ particle in a gravitational field would be exactly the opposite to that of a K^0; the $\bar{K}^0$ would "fall upwards". For U, use in turn the gravitational potential of the Earth and that of the Sun and determine the limits for η from the observations on the K^0–$\bar{K}^0$ system.

26.3 Show that left-handed neutrino fields remain left-handed under the transformation (26.68).

26.4 Consider scalar currents $\mathbf{S}(x)$, vector currents $\mathcal{J}_\lambda(x)$ and tensor currents $\mathbf{T}_{\lambda\rho}(x)$, $\mathbf{M}_{\lambda\rho\tau}(x)$, etc., which are constructed from the fundamental fields and their derivatives at the point x, e.g., $\mathbf{S}(x) = \bar{\psi}(x)\psi(x)$, $\bar{\psi}(x)\gamma_5\psi(x)$, $\mathrm{Tr}(\mathbf{G}_{\mu\nu}(x)\mathbf{G}^{\mu\nu}(x))$, $\mathcal{J}_\lambda(x) = \bar{\psi}(x)\gamma_\lambda\psi(x)$, $\bar{\psi}(x)\gamma_\lambda\gamma_5\psi(x)$, $\mathrm{Tr}(\mathbf{G}_{\mu\nu}\partial_\lambda\mathbf{G}^{\mu\nu}(x))$, etc. Prove (26.91) for such currents and discuss the transformation behavior of tensor currents of higher rank.

26.5 According to the results of Exercise 3.8, the angular momentum operators can be represented as integrals over tensor currents of rank three. What does this imply for their behavior under the CPT transformation? Derive from this, and from (26.98), how the CPT transformation acts on one-particle states and compare with the product of the C, P, and T-transformations as given in Sect. 16.2.

CHAPTER 27

Order and Disorder in Elementary Particle Physics

In the previous chapters we have obtained a detailed view of the principles of the standard model and of its successes in comparison with experiment. All presently available experimental results seem to be compatible with the standard model, which thus seems to provide a correct description of particle physics, at least up to center of mass energies of about 100 GeV. What can we learn from this and what are the open questions?

The standard model brought order into the description of the *forces* that act between particles. An essential discovery was that the strong, electromagnetic, and weak forces can be derived from the postulate of gauge symmetries. The corresponding gauge group $\mathscr{F} = \mathrm{SU}(3) \times \mathrm{SU}(2) \times \mathrm{U}(1)$ is recognized as a fundamental symmetry group of Nature (see Sect. 22.5). Three coupling constants, corresponding to the three factors of $\mathscr{F}$, determine the strengths of these forces:

$$\begin{aligned} \mathrm{SU}(3)&: \quad g_3 \equiv g_s, \\ \mathrm{SU}(2)&: \quad g_2 \equiv g = e/\sin \vartheta_W, \\ \mathrm{U}(1)&: \quad g_1 \equiv g' = e/\cos \vartheta_W. \end{aligned} \tag{27.1}$$

Experimentally, we found at an energy scale of about 5 GeV [see (20.34), (24.35)]:

$$\begin{aligned} \frac{g_3^2}{4\pi} &\cong 0.16, \\ \frac{g_2^2}{4\pi} &= \frac{\alpha}{\sin^2 \vartheta_W} \cong 0.032, \\ \frac{g_1^2}{4\pi} &= \frac{\alpha}{\cos^2 \vartheta_W} \cong 0.009. \end{aligned} \tag{27.2}$$

One of the main questions of contemporary elementary particle physics concerns the origin of the symmetry group $\mathscr{F}$. Is $\mathscr{F}$ really a fundamental symmetry group that cannot be derived from simpler principles? If this is so, then the observed values of g_1, g_2, and g_3 would have no further explanation. But perhaps $\mathscr{F}$ is part of a larger symmetry group that becomes visible at higher energies and within which g_1, g_2, and g_3 are related to one another. In general, we might ask about the symmetries that govern physics at center of mass energies much larger than 100 GeV. Will it eventually be possible to include the gravitational force, too, in a unified scheme? On what energy scale does gravity begin to play a role in the interactions between individual elementary particles? We will discuss this last question in somewhat more detail.

We set ourselves the task of resolving structures of a given linear dimension Δx in an elementary particle, for example, the proton. For this purpose we must expose the particle to a probe whose wavelength is smaller than or equal to Δx. As a concrete example we might think of the deep inelastic electron–proton scattering (Chap. 18). For the momentum p and energy E of the particle-probe we then have

$$E \cong p \gtrsim \frac{1}{\Delta x}. \tag{27.3}$$

We work here in the extreme relativistic limit since we are interested in very small lengths Δx, corresponding to very large momenta. However, such a particle-probe also produces a gravitational potential U of magnitude

$$U \sim \frac{G_N E}{\Delta x} \gtrsim \frac{G_N}{(\Delta x)^2}. \tag{27.4}$$

Here G_N is Newton's gravitational constant. According to general relativity this causes the metric in the space–time continuum to be changed. For example, for the component g_{00} of the metric tensor (see, e.g., Weinberg 1972b, Straumann 1984) one has

$$g_{00} \cong 1 + 2U. \tag{27.5}$$

The particle-probe will influence the metric at the position of the target particle when U is of the order of 1, i.e., when

$$U \sim \frac{G_N}{(\Delta x)^2} \sim 1, \qquad \Delta x \sim (G_N)^{1/2}. \tag{27.6}$$

Using the gravitational constant, one thus defines a length and a mass known as the Planck length and Planck mass:

$$l_{\text{Planck}} = (m_{\text{Planck}})^{-1} = (G_N)^{1/2}. \tag{27.7}$$

From the numerical value of G_N, which in conventional units is

$$G_N = 6.67 \cdot 10^{-8}\ \text{cm}^3\ \text{g}^{-1}\ \text{s}^{-2} \tag{27.8}$$

one obtains, after a simple conversion into our present units (see Chap. 1)

$$\begin{aligned} l_{\text{Planck}} &= 1.61 \cdot 10^{-33}\,\text{cm}, \\ m_{\text{Planck}} &= 1.22 \cdot 10^{19}\,\text{GeV}. \end{aligned} \tag{27.9}$$

At these distances and energies, the effect of gravitation on elementary particles must certainly be taken into account.

We return now to the standard model. For the forces, a very extensive though not yet completely satisfactory "order" or "degree of understanding" has been established: The forces can be "derived" from a physical principle, the gauge principle. For the spectrum of the leptons and quarks and for their properties, it is quite a different matter, not to mention the Higgs sector where our theoretical understanding is least complete. Many questions remain unsolved here: Why do both leptons *and* quarks exist and why are there three families? How can we understand the hypercharge Y, which in Chap. 22, had to be adjusted by hand in order to obtain the observed electrical charges Q? The standard model yields, for example, no explanation for the fact that electron and proton have charges of equal magnitude. Experimentally, this has been verified to an accuracy of 1 in 10^{21} (Dylla 1973). Can the observed masses of the leptons and quarks and the values of the Kobayashi–Maskawa matrix elements be derived theoretically? Here, the standard model has almost no predictive power.

The standard model also yields a consistency condition for the quantum numbers of the fermions, as will now be discussed (Bouchiat 1972; Korthals Altes 1972; Gross 1972). We take as an example the $Z\gamma\gamma$ vertex. According to the Feynman rules of QFD (Appendix G), there is no elementary vertex for this. But one needs only diagrams with a single loop, so-called triangle diagrams (Fig. 27.1), in order to obtain such a coupling. These triangle diagrams have "peculiar" properties; they lead to so-called anomalies (Adler 1969; Bell 1969). For our problem, we can argue qualitatively as follows: The coupling of three gauge bosons to one another is fixed by the gauge symmetry (Chap. 22). The diagram of Fig. 27.1 actually vanishes for a real Z and real photons since the Z, as a spin-1 particle, cannot decay into two real photons (see Chap. 13). However, as soon as we consider the decay of a virtual Z-boson into two real photons, the very same diagram yields a nonvanishing contribution that stems from the *axial-vector* coupling of the Z. It turns out that this leads to a serious problem for the theory, namely that its renormalizability

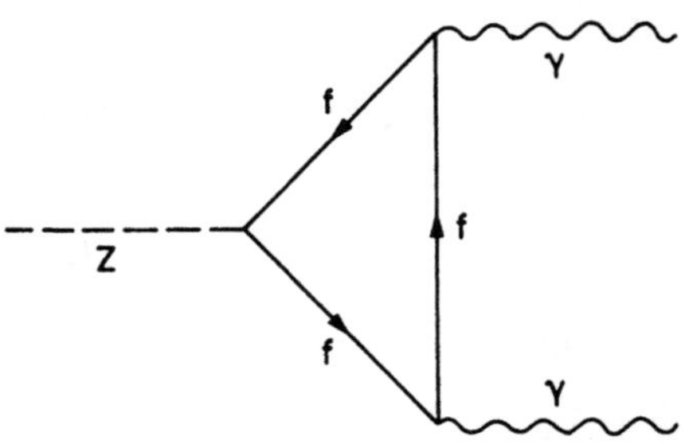

Figure 27.1 The triangle diagram that leads to a $Z_{\gamma\gamma}$ coupling. All fundamental fermions contribute in the loop.

is lost. A nonvanishing contribution from diagrams of the type shown in Fig. 27.1 presents us, essentially, with a $Z\gamma\gamma$ vertex which would be incompatible with the gauge symmetry and which—if used again in other diagrams—would lead to one new infinity after another. The only escape from this that has been suggested to date is to demand that the *sum* of all contributions of the fermions f in the triangle diagram disappears. If, for the coupling of the fermions to the Z-boson, we use the notation of (24.40), then this condition can be shown to read

$$\sum_{\mathrm{f}} g_{\mathrm{A}}^{\mathrm{f}} Q_{\mathrm{f}}^2 N_{\mathrm{c}}^{\mathrm{f}} = 0, \tag{27.10}$$

where $N_{\mathrm{c}}^{\mathrm{f}}$ is the number of colors ($N_{\mathrm{c}}^{\mathrm{f}} = 1$ for fermions and $N_{\mathrm{c}}^{\mathrm{f}} = 3$ for quarks). We now put, as in the standard model, $g_{\mathrm{A}}^{\mathrm{f}} = T_{3\mathrm{L}}^{\mathrm{f}}$ and $Q_{\mathrm{f}} = T_{3\mathrm{L}}^{\mathrm{f}} + Y_{\mathrm{L}}^{\mathrm{f}}$, where $T_{3\mathrm{L}}^{\mathrm{f}}$ and $Y_{\mathrm{L}}^{\mathrm{f}}$ are the eigenvalues of the third isospin component and of the hypercharge for the left-handed fermions. With this we obtain

$$\begin{aligned} \sum_{\mathrm{f}} g_{\mathrm{A}}^{\mathrm{f}} Q_{\mathrm{f}}^2 N_{\mathrm{c}}^{\mathrm{f}} &= \sum_{\mathrm{f}} T_{3\mathrm{L}}^{\mathrm{f}} (T_{3\mathrm{L}}^{\mathrm{f}} + Y_{\mathrm{L}}^{\mathrm{f}})^2 N_{\mathrm{c}}^{\mathrm{f}} \\ &= \sum_{\mathrm{f}} 2 (T_{3\mathrm{L}}^{\mathrm{f}})^2 Q_{\mathrm{f}} N_{\mathrm{c}}^{\mathrm{f}} = 0. \end{aligned} \tag{27.11}$$

Here we use the facts that $Y_{\mathrm{L}}^{\mathrm{f}}$ has a fixed value for every isomultiplet and that the sums over $T_{3\mathrm{L}}^{\mathrm{f}}$ and $(T_{3\mathrm{L}}^{\mathrm{f}})^3$ vanish. In the standard model, all left-handed fermions are arranged in isodoublets and thus $T_{3\mathrm{L}}^{\mathrm{f}} = \pm\frac{1}{2}$. Equation (27.11) can therefore be written in the simple form

$$\sum_{\mathrm{f}} Q_{\mathrm{f}} N_{\mathrm{c}}^{\mathrm{f}} = 0. \tag{27.12}$$

The sum of the charges of the fermions, weighted by their multiplicities in color space, must give zero. As can easily be checked, (27.12) is satisfied separately by each family of fermions. Here we consider the first family ν_{e}, e, u, d (Table 22.1). We find

$$Q_{\nu_{\mathrm{e}}} + Q_{\mathrm{e}} + 3(Q_{\mathrm{u}} + Q_{\mathrm{d}}) = 0 + (-1) + 3(\tfrac{2}{3} - \tfrac{1}{3}) = 0.$$

Within the standard model the family-wise compensation looks like a "coincidence" since (27.12) must only be valid upon summation over the fermions of *all* families.

In the following we will briefly discuss a few theories that have been suggested as extensions of the standard model. These theories have the common feature that at center of mass energies $\sqrt{s} \lesssim 100$ GeV, they reproduce the predictions of the standard model sufficiently accurately to be compatible with current experimental results. At higher energies, however, the individual theories produce radically differing predictions.

27.1 Grand Unification

One assumes here that the symmetry group $\mathscr{F}$ of the standard model is part of a larger group $\mathscr{G}$, which only becomes manifest at very high energies—in most theories, only at energies of 10^{15}–10^{19} GeV. In the ideal case, the group

$\mathcal{G}$ should allow only *one* independent coupling constant when used as a gauge group. This is the idea behind the grand unified theories (GUT) (Pati 1973; Georgi 1974). The large group $\mathcal{G}$ should reduce to the group $\mathcal{F}$ through spontaneous symmetry breaking at a mass scale m_X. The group $\mathcal{F}$ is, in turn, broken down—as in the standard model—at a mass scale $m_Z \cong 100$ GeV to the group $SU(3) \times U_{em}(1)$ (cf. Sect. 22.5):

$$\begin{array}{l} \mathcal{G} \\ \big\downarrow {\scriptstyle m_X \gtrsim 10^{15}\,\mathrm{GeV}} \\ \mathcal{F} = SU(3) \times SU(2) \times U(1) \\ \big\downarrow {\scriptstyle m_Z \cong 10^{2}\,\mathrm{GeV}} \\ SU(3) \times U_{em}(1) \end{array}$$

In this scenario the group $\mathcal{F}$ would be the manifest symmetry group from $m_Z \cong 10^2$ GeV up to $m_X \cong 10^{15}$ GeV. In this range there should be no new physics—it would be a "desert".

The simplest and most elegant model of this type is the SU(5) model of Georgi and Glashow (Georgi 1974), which initially seemed very promising. It was possible, for example, to reduce the three coupling constants g_1, g_2, and g_3 to a single constant, and a number of ratios between quark and lepton masses could be correctly calculated. For the coupling constants, the argument runs approximately as follows: Let us assume that at energies $E \gtrsim m_X$ all forces are united and characterized by *one* coupling constant g_{GUT}. At m_X, spontaneous symmetry breaking should begin for which we postulate a Higgs mechanism in analogy to that in the standard model. Below m_X the manifest symmetry group is $\mathcal{F}$ and the strengths of the forces are characterized by three parameters g_1, g_2, and g_3. We have already seen in Chap. 19 that g_3 should be taken as an effective parameter that depends on the energy (19.25). In exact analogy, one finds that g_1 and g_2, too, must be treated as energy-dependent effective coupling parameters whose expansion as a function of energy can be calculated from perturbation theory. At an energy scale m_X, all three coupling strengths g_1, g_2, and g_3 should be expressible in terms of g_{GUT} where, of course, the exact relations are dependent on the form assumed for the large group $\mathcal{G}$. In the SU(5) model the relations read

$$\frac{5}{3}\frac{g_1^2(m_X^2)}{4\pi} = \frac{g_2^2(m_X^2)}{4\pi} = \frac{g_3^2(m_X^2)}{4\pi} = \frac{g_{GUT}^2(m_X^2)}{4\pi}. \tag{27.13}$$

One must now try to find an energy scale m_X and a value g_{GUT} for which the observed values (27.2) of $g_i^2/4\pi$ $(i = 1, 2, 3)$ are reproduced. In the SU(5) model this is indeed possible for $m_X \cong 10^{14}$–10^{15} GeV and $g_{GUT}^2(m_X^2)/4\pi \cong 1/45$. The corresponding evolution of the coupling parameters is shown in Fig. 27.2. It should be noted that g_3 and g_2 decrease with increasing energy—the corresponding theories are asymptotically free—whereas g_1 increases. This yields a natural explanation for the observed order of the coupling strengths in (27.2). The large value of m_X can be attributed to the fact that the coupling

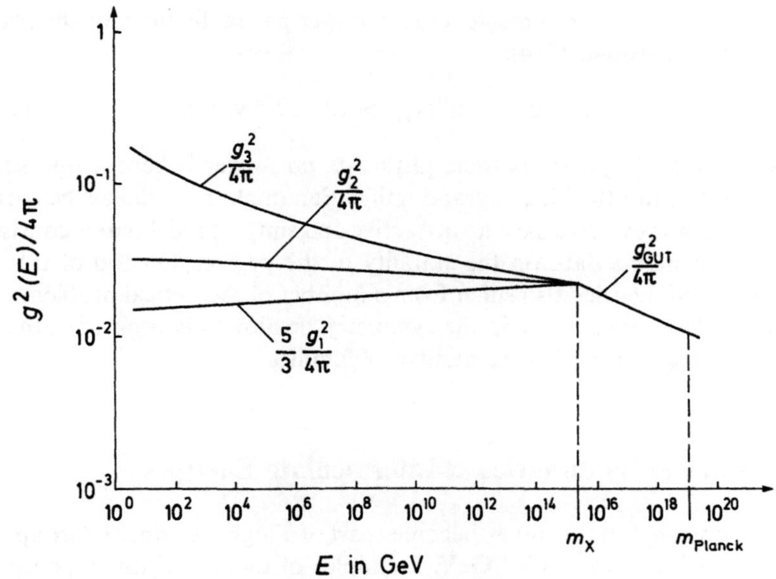

Figure 27.2 The effective coupling parameters $g_i^2(E^2)/4\pi$ ($i = 1, 2, 3$) and $g_{GUT}^2(E^2)/4\pi$ as functions of the energy E according to the SU(5) model (schematic).

parameters only vary logarithmically [see (19.25)], so that the strong and electroweak forces have the same strength only at very high energies.

Within this theory g_1, g_2, and g_3 are thus described by two parameters (m_X and g_{GUT}) and we therefore obtain *one* relation between them as a prediction of the theory. For $\sin^2 \vartheta_W$ this relation yields (see Ross 1984 and references therein)

$$\sin^2 \vartheta_W = 0.214 \begin{cases} +0.006 \\ -0.004 \end{cases}. \tag{27.14}$$

This agrees amazingly well with the experimental value (24.35) and is regarded as a great triumph of the SU(5) model.

In the SU(5) model leptons and quarks are jointly grouped into SU(5) multiplets. The gauging of the SU(5) group with the subsequent implementation of spontaneous symmetry breaking leads to gauge bosons of a mass of approximately m_X whose emission causes a quark to change into a lepton. In this way, the most spectacular of the predictions of the SU(5) model, that of the instability of the proton, is obtained. In particular, it was possible to reliably estimate the decay rate for the channel

$$p \to e^+ + \pi^0$$

(see Langacker 1981; Berezinsky 1981; Brodsky 1984) to obtain

$$\Gamma^{-1}(p \to e^+ \pi^0)|_{\text{theor.}} \cong 10^{28 \pm 1.3} \text{ years.} \tag{27.15}$$

Experiments, however, indicate a much larger partial lifetime for the proton (Jones 1984; Koshiba 1984):

$$\Gamma^{-1}(\mathrm{p} \to \mathrm{e}^+ \pi^0)|_{\mathrm{exp.}} > 1.5 \cdot 10^{32} \text{ years.} \tag{27.16}$$

As a result of this, most particle physicists no longer believe in the simple SU(5) model. But the idea of grand unification might nonetheless be correct. At present, however, we have no attractive and simple model that is consistent with the available data on the stability of the proton. On top of this, the grand unified theories also suffer from a number of theoretical problems. For example, they cannot explain the symmetry breaking hierarchy, i.e., the fact that $m_Z \ll m_X$, nor predict the number of families.

27.2 Further Symmetries at Intermediate Energies

It is possible that the group $\mathscr{F}$ becomes part of a higher symmetry group even at energies lower than 10^{15} GeV. Examples of models of this type are the so-called left–right symmetric models in which the symmetry breaking evolves through the following stages (Pati 1975; Mohapatra 1975; Senjanovic 1975)

$$\begin{array}{l} \downarrow {\scriptstyle ?} \\ \mathrm{SU}(3) \times \mathrm{SU_R}(2) \times \mathrm{SU_L}(2) \times \mathrm{U}(1) \\ \downarrow {\scriptstyle m_{\mathrm{W_R}} \cong 10^4\ \mathrm{GeV}} \\ \mathscr{F} = \mathrm{SU}(3) \times \mathrm{SU_L}(2) \times \mathrm{U}(1) \\ \downarrow {\scriptstyle m_{\mathrm{W_L}} \cong m_Z \cong 10^2\ \mathrm{GeV}} \\ \mathrm{SU}(3) \times \mathrm{U_{em}}(1). \end{array}$$

The isospin group and the W-bosons of the standard model are denoted by $\mathrm{SU_L}(2)$ and $\mathrm{W_L}$, respectively, since, here only the left-handed fermions carry nonvanishing isospin and the ordinary W-bosons couple to a purely left-handed current. In the left-right symmetric models one assumes the existence of a group $\mathrm{SU_R}(2)$ with respect to which the right-handed fermions have nonvanishing charge. Correspondingly, one has bosons $\mathrm{W_R}$ that couple to right-handed charged currents. The group $\mathrm{SU_R}(2)$ is supposed to be spontaneously broken at a scale of the order of the mass $m_{\mathrm{W_R}}$ of the $\mathrm{W_R}$-bosons. Usually, one again postulates a suitable Higgs mechanism. The mass $m_{\mathrm{W_R}}$ must be large enough to ensure that the corresponding Fermi constant $G_\mathrm{R} \sim \alpha/m^2_{\mathrm{W_R}} \sim G m^2_{\mathrm{W_L}}/m^2_{\mathrm{W_R}}$ is sufficiently small to be compatible with the experimental upper bound for right-handed charged currents.

Let us consider, for example, μ-decay (Sect. 23.1). If right-handed currents were present we would typically (although not necessarily) expect the Michel parameter ρ to deviate from the value $\frac{3}{4}$:

$$\rho = \tfrac{3}{4} + \mathrm{O}\left(\frac{G_\mathrm{R}^2}{G^2}\right). \tag{27.17}$$

Experimentally one finds (23.8):

$$|\rho - \tfrac{3}{4}| \leq 0.005. \tag{27.18}$$

From this we conclude

$$\frac{G_R^2}{G^2} \cong \left(\frac{m_{W_L}}{m_{W_R}}\right)^4 \lesssim 0.005,$$
$$m_{W_R} \gtrsim 3.8\, m_{W_L} \cong 300 \text{ GeV}. \tag{27.19}$$

A more precise analysis of many experimental data has confirmed this lower bound (see Bég 1982 for a review). A significantly higher limit of

$$m_{W_R} \gtrsim 21 m_{W_L} \cong 1700 \text{ GeV}$$

was suggested by an analysis of the K^0–$\bar{K}^0$ system in the framework of left–right symmetric models (Beall 1982).

A noteworthy feature of the left–right symmetric models is their capability of spontaneously producing the observed parity violation. From an aesthetic point of view this is most satisfying. The basic equations would be fully symmetric and only the vacuum state—due to the nonvanishing vacuum expectation value of a suitable Higgs field—singles out right from left. Even C and CP violation could be spontaneously produced in this way. Models of this type have also led to interesting predictions for the electric dipole moment of the neutron (Ecker 1983).

Another category of theories with new symmetries at intermediate energies are models in which the quarks and leptons are composite. The idea behind these is that we now know of too many leptons and quarks for them all to be truly fundamental particles. It is thus tempting to postulate yet another level of *subparticles* from which the quarks and leptons are formed as bound states. An example of a detailed model of this type is the Harari–Shupe model (Harari 1979, 1981; Shupe 1979). The subparticles are assumed to be bound together by a completely new strong interaction. The new gauge group corresponding to this is usually termed the "metacolor" or "hypercolor" group. The scale of this interaction might even lie in the region 200–300 GeV. In such theories, the W- and Z-bosons are often also composite objects. The gauge group $SU_L(2) \times U(1)$ of the electroweak interaction is then simply an approximation and not a fundamental symmetry group. Such models predict deviations—albeit only slight ones—in the properties of the W- and the Z-bosons from those expected in the standard model (see Harari 1984 for a review). The results of the experiments that will be performed at the new storage rings SLC and LEP are expected to shed some light on the validity of such concepts.

27.3 Supersymmetry, Strings, and Superstrings

Supersymmetry (SUSY) was introduced into particle physics by Akulov, Volkov, Wess, and Zumino (Akulov 1972; Wess 1974a, b). SUSY is a symmetry relating bosons and fermions, which are grouped into supermultiplets.

In Nature we know of a whole series of fundamental fermions and bosons (Table 22.1). It would be very satisfactory if SUSY enabled us to combine a few of these to form a supermultiplet. Unfortunately, this is not the case. Within the present SUSY models the number of particles must, roughly speaking, be *doubled*. For the known fundamental fermions (leptons or quarks) there should be as yet undiscovered spin-zero partners, the so-called sfermions (sleptons or squarks). The known bosons (photons, gluons, W-, Z-, and Higgs particles) should, in analogy, be accompanied by spin-$\frac{1}{2}$ partners, the so-called "bosinos" (photinos, gluinos, winos, zinos, higgsinos). If SUSY were an exact spectrum symmetry, the particles and their respective superpartners would have to have the same mass. This is certainly not the case in Nature and, therefore, SUSY most be broken. The mass scale m_{SUSY}, at which SUSY is broken should be a rough guide to the expected masses of the superparticles. Unfortunately, the theory has not yet succeeded in determining m_{SUSY}. We must rely here on experiments, but these, as yet, provide no evidence in support of SUSY.

Although it is unclear if, and when, evidence for SUSY will be found experimentally, SUSY models have been studied extensively. Such models have many attractive features. For example, they often lead to better high energy behavior and renormalization properties than conventional theories. Eventually, it might even be possible to construct field theories with no divergences. Finally, SUSY gauge theories open up the possibility of uniting gravitation with the other interactions. The final goal of such a supergravitation theory would be to derive all phenomena from one theory with only a single parameter—the Planck mass (27.7). This goal, however, would not yet seem to be in sight. For further information about SUSY and its applications we refer the reader to Wess (1983), Haber (1985), Ross (1984) and the references therein.

Currently very popular are string, and especially superstring, theories. In these, one makes a radical step and considers as basic objects, instead of space–time points and field variables defined at points, *extended* objects, i.e., strings. The length of the strings is, however, thought to be only of the order of the Planck length (27.9), and for more accessible distances, one recovers with sufficient precision the ordinary field theories. The hope in this approach is that the strings existing at the Planck length will allow a successful marriage of quantum theory and gravitation and, at the same time, determine in a unique way, in the sense of a "low" energy approximation, the field theory of elementary particles relevant at our present energies. In this approach, the field theory is expected to be a supergravity theory. For further reading we refer to Green (1987).

27.4 Order Out of Chaos

In discussing the various possibilities for extending the standard model, we have assumed until now that the effective symmetry *increases* at higher energies. But it is not impossible to imagine the opposite, namely, that at high energies completely chaotic behavior is the rule, and that the symmetries observed at low energy are little more than an illusion. We are familiar with a similar situation in solid state physics. Think, for example, of an amorphous piece of material such as glass. On an atomic scale, this displays neither translational nor rotational symmetry. On a macroscopic scale, however, glass appears to be homogeneous and isotropic—symmetries appear as "infrared phenomena."

In the particle world things could be similar. The idea of understanding the symmetries of particle physics as infrared phenomena is not a particularly new one (see Wilson 1974). This direction of thought has been further developed by H.B. Nielsen (Nielsen 1985). He has, in fact, been able to show that certain models do possess the properties outlined above. The eventual aim of this approach would be to *derive* all observed symmetries, including the Lorentz invariance of special relativity. The basic postulate upon which the entire theory could be constructed might read: *In the beginning there was chaos.*

The further development of elementary particle physics is thus far from predictable, as we hope to have shown here. It might lead to a "desert", to new symmetries, or even to chaos. It will be very interesting to see what new phenomena Nature will display as we investigate her elementary particles at ever higher energies and ever shorter distances.

APPENDIX A

Dirac Matrices and Spinors

Defining relation:

$$\gamma^\mu \gamma^\nu + \gamma^\nu \gamma^\mu = 2g^{\mu\nu}. \tag{A.1}$$

Definitions:

$$\not{a} = a^\mu \gamma_\mu, \tag{A.2}$$

$$\gamma_5 = \mathrm{i}\gamma^0\gamma^1\gamma^2\gamma^3, \tag{A.3}$$

$$\sigma^{\mu\nu} = \frac{\mathrm{i}}{2}(\gamma^\mu\gamma^\nu - \gamma^\nu\gamma^\mu). \tag{A.4}$$

Relations:

$$\gamma^\mu\gamma_\mu = 4, \tag{A.5}$$

$$\gamma^\mu \not{a} \gamma_\mu = -2\not{a}, \tag{A.6}$$

$$\gamma^\mu \not{a}\not{b} \gamma_\mu = 4(a, b), \tag{A.7}$$

$$\gamma^\mu \not{a}\not{b}\not{c} \gamma_\mu = -2\not{c}\not{b}\not{a}, \tag{A.8}$$

$$\gamma^\mu\gamma_5 + \gamma_5\gamma^\mu = 0, \tag{A.9}$$

$$\gamma_5\gamma_5 = 1. \tag{A.10}$$

From (A.1) it immediately follows for the trace that

$$\mathrm{Tr}(\gamma^\mu\gamma^\nu) = 4g^{\mu\nu}. \tag{A.11}$$

The trace of an odd number of γ-matrices vanishes:

$$\mathrm{Tr}(\not{a}_1\not{a}_2 \dots \not{a}_{2n+1}) = 0. \tag{A.12}$$

To prove this we use the matrix γ_5, for which from (A.9), (A.10) we have

$$\begin{aligned} \mathrm{Tr}(\not{a}_1\not{a}_2 \ldots \not{a}_{2n+1}) &= \mathrm{Tr}(\gamma_5\not{a}_1\gamma_5\gamma_5\not{a}_2\gamma_5 \ldots \gamma_5\not{a}_{2n+1}\gamma_5) \\ &= (-1)^{2n+1}\,\mathrm{Tr}(\not{a}_1\not{a}_2 \ldots \not{a}_{2n+1}). \end{aligned} \tag{A.13}$$

A further relationship is

$$\mathrm{Tr}(\not{a}_1 \ldots \not{a}_n) = \mathrm{Tr}(\not{a}_n \ldots \not{a}_1) \tag{A.14}$$

for which the proof is similar to that of (A.13), but using the matrix **S**(C) (4.124) instead of γ_5.

Relations concerning γ_5:

$$\gamma_\mu\gamma_5 = \frac{\mathrm{i}}{6}\varepsilon_{\mu\nu\rho\sigma}\gamma^\nu\gamma^\rho\gamma^\sigma, \tag{A.15}$$

$$\mathrm{Tr}\,\gamma_5 = \mathrm{Tr}(\gamma^\mu\gamma_5) = \mathrm{Tr}(\gamma^\mu\gamma^\nu\gamma_5) = \mathrm{Tr}(\gamma^\mu\gamma^\nu\gamma^\rho\gamma_5) = 0, \tag{A.16}$$

$$\mathrm{Tr}(\gamma^\mu\gamma^\nu\gamma^\rho\gamma^\sigma\gamma_5) = 4\mathrm{i}\varepsilon^{\mu\nu\rho\sigma}. \tag{A.17}$$

The trace of the product of an even number ($2n$) of γ-matrices can be expressed in terms of traces over $2(n-1)$ γ-matrices:

$$\mathrm{Tr}(\not{a}_1\not{a}_2 \ldots \not{a}_{2n}) = \sum_{k=2}^{2n}(-1)^k(a_1, a_k)\,\mathrm{Tr}(\not{a}_2 \ldots \not{a}_{k-1}\not{a}_{k+1} \ldots \not{a}_{2n}). \tag{A.18}$$

To prove this $\not{a}_1$ is moved to the right by repeated anticommutation:

$$\begin{aligned} \mathrm{Tr}(\not{a}_1\not{a}_2 \ldots \not{a}_{2n}) &= \mathrm{Tr}((\{\not{a}_1, \not{a}_2\} - \not{a}_2\not{a}_1)\not{a}_3 \ldots \not{a}_{2n}) \\ &= 2(a_1, a_2)\,\mathrm{Tr}(\not{a}_3 \ldots \not{a}_{2n}) - \mathrm{Tr}(\not{a}_2\not{a}_1\not{a}_3 \ldots \not{a}_{2n}) \\ &= 2\sum_{k=2}^{2n}(-1)^k(a_1, a_k)\,\mathrm{Tr}(\not{a}_2 \ldots \not{a}_{k-1}\not{a}_{k+1} \ldots \not{a}_{2n}) \\ &\quad + (-1)^{2n-1}\,\mathrm{Tr}(\not{a}_2\not{a}_3 \ldots \not{a}_{2n}\not{a}_1). \end{aligned} \tag{A.19}$$

Since the matrices in a trace can be cyclically permuted, (A.18) follows.

For the trace of four γ-matrices we obtain from (A.18)

$$\mathrm{Tr}(\gamma^\lambda\gamma^\mu\gamma^\nu\gamma^\rho) = 4\{g^{\lambda\mu}g^{\nu\rho} + g^{\lambda\rho}g^{\mu\nu} - g^{\lambda\nu}g^{\mu\rho}\}. \tag{A.20}$$

The following are some important relations satisfied by the solutions of the Dirac equation (4.38), (4.41), where $s, r = \pm\frac{1}{2}$ are the spin indices:

$$\begin{aligned} (\not{p} - m)u_s(p) &= 0, \\ (\not{p} + m)v_s(p) &= 0, \end{aligned} \tag{A.21}$$

$$\begin{aligned} \sum_{s=\pm 1/2} u_s(p)\bar{u}_s(p) &= \not{p} + m, \\ \sum_{s=\pm 1/2} v_s(p)\bar{v}_s(p) &= \not{p} - m, \end{aligned} \tag{A.22}$$

$$\begin{aligned} \bar{u}_r(p)u_s(p) &= 2m\delta_{rs}, \\ \bar{v}_r(p)v_s(p) &= -2m\delta_{rs}, \end{aligned} \tag{A.23}$$

$$\begin{aligned} \bar{u}_r(p)\gamma^\mu u_s(p) &= 2p^\mu\delta_{rs}, \\ \bar{v}_r(p)\gamma^\mu v_s(p) &= 2p^\mu\delta_{rs}, \end{aligned} \tag{A.24}$$

$$\begin{aligned} \bar{u}_r(p')\{2m\gamma^\mu - (p'+p)^\mu - \mathrm{i}\sigma^{\mu\nu}(p'-p)_\nu\}u_s(p) = 0, \\ \bar{v}_r(p')\{2m\gamma^\mu + (p'+p)^\mu + \mathrm{i}\sigma^{\mu\nu}(p'-p)_\nu\}v_s(p) = 0. \end{aligned} \tag{A.25}$$

APPENDIX B

The Feynman Rules of QED

We shall make use of the following correspondence between physical situations, the elements of diagrams, and analytical expressions. Spin indices are not written explicitly.

electron in initial state — incoming electron line

$e^-(p)$

$u(p)$ 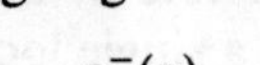

electron in final state — outgoing electron line

$e^-(p)$

$\bar{u}(p)$ 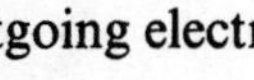

positron in initial state — outgoing electron line

$e^+(q)$

$\bar{v}(q)$ 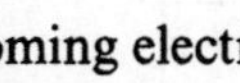

positron in final state — incoming electron line

$e^+(q)$

$v(q)$

photon in initial state — incoming photon line

$\gamma(k)$

ε^μ 

photon in final state — outgoing photon line

$\varepsilon^{\mu *}$

virtual photon — internal photon line

$$\frac{-\mathrm{i}g_{\mu\nu}}{k^2 + \mathrm{i}\varepsilon}$$

virtual electron — internal electron line

$$\mathrm{i}\frac{\not{p} + m}{p^2 - m^2 + \mathrm{i}\varepsilon}$$

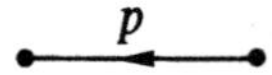

elementary process — vertex

$\mathrm{i}e\gamma^{\mu}$

The transition amplitude for a particular reaction is given by the sum of all the diagrams with the appropriate incoming and outgoing lines. Within the diagrams, any number of vertices are allowed. At each vertex, four-momentum must be conserved. This also guarantees overall four-momentum conservation, i.e., the sum of the incoming four-momenta is equal to the sum of the outgoing four-momenta. The four-momenta flow through the diagrams like electrical currents through a network of electrical conductors. Internal currents correspond to loop momenta. Diagrams with no loops are called *tree diagrams.* Examples are the diagrams of Fig. 10.2.

In Fig. B.1 we show an example of a diagram with one loop. The reaction is $\gamma(k_1) + \gamma(k_2) \to \gamma(k_3) + \gamma(k_4)$ and the loop momentum is l.

It is implicitly understood that loop momenta are to be integrated over with a measure given for a single loop by

$$\int \frac{\mathrm{d}l}{(2\pi)^4}. \tag{B.1}$$

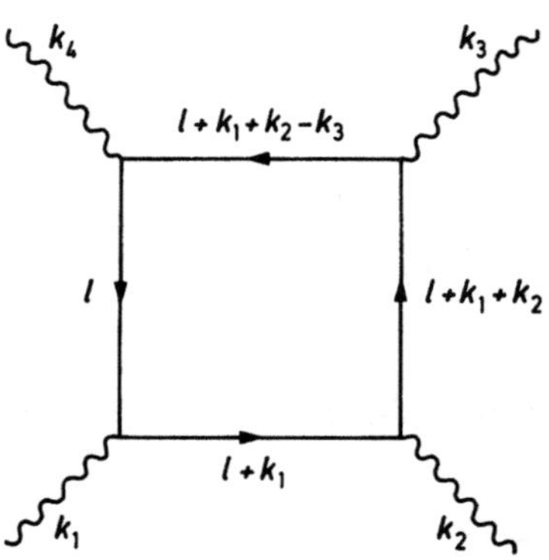

Figure B.1 A diagram for photon–photon scattering with one loop.

Figure B.2 Examples of diagrams for electron–electron scattering. The diagrams (b) and (c) receive a minus sign relative to (a).

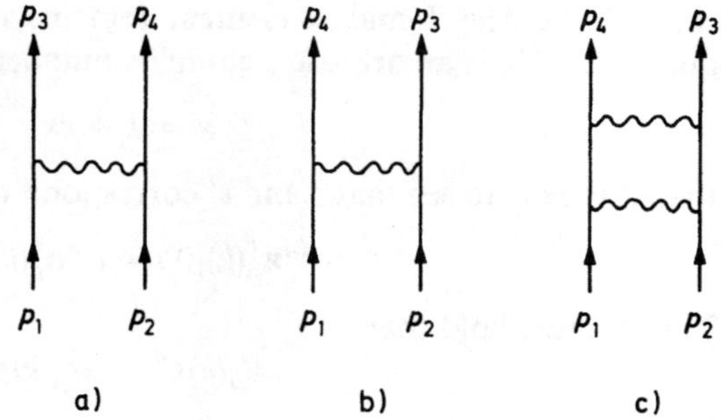

The S-matrix element is obtained from the transition amplitude by multiplication with the δ-function from the energy-momentum conservation:

$$(2\pi)^4\delta\left(\sum_f p_f - \sum_i p_i\right), \tag{B.2}$$

where $p_i(p_f)$ are the momenta of the incoming (outgoing) particles.

A few other rules should be noted. For every closed fermion loop a factor of (-1) appears. Diagrams differing from one another only through a permutation of the labels of the external fermion lines receive relative to each other as factors the sign of the corresponding permutation. (An overall phase factor for the complete amplitude remains, of course, arbitrary.) Diagrams which are identical as far as the fermion lines and their labeling are concerned get no extra sign factor relative to each other. Some examples are shown in Fig. B.2.

To the interaction with an external electromagnetic field there corresponds the following analytical expression and vertex:

$$\mathrm{i}e\gamma^\mu \int \mathrm{d}x\, \mathrm{e}^{\mathrm{i}(p'-p)x} A_\mu^{\mathrm{ext}}(x)$$

$$= -\mathrm{i}e\gamma^\mu \frac{1}{(p'-p)^2} \int \mathrm{d}x\, \mathrm{e}^{\mathrm{i}(p'-p)x} j_\mu^{\mathrm{ext}}(x).$$

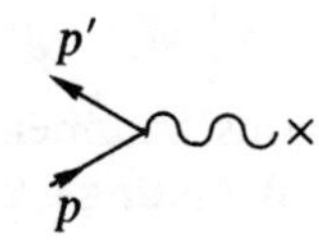

If such a vertex appears in a diagram, then the δ-function of the energy-momentum conservation (B.2) is to be omitted in the S-matrix element.

For calculating scattering cross-sections for unpolarized particles in the initial (final) state one must average (sum) over the corresponding spin orientations. The spin sums for Dirac spinors are given in (A.22). The spin sums for photons require some further explanation.

Let us consider a reaction in which an unpolarized photon with four-momentum k is incident. The Feynman rules yield as a result the amplitude for a photon with polarization vector ε where the dependence on ε is of the general form

$$\langle \ldots | \mathrm{T} | \ldots \gamma(k, \varepsilon)\rangle = \mathscr{M}_\mu(k)\varepsilon^\mu. \tag{B.3}$$

As discussed in Chap. 7, we can replace the photon state by an equivalent state (7.31). The T-matrix element remains unchanged, as do all physical quantities. If c is an arbitrary complex number and

$$\varepsilon' = \varepsilon + ck, \tag{B.4}$$

then it is easy to see that ε and ε' correspond to equivalent photon states:

$$\varepsilon^\mu \mathbf{a}_\mu^\dagger(k)|0\rangle \sim \varepsilon'^\mu \mathbf{a}_\mu^\dagger(k)|0\rangle. \tag{B.5}$$

Thus it must hold that

$$\mathcal{M}_\mu(k)\varepsilon^\mu = \mathcal{M}_\mu(k)\varepsilon'^\mu, \tag{B.6}$$

from which it follows that

$$\mathcal{M}_\mu(k)k^\mu = 0. \tag{B.7}$$

In calculating the cross-section σ for an unpolarized photon we must average over the physical polarization directions of the photon in the initial state:

$$\sigma \propto \sum_{\text{phys.}\,\varepsilon}' |\mathcal{M}_\mu(k)\varepsilon^\mu|^2. \tag{B.8}$$

To evaluate this sum we make use of the coordinate system defined in (7.24). As physical polarization vectors we can choose $\varepsilon_1 = (0, \boldsymbol{e}_1)$ and $\varepsilon_2 = (0, \boldsymbol{e}_2)$. Equation (B.7) yields

$$\begin{aligned} \mathcal{M}_\mu(k)k^\mu &= k^0(\mathcal{M}_0(k) - \vec{\mathcal{M}}(k)\boldsymbol{e}_3) = 0, \\ \mathcal{M}_0(k) &= \vec{\mathcal{M}}(k)\boldsymbol{e}_3. \end{aligned} \tag{B.9}$$

For the cross-section in (B.8) we then obtain

$$\begin{aligned} \sigma \propto \sum_{\text{phys.}\,\varepsilon}' |\mathcal{M}_\mu\varepsilon^\mu|^2 &= \tfrac{1}{2}\sum_{i=1}^{2} |\mathcal{M}_\mu\varepsilon_i^\mu|^2 = \tfrac{1}{2}\{|\vec{\mathcal{M}}\boldsymbol{e}_1|^2 + |\vec{\mathcal{M}}\boldsymbol{e}_2|^2\} \\ &= \tfrac{1}{2}\{|\vec{\mathcal{M}}\boldsymbol{e}_1|^2 + |\vec{\mathcal{M}}\boldsymbol{e}_2|^2 + |\vec{\mathcal{M}}\boldsymbol{e}_3|^2 - |\mathcal{M}_0|^2\} \\ &= \tfrac{1}{2}\mathcal{M}_\mu(-g^{\mu\nu})\mathcal{M}_\nu^*. \end{aligned} \tag{B.10}$$

An analogous argument holds for an unpolarized photon in the final state. Thus for calculating cross-sections in QED we can make the replacement

$$\sum_{\text{phys.}\,\varepsilon} \varepsilon^\mu\varepsilon^{\nu*} \to -g^{\mu\nu}. \tag{B.11}$$

A useful formula regarding a type of phase-space integral that frequently occurs in the calculation of cross-sections is

$$\begin{aligned} &\int \mathrm{d}q_1 \int \mathrm{d}q_2\, \delta_+(q_1^2 - m_1^2)\delta_+(q_2^2 - m_2^2)\delta(p - q_1 - q_2) \\ &\quad \cdot\{a + b^\mu(q_1 - q_2)_\mu + c^{\mu\nu}(q_1 - q_2)_\mu(q_1 - q_2)_\nu\} \\ &= \theta(p^0)\theta(p^2 - (m_1 + m_2)^2)\frac{\pi}{2p^2}w(p^2, m_1^2, m_2^2)\left\{a + b^\mu p_\mu\frac{m_1^2 - m_2^2}{p^2}\right. \\ &\quad \left. + c^{\mu\nu}\left[(4p_\mu p_\nu - g_{\mu\nu}p^2)\frac{1}{3(p^2)^2}w^2(p^2, m_1^2, m_2^2) - p_\mu p_\nu\frac{1}{p^2}(p^2 - 2m_1^2 - 2m_2^2)\right]\right\}. \end{aligned} \tag{B.12}$$

Here p is an external momentum, $m_1 \geq 0$ and $m_2 \geq 0$ are masses, and a, b^μ, $c^{\mu\nu}$ are arbitrary constants. The function w is defined in (2.72) and

$$\delta_+(q^2 - m^2) \equiv \theta(q^0)\delta(q^2 - m^2) = \frac{1}{2q^0}\delta(q_0 - \sqrt{\boldsymbol{q}^2 + m^2}). \qquad \text{(B.13)}$$

In order to prove (B.12) one must first realize that the integral is only nonzero for $p^2 \geq 0$ and $p^0 \geq 0$ and that the individual parts are invariant or covariant under Lorentz transformations. For $p^2 > 0$ the integral is easily calculated in a frame where the spatial part of p is zero. For $p^2 = 0$ the integral gives zero except when $m_1 = m_2 = 0$. In the latter case, we *define* it as the limiting value for $p^2 \to +0$.

APPENDIX C

The Groups SU(2) and SU(3)

C.1 The Group SU(2)

The group SU(2) is the group of all unitary 2 × 2 matrices $\mathbf{U}$ with determinant 1:

$$\begin{aligned} \mathbf{U}\mathbf{U}^\dagger &= 1, \\ \det \mathbf{U} &= 1. \end{aligned} \tag{C.1}$$

This group plays a very important role in particle physics since both the rotations in space and also the groups of the strong and weak isospin have the mathematical structure of the SU(2) group.

Infinitesimal transformations

$$\mathbf{U} = 1 + \mathrm{i}\delta\varphi^j \tfrac{1}{2}\sigma^j. \tag{C.2}$$

Here $\delta\varphi^j$ are infinitesimal parameters and the σ^j ($j = 1, 2, 3$) are the Pauli matrices:[1]

$$\sigma^1 = \begin{pmatrix} 0 & 1 \\ 1 & 0 \end{pmatrix}, \quad \sigma^2 = \begin{pmatrix} 0 & -\mathrm{i} \\ \mathrm{i} & 0 \end{pmatrix}, \quad \sigma^3 = \begin{pmatrix} 1 & 0 \\ 0 & -1 \end{pmatrix}. \tag{C.3}$$

[1] The notation σ^j for the Pauli matrices is conventionally used when considering rotations in ordinary space. When referring to rotations in an isospin space we denote the Pauli matrices by τ_a ($a = 1, 2, 3$).

Relations for Pauli matrices:

$$[\sigma^i, \sigma^j] = 2\mathrm{i}\varepsilon^{ijk}\sigma^k, \tag{C.4}$$

$$\{\sigma^i, \sigma^j\} = 2\delta^{ij}, \tag{C.5}$$

$$\mathrm{Tr}(\sigma^i\sigma^j) = 2\delta^{ij}, \tag{C.6}$$

$$(\boldsymbol{a\sigma})(\boldsymbol{b\sigma}) = (\boldsymbol{a}, \boldsymbol{b}) + \mathrm{i}(\boldsymbol{a} \times \boldsymbol{b})\boldsymbol{\sigma}, \tag{C.7}$$

$$\sigma^j\sigma^k\sigma^j = -\sigma^k, \tag{C.8}$$

$$\sigma^i_{sr}\sigma^i_{s'r'} = 2\delta_{sr'}\delta_{s'r} - \delta_{sr}\delta_{s'r'}, \tag{C.9}$$

$$\varepsilon\sigma^i\varepsilon^{\mathrm{T}} = -\sigma^{i\mathrm{T}}. \tag{C.10}$$

Here $\boldsymbol{a}$ and $\boldsymbol{b}$ are any two vectors and ε in (C.10) is the antisymmetric 2×2 matrix

$$\varepsilon = \begin{pmatrix} 0 & 1 \\ -1 & 0 \end{pmatrix}. \tag{C.11}$$

A unitary representation of a general group G with elements $\mathbf{U}$, $\mathbf{V}$, ... is understood to mean a mapping of G into unitary matrices $\mathbf{D}(\mathbf{U})$, $\mathbf{D}(\mathbf{V})$, ... over a space R of dimension n, such that the group property is retained:

$$\begin{aligned} &\mathbf{U} \to \mathbf{D}(\mathbf{U}), \\ &\mathbf{D}(\mathbf{U})\mathbf{D}^\dagger(\mathbf{U}) = 1, \\ &\mathbf{D}(\mathbf{U})\mathbf{D}(\mathbf{V}) = \mathbf{D}(\mathbf{U}\cdot\mathbf{V}). \end{aligned} \tag{C.12}$$

A representation is called *reducible* if, for a suitable choice of basis, all matrices $\mathbf{D}(\mathbf{U})$ can be *simultaneously* brought into block diagonal form:

$$\mathbf{D}(\mathbf{U}) = \left(\begin{array}{c|c} \mathbf{D}_1(\mathbf{U}) & 0 \\ \hline 0 & \mathbf{D}_2(\mathbf{U}) \end{array}\right). \tag{C.13}$$

Here

$$\begin{aligned} &\mathbf{D}(\mathbf{U}) \text{ are } n \times n \text{ matrices,} \\ &\mathbf{D}_1(\mathbf{U}) \text{ are } n_1 \times n_1 \text{ matrices,} \\ &\mathbf{D}_2(\mathbf{U}) \text{ are } n_2 \times n_2 \text{ matrices,} \end{aligned}$$

and

$$n_1 + n_2 = n, \qquad n_1 \geq 1, \quad n_2 \geq 1. \tag{C.14}$$

A representation is called *irreducible* if this form for the matrices $\mathbf{D}(\mathbf{U})$ is not possible for any choice of basis. As can easily be shown, a representation is irreducible if and only if for every state $|\alpha\rangle \neq 0$ in the representation space R, the states obtained by applying the representation matrices $\mathbf{D}(\mathbf{U})$ to $|\alpha\rangle$, span the space R.

Two representations

$$\begin{aligned} &\mathbf{U} \to \mathbf{D}_1(\mathbf{U}), \\ &\mathbf{U} \to \mathbf{D}_2(\mathbf{U}), \end{aligned} \tag{C.15}$$

are said to be *equivalent* if a nonsingular matrix $\mathbf{S}$ exists such that

$$\mathbf{S}^{-1}\mathbf{D}_1(\mathbf{U})\mathbf{S} = \mathbf{D}_2(\mathbf{U}). \tag{C.16}$$

Important physical implications follow from *Schur's lemma*. Let

$$\begin{aligned} \mathbf{U} &\to \mathbf{D}_1(\mathbf{U}),\\ \mathbf{U} &\to \mathbf{D}_2(\mathbf{U}), \end{aligned} \tag{C.17}$$

be two irreducible representations of a group G in spaces of dimensions n_1 and n_2, respectively. If for all $\mathbf{U} \in G$ we have

$$\mathbf{D}_1(\mathbf{U})\mathbf{S} = \mathbf{S}\mathbf{D}_2(\mathbf{U}). \tag{C.18}$$

then either $\mathbf{S} \equiv 0$ or $n_1 = n_2$ and det $\mathbf{S} \neq 0$ so that the two representations are equivalent (for a proof see, e.g., Weyl 1931). As an important corollary one obtains that a matrix $\mathbf{S}$ which is invariant under the transformations of an *irreducible* representation $\mathbf{U} \to \mathbf{D}(\mathbf{U})$ must be a multiple of the unit matrix. Thus if for all $\mathbf{U} \in G$

$$\mathbf{D}^{-1}(\mathbf{U})\mathbf{S}\mathbf{D}(\mathbf{U}) = \mathbf{S}, \tag{C.19}$$

then with λ a complex number it follows that

$$\mathbf{S} = \lambda \cdot 1. \tag{C.20}$$

This statement can be proved by elementary means. Every matrix $\mathbf{S}$ has at least one eigenvalue λ and one eigenvector $|\lambda\rangle \neq 0$:

$$\mathbf{S}|\lambda\rangle = \lambda|\lambda\rangle. \tag{C.21}$$

According to (C.19) we have

$$\mathbf{S}\mathbf{D}(\mathbf{U})|\lambda\rangle = \mathbf{D}(\mathbf{U})\mathbf{S}|\lambda\rangle = \lambda\mathbf{D}(\mathbf{U})|\lambda\rangle. \tag{C.22}$$

Since, for an irreducible representation, the vectors $\mathbf{D}(\mathbf{U})|\lambda\rangle$ with $\mathbf{U} \in G$ span the entire space, (C.20) follows.

We now discuss representations of the group SU(2) (see, e.g., Edmonds 1957). If

$$\mathbf{U} \to \mathbf{D}(\mathbf{U})$$

is a representation, then we define the generating operators $\mathbf{J}^i$ ($i = 1, 2, 3$), i.e., the angular momentum operators, by the transformations representing the infinitesimal group elements (C.2)

$$\mathbf{D}(1 + \mathrm{i}\delta\varphi^i \tfrac{1}{2}\sigma^i) = 1 + \mathrm{i}\delta\varphi^i\mathbf{J}^i. \tag{C.23}$$

Furthermore, define

$$\vec{\mathbf{J}}^2 = (\mathbf{J}^1)^2 + (\mathbf{J}^2)^2 + (\mathbf{J}^3)^2. \tag{C.24}$$

The irreducible representations D(j) of the group SU(2) are characterized by numbers $j = 0, \frac{1}{2}, 1, \ldots$. The representation D($j$) is realized in a space of dimension $2j + 1$ with orthonormal basis vectors

$$|j, j_3\rangle \qquad (j_3 = -j, \ldots, j).$$

with a suitable (canonical) choice of phases for this basis the generating operators are represented by

$$\mathbf{J}^3 |j, j_3\rangle = j_3 |j, j_3\rangle,$$
$$(\mathbf{J}^1 \pm \mathrm{i}\mathbf{J}^2)|j, j_3\rangle = \sqrt{(j \mp j_3)(j \pm j_3 + 1)}\, |j, j_3 \pm 1\rangle, \tag{C.25}$$
$$\vec{\mathbf{J}}^2 |j, j_3\rangle = j(j+1)|j, j_3\rangle. \tag{C.26}$$

Equation (C.25) also defines the matrices $\mathbf{S}^i$ $(i = 1, 2, 3)$ of (16.22) for general spin s:

$$\mathbf{S}^i_{s'_3, s_3} = \langle s, s'_3 | \mathbf{J}^i | s, s_3 \rangle \qquad (s'_3, s_3 = -s, \ldots, s). \tag{C.27}$$

The product of two irreducible representations D(k) and D(s) can be reduced out, that is, decomposed into irreducible parts, by means of the Clebsch–Gordan series

$$\mathrm{D}(k) \otimes \mathrm{D}(s) = \sum_{j=|k-s|}^{k+s} \mathrm{D}(j). \tag{C.28}$$

The construction of the basis vectors for D(j) in the product space proceeds with the help of the Clebsch–Gordan coefficients (see Edmonds 1957)

$$|j, j_3\rangle = \sum_{k_3, s_3} (k, k_3; s, s_3 | j, j_3) |k, k_3\rangle |s, s_3\rangle. \tag{C.29}$$

The following relations are valid

$$\sum_{j, j_3} (k, k_3; s, s_3 | j, j_3)(k, k'_3; s, s'_3 | j, j_3) = \delta_{k_3 k'_3} \delta_{s_3 s'_3}, \tag{C.30}$$

$$\sum_{k_3, s_3} (k, k_3; s, s_3 | j, j_3)(k, k_3; s, s_3 | j', j'_3) = \delta_{jj'} \delta_{j_3 j'_3}. \tag{C.31}$$

For representations D(l) with integer l $(l = 0, 1, 2, \ldots)$ the spherical harmonics can be used as a basis. We now summarize the most important properties of these functions for which the generating operators will be denoted by $\vec{\mathbf{L}}$ instead of $\vec{\mathbf{J}}$.

Legendre Polynomials:

$$P_l(z) = \frac{1}{2^l l!} \frac{\mathrm{d}^l}{\mathrm{d}z^l} (z^2 - 1)^l \qquad (l = 0, 1, 2, \ldots). \tag{C.32}$$

Unit vector:

$$\hat{\mathfrak{x}} = \begin{pmatrix} \sin\vartheta \cos\varphi \\ \sin\vartheta \sin\varphi \\ \cos\vartheta \end{pmatrix} \qquad (0 \le \vartheta \le \pi, 0 \le \varphi \le 2\pi). \tag{C.33}$$

Spherical harmonics:

$$Y_{l, l_3}(\hat{\mathfrak{x}}) = \left(\frac{2l+1}{4\pi}\right)^{1/2} \left(\frac{(l-|l_3|)!}{(l+|l_3|)!}\right)^{1/2} (-1)^{1/2(l_3 + |l_3|)}$$
$$\cdot\, \mathrm{e}^{\mathrm{i} l_3 \varphi} (\sin\vartheta)^{|l_3|} \left(\frac{d}{d\cos\vartheta}\right)^{|l_3|} P_l(\cos\vartheta) \quad (l = 0, 1, 2, \ldots; l_3 = -l, \ldots, l) \tag{C.36}$$

Relations:

$$\begin{aligned} Y_{l,l_3}(-\hat{x}) &= (-1)^l Y_{l,l_3}(\hat{x}), \\ Y^*_{l,l_3}(\hat{x}) &= (-1)^{l_3} Y_{l,-l_3}(\hat{x}). \end{aligned} \tag{C.37}$$

$$x \times \frac{1}{\mathrm{i}} \nabla Y_{l,l_3}(\hat{x}) = \sum_{l_3'} Y_{l,l_3'}(\hat{x}) \langle l, l_3' | \vec{\mathbf{L}} | l, l_3 \rangle. \tag{C.38}$$

Here the matrix elements of $\vec{\mathbf{L}}$ are given by (C.25).

Addition theorem:

$$\sum_{l_3=-l}^{l} Y_{l,l_3}(\hat{x}) Y^*_{l,l_3}(\hat{x}') = \frac{2l+1}{4\pi} P_l(\hat{x}\hat{x}'). \tag{C.39}$$

We now calculate the sum of Clebsch–Gordan coefficients that is needed in Sect. 16.3. Here $l = 0, 1, 2, \ldots, j = l \pm \frac{1}{2}$ and

$$\begin{aligned} \vec{\mathbf{J}} &= \vec{\mathbf{L}} + \tfrac{1}{2}\boldsymbol{\sigma}, \\ j' &= l \mp \tfrac{1}{2} \qquad \text{for} \quad j = l \pm \tfrac{1}{2}. \end{aligned} \tag{C.40}$$

We find

$$\begin{aligned} &\sum_{j_3} (l, l_3'; \tfrac{1}{2}, s_3' | j, j_3)(l, l_3; \tfrac{1}{2}, s_3 | j, j_3) \\ &= (\langle l, l_3' | \langle \tfrac{1}{2}, s_3' |) \left(\sum_{j_3} | j, j_3 \rangle \langle j, j_3 | \right) (| l, l_3 \rangle | \tfrac{1}{2}, s_3 \rangle) \\ &= (\langle l, l_3' | \langle \tfrac{1}{2}, s_3' |) \left(\frac{\vec{\mathbf{J}}^2 - j'(j'+1)}{j(j+1) - j'(j'+1)} \right) (| l, l_3 \rangle | \tfrac{1}{2}, s_3 \rangle) \\ &= (j(j+1) - j'(j'+1))^{-1} (\langle l, l_3' | \langle \tfrac{1}{2}, s_3' |) \\ &\quad \cdot \{ \vec{\mathbf{L}}^2 + \vec{\mathbf{L}}\boldsymbol{\sigma} + \tfrac{3}{4} - j'(j'+1) \} (| l, l_3 \rangle | \tfrac{1}{2}, s_3 \rangle) \\ &= (j(j+1) - j'(j'+1))^{-1} \{ (l(l+1) + \tfrac{3}{4} - j'(j'+1)) \delta_{l_3' l_3} \delta_{s_3' s_3} \\ &\quad + \langle l, l_3' | \vec{\mathbf{L}} | l, l_3 \rangle \boldsymbol{\sigma}_{s_3', s_3} \} \\ &= (2l+1)^{-1} \{ (j + \tfrac{1}{2}) \delta_{l_3' l_3} \delta_{s_3' s_3} + 2(j-l) \langle l, l_3' | \vec{\mathbf{L}} | l, l_3 \rangle \boldsymbol{\sigma}_{s_3', s_3} \} \end{aligned} \tag{C.41}$$

C.2 The Group SU(3)

The group SU(3) is the group of all unitary 3×3 matrices $\mathbf{U}$ with determinant 1:

$$\begin{aligned} \mathbf{U}\mathbf{U}^\dagger &= 1, \\ \det \mathbf{U} &= 1. \end{aligned} \tag{C.42}$$

Infinitesimal transformations:

$$\mathbf{U} = 1 + \mathrm{i}\delta\varphi_a \tfrac{1}{2}\lambda_a. \tag{C.43}$$

where we make use of the summation convention. The $\delta\varphi_a$ are infinitesimal parameters and the λ_a ($a = 1, \dots, 8$) are the Gell-Mann matrices:

$$\lambda_1 = \begin{pmatrix} 0 & 1 & 0 \\ 1 & 0 & 0 \\ 0 & 0 & 0 \end{pmatrix}, \quad \lambda_2 = \begin{pmatrix} 0 & -\mathrm{i} & 0 \\ \mathrm{i} & 0 & 0 \\ 0 & 0 & 0 \end{pmatrix}, \quad \lambda_3 = \begin{pmatrix} 1 & 0 & 0 \\ 0 & -1 & 0 \\ 0 & 0 & 0 \end{pmatrix},$$

$$\lambda_4 = \begin{pmatrix} 0 & 0 & 1 \\ 0 & 0 & 0 \\ 1 & 0 & 0 \end{pmatrix}, \quad \lambda_5 = \begin{pmatrix} 0 & 0 & -\mathrm{i} \\ 0 & 0 & 0 \\ \mathrm{i} & 0 & 0 \end{pmatrix}, \quad \lambda_6 = \begin{pmatrix} 0 & 0 & 0 \\ 0 & 0 & 1 \\ 0 & 1 & 0 \end{pmatrix},$$

$$\lambda_7 = \begin{pmatrix} 0 & 0 & 0 \\ 0 & 0 & -\mathrm{i} \\ 0 & \mathrm{i} & 0 \end{pmatrix}, \quad \lambda = \frac{1}{\sqrt{3}} \begin{pmatrix} 1 & 0 & 0 \\ 0 & 1 & 0 \\ 0 & 0 & -2 \end{pmatrix}. \tag{C.44}$$

Relations:

$$\mathrm{Tr}\, \lambda_a = 0, \tag{C.45}$$

$$\mathrm{Tr}(\lambda_a \lambda_b) = 2\delta_{ab}, \tag{C.46}$$

$$[\lambda_a, \lambda_b] = 2\mathrm{i} f_{abc} \lambda_c, \tag{C.47}$$

$$\{\lambda_a, \lambda_b\} = \tfrac{4}{3}\delta_{ab} + 2 d_{abc} \lambda_c, \tag{C.48}$$

where f_{abc} is totally antisymmetric, and d_{abc} is totally symmetric with values for the independent components as given in Table C.1.

Table C.1 The independent, nonvanishing components of f_{abc} and d_{abc}.

(a, b, c)	f_{abc}	(a, b, c)	d_{abc}
123	1	118	$1/\sqrt{3}$
147	1/2	146	1/2
156	−1/2	157	1/2
246	1/2	228	$1/\sqrt{3}$
257	1/2	247	−1/2
345	1/2	256	1/2
367	−1/2	338	$1/\sqrt{3}$
458	$\sqrt{3}/2$	344	1/2
678	$\sqrt{3}/2$	355	1/2
		366	−1/2
		377	−1/2
		448	$-1/(2\sqrt{3})$
		558	$-1/(2\sqrt{3})$
		668	$-1/(2\sqrt{3})$
		778	$-1/(2\sqrt{3})$
		888	$-1/\sqrt{3}$

$$f_{abr}f_{rcs} + f_{bcr}f_{ras} + f_{car}f_{rbs} = 0, \tag{C.49}$$

$$f_{abr}d_{rcs} + f_{cbr}d_{ras} = d_{acr}f_{rbs}, \tag{C.50}$$

$$f_{ars}f_{brs} = 3\delta_{ab}, \tag{C.51}$$

$$d_{aab} = 0, \tag{C.52}$$

$$d_{ars}d_{brs} = \tfrac{5}{3}\delta_{ab}, \tag{C.53}$$

$$f_{abc}\lambda_b\lambda_c = 3\mathrm{i}\lambda_a, \tag{C.54}$$

$$\lambda_a\lambda_a = \tfrac{16}{3}, \tag{C.55}$$

$$\lambda_a\lambda_b\lambda_a = -\tfrac{2}{3}\lambda_b, \tag{C.56}$$

$$(\lambda_a)_{\alpha\beta}(\lambda_a)_{\alpha'\beta'} = -\tfrac{2}{3}\delta_{\alpha\beta}\delta_{\alpha'\beta'} + 2\delta_{\alpha\beta'}\delta_{\alpha'\beta}. \tag{C.57}$$

The most important representations of the SU(3) group are discussed in Sect. 17.1.

APPENDIX D

The Feynman Rules of QCD

In this Appendix we present the Feynman rules of QCD. The derivation of these rules from the Lagrange density (19.22) is best achieved using the path-integral formalism. Here we restrict ourselves to giving only the results.

The Feynman diagrams of QCD are constructed from three types of lines: quark lines, gluon lines, and so-called Faddev–Popov ghost lines. Within QCD it is not possible that lines corresponding to different quark flavors are joined together. The ghosts are to be treated like a complex scalar particle with zero mass. Thus ghost and antighost are distinct. The corresponding lines, like the quark lines, are given an arrow. The ghosts transform according to the adjoint representation of the color group SU(3) and thus carry a color index a ($a = 1, \ldots, 8$).

We shall make use of the following scheme to relate physical situations, parts of diagrams, and analytical expressions:

quark in initial state	incoming quark line
$u(p)$	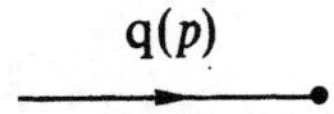
quark in final state	outgoing quark line
$\bar{u}(p)$	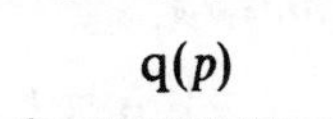

antiquark in initial state — outgoing quark line

$\bar{v}(p)$

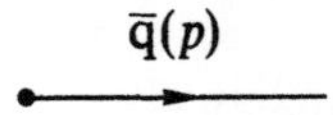

antiquark in final state — incoming quark line

$v(p)$

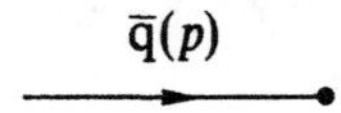

ghost in initial or antighost in final state — incoming ghost line

1

ghost in final or antighost in initial state — outgoing ghost line

1

virtual quark with flavour j — internal quark line

$$\frac{\mathrm{i}}{\not p - m_j + \mathrm{i}\varepsilon}$$

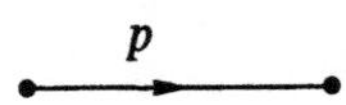

gluon in initial state — incoming gluon line

ε^{μ}

gluon in final state — outgoing gluon line

$\varepsilon^{*\mu}$

virtual gluon — internal gluon line

$$\mathrm{i}\delta^{ab}\left\{\frac{-g_{\mu\nu}}{k^2 + \mathrm{i}\varepsilon} + \frac{(1-\xi)k_\mu k_\nu}{(k^2 + \mathrm{i}\varepsilon)^2}\right\}$$

virtual ghost — internal ghost line

$$\mathrm{i}\frac{\delta^{ab}}{p^2 + \mathrm{i}\varepsilon}$$

The elementary processes of QCD correspond to the following vertices:

3-gluon vertex $\left(\sum_{j=1}^{3} k_j = 0, \text{ all momenta incoming}\right)$

$$-g_s f_{a_1 a_2 a_3} \cdot \{(k_1 - k_2)_{\mu_3} g_{\mu_1\mu_2} + (k_2 - k_3)_{\mu_1} g_{\mu_2\mu_3} + (k_3 - k_1)_{\mu_2} g_{\mu_3\mu_1}\}$$

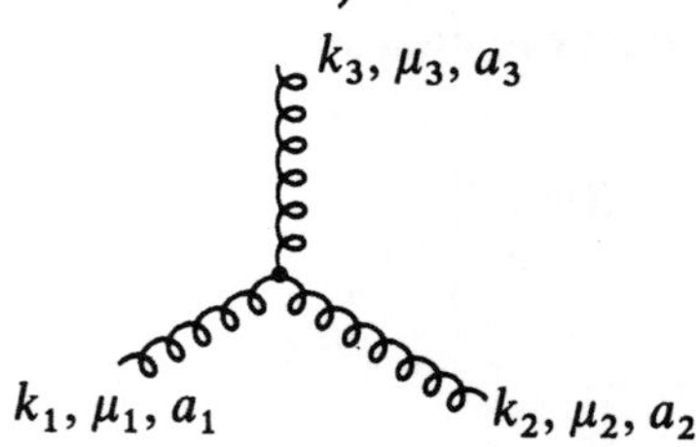

4-gluon vertex $\left(\sum_{j=1}^{4} k_j = 0, \text{ all momenta incoming}\right)$

$$\mathrm{i}g_s^2 f_{a_1 a_2 b} f_{a_3 a_4 b}$$
$$\cdot (g_{\mu_2\mu_3} g_{\mu_1\mu_4} - g_{\mu_1\mu_3} g_{\mu_2\mu_4})$$
$$+ \text{cyclic perms. of } 1, 2, 3$$

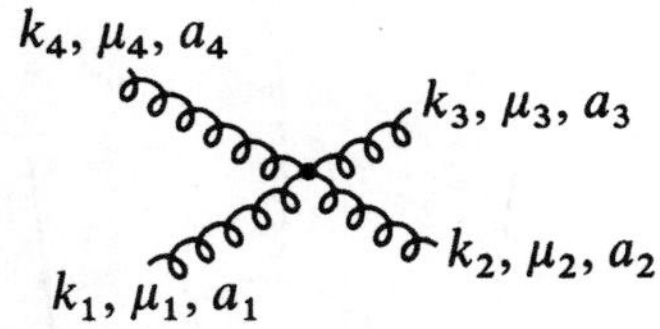

ghost–gluon vertex (p, k incoming; p' outgoing; $p + k = p'$)

$$g_s f_{abc} p'_\mu$$

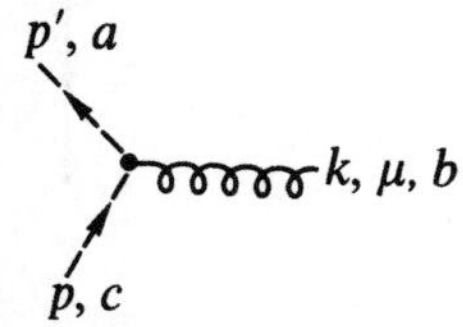

quark–gluon vertex (p, k incoming; p' outgoing; $p + k = p'$)

$$-\mathrm{i}g_s \gamma_\mu \tfrac{1}{2}\lambda_a$$

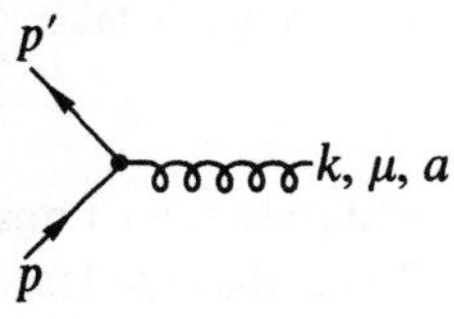

The parameters that appear in the above Feynman rules are the coupling constant g_s, the quark masses m_j ($j = 1, \ldots, f$), and the arbitrary gauge parameter ξ which must drop out in the expressions obtained for observable quantities. It is usually most convenient to set $\xi = 1$; this is the so-called Feynman gauge. We used this gauge throughout our discussion of QED where an analogous gauge parameter can be introduced.

The rules for loops and for the δ-function of energy-momentum conservation are the same as in QED (Appendix B). For every closed fermion loop and every ghost loop a factor of (-1) must be included. For closed gluon loops one must take the so-called statistical factor into account. This is obtained by counting up all possible contractions of field operators in the perturbation expansion, which, for QCD too, is as given in (9.14). The diagram in Fig. D.1(a), for example, has the statistical factor 1, and that of Fig. D.1(b) has the factor $\frac{1}{2}$. In order to see this we write the three-gluon coupling schematically as

$$\frac{1}{3!} :\mathbf{G}^3(x):,$$

where $\mathbf{G}(x)$ is the gluon field operator. The diagram of Fig. D.1(a) then corresponds to a certain Fourier transform of the vacuum expectation value

$$\langle 0 | \frac{1}{3!} :\mathbf{G}^3(x): \mathbf{G}(x_1)\mathbf{G}(x_2)\mathbf{G}(x_3) | 0 \rangle \propto 1. \tag{D.1}$$

The evaluation is carried out using Wick's theorem from Sect. 8.2. $\mathbf{G}(x_1)$ must be contracted in turn with each of the three operators from $:\mathbf{G}^3(x):$. For $\mathbf{G}(x_2)$

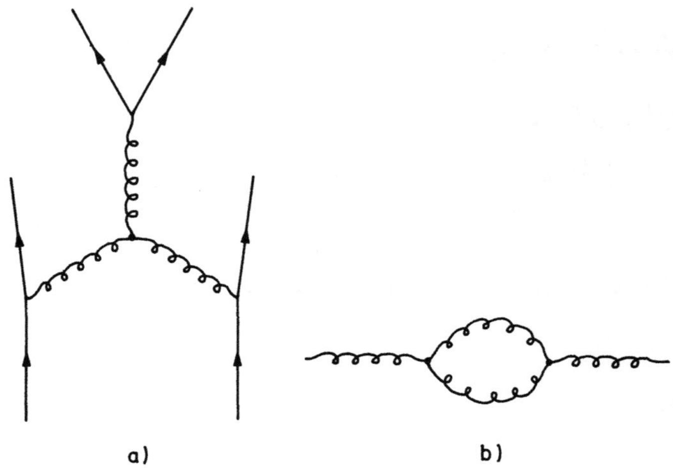

Figure D.1 A diagram with a three-gluon vertex in which no loop occurs, thus giving a statistical factor of 1(a). A self-energy diagram for the gluon propagator. In this a statistical factor of $\frac{1}{2}$ must be taken (b).

only two possibilities then remain, and for $\mathbf{G}(x_3)$ only one. The net result indicated in (D.1) is that the factor 1/3! is cancelled. If we now move on to the diagram of Fig. D.1(b) we find, with the factor 1/2! from the exponential series in (9.14), a corresponding vacuum expectation value of

$$\frac{1}{2!}\langle 0|\frac{1}{3!}:\mathbf{G}^3(x):\frac{1}{3!}:\mathbf{G}^3(y):\mathbf{G}(x_1)\mathbf{G}(x_2)|0\rangle \propto \tfrac{1}{2}. \tag{D.2}$$

This statistical weight of $\frac{1}{2}$ can also be interpreted as phase-space factor for the two gluons in the intermediate state since these are indistinguishable.

When calculating the polarization sums for gluons in the initial or final states there are likewise a few points that should be noted. It is always possible to do the calculation noncovariantly, i.e., to consider only the transverse polarization degrees of freedom of the gluons (cf. Appendix E). Alternatively, one can work with covariant quantities, i.e., replace the polarization sum of a gluon in the initial or final state by

$$\sum_{\text{Spins}} \varepsilon_\mu \varepsilon_\nu^* = -g_{\mu\nu}. \tag{D.3}$$

In this case, however, it is also necessary to take into account processes of the same order with incoming and outgoing ghost lines, treating ghosts and antighosts formally as if they were extra degrees of freedom for the gluons to be considered in their spin sums. (Of course only two external gluon lines can be replaced at a time by ghost lines since an odd number of external ghost lines is impossible, as is an odd number of external fermion lines.) Amplitudes corresponding to diagrams which are identical except for permutations of the labels of external ghost lines receive extra sign factors relative to each other in the same way as for fermions (cf. Appendix B). In calculating transition probabilities one has furthermore to include a factor $(-1)^{n/2}$ where n is the number of external ghost lines in the corresponding diagram. Thus some ghost processes, for instance, the production of a single ghost-antighost pair ($n = 2$), get assigned a negative probability (cf. Cutler 1978).

These rules are sufficient for the calculation of all tree diagrams, i.e., diagrams without loops. For the regularization and renormalization procedures to be applied beyond this level the reader is referred to the available textbooks, e.g., Itzykson (1980).

APPENDIX E

The Q^2-Evolution of Quark and Gluon Distribution Functions of the Nucleons in QCD

We will begin by deriving a Hamiltonian for QCD. For this purpose it is convenient to add a total divergence to the Lagrange density given in (19.22); this is always possible. Thus we start out with the following Lagrange density:

$$\mathscr{L} = -\tfrac{1}{4} G^a_{\lambda\rho} G^{\lambda\rho a} + \sum_{j=1}^{f} \bar{\mathbf{q}}^j (\mathrm{i}\gamma^\lambda \mathbf{D}_\lambda - m_j)\mathbf{q}^j$$

$$+ \partial_\lambda \left\{ \tfrac{1}{2}(G^{\lambda a}\partial_\rho G^{\rho a} - G^{\rho a}\partial_\rho G^{\lambda a}) - \sum_{j=1}^{f} \bar{\mathbf{q}}^j \frac{\mathrm{i}}{2}\gamma^\lambda \mathbf{q}^j \right\}, \qquad \text{(E.1)}$$

where we initially regard the fields as classical quantities. We make use of the gauge freedom to set the time components of the gluon fields equal to zero:

$$G^a_0 = 0 \qquad (a = 1, \dots, 8). \qquad \text{(E.2)}$$

With this gauge condition the Lagrange function reads

$$L = \int \mathrm{d}^3x\, \mathscr{L} = \int \mathrm{d}^3x \left\{ \tfrac{1}{2}\dot{\boldsymbol{G}}^a \dot{\boldsymbol{G}}^a - \tfrac{1}{2}\nabla_\alpha G^{\beta a}\nabla_\alpha G^{\beta a} \right.$$

$$+ \tfrac{1}{2}(\boldsymbol{\nabla}\boldsymbol{G}^a)(\boldsymbol{\nabla}\boldsymbol{G}^a) - g_s f_{abc} G^{\alpha a} G^{\beta b}\nabla_\alpha G^{\beta c} - \frac{g_s^2}{4} f_{abc} f_{ars} G^{\alpha b} G^{\beta c} G^{\alpha r} G^{\beta s}$$

$$\left. + \sum_{j=1}^{f} \left[\bar{\mathbf{q}}^j \frac{\mathrm{i}}{2}\gamma^0 \dot{\mathbf{q}}^j - \dot{\bar{\mathbf{q}}}^j \frac{\mathrm{i}}{2}\gamma^0 \mathbf{q}^j + \bar{\mathbf{q}}^j \left(\frac{\mathrm{i}}{2}\boldsymbol{\gamma}\overrightarrow{\boldsymbol{\nabla}} - \frac{\mathrm{i}}{2}\boldsymbol{\gamma}\overleftarrow{\boldsymbol{\nabla}} - m_j + g_s \boldsymbol{\gamma}\boldsymbol{G}^a \frac{\lambda_a}{2} \right) \mathbf{q}^j \right] \right\}. \qquad \text{(E.3)}$$

Here α and β are spatial indices ($\alpha, \beta = 1, 2, 3$). According to (3.105) we obtain the canonical conjugate momenta $\Pi^{\alpha a}$ corresponding to the gluon fields $G^{\alpha a}$ by taking the functional derivative of L with respect to $\dot{G}^{\alpha a}$:

$$\Pi^{\alpha a} = \frac{\delta L}{\delta \dot{G}^{\alpha a}} = \dot{G}^{\alpha a}. \tag{E.4}$$

We further find

$$\frac{\delta L}{\delta \dot{\mathbf{q}}^j} = \bar{\mathbf{q}}^j \frac{\mathrm{i}}{2}\gamma^0, \qquad \frac{\delta L}{\delta \dot{\bar{\mathbf{q}}}^j} = -\frac{\mathrm{i}}{2}\gamma^0 \mathbf{q}^j. \tag{E.5}$$

The Hamiltonian is obtained from the general rule of (3.107) as

$$\begin{aligned} H &= \int \mathrm{d}^3x \left\{ \frac{\delta L}{\delta \dot{G}^{\alpha a}} \dot{G}^{\alpha a} + \sum_{j=1}^{f} \left(\frac{\delta L}{\delta \dot{\mathbf{q}}^j} \dot{\mathbf{q}}^j + \dot{\bar{\mathbf{q}}}^j \frac{\delta L}{\delta \dot{\bar{\mathbf{q}}}^j} \right) \right\} - L \\ &= \int \mathrm{d}^3x \left\{ \tfrac{1}{2}\Pi^a\Pi^a + \tfrac{1}{2}\nabla_\beta G^{\alpha a}\nabla_\beta G^{\alpha a} - \tfrac{1}{2}(\nabla G^a)(\nabla G^a) \right. \\ &\quad + g_s f_{abc} G^{\alpha a} G^{\beta b} \nabla_\alpha G^{\beta c} + \frac{g_s^2}{4} f_{abc} f_{ars} G^{\alpha b} G^{\beta c} G^{\alpha r} G^{\beta s} \\ &\quad \left. + \sum_{j=1}^{f} \left[\bar{\mathbf{q}}^j \left(-\frac{\mathrm{i}}{2}\gamma\overrightarrow{\nabla} + \frac{\mathrm{i}}{2}\gamma\overleftarrow{\nabla} + m_j - g_s \gamma G^a \frac{\lambda_a}{2} \right) \mathbf{q}^j \right] \right\}. \end{aligned} \tag{E.6}$$

We can now move over to quantum field theory and impose the canonical commutation rules of Sect. 3.4 for the gluon fields. The choice of gauge (E.2) is quite essential here, since with this gauge, none of the canonical momenta $\Pi^{\alpha a}$ is identically zero (cf. Exercise 9.3). For the quark fields we demand canonical anticommutation rules as in (4.92). The Hamilton function (E.6) then becomes the Hamilton operator of the quantized theory.

If we only consider the flavors u, d, and s ($f = 3$) then we see from (E6) that the Hamilton operator is flavor-SU(3) invariant apart from the mass term. Assuming the validity of (17.56), the mass term can be decomposed into a flavor-singlet and a flavor-octet part exactly as in (17.57). Thus (17.58) has explicitly been verified in the framework of QCD.

We now calculate the expansion coefficients $C(\boldsymbol{k}, \boldsymbol{p}')$ occurring in (19.40). We normalize our states to 1 in a normalization volume V. With the appropriate quark and gluon wavefunctions we then obtain from (19.39)

$$\begin{aligned} &\langle G^a_{K'}(\boldsymbol{k}, \boldsymbol{\varepsilon}), \mathrm{q}_{K'}(\boldsymbol{p}') | \mathbf{H}'_{\Delta K} | \mathrm{q}_K(\boldsymbol{p}) \rangle \\ &= -g_s (2\pi)^3 \delta^3(\boldsymbol{k} + \boldsymbol{p}' - \boldsymbol{p}) \frac{1}{\sqrt{2Vk_0 2Vp'_0 2Vp_0}} \bar{u}(p')(\boldsymbol{\varepsilon}^* \cdot \boldsymbol{\gamma}) \frac{\lambda_a}{2} u(p). \end{aligned} \tag{E.7}$$

For the evaluation of the phase-space integral (19.41) the following points are to be noted.

In a noncovariant perturbation theory only momentum is conserved at the vertices of Fig. 19.6 but not the energy. For the momenta we thus write

$$\boldsymbol{p} = \begin{pmatrix} 0 \\ 0 \\ |\boldsymbol{p}| \end{pmatrix}, \qquad \boldsymbol{k} = \begin{pmatrix} \boldsymbol{k}_\perp \\ x|\boldsymbol{p}| \end{pmatrix}, \qquad \boldsymbol{p}' = \begin{pmatrix} -\boldsymbol{k}_\perp \\ (1-x)|\boldsymbol{p}| \end{pmatrix}, \tag{E.8}$$

where in accordance with (19.35) we always demand

$$|\boldsymbol{p}|, \qquad x|\boldsymbol{p}|, \qquad (1-x)|\boldsymbol{p}| \gg |\boldsymbol{k}_\perp|. \tag{E.9}$$

As a consequence we have, for example, the approximate relation

$$|\boldsymbol{k}| + |\boldsymbol{p}'| - |\boldsymbol{p}| \cong \frac{\boldsymbol{k}_\perp^2}{2|\boldsymbol{p}|}\frac{1}{x(1-x)}. \tag{E.10}$$

The sum over momenta in (19.41) is always to be performed in the limit of infinite normalization volume, $V \to \infty$, where the following replacements can be made:

$$\frac{1}{V}\sum_{p'} \to \int \frac{\mathrm{d}^3p'}{(2\pi)^3}, \qquad \frac{1}{V}\sum_{k} \to \int \frac{\mathrm{d}^3k}{(2\pi)^3}. \tag{E.11}$$

These replacements can be justified by the enumeration of stationary states in a cube of volume V using periodic boundary conditions.

If we now put (E.7) into (19.41) and apply the approximation (E.9) throughout, then we obtain

$$\begin{aligned} N_G(x, Q'^2) &= g_s^2 \int_{K \le |\boldsymbol{k}_\perp| \le K'} \frac{\mathrm{d}^3k\,\mathrm{d}^3p'}{(2\pi)^3}\frac{1}{8k_0p_0'p_0}\delta\left(x - \frac{\boldsymbol{k}\cdot\boldsymbol{p}}{|\boldsymbol{p}|^2}\right)\delta^3(\boldsymbol{k}+\boldsymbol{p}'-\boldsymbol{p}) \\ &\quad \cdot \frac{4|\boldsymbol{p}|^2x^2(1-x)^2}{|\boldsymbol{k}_\perp|^4} \sum_{\text{Spins, color}}{}' |\bar{u}(p')\boldsymbol{\varepsilon}^*\cdot\boldsymbol{\gamma}\frac{\lambda_a}{2}u(p)|^2 \\ &= \frac{g_s^2}{(2\pi)^3}\frac{1}{2}\int_{K \le |\boldsymbol{k}_\perp| \le K'} \frac{\mathrm{d}^2k_\perp}{|\boldsymbol{k}_\perp|^4}x(1-x)\sum_{\text{Spins, color}}{}' |\bar{u}(p')\boldsymbol{\varepsilon}^*\cdot\boldsymbol{\gamma}\frac{\lambda_a}{2}u(p)|^2. \end{aligned} \tag{E.12}$$

In (E.12) we have to carry out a sum over gluon polarizations. This sum only extends over transverse gluons since the other gluon polarizations are only gauge degrees of freedom. Thus we have to make the substitution

$$\sum_{\text{Spins}} \varepsilon^\alpha \varepsilon^{*\beta} = \delta^{\alpha\beta} - \frac{k^\alpha k^\beta}{|\boldsymbol{k}|^2}. \tag{E.13}$$

After averaging over the spin and color of the quark in the initial state, and summing over the spin and color of quark and gluon in the final state, we obtain

$$\begin{aligned} \sum_{\text{Spins, color}}{}' |\bar{u}(p')\boldsymbol{\varepsilon}^*\cdot\boldsymbol{\gamma}\frac{\lambda_a}{2}u(p)|^2 &= \frac{16}{3|\boldsymbol{k}|^2}\{|\boldsymbol{k}|^2|\boldsymbol{p}||\boldsymbol{p}'| - (\boldsymbol{p}\cdot\boldsymbol{k})(\boldsymbol{p}'\cdot\boldsymbol{k})\} \\ &\cong \frac{16}{3x^2}\boldsymbol{k}_\perp^2\left\{1 + \frac{x^2}{2(1-x)}\right\}. \end{aligned} \tag{E.14}$$

Inserting this result into (E.12) now yields

$$N_G(x, Q'^2) = \frac{g_s^2}{(2\pi)^3} \int_{K \le |\boldsymbol{k}_\perp| \le K'} \frac{d^2k_\perp}{|\boldsymbol{k}_\perp|^2} \frac{4}{3} \frac{1 + (1-x)^2}{x}$$
$$= \frac{g_s^2}{8\pi^2} \frac{4}{3} \frac{1 + (1-x)^2}{x} \Delta \ln K^2. \tag{E.15}$$

According to (19.35) we have for constant ε

$$\Delta \ln K^2 = \Delta \ln Q^2. \tag{E.16}$$

Therefore (E.15) has the form anticipated in (19.42) with $P_{Gq}(x)$ given by (19.43).

As another topic we discuss in this Appendix the explicit parametrization of the distribution functions of the proton suggested by Duke and Owens (1984). These authors made a general fit to the data on deep inelastic lepton–nucleon scattering and related processes in the interval $4 \le Q^2 \le 200$ GeV^2 treating the Q^2-evolution of the distribution functions according to the Altarelli–Parisi equation. The resulting functions were approximated by a simple functional form. Further approximations were the use of only four quarks, u, d, s, and c, which were taken to be massless, and the assumptions:

$$\begin{aligned} N_{\bar{u}}^p(x) = N_{\bar{d}}^p(x) = N_{\bar{s}}^p(x) = N_s^p(x), \\ N_c^p(x) = N_{\bar{c}}^p(x). \end{aligned} \tag{E.17}$$

As a starting point for the Q^2-evolution $Q_0^2 = 4$ GeV^2 was chosen and, for this value, the distribution of c-quarks in the proton was set identically to zero. One of the parametrizations thus obtained for the distribution functions of the proton is the following:

$$\begin{aligned} xV_{ud}(x) &\equiv x[N_u^p(x) + N_d^p(x) - N_{\bar{u}}^p(x) - N_{\bar{d}}^p(x)] \\ &= \frac{3x^{\eta_1}(1-x)^{\eta_2}(1+\gamma_{ud}x)}{[1 + \gamma_{ud}\eta_1(1+\eta_1+\eta_2)^{-1}]B(\eta_1, \eta_2+1)}, \end{aligned} \tag{E.18}$$

$$\begin{aligned} xV_d(x) &\equiv x[N_d^p(x) - N_{\bar{d}}^p(x)] \\ &= \frac{x^{\eta_3}(1-x)^{\eta_4}(1+\gamma_d x)}{[1 + \gamma_d\eta_3(1+\eta_3+\eta_4)^{-1}]B(\eta_3, \eta_4+1)}, \end{aligned} \tag{E.19}$$

$$\begin{aligned} xS(x) &\equiv 2x[N_{\bar{u}}^p(x) + N_{\bar{d}}^p(x) + N_{\bar{s}}^p(x)] \\ &= A_S x^{a_S}(1-x)^{b_S}(1 + \alpha_S x + \beta_S x^2 + \gamma_S x^3), \end{aligned} \tag{E.20}$$

$$xN_G(x) = A_G x^{a_G}(1-x)^{b_G}(1 + \alpha_G x + \beta_G x^2 + \gamma_G x^3), \tag{E.21}$$

$$x[N_c^p(x) + N_{\bar{c}}^p(x)] = A_c x^{a_c}(1-x)^{b_c}(1 + \alpha_c x + \beta_c x^2 + \gamma_c x^3). \tag{E.22}$$

The parameters η_1, η_2, etc., appearing here are given in Table E.1. They are

Table E.1 The functional form of the parameters η_1, η_2, etc., from (E.18)–(E.22).

$\eta_1 = 0.419 + 0.004\,s - 0.007\,s^2$	$\eta_3 = 0.763 - 0.237\,s + 0.026\,s^2$
$\eta_2 = 3.46 + 0.724\,s - 0.066\,s^2$	$\eta_4 = 4.00 + 0.627\,s - 0.019\,s^2$
$\gamma_{\mathrm{ud}} = 4.40 - 4.86\,s + 1.33\,s^2$	$\gamma_{\mathrm{d}} = -0.421\,s + 0.033\,s^2$
$A_S = 1.265 - 1.132\,s + 0.293\,s^2$	$\alpha_S = 6.31\,s - 0.273\,s^2$
$a_S = -0.372\,s - 0.029\,s^2$	$\beta_S = -10.5\,s - 3.17\,s^2$
$b_S = 8.05 + 1.59\,s - 0.153\,s^2$	$\gamma_S = 14.7\,s + 9.80\,s^2$
$A_{\mathrm{c}} = 0.135\,s - 0.075\,s^2$	$\alpha_{\mathrm{c}} = -3.03\,s + 1.50\,s^2$
$a_{\mathrm{c}} = -0.036 - 0.222\,s - 0.058\,s^2$	$\beta_{\mathrm{c}} = 17.4\,s - 11.3\,s^2$
$b_{\mathrm{c}} = 6.35 + 3.26\,s - 0.909\,s^2$	$\gamma_{\mathrm{c}} = -17.9\,s + 15.6\,s^2$
$A_G = 1.56 - 1.71\,s + 0.638\,s^2$	$\alpha_G = 9.0 - 7.19\,s + 0.255\,s^2$
$a_G = -0.949\,s + 0.325\,s^2$	$\beta_G = -16.5\,s + 10.9\,s^2$
$b_G = 6.0 + 1.44\,s - 1.05\,s^2$	$\gamma_G = 15.3\,s - 10.1\,s^2$

functions of the variable

$$s = \ln \frac{\ln \dfrac{Q^2}{\Lambda^2}}{\ln \dfrac{Q_0^2}{\Lambda^2}}, \tag{E.23}$$

where $Q_0^2 = 4\ \mathrm{GeV}^2$ and $\Lambda = 200$ MeV. In this way the Q^2-dependence of the distribution functions is taken into account. The function $B(\eta, \eta')$ is Euler's beta function

$$B(\eta, \eta') = \frac{\Gamma(\eta)\Gamma(\eta')}{\Gamma(\eta + \eta')}. \tag{E.24}$$

The above parametrization is reliable (error approx. $\pm 1\%$) for $Q_0^2 \leq Q^2 \leq 10^6\ \mathrm{GeV}^2$ in the region $x \geq 0.05$. The resulting distribution functions for $Q^2 = 4\ \mathrm{GeV}^2$ and $Q^2 = 10^4\ \mathrm{GeV}^2$ are shown in Fig. 19.11.

Table E.2 The momentum fractions carried by the individual partons in the proton according to the parametrization of (E.18)–(E.22). The momentum sum rule (18.63) is not exactly satisfied. (The region of small x values is not sufficiently well described by the parametrization!.)

	$\int_0^1 dx\, xN_j(x)$	
Parton j	$Q^2 = 4\ \mathrm{GeV}^2$	$Q^2 = 10^4\ \mathrm{GeV}^2$
u	0.28	0.20
d	0.16	0.12
G	0.48	0.59
$\bar{\mathrm{u}} + \bar{\mathrm{d}} + \bar{\mathrm{s}} + \mathrm{s}$	0.09	0.13
$\mathrm{c} + \bar{\mathrm{c}}$	0	0.02
Total	1.01	1.06

The functions $V_{\mathrm{ud}}(x)$ and $V_{\mathrm{d}}(x)$ of (E.18), (E.19) are the so-called valence-quark distributions for u- and d-quarks and for the d-quark, respectively. The normalization of these distribution functions is determined by the sum rules for the baryon number and the electrical charge (18.65), (18.66) and reads

$$\int_0^1 \mathrm{d}x\, V_{\mathrm{ud}}(x) = 3, \tag{E.25}$$

$$\int_0^1 \mathrm{d}x\, V_{\mathrm{d}}(x) = 1. \tag{E.26}$$

These relations are satisfied by the explicit expressions given in (E.18), (E.19). Also of interest are the momentum fractions carried by the individual parton types. These are given in Table E.2 for $Q^2 = 4\ \mathrm{GeV}^2$ and $Q^2 = 10^4\ \mathrm{GeV}^2$.

APPENDIX F

The Fierz Transformation

Let ψ_1, ψ_2, ψ_3, and ψ_4 be four arbitrary Dirac field operators. In Table F.1 we consider a set of five Lorentz invariant Lagrange densities $\mathscr{L}_i$ constructed from these fields, that are true scalars also with respect to reflections in space. We assert that the Lagrange densities that are obtained from these by the exchange of ψ_2 and ψ_4 or ψ_1 and ψ_3, can be represented as linear combinations of the original densities $\mathscr{L}_i$,

$$\mathscr{L}_i(\bar{\psi}_1, \psi_4, \bar{\psi}_3, \psi_2) = \sum_j C_{ij} \mathscr{L}_j(\bar{\psi}_1, \psi_2, \bar{\psi}_3, \psi_4), \tag{F.1}$$

with C_{ij} as given in Table F.2. It further holds that

$$\sum_j C_{ij} C_{jk} = \delta_{ik}. \tag{F.2}$$

In order to prove this claim we consider, for example, the Lagrange density $i = \mathrm{S}$ in (F.1) and write out the Dirac indices $(\alpha, \ldots, \delta)$ explicitly

$$\begin{aligned} :(\bar{\psi}_1 \psi_4)(\bar{\psi}_3 \psi_2): &= :\bar{\psi}_{1\alpha} \delta_{\alpha\delta} \psi_{4\delta} \, \bar{\psi}_{3\beta} \delta_{\beta\gamma} \psi_{2\gamma}: \\ &= :\bar{\psi}_{1\alpha} \bar{\psi}_{3\beta} M_{\alpha\beta,\gamma\delta} \psi_{2\gamma} \psi_{4\delta}:, \end{aligned} \tag{F.3}$$

where

$$M_{\alpha\beta,\gamma\delta} = \delta_{\alpha\delta} \delta_{\beta\gamma}. \tag{F.4}$$

For fixed indices α and γ, $M_{\alpha\beta,\gamma\delta}$ is a 4×4 matrix in the indices β and δ. As such it can be expanded in the basis formed by the 4×4 matrices

$$\mathbb{1}, \quad \gamma^\mu, \quad \sigma^{\mu\nu}, \quad \gamma^\mu \gamma_5, \quad \gamma_5 \tag{F.5}$$

Table F.1 Parity conserving scalar Lagrange densities formed from four Dirac field operators. The colons symbolize normal ordering.

i	$\mathscr{L}_i(\bar{\psi}_1, \psi_2, \bar{\psi}_3, \psi_4)$
S	$:(\bar{\psi}_1\psi_2)(\bar{\psi}_3\psi_4):$
V	$:(\bar{\psi}_1\gamma_\mu\psi_2)(\bar{\psi}_3\gamma^\mu\psi_4):$
T	$:(\bar{\psi}_1\sigma_{\mu\nu}\psi_2)(\bar{\psi}_3\sigma^{\mu\nu}\psi_4):$
A	$:(\bar{\psi}_1\gamma_\mu\gamma_5\psi_2)(\bar{\psi}_3\gamma^\mu\gamma_5\psi_4):$
P	$:(\bar{\psi}_1\gamma_5\psi_2)(\bar{\psi}_3\gamma_5\psi_4):$

with coefficients A^i

$$\begin{aligned} M_{\alpha\beta,\gamma\delta} = {} & A^1_{\alpha\gamma}\cdot(\mathbb{1})_{\beta\delta} + (A^2_\mu)_{\alpha\gamma}\cdot(\gamma^\mu)_{\beta\delta} \\ & + (A^3_{\mu\nu})_{\alpha\gamma}\cdot(\sigma^{\mu\nu})_{\beta\delta} + (A^4_\mu)_{\alpha\gamma}\cdot(\gamma^\mu\gamma_5)_{\beta\delta} + A^5_{\alpha\gamma}\cdot(\gamma_5)_{\beta\delta} \end{aligned} \tag{F.6}$$

that remain to be determined. To do so we multiply (F.6) from the right with an element of the basis (F.5) and take the trace. The indices α and γ remain fixed. As is easily seen, a corresponding coefficient A^i is thereby projected out from the right-hand side of (F.6). We obtain, for example,

$$\begin{aligned} M_{\alpha\beta,\gamma\delta}(\mathbb{1})_{\delta\beta} &= A^1_{\alpha\gamma}(\mathbb{1})_{\beta\delta}(\mathbb{1})_{\delta\beta} = A^1_{\alpha\gamma}\cdot 4, \\ A^1_{\alpha\gamma} &= \tfrac{1}{4}M_{\alpha\beta,\gamma\delta}\delta_{\delta\beta} = \tfrac{1}{4}\delta_{\alpha\delta}\delta_{\beta\gamma}\delta_{\delta\beta} = \tfrac{1}{4}\delta_{\alpha\gamma}. \end{aligned} \tag{F.7}$$

$$\begin{aligned} M_{\alpha\beta,\gamma\delta}(\gamma_\nu)_{\delta\beta} &= (A^2_\mu)_{\alpha\gamma}\mathrm{Sp}(\gamma^\mu\gamma_\nu) = (A^2_\nu)_{\alpha\gamma}\cdot 4, \\ (A^2_\nu)_{\alpha\gamma} &= \tfrac{1}{4}(\gamma_\nu)_{\alpha\gamma}. \end{aligned} \tag{F.8}$$

Taking account of the anticommutation rules for Dirac field operators we therefore get

$$\begin{aligned} :(\bar{\psi}_1\psi_4)(\bar{\psi}_3\psi_2): &= :\bar{\psi}_{1\alpha}\bar{\psi}_{3\beta}\{\tfrac{1}{4}\delta_{\alpha\gamma}\delta_{\beta\delta} + \tfrac{1}{4}(\gamma_\mu)_{\alpha\gamma}(\gamma^\mu)_{\beta\delta} + \cdots\}\psi_{2\gamma}\psi_{4\delta}: \\ &= -\tfrac{1}{4}:(\bar{\psi}_1\psi_2)(\bar{\psi}_3\psi_4): - \tfrac{1}{4}:(\bar{\psi}_1\gamma_\mu\psi_2)(\bar{\psi}_3\gamma^\mu\psi_4): + \cdots . \end{aligned} \tag{F.9}$$

Table F.2 The values of the coefficients C_{ij} in (F.1).

i \ j	S	V	T	A	P
S	−1/4	−1/4	−1/8	1/4	−1/4
V	−1	1/2	0	1/2	1
T	−3	0	1/2	0	−3
A	1	1/2	0	1/2	−1
P	−1/4	1/4	−1/8	−1/4	−1/4

We have thus reproduced the coefficients C_{SS} and C_{SV} of Table F.2. All other coefficients can be determined in an analogous way. Equation (F.2) is easy to check explicitly. It is necessarily valid since, by twice exchanging ψ_2 and ψ_4, we of course retrieve the original Lagrange density.

The pseudoscalar Lagrange densities are obtained from those of Table F.1 by multiplying one of the field operators by γ_5. For example, we can make the substitution

$$\psi_4 \to \gamma_5 \psi_4 . \tag{F.10}$$

With this substitution (F.1) also yields the formulas of the Fierz transformation for pseudoscalar Lagrange densities.

APPENDIX G

The Feynman Rules for the Standard Model in the Unitary Gauge

We begin by considering a free massive vector boson, e.g., the Z-boson. The appropriate Lagrange density $\mathscr{L}_Z^{(0)}$ is:

$$\mathscr{L}_Z^{(0)} = -\tfrac{1}{4}(\partial_\mu Z_\nu - \partial_\nu Z_\mu)(\partial^\mu Z^\nu - \partial^\nu Z^\mu) + \tfrac{1}{2} m_Z^2 Z_\mu Z^\mu. \tag{G.1}$$

Here $Z_\mu = Z_\mu^*$ are initially classical real fields and $m_Z \neq 0$. The equations of motion then read:

$$\{g^{\mu\nu}(\Box + m_Z^2) - \partial^\mu \partial^\nu\} Z_\nu(x) = 0. \tag{G.2}$$

If we form the divergence, it follows that

$$m_Z^2 \partial^\mu Z_\mu(x) = 0 \tag{G.3}$$

and thus, because $m_Z \neq 0$

$$\partial^\mu Z_\mu(x) = 0. \tag{G.4}$$

An equivalent form for the equations of motion is therefore

$$\begin{aligned} (\Box + m_Z^2) Z_\mu(x) &= 0, \\ \partial^\mu Z_\mu(x) &= 0. \end{aligned} \tag{G.5}$$

The solutions of negative frequency, which correspond to plane waves, are of the form

$$Z_\mu(x) \propto \mathrm{e}^{-ikx}\, \varepsilon^\mu(k), \tag{G.6}$$

where

$$k = \begin{pmatrix} k^0 \\ \boldsymbol{k} \end{pmatrix}, \qquad k^0 = \omega \equiv +\sqrt{m_Z^2 + \boldsymbol{k}^2},$$

and where $\varepsilon(k)$ is the polarization vector that must satisfy the condition

$$k^\mu \varepsilon_\mu(k) = 0. \tag{G.7}$$

A basis $\varepsilon_1, \varepsilon_2, \varepsilon_3$ for the polarization vectors to a given k value is obtained as follows. We choose three unit vectors $\boldsymbol{e}_1, \boldsymbol{e}_2, \boldsymbol{e}_3$, where $\boldsymbol{e}_3 = \hat{\boldsymbol{k}} \equiv \boldsymbol{k}/|\boldsymbol{k}|$ and

$$\boldsymbol{e}_i^* \cdot \boldsymbol{e}_j = \delta_{ij} \qquad (1 \le i, j \le 3). \tag{G.8}$$

We then write

$$\varepsilon_1(k) = \begin{pmatrix} 0 \\ \boldsymbol{e}_1 \end{pmatrix}, \qquad \varepsilon_2(k) = \begin{pmatrix} 0 \\ \boldsymbol{e}_2 \end{pmatrix}, \qquad \varepsilon_3(k) = \frac{1}{m_Z}\begin{pmatrix} |\boldsymbol{k}| \\ \omega\hat{\boldsymbol{k}} \end{pmatrix} \tag{G.9}$$

and obtain

$$\begin{aligned} \varepsilon_i^{\mu *}(k)\varepsilon_{j\mu}(k) &= -\delta_{ij}, \\ k^\mu \varepsilon_{i\mu}(k) &= 0, \\ \sum_{i=1}^{3} \varepsilon_i^\mu(k)\varepsilon_i^{\nu *}(k) &= -g^{\mu\nu} + \frac{k^\mu k^\nu}{m_Z^2}. \end{aligned} \tag{G.10}$$

An arbitrary polarization vector $\varepsilon(k)$ of the Z-boson, i.e., a general solution of (G.7), has the form

$$\varepsilon(k) = \sum_{i=1}^{3} c_i \varepsilon_i^\mu(k), \tag{G.11}$$

where the c_i are any complex numbers. We will demand the normalization condition

$$\varepsilon^*(k)\varepsilon(k) = -\sum_{i=1}^{3} |c_i|^2 = -1. \tag{G.12}$$

We now go over to quantum field theory and, instead of the classical real fields $Z_\mu = Z_\mu^*$, we consider Hermitian field operators $\mathbf{Z}_\mu = \mathbf{Z}_\mu^\dagger$, satisfying (G.5). Following the same methods as used in Chap. 3, we obtain an expansion of the field operators $\mathbf{Z}_\mu(x)$ in terms of creation and annihilation operators $\mathbf{a}_i^\dagger(\boldsymbol{k})$ and $\mathbf{a}_i(\boldsymbol{k})$, and the commutation rules for the latter:

$$\mathbf{Z}_\mu(x) = \int \frac{\mathrm{d}^3 k}{(2\pi)^3 2\omega} \sum_{j=1}^{3} \{\mathrm{e}^{\mathrm{i}kx}\, \varepsilon_{j\mu}^*(k)\mathbf{a}_j^\dagger(\boldsymbol{k}) + \mathrm{e}^{-\mathrm{i}kx}\, \varepsilon_{j\mu}(k)\mathbf{a}_j(\boldsymbol{k})\}, \tag{G.13}$$

$$[\mathbf{a}_i(\boldsymbol{k}), \mathbf{a}_j^\dagger(\boldsymbol{k}')] = \delta_{ij}(2\pi)^3 2\omega\delta^3(\boldsymbol{k} - \boldsymbol{k}'). \tag{G.14}$$

The vacuum expectation value of the T-product of two field operators yields the Feynman propagator (Sect. 7.2). In the present case the T-product by itself does not lead to covariant results and it is convenient to introduce a modified time-ordered product, the T*-product, that has been made covariant. The Feynman propagator is then *defined* as (see Weinberg 1973)

$$\langle 0|\mathrm{T}^*(\mathbf{Z}_\mu(x)\mathbf{Z}_\nu(y))|0\rangle = \int \frac{\mathrm{d}k}{(2\pi)^4}\mathrm{e}^{-\mathrm{i}k(x-y)} \frac{\mathrm{i}\left(-g_{\mu\nu} + \dfrac{k_\mu k_\nu}{m_Z^2}\right)}{k^2 - m_Z^2 + \mathrm{i}\varepsilon}. \tag{G.15}$$

For the W-bosons the formulas are similar with modifications arising from the fact that the corresponding field operators $\mathbf{W}_\mu^\pm$ are non-Hermitian but satisfy the condition

$$(\mathbf{W}_\mu^-)^\dagger = \mathbf{W}_\mu^+ . \tag{G.16}$$

The expansion of $\mathbf{W}_\mu^-$ reads

$$\mathbf{W}_\mu^-(x) = \int \frac{d^3k}{(2\pi)^3 2\omega} \sum_{j=1}^{3} \{e^{ikx}\, \varepsilon_{j\mu}^*(k)\mathbf{b}_j^\dagger(\boldsymbol{k}) + e^{-ikx}\, \varepsilon_{j\mu}(k)\mathbf{a}_j(\boldsymbol{k})\}, \tag{G.17}$$

where we now have $k^0 = \omega \equiv +\sqrt{m_W^2 + \boldsymbol{k}^2}$. Furthermore, the annihilation operators $\mathbf{a}_j(\boldsymbol{k})$ and $\mathbf{b}_j(\boldsymbol{k})$ of W^- and W^+ particles satisfy the commutation rules:

$$[\mathbf{a}_i(\boldsymbol{k}), \mathbf{a}_j^\dagger(\boldsymbol{k}')] = [\mathbf{b}_i(\boldsymbol{k}), \mathbf{b}_j^\dagger(\boldsymbol{k}')] = \delta_{ij}(2\pi)^3 2\omega\delta^3(\boldsymbol{k} - \boldsymbol{k}'). \tag{G.18}$$

The starting point for a derivation of the Feynman rules for the standard model in the unitary gauge is the Lagrange density of (22.123). In order to obtain the desired rules one can apply either the canonical formalism (see Weinberg 1973) or the path-integral method (see, e.g., Itzykson 1980). Here we will simply give the result.

The rules for incoming, outgoing, and virtual fermions, photons, and gluons remain the same as in QED and QCD (Appendices B and D). It should be noted, however, that in the standard model only left-handed neutrino fields appear. This means that only u_- occurs as Dirac spinor in the wave function of an external neutrino, and only v_- for antineutrinos (21.20), (21.21).

All vertices and rules concerning gluons remain as in QCD.

The following rules—specific to QFD—represent the additions.

W^- initial state — incoming W line

$\varepsilon(k)$

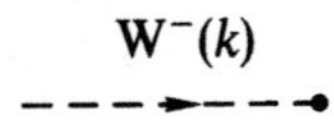

W^- in final state — outgoing W line

$\varepsilon^*(k)$

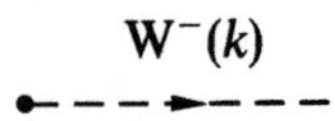

W^+ in initial state — outgoing W line

$\varepsilon(k)$

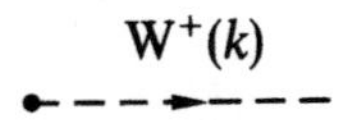

W^+ in final state — incoming W line

$\varepsilon^*(k)$

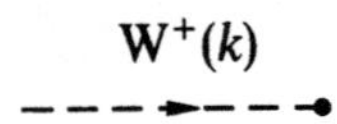

Z in initial (final) state	external Z line
$\varepsilon(k)(\varepsilon^*(k))$	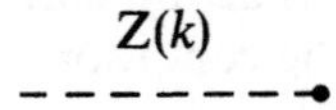
Higgs particle in initial (final) state	external ρ' line
1	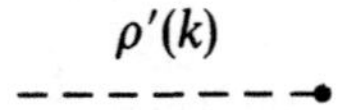
virtual W boson	internal W line
$\dfrac{\mathrm{i}\left(-g^{\mu\nu}+\dfrac{k^\mu k^\nu}{m_W^2}\right)}{k^2-m_W^2+\mathrm{i}\varepsilon}$	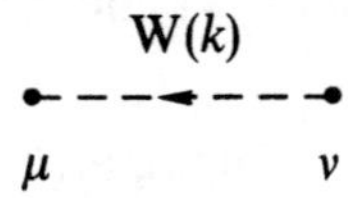
virtual Z boson	internal Z line
$\dfrac{\mathrm{i}\left(-g^{\mu\nu}+\dfrac{k^\mu k^\nu}{m_Z^2}\right)}{k^2-m_Z^2+\mathrm{i}\varepsilon}$	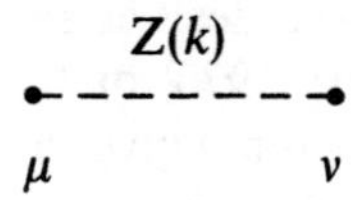
virtual Higgs particle	internal ρ' line
$\dfrac{\mathrm{i}}{k^2-m_{\rho'}^2+\mathrm{i}\varepsilon}$	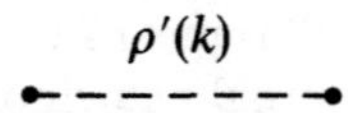

Vertices with 3 vector bosons $\left(\sum_{j=1}^{3} k_j = 0, \text{ all momenta incoming}\right)$:

$$\mathrm{i}e\{(k_1-k_2)_{\mu_3}g_{\mu_1\mu_2} + (k_2-k_3)_{\mu_1}g_{\mu_2\mu_3} + (k_3-k_1)_{\mu_2}g_{\mu_3\mu_1}\}$$

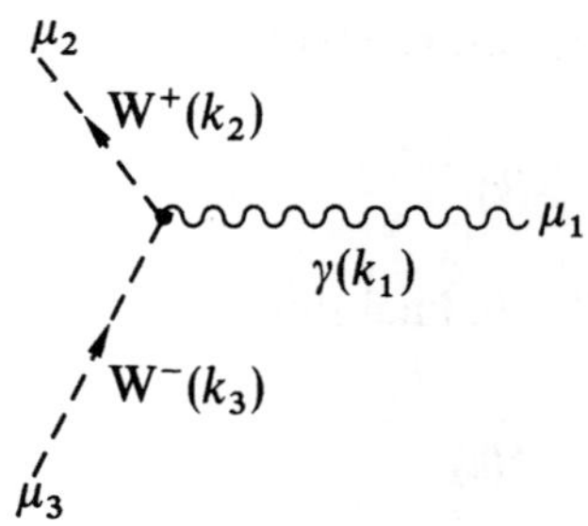

$$\mathrm{i}e\frac{\cos\vartheta_W}{\sin\vartheta_W}\{\text{as above}\}$$

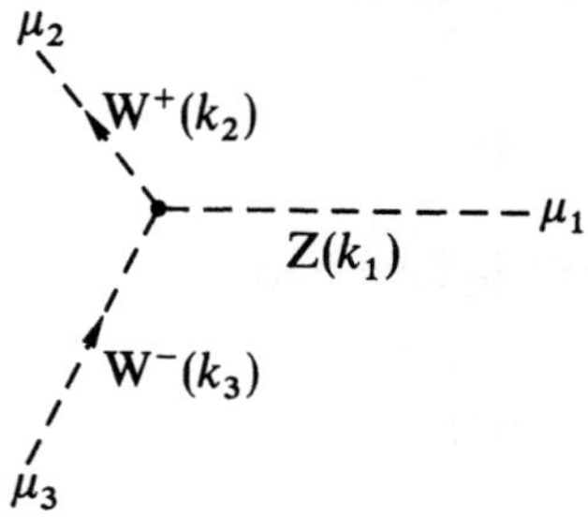

Vertices with 4 vector bosons $\left(\sum_{j=1}^{4} k_j = 0, \text{ all momenta incoming} \right)$:

$$ie^2\{g_{\mu_1\mu_3}g_{\mu_2\mu_4} + g_{\mu_1\mu_4}g_{\mu_2\mu_3} - 2g_{\mu_1\mu_2}g_{\mu_3\mu_4}\}$$

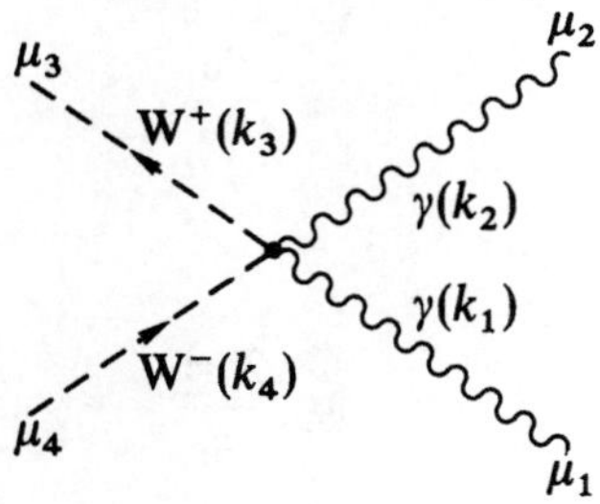

$$ie^2 \frac{\cos \vartheta_W}{\sin \vartheta_W} \{\text{as above}\}$$

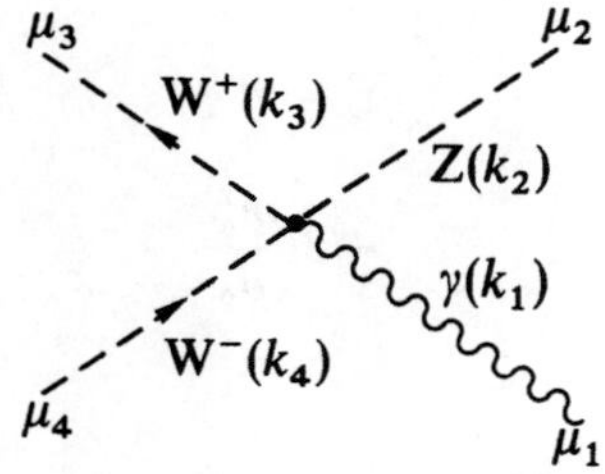

$$ie^2 \frac{\cos^2 \vartheta_W}{\sin^2 \vartheta_W} \{\text{as above}\}$$

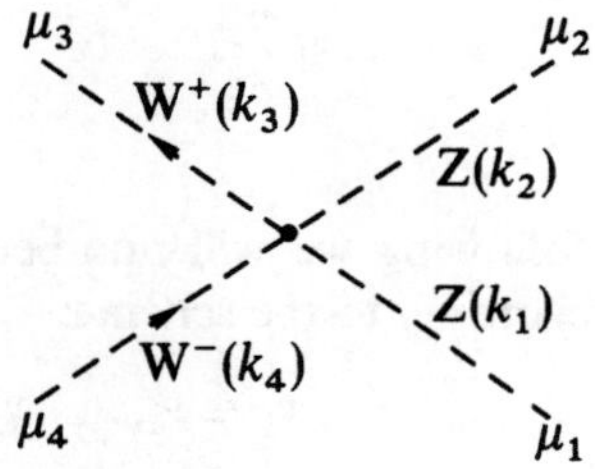

$$-ie^2 \frac{1}{\sin^2 \vartheta_W} \{\text{as above}\}$$

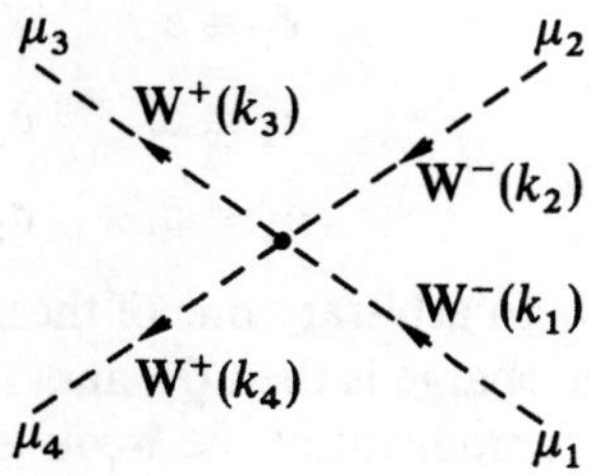

Vertices with Higgs bosons:

$$ig_{\mu\nu} \frac{2m_W^2}{\rho_0}$$

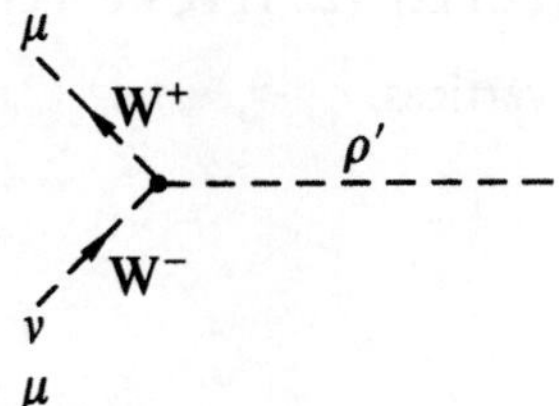

$$ig_{\mu\nu} \frac{2m_W^2}{\rho_0^2}$$

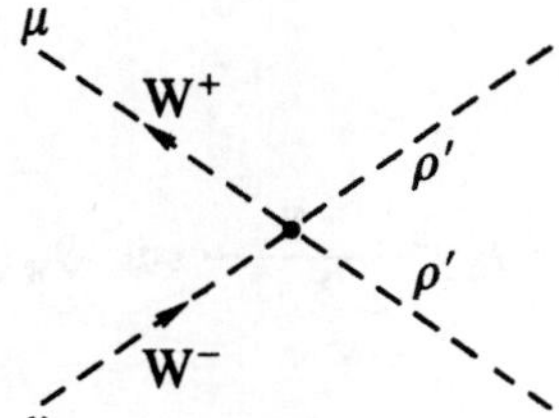

$$\mathrm{i}g_{\mu\nu}\frac{2m_Z^2}{\rho_0}$$

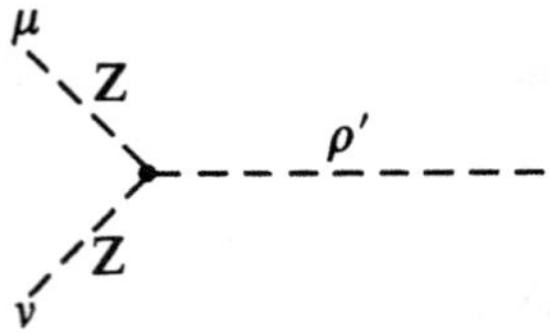

$$\mathrm{i}g_{\mu\nu}\frac{2m_Z^2}{\rho_0^2}$$

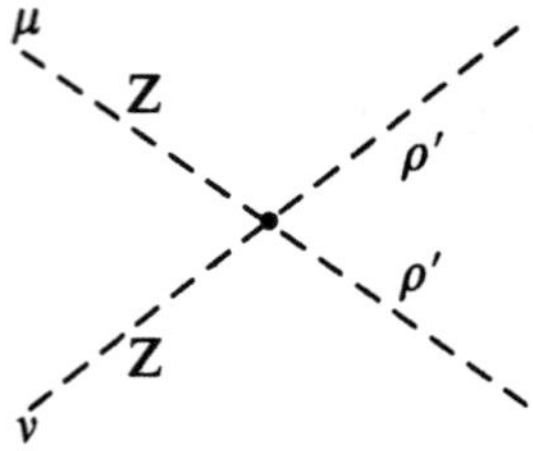

$$-3\mathrm{i}\frac{m_{\rho'}^2}{\rho_0}$$

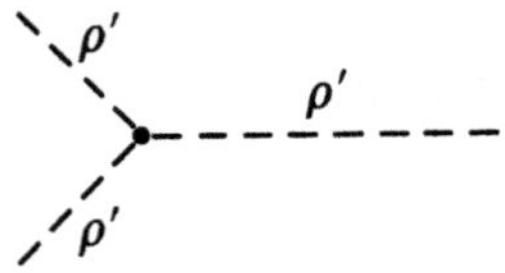

$$-3\mathrm{i}\frac{m_{\rho'}^2}{\rho_0^2}$$

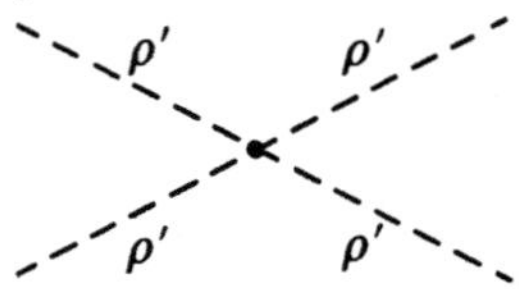

In the following we will number the leptons and quarks of the various families according to the scheme:

$$\begin{array}{lll} \nu_1 \equiv \nu_e, & \nu_2 \equiv \nu_\mu, & \nu_3 \equiv \nu_\tau; \\ \ell_1 \equiv \mathrm{e}, & \ell_2 \equiv \mu, & \ell_3 \equiv \tau; \\ \mathrm{u}_1 \equiv \mathrm{u}, & \mathrm{u}_2 \equiv \mathrm{c}, & \mathrm{u}_3 \equiv \mathrm{t}; \\ \mathrm{d}_1 \equiv \mathrm{d}, & \mathrm{d}_2 \equiv \mathrm{s}, & \mathrm{d}_3 \equiv \mathrm{b}. \end{array}$$

To refer to an arbitrary one of these particles we will use the generic label f; the electric charge is then Q_f, and T_3^f is the eigenvalue obtained by applying the third component of the weak isospin to the *left-handed* part of f. From (22.123), (22.77), (22.112)–(22.114), we then obtain the following vertices:

Fermion–Boson vertices:

$$-\mathrm{i}eQ_f\gamma^\mu$$

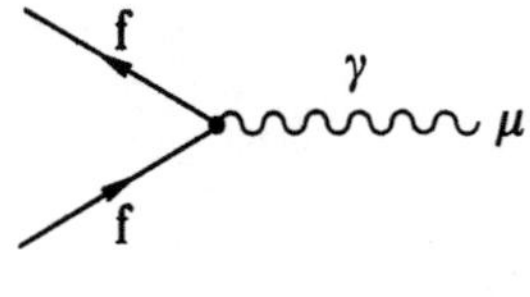

$$-\mathrm{i}\frac{e}{\sin\vartheta_W\cos\vartheta_W}\left\{T_3^f\gamma^\mu\frac{1-\gamma_5}{2}-\sin^2\vartheta_W Q_f\gamma^\mu\right\}$$

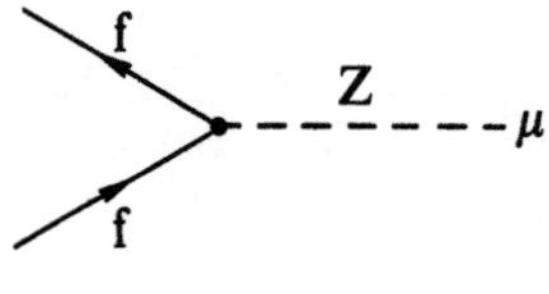

$$-\mathrm{i}\frac{e}{\sqrt{2}\sin\vartheta_{\mathrm{W}}}\gamma^{\mu}\frac{1-\gamma_5}{2}$$

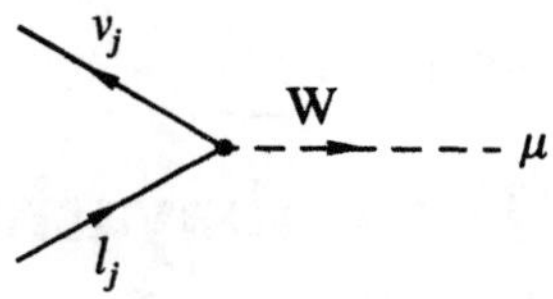

$$-\mathrm{i}\frac{e}{\sqrt{2}\sin\vartheta_{\mathrm{W}}}\gamma^{\mu}\frac{1-\gamma_5}{2}$$

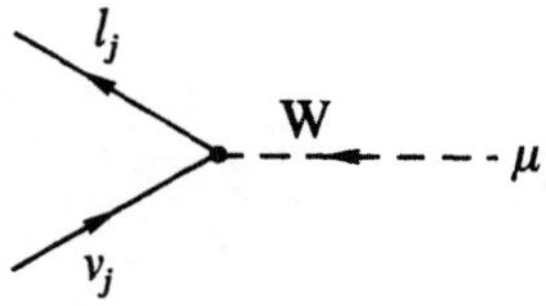

$$-\mathrm{i}\frac{e}{\sqrt{2}\sin\vartheta_{\mathrm{W}}}V_{ij}\,\gamma^{\mu}\frac{1-\gamma_5}{2}$$

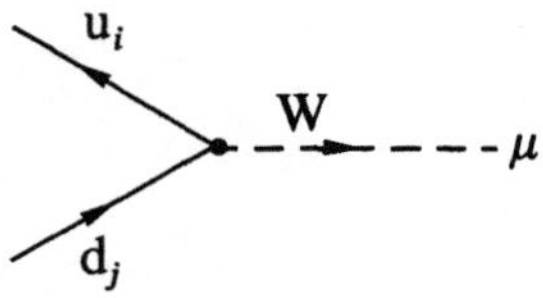

$$-\mathrm{i}\frac{e}{\sqrt{2}\sin\vartheta_{\mathrm{W}}}V_{ji}^{*}\,\gamma^{\mu}\frac{1-\gamma_5}{2}$$

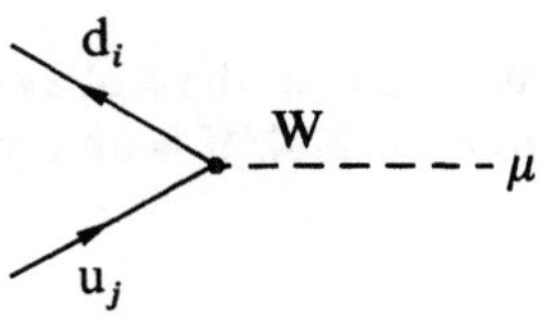

$$-\mathrm{i}\frac{m_{\mathrm{f}}}{\rho_0}$$

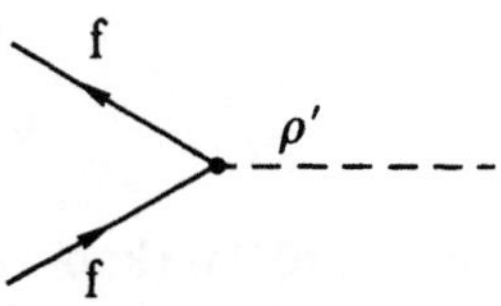

With the aid of these rules one can calculate all tree diagrams in the standard model. For diagrams that include loops one must—even in the unitary gauge—also consider Fadeev–Popov ghosts, which couple to the Higgs field ρ', or corresponding contact terms (see Weinberg 1973). However, for the calculation of diagrams with loops, the unitary gauge is not particularly well suited since its individual diagrams—as in a *nonrenormalizable* theory—are highly divergent, although the sum of the diagrams of a particular order shows only the divergence of a *renormalizable* theory. Thus for carrying out such calculations, one usually chooses a gauge in which the individual diagrams themselves only possess the divergences of a renormalizable theory, e.g., the Feynman–'t Hooft gauge (see, e.g., Itzkyson 1980; Becher 1984).

APPENDIX H

The Kobayashi–Maskawa Matrix for Three Families

We consider the Kobayashi–Maskawa matrix **V** for three families. According to Sect. 22.4, **V** is unitary:

$$\mathbf{V} = \begin{pmatrix} V_{11} & V_{12} & V_{13} \\ V_{21} & V_{22} & V_{23} \\ V_{31} & V_{32} & V_{33} \end{pmatrix}, \tag{H.1}$$

$$\mathbf{V}\mathbf{V}^{+} = \mathbb{1}. \tag{H.2}$$

From (22.102) we know that we have the freedom to multiply **V** from the right and the left by unitary diagonal matrices:

$$\begin{aligned}\mathbf{V} \to & \begin{pmatrix} e^{-i\varphi_1} & 0 & 0 \\ 0 & e^{-i\varphi_2} & 0 \\ 0 & 0 & e^{-i\varphi_3} \end{pmatrix} \mathbf{V} \begin{pmatrix} e^{i\chi_1} & 0 & 0 \\ 0 & e^{i\chi_2} & 0 \\ 0 & 0 & e^{i\chi_3} \end{pmatrix} \\ = & \begin{pmatrix} e^{-i(\varphi_1-\chi_1)} V_{11} & e^{-i(\varphi_1-\chi_2)} V_{12} & e^{-i(\varphi_1-\chi_3)} V_{13} \\ e^{-i(\varphi_2-\chi_1)} V_{21} & e^{-i(\varphi_2-\chi_2)} V_{22} & e^{-i(\varphi_2-\chi_3)} V_{23} \\ e^{-i(\varphi_3-\chi_1)} V_{31} & e^{-i(\varphi_3-\chi_2)} V_{32} & e^{-i(\varphi_3-\chi_3)} V_{33} \end{pmatrix}. \end{aligned} \tag{H.3}$$

Of the phase differences $\varphi_i - \chi_j$, five are linearly independent and can be freely chosen. We can thus prescribe the phases of five of the matrix elements V_{ij}. As a canonical convention we impose

$$\begin{aligned} V_{11} &\geq 0, \\ V_{12} &\geq 0, \\ V_{13} &\geq 0, \\ V_{21} &\leq 0, \\ V_{31} &\leq 0. \end{aligned} \tag{H.4}$$

Our task now is to find a suitable parametrization for unitary matrices $\mathbf{V}$ that satisfy (H.4). Since $\mathbf{V}$ is unitary, we have

$$\begin{aligned} (V_{11})^2 + (V_{12})^2 + (V_{13})^2 &= 1, \\ (V_{11})^2 + (V_{21})^2 + (V_{31})^2 &= 1. \end{aligned} \tag{H.5}$$

These equations and the inequalities (H.4) are identically satisfied if we introduce the angles $\vartheta_1, \vartheta_2, \vartheta_3$ with

$$0 \leq \vartheta_i \leq \frac{\pi}{2} \qquad (i = 1, 2, 3) \tag{H.6}$$

and set

$$\begin{aligned} V_{11} &= c_1, \\ V_{12} &= s_1 c_3, \\ V_{13} &= s_1 s_3, \\ V_{21} &= -s_1 c_2, \\ V_{31} &= -s_1 s_2, \end{aligned} \tag{H.7}$$

where $c_i \equiv \cos \vartheta_i$ and $s_i \equiv \sin \vartheta_i$. In addition, we have to satisfy the unitarity relations

$$V_{11} V_{21} + V_{12} V_{22} + V_{13} V_{23} = 0, \tag{H.8}$$

$$(V_{21})^2 + |V_{22}|^2 + |V_{23}|^2 = 1. \tag{H.9}$$

From (H.8) it follows that

$$c_3 V_{22} + s_3 V_{23} = c_1 c_2 \tag{H.10}$$

with the general solution

$$\begin{aligned} V_{22} &= c_1 c_2 c_3 + \lambda s_2 s_3, \\ V_{23} &= c_1 c_2 s_3 - \lambda s_2 c_3, \end{aligned} \tag{H.11}$$

where λ is complex. Equation (H.9) implies $|\lambda| = 1$, i.e., we can write

$$\lambda = -e^{i\delta}, \tag{H.12}$$

where $0 \leq \delta \leq 2\pi$. From the relations

$$\begin{aligned} V_{11} V_{31} + V_{12} V_{32} + V_{13} V_{33} &= 0, \\ (V_{31})^2 + |V_{32}|^2 + |V_{33}|^2 &= 1, \end{aligned} \tag{H.13}$$

it follows analogously that

$$\begin{aligned} V_{32} &= c_1 s_2 c_3 - \lambda' c_2 s_3, \\ V_{33} &= c_1 s_2 s_3 + \lambda' c_2 c_3, \\ |\lambda'| &= 1. \end{aligned} \tag{H.14}$$

The relation

$$V_{21}^* V_{31} + V_{22}^* V_{32} + V_{23}^* V_{33} = 0, \tag{H.15}$$

finally, yields

$$\lambda = \lambda'. \tag{H.16}$$

Collecting all this information together we obtain the following canonical form for $\mathbf{V}$:

$$\mathbf{V} = \begin{pmatrix} c_1 & s_1 c_3 & s_1 s_3 \\ -s_1 c_2 & c_1 c_2 c_3 - e^{i\delta} s_2 s_3 & c_1 c_2 s_3 + e^{i\delta} s_2 c_3 \\ -s_1 s_2 & c_1 s_2 c_3 + e^{i\delta} c_2 s_3 & c_1 s_2 s_3 - e^{i\delta} c_2 c_3 \end{pmatrix}, \tag{H.17}$$

where

$$\begin{aligned} c_i &\equiv \cos \vartheta_i, \\ s_i &\equiv \sin \vartheta_i, \\ 0 \le \vartheta_i &\le \frac{\pi}{2}, \\ 0 \le \delta &< 2\pi. \end{aligned} \tag{H.18}$$

Another commonly used convention for the matrix $\mathbf{V}$ is given by Maiani (1977).

APPENDIX I

The Wigner–Weisskopf Approximation for the Description of the Decay of Unstable Particles and the K^0–$\bar{K}^0$ System

I.1 General Formalism

The problem to be treated here is the calculation of the time evolution of the wavefunctions of unstable states such as the mesons K^0 and $\bar{K}^0$.

We consider a system that is described by a Hamilton operator $\mathbf{H}$:

$$\mathbf{H} = \mathbf{H}_0 + \mathbf{H}', \tag{I.1}$$

where $\mathbf{H}_0$ is the unperturbed Hamilton operator and $\mathbf{H}'$ is a small perturbation. In the case of the K-mesons, we let $\mathbf{H}_0$ be the Hamiltonian of the strong interaction and $\mathbf{H}'$ that of the weak interaction. The eigenstates of $\mathbf{H}_0$ are assumed to consist of n degenerate discrete states $|\alpha\rangle$ and a series of continuum states $|\beta\rangle$:

$$\mathbf{H}_0|\alpha\rangle = E_0|\alpha\rangle \qquad (\alpha = 1, \ldots, n), \tag{I.2}$$

$$\mathbf{H}_0|\beta\rangle = E_\beta|\beta\rangle. \tag{I.3}$$

When $\mathbf{H}'$ is switched on, it should be possible for the states $|\alpha\rangle$ to decay into the continuum states $|\beta\rangle$.

To begin with we shall work in the Schrödinger picture. Every state vector $|t\rangle_S$ at time t can be expanded in terms of the eigenstates $|\alpha\rangle$ and $|\beta\rangle$ of $\mathbf{H}_0$:

$$|t\rangle_S = \sum_{\alpha=1}^{n} \psi_\alpha(t)|\alpha\rangle + \sum_\beta c_\beta(t)|\beta\rangle. \tag{I.4}$$

As an initial state at $t = 0$ we consider a superposition of the discrete states $|\alpha\rangle$:

$$|t = 0\rangle_S = \sum_{\alpha=1}^{n} \psi_\alpha^{(0)}|\alpha\rangle. \tag{I.5}$$

The problem is the calculation of the time evolution of such a state. We expect an exponential decay of the amplitudes of the discrete states $|\alpha\rangle$ and a corresponding growth of the amplitudes of the continuum states $|\beta\rangle$. The Wigner–Weisskopf method shows that with certain approximations this is actually the case.

The time evolution of the state vectors is described by the Schrödinger equation:

$$\mathrm{i}\frac{\partial}{\partial t}|t\rangle_S = \mathbf{H}|t\rangle_S. \tag{I.6}$$

We now switch to the interaction picture which turns out to be very convenient here. We write

$$\begin{aligned} |t = 0\rangle_W &= |t = 0\rangle_S, \\ |t\rangle_W &= \mathrm{e}^{\mathrm{i}\mathbf{H}_0 t}\,|t\rangle_S = \sum_{\alpha=1}^{n} a_\alpha(t)|\alpha\rangle + \sum_\beta b_\beta(t)|\beta\rangle. \end{aligned} \tag{I.7}$$

The equation of motion then becomes

$$\mathrm{i}\frac{\partial}{\partial t}|t\rangle_W = \mathbf{H}'(t)|t\rangle_W, \tag{I.8}$$

where

$$\mathbf{H}'(t) = \mathrm{e}^{\mathrm{i}\mathbf{H}_0 t}\,\mathbf{H}'\,\mathrm{e}^{-\mathrm{i}\mathbf{H}_0 t}. \tag{I.9}$$

In terms of amplitudes (I.8) reads:

$$\mathrm{i}\frac{\partial a_\alpha(t)}{\partial t} = \sum_{\alpha'} \langle\alpha|\mathbf{H}'|\alpha'\rangle a_{\alpha'}(t) + \sum_\beta \mathrm{e}^{\mathrm{i}(E_0 - E_\beta)t}\,\langle\alpha|\mathbf{H}'|\beta\rangle b_\beta(t), \tag{I.10}$$

$$\begin{aligned} \mathrm{i}\frac{\partial b_\beta(t)}{\partial t} &= \sum_{\alpha'} \mathrm{e}^{\mathrm{i}(E_\beta - E_0)t}\,\langle\beta|\mathbf{H}'|\alpha'\rangle a_{\alpha'}(t) \\ &\quad + \sum_{\beta'} \mathrm{e}^{\mathrm{i}(E_\beta - E_{\beta'})t}\,\langle\beta|\mathbf{H}'|\beta'\rangle b_{\beta'}(t). \end{aligned} \tag{I.11}$$

The first approximation that is introduced in the Wigner–Weisskopf approach is to neglect the second sum on the right-hand side of (I.11). For our example of K-meson decay this amounts to the neglect of the weak interaction for the particles into which the K-mesons decay. In particular, this means that we consider the π-mesons and muons that occur in K-decay as stable particles.

With this approximation, equations (I.10), (I.11) can be solved. With the initial conditions

$$\begin{aligned} a_\alpha(0) &= \psi_\alpha^{(0)}, \\ b_\beta(0) &= 0, \end{aligned} \tag{I.12}$$

we first obtain by integration

$$b_\beta(t) = -\mathrm{i}\sum_{\alpha'}\int_0^t \mathrm{d}t'\, \mathrm{e}^{\mathrm{i}(E_\beta - E_0)t'} \langle\beta|\mathbf{H}'|\alpha'\rangle a_{\alpha'}(t'), \tag{I.13}$$

$$\begin{aligned} a_\alpha(t) = \psi_\alpha^{(0)} - \mathrm{i}\int_0^t \mathrm{d}t' \sum_{\alpha'} \langle\alpha|\mathbf{H}'|\alpha'\rangle a_{\alpha'}(t') \\ - \sum_{\alpha',\beta}\int_0^t \mathrm{d}t' \int_0^{t'} \mathrm{d}t''\, \mathrm{e}^{\mathrm{i}(E_0 - E_\beta)(t'-t'')} \langle\alpha|\mathbf{H}'|\beta\rangle\langle\beta|\mathbf{H}'|\alpha'\rangle a_{\alpha'}(t''). \end{aligned} \tag{I.14}$$

In contrast to (I.10), (I.11), where the amplitudes a_α and b_β are mutually coupled (I.14) involves the amplitudes a_α only which are *finite* in number. We can solve (I.14) by means of a Laplace transform. We write

$$\tilde{a}_\alpha(\sigma) = \int_0^\infty \mathrm{d}t\, \mathrm{e}^{-\sigma t}\, a_\alpha(t) \tag{I.15}$$

and obtain

$$\tilde{a}_\alpha(\sigma) = \frac{1}{\sigma}\psi_\alpha^{(0)} - \frac{\mathrm{i}}{\sigma}\sum_{\alpha'} W_{\alpha\alpha'}(\sigma)\tilde{a}_{\alpha'}(\sigma), \tag{I.16}$$

where

$$W_{\alpha\alpha'}(\sigma) = \langle\alpha|\mathbf{H}'|\alpha'\rangle + \sum_\beta \frac{\langle\alpha|\mathbf{H}'|\beta\rangle\langle\beta|\mathbf{H}'|\alpha'\rangle}{E_0 - E_\beta + \mathrm{i}\sigma}. \tag{I.17}$$

We go over to matrix notation and set

$$\psi^{(0)} = \begin{pmatrix} \psi_1^{(0)} \\ \vdots \\ \psi_n^{(0)} \end{pmatrix}, \quad a(t) = \begin{pmatrix} a_1(t) \\ \vdots \\ a_n(t) \end{pmatrix}, \quad \tilde{a}(\sigma) = \begin{pmatrix} \tilde{a}_1(\sigma) \\ \vdots \\ \tilde{a}_n(\sigma) \end{pmatrix}, \tag{I.18}$$

$$\mathbf{W}(\sigma) = (W_{\alpha\alpha'}(\sigma)).$$

With this we obtain from (I.16)

$$(\sigma + \mathrm{i}\mathbf{W}(\sigma))\cdot\tilde{a}(\sigma) = \psi^{(0)}, \tag{I.19}$$

$$\tilde{a}(\sigma) = (\sigma + \mathrm{i}\mathbf{W}(\sigma))^{-1}\cdot\psi^{(0)}. \tag{I.20}$$

The inverse Laplace transform may then be applied to yield

$$a(t) = \frac{1}{2\pi\mathrm{i}}\int_{\sigma_0 - \mathrm{i}\infty}^{\sigma_0 + \mathrm{i}\infty} \mathrm{d}\sigma\, \mathrm{e}^{\sigma t}\, \tilde{a}(\sigma) = \frac{1}{2\pi\mathrm{i}}\int_{\sigma_0 - \mathrm{i}\infty}^{\sigma_0 + \mathrm{i}\infty} \mathrm{d}\sigma\, \mathrm{e}^{\sigma t}\, (\sigma + \mathrm{i}\mathbf{W}(\sigma))^{-1}\psi^{(0)}. \tag{I.21}$$

The value of σ_0 must be so chosen that the integration path in the complex σ-plane lies to the right of all singularities of $(\sigma + \mathrm{i}\mathbf{W}(\sigma))^{-1}$; see Fig. I.1.

We now discuss the properties of the matrix function $\mathbf{W}(\sigma)$ in the complex σ-plane. According to (I.17) $\mathbf{W}(\sigma)$ is analytic except at those points where

$$E_0 - E_\beta + \mathrm{i}\sigma = 0 \tag{I.22}$$

is satisfied for some β. For this it is necessary to assume that the integral over

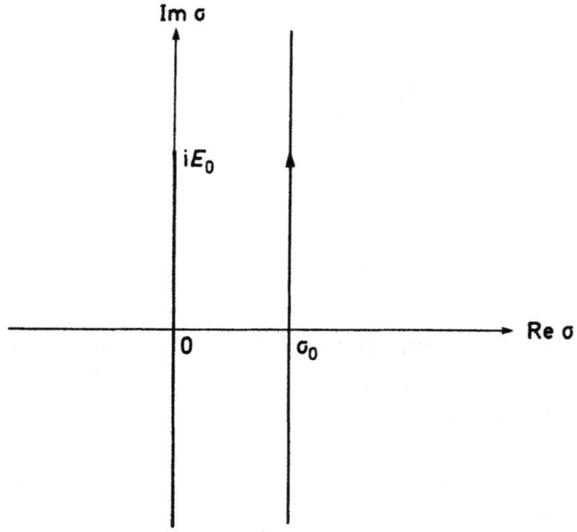

Figure I.1 The integration path in the complex σ-plane for the integral in (I.21). The position of the cut of $\mathbf{W}(\sigma)$ and $[\sigma + i\mathbf{W}(\sigma)]^{-1}$ is indicated by the thick line.

the continuum states $|\beta\rangle$ is sufficiently convergent. We now set

$$E_0' = E_0 - \min_{\beta} E_\beta .$$

As can easily be seen, the matrix valued function $\mathbf{W}(\sigma)$ has a cut from $-i\infty$ to iE_0' along the imaginary axis (Fig. I.1). We will now demonstrate that $\sigma + i\mathbf{W}(\sigma)$ has a nonvanishing determinant for Re $\sigma \neq 0$, and thus, an inverse. We shall use the notation

$$\begin{aligned} \sigma_1 &= \mathrm{Re}\,\sigma, \\ \sigma_2 &= \mathrm{Im}\,\sigma. \end{aligned} \tag{I.23}$$

According to (I.17) we can write

$$\sigma + i\mathbf{W}(\sigma) = \mathbf{X} + i\mathbf{Y}, \tag{I.24}$$

where $\mathbf{X} = (X_{\alpha\alpha'})$, $\mathbf{Y} = (Y_{\alpha\alpha'})$ and

$$\begin{aligned} X_{\alpha\alpha'} &= \sigma_1 \left[\delta_{\alpha\alpha'} + \sum_\beta \frac{\langle\alpha|\mathbf{H}'|\beta\rangle\langle\beta|\mathbf{H}'|\alpha'\rangle}{(E_0 - E_\beta - \sigma_2)^2 + \sigma_1^2} \right], \\ Y_{\alpha\alpha'} &= \sigma_2\delta_{\alpha\alpha'} + \langle\alpha|\mathbf{H}'|\alpha'\rangle + \sum_\beta \langle\alpha|\mathbf{H}'|\beta\rangle\langle\beta|\mathbf{H}'|\alpha'\rangle \frac{E_0 - E_\beta - \sigma_2}{(E_0 - E_\beta - \sigma_2)^2 + \sigma_1^2}. \end{aligned} \tag{I.25}$$

It clearly holds that

$$\begin{aligned} \mathbf{X} &= \mathbf{X}^\dagger, \\ \mathbf{Y} &= \mathbf{Y}^\dagger. \end{aligned} \tag{I.26}$$

We first consider the case $\sigma_1 > 0$. It then follows immediately from (I.25) that $\mathbf{X}$ is a positive definite matrix:

$$\mathbf{X} > 0. \tag{I.27}$$

Thus $\mathbf{X}$ has an inverse and

$$(\mathbf{X} + \mathrm{i}\mathbf{Y})\mathbf{X}^{-1}(\mathbf{X} - \mathrm{i}\mathbf{Y}) = \mathbf{X} + \mathbf{Y}\mathbf{X}^{-1}\mathbf{Y} > 0. \tag{I.28}$$

From this it follows that

$$\begin{aligned} |\det(\mathbf{X} + \mathrm{i}\mathbf{Y})|^2(\det \mathbf{X})^{-1} &> 0, \\ \det(\mathbf{X} + \mathrm{i}\mathbf{Y}) &\neq 0. \end{aligned} \tag{I.29}$$

We have thus shown that for $\sigma_1 > 0$ the matrix $\mathbf{X} + \mathrm{i}\mathbf{Y}$ has an inverse. For $\sigma_1 < 0$ the proof is analogous. From (I.24) the matrix function $(\sigma + \mathrm{i}\mathbf{W}(\sigma))^{-1}$ is thus regular for $\operatorname{Re}\sigma \neq 0$. Its singularities lie on the imaginary axis and above $\mathrm{i}E'_0$ it is clear that only poles can occur. Below $\mathrm{i}E'_0$, both $(\sigma + \mathrm{i}\mathbf{W}(\sigma))^{-1}$ and $\mathbf{W}(\sigma)$ generally have a cut.

We can now move the integration contour in (I.21) almost onto the imaginary axis from the right (Fig. I.1). The question is: From where do we expect the main contribution to the integral?

For $\mathbf{H}' = 0$ we have $\mathbf{W}(\sigma) = 0$ and therefore as singularity only a pole at $\sigma = 0$ occurs in the integrand of (I.21). In this case we immediately obtain

$$a(t) = \psi^{(0)} \qquad \text{for} \quad t \geq 0, \tag{I.30}$$

as is necessary since the states $|\alpha\rangle$ are then stable. Here the main contribution to the integral (I.21) naturally comes from the region of the pole at $\sigma = 0$. When $\mathbf{H}'$ is switched on we expect that this pole will move through the cut onto the second sheet, but for small perturbations we expect it to stay close to the imaginary axis. This pole should continue to yield the main contribution to the integral (I.21). The second approximation of the Wigner–Weisskopf approach is to consider only the contribution from this pole, that is, to replace $\mathbf{W}(\sigma)$ in the vicinity of $\sigma = 0$, $\operatorname{Re}\sigma > 0$, by a constant:

$$\mathbf{W}(\sigma) \to \mathbf{W} \equiv \lim_{\sigma \to +0} \mathbf{W}(\sigma). \tag{I.31}$$

From (I.17) this constant is (P = principal value prescription)

$$\begin{aligned} W_{\alpha\alpha'} = \langle\alpha|\mathbf{H}'|\alpha'\rangle &+ \sum_\beta \frac{\langle\alpha|\mathbf{H}'|\beta\rangle\langle\beta|\mathbf{H}'|\alpha'\rangle}{(E_0 - E_\beta)_\mathrm{P}} \\ &- \mathrm{i}\pi \sum_\beta \delta(E_0 - E_\beta)\langle\alpha|\mathbf{H}'|\beta\rangle\langle\beta|\mathbf{H}'|\alpha'\rangle. \end{aligned} \tag{I.32}$$

Using the residue theorem to evaluate (I.21), we obtain for $t > 0$

$$a(t) = \frac{1}{2\pi\mathrm{i}} \oint \mathrm{d}\sigma\, \mathrm{e}^{\sigma t} \frac{1}{\sigma + \mathrm{i}\mathbf{W}} \psi^{(0)} = \mathrm{e}^{-\mathrm{i}\mathbf{W}t}\, \psi^{(0)}. \tag{I.33}$$

Returning to the Schrödinger picture, we find from (I.7) that the time evolution of the amplitudes $\psi_\alpha(t)$ of the discrete states in (I.4) is given by

$$\psi(t) \equiv (\psi_\alpha(t)) = \mathrm{e}^{-\mathrm{i}\mathcal{M}t}\, \psi^{(0)}, \tag{I.34}$$

where

$$\mathcal{M} = E_0 + \mathbf{W}. \tag{I.35}$$

The matrix $\mathcal{M}$ is known as the non-Hermitian mass- or energy-matrix. We define the Hermitian part $\mathbf{M} = \mathbf{M}^\dagger$ of the mass-matrix and the decay matrix $\Gamma = \Gamma^\dagger$ by

$$\mathbf{M} = \tfrac{1}{2}(\mathcal{M} + \mathcal{M}^\dagger), \qquad \Gamma = \mathrm{i}(\mathcal{M} - \mathcal{M}^\dagger). \tag{I.36}$$

We then have

$$\mathcal{M} = \mathbf{M} - \frac{\mathrm{i}}{2}\Gamma \tag{I.37}$$

with $\mathbf{M}$ and Γ given according to (I.32) by

$$M_{\alpha\alpha'} = E_0\delta_{\alpha\alpha'} + \langle\alpha|\mathbf{H}'|\alpha'\rangle + \sum_\beta \frac{\langle\alpha|\mathbf{H}'|\beta\rangle\langle\beta|\mathbf{H}'|\alpha'\rangle}{(E_0 - E_\beta)_\mathrm{P}},$$
$$\Gamma_{\alpha\alpha'} = \sum_\beta 2\pi\delta(E_0 - E_\beta)\langle\alpha|\mathbf{H}'|\beta\rangle\langle\beta|\mathbf{H}'|\alpha'\rangle. \tag{I.38}$$

Thus the time evolution of the amplitudes describing the persistence of the initial particles has been determined. We now show that (I.34) implies that $\psi(t)$ decays exponentially with time t. For this we introduce the right and left eigenvectors $\varphi_j = (\varphi_{j\alpha})$ and $\tilde{\varphi}_j = (\tilde{\varphi}_{j\alpha})$ $(\alpha = 1,\ldots,n; j = 1,\ldots,n)$ of the generally non-Hermitian matrix $\mathcal{M}$. As can easily be seen, the eigenvalues M_j for the right and left eigenvectors are identical and we have

$$\mathcal{M}\varphi_j = M_j\varphi_j,$$
$$\tilde{\varphi}_j^\dagger\mathcal{M} = \tilde{\varphi}_j^\dagger M_j \qquad (j = 1,\ldots n;\ \text{no summation over } j). \tag{I.39}$$

We assume here that $\mathcal{M}$ has n linearly independent eigenvectors, as is the case, for example, when all eigenvalues M_j $(j = 1, \ldots, n)$ are different. It is then evident that one can demand the following normalization for the eigenvectors:

$$\tilde{\varphi}_j^\dagger\varphi_k = \delta_{jk}. \tag{I.40}$$

We write the (generally complex) eigenvalues as

$$M_j = m_j - \frac{\mathrm{i}}{2}\Gamma_j. \tag{I.41}$$

where m_j and Γ_j are real. It is easy to show that

$$\Gamma_j \geq 0. \tag{I.42}$$

The proof starts from (I.38) which implies that the decay matrix Γ is at least positive semidefinite, $\Gamma \geq 0$. With (I.36) it then follows that

$$\varphi_j^\dagger\Gamma\varphi_j = \varphi_j^\dagger\mathrm{i}(\mathcal{M} - \mathcal{M}^\dagger)\varphi_j = \Gamma_j\varphi_j^\dagger\varphi_j \geq 0 \tag{I.43}$$

and thus (I.42) is valid.

If we have an arbitrary initial state vector $\psi^{(0)}$, we can expand it in terms of the basis vectors φ_j:

$$\psi^{(0)} = \sum_{j=1}^n h_j\varphi_j. \tag{I.44}$$

where

$$h_j = \tilde{\varphi}_j^\dagger \psi^{(0)}. \tag{I.45}$$

For the wavefunction (I.34) one then obtains

$$\psi(t) = \mathrm{e}^{-\mathrm{i}\mathcal{M}t}\,\psi^{(0)} = \sum_{j=1}^{n} \mathrm{e}^{(-\mathrm{i}m_j t-(1/2)\Gamma_j t)}\,h_j \varphi_j. \tag{I.46}$$

The amplitude for finding the particle undecayed thus decreases like a superposition of exponentials. The states with a pure exponential decay are the eigenstates φ_j of the non-Hermitian mass matrix $\mathcal{M}$. If

$$\psi^{(0)} = \varphi_j, \tag{I.47}$$

then we have

$$\psi(t) = \mathrm{e}^{(-\mathrm{i}m_j t-(1/2)\Gamma_j t)}\,\varphi_j. \tag{I.48}$$

In this case, the probability of finding the particle undecayed decreases with t according to

$$|\psi(t)|^2 = \mathrm{e}^{-\Gamma_j t}\,\varphi_j^\dagger \varphi_j. \tag{I.49}$$

The quantities Γ_j are therefore the decay rates of the eigenstates φ_j of $\mathcal{M}$.

Using (I.13) it is now straightforward to calculate the time development of the continuum amplitudes $b_\beta(t)$ in (I.7) or equivalently $c_\beta(t)$ in (I.4). The result for c_β is

$$c_\beta(t) = \sum_{\alpha,\alpha'} \langle\beta|\mathbf{H}'|\alpha\rangle\,[(\mathrm{e}^{-\mathrm{i}\mathcal{M}t} - \mathrm{e}^{-\mathrm{i}E_\beta t})(\mathcal{M} - E_\beta)^{-1}]_{\alpha\alpha'}\psi_{\alpha'}^{(0)}. \tag{I.50}$$

The exponential decay law for $\psi(t)$ (I.46) only represents an approximation. It can be shown that this approximation fails both for very small and for very long times if the spectrum of the Hamilton operator $\mathbf{H}$ is bounded from below (Khalfin 1957, 1968; Fonda 1978). In passing we mention that such deviations from an exponential decay behavior have been discussed in connection with proton decay and double β-decay (Khalfin 1982; Fleming 1983; Chiu 1982; Klapdor 1984).

The n-dimensional vectors $\psi(t)$, φ_j, and $\tilde{\varphi}_j$ can be identified with state vectors in the subspace $\mathcal{H}_n$ that is spanned by the states $|\alpha\rangle$ ($\alpha = 1, \ldots, n$) by setting

$$\begin{aligned} |\psi(t)\rangle &= \sum_{\alpha=1}^{n} \psi_\alpha(t)|\alpha\rangle, \\ |\varphi_j\rangle &= \sum_{\alpha=1}^{n} \varphi_{j\alpha}|\alpha\rangle, \\ |\tilde{\varphi}_j\rangle &= \sum_{\alpha=1}^{n} \tilde{\varphi}_{j\alpha}|\alpha\rangle. \end{aligned} \tag{I.51}$$

In an analogous way we can write $\mathcal{M}$ as an operator in $\mathcal{H}_n$. It is readily demonstrated that $\mathcal{M}$ can be written in the form

$$\mathcal{M} = \sum_{j=1}^{n} M_j |\varphi_j\rangle\langle\tilde{\varphi}_j|. \tag{I.52}$$

It is important to emphasize here that $|\psi(t)\rangle$ (I.51) is only the projection of the total state vector $|t\rangle_S$ of (I.4) onto the subspace $\mathscr{H}_n$. The norm of the total state vector $|t\rangle_S$ in (I.4) is conserved. The amount of norm that disappears from the subspace $\mathscr{H}_n$ of discrete states due to the decrease in the amplitudes $\psi_\alpha(t)$, in fact, reappears as the norm in the subspace of continuum states $|\beta\rangle$.

Up to this point our discussion has made no recourse to any possible discrete or continuous symmetries of the Hamilton operator $\mathbf{H}$ in (I.1). We will now assume CPT invariance to hold as is the case for elementary particle systems that are not subjected to any external fields (Sect. 26.2.2). From (26.94) with $\mathbf{H} \equiv \mathbf{P}_0$ we have:

$$(\mathbf{V}(\Theta)\mathbf{H}\mathbf{V}^{-1}(\Theta))^\dagger = \mathbf{H}. \tag{I.53}$$

It follows that

$$(\mathbf{V}(\Theta)\, \mathrm{e}^{-\mathrm{i}\mathbf{H}t}\, \mathbf{V}^{-1}(\Theta))^\dagger = \mathrm{e}^{-\mathrm{i}\mathbf{H}t}. \tag{I.54}$$

In a similar way one can show that in the case of CPT invariance, the S-matrix satisfies

$$(\mathbf{V}(\Theta)\mathbf{S}\mathbf{V}^{-1}(\Theta))^\dagger = \mathbf{S}. \tag{I.55}$$

To prove (I.55) we insert on the left-hand side the exact solution (9.14) for the S-matrix in terms of the interaction Hamiltonian $\mathbf{H}'(t)$ and use

$$(\mathbf{V}(\Theta)\mathbf{H}'(t)\mathbf{V}^{-1}(\Theta))^\dagger = \mathbf{H}'(-t),$$

which, as is easily seen, holds in the case of CPT invariance.

The main result of the above discussion is that under the stated approximations the Schrödinger equation is solved by means of a mass matrix. For (discrete) states $|\alpha_1\rangle$, $|\alpha_2\rangle$ from $\mathscr{H}_n$ we thus have the approximate equation

$$\langle\alpha_1|\mathrm{e}^{-\mathrm{i}\mathbf{H}t}|\alpha_2\rangle = \langle\alpha_1|\mathrm{e}^{-\mathrm{i}\mathscr{M}t}|\alpha_2\rangle. \tag{I.56}$$

We now consider the CPT-transformed states corresponding to $|\alpha\rangle$ and set

$$|\alpha'\rangle = \mathbf{V}(\Theta)|\alpha\rangle \qquad (\alpha = 1, \ldots, n). \tag{I.57}$$

The states $|\alpha'\rangle$ are *a priori* not necessarily elements of the subspace $\mathscr{H}_n$. However, we can always achieve this by taking as new states $|\alpha\rangle$ the set formed by the union of the old states $|\alpha\rangle$ and the states $|\alpha'\rangle$. On account of CPT invariance we then conclude from (I.54):

$$\begin{aligned} \langle\alpha_1|\mathrm{e}^{-\mathrm{i}\mathbf{H}t}|\alpha_2\rangle &= \langle\alpha_2'|\mathrm{e}^{-\mathrm{i}\mathbf{H}t}|\alpha_1'\rangle, \\ \langle\alpha_1|\mathscr{M}|\alpha_2\rangle &= \langle\alpha_2'|\mathscr{M}|\alpha_1'\rangle. \end{aligned} \tag{I.58}$$

Given CPT invariance, $\mathscr{M}$ must satisfy these symmetry conditions. In particular, if we put $|\alpha_1\rangle = |\alpha_2\rangle = |\alpha\rangle$, then using (I.37) it follows that:

$$\begin{aligned} \langle\alpha|\mathscr{M}|\alpha\rangle &= \langle\alpha'|\mathscr{M}|\alpha'\rangle, \\ \langle\alpha|\mathbf{M}|\alpha\rangle &= \langle\alpha'|\mathbf{M}|\alpha'\rangle, \\ \langle\alpha|\Gamma|\alpha\rangle &= \langle\alpha'|\Gamma|\alpha'\rangle. \end{aligned} \tag{I.59}$$

Since CPT transforms particles into antiparticles, we can conclude from this that they have the same (real) mass and the same total decay width [see (16.50)].

I.2 Application to the System of Neutral K-Mesons

We now apply the formalism discussed in the previous section to the decay of neutral K-mesons. In this, the operators $\mathbf{H}_0$ and $\mathbf{H}'$ in (I.1) are taken as the Hamiltonians of the strong and the weak interaction, respectively. As unperturbed discrete states $|\alpha\rangle$ in (I.2) we consider the CP eigenstates $|K_1^0\rangle$ and $|K_2^0\rangle$ of (26.27). Since there is no strictly conserved CP operator a convention has to be made. The choice to which we adhere is defined by (26.79). (For an alternative convention see Lee 1966.) When the weak interaction is "switched on", the K-mesons decay into continuum states $|\beta\rangle = |\pi\pi\rangle, |\pi\pi\pi\rangle, |\pi\ell\nu\rangle$, etc. The mass matrix $\mathcal{M}$ is a 2×2 matrix (26.55) which is an operator in the subspace spanned by the states $|\mathrm{K}_1^0\rangle$ and $|\mathrm{K}_2^0\rangle$. The eigenvalues of $\mathcal{M}$ are denoted by $M_{\mathrm{L,S}}$ and the *right* eigenvectors by $|\mathrm{K}_{\mathrm{L,S}}\rangle$ (26.19). Here we identify state vectors and eigenvectors of $\mathcal{M}$ in the sense of (I.51). Taking note of the normalization conditions (26.21), we can write, quite generally

$$\begin{aligned} |\mathrm{K_S}\rangle &= (1 + |\varepsilon_\mathrm{S}|^2)^{-1/2}(|\mathrm{K}_1^0\rangle + \varepsilon_\mathrm{S}|\mathrm{K}_2^0\rangle), \\ |\mathrm{K_L}\rangle &= \mathrm{e}^{\mathrm{i}\varphi}\,(1 + |\varepsilon_\mathrm{L}|^2)^{-1/2}(|\mathrm{K}_2^0\rangle + \varepsilon_\mathrm{L}|\mathrm{K}_1^0\rangle), \end{aligned} \tag{I.60}$$

where ε_S and ε_L are complex numbers and φ is a phase. The inequality (26.22) yields the condition

$$\mathrm{e}^{\mathrm{i}\varphi}\,(\varepsilon_\mathrm{L} + \varepsilon_\mathrm{S}^*) \geq 0. \tag{I.61}$$

The *left* eigenvectors of $\mathcal{M}$ are denoted by $|\tilde{\mathrm{K}}_{\mathrm{S,L}}\rangle$. According to (I.40) they must satisfy the following equations:

$$\begin{aligned} \langle\tilde{\mathrm{K}}_\mathrm{S}|\mathrm{K_S}\rangle &= \langle\tilde{\mathrm{K}}_\mathrm{L}|\mathrm{K_L}\rangle = 1, \\ \langle\tilde{\mathrm{K}}_\mathrm{S}|\mathrm{K_L}\rangle &= \langle\tilde{\mathrm{K}}_\mathrm{L}|\mathrm{K_S}\rangle = 0. \end{aligned} \tag{I.62}$$

From this we find

$$\begin{aligned} |\tilde{\mathrm{K}}_\mathrm{S}\rangle &= \frac{(1 + |\varepsilon_\mathrm{S}|^2)^{1/2}}{1 - \varepsilon_\mathrm{L}^*\varepsilon_\mathrm{S}^*}(|\mathrm{K}_1^0\rangle - \varepsilon_\mathrm{L}^*|\mathrm{K}_2^0\rangle), \\ |\tilde{\mathrm{K}}_\mathrm{L}\rangle &= \mathrm{e}^{\mathrm{i}\varphi}\frac{(1 + |\varepsilon_\mathrm{L}|^2)^{1/2}}{1 - \varepsilon_\mathrm{L}^*\varepsilon_\mathrm{S}^*}(|\mathrm{K}_2^0\rangle - \varepsilon_\mathrm{S}^*|\mathrm{K}_1^0\rangle). \end{aligned} \tag{I.63}$$

Furthermore, from (I.52) we have

$$\mathcal{M} = M_\mathrm{S}|\mathrm{K_S}\rangle\langle\tilde{\mathrm{K}}_\mathrm{S}| + M_\mathrm{L}|\mathrm{K_L}\rangle\langle\tilde{\mathrm{K}}_\mathrm{L}|. \tag{I.64}$$

We now derive some consequences of CPT invariance. The states $|\mathrm{K}^0\rangle$ and $|\bar{\mathrm{K}}^0\rangle$ are defined as eigenstates of the strangeness operator $\mathbf{S}$, which can be considered as a charge operator in the sense of (26.92). According to the arguments of Sect. 26.2.2, it then follows that the CPT transformation converts $|\mathrm{K}^0\rangle$ into $|\bar{\mathrm{K}}^0\rangle$. By suitably choosing the phases of the state vectors we can arrange to have

$$\begin{aligned} \mathbf{V}(\Theta)|\mathrm{K}^0\rangle &= |\bar{\mathrm{K}}^0\rangle, \\ \mathbf{V}(\Theta)|\bar{\mathrm{K}}^0\rangle &= |\mathrm{K}^0\rangle. \end{aligned} \tag{I.65}$$

For the mass matrix we then obtain from (I.57) as a consequence of CPT invariance

$$\langle \mathrm{K}^0|\mathcal{M}|\mathrm{K}^0\rangle = \langle \bar{\mathrm{K}}^0|\mathcal{M}|\bar{\mathrm{K}}^0\rangle, \tag{I.66a}$$

or in the basis $|\mathrm{K}^0_{1,2}\rangle$

$$\langle \mathrm{K}^0_1|\mathcal{M}|\mathrm{K}^0_2\rangle + \langle \mathrm{K}^0_2|\mathcal{M}|\mathrm{K}^0_1\rangle = 0, \tag{I.66b}$$

which has already been used in (26.58). If we substitute the expression for the mass matrix (I.64) into (I.66b) then we find that CPT invariance implies for the states $|\mathrm{K}_\mathrm{S}\rangle$, $|\mathrm{K}_\mathrm{L}\rangle$ that

$$\varepsilon_\mathrm{S} = \varepsilon_\mathrm{L}. \tag{I.67}$$

In order to determine the phase φ in (I.60) we consider the semileptonic decays of K_L (26.36). In the standard model this decay proceeds—to leading order in the weak interaction—via the following decay of an s-quark or an $\bar{\mathrm{s}}$-antiquark in the K-meson

$$\begin{aligned} \mathrm{s} &\to \mathrm{u} + \ell^- + \bar{\nu}_\ell, \\ \bar{\mathrm{s}} &\to \bar{\mathrm{u}} + \ell^+ + \nu_\ell. \end{aligned} \tag{I.68}$$

In this process the charge and strangeness of the quarks both change in the same direction by one unit (the $\Delta S = \Delta Q$ rule). Thus, to leading order in the weak interaction, a K^0-meson can only decay into ℓ^+-leptons, and a $\bar{\mathrm{K}}^0$-meson only into ℓ^--leptons. The opposite decay processes are forbidden:

$$\begin{aligned} \mathrm{K}^0 &\not\to \pi^+\ell^-\bar{\nu}_\ell, \\ \bar{\mathrm{K}}^0 &\not\to \pi^-\ell^+\nu_\ell. \end{aligned} \tag{I.69}$$

As a result of the CPT invariance of the S-matrix (I.55) it further holds that

$$|\langle \ell^-\bar{\nu}_\ell\pi^+|\mathbf{T}|\mathrm{K}^0\rangle| = |\langle \bar{\mathrm{K}}^0|\mathbf{T}|\ell^+\nu_\ell\pi^-\rangle| = |\langle \ell^+\nu_\ell\pi^-|\mathbf{T}^\dagger|\bar{\mathrm{K}}^0\rangle|. \tag{I.70}$$

From the unitarity relation (5.65), and setting there $|\mathrm{i}\rangle = |\bar{\mathrm{K}}^0\rangle$ and $|\mathrm{i}'\rangle = |\ell^+\nu_\ell\pi^-\rangle$, we now see that

$$\langle \ell^+\nu_\ell\pi^-|\mathbf{T}|\bar{\mathrm{K}}^0\rangle - \langle \ell^+\nu_\ell\pi^-|\mathbf{T}^\dagger|\bar{\mathrm{K}}^0\rangle = \mathrm{i}\sum_f (2\pi)^4\delta(P_f - P_i) \cdot \langle f|\mathbf{T}|\ell^+\nu_\ell\pi^-\rangle^*\langle f|\mathbf{T}|\bar{\mathrm{K}}^0\rangle \tag{I.70a}$$

But the matrix elements on the right-hand side of (I.70a) are nonzero only because of electroweak interactions. (In their absence, $\bar{\mathrm{K}}^0$ is stable and ℓ^+, ν_ℓ, and π^- do not interact.) Thus the right-hand side of (I.70a) is of second order in the electroweak transition matrix elements and, to a good approximation, we can set it to zero. In this way we find equality for the amplitudes on the left-hand side of (I.70a) and using (I.70) we get

$$\Gamma(\mathrm{K}^0 \to \ell^- \bar{\nu}_\ell \pi^+) = \Gamma(\bar{\mathrm{K}}^0 \to \ell^+ \nu_\ell \pi^-). \tag{I.71}$$

With this we obtain

$$\begin{aligned} \Gamma(\mathrm{K_L} \to \ell^- \bar{\nu}_\ell \pi^+) &= \frac{1}{2}\frac{|1-\varepsilon_\mathrm{L}|^2}{1+|\varepsilon_\mathrm{L}|^2}\Gamma(\bar{\mathrm{K}}^0 \to \ell^- \bar{\nu}_\ell \pi^+), \\ \Gamma(\mathrm{K_L} \to \ell^+ \nu_\ell \pi^-) &= \frac{1}{2}\frac{|1+\varepsilon_\mathrm{L}|^2}{1+|\varepsilon_\mathrm{L}|^2}\Gamma(\mathrm{K}^0 \to \ell^+ \nu_\ell \pi^-) \end{aligned} \tag{I.72}$$

and for the asymmetry (26.38)

$$\delta = \frac{2 \operatorname{Re} \varepsilon_\mathrm{L}}{1+|\varepsilon_\mathrm{L}|^2}. \tag{I.73}$$

Since experimentally one finds $\delta > 0$, it follows that $\operatorname{Re} \varepsilon_\mathrm{L} > 0$ and, from (I.67), (I.61)

$$e^{i\varphi} = 1, \tag{I.74}$$

thus determining the phase of $|\mathrm{K_L}\rangle$ in (I.60).

We next discuss the unitarity relation of Bell and Steinberger (Bell 1966), by which the phase of ε (26.51) is determined. From (I.36), (I.38) it follows that

$$\begin{aligned} i\langle \mathrm{K_S}|\mathcal{M} - \mathcal{M}^\dagger|\mathrm{K_L}\rangle &= \langle \mathrm{K_S}|\Gamma|\mathrm{K_L}\rangle, \\ i(M_\mathrm{L} - M_\mathrm{S}^*)\langle \mathrm{K_S}|\mathrm{K_L}\rangle &= \sum_f \langle f|\mathbf{T}|\mathrm{K_S}\rangle^* \langle f|\mathbf{T}|\mathrm{K_L}\rangle. \end{aligned} \tag{I.75}$$

The matrix elements $\langle\beta|\mathbf{H}'|\alpha'\rangle$ that appear in (I.38) should be identified with the T-matrix elements $\langle f|\mathbf{T}|\mathrm{K_L}\rangle$, etc. In Sect. I.1, $\mathbf{H}_0$ was the Hamiltonian of the total strong interaction and *not* that of free particles. From the viewpoint of the interaction picture, in the amplitude $\langle\beta|\mathbf{H}'|\alpha'\rangle$, one must consider a single weak elementary process accompanied by arbitrarily many strong processes, i.e., we are dealing with the T-matrix element to first order in the weak interaction.

The $\mathrm{K_S}$-meson mainly decays into the $\pi\pi$ state with $I = 0$ ($\Delta I = \frac{1}{2}$ rule; see below). We thus expect that the state $|f\rangle = |\pi\pi, I = 0\rangle$ will yield an important contribution to the right-hand side of (I.75). If we consider *only* this state, then (I.75) gives

$$i(M_\mathrm{L} - M_\mathrm{S}^*)\langle \mathrm{K_S}|\mathrm{K_L}\rangle = \langle \pi\pi, 0|\mathbf{T}|\mathrm{K_S}\rangle^* \langle \pi\pi, 0|\mathbf{T}|\mathrm{K_L}\rangle = \varepsilon\Gamma(\mathrm{K_S} \to \pi\pi(I=0)). \tag{I.76}$$

The inclusion of the other final states changes this relation by at most $\pm 10\%$ (Schubert 1970). To this accuracy it follows that

$$\varepsilon = i\frac{m_\mathrm{L} - m_\mathrm{S} - \frac{i}{2}(\Gamma_\mathrm{L} + \Gamma_\mathrm{S})}{\Gamma(\mathrm{K_S} \to \pi\pi(I=0))}\langle \mathrm{K_S}|\mathrm{K_L}\rangle. \tag{I.77}$$

Experimentally, one finds $\Gamma_\mathrm{L} \ll \Gamma_\mathrm{S} \cong \Gamma(\mathrm{K_S} \to \pi\pi(I=0))$ and $m_\mathrm{L} - m_\mathrm{S} \cong \Gamma_\mathrm{S}/2$.

With these results (I.77) becomes

$$\varepsilon \cong e^{i(\pi/4)} \frac{1}{\sqrt{2}} \langle K_S | K_L \rangle, \tag{I.78}$$

and thus, with (26.22), the phase of ε is determined to within $\pm 10\%$.

We will now make use of CPT invariance and the generalized optical theorem (5.65) to determine the phase of ε' modulo π. From (I.65) we have

$$\begin{aligned} \mathbf{V}(\Theta)|K_1^0\rangle &= -|K_1^0\rangle, \\ \mathbf{V}(\Theta)|K_2^0\rangle &= |K_2^0\rangle. \end{aligned} \tag{I.79}$$

We can choose the phases of the $\pi\pi$ states such that

$$\mathbf{V}(\Theta)|\pi\pi, I\rangle = |\pi\pi, I\rangle. \tag{I.80}$$

It then follows from CPT invariance that

$$\langle K_{1,2}^0 | \mathbf{T} | \pi\pi, I\rangle = \mp \langle \pi\pi, I | \mathbf{T} | K_{1,2}^0 \rangle. \tag{I.81}$$

The unitarity relation (5.65) further yields

$$\begin{aligned} &\frac{1}{2i}\{\langle \pi\pi, I|\mathbf{T}|K_{1,2}^0\rangle - \langle K_{1,2}^0|\mathbf{T}|\pi\pi, I\rangle^*\} \\ &\quad = \frac{1}{2}\sum_f (2\pi)^4 \delta(P_f - P_i) \langle f|\mathbf{T}|\pi\pi, I\rangle^* \langle f|\mathbf{T}|K_{1,2}^0\rangle \\ &\quad = -\frac{1}{2i}\sum_f \langle f|(\mathbf{S} - 1)|\pi\pi, I\rangle^* \langle f|\mathbf{T}|K_{1,2}^0\rangle. \end{aligned} \tag{I.82}$$

The K-mesons are stable with respect to the strong and electromagnetic interactions and thus $\langle f|\mathbf{T}|K_{1,2}^0\rangle$ and $\langle \pi\pi, I|\mathbf{T}|K_{1,2}^0\rangle$ are transition matrix elements of the weak interaction. If in (I.82) we neglect quadratic terms in the weak transition amplitudes and also electromagnetic corrections, then we can regard $\langle f|\mathbf{S} - 1|\pi\pi, I\rangle$ as strong amplitudes, i.e., we can regard $\mathbf{S}$ as S-operator of the strong interaction. Two pions of center of mass energy $\sqrt{s} = m_K$ can only scatter elastically. Using the methods presented in Sect. 16.3 we thus obtain:

$$\mathbf{S}|\pi\pi, I\rangle = e^{2i\delta_I} |\pi\pi, I\rangle, \tag{I.83}$$

where δ_I ($I = 0, 2$) are the S-wave scattering phases for center of mass energy $\sqrt{s} = m_K$. Inserting this in (I.82) yields after some calculation

$$\begin{aligned} \langle \pi\pi, I|\mathbf{T}|K_1^0\rangle &= e^{i\delta_I} A_1^I, \\ \langle \pi\pi, I|\mathbf{T}|K_2^0\rangle &= e^{i(\delta_I + (1/2)\pi)} A_2^I. \end{aligned} \tag{I.84}$$

Here $A_{1,2}^I$ are *real* amplitudes of undetermined sign.[1] If we now calculate

[1] In the literature $A^I \equiv (A_1^I + iA_2^I)/\sqrt{2}$ is often used. One then has $\langle \pi\pi, I|\mathbf{T}|K^0\rangle = e^{i\delta_I} A^I$ and $\langle \pi\pi, I|\mathbf{T}|\bar{K}^0\rangle = -e^{i\delta_I} (A^I)^*$.

ε and ε', then neglecting terms of order ε_L^2 we obtain:

$$\varepsilon = \varepsilon_L + i\frac{A_2^0}{A_1^0},$$
$$\varepsilon' = e^{i(\delta_2 - \delta_0 + (1/2)\pi)} \frac{A_1^0 A_2^2 - A_2^0 A_1^2}{\sqrt{2}\,(A_1^0)^2}. \tag{I.85}$$

From the measured scattering phases (see Martin 1976)

$$\delta_2 - \delta_0 = -(0.29 \pm 0.03)\pi, \tag{I.86}$$

and from (I.78) one obtains

$$\arg\frac{\varepsilon'}{\varepsilon} \cong -0.05\pi \quad \text{or} \quad (1 - 0.05)\pi. \tag{I.87}$$

Thus $\exp(i0.05\pi)\varepsilon'/\varepsilon$ is a real quantity whose sign is not fixed by the above arguments but is experimentally measurable.

We turn finally to the determination of ω (26.46). Here the CP violation is of no importance. Thus we write $|K_S\rangle \cong |K_1^0\rangle$. It then follows from (I.84) that

$$\omega = e^{i(\delta_2 - \delta_0)} \frac{A_1^2}{A_1^0}. \tag{I.88}$$

Therefore, to modulo π, the phase of ω is fixed. A simple calculation yields

$$\frac{\Gamma(K_S \to \pi^+\pi^-)}{\Gamma(K_S \to \pi^0\pi^0)} = 2\left|\frac{1 + \frac{1}{\sqrt{2}}\omega}{1 - \sqrt{2}\,\omega}\right|^2. \tag{I.89}$$

From the experimental value for this ratio, which is close to two (see Table 26.1), (I.88) and (I.89) yield two solutions for ω. The solution with small $|\omega|$ is given in Table 26.1. From a comparison with the decay processes $K^\pm \to \pi^\pm\pi^0$ it is possible to exclude the solution with large $|\omega|$ if one assumes that the weak interaction produces no transitions in which the strong isospin changes by $\Delta I = \frac{5}{2}$. In the standard model this is correct to first order in the weak interaction.

APPENDIX J

Solutions to Selected Exercises

Solutions to exercises that are already given in the text, or those that follow in a straightforward manner from the text, are not included in the following.

2.1 If one considers how a Lorentz transformation Λ which satisfies the conditions (2.39) affects the four-vector x of (2.36), then one readily finds the representation $\Lambda = \Lambda(\mathbf{R})\Lambda(v)\Lambda(\mathbf{R}')$. Here $\Lambda(\mathbf{R})$ and $\Lambda(\mathbf{R}')$ are proper rotations (as in (2.32) with $\det \mathbf{R} = 1$ and $\det \mathbf{R}' = 1$) and $\Lambda(v)$ is a special Lorentz transformation into the x^1-direction (2.33). By means of continuous variations of $\mathbf{R}$, v, $\mathbf{R}'$, Λ can be connected to the identity transformation.

2.2 $E_{\mathrm{L}} = 1496.40$ GeV; $u = -2804.14\ \mathrm{GeV}^3$; $\vartheta_{\mathrm{L}} = 2.66'$.

3.2 Let $\varphi(x, t)$ be any real solution of (3.14). We choose an arbitrary time t_0 and set

$$\alpha(\boldsymbol{k}) = \mathrm{e}^{\mathrm{i}\omega t_0} \int \mathrm{d}^3x\, \mathrm{e}^{-\mathrm{i}\boldsymbol{k}\boldsymbol{x}}\, [\omega\varphi(\boldsymbol{x}, t_0) + \mathrm{i}\dot{\varphi}(\boldsymbol{x}, t_0)],$$

$$\tilde{\varphi}(\boldsymbol{x}, t) = \int \frac{\mathrm{d}^3k}{(2\pi)^3} \frac{1}{2\omega} [\mathrm{e}^{\mathrm{i}kx}\, \alpha^*(\boldsymbol{k}) + \mathrm{e}^{-\mathrm{i}kx}\, \alpha(\boldsymbol{k})].$$

$\tilde{\varphi}(\boldsymbol{x}, t)$ is a solution of (3.14) and we have: $\tilde{\varphi}(\boldsymbol{x}, t_0) = \varphi(\boldsymbol{x}, t_0)$, $\dot{\tilde{\varphi}}(\boldsymbol{x}, t_0) = \dot{\varphi}(\boldsymbol{x}, t_0)$. From Cauchy's existence and uniqueness theorem for differential equations of the hyperbolic type it then follows that $\tilde{\varphi}(\boldsymbol{x}, t) = \varphi(\boldsymbol{x}, t)$ for all t.

3.3 For $|\boldsymbol{x}| > t$ we have

$$\psi(\boldsymbol{x}, t) = \frac{\mathrm{i}m^4t}{2\pi^2z^2} K_2(z) \xrightarrow[z \gg 1]{} \frac{\mathrm{i}m^4t}{2\pi^2z^2}\left(\frac{\pi}{2z}\right)^{1/2} \mathrm{e}^{-z},$$

where $z = m(x^2 - t^2)^{1/2}$ and $K_2(z)$ is a modified Bessel function (see Bateman 1953).

3.4 Let $m \geq 0$, $k^0 = \omega = +\sqrt{\boldsymbol{k}^2 + m^2}$, $k = (k^\mu)$. We subject k to a Lorentz transformation $\Lambda = (\Lambda^\mu{}_\nu)$, i.e., we set $k^\mu = \Lambda^\mu{}_\nu \bar{k}^\nu$. From (2.26) it then follows that

$$\det[\Lambda^i{}_j + \Lambda^i{}_0 \bar{k}^j(\bar{k}^0)^{-1}] = (\det \Lambda) \cdot [\Lambda^0{}_0 + \Lambda^0{}_j \bar{k}^j(\bar{k}^0)^{-1}].$$

With the help of this relation one readily obtains the desired proof.

3.5 Let $x' = \Lambda x + a$ be a Poincaré transformation where Λ is from the identity component. If the states in Hilbert space are kept unchanged, the field operators in the primed and unprimed systems obey $\Phi'(x') = \Phi(x)$. In order to be truly equivalent, the total set of states and observables must be the same for both observer systems. There should thus be a unitary transformation in the space of states $\mathbf{U}(\Lambda, a)$ such that the effect of the original field operator Φ with respect to the transformed states is the same as that of the field operator Φ' with respect to the untransformed ones:

$$\mathbf{U}^{-1}(\Lambda, a)\Phi(x')\mathbf{U}(\Lambda, a) = \Phi'(x') = \Phi(x) = \Phi(\Lambda^{-1}(x' - a)).$$

This transformation does indeed exist. It is defined by

$$\mathbf{U}(\Lambda, a)\mathbf{a}^\dagger(\boldsymbol{k})\mathbf{U}^{-1}(\Lambda, a) = e^{ik'a}\,\mathbf{a}^\dagger(\boldsymbol{k}'),$$

$$\mathbf{U}(\Lambda, a)|0\rangle = |0\rangle,$$

where $k' = \Lambda k$. For the one-meson states of (3.68) we then have

$$\mathbf{U}(\Lambda, a)|\boldsymbol{k}\rangle = \exp(ik'a)|\boldsymbol{k}'\rangle.$$

3.6

$$L[\varphi, \dot{\varphi}] = \int d^3x\, \tfrac{1}{2}\{(\dot{\varphi}(\boldsymbol{x}, t))^2 - (\nabla\varphi(\boldsymbol{x}, t))^2 - m^2(\varphi(\boldsymbol{x}, t))^2\},$$

$$\Pi(\boldsymbol{x}, t) = \dot{\varphi}(\boldsymbol{x}, t),$$

$$H[\varphi, \Pi] = \int d^3x\, \tfrac{1}{2}\{(\Pi(\boldsymbol{x}, t))^2 + (\nabla\varphi(\boldsymbol{x}, t))^2 + m^2(\varphi(\boldsymbol{x}, t))^2\}.$$

3.8 Let the Lagrange density $\mathscr{L}$ be that of (3.82). We assume rotational invariance, i.e., that $\mathscr{L}$ is invariant under the infinitesimal transformations:

$$t \to t' = t,$$

$$x^j \to x'^j = x^j + \varepsilon^{jkl} x^k \delta\eta^l,$$

$$\varphi_r(x) \to \varphi_r'(x') = \varphi_r(x)$$

Here $\delta\boldsymbol{\eta} = (\delta\eta^l)$ is the infinitesimal rotation vector. The conserved current (3.90) yields the angular momentum tensor $M^\lambda{}_{jk} = T^\lambda{}_j x_k - T^\lambda{}_k x_j$, whose integral gives the conserved angular momentum vector $\boldsymbol{J}$ where

$$J^l = -\tfrac{1}{2}\varepsilon^{ljk} \int d^3x\, M^0{}_{jk}.$$

Rotational invariance of the Lagrange density thus yields the conservation of angular momentum.

4.2 The proper rotations $\Lambda(\mathbf{R})$ (with det $\mathbf{R} = +1$) and the special Lorentz transformations $\Lambda(v)$ (2.33) can readily be written in the form $\Lambda = \exp(\mathbf{h}\tau)$. For a Λ such as this, $\mathbf{S}(\Lambda) = \exp(-\frac{1}{4}\mathrm{i}\tau h^{\mu\nu}\sigma_{\mu\nu})$ satisfies (4.48).

4.3 For $\Lambda = \exp(\mathbf{h}\tau)$ it follows from Exercise 4.2 that $\mathbf{S}^{-1}(\Lambda) = \exp(\frac{1}{4}\mathrm{i}\tau h^{\mu\nu}\sigma_{\mu\nu})$ and $\bar{\mathbf{S}}(\Lambda) = \exp(\frac{1}{4}\mathrm{i}\tau h^{\mu\nu}\bar{\sigma}_{\mu\nu}) = \mathbf{S}^{-1}(\Lambda)$ since $\bar{\sigma}_{\mu\nu} = \sigma_{\mu\nu}$. The extension of this result to all Λ in the identity component proceeds as in Exercise 4.2.

4.5 The following relations are valid

$$(\mathbf{V}(\Theta)\psi(x)\mathbf{V}^{-1}(\Theta))^\dagger = -\gamma_5\psi(-x),$$

$$-\gamma_5 u_s(p) = (-1)^{s-1/2} v_{-s}(p),$$

$$-\gamma_5 v_s(p) = (-1)^{s+1/2} u_{-s}(p),$$

$$\mathbf{V}(\Theta)\mathbf{a}_s^\dagger(p)\mathbf{V}^{-1}(\Theta) = (-1)^{s-1/2}\mathbf{b}_{-s}^\dagger(p),$$

$$\mathbf{V}(\Theta)\mathbf{b}_s^\dagger(p)\mathbf{V}^{-1}(\Theta) = (-1)^{s+1/2}\mathbf{a}_{-s}^\dagger(p),$$

where $\mathbf{V}(\Theta)$ is antiunitary and $s = \pm\frac{1}{2}$.

5.2
$$\Gamma(\mathrm{h} \to \mathrm{e}^+\mathrm{e}^-) = \frac{m_\mathrm{h}}{8\pi}\left(\frac{m_\mathrm{e}}{\rho_0}\right)^2\left(1 - \frac{4m_\mathrm{e}^2}{m_\mathrm{h}^2}\right)^{3/2}.$$

5.3 We make the ansatz

$$|i\rangle = \int \frac{\mathrm{d}^3\tilde{p}}{(2\pi)^3\sqrt{2\tilde{p}_0}} f_p(\tilde{p})|\mathrm{e}^-(\tilde{p}, s)\rangle,$$

$$|f\rangle = \int \frac{\mathrm{d}^3\tilde{p}'}{(2\pi)^3\sqrt{2\tilde{p}_0'}} f'_{p'}(\tilde{p}')|\mathrm{e}^-(\tilde{p}', s')\rangle,$$

with

$$\langle i|i\rangle = \int \frac{\mathrm{d}^3\tilde{p}}{(2\pi)^3}|f_p(\tilde{p})|^2 = 1,$$

$$\langle f|f\rangle = \int \frac{\mathrm{d}^3\tilde{p}'}{(2\pi)^3}|f'_{p'}(\tilde{p}')|^2 = 1.$$

We further assume that $|f_p(\tilde{p})|(|f'_{p'}(p')|)$ has its maximum at p ($p' \neq p$) and, for simplicity, that it only depends on $|\tilde{p} - p|$ ($|\tilde{p}' - p'|$). Narrow wave packets can then (in general) be approximated by Gaussian wave packets. To see this we write

$$f_p(\tilde{p}) = |f_p(\tilde{p})|\,\mathrm{e}^{\mathrm{i}\varphi(\tilde{p})}$$

and expand $\ln|f_p(\tilde{p})|$ and $\varphi(\tilde{p})$ around $\tilde{p} = p$ in powers of $(\tilde{p} - p)$ to get, keeping in mind the maximum conditions,

$$\ln|f_p(\tilde{p})| = \ln|f_p(p)| - \sigma_i^2(\tilde{p} - p)^2 + \cdots,$$

$$\varphi(\tilde{p}) = \varphi_0 - \mathrm{i}(\tilde{p} - p)\cdot x_i^{(0)} + \cdots,$$

where $\sigma_i \geq 0$ and $x_i^{(0)}$ are defined as expansion coefficients. Assuming $\sigma_i > 0$, i.e., a nonvanishing second order coefficient of $\ln|f_p(\tilde{p})|$, neglecting higher order terms and the irrelevant phase φ_0, and taking into account the normalization condition for $f_p(\tilde{p})$ we find:

$$f_p(\tilde{p}) \cong (2\pi)^{3/4}(2\sigma_i)^{3/2} \exp[-i(\tilde{p}-p)x_i^0 - (\tilde{p}-p)^2\sigma_i^2].$$

In an analogous way we obtain

$$f'_{p'}(\tilde{p}') \cong (2\pi)^{3/4}(2\sigma_f)^{3/2} \exp[-i(\tilde{p}'-p')x_f^0 - (\tilde{p}'-p')^2\sigma_f^2]$$

where $\sigma_f > 0$, and x_f^0 are again constants. In coordinate space this leads to

$$\langle 0|\psi(x)|i\rangle$$
$$\cong (2\pi)^{-3/4}\sigma_i^{-3/2}(2p_0)^{-1/2}u_s(p)\exp\left[i(px - p_0t) - \tfrac{1}{4}\sigma_i^{-2}\left(x - x_i^0 - \frac{p}{p_0}t\right)^2\right].$$

The expectation values of observable quantities such as $\bar{\psi}(x)\gamma^\mu\psi(x)$ then behave as follows:

$$\langle i|\bar{\psi}(x)\gamma^\mu\psi(x)|i\rangle = \langle i|\bar{\psi}(x)|0\rangle\gamma^\mu\langle 0|\psi(x)|i\rangle$$
$$\propto \exp\left[-\tfrac{1}{2}\sigma_i^{-2}\left(x - x^0 - \frac{p}{p_0}t\right)^2\right].$$

Thus $|i\rangle$ represents a Gaussian wave packet of linear extent σ_i which moves with the velocity p/p_0. $|f\rangle$ can be described analogously. We now assume that the center of the incoming wave packet arrives at the scattering center $x = 0$ at a time t_0. The center of the outgoing wave packet, whose form and temporal extent are determined by the measuring arrangement, is at $x = 0$ at the time t'_0. We set $T_i = \sigma_i p_0/|p|$, $T_f = \sigma_f p'_0/|p'|$ and make the assumptions:

$$t_0 - \tfrac{1}{2}T_i \ll t'_0 - \tfrac{1}{2}T_f \ll t'_0 + \tfrac{1}{2}T_f \ll t_0 + \tfrac{1}{2}T_i,$$

i.e., the temporal extent of the incoming wave is taken to be much longer than the time of observation of the scattered wave. For $\tilde{p}' \neq \tilde{p}$ we write the S-matrix element as

$$\langle e^-(\tilde{p}', s')|\mathbf{S}|e^-(\tilde{p}, s)\rangle = 2\pi i\delta(\tilde{p}'_0 - \tilde{p}_0)\tilde{T}_{s's}(\tilde{p}', \tilde{p}).$$

For $\sigma_i \to \infty$, $\sigma_f \to \infty$, $\sigma_i \gg \sigma_f$ we then obtain the following approximation for the S-matrix element:

$$\langle f|\mathbf{S}|i\rangle \cong (2\pi)^{-1}\sigma_f^{-3/2}\sigma_i^{-3/2}(2p'_0 2p_0)^{-1/2} i\tilde{T}_{s's}(p', p)\sqrt{2}\,T_f$$
$$\cdot \exp[i(p'_0 - p_0)t'_0 - (p'_0 - p_0)^2 T_f^2],$$

for the transition probability w_{fi}:

$$w_{fi} \cong (2\pi)^{-2}\sigma_f^{-3}\sigma_i^{-3}(2p'_0 2p_0)^{-1}|\tilde{T}_{s's}(p', p)|^2 2T_f^2 \exp[-2(p'_0 - p_0)^2 T_f^2],$$

for the flux of incoming particles:

$$\Phi \cong (2\pi)^{-3/2}\sigma_i^{-3}|p|p_0^{-1}$$

and for the outgoing state:

$$|f\rangle \xrightarrow[\sigma_f \to \infty]{} (2\pi)^{-3/4}(2p'_0)^{-1/2}\sigma_f^{-3/2}|e^-(p', s')\rangle.$$

If t_1 is the time of arrival of the center of the scattered wave packet at the measuring device, and $\Gamma_{fi}(t_1)$ the corresponding transition *rate*, then it must be true that

$$w_{fi} \cong \int dt\, \Gamma_{fi}(t_1)\exp[-\tfrac{1}{2}T_f^{-2}(t-t_1)^2].$$

Thus we obtain for the scattering cross-section:

$$d\sigma = \frac{d^3p'}{(2\pi)^3 2p'_0} \lim_{\sigma_f\to\infty} (2\pi)^{3/2} 2p'_0 \sigma_f^3 \Phi^{-1}\Gamma_{fi}(t_1)$$

in agreement with (5.26).

7.1 Let $F[A_\mu]$ be an initially classical functional of the classical fields $A_\mu(x)$, expandable in the general form

$$F[A_\mu] = \sum_{n=0}^{\infty} \int dx_1 \ldots \int dx_n\, f_n^{\mu_1\ldots\mu_n}(x_1,\ldots,x_n) A_{\mu_1}(x_1)\ldots A_{\mu_n}(x_n).$$

Without loss of generality we can insist that the f_n are symmetric with respect to arbitrary exchanges $(\mu_i, x_i) \leftrightarrow (\mu_j, x_j)$. Equation (7.61) yields the condition

$$\frac{\partial}{\partial x_1^{\mu_1}} f_n^{\mu_1\ldots\mu_n}(x_1,\ldots,x_n) = 0$$

for $n \geq 1$. In the quantum theory we write:

$$\mathbf{F}[\mathbf{A}_\mu] = \sum_{n=0}^{\infty} \int dx_1 \ldots \int dx_n\, f_n^{\mu_1\ldots\mu_n}(x_1,\ldots,x_n) :\mathbf{A}_{\mu_1}(x_1)\ldots \mathbf{A}_{\mu_n}(x_n):.$$

If one substitutes the expansion of $\mathbf{A}_\mu(x)$ (7.11) into this equation, and expresses everything in terms of the operators α (7.25), then it is found that $\mathbf{F}[\mathbf{A}_\mu]$ contains no operators α_3 or $\alpha_3^\dagger$. Thus from (7.26) it must hold that

$$[\alpha_0(\boldsymbol{k}), \mathbf{F}[\mathbf{A}_\mu]] = [\alpha_0(\boldsymbol{k}), \mathbf{F}^\dagger[\mathbf{A}_\mu]] = 0.$$

From this and from (7.27) it follows for a physical state vector $|Z\rangle$, that $\mathbf{F}|Z\rangle$ and $\mathbf{F}^\dagger|Z\rangle$ are also physical state vectors. If $|Z_1\rangle$ and $|Z_2\rangle$ are two equivalent state vectors, then for all physical state vectors $|Z\rangle$ we have

$$\langle Z|(|Z_1\rangle - |Z_2\rangle) = 0,$$

and thus with the substitution $\langle Z| \to \langle Z|\mathbf{F}$:

$$\langle Z|\mathbf{F}(|Z_1\rangle - |Z_2\rangle) = 0$$

Therefore $\mathbf{F}|Z_1\rangle$ and $\mathbf{F}|Z_2\rangle$ are equivalent, which completes the required proof.

7.3 If (Λ, a) is a Poincaré transformation as in (2.28) with $\Lambda^0{}_0 > 0$, then we define a unitary transformation $\mathbf{U}(\Lambda, a)$ in the space of states by

$$\mathbf{U}(\Lambda, a)\mathbf{a}_\mu^\dagger(\boldsymbol{k})\mathbf{U}^{-1}(\Lambda, a) = \exp(ik'a)\mathbf{a}_\nu^\dagger(\boldsymbol{k}')\Lambda^\nu{}_\mu,$$

$$\mathbf{U}(\Lambda, a)|0\rangle = |0\rangle,$$

where $k' = \Lambda k$. Thus $\mathbf{A}^\mu(x)$ obeys

$$\mathbf{U}^{-1}(\Lambda, a)\mathbf{A}^\mu(x)\mathbf{U}(\Lambda, a) = \Lambda^\mu{}_\nu \mathbf{A}^\nu(\Lambda^{-1}(x-a)),$$

and for the states of (7.33) we have

$$\mathbf{U}(\Lambda, a)|\boldsymbol{k}, \varepsilon\rangle = \exp(\mathrm{i}k'a)|\boldsymbol{k}', \varepsilon'\rangle,$$

where $\varepsilon' = \Lambda\varepsilon$.

9.2 The Lagrange density $\mathscr{L}$ in (9.5) yields the following equations of motion:

$$\partial_\lambda F^{\lambda\mu}(x) = -e\bar{\psi}(x)\gamma^\mu\psi(x),$$
$$(\mathrm{i}\gamma^\mu\partial_\mu + e\,\gamma^\mu A_\mu - m)\psi(x) = 0.$$

9.3 $\tilde{\mathscr{L}} = -\frac{1}{2}(\partial_\mu A_\nu)(\partial^\mu A^\nu) - \frac{1}{2}(\partial_\mu\bar{\psi})\mathrm{i}\gamma^\mu\psi + \frac{1}{2}\bar{\psi}\mathrm{i}\gamma^\mu\partial_\mu\psi - m\bar{\psi}\psi + e\bar{\psi}\gamma^\mu A_\mu\psi,$

$$\tilde{L} = \int \mathrm{d}^3x\, \tilde{\mathscr{L}},$$

$$\Pi^0 = \frac{\delta\tilde{L}}{\delta\dot{A}^0} = -\dot{A}^0,$$

$$\Pi^j = \frac{\delta\tilde{L}}{\delta\dot{A}^j} = \dot{A}^j,$$

$$\begin{aligned} H &= \int \mathrm{d}^3x\left\{\Pi^0\dot{A}^0 + \Pi^j\dot{A}^j + \dot{\bar{\psi}}\frac{\delta L}{\delta\dot{\bar{\psi}}} + \frac{\delta L}{\delta\dot{\psi}}\dot{\psi}\right\} - \tilde{L} \\ &= \int \mathrm{d}^3x\{-\tfrac{1}{2}\Pi^0\Pi^0 - \tfrac{1}{2}(\nabla A^0)(\nabla A^0) + \tfrac{1}{2}\Pi^j\Pi^j + \tfrac{1}{2}(\nabla A^j)(\nabla A^j) \\ &\quad + \tfrac{1}{2}(\nabla\bar{\psi})\mathrm{i}\gamma\psi - \tfrac{1}{2}\bar{\psi}\mathrm{i}\gamma\nabla\psi + m\bar{\psi}\psi - e\bar{\psi}\gamma^\mu A_\mu\psi\}. \end{aligned}$$

10.2 From the diagrams in Fig. 10.3 one obtains the invariant cross-section for Bhabha scattering as

$$\frac{\mathrm{d}\sigma}{\mathrm{d}t} = \frac{4\pi\alpha^2}{s(s-4m^2)}\frac{1}{t^2s^2}\{(2m^2-u)^2(s^2+t^2) + st(-4m^2u + 12m^4 + st)\}.$$

One should note that the square of the invariant matrix element is obtained from that of Møller scattering by making the replacement $s \leftrightarrow u$, as must be the case from crossing symmetry.

10.4 The differential cross-section for the reaction

$$\mathrm{e}^-(p_1) + \mathrm{e}^+(p_2) \to \gamma(k_1) + \gamma(k_2)$$

is given for unpolarized particles to leading order in α and under the assumptions $s, t, u \gg m_e^2$, by

$$\frac{\mathrm{d}\sigma}{\mathrm{d}\Omega} = \frac{\alpha^2}{2s}\frac{t^2+u^2}{tu} = \frac{\alpha^2}{s}\frac{1+\cos^2\vartheta}{\sin^2\vartheta}.$$

Here $s = (p_1+p_2)^2$, $t = (p_1-k_1)^2$, $u = (p_1-k_2)^2$; ϑ is the scattering angle [$\cos\vartheta = \boldsymbol{p}_1\boldsymbol{k}_1/(|\boldsymbol{p}_1||\boldsymbol{k}_1|)$] and $\mathrm{d}\Omega$ is the solid angle element of $\boldsymbol{k}_1$, in each case with reference to the center of mass system. Since the two photons in the final state are indistinguishable, the physical range of ϑ is given by $0 \le \vartheta \le \pi/2$.

11.1 If all the mass terms in (11.13) are taken into account one has

$$\frac{d\sigma}{d\Omega} = \frac{\alpha^2}{4s}\left(1 - \frac{4m_e^2}{s}\right)^{-1/2}\left(1 - \frac{4m_\mu^2}{s}\right)^{1/2}\left\{1 + 4\frac{m_e^2 + m_\mu^2}{s} + \left(1 - \frac{4m_e^2}{s}\right)\left(1 - \frac{4m_\mu^2}{s}\right)\cos^2 \vartheta\right\}$$

$$\sigma_{\text{tot}}(e^-e^+ \to \mu^-\mu^+) = \frac{4\pi\alpha^2}{3s}\left(1 - \frac{4m_e^2}{s}\right)^{-1/2}\left(1 - \frac{4m_\mu^2}{s}\right)^{-1/2} \cdot\left\{1 + 2\frac{m_e^2 + m_\mu^2}{s} + 4\frac{m_e^2 m_\mu^2}{s^2}\right\}.$$

11.2 Let $s = (p_1 + p_2)^2$ and let ϑ be the scattering angle in the center of mass system $[\cos \vartheta = \boldsymbol{p}_1\boldsymbol{p}_3/(|\boldsymbol{p}_1||\boldsymbol{p}_3|)]$. Neglecting terms of order m_e^2/s, one obtains for the differential and total cross-sections:

$$\frac{d\sigma}{d\Omega}(e^-e^+ \to \pi^-\pi^+) = \frac{\alpha^2}{8s}\left(1 - \frac{4m_\pi^2}{s}\right)^{3/2} \sin^2 \vartheta |F_\pi(s)|^2,$$

$$\sigma_{\text{tot}}(e^-e^+ \to \pi^-\pi^+) = \frac{4\pi\alpha^2}{3s}\frac{1}{4}\left(1 - \frac{4m_\pi^2}{s}\right)^{3/2} |F_\pi(s)|^2.$$

As a result of the crossing symmetry it follows from (11.22) that

$$\langle\pi^+(p_4)|\mathbf{j}_\mu(0)|\pi^+(p_3)\rangle = e(p_3 + p_4)_\mu F_\pi((p_3 - p_4)^2).$$

On the other hand, with $\mathbf{Q}$ the charge operator, we have

$$\begin{aligned}\langle\pi^+(p_4)|\mathbf{Q}|\pi^+(p_3)\rangle &= e\langle\pi^+(p_4)|\pi^+(p_3)\rangle = \int_{t=\text{const.}} d^3x\,\langle\pi^+(p_4)|\mathbf{j}^0(x)|\pi^+(p_3)\rangle \\ &= \int_{t=\text{const.}} d^3x \exp[i(p_4 - p_3)x]\langle\pi^+(p_4)|\mathbf{j}^0(0)|\pi^+(p_3)\rangle \\ &= eF_\pi(0)2p_3^0(2\pi)^3\delta^3(\boldsymbol{p}_4 - \boldsymbol{p}_3).\end{aligned}$$

If we take account of the normalization of the states (3.69) then it follows that $F_\pi(0) = 1$. Without proof we further note that $F_\pi(s)$ is an analytical function in the complex s-plane with a cut along the positive real axis for $s \geq 4m_\pi^2$. In the reaction $e^+e^- \to \pi^-\pi^+$ one measures the boundary value of $F_\pi(s)$ obtained by approaching the cut from above.

13.1 The state vector (normalized to one) of the $1\,^3S_1$ positronium at rest with polarization vector $\boldsymbol{\lambda}$ ($|\boldsymbol{\lambda}| = 1$) is

$$|1\,^3S_1, \boldsymbol{\lambda}\rangle = \frac{1}{\sqrt{V}}\int \frac{d^3p}{(2\pi)^3 2p^0} f(\boldsymbol{p})2^{-1/2}(\boldsymbol{\lambda\sigma}\varepsilon)_{ss'}\mathbf{a}_s^\dagger(\boldsymbol{p})\mathbf{b}_{s'}^\dagger(-\boldsymbol{p})|0\rangle,$$

where $f(\boldsymbol{p})$ is as in (13.19) and ε as in (C.11). To leading order in α the S-matrix element for the annihilation into three photons is given by

$$\langle \gamma(k_1, \varepsilon_1), \gamma(k_2, \varepsilon_2), \gamma(k_3, \varepsilon_3)|\mathbf{S}|1\,^3S_1, \boldsymbol{\lambda}\rangle$$

$$= -ie^3 \frac{1}{\sqrt{2V}} \frac{\psi(0)}{m_e^2} (2\pi)^4 \delta^3(\boldsymbol{k}_1 + \boldsymbol{k}_2 + \boldsymbol{k}_3)\delta(2m_e - |\boldsymbol{k}_1| - |\boldsymbol{k}_2| - |\boldsymbol{k}_3|)$$

$$\cdot\{(\boldsymbol{\varepsilon}_1^* \cdot \boldsymbol{\varepsilon}_2^*)(\boldsymbol{\varepsilon}_3^* \cdot \boldsymbol{\lambda}) - (\hat{\boldsymbol{k}}_1 \times \boldsymbol{\varepsilon}_1^*)\cdot(\hat{\boldsymbol{k}}_2 \times \boldsymbol{\varepsilon}_2^*)(\boldsymbol{\varepsilon}_3^* \cdot \boldsymbol{\lambda}) + (\hat{\boldsymbol{k}}_1 \times \boldsymbol{\varepsilon}_1^* \cdot \boldsymbol{\varepsilon}_2^*)(\hat{\boldsymbol{k}}_3 \times \boldsymbol{\varepsilon}_3^* \cdot \boldsymbol{\lambda})$$

$$+ (\hat{\boldsymbol{k}}_2 \times \boldsymbol{\varepsilon}_2^* \cdot \boldsymbol{\varepsilon}_1^*)(\hat{\boldsymbol{k}}_3 \times \boldsymbol{\varepsilon}_3^* \cdot \boldsymbol{\lambda}) + \text{cyclic permutations.}$$

Here $\hat{\boldsymbol{k}}_i = \boldsymbol{k}_i/|\boldsymbol{k}_i|$. By taking the square and integrating over the phase space of the three photons one arrives at (13.37).

13.2 A basis for the vector functions $\boldsymbol{A}$ can be constructed from $\boldsymbol{\varepsilon}_1, \boldsymbol{\varepsilon}_2$ and the spherical harmonics $Y_{l,l_3}(\hat{\boldsymbol{k}})$ where $\hat{\boldsymbol{k}} = \boldsymbol{k}/|\boldsymbol{k}|$ (cf. Appendix C). We thus have to ask: In how many ways can we get the representation $D(1)$ (for $\boldsymbol{A}$) from the product representation $D(1) \otimes D(1) \otimes D(l)$, where $l = 0, 1, 2 \ldots$ and the first two factors stem from $\boldsymbol{\varepsilon}_{1,2}$? Using the Clebsch–Gordan series (C.28) we combine first $\boldsymbol{\varepsilon}_1$ and $\boldsymbol{\varepsilon}_2$ to form a scalar, vector, and symmetric second-rank tensor

$$D(1) \otimes D(1) \otimes D(l) = \sum_{n=0}^{2} D(n) \otimes D(l).$$

Again using (C.28) we get the representation $D(1)$ for $(n, l) = (0, 1), (1, 0), (1, 1), (1, 2), (2, 1), (2, 2), (2, 3)$ corresponding to the amplitudes (a)–(g) of (13.12).

16.3 Let (Λ', a') be a Poincaré transformation with Λ' from the identity component and furthermore let $p' = \Lambda' p$, $\Lambda_R = \Lambda_{p'}^{-1}\Lambda'\Lambda_p$. The transformation Λ_R is a pure rotation. The corresponding rotation matrix is denoted by $\mathbf{R}(\Lambda', p)$. We have

$$\mathbf{U}(\Lambda', a')|\mathrm{a}(p, s_3)\rangle = \exp(\mathrm{i}p'a')|\mathrm{a}(p', s_3')\rangle D_{s_3's_3}(\mathbf{R}(\Lambda', p)),$$

where $\mathbf{D}(\mathbf{R}) = (D_{s_3's_3}(\mathbf{R}))$ is a matrix from the spin s representation of the rotation group (see Edmonds 1957). The normalization of the states is deduced from the following argument. On account of (16.26) the state vectors must satisfy a continuum normalization of the form

$$\langle \mathrm{a}(\tilde{p}, r_3)|\mathrm{a}(p, s_3)\rangle = N_{r_3s_3}(p)(2\pi)^3 2p_0 \delta^3(\tilde{\boldsymbol{p}} - \boldsymbol{p}),$$

where $\mathbf{N}(p) = (N_{r_3s_3}(p))$ is a matrix that has to be determined. By applying a transformation $\mathbf{U}(\Lambda', 0)$ we find

$$\mathbf{D}^\dagger(\mathbf{R}(\Lambda', p))\mathbf{N}(p')\mathbf{D}(\mathbf{R}(\Lambda', p)) = \mathbf{N}(p),$$

where $p' = \Lambda p$. If Λ' is chosen as a pure rotation $\Lambda(\mathbf{R}_0)$ then $\mathbf{R}(\Lambda', \boldsymbol{p} = 0) = \mathbf{R}_0$ and the irreducibility of the representation of the rotation group for spin s implies, that $\mathbf{N}(\boldsymbol{p} = 0)$ is a multiple of the unit matrix: $\mathbf{N}(\boldsymbol{p} = 0) = \lambda\mathbb{1}$. Since the norm of the states must be positive it follows that $\lambda > 0$ and by rescaling we can arrange that $\lambda = 1$. Using $\Lambda' = \Lambda_p^{-1}$ we then conclude that $\mathbf{N}(p) = \mathbb{1}$ for all p.

16.6 As an example we consider a resonance at $s = s_0$ with $j^P = \frac{3}{2}^+$. According to Table 16.1, we must then have $l = 1$ and the resonance states produced are $|s_0; 1; \frac{3}{2}; \pm\frac{1}{2}\rangle$. If (16.60), (16.58) are used to construct these states from the π^+p final states then we find

$$|s_0; 1; \tfrac{3}{2}, \pm\tfrac{1}{2}\rangle = \sum_{l_3, r} (1, l_3; \tfrac{1}{2}, r | \tfrac{3}{2}, \pm\tfrac{1}{2}) \int d\Omega_{p'}\, Y_{1,l_3}(\hat{\boldsymbol{p}}')|\pi^+(\boldsymbol{p}'), p(-\boldsymbol{p}', r)\rangle$$

$$= \int d\Omega_{p'} \{\sqrt{\tfrac{2}{3}}\, Y_{1,0}(\hat{\boldsymbol{p}}')|\pi^+(\boldsymbol{p}'), p(-\boldsymbol{p}', \pm\tfrac{1}{2})\rangle$$

$$+ \sqrt{\tfrac{1}{3}}\, Y_{1,\pm 1}(\hat{\boldsymbol{p}}')|\pi^+(\boldsymbol{p}'), p(-\boldsymbol{p}', \mp\tfrac{1}{2})\rangle\},$$

$$\frac{d\sigma}{d\Omega} \propto \tfrac{2}{3}|Y_{1,0}(\hat{\boldsymbol{p}}')|^2 + \tfrac{1}{3}|Y_{1,\pm 1}(\hat{\boldsymbol{p}}')|^2 \propto 1 + 3\cos^2\vartheta.$$

This is in agreement with (16.81).

16.7

$$a_{1,3/2}(s) = -\frac{m_\Delta \Gamma_\Delta}{s - m_\Delta^2 + i m_\Delta \Gamma_\Delta},$$

$$\delta_{1,3/2}(s) = \arctan\left(\frac{m_\Delta \Gamma_\Delta}{m_\Delta^2 - s}\right).$$

16.8 We take the collision axis as the x^1-axis and, for an arbitrary reference system of the class considered, we define the rapidity y_c of a particle c with four-momentum p_c as

$$y_c = \tfrac{1}{2}\ln\frac{p_c^0 + p_c^1}{p_c^0 - p_c^1}.$$

Rotations around the collision axis do not change y_c. The Lorentz boosts along the collision axis have the form given by $\Lambda(v)$ (2.33). If we set $p_c' = \Lambda(v)p_c$ then we find

$$y_c' = y_c + \tfrac{1}{2}\ln\frac{1+v}{1-v}.$$

Thus a Lorentz transformation along the collision axis has the effect of shifting all rapidities by a constant amount.

17.2 The proof, that (17.21) follows from (17.16) proceeds in a manner similar to the one used in Exercise 16.1. Here we deal with the proof of the reverse conclusion. We begin by stating the following subsidiary relation, valid for any $n \times n$ matrices $\mathbf{A}$, $\mathbf{B}$:

$$\exp(\mathbf{A} + \varepsilon\mathbf{B})\cdot\exp(-\mathbf{A})$$

$$= 1 + \varepsilon\left\{\frac{1}{1!}\mathbf{B} + \frac{1}{2!}[\mathbf{A}, \mathbf{B}] + \frac{1}{3!}[\mathbf{A}, [\mathbf{A}, \mathbf{B}]] + \cdots\right\} + O(\varepsilon^2).$$

To prove this we expand $\mathbf{F}(t, \varepsilon) = \exp(t(\mathbf{A} + \varepsilon\mathbf{B}))\cdot\exp(-t\mathbf{A})$, where t is any complex number, in a power series in t:

$$\mathbf{F}(t, \varepsilon) = \sum_{n=0}^{\infty} \frac{t^n}{n!}\mathbf{C}_n(\varepsilon).$$

We have $\mathbf{C}_0(\varepsilon) = 1$ and

$$\mathbf{C}_n(\varepsilon) = \left(\frac{\mathrm{d}}{\mathrm{d}t}\right)^n \exp(t(\mathbf{A} + \varepsilon\mathbf{B}))\cdot\exp(-t\mathbf{A})|_{t=0}$$

$$= \frac{\mathrm{d}}{\mathrm{d}t}\exp(t(\mathbf{A} + \varepsilon\mathbf{B}))\cdot\mathbf{C}_{n-1}(\varepsilon)\cdot\exp(-t\mathbf{A})|_{t=0}$$

$$= (\mathbf{A} + \varepsilon\mathbf{B})\mathbf{C}_{n-1}(\varepsilon) - \mathbf{C}_{n-1}(\varepsilon)\mathbf{A} = [\mathbf{A}, \mathbf{C}_{n-1}(\varepsilon)] + \varepsilon\mathbf{B}\mathbf{C}_{n-1}(\varepsilon)$$

for $n \geq 1$. The validity of the subsidiary relation then follows immediately. We now note that every $\mathbf{U} \in \mathrm{SU}(3)$ can be represented in the form $\mathbf{U} = \exp(\mathrm{i}\varphi_a\frac{1}{2}\lambda_a)$ with real φ_a. Let $\mathbf{U} = \exp(\mathrm{i}\varphi_a\frac{1}{2}\lambda_a)$ and $\mathbf{V} = \exp(\mathrm{i}\psi_a\frac{1}{2}\lambda_a)$ be any matrices from SU(3). We write

$$\mathbf{U}(\tau) = \exp(\mathrm{i}\tau\varphi_a\tfrac{1}{2}\lambda_a), \qquad \mathbf{W}(\tau) = \mathbf{U}(\tau)\mathbf{V} = \exp(\mathrm{i}\chi_a(\tau)\tfrac{1}{2}\lambda_a),$$

where $0 \leq \tau \leq 1$. We then have

$$\frac{\mathrm{d}}{\mathrm{d}\tau}\mathbf{W}(\tau) = \mathrm{i}\varphi_a\tfrac{1}{2}\lambda_a\mathbf{W}(\tau).$$

Let $\mathbf{F}_a$ be Hermitian matrices that satisfy (17.21). We set

$$\mathbf{D}(\mathbf{U}(\tau)) = \exp(\mathrm{i}\tau\varphi_a\mathbf{F}_a), \qquad \mathbf{D}(\mathbf{V}) = \exp(\mathrm{i}\psi_a\mathbf{F}_a),$$

$$\mathbf{D}(\mathbf{W}(\tau)) = \exp(\mathrm{i}\chi_a(\tau)\mathbf{F}_a).$$

It obviously holds that

$$\frac{\mathrm{d}}{\mathrm{d}\tau}\mathbf{D}(\mathbf{U}(\tau))\mathbf{D}(\mathbf{V}) = \mathrm{i}\varphi_a\mathbf{F}_a\mathbf{D}(\mathbf{U}(\tau))\mathbf{D}(\mathbf{V}),$$

$$\mathbf{D}(\mathbf{U}(\tau))\mathbf{D}(\mathbf{V})|_{\tau=0} = \mathbf{D}(\mathbf{W}(\tau))|_{\tau=0} = \mathbf{D}(\mathbf{V}).$$

For the calculation of $(\mathrm{d}\mathbf{W}(\tau)/\mathrm{d}\tau)\mathbf{W}^{-1}(\tau)$ and $(\mathrm{d}\mathbf{D}(\mathbf{W}(\tau))/\mathrm{d}\tau)\mathbf{D}^{-1}(\mathbf{W}(\tau))$ we simply need, according to our subsidiary relation, to know the commutation rules of the matrices $\frac{1}{2}\lambda_a$ and $\mathbf{F}_a$, respectively. By assumption these are identical. It follows that

$$\frac{\mathrm{d}}{\mathrm{d}\tau}\mathbf{D}(\mathbf{W}(\tau)) = \mathrm{i}\varphi_a\mathbf{F}_a\mathbf{D}(\mathbf{W}(\tau)),$$

which in turn implies the equality of $\mathbf{D}(\mathbf{U}(\tau))\cdot\mathbf{D}(\mathbf{V})$ and $\mathbf{D}(\mathbf{W}(\tau))$ since they satisfy the same differential equation and the same initial conditions. Thus the group property (17.16) is proven.

17.4 The operators $\mathbf{I}_\pm$, $\mathbf{I}_3 = \mathbf{F}_3$; $\mathbf{U}_\pm$, $\mathbf{U}_3 = \frac{1}{2}(\sqrt{3}\mathbf{F}_8 - \mathbf{F}_3)$; and $\mathbf{V}_\pm$, $\mathbf{V}_3 = \frac{1}{2}(\sqrt{3}\mathbf{F}_8 + \mathbf{F}_3)$ are the generators of three different SU(2) subgroups of SU(3): the isospin, the U-spin, and the V-spin group. With respect to isospin, the u- and d-quarks form an isodoublet and the s-quark a singlet; with respect to the U-spin, d and s form a doublet and u a singlet and with respect to the V-spin, u and s form a doublet and d a singlet. Particles that transform according to a general representation of the SU(3) group can be classified according to isospin, U-spin or V-spin multiplets. In the corresponding weight diagram, these lie on straight lines parallel to the u–d, d–s, and u–s lines, respectively.

17.6 We consider for any of the representations in question the basis formed by the common eigenstates of $\mathbf{I}_3, \overrightarrow{\mathbf{I}}^2$, and $\mathbf{Y}$. To visualize this, mark the corresponding points in the appropriate weight diagram. One first demonstrates that it is possible, by repeated application of the operators $\mathbf{I}_\pm$, $\mathbf{U}_\pm$, $\mathbf{V}_\pm$ to the state with maximum eigenvalue of $\mathbf{I}_3$, to generate all the basis states. Second, one shows that an arbitrary state (not necessarily a state from the basis) can be converted into a multiple ($\neq 0$) of the state with maximum eigenvalue of $\mathbf{I}_3$ by suitable repeated application of $\mathbf{I}_\pm$, $\mathbf{U}_\pm$ and $\mathbf{V}_\pm$. Thus the states that one can obtain by repeated application of the generators of the SU(3) group to an arbitrary state always span the entire representation space. The representations considered are therefore irreducible.

17.7 For the 0^--mesons one finds $\bar{m}_8 = \frac{4}{3}m_K - \frac{1}{3}m_\pi$, $m_1 = m_\eta + m_{\eta'} - \bar{m}_8$, $\tan\vartheta = (m_\eta - \bar{m}_8)(\bar{m}_8 m_1 - m_\eta m_{\eta'})^{-1/2}$. If the mass values from Table 15.1 are taken, then (17.77) is obtained. The same procedure applied to the 1^--mesons ρ, K^*, ω, ϕ leads to $\tan\vartheta = -1.31$, $\vartheta = -52.6°$. "Ideal" mixing (17.78) corresponds to $\tan\vartheta = -\sqrt{2}$.

17.8 We will use the generic labels a, b ... to refer to one of the states $\overset{\uparrow}{\mathrm{u}}, \ldots, \overset{\downarrow}{\mathrm{s}}$. The linearly independent symmetric combinations of three such states can be enumerated as follows: There are 6 states of the type aaa; $6 \times 5 = 30$ states of the type aab with $a \neq b$; $6 \times 5 \times 4/3! = 20$ states of the type abc with $a \neq b \neq c \neq a$. If all quarks are put into spatial S-wave states, then it is clear that this 56-representation of the SU(6) group only contains states with total angular momentum $\frac{1}{2}$ and $\frac{3}{2}$. According to (17.55), only the SU(3) representations 8 and 10 occur. (The 1-representation cannot arise, as explained in the text.) The decuplet with spin $\frac{3}{2}$ has a totally symmetric wavefunction in flavor space and thus certainly belongs to the 56-representation of SU(6); it yields $10 \times 4 = 40$ states. The remaining 16 states must therefore form an octet with spin $\frac{1}{2}$; hence (17.80) is proven.

18.4 If $N_{p'}(x)$, etc., are the distribution functions averaged over proton and neutron, then in the model with bare nucleons as partons one finds

$$F_2^{(eN)}(x) = x(N_{p'}(x) + N_{\bar{p}'}(x)),$$

$$F_2^{(\nu N)}(x) = 2x(N_{n'}(x) + N_{\bar{p}'}(x)).$$

On account of isospin symmetry $N_{p'}(x) = N_{n'}(x)$ and thus $F_2^{(\nu N)}(x)/F_2^{(eN)}(x) = 2$.

18.5

$$\mathbf{P}_I|\mathrm{p}\rangle = -|\mathrm{n}\rangle, \qquad \mathbf{P}_I|\mathrm{n}\rangle = |\mathrm{p}\rangle,$$

$$\mathbf{P}_I|\mathrm{u}\rangle = -|\mathrm{d}\rangle, \qquad \mathbf{P}_I|\mathrm{d}\rangle = |\mathrm{u}\rangle.$$

18.6 With $G = 1.16 \cdot 10^{-5}\ \mathrm{GeV}^{-2}$ we have $G^2M/\pi = 1.56 \cdot 10^{-38}\ \mathrm{cm}^2\ \mathrm{GeV}^{-1}$ and from Fig. 18.14 it follows that $\langle q \rangle = 0.40 \pm 0.02$; $\langle \bar{q} \rangle = 0.08 \pm 0.02$, and $I \cong \langle q \rangle + \langle \bar{q} \rangle = 0.48 \pm 0.02$.

18.7 With the help of isospin symmetry one can derive the following inequalities for the distribution functions of the *proton*:

$$2N_u(x) - N_d(x) \geq 0, \qquad N_d(x) \geq 0, \qquad N_s(x) \geq 0,$$

$$2N_{\bar{d}}(x) - N_{\bar{u}}(x) \geq 0, \qquad N_{\bar{u}}(x) \geq 0, \qquad N_{\bar{s}}(x) \geq 0.$$

To determine the extremal values of $F_2^{(en)}(x)/F_2^{(ep)}(x)$ one only needs to consider the cases where all but one of the above inequalities become equalities. In this way one sees that the minimum of $F_2^{(en)}(x)/F_2^{(ep)}(x)$ is equal to $\frac{1}{4}$, and that this lower limit can only be reached for those x values where the distribution functions $N_d(x)$, $N_s(x)$, $N_{\bar{u}}(x)$, $N_{\bar{d}}(x)$, and $N_{\bar{s}}(x)$ vanish relative to $N_u(x)$.

18.8 The result is Adler's sum rule:

$$\int_0^\infty d\nu\,[W_2^{(\bar{\nu}p)}(\nu, Q^2) - W_2^{(\nu p)}(\nu, Q^2)] = 2.$$

18.9 The result is the Gross–Llewellyn–Smith sum rule:

$$\int_0^1 dx\,[F_3^{(\bar{\nu}p)}(x) + F_3^{(\nu p)}(x)] = -6.$$

20.1 The minimum value of T for planar events is $T = 2/\pi$ and this occurs for planar isotropic events. For events with only three massless hadrons $T_{\min} = \frac{2}{3}$.

20.3 We consider hypothetical scalar gluons $\tilde{G}$, and for the Lagrange density of their interaction with the quarks we set $\mathscr{L}'(x) = \tilde{g}\bar{q}(x)q(x)\tilde{G}(x)$. Here $\tilde{g}$ is the coupling constant. To leading order in $\tilde{g}$ we then obtain (with $\frac{2}{3} \le T \le 1$):

$$\frac{1}{\sigma_{\mathrm{tot}}}\frac{d\sigma}{dT}(e^+e^- \to q\bar{q}\tilde{G}) = \left(\frac{\tilde{g}}{4\pi}\right)^2 [(1-T)^{-1} - 9(1-T) - 2\ln(1-T) + 2\ln(2T-1)],$$

$$\langle 1 - T\rangle = \left(\frac{\tilde{g}}{4\pi}\right)^2 \frac{1}{36}(2 + 9\ln 3).$$

20.6 We consider the decay $J/\psi \to G^a G^b G^c$, where a, b, c are the color indices of the gluons. The J/ψ is a color singlet with charge conjugation $C = -1$ (Table 13.1). Because of color conservation, the wave function of the three gluons in the final state can only be proportional to f_{abc} or d_{abc}. By means of an explicit calculation or by making use of the charge conjugation invariance of QCD, one sees that only d_{abc} is actually possible. The charge conjugation transformation, C, has the following effect on the quark field operators [cf. (4.126)]:

$$C\colon \mathbf{q}(x) \to \mathbf{S}(C)\bar{\mathbf{q}}^{\mathrm{T}}(x).$$

If we supplement this with a transformation of the gluon fields (Chap. 19)

$$C\colon \mathbf{G}_\lambda(x) \to -\mathbf{G}_\lambda^{\mathrm{T}}(x),$$

then $\mathscr{L}_{\mathrm{QCD}}$ (19.22) remains invariant. A wavefunction for three gluons proportional to d_{abc} is thus odd with respect to C:

$$C\colon d_{abc} = \tfrac{1}{4}\,\mathrm{Tr}\{\lambda_a, \lambda_b\}\lambda_c \to \tfrac{1}{4}\,\mathrm{Tr}\{-\lambda_a^{\mathrm{T}} - \lambda_b^{\mathrm{T}}\}(-\lambda_c^{\mathrm{T}}) = -d_{abc}.$$

Analogously, one sees that f_{abc} is even with respect to C and therefore, in the decay of the J/ψ with $C = -1$, it cannot appear. The amplitude for the decay $J/\psi \to G^a G^b G^c$ can thus be obtained from that of orthopositronium decay (Exercise 13.1) with the substitutions

$$e^3 \to g_s^3 \frac{1}{\sqrt{3}}\,\mathrm{Tr}\left(\frac{\lambda_a}{2}\frac{\lambda_b}{2}\frac{\lambda_c}{2}\right)\Bigg|_{\mathrm{symm.\,part}} = g_s^3 \frac{1}{4\sqrt{3}}\, d_{abc},$$

$$m_e \to m_c \cong \tfrac{1}{2} m_{J/\psi}.$$

After summation over the color degrees of freedom of the gluons, (13.37) then yields (20.31).

21.1 Parity invariance demands $g_j' = 0$; charge conjugation invariance demands $g_j = g_j^*$, $g_j' = -g_j'^*$; time reversal invariance demands $g_j = g^*$, $g_j = g_j'^*$, where in each case $j = 1, \ldots, 5$.

23.1 Neglecting the mass of the electron and using the Lagrange density from (23.69), we still obtain (23.2a) for the total decay rate of the muon; however the Michel parameter is found to be $\rho = \frac{3}{8}\,\mathrm{Re}(1 + ab^*)$.

23.2 In this case the Noether currents are

$$\mathbf{J}_a^\mu = (\bar{\mathsf{u}}(x), \bar{\mathsf{d}}(x))\tfrac{1}{2}\tau_a\gamma^\mu\begin{pmatrix}\mathsf{u}(x)\\ \mathsf{d}(x)\end{pmatrix},$$

where $a = 1, 2, 3$.

23.5 From the Lorentz invariance it follows that

$$\langle 0|\mathbf{J}_\lambda(0)|\pi^-(p)\rangle = \mathrm{i}p_\lambda f_\pi' \qquad \text{and} \qquad \langle 0|\mathbf{J}(0)|\pi^-(p)\rangle = g_\pi,$$

where f_π' and g_π are constants. The leptonic decay rates are then obtained from (23.49) by the replacements $V_{11}f_\pi \to |f_\pi'|$ for case (i), and $V_{11}f_\pi \to |g_\pi|/m_\ell$ for case (ii).

23.6 We use a simple spectator model by neglecting the motion of the b-quark in the B-meson and setting $m_\mathrm{b} \cong m_\mathrm{B}$. Let E be the energy of the lepton ℓ^- in the rest frame of the B-meson and $x = 2E/m_\mathrm{B}$. For the differential decay rate in the semileptonic decay one then obtains

$$\frac{\mathrm{d}\Gamma}{\mathrm{d}x} = \frac{\tilde{G}^2 m_\mathrm{B}^5}{16\pi^3}x^2(1 - x - \eta)^2(1 - x)^{-3}\left[(1 - x)\left(\frac{1}{2} - \frac{x}{3}\right) + \left(\frac{1}{2} - \frac{x}{6}\right)\eta\right],$$

with $0 \leq x \leq 1 - \eta$. For the transition $\mathrm{b} \to \mathrm{c}\ell^-\bar{\nu}_\ell$ we have $\tilde{G} = GV_{23}$ and $\eta = m_\mathrm{c}^2/m_\mathrm{b}^2$, whereas for $b \to \mathrm{u}\ell^-\bar{\nu}_\ell$ one must use $\tilde{G} = GV_{13}$ and $\eta = 0$.

24.2 We write

$$\tilde{Q}_\mathrm{f}(s) = Q_\mathrm{f}\left[1 + C\frac{s\, m_{Z'}^2}{(\sin\vartheta_\mathrm{W}\cos\vartheta_\mathrm{W} m_\mathrm{Z})^2(s - m_{Z'}^2 + \mathrm{i}m_{Z'}\Gamma_{Z'})}\right].$$

The differential cross-section for the reaction $\mathrm{e}^+\mathrm{e}^- \to \bar{\mathrm{f}}\,\mathrm{f}$ is as in (24.43) but with

$$R_\mathrm{f}(s) = N_\mathrm{c}^\mathrm{f}\{|\tilde{Q}_\mathrm{f}(s)|^2 + 2g_\mathrm{V}^\mathrm{e}g_\mathrm{V}^\mathrm{f}\,\mathrm{Re}[K(s)\tilde{Q}_\mathrm{f}^*(s)] + [(g_\mathrm{V}^\mathrm{e})^2 + (g_\mathrm{A}^\mathrm{e})^2][(g_\mathrm{V}^\mathrm{f})^2 + (g_\mathrm{A}^\mathrm{f})^2]|K(s)|^2\},$$

$$I_\mathrm{f}(s) = N_\mathrm{c}^\mathrm{f}\{4g_\mathrm{A}^\mathrm{e}g_\mathrm{A}^\mathrm{f}\,\mathrm{Re}[K(s)\tilde{Q}_\mathrm{f}^*(s)] + 8g_\mathrm{V}^\mathrm{e}g_\mathrm{A}^\mathrm{e}g_\mathrm{V}^\mathrm{f}g_\mathrm{A}^\mathrm{f}|K(s)|^2\}.$$

For $s \ll m_\mathrm{Z}^2, m_{Z'}^2$, this leads to $A_\mathrm{f}(s)$ as in (24.54). For $\delta R_\mathrm{f}(s)$, on the other hand, one obtains

$$\delta R_\mathrm{f}(s) = N_\mathrm{c}^\mathrm{f}[-Q_\mathrm{f}^2 8C + 2Q_\mathrm{f}g_\mathrm{V}^\mathrm{e}g_\mathrm{V}^\mathrm{f}]\frac{Gs}{2\sqrt{2}\pi\alpha}.$$

With this one obtains from (24.48), (24.55) the experimental upper limit $C \lesssim 0.035$. A more detailed analysis of the data has confirmed this limit (cf. Wu 1984).

24.3 We consider the reaction $e^+(p_1) + e^-(p_2) \to e^+(p_3) + e^-(p_4)$ and set $s = (p_1 + p_2)^2, t = (p_1 - p_3)^2, u = (p_1 - p_4)^2$. On inclusion of all diagrams with the exchange of one boson, the differential cross-section is found to be

$$\frac{d\sigma}{dt} = \frac{\pi\alpha^2}{s^2}\{(t^2 + u^2)A(s, t) + 2s^2B(s, t) + 2s(t - u)C(s, t)\},$$

where (with $\eta = \pm 1$)

$$A(s, t) = \sum_\eta (|C_V^\eta|^2 + |C_A^\eta|^2),$$

$$B(s, t) = t^{-2}|1 - K(t)[(g_V^e)^2 + (g_A^e)^2]|^2,$$

$$C(s, t) = \sum_\eta \mathrm{Re}(C_V^\eta C_A^{\eta *}),$$

$$C_V^\eta = (2t)^{-1} + s^{-1} - (2t)^{-1}K(t)(g_V^e - \eta g_A^e)^2 - s^{-1}K(s)(g_V^e - \eta g_A^e)g_V^e,$$

$$C_A^\eta = (2t)^{-1} - (2t)^{-1}K(t)(g_V^e - \eta g_A^e)^2 + s^{-1}K(s)(g_V^e - \eta g_A^e)\eta g_A^e.$$

The function K is given by (24.45).

25.2 We consider e^+e^- states in the center of mass frame and define states with total spin S and angular momentum l with the help of the Clebsch–Gordan coefficients by

$$|e^+e^-; l, l_3; S, S_3\rangle = \sum_{r,r'} \int d\Omega_p\, Y_{l,l_3}(\hat{p})(\tfrac{1}{2}, r; \tfrac{1}{2}r'|S, S_3)\, |e^+(p, r), e^-(-p, r')\rangle.$$

Here $S = 0, 1$ and $l = 0, 1, 2, \ldots$. A suitable set of eigenstates with total angular momentum j is then

$$|e^+e^-; S; l; j, j_3\rangle = \sum_{l_3, S_3} (l, l_3; S, S_3|j, j_3)\, |e^+e^-; l, l_3; S, S_3\rangle.$$

The Z-boson is produced via the vector coupling ($g_V^{e,\mu}$) in the partial waves with $j = 1, S = 1, l = 0$ and 2 and via the axial-vector coupling ($g_A^{e,\mu}$) in the partial wave with $j = 1, S = 1, l = 1$.

25.3 We consider the reaction $e^+(p_1) + e^-(p_2) \to \gamma(k_1) + Z(k_2)$ in the center of mass frame and set

$$s = (p_1 + p_2)^2, \qquad \cos\vartheta = \boldsymbol{p}_1\boldsymbol{k}_1/(|\boldsymbol{p}_1||\boldsymbol{k}_1|), \qquad E_\gamma = k_1^0 = (s - m_Z^2)/(2\sqrt{s}).$$

If the mass of the electron is neglected, the differential cross-section is given by

$$\frac{d\sigma}{d\Omega}(e^+e^- \to \gamma Z) = \frac{1}{2s}\left(\frac{\alpha}{\sin\vartheta_W \cos\vartheta_W}\right)^2 [(g_V^e)^2 + (g_A^e)^2]\frac{(\sqrt{s} - E_\gamma)^2 + E_\gamma^2\cos^2\vartheta}{\sqrt{s}\,E_\gamma \sin^2\vartheta},$$

where $d\Omega$ is the element of solid angle for $\boldsymbol{k}_1$. For the reaction (25.21), one must then multiply the above result by the branching ratio $\Gamma(Z \to \text{neutrinos})/\Gamma_Z$.

25.5 When the finite width of the Z is taken into account, one finds for the reaction (25.90)

$$\sigma = \frac{4\pi s}{3m_Z^2}\sum_{q=u,d}\frac{\Gamma(Z \to q\bar{q})}{m_Z}\int_0^1 dx_1\, x_1 N_q^p(x_1)\int_0^1 dx_2\, x_2 N_{\bar{q}}^{\bar{p}}(x_2) \cdot \frac{m_Z\Gamma(Z \to X')}{(x_1x_2s - m_Z^2)^2 + m_Z^2\Gamma_Z^2}.$$

Numerically this leads to a total Z-production cross-section at 540 GeV which differs by +1% from that given in (25.49).

26.2 For K-mesons *at rest* in a gravitational field, the assumed effect would lead to the following change in the mass matrix:

$$\mathcal{M} \to \mathcal{M} + (1 - \tfrac{1}{2}\eta) m_K U \cdot \mathbb{1} + \mathcal{M}',$$

where

$$\mathcal{M}' = \tfrac{1}{2}\eta m_K U(|K^0\rangle\langle K^0| - |\bar{K}^0\rangle\langle\bar{K}^0|).$$

The term $\mathcal{M}'$ would violate CP and CPT, but not T. If the source of the CP violation were merely in $\mathcal{M}'$, then we would conclude

$$\eta_{+-} = \eta_{00} \cong \eta m_K U 2^{-1/2} \Gamma_S^{-1} \exp(-i\tfrac{1}{4}\pi).$$

However, such a phase for η_{+-} or η_{00} is not compatible with experiment (see Table 26.1). If we allow a maximum contribution of 20% from terms of the "wrong" phase, then we obtain $|\eta| \lesssim 6.10^{-4}\Gamma_S/|m_K U|$. (The more detailed analysis of possible CPT violating terms of Schubert (1970) gives the same order of magnitude; see Kleinknecht (1976).) If we insert for U the gravitational potential of the Earth, then we find:

$$|m_K U| \cong 0.4 \text{ eV} \cong 5 \cdot 10^4 \Gamma_S, \qquad |\eta| \lesssim 1.4 \cdot 10^{-8}.$$

For the potential of the Sun one obtains analogously:

$$|m_K U| \cong 6 \text{ eV} \cong 8 \cdot 10^5 \Gamma_S, \qquad |\eta| \lesssim 9 \cdot 10^{-10}.$$

For K-mesons that are moving in a gravitational field—for example, a K-meson beam in the laboratory—the above analysis is not directly applicable. One would expect in general a velocity dependence of the parameters of CP violation under the influence of cosmological fields, such as the gravitational field (see Nachtmann 1969; Fischbach 1982).

References*

Abela, R. et al., *Nucl. Phys.* **A395**, 413 (1983)

Abers, E.S. and B.W. Lee, *Phys. Rep.* **9C**, 1 (1973)

Abramowicz, H. et al. (CDHS Coll.), *Z. Phys.* **C17**, 283 (1983)

Adler, S.L., *Phys. Rev.* **140**, B736 (1965)

Adler, S.L., *Phys. Rev.* **143**, 1144 (1966)

Adler, S.L. and R.F. Dashen, *Current Algebras* (W.A. Benjamin, New York and Amsterdam, 1968)

Adler, S.L., *Phys. Rev.* **177**, 2426 (1969)

Åkesson, T. et al. (AFS Coll.), *Z. Phys.* **C25**, 13 (1984)

Akhundov, A.A. D. Yu. Bardin, and T. Riemann, *Nucl. Phys.* **B276**, 1 (1986)

Akulov, D.V. and V.P. Volkov, *JETP Lett.* **16**, 438 (1972)

Albajar, C. et al. (UA1 Coll.), *Phys. Lett.* **186B**, 247 (1987)

Albert, D., W. Marciano, D. Wyler, and Z. Parsa, *Nucl. Phys.* **B166**, 460 (1980)

Albrecht, H. et al. (ARGUS Coll.), *Phys. Lett.* **192B**, 245 (1987)

Alpgård, K. et al. (UA5 Coll.), *Phys. Lett.* **121B**, 209 (1983)

Altarelli, G. and G. Parisi, *Nucl. Phys.* **B126**, 298 (1977)

Altarelli, G., R.K. Ellis, and G. Martinelli, *Nucl. Phys.* **B157**, 461 (1979)

Altarelli, G., *Phys. Rep.* **C81**, 1 (1982)

* Letters added to the year of publication serve to discriminate between references in the text; they have no bibliographical meaning.

Altarelli, G., N. Cabibbo, G. Corbo, L. Maiani, and G. Martinelli, *Nucl. Phys.* **B208**, 365 (1982a)

Altarelli, G., R.K. Ellis, M. Greco, and G. Martinelli, *Nucl. Phys.* **B246**, 12 (1984)

Altarev, I.S. et al., *Phys. Lett.* **102B**, 13 (1981)

Althoff, M. et al. (TASSO Coll.), *Z. Phys.* **C22**, 13 (1984)

Althoff, M. et al. (TASSO Coll.), *Z. Phys.* **C22**, 307 (1984a)

Anderson, C.D., *Science* **76**, 238 (1932)

Anderson, C.D., *Phys. Rev.* **43**, 491 (1933)

Anderson, C.D., *Phys. Rev.* **51**, 884 (1937)

Anderson, C.D. and S. Neddermeyer, *Phys. Rev.* **54**, 88 (1938)

Anderson, H.L. et al., *Phys. Rev. Lett.* **38**, 1450 (1977)

Andersson, B., G. Gustafson, and T. Sjöstrand, *Z. Phys.* **C6**, 235 (1980)

Andersson, B., G. Gustafson, G. Ingelman, and T. Sjöstrand, *Phys. Rep.* **C97**, 31 (1983)

Appelquist, Th. and H. Georgi, *Phys. Rev.* **D8**, 4000 (1973)

Appelquist, Th. and H.D. Politzer, *Phys. Rev. Lett.* **34**, 43 (1975)

Argento, W. et al., *Phys. Lett.* **120B**, 245 (1983)

Arnison, G. et al. (UA1 Coll.), *Phys. Lett.* **122B**, 103 (1983a)

Arnison, G. et al. (UA1 Coll.), *Phys. Lett.* **123B**, 108 (1983b)

Arnison, G. et al. (UA1 Coll.), *Phys. Lett.* **126B**, 398 (1983c)

Arnison, G. et al. (UA1 Coll.), *Phys. Lett.* **128B**, 336 (1983d)

Arnison, G. et al. (UA1 Coll.), *Phys. Lett.* **129B**, 273 (1983e)

Arnison, G. et al. (UA1 Coll.), *Phys. Lett.* **132B**, 214 (1983f)

Arnison, G. et al. (UA1 Coll.), *Phys. Lett.* **132B**, 223 (1983g)

Arnison, G. et al. (UA1 Coll.), *Phys. Lett.* **134B**, 469 (1984a)

Arnison, G. et al. (UA1 Coll.), *Phys. Lett.* **136B**, 423 (1984b)

Arnison, G. et al. (UA1 Coll.), *Phys. Lett.* **139B**, 115 (1984c)

Arnison, G. et al. (UA1 Coll.), *Phys. Lett.* **147B**, 241 (1984d)

Arnison, G. et al. (UA1 Coll.), *Phys. Lett.* **147B**, 493 (1984e)

Ashkin, A. et al., *Phys. Rev.* **94**, 357 (1954)

Aubert, J.J. et al., *Phys. Rev. Lett.* **33**, 1404 (1974)

Aubert, J.J. et al. (European Muon Collaboration), *Phys. Lett.* **123B**, 275 (1983)

Augustin, J.E. et al., *Phys. Rev. Lett.* **33**, 1406 (1974)

Aveda, B. et al. (MARK J. Coll.), *Phys. Rev. Lett.* **48**, 1701 (1982)

Bacci, C. et al., *Phys. Lett.* **86B**, 234 (1979)

Baće, M., *Phys. Lett.* **78B**, 132 (1978)

Bagnaia, P. et al. (UA2 Coll.), *Phys. Lett.* **129B**, 130 (1983)

Bagnaia, P. et al. (UA2 Coll.), *Phys. Lett.* **138B**, 430 (1984a)

Bagnaia, P. et al. (UA2 Coll.), *Z. Phys.* **C20**, 117 (1984b)

Bagnaia, P. et al. (UA2 Coll.), *Phys. Lett.* **144B**, 283 (1984c)

Bagnaia, P. et al. (UA2 Coll.), *Phys. Lett.* **144B**, 291 (1984d)

Bagnaia, P. et al. (UA2 Coll.), *Z. Phys.* **C24**, 1 (1984e)

Baluni, V., *Phys. Rev.* **D19**, 2227 (1979)

Bander, M., D. Silverman, and A. Soni, *Phys. Rev. Lett.* **44**, 7 and E. 962 (1980)

Banner, M. et al. (UA2 Coll.), *Phys. Lett.* **118B**, 203 (1982)

Banner, M. et al. (UA2 Coll.), *Phys. Lett.* **122B**, 476 (1983)

Barber, D.P. et al., (MARK J Coll.), *Phys. Rev. Lett.* **43**, 830 (1979)

Barber, W.C. et al., *Phys. Rev. Lett.* **16**, 1127 (1966)

Barber, W.C. et al., *Phys. Rev.* **D3**, 2796 (1971)

Bardeen, W.A., A.J. Buras, D.W. Duke, and T. Muta, *Phys. Rev.* **D18**, 3998 (1978)

Barnes, V.E. et al., *Phys. Rev. Lett.* **12**, 204 (1964)

Bartel, W. et al. (JADE Coll.), *Phys. Lett.* **88B**, 171 (1979)

Bartel, W. et al. (JADE Coll.), *Phys. Lett.* **91B**, 142 (1980)

Bartel, W. et al. (JADE Coll.), *Phys. Lett.* **108B**, 140 (1982)

Bartel, W. et al. (JADE Coll.), *Z. Phys.* **C19**, 197 (1983)

Bartel, W. et al. (JADE Coll.), *Z. Phys.* **C26**, 507 (1985)

Bateman Manuscript Project, *Higher Transcendental Functions*, Vol. 2 (McGraw-Hill, New York, 1953)

Bauer, M., B. Stech, and M. Wirbel, *Z. Phys.* **C34**, 103 (1987)

Beall, G., M. Bander, and A. Soni, *Phys. Rev. Lett.* **48**, 848 (1982)

Becher, P., M. Böhm, and H. Joos, *Gauge Theories of Strong and Electroweak Interactions* (J. Wiley, New York, 1984)

Becquerel, H., *C.R. Acad. Sci.* (*Paris*) **122**, 501 (1896)

Bég, M.A.B. and A. Sirlin, *Phys. Rep.* **88**, 1 (1982)

Behrend, H.J. et al. (CELLO Coll.), *Z. Phys.* **C14**, 283 (1982)

Bell, J.S. and J. Steinberger, in: *Proc. Int. Conf. Elementary Particles, Oxford, 1965*, ed. by T.R. Walsh (Rutherford Appleton Laboratory, Didcot, 1966)

Bell, J.S. and R. Jackiw, *Nuovo Cim.* **51**, 47 (1969)

Berezinsky, V.S., B.L. Joffe, and Ya.I. Kogan, *Phys. Lett.* **105B**, 33 (1981)

Berger, Ch. et al. (PLUTO Coll.), *Phys. Lett.* **78B**, 176 (1978)

Berger, Ch. et al. (PLUTO Coll.), *Phys. Lett.* **86B**, 418 (1979)

Berger, Ch. et al. (PLUTO Coll.), *Z. Phys.* **C21**, 53 (1983)

Bergsma, F. et al. (CHARM Coll.), *Phys. Lett.* **123B**, 269 (1983)

Bergsma, F. et al. (CHARM Coll.), *Phys. Lett.* **147B**, 481 (1984)

Berkelman, K., *Phys. Rep.* **C98**, 145 (1983)

Berko, S. and H.N. Pendleton, *Ann. Rev. Nucl. Part. Sci.* **30**, 543 (1980)

Bernreuther, W., O. Nachtmann, and B. Stech, *Z. Phys.* **C4**, 257 (1980)

Bernreuther, W. and W. Wetzel, *Nucl. Phys.* **B197**, 228 (1982)

Bernstein, A. and A.K. Mann, *Amer. J. Phys.* **24**, 445 (1956)

Bethe, H. and W. Heitler, *Proc. Roy. Soc.* **A146**, 83 (1934)

Bethe, H., *Phys. Rev.* **72**, 339 (1947)

Bjorken, J.D. and S.D. Drell, *Relativistic Quantum Fields* (McGraw-Hill, New York, 1965)

Bjorken, J.D., *Phys. Rev. Lett.* **16**, 408 (1966)

Bjorken, J.D., *Phys. Rev.* **179**, 1547 (1969)

Bjorken, J.D. and E.A. Paschos, *Phys. Rev.* **185**, 1975 (1969a)

Bjorkland, R. et al., *Phys. Rev.* **77**, 213 (1950)

Black, J.K. et al., *Phys. Rev. Lett.* **54**, 1628 (1985)

Blair, R. et al., *Phys. Rev. Lett.* **51**, 343 (1983)

Bleuler, K., *Helv. Phys. Acta* **23**, 567 (1950)

Bloch, F. and A. Nordsiek, *Phys. Rev.* **52**, 54 (1937)

Bodek, A. et al., *Phys. Rev. Lett.* **30**, 1087 (1973)

Böhm, A., "Testing the Standard Model, a Summary of the Moriond Conference", in: *Proc. 19th Rencontre de Moriond, Leptonic session*, ed. by J. Tran Trank Van (Editions Frontieres, Paris), Vol. 1

Böhm, M. and W. Hollik, *Z. Phys.* **C23**, 31 (1984)

Bogoliubov, N.N. and D.V. Shirkov, *Introduction to the Theory of Quantized Fields* (Interscience, New York, 1959)

Bohr, N., *Phil. Mag.* **26**, 476 (1913)

Bohr, N. and L. Rosenfeld, *Dansk. Vid. Sels. Mat.-fys. Med.* **12**, 1 (1933)

Bosetti, B.C. et al., *Nucl. Phys.* **B142**, 1 (1978)

Bouchiat, C. and L. Michel, *Phys. Rev.* **106**, 170 (1957)

Bouchiat, C., J. Iliopoulos, and Ph. Meyer, *Phys. Lett.* **38B**, 519 (1972)

Bourquin, M. et al., *Z. Phys.* **C21**, 27 (1983)

Bowdery, C.K., "New Lepton Asymmetry Results from PETRA", in: *Proc. 22nd Int. Conf. High Energy Physics, Leipzig, 1984*, ed. by A. Meyer and E. Wieczorek (Akademie Wiss. DDR, Zeuthen, 1984)

Brandelik, R. et al. (TASSO Coll.), *Phys. Lett.* **86B**, 243 (1979)

Brandelik, R. et al. (TASSO Coll.), *Phys. Lett.* **110B**, 173 (1982)

Brandt, S., Ch. Peyrou, R. Sosnowski, and A. Wroblewski, *Phys. Lett.* **12**, 57 (1964)

Breakstone, A. et al., *Phys. Rev.* **D30**, 528 (1984)

Breakstone, A. et al., *Z. Phys.* **C23**, 9; **C25**, 21 (1984a)

Brodsky, S.J., J. Ellis, J.S. Hagelin, and C.T. Sachrajda, *Nucl. Phys.* **B238**, 561 (1984)

Buchmüller, W. and S.-H.H. Tye, *Phys. Rev.* **D24**, 132 (1981)

Buras, A.J., "CP-Violation in the Standard Model and Beyond", in: *Proc. Workshop on the Future of Intermediate Energy Physics in Europe*, ed. by H. Koch and F. Scheck (Kernforschungsgesellschaft, Karlsruhe, 1984)

Burkhardt, H. et al., *Phys. Lett.* **B206**, 169 (1988)

Burmester, J. et al., *Phys. Lett.* **68B**, 297 (1977)

Burmester, J. et al., *Phys. Lett.* **68B**, 301 (1977a)

Bussey, P.J. et al., *Nucl. Phys.* **B58**, 363 (1973)

Cabibbo, N., *Phys. Rev. Lett.* **10**, 531 (1963)

Cabibbo, N., G. Parisi, and M. Testa, *Lett. Nuovo Cim.* **4**, 35 (1970)

Callan, C.G. Jr. and D.J. Gross, *Phys. Rev. Lett.* **22**, 156 (1969)

Callan, C.G. Jr. and D.J. Gross, *Phys. Rev.* **D8**, 4383 (1973)

Callan, C.G. Jr., R.F. Dashen, and D.J. Gross, *Phys. Lett.* **63B**, 334 (1976)

Carlson, A.G. et al., *Phil. Mag.* **41**, 701 (1950)

Casella, R., *Phys. Rev. Lett.* **22**, 554 (1969)

Cazzoli, E.G. et al., *Phys. Rev. Lett.* **34**, 1125 (1975)

Celmaster, W. and R.J. Gonsalves, *Phys. Rev. Lett.* **44**, 560 (1979)

Chadwick, J., *Verh. Dtsch. Phys. Ges.* **16**, 383 (1914)

Chadwick, J., *Proc. Roy. Soc. London* **A136**, 692 (1932)

Chadwick, G., in: *Proc. Int. Europhy. Conf. on High Energy Physics, Brighton, 1983*, ed. by J. Guy and C. Costain (Rutherford Appleton Laboratory, Didcot, 1983)

Chetyrkin, K.G., A.L. Kataev, and F.V. Tkachev, *Phys. Lett.* **85B**, 277 (1979)

Chiu, C.B., B. Misra, and E.C.G. Sudarshan, *Phys. Lett.* **117B**, 34 (1982)

Christ, N., B. Hasslacher, and A.H. Mueller, *Phys. Rev.* **D6**, 3543 (1972)

Christensen, J.H., J.W. Cronin, V.L. Fitch, and R. Turlay, *Phys. Rev. Lett.* **13**, 138 (1964)

Christensen, J.H., J.W. Cronin, V.L. Fitch, and R. Turlay, *Phys. Rev.* **140**, B74 (1965)

Coleman, S. and D.J. Gross, *Phys. Rev. Lett.* **31**, 851 (1973)

Collins, P.D.B., *Regge Theory and High Energy Physics* (Cambridge University Press, Cambridge, 1977)

Combridge, B.L., J. Kripfganz, and J. Ranft, *Phys. Lett.* **70B**, 234 (1977)

Compton, A.H., *Phys. Rev.* **21**, 483 (1923)

Cox, B., *J. Physique, Colloques* **43**, C3-140 (1982)

Crewther, R., P. Di Vecchia, G. Veneziano, and E. Witten, *Phys. Lett.* **88B**, 123 (1979), E. **91B**, 487 (1980)

Cronin, J.W., *Rev. Mod. Phys.* **53**, 373 (1981)

Cutler, R. and D. Sivers, *Phys. Rev.* **D17**, 196 (1978)

Danby, G., J.M. Gaillard, K. Goulianos, L.M. Lederman, N. Mistry, M. Schwartz, and J. Steinberger, *Phys. Rev. Lett.* **9**, 36 (1962)

Darwin, C.G., *Proc. Roy. Soc. London* **A118**, 654 (1928)

Davisson, C.J. and L.H. Germer, *Phys. Rev.* **30**, 705 (1927)

Deden, H. et al., *Nucl. Phys.* **B85**, 269 (1975)

de Groot, J.G.H. et al., *Phys. Lett.* **82B**, 292 (1979)

de Groot, E.H., G.J. Gounaris, and D. Schildknecht, *Z. Phys.* **C5**, 127 (1980)

de Kerret, H. et al., *Phys. Lett.* **68B**, 374 (1977)

De Rujula, A., H. Georgi, and S.L. Glashow, *Phys. Rev.* **D12**, 147 (1975)

De Rujula, A., H. Georgi, and H.D. Politzer, *Ann. Phys. (N.Y.)* **103**, 315 (1977)

De Rujula, A., J. Ellis, E.G. Floratos, and M.K. Gaillard, *Nucl. Phys.* **B138**, 387 (1978)

Dine, M. and J. Sapirstein, *Phys. Rev. Lett.* **43**, 668 (1979)

Dine, M., W. Fischler, and M. Srednicki, *Phys. Lett.* **104B**, 199 (1981)

Dirac, P.A.M., *Proc. Roy. Soc. London* **A114**, 243, 710 (1927)

Dirac, P.A.M., *Proc. Roy. Soc. London* **A117**, 610 (1928)

Dirac, P.A.M., *Proc. Roy. Soc. London* **A126**, 360 (1930)

Dirac, P.A.M., *The Principles of Quantum Mechanics*, 4th ed. (Oxford University Press, Oxford, 1958)

Dittmann, P. and V. Hepp, *Z. Phys.* **C10**, 283 (1981)

Drees, J. and H.E. Montgomery, *Ann. Rev. Nucl. Part. Sci.* **33**, 383 (1983)

Drell, S.D., D.J. Levy, and T.M. Yan, *Phys. Rev.* **187**, 2159 (1969)

Drell, S.D. and T.M. Yan, *Phys. Rev. Lett.* **25**, 316 (1970)

Drell, S.D., D.J. Levy, and T.M. Yan, *Phys. Rev.* **D1**, 1617 (1970a)

Drell, S.D. and T.M. Yan, *Ann. Phys.* (*N.Y.*) **66**, 578 (1971) and references therein

Drell, S.D., *Physica* **A96**, 3 (1979)

Duke, D.W. and J.F. Owens, *Phys. Rev.* **D30**, 49 (1984)

Dumbrajs, O. et al., *Nucl. Phys.* **B216**, 277 (1983)

Dydak, F., "Experimental Results from Lepton–Hadron and Photon–Hadron Scattering: Structure Functions and Final States", in *Proc. 1983 Int. Symp. on Lepton and Photon Interactions at High Energies*, ed. by D.G. Cassel and D.L. Kreinick (Cornell Univ., Ithaca, New York 1983)

Dylla, H.F. and J.G. King, *Phys. Rev.* **A7**, 1224 (1973)

Dyson, F.J., *Symmetry Groups in Nuclear and Particle Physics* (W.A. Benjamin, New York and Amsterdam, 1966)

Ecker, G., W. Grimus, and H. Neufeld, *Nucl. Phys.* **B229**, 421 (1983)

Edmonds, A.R., *Angular Momentum in Quantum Mechanics* (Princeton University Press, Princeton, 1957)

Eggert, K., "W, Z^0 and Then", Invited paper given at *Int. Symp. on Cosmic Ray and Particle Physics, Tokyo, 1984*

Einstein, A., *Ann. Phys.* (*Leipzig*) **17**, 132 (1905)

Einstein, A., *Ann. Phys.* (*Leipzig*) **17**, 891 (1905a)

Einstein, A., *Ann. Phys.* (*Leipzig*) **18**, 639 (1905b)

Einstein, A., *Ann. d. Phys.* (*Leipzig*) **49**, 769 (1916)

Einstein, A., *Phys. Z.* **18**, 121 (1917)

Eisele, F., "Structure Functions", *J. Physique, Colloques* **43**, C3-337 (1982)

Ellis, C.D. and W.A. Wooster, *Proc. Roy. Soc. London* **A117**, 109 (1927)

Ellis, J., M.K. Gaillard, and D.V. Nanopoulos, *Nucl. Phys.* **B100**, 313 (1975)

Ellis, J., M.K. Gaillard, and D.V. Nanopoulos, *Nucl. Phys.* **B106**, 292 (1976)

Ellis, J., M.K. Gaillard, and D.V. Nanopoulos, *Nucl. Phys.* **B109**, 213 (1976a)

Ellis, R.K., D.A. Ross, and A.E. Terrano, *Nucl. Phys.* **B178**, 421 (1981)

Englert, F. and R. Brout, *Phys. Rev. Lett.* **13**, 321 (1964)

Estermann, I. and O. Stern, *Z. Phys.* **85**, 17 (1933)

Evans, E.D., *Handbuch der Physik*, Vol. 34, 2, ed. by S. Flügge (Springer-Verlag, Berlin, 1958)

Fabrizius, K., I. Schmitt, G. Kramer, and G. Schierholz, *Z. Phys.* **C6**, 315 (1981)

Farhi, E., *Phys. Rev. Lett.* **39**, 1587 (1977)

Feldman, G.J. et al., *Phys. Rev. Lett.* **38**, 117 (1976)

Fermi, E., *Ricera Scient.* **2**, issue 12 (1933)

Fermi, E., *Z. Phys.* **88**, 161 (1934)

Fernandez, E. et al. (MAC Coll.), *Phys. Rev. Lett.* **50**, 1238 (1983)

Fetscher, W., *Phys. Lett.* **140B**, 117 (1984)

Feynman, R.P. and M. Gell-Mann, *Phys. Rev.* **109**, 193 and **111**, 362 (1958)

Feynman, R.P., *Phys. Rev. Lett.* **23**, 1415 (1969)

Feynman, R.P., *Photon–Hadron Interactions* (W.A. Benjamin, New York and Amsterdam, 1972)

Field, R.D. and R.P. Feynman, *Nucl. Phys.* **B136**, 1 (1978)

Fierz, M., *Z. Phys.* **104**, 553 (1937)

Fischbach, E. et al., *Phys. Lett.* **116B**, 73 (1982)

Fleming, G.N., *Phys. Lett.* **125B**, 287 (1983)

Fock, V., *Z. Physik* **75**, 622 (1932)

Fonda, L., G.C. Ghirardi, and A. Rimini, *Rep. Progr. Phys.* **41**, 587 (1978)

Frampton, P.H. and P. Vogel, *Phys. Rep.* **C82**, 339 (1982)

Frisch, R. and O. Stern, *Z. Phys.* **85**, 4 (1933)

Fritzsch, H. and M. Gell-Mann, in: *Proc. XVth Intern. Conf. on High Energy Physics, Chicago-Batavia, 1973* (NAL, Batavia, 1973)

Fritzsch, H., M. Gell-Mann, and H. Leutwyler, *Phys. Lett.* **B47**, 365 (1973a)

Fritzsch, H. and P. Minkowski, *Phys. Lett.* **90B**, 455 (1980)

Froissart, M., *Phys. Rev.* **123**, 1053 (1961)

Gaillard, M.K. and B.W. Lee, *Phys. Rev.* **D10**, 897 (1974)

Gaillard, M.K., B.W. Lee, and J.L. Rosner, *Rev. Mod. Phys.* **47**, 277 (1975)

Gaillard, M.K. and L. Maiani, in: *Proc. Summer Inst. on Quarks and Leptons, Cargèse, 1979* (Plenum, New York, 1980)

Gamow, G. and E. Teller, *Phys. Rev.* **49**, 895 (1936)

Garwin, R.L., L.M. Lederman, and M. Weinrich, *Phys. Rev.* **105**, 1415 (1957)

Gasser, J. and H. Leutwyler, *Phys. Rep.* **87**, 77 (1982)

Gell-Mann, M., *Phys. Rev.* **92**, 833 (1953)

Gell-Mann, M. and A. Pais, *Phys. Rev.* **97**, 1387 (1955)

Gell-Mann, M., "The Eightfold Way: A Theory of Strong Interaction Symmetry", California Institute of Technology Synchroton Laboratory Report CTSL-20 (1961) (unpublished)

Gell-Mann, M. und Y. Ne'eman, *The Eightfold Way* (W.A. Benjamin, New York and Amsterdam, 1964)

Gell-Mann, M., *Phys. Lett.* **8**, 214 (1964a)

Gell-Mann, M., *Physics* **1**, 63 (1964b)

Gell-Mann, M., *Acta Physics Austriaca*, Suppl. **9**, 733 (1972)

Georgi, H. and S.L. Glashow, *Phys. Rev. Lett.* **32**, 438 (1974)

Gershtein, S.S. and Ya.B. Zeldovich, *Sov. Phys. J.E.T.P.* **2**, 596 (1956)

Geweniger, C., "Measurement of $\sin^2 \vartheta_W$ from Neutrino–Nucleon Scattering", in: *Proc. 11th Int. Conf. on Neutrino Physics and Astrophysics, Nordkirchen, 1984*, ed. by K. Kleinknecht and E.A. Paschos (World Scientific, Singapore, 1984)

Giacomelli, G. and M. Jacob, *Phys. Rep.* **C55**, 1 (1979)

Gibbons, G.W., S.W. Hawking, and S. Siklos (eds.), *The Very Early Universe* (Cambridge University Press, Cambridge, 1983)

Ginsparg, P.H. and M.B. Wise, *Phys. Lett.* **127B**, 265 (1983)

Glashow, S.L., *Nucl. Phys.* **22**, 579 (1961)

Glashow, S.L., J. Iliopoulos, and L. Maiani, *Phys. Rev.* **D2**, 1285 (1970)

Goldhaber, G. et al., *Phys. Rev. Lett.* **37**, 255 (1976)

Goldman, T. and W. Wilson, *Phys. Rev.* **D15**, 709 (1977)

Goldstein, H., *Classical Mechanics*, 2nd ed. (Addison-Wesley, Reading, Ma., 1980)

Good, M.L., *Phys. Rev.* **121**, 311 (1961)

Gordon, W., *Z. Physik* **40**, 117 (1927)

Gordon, W., *Z. Phys.* **48**, 11 (1928)

Gorishny, S.G., A.L. Kataev, S.A. Larin, "Next-Next-to Leading $O(\alpha_s^3)$ QCD Corrections to $\sigma_{tot}(e^+e^- \to \text{Hadrons})$", in *Proc. Conf. Hadron Structure '87, Smolenice*, ed. by D. Krupa (Institute of Physics, Bratislava, 1988)

Gottfried, K., *Quantum Mechanics*, Vol. I (W.A. Benjamin, Reading, Ma., 1966)

Gottfried, K., "Hadronic Spectroscopy", in: *Proc. Int. Europhys. Conf. on High Energy Physics, Brighton, 1983*, ed. by J. Guy and C. Costain (Rutherford Appleton Laboratory, Didcot, 1983)

Gourdin, M., *Unitary Symmetries* (North-Holland, Amsterdam, 1967)

Green, M.B., J.H. Schwarz, and E. Witten, *Superstring Theory*, Vols. 1, 2 (Cambridge University Press, Cambridge, 1987)

Greenberg, O.W., *Phys. Rev. Lett.* **13**, 598 (1964)

Gromes, D., "Hyperfine Interactions in Baryons", in: *Proc. IVth Int. Conf. on Baryon Resonances* (University of Toronto, 1980)

Gross, D.J. and C.H. Llewellyn Smith, *Nucl. Phys.* **B14**, 337 (1969)

Gross, D.J. and R. Jackiw, *Phys. Rev.* **D6**, 477 (1972)

Gross, D.J. and F. Wilczek, *Phys. Rev. Lett.* **30**, 1343 (1973)

Gross, D.J. and F. Wilczek, *Phys. Rev.* **D8**, 3633 (1973a)

Gross, D.J. and F. Wilczek, *Phys. Rev.* **D9**, 980 (1974)

Guberina, B., J.J. Kühn, R.D. Peccei, and R. Rückl, *Nucl. Phys.* **B174**, 317 (1980)

Gupta, S.N., *Proc. Phys. Soc. London* **A63**, 681 (1950)

Guralnik, G.S., C.R. Hagen, and T.W.B. Kibble, *Phys. Rev. Lett.* **13**, 585 (1964)

Gürsey, F. and L.A. Radicati, *Phys. Rev. Lett.* **13**, 173 (1964)

Haber, H.E. and G.L. Kane, *Phys. Rep.* **C117**, 75 (1985)

Hagedorn, R. and J. Ranft, *Suppl. Nuovo Cim.* **6**, 169 (1968)

Hagedorn, R., "Multiplicities, p_T Distributions and the Expected Hadron–Quark–Gluon Phase Transition", Report CERN-TH-3684 (1983)

Hagelin, J.S., *Nucl. Phys.* **B193**, 123 (1981)

Haidt, D., "Experimental Tests of Gauge Theories," DESY Report 84-108 (1984)

Hanson, G., in: *Proc. Int. Europhys. Conf. on High Energy Physics, Brighton, 1983*, ed. by J. Guy and C. Costain (Rutherford Appleton Laboratory, Didcot, 1983)

Harari, H., *Phys. Lett.* **86B**, 83 (1979)

Harari, H. and N. Seiberg, *Phys. Lett.* **98B**, 269 (1981)

Harari, H., *Phys. Rep.* **C104**, 159 (1984)

Hasert, F.J. et al., *Phys. Lett.* **46B**, 121 (1973a)

Hasert, F.J. et al., *Phys. Lett.* **46B**, 138 (1973b)

Hasert, F.J. et al., *Nucl. Phys.* **B73** (1974)

Hegerfeld, G.C., *Phys. Rev.* **D10**, 3320 (1974)

Heisenberg, W. and W. Pauli, *Z. Phys.* **56**, 1 (1929)

Heisenberg, W. and W. Pauli, *Z. Phys.* **59**, 169 (1930)

Heisenberg, W., *Z. Phys.* **77**, 1 (1932)

Heisenberg, W., *Z. Phys.* **120**, 513 and 673 (1943)

Heisenberg, W., *Z. Phys.* **123**, 93 (1944)

Heisenberg, W., *Z. Phys.* **133**, 65 (1952)

Heitler, W., *The Quantum Theory of Radiation*, 3rd. ed. (Oxford University Press, Oxford, 1970)

Hepp, V., in: "*Present Status and Aims of Quantum Electrodynamics*", Lecture Notes in Physics, Vol. 143, ed. by G. Gräff, E. Klempt, G. Werth (Springer-Verlag, Berlin, Heidelberg, 1981)

Herb, S.W. et al., *Phys. Rev. Lett.* **39**, 252 (1977)

Higgs, P.W., *Phys. Lett.* **12**, 132 (1964)

Higgs, P.W., *Phys. Rev. Lett.* **13**, 508 (1964a)

Higgs, P.W., *Phys. Rev.* **145**, 1156 (1966)

Hofstadter, R. and J.A. McIntyre, *Phys. Rev.* **76**, 1269 (1949)

Höhler, G., *Pion–Nucleon Scattering*, Landolt-Börnstein, Vol. I/9b, ed. by H. Schopper (Springer-Verlag, Berlin, Heidelberg, 1983)

Hohn, F.E., *Elementary Matrix Algebra* (Macmillan, New York, 1958)

Horgan, R.R. and M. Jacob, *Nucl. Phys.* **B179**, 441 (1981)

Horgan, R.R. and P.V. Landshoff, *Phys. Lett.* **110B**, 493 (1982)

Hoyer, P., P. Osland, H.G. Sander, T.F. Walsh, and P.M. Zerwas, *Nucl. Phys.* **B161**, 349 (1979)

Hung, P.Q. and J.J. Sakurai, *Ann. Rev. Nucl. Part. Sci.* **31**, 375 (1981)

Isgur, N. and G. Karl, *Phys. Rev.* **D18**, 4187 (1978)

Isgur, N. and G. Karl, *Phys. Rev.* **D19**, 2653 (1979)

Itzykson, Cl. and J.-B. Zuber, *Quantum Field Theory* (McGraw-Hill, New York, 1980)

Jackiw, R. and C. Rebbi, *Phys. Rev. Lett.* **37**, 172 (1976)

Jackson, J.D., *Classical Electrodynamics*, 2nd ed. (J. Wiley, New York, 1975)

Jacob, M., *Nuovo Cim.* **9**, 826 (1958)

Jacob, M. and G.C. Wick, *Ann. Phys.* (*N.Y.*) **7**, 404 (1959)

Jauch, J.M. and F. Rohrlich, *The Theory of Photons and Electrons* (Addison-Wesley, Cambridge, Ma., 1955)

Jones, L.M., K.E. Lassila, U. Sukhatme and D.E. Willen, *Phys. Rev.* **D23**, 717 (1981)

Jones, T.W., "Results from the IMB Detector", in: *Proc. 11th Int. Conf. on Neutrino Physics and Astrophysics, Nordkirchen, 1984*, ed. by K. Kleinknecht and E.A. Paschos (World Scientific, Singapore, 1984)

Jonker, M. et al., *Phys. Lett.* **99B**, 265 (1981)

Jordan, P., *Z. Physik* **44**, 473 (1927)

Jordan, P. and O. Klein, *Z. Physik* **45**, 751 (1927a)

Jordan, P., *Z. Physik* **45**, 766 (1927b)

Jordan, P. and W. Pauli, *Z. Phys.* **47**, 151 (1928)

Jordan, P. and E. Wigner, *Z. Physik* **47**, 631 (1928a)

Källen, G., *Elementary Particle Physics* (Addison-Wesley, Reading, Ma., 1964)

Kemmer, N., *Proc. Cambridge Phil. Soc.* **34**, 354 (1938)

Khalfin, L.A., *Dokl. Akad. Nauk. USSR* **115**, 277 (1957); Engl. transl.: *Soviet Phys. Dokl.* **2**, 340 (1957)

Khalfin, L.A., *Pis'ma Zh. Eksp. Teor. Fiz.* **8**, 106 (1968)

Khalfin, L.A., *Phys. Lett.* **112B**, 223 (1982)

Kibble, T.W., *Phys. Rev.* **155**, 1554 (1967)

Kim, J.E., *Phys. Rev. Lett.* **43**, 103 (1979)

Kim, J.E., P. Langacker, M. Levine, and H.H. Williams, *Rev. Mod. Phys.* **53**, 211 (1981)

Kinoshita, T. and W.B. Lindquist, *Phys. Rev. Lett.* **47**, 1573 (1981)

Klapdor, H.V. and K. Grotz, *Phys. Rev.* **C30**, 2098 (1984)

Klein, O., *Z. Physik* **41**, 407 (1927)

Klein, O. und Y. Nishina, *Z. Phys.* **52**, 853 (1929)

Klein, O., *New Theories in Physics* (Intern. Institute of Intellectual Cooperation, Paris, 1939)

Kleinknecht, K., *Ann. Rev. Nucl. Sci.* **26**, 1 (1976)

Kleinknecht, K. and B. Renk, *Phys. Lett.* **130B**, 459 (1983)

Kleinknecht, K., *Comm. Nucl. Part. Phys.* **13**, 219 (1984)

Kobayashi, M. and T. Maskawa, *Progr. Theor. Phys.* **49**, 652 (1973)

Kogut, J. and L. Susskind, *Phys. Rev.* **D9**, 697, 3391 (1974)

Körner, J.G., G. Schierholz, and J. Willrodt, *Nucl. Phys.* **B185**, 365 (1981)

Korthals Altes, C.P. and M. Perrottet, *Phys. Lett.* **39B**, 546 (1972)

Koshiba, M., "Experiments on Baryon Number Nonconservation", in: *Proc. XXII Int. Conf. on High Energy Physics, Leipzig, 1984*, ed. by A. Meyer and E. Wieczorek (Akademie Wiss. DDR, Zeuthen, 1984)

Kubar-André, J. and F.E. Paige, *Phys. Rev.* **D19**, 221 (1979)

Kugel, H.W. and D.E. Murnick, *Rep. Prog. Phys.* **40**, 297 (1977)

Kühn, J.H. and P.M. Zerwas, *Phys. Lett.* **154B**, 448 (1985)

Kühnelt, H. and O. Nachtmann, *Nucl. Phys.* **B88**, 41 (1975)

Kunszt, Z. and E. Pietarinen, *Phys. Lett.* **132B**, 453 (1983)

La Rue, G.S., W.M. Fairbank, and J.D. Phillips, *Phys. Rev. Lett.* **42**, 142, 1019 (E) (1979)

Lamb, W.E. and R.C. Retherford, *Phys. Rev.* **72**, 241 (1947)

Landau, L.D., *Dokl. Akad. Nauk USSR* **60**, 207 (1948)

Landau, L.D., *JETP* **32**, 405 (1956a)

Landau, L.D., *JETP* **32**, 407 (1956b)

Landau, L.D. and E.M. Lifschitz, *The Classical Theory of Fields*, 4th ed. (Pergamon Press, Oxford, 1975)

Landé, E., E.T. Booth, J. Impeduglia, L. Lederman, and W. Chinowsky, *Phys. Rev.* **103**, 1901 (1956)

Landshoff, P.V., J.C. Polkinghorne, and R.D. Short, *Nucl. Phys.* **B28**, 225 (1971)

Landshoff, P.V. and J.C. Polkinghorne, *Nucl. Phys.* **B28**, 240 (1971a)

Langacker, P., *Phys. Rep.* **C72**, 186 (1981)

Langacker, P., "Neutrino Mass and Oscillations: Theory and Experiment", Report, Penn. U. UPR-0386T (1989)

Lattes, C.M.G., H. Muirhead, C.F. Powell, and G.P. Occhialini, *Nature* **159**, 694 (1947)

v. Laue, M., *Die Relativitätstheorie*, Vol. 1 (Vieweg, Braunschweig, 1952)

Lee, T.D. and C.N. Yang, *Phys. Rev.* **104,** 254 (1956)

Lee, T.D. and C.N. Yang, *Phys. Rev.* **105**, 1671 (1957)

Lee, T.D. and C.S. Wu, *Ann. Rev. Nucl. Sci.* **16**, 511 (1966)

Lee, B.W., C. Quigg, and H.B. Thacker, *Phys. Rev.* **D16**, 1519 (1977)

Lee-Franzini, J., *Nucl. Phys.* B (Proc. Suppl.) **3**, 139 (1988)

Leutwyler, H. and M. Roos, *Z. Phys.* **C25**, 91 (1984)

Linde, A.D., *JETP Lett.* **23**, 64 (1976)

Linde, A.D., "Elementary Particles and Cosmology", in: *Proc. 22nd Int. Conf. on High Energy Physics, Leipzig, 1984*, ed. by A. Meyer and E. Wieczorek (Akademie Wiss. DDR, Zeuthen, 1984)

Lipkin, H.J., *Lie Groups for Pedestrians*, 2nd. ed. (North-Holland, Amsterdam, 1966)

Llewellyn Smith, C.H., and J.F. Wheater, *Nucl. Phys.* **B208**, 27 (1982)

Lüders, G., *Dan. Mat. Fys. Medd.* **28**, 5 (1954)

Lüders, G., *Ann. Phys.* (*N.Y.*) **2**, 1 (1957)

Mac Farlane, D.B. et al., *Z. Phys.* **C26**, 1 (1984)

Mackenzie, P.B. and G.P. Lepage, *Phys. Rev. Lett.* **47**, 1244 (1981)

Maiani, L., "New Currents", in: *Proc. 1977 Int. Symp. on Lepton and Photon Int. at High Energies*, ed. by F. Gutbrod (DESY, Hamburg, 1977)

Majumdar, D.P., *Phys. Rev.* **D3**, 2869 (1971)

Mandelstam, S., *Phys. Rev.* **112**, 1344 (1958)

Marciano, W. and A. Sirlin, *Phys. Rev. Lett.* **36**, 1425 (1976)

Marciano, W. and H. Pagels, *Phys. Rep.* **36C**, 137 (1978)

Marciano, W. and L. Wyler, *Z. Phys.* **C3**, 181 (1979)

Marciano, W.J. and A. Sirlin, *Phys. Rev.* **D22**, 2695 (1980)

Marciano, W.J. and A. Sirlin, *Nucl. Phys.* **B189**, 442 (1981)

Marciano, W.J. and A. Sirlin, *Phys. Rev.* **D27**, 52 (1983)

Martin, A., *Phys. Rev.* **129**, 1432 (1963)

Martin, B.R., D. Morgan and G. Shaw, *Pion–Pion Interaction in Particle Physics* (Academic Press, New York, 1976)

Michelson, A.A., *Sill. J.* **22**, 120 (1881)

Michelson, A.A. and E.W. Morley, *Am. Jour. Sci.* **34**, 333 (1887)

Michel, L., *Proc. Phys. Soc. London* **A63**, 514 (1950)

Miller, G. et al., *Phys. Rev.* **D5**, 528 (1972)

Minkowski, H., *Phys. Zs.* **10**, 104 (1909)

Mohapatra, R.N. and J.C. Pati, *Phys. Rev.* **D11**, 566, 2558 (1975)

Möhl, D., G. Petrucci, L. Thorndahl, and S. van der Meer, *Phys. Rep.* **58**, 73 (1980)

Møller, C., *Ann. Phys.* (*Leipzig*) **14**, 531 (1932)

Morpurgo, G., *Acta Physica Austriaca, Suppl.* **21**, 5 (1979)

Mott, N.F., *Proc. Roy. Soc. London* **A126**, 259 (1930)

Mott, N.F. and H.S.W. Massey, *Theory of Atomic Collisions*, 3rd ed., pp. 302 and 818 (Oxford University Press, Oxford, 1965)

Nachtmann, O., *Acta Phys. Austriaca Suppl.* **VI**, 485 (1969)

Nachtmann, O., *Nucl. Phys.* **B38**, 397 (1972)

Nachtmann, O., *Nucl. Phys.* **B63**, 237 (1973)

Nachtmann, O., in: *Weak Interactions*, ed. by M.K. Gaillard and M. Nicolic (Institut National de Physique Nucleaire et de Physique de Particules, Paris, 1977)

Nachtmann, O. and A. Reiter, *Z. Phys.* **C16**, 45 (1982)

Nachtmann, O. and A. Reiter, *Z. Phys.* **C24**, 283 (1984)

Naroska, B., in: *Proc. 1983 Int. Symp. on Lepton and Photon Int. at High Energies*, ed. by D.G. Cassel and D.L. Kreinick (Cornell Univ., Ithaca, New York 1983)

Naroska, B., *Phys. Rep.* **C148**, 67 (1987)

Ne'eman, Y., *Nucl. Phys.* **26**, 222 (1961)

Nielsen, H.B., D.L. Bennet, and N. Brene, "The Random Dynamics Project" in: Niels Bohr Centennial Conf. 1985, *Recent Developments in Quantum Field Theory* (North-Holland, Amsterdam, 1985)

Nishijima, K., *Prog. Theor. Phys.* **13**, 285 (1955)

Noether, E., Kgl. Ges. d. Wiss. Nachrichten, Math.-phys. Klasse, Göttingen, p. 235 (1918)

Novikov, V.A., L.B. Okun, M.A. Shifman, A.I. Vainshtein, M.B. Voloshin, and V.I. Zakharov, *Phys. Rep.* **C41**, 1 (1978)

Oppenheimer, J.R., *Phys. Rev.* **35**, 939 (1930)

Ore, A. and J.L. Powell, *Phys. Rev.* **75**, 1696 (1949)

Pais, A., *Phys. Rev.* **86**, 663 (1952)

Pais, A., *Rev. Mod. Phys.* **49**, 925 (1977)

Pais, A., *Rev. Mod. Phys.* **51**, 863 (1979)

Panman, J., in: *Proc. 11th Int. Conf. on Neutrino Physics and Astrophysics, Nordkirchen, 1984*, ed. by K. Kleinknecht and E.A. Paschos (World Scientific, Singapore, 1984)

Parisi, G., *Nucl. Phys.* **B59**, 641 (1973)

Particle Data Group, *Rev. Mod. Phys.* **52**, 2, Part II (1980)

Particle Data Group, *Rev. Mod. Phys.* **56**, No. 2, Part II (1984)

Particle Data Group, *Phys. Lett.* **170B**, 1 (1986)

Paschos, E.A. and M. Wirbel, *Nucl. Phys.* **B194**, 189 (1982)

Paschos, E.A. and U. Türke, *Phys. Lett.* **116B**, 360 (1982a)

Paschos, E.A., B. Stech, and U. Türke, *Phys. Lett.* **128B**, 240 (1983)

Paschos, E.A. and U. Türke, *Nucl. Phys.* **B243**, 29 (1984)

Pati, J.C. and A. Salam, *Phys. Rev.* **D8**, 1240 (1973)

Pati, J.C. and A. Salam, *Phys. Rev.* **D10**, 275 (1975)

Pauli, W., *Z. Physik* **43**, 601 (1927)

Pauli, W., *Ann. Inst. Henri Poincaré* **6**, 137 (1936)

Pauli, W., *Phys. Rev.* **58**, 716 (1940)

Pauli, W., in: *Niels Bohr and the Development of Physics*, ed. by W. Pauli, L. Rosenfeld and V. Weisskopf (McGraw-Hill, New York and Pergamon Press, London, 1955)

Pauli, W., *Collected Scientific Papers*, Vol. 2, p. 1313 (Interscience, New York, 1964)

Pauli, W., "Zur älteren und neueren Geschichte des Neutrinos", in: *W. Pauli, Physik und Erkenntnistheorie*, ed. by R.U. Sexl (Vieweg, Braunschweig, 1984)

Peccei, R.D. and H.R. Quinn, *Phys. Rev. Lett.* **38**, 1440 (1977)

Perkins, D.H., *Introduction to High Energy Physics*, 2nd ed. (Addison-Wesley, Reading, Ma., 1982)

Perl, M.L., *High Energy Hadron Physics* (Wiley, New York, 1974)

Perl, M.L. et al., *Phys. Rev. Lett.* **35**, 1489 (1975)

Peterman, A., *Phys. Rep.* **53C**, 157 (1979)

Pietschmann, H., H. Rupertsberger, and K. Svozil, *Z. Phys.* **C12**, 367 (1982)

Piketty, C.A., "Parity Violation in Atoms", in: *Proc. 11th Int. Conf. on Neutrino Physics and Astrophysics, Nordkirchen, 1984*, ed. by K. Kleinknecht and E.A. Paschos (World Scientific, Singapore, 1984)

Politzer, H.D., *Phys. Rev. Lett.* **30**, 1346 (1973)

Politzer, H.D., *Phys. Rep.* **14C**, 129 (1974)

Polyakov, A.M., *Soviet Phys. JETP* **32**, 296 (1971)

Pomeranchuk, I.J., *Soviet Phys. JETP* **7**, 499 (1958)

Prescott, Ch. et al., *Phys. Lett.* **77B**, 347 (1978)

Prescott, Ch. et al., *Phys. Lett.* **84B**, 524 (1979)

Puppi, G., *Nuovo Cim.* **5**, 587 (1948)

Quenzer, A. et al., *Phys. Lett.* **76B**, 512 (1978)

Rabi, I.I., Paper presented at EPS Int. Conf. on High Energy Physics, Palermo, June, 1975

Racah, G., *Nuovo Cimento* **11**, 7 (1934)

Racah, G., *Nuovo Cimento* **13**, 69 (1936)

Radermacher, E., "The Experimental Discovery of the Intermediate Vector Bosons W^+, W^- and Z^0 at the CERN $p\bar{p}$ Collider", in: *Progress in Particle and Nuclear Physics*, Vol. 14, ed. by A. Faessler (Pergamon Press, 1985)

Rauch, H. et al., *Phys. Lett.* **54A**, 425 (1975)

Rebbi, C., *Lattice Gauge Theories and Monte Carlo Simulations* (World Scientific, Singapore, 1983)

Reinders, L.J., H.R. Rubinstein, and S. Yazaki, *Nucl. Phys.* **B186**, 109 (1981)

Reinders, L.J., H.R. Rubinstein, and S. Yazaki, *Phys. Rep.* **C127**, 1 (1985)

Reines, F. and C.L. Cowan, Jr., *Phys. Rev.* **92**, 830 (1953)

Reines, F., C.L. Cowan, Jr., F.B. Harrison, A.D. McGuire, and H.W. Kruse, *Science* **124**, 103 (1956)

Reines, F. and C.L. Cowan, Jr., *Phys. Rev.* **113**, 273 (1959)

Reines, F., H.S. Gurr, and H.W. Sobel, *Phys. Rev. Lett.* **37**, 315 (1976)

Rich, A., *Rev. Mod. Phys.* **53**, 127 (1981)

Richardson, J.L., *Phys. Lett.* **82B**, 272 (1979)

Rith, K., "Review of Present Structure Function Data", in: *Proc. Int. Europhys. Conf. on High Energy Physics*, Brighton, 1983, ed. by J. Guy and C. Costain (Rutherford Appleton Laboratory, Didcot, 1983)

Rochester, G.D. and C.C. Butler, *Nature* **160**, 855 (1947)

Roesch, L.Ph. et al., *Helv. Phys. Acta* **55**, 74 (1982)

Rosen, S.P., *Phys. Rev. Lett.* **44**, 4 (1980)

Rosner, J.L., *Phys. Rep.* **11C**, 189 (1974)

Ross, G.G., in: *Proc. 11th Int. Conf. on Neutrino Physics and Astrophysics, Nordkirchen*, 1984. ed. by K. Kleinknecht and E.A. Paschos (World Scientific, Singapore, 1984)

Roy, S.M., *Phys. Rep.* **C5**, 125 (1972)

Rubbia, C., P. McIntyre and D. Cline, in: *Proc. Int. Neutrino Conf., Aachen*, 1976, ed. by H. Faissner, H. Reithler, P. Zerwas (Vieweg, Braunschweig, 1977)

Rubbia, C., "Physics-Results of the UA1 Collaboration at the CERN Proton-Antiproton Collider", in: *Proc. 11th. Int. Conf. on Neutrino Physics and Astrophysics, Nordkirchen*, 1984, ed. by K. Kleinknecht and E.A. Paschos (World Scientific, Singapore, 1984)

Rutherford, E., *Phil. Mag.* **47**, 109 (1899)

Rutherford, E., *Phil Mag.* **21**, 669 (1911)

Rutherford, E., *Proc. Manchester Lit. Philos. Soc.* **55**, 18 (1911a)

Rutherford, E., *Phil. Mag.* **37**, 581 (1919)

Sakata, S., *Progr. Theor. Phys.* **16**, 686 (1956)

Sakita, B., *Phys. Rev.* **136B**, 1756 (1964)

Salam, A., *Nuovo Cim.* **5**, 299 (1957)

Salam, A., in: *Proc. 8th Nobel Symposium*, ed. by N. Svartholm (Almquist and Wiskell, Stockholm, 1968)

Scharre, D.L., "SPEAR Results 1981", in: *Proc. 1981 Int. Symp. on Lepton and Photon Interactions at High Energies*, ed. by W. Pfeil (University Bonn, 1981)

Scheck, F., *Phys. Rep.* **44**, 187 (1978)

Schiff L.I., *Quantum Mechanics* (McGraw-Hill, New York, 1955)

Schrödinger, E., *Ann. Phys.* (*Leipzig*) **81**, 109 (1926)

Schubert, K.R. et al., *Phys. Lett.* **31B**, 662 (1970)

Schubert, K.R., "Weak Decays of Heavy Mesons", in: *Proc. 11th Int. Conf. on Neutrino Physics and Astrophysics, Nordkirchen* 1984, ed. by K. Kleinknecht and E.A. Paschos (World Scientific, Singapore, 1984)

Schwinger, J., *Phys. Rev.* **75**, 898 (1949)

Schwinger, J. (ed.), "*Quantum Electrodynamics*" (Dover, New York, 1958)

Senjanovic, G. and R.N. Mohapatra, *Phys. Rev.* **D12**, 1502 (1975)

Shaw, R., Ph.D. Thesis, Cambridge, unpublished (1954)

Shifman, M.A., A.I. Vainshtein, and V.I. Zakharov, *Nucl. Phys.* **B166**, 493 (1980)

Shifman, M.A., "Theory of Heavy Quark Antiquark States", in: *Proc. 1981 Int. Symp. on Lepton and Photon Interactions at High Energies*, ed. by W. Pfeil (University Bonn, 1981)

Shifman, M.A., *Nucl. Phys.* B (Proc. Suppl.) **3**, 289 (1988)

Shupe, M.A., *Phys. Lett.* **86B**, 87 (1979)

Sirlin, A., *Phys. Rev.* **D22**, 971 (1980)

Sirlin, A. and W.J. Marciano, *Nucl. Phys.* **B189**, 442 (1981)

Sommerfeld, A., *Mechanics* (Academic Press, New York, 1952)

Stech, B. and J.H.D. Jensen, *Z. Phys.* **141**, 175 (1955)

Sterman, G. and S. Weinberg, *Phys. Rev. Lett.* **39**, 1436 (1977)

Straumann, N., *General Relativity and Relativistic Astrophysics* (Springer, Heidelberg, 1984)

Streater, R.F. and A.S. Wightman, *PCT, Spin and Statistics and All That* (W.A. Benjamin, New York and Amsterdam, 1964)

Street, J.C. and E.C. Stevenson, *Phys. Rev.* **52**, 1003 (1937)

Stueckelberg, E.C.G., *Helv. Phys. Acta* **14**, 322, 588 (1941)

Sudarshan, E.C.G. and R.E. Marshak, in: *Proc. Padua Conf. on Mesons and Recently Discovered Particles*, p. 14 (1957)

Takasaki, F., *Nucl. Phys.* **B** (Proc. Suppl.) **3**, 17 (1988)

Theis, W.R., *Z. Phys.* **150**, 590 (1958)

Theriot, E.D. Jr., et al., *Phys. Rev.* **A2**, 707 (1970)

Thirring, W., *Acta Physica Austriaca*, Suppl. II, 205 (1965)

Thomson, J.J., *Phil. Mag.* **44**, 269 (1897)

Thomson, G.P. and A. Reid, *Nature* **119**, 890 (1927)

Thomé, W. et al., *Nucl. Phys.* **B129**, 365 (1977)

't Hooft, G., *Nucl. Phys.* **B33**, 173 (1971)

't Hooft, G., *Phys. Lett.* **37B**, 195 (1971a)

't Hooft, G., unpublished comments at a conference in Marseille (1972)

't Hooft, G. and M. Veltman, Diagrammar, CERN-Report 73-9 (1973)

't Hooft, G., *Phys. Rev. Lett.* **37**, 8 (1976a)

't Hooft, G., *Phys. Rev.* **D14**, 3432 (1976b)

Tsai, Y.S., *Rev. Mod. Phys.* **46**, 815 (1974)

van der Meer, S., Report CERN ISR Po/72-31 (1972)

van Dyck, R.S. Jr., P.B. Schwinberg, and H.G. Dehmelt, *Bull. Amer. Phys. Soc.* **24**, 758 (1979)

Veltman, M., "Gauge Field Theories", in: *Proc. 6th Int. Symp. on Electron and Photon Interactions at High Energies, Bonn, 1973*, ed. by H. Rollnick and W. Pfeil (North-Holland, Amsterdam 1974)

Veltman, M., *Acta Phys. Polon.* **B8**, 475 (1977)

Vermaseren, J., J. Gaemers, and S. Oldham, *Nucl. Phys.* **B187**, 301 (1981)

Wandzura, S., *Nucl. Phys.* **B122**, 412 (1977)

Webber, B.R., *Nucl. Phys.* **B238**, 492 (1984)

Weinberg, S., *Phys. Rev. Lett.* **19**, 1264 (1967)

Weinberg, S., *Phys. Rev.* **D5**, 1412 (1972a)

Weinberg, S., *Gravitation and Cosmology* (J. Wiley, New York, 1972b)

Weinberg, S., *Phys. Rev.* **D7**, 1068 (1973)

Weinberg, S., *Phys. Rev. Lett.* **36**, 294 (1976)

Weinberg, S., *Phys. Rev. Lett.* **40**, 223 (1978)

Weinberg, S., *Phys. Lett.* **91B**, 51 (1980)

Weisberger, W.I., *Phys. Rev.* **143**, 1302 (1966)

Weisskopf, V.F. and E.P. Wigner, *Z. Phys.* **63**, 54 and **65**, 18 (1930)

Werner, S.A. et al., *Phys. Rev. Lett.* **35**, 1053 (1975)

Wess, J. and B. Zumino, *Phys. Lett.* **49B**, 52 (1974a)

Wess, J. and B. Zumino, *Nucl. Phys.* **B70**, 39 (1974b)

Wess, J. and J. Bagger, *Supersymmetry and Supergravity* (Princeton University Press, Princeton, N.J., 1983)

Wetzel, W., *Nucl. Phys.* **B227**, 1 (1983)

Wetzel, W., "Electroweak Radiative Corrections for $e^+e^- \to \mu^+\mu^-$ at PETRA Energies", Report, University of Heidelberg, HD-THEP-83-9 (1983a)

Weyl, H., *Z. Phys.* **56**, 330 (1929)

Weyl, H., *The Theory of Groups and Quantum Mechanics* (Dover, New York, 1931)

Wheeler, J.A., *Phys. Rev.* **52**, 1107 (1937)

Wick, G.C., *Phys. Rev.* **80**, 268 (1950)

Wicklund, E., "Weak-Electromagnetic Interference Effects in e^+e^- Annihilation to Hadronic Final States at PETRA", in: *Proc. 22nd Int. Conf. on High Energy Physics, Leipzig, 1984*, ed. by A. Meyer and E. Wieczorek (Akademie Wiss. DDR, Zeuthen, 1984)

Wigner, E.P., *Gött. Nach.* **31**, 546 (1932)

Wigner, E.P., *Ann. Math.* **40**, 149 (1939)

Wilczek, F., *Phys. Rev. Lett.* **39**, 1304 (1977)

Wilczek, F., *Phys. Rev. Lett.* **40**, 279 (1978)

Wilkinson, D.H., in: *Symmetries and Nuclei, Nuclear Physics with Heavy Ions and Mesons*, Vol. II, ed. by R. Balian, M. Rho, and G. Ripka (North-Holland, Amsterdam, 1978)

Wilkinson, D.H., *Progr. in Particle and Nuclear Physics* **6**, 325 (1980)

Wilkinson, D.H., *Nucl. Phys.* **A377**, 474 (1982)

Wilson, K.G., *Phys. Rev.* **179**, 1499 (1969)

Wilson, K.G., "Products of Currents", in: *Proc. 1971 Int. Symp. on Electron and Photon Interactions at High Energy*, ed. by N.B. Mistry (Cornell Univ., Ithaca, New York 1972)

Wilson, K.G., *Phys. Rev.* **D10**, 2445 (1974a)

Wilson, K.G. and J. Kogut, *Phys. Rep.* **C12**, 75 (1974b)

Wilson, K.G., *Rev. Mod. Phys.* **47**, 773 (1975)

Wise, M.B., H. Georgi, and S.L. Glashow, *Phys. Rev. Lett.* **47**, 402 (1981)

Wolf, G., *J. Physique, Colloques* **43**, C3-525 (1982)

Wolfenstein, L., *Phys. Rev. Lett.* **13**, 562 (1964)

Wolfenstein, L., *Nucl. Phys.* **B246**, 45 (1984)

Wu, C.S., E. Ambler, R. Hayward, D. Hoppes, and R. Hudson, *Phys. Rev.* **105**, 1413 (1957)

Wu, S.L., *Phys. Rep.* **C107**, 59 (1984)

Yan, T.M., in: *Lepton Pair Production*, ed. by J. Tranh Than Van (Editions Frontieres, Paris, 1981), pp. 369 ff

Yang, C.N., *Phys. Rev.* **77**, 242 (1950)

Yang, C.N. and R. Mills, *Phys. Rev.* **96**, 191 (1954)

Yennie, D., S. Frautschi, and H. Suura, *Ann. Phys.* (*N.Y.*) **13**, 379 (1961)

Yukawa, H., *Proc. Phys. Math. Soc. Japan* **17**, 48 (1935)

Zee, A., *Phys. Rev.* **D8**, 4038 (1973)

Zweig, G., CERN-Report TH-401 (1964) and TH-412 (1964), published in: *Developments in the Quark Theory of Hadrons, A Reprint Collection*, Vol. I: 1964–1978, ed. by D.B. Lichtenberg and S.P. Rosen (Hadronic Press, Nonamtum, Ma., 1980)

Index

GPSR Compliance

The European Union's (EU) General Product Safety Regulation (GPSR) is a set of rules that requires consumer products to be safe and our obligations to ensure this.

If you have any concerns about our products, you can contact us on ProductSafety@springernature.com

In case Publisher is established outside the EU, the EU authorized representative is:

Springer Nature Customer Service Center GmbH
Europaplatz 3
69115 Heidelberg, Germany

Batch number: 10704711

Printed by Printforce, the Netherlands